div D = 4πρ
div E = 4πρ'

THE NATURE OF PHYSICS

PETER J. BRANCAZIO

BROOKLYN COLLEGE

CITY UNIVERSITY OF NEW YORK

MACMILLAN PUBLISHING CO., INC.

NEW YORK

COLLIER MACMILLAN PUBLISHERS

LONDON

Printed in the United States of America

Macmillan Publishing Co., Inc.
866 Third Avenue
New York, New York 10022

Collier-Macmillan Canada, Ltd.

Library of Congress Cataloging in Publication Data

Brancazio, Peter J.
The nature of physics.

Bibliography: p.
1. Physics. I. Title.
QC21.2.B725 530 73-19495
ISBN 0-02-313500-X

Printing: 1 2 3 4 5 6 7 8 Year: 5 6 7 8 9 0

for Davy and Larry

without whose cheerful assistance this book would have been finished much sooner

PREFACE

This book developed out of my eleven years of teaching an introductory course in physics or physical science to nonscience majors. Traditionally, many students have entered such a course with great reluctance and trepidation because they lack mathematical proficiency and because they believe the subject matter is uninteresting or too difficult for them. To some extent their fears are justified, for many introductory physics courses are organized around problem-solving techniques, with insufficient emphasis on the understanding of the fundamental concepts of physics. My aim was to find a way to teach the principles of physics that would get across the essence of physics without undue emphasis on its mathematical aspects and would be, above all, interesting and enjoyable for the student.

In order to meet this objective, I have tried to write a book that aligns itself with the strengths and weaknesses of the liberal arts student. It is my belief that a student who is capable of grasping sophisticated concepts in philosophy, psychology, or sociology certainly ought to be able to understand the basic concepts of physics as well, if he is not unduly burdened with mathematical details. Moreover, a student with some background in philosophy and history can readily appreciate the philosophical and historical evolution of ideas in physics. I have therefore placed considerable emphasis on the historical development and philosophical content of the major physical concepts and theories, and played down the teaching of problem-solving techniques. My overall purpose has been to present physics not as a monolithic collection of facts and formulas, but as an organic body of concepts that have evolved from man's ongoing attempts to understand the universe. I have tried to point out the *continuing* evolution of physics. By making numerous references to theories that were popular in the past and are now regarded as incorrect, such as the geocentric theory and the ether hypothesis, and by mentioning some still-unsolved problems, I have attempted to give the student an appreciation for the fact that modern-day physics is by no means the last word.

In adopting a historical-philosophical approach to a subject, one runs the risk of neglecting content. I have tried to avoid the obvious pitfalls of this approach. The major areas of physics—mechanics, light and electromagnetism, relativity, quantum theory, and atomic and nuclear physics—are covered in the text. Although the development of each of these topics is along historical lines, the main emphasis is placed on an understanding of the significant concepts and theories. Mathematical formulas have been used wherever they are needed to illuminate the physics, and sample problems have been included in the text to help the student understand how to use these formulas. The questions at the end of each chapter are divided into two parts: "Questions for Review," which help the student to summarize and understand the main ideas presented in the chapter, and "Exercises," which include numerical problems as well as some thought-provoking questions whose answers cannot be found directly in the text.

A few particular features of the organization of the text are worth noting. Chapter 1—"What Is Science?"—deals with the philosophical assumptions of science and outlines the scientific method. The instructor should avoid the temptation to skip over this chapter, since it sets the tone for the rest of the book. In particular, the student should have some introduction to the scientific method, for the relationship between observations, laws, and theories is stressed repeatedly throughout the book. The second chapter deals with the development of astronomy from ancient times to the Renaissance. Most introductory physics courses do not deal with astronomy in much depth. However, there are strong historical and pedagogical reasons for beginning a physics book with a section on astronomy. First of all, it should be realized that the study of astronomy was the core of physics for more than two thousand years and was the source of the seminal contributions to science made by Kepler, Galileo, and Newton. Moreover, astronomy is the area of physics that is most accessible and interesting to the nonscience student. As such, it is an excellent starting point for the study of physics.

Three full chapters have been devoted to the theory of relativity, a subject that is of particular interest to the nonscientist. In these chapters I have stressed the conceptual basis of relativity, rather than its consequences and applications. I have attempted to trace the main sources of the theory, which include the personality of Einstein himself. Some instructors may find my approach to the subject a bit unorthodox, but I can attest to its success with several classes of students at Brooklyn College. I have found the chapter on general relativity to be a particularly successful one. It has been one of the greatest pleasures of my teaching career to watch students who have enrolled in a physics course with considerable misgivings discover after a few months that they are capable of understanding the basic principles of general relativity.

There are a great many people whose assistance in one form or another merits acknowledgment, beginning with my parents, who had the good sense to send me to college. I am greatly indebted to the hundreds of Brooklyn College students who, over the years, really forced me to understand physics. They, more than anyone else, are responsible for the form this book has taken. I owe a special thanks to a remarkable group of philosophy students at Brooklyn College who befriended me in the winter of 1967 and introduced me to the philosophy of science. In particular, Jay Newman and Jules Coleman were most responsible for my education in this area. My editors at Macmillan—Michael Hamilton, Wallace Shows, and especially John Beck—have my appreciation for their encouragement and assistance, and above all for their tolerance of my indolence. I would also like to thank the anonymous reviewers, assembled by my editors, who reviewed portions of the manuscript and suggested useful corrections and revisions. I am grateful to Professor Arthur E. Walters of Rutgers University for his very careful reading of the entire manuscript.

I would like to thank Lydia Fazio for her help in typing and proofreading the manuscript and Sherry Olynyk for her typing and editorial assistance. I am particularly indebted to Elisabeth H. Belfer of Macmillan for the highly competent manner in which she has steered this manuscript through production.

Finally, I wish to express my appreciation to my family for tolerating my seeming indifference and inability to indulge in normal conversation while I was writing or thinking about the next chapter. My deepest gratitude is reserved for my wife, Ronnie, who read and reread, typed and retyped, edited, suggested, inspired, pushed, and allowed me to ignore my obligations as a father and homeowner without really complaining too much.

P. J. B.

CONTENTS

INTRODUCTION

PART ONE
FROM THE CAVE TO THE RENAISSANCE

PART THREE

LIGHT, ELECTROMAGNETISM, AND THE ETHER HYPOTHESIS

PART FOUR

THE EINSTEINIAN REVOLUTION

INTRODUCTION

WHAT IS SCIENCE?

1

The eternal mystery of the world is its comprehensibility.

Albert Einstein

WE LIVE IN the Age of Science. Virtually every aspect of our everyday lives is affected in some way by the scientific and technological developments of the past hundred years. The availability of electric power has enabled us to fill our homes with all sorts of conveniences and labor-saving devices—from the light bulb to the washing machine to the electric toothbrush. The automobile and the airplane provide us with such mobility that we can go from almost any point on earth to any other point in a day or two. As a result of advances in biology and medicine, many dreaded diseases have been eradicated, and we can look forward to healthier and longer lives. The field of communications has developed to such an extent that we can pick up a telephone and be connected within minutes to almost any city on earth. An even more amazing development is our ability to land men on the moon and to watch them on live color television as they perform their experiments.

Even in the arts, science and technology have affected developments to a significant degree. Technological advances in the printing industry have made it possible to produce books at low cost, so that the great works of literature are available to everyone. Classical music has been brought out of the concert hall and into the home by way of the long-playing phonograph record, the hi-fi, and the stereo tape recorder. The memory of what it was like to hear Beethoven conducting his own compositions was lost forever with the death of the last surviving member of the audience, but in our century we have found a way to preserve for future generations the genius of Toscanini, Stravinsky conducting his own works, and even the songs of the Beatles. One wonders how enjoyable a rock concert would be without the sophisticated electronic amplification equipment that usually fills half the stage.

But there is another side to the story. Science has provided mankind with new standards of health and physical comfort; yet at the same time it has created grave and ominous social problems. As a result of our ability to control disease, we are faced with the problem of the overpopulation of our planet. Uncontrolled industrialization has led to the pollution of our waters and our atmosphere. The privacy of the individual is being eroded through the use of computers and electronic listening devices. Throughout history scientific and technological advances have been employed in the development of more efficient weapons of war—and today weapons are available that can wipe out whole cities in an instant and destroy all life on the earth. Thus, through science, man has devised the means for his own annihilation.

There are many people who believe that science and technology have progressed *too* rapidly, that mankind is unable or unwilling to keep science under control, and that scientists themselves are in a way responsible because they have proceeded with their research without taking into consideration the consequences of their discoveries. These considerations, as well as many other questions that have been raised about the role of science in our society, are extremely complex ones, and if we are to deal with them intelligently, we must have some familiarity with the workings of science.

The British author C. P. Snow has warned of the existence of "two cultures." In his view the intellectual community is divided into two camps—the sciences and the humanities—with very little communication between them. Each side is contemptuous and somewhat distrustful of the other. In the extreme, the scientist looks upon the person in the humanities as one who is totally ignorant (probably as a result of lesser intelligence) of the ways of science. The person in the humanities considers the scientist as a mechanical, uncultured individual with no sense of moral responsibility. These are, of course, extreme stereotypes. Nevertheless, a gap does exist between the two cultures, and it is the responsibility of every educated person to learn how to view the world from both sides of the intellectual fence. For this reason college students who are science majors are encouraged to take courses in literature, history, music, and art, and students who are majoring in liberal arts are often required to take basic courses in the physical, biological, and social sciences.

There are other reasons why the liberal arts student should study science. The scientific method represents a particular philosophical outlook, a certain way of viewing the world. The scientist has an abiding faith in the orderliness of the natural world; rather than accept things as they are, he tends to look for the reasons behind every phenomenon. Such an approach is useful even for the nonscientist, for when a person adopts the scientific outlook, he not only gains some understanding of the natural and mechanical objects of his world, but he also develops a facility for thinking analytically and objectively. In this respect the aims of this textbook are twofold: (1) to present the modern-day picture of the nature of the universe by analyzing in detail the major physical theories and (2) to illustrate, by emphasizing the historical development of these theories, the way in which scientists operate.

We will begin our survey by examining the philosophical underpinnings of the scientific outlook.

1–1

PHILOSOPHICAL FOUNDATIONS

At first glance the philosophical basis of science seems rather straightforward. We observe that we are part of a vast universe that appears to run in a very orderly manner. We describe this order by saying that the universe is governed by a set of principles that we call the "laws of nature." The role of the scientist is to make systematic observations, perform experiments, and formulate hypotheses that will enable him to identify these laws. As scientists refine their observations and experiments, more "laws of nature" are discovered, and in many instances the older laws are overthrown and replaced by more accurate ones. The ultimate goal of this process is to explain every phenomenon in the universe in terms of one law or another.

However, there are several hidden assumptions in this approach. First of all, let us consider how we learn about the world around us. *All our information about the world is obtained through our five senses*—sight, hearing, touch, smell, and taste. Yet we know that our senses are limited. We cannot see infrared or ultraviolet light. We know that dogs can hear high-pitched sounds that humans cannot hear. Most animals have a much more highly developed sense of smell than we do. Thus it is evident that the picture of the external world that our minds create is an incomplete one. Moreover, our senses often provide us with misleading information. There are many common optical illusions that cause us to make incorrect judgments as to the size and shape of, or distance to, a given object. Unusual weather conditions can make us see mirages. We can be fooled very convincingly by magicians and ventriloquists, and some people's senses can be controlled to a considerable degree by a competent hypnotist. The effect on the mind of various hallucinogenic drugs is also well known. Given the limitations of our senses and the imperfections in the way in which our minds interpret sensory information, we cannot be certain that the picture we have of the world corresponds to reality. If we cannot trust the world of our senses, is there any way in which we can determine the nature of the "real" world?

Virtually all of the great philosophers have attempted to answer questions of this nature. Plato believed man could determine the nature of reality by reason alone. The world of our senses is but an approximation to the real world, Plato claimed; hence our reason should take precedence over our observations. Descartes' famous statement *Cogito ergo sum* ("I think, therefore I am") signified his belief that the only thing he could be certain of was the existence of his own consciousness. Our senses, he concluded, cannot supply us with unequivocal knowledge about a world external to the mind. The eighteenth-century philosopher Berkeley went so far as to deny the existence of an external world entirely. In his view reality consisted solely of minds and their ideas. What appear to be objects in an "external world" are only ideas in our minds. Another eighteenth-century philosopher, David Hume, adopted an even more skeptical attitude. He claimed that we can neither affirm nor deny the existence of the external world; he considered it to be a hypothesis that we cannot prove or disprove.

If philosophers tell us that we cannot even be certain of the existence of a world external to our senses, what is the point of conducting scientific inquiry? The scientist, lacking formal proof of the existence of an external world, must simply *assume* its existence. Hence *a belief in the existence of an external world is one of the fundamental assumptions of science.*

Having accepted the existence of an external world, the scientist is faced with a second problem: What is the relationship between the external world and the limited and possibly misleading information provided by our senses? It is clear that *scientists are restricted to the investigation of sensory information,* and they cannot rightfully claim to be describing reality. To be sure, scientists may devise measuring instruments, such as infrared detectors and telescopes, to extend the range and accuracy of their own senses, but the data (meter readings, photographs, etc.) provided by these instruments must also be classified as sensory information; that is, we can learn about the results of our instrumental measurements only through our senses.

Thus we find that *the subject matter of scientific inquiry is sensory information.* The role of the scientist is to devise rational and orderly schemes for describing and correlating the sensory information that enters his consciousness (including sensory information that other observers have obtained and communicated to him). Any scientist is free to speculate as to the connection between sensory information and the "real" world, but in doing so, the scientist is acting as a philosopher rather than a scientist, for his speculations will not be verifiable by the methods of science.

From this point on objects and phenomena in the external world will be spoken about directly, mainly to simplify the terminology. Nevertheless, we should remember that science deals with the sensory information received from external objects and not with external objects themselves.

Another major philosophical problem underlying science involves the apparent *orderliness of the universe.* We observe that events take place in the world in a very systematic way. A heavy object always falls in exactly the same way when it is dropped from a given height. The sun, moon, and stars appear to move in smooth and regularly repeating patterns. We have come to expect the behavior of inanimate objects to be orderly, predictable, and uncapricious. Virtually every effect or phenomenon seems to be attributable to some cause, and in fact, we begin to learn about cause-and-effect relationships even as infants. A baby quickly realizes that if he cries, he will get attention from his mother, or if he touches the oven, he will get burned. As we grow older, we make our way through life by relying on our accumulated knowledge of the patterns of cause and effect that appear in the world.

Philosophers, however, have shown that this approach is not quite as rational as it appears to be. Hume pointed out that a certain event's having occurred with great regularity in the past does not *logically* require that it will occur in the future. In other words, the observation that the sun has risen every day for as long as man can remember does not *prove* that it will rise tomorrow. We can only anticipate the rising of the sun with considerable confidence because we know that this policy has been successful many, many times before without a single failure.

Hume came to similar conclusions with respect to cause and effect. We may believe that event A *caused* event B, but all we are actually certain of is that *event A preceded event B in time.* That is, we do not directly observe a causal action; we see only a sequence of events. We cannot prove with logical rigor that B was a direct consequence of A. Hume recognized, however, that it would be foolish not to accept the idea of cause and effect. He suggested, therefore, that we adopt the concept of causality as a useful procedure for predicting future events, even though we cannot be absolutely certain that the procedure will succeed.

Thus the scientist believes that events occur in the universe in a systematic and orderly way. He believes that every effect has a cause—or in other words, that every phenomenon or event can be explained in terms of some antecedent condition. Moreover, the scientist believes that under identical conditions the same cause will always produce the same effect. These beliefs are based on our observations of the way that the universe has behaved in the past and the assumption that the universe will continue to behave in exactly the same way in the future. It is, then, a basic assumption of science that the universe is—and will continue to be—inherently orderly.

It is of particular interest to the scientist that the orderliness of the universe seems to be mathematically based; that is, many of the cause-and-effect relationships can be stated in the form of mathematical equations. Numerous philosophers and scientists throughout history have come to believe that the universe was *designed* according to a mathematical scheme—a belief that implies that mathematics existed before the physical universe came into being. Others have taken the point of view that man developed mathematics as a means of describing the universe, and that the universe is not inherently mathematical. Questions as to why the universe exhibits a mathematical order, or whether the universe was created or has always existed, or why there is a universe at all are sources of endless philosophical speculation. One could believe, as many scientists do, that the universe was created by a Supreme Being, and that order was imposed on the universe at its creation. But this is also an assumption, which each individual accepts or rejects as a matter of faith.

We see, then, that belief in the existence of a physical universe external to our minds and the belief that this universe is inherently orderly are *fundamental assumptions* of science. Although we cannot demonstrate the truth of these beliefs with any logical finality, we accept them because it is useful and profitable to do so. Without these assumptions scientific research becomes a meaningless enterprise. Having thus established the ground rules, we will now proceed to a description of the way in which scientific research is conducted.

1–2

THE SCIENTIFIC METHOD

Why is it that such diverse disciplines as sociology, biology, and atomic physics can all be lumped under the general title "science"? The connecting link is that they all share the same methodology. Every science selects a particular class of phenomena (such as human behavior or the motions of celestial bodies), attempts to describe these phenomena in a systematic way, and then tries to explain these phenomena in terms of basic principles.

Let us examine this procedure in more detail. The first step in any scientific study is to collect data relevant to the phenomenon under study. This is done either by making careful observations of some natural phenomenon or by performing an experiment in which a phenomenon is studied under predetermined conditions. In an experiment the various controlling factors can usually be isolated and varied at will. For example, if we are studying the motion of a pendulum, we can vary the length of the pendulum, the weight of the pendulum bob, and so on. Consequently, the experimental approach is the preferable way of collecting data. However, if one is interested in the motions of the moon or possibly the sexual habits of Bengal tigers, it is difficult if not impossible to perform an experiment, and one must be content with being a passive observer.

Since scientific relationships are preferably expressed in mathematical form, scientific data should be quantitative rather than qualitative. That is, whenever possible, scientific data should be in the form of numerical measurements of some physical quantity. It is not enough to state that "a heavy object falls when dropped"; such quantities as the amount of time it took the object to fall, the initial distance of the object from the ground, and the size and weight of the object should be specified numerically if the observation is to have scientific value. In many areas of scientific research, however, it is difficult, if not impossible, to make precise quantitative measurements. For example, such terms as "creativity," "anxiety," and "social adjustment" are not easily describable in numerical form. However, it is the goal of scientists to devise numerical scales of measurement for such abstract and difficult-to-define concepts.

It is essential for scientific data to be *reproducible.* That is, a measurement or observation should be made in such a way that any observer can obtain the same results (within the limits of experimental accuracy) under the same conditions. One of the main reasons most scientists have been reluctant to accept the idea of extrasensory perception (ESP) is the almost complete lack of *reproducible* data in support of its existence. Subjects who seem to exhibit ESP ability on one occasion may show no such ability at another time or with a different experimenter. Unless a body of reproducible scientific data has been established, a scientifically meaningful investigation cannot proceed.

SCIENTIFIC LAWS

As the scientist collects more and more data, he begins to look for patterns or regularities in the data. These regularities normally consist of systematic relationships between two or more of the quantities being measured. We might find, for example, that the period (or time for one complete swing) of a pendulum increases as the length of the pendulum increases, or that as we decrease the pressure on a sample of gas maintained at a constant temperature, the volume of the gas increases. If possible, the relationship is stated in precise mathematical form. For example, careful measurements show that when the pressure and volume of a fixed quantity of gas are changed, the product of the pressure (P) and the volume (V) always has the same numerical value. This observation can be described in mathematical form by the equation $PV = k$, where k is a numerical constant.

A relationship of this sort is known as a ***scientific*** or ***natural law.*** A scientific law is a *description* of a systematic natural relationship or recurring natural phenomenon. There is no attempt to give a reason for the relationship, or to explain *why* the phenomenon occurs. A law is simply a statement of *how* nature appears to behave. In the less exact sciences a law may simply describe a general tendency, such as "blood pressure increases with age," or "aspirin reduces pain." However, the scientist's aim is to state a law as an exact mathematical relationship that is rigidly adhered to in the phenomenon under study.

The validity of a law is limited in at least two ways. First, most laws are only valid under a specific set of conditions. (For example, the pressure–volume relationship $PV = k$ does not apply if the gas temperature is not constant, and the law is considerably less accurate at lower temperatures than at higher temperatures.) Second, the accuracy of a law is generally only as good as the accuracy of the data from which it was formulated. Thus the validity of a law may be weakened when more accurate measurements are made, and the law must then be revised or even rejected outright. When this happens, it is sometimes reported that "a law of nature has been overthrown," as if to imply that scientists have somehow managed to circumvent or alter the ways of nature, but in fact the "overthrow" of a law is simply the replacement of a less accurate description of nature with a more accurate one. A scientist considers a law to be a *provisional* statement based on the best data available at the time. Hence the revision of a law on the basis of more accurate data represents scientific progress.

As a case in point, let us consider the development of one of the laws of planetary motion (to be discussed more fully in the following chapter). For over two thousand years it was believed that planets traveled in circular orbits. This belief was based on the most accurate astronomical observations available at the time. Near the end of the sixteenth century more accurate data were obtained, and it was then found that the planetary orbits were elliptical rather than circular. A few centuries later even more accurate observations were made with high-powered telescopes, and it was discovered that planetary orbits were not exactly elliptical. Once again, the laws of planetary motion had to be revised. Thus astronomers have obtained a better and better description of planetary motion. It would not be surprising if future observations (made, perhaps, from orbiting space stations) necessitated further revisions.

In spite of these limitations, scientific laws are of great value in that they enable us to make predictions about the future. Our belief in the orderliness of the universe implies that laws that have been operating in the past will continue to operate in the future. Consequently, we can confidently predict from the law of falling bodies that the next time we hold a heavy object in midair and then let go, it will fall to the ground. We can even use the mathematical formulation of the law to predict the amount of time it takes the body to fall. Similarly, on the basis of past observations, we can formulate a law that will allow us to predict with great accuracy the time and place of the next solar eclipse. As a result, when there is a solar eclipse, men no longer fear (as they did in the past) that the sun has been swallowed by a demon or that the world is coming to an end. A solar eclipse is still one of the most spectacular events in nature, but it loses a great deal of its awesomeness when the time of its occurrence can be predicted to the second. In this respect the ability to make predictions about natural phenomena is of great benefit to us. Man no longer lives in fear of the natural world; he is able to predict its behavior and ultimately to control and change it.

SCIENTIFIC THEORIES

We have learned that scientific laws attempt to describe *how* nature behaves under certain conditions. However, the scientist wants to have more than a mere description of nature; he would like to know *why* nature behaves the way it does. For this purpose scientists propose ***theories.*** The formulation of a scientific theory is the final and most important part of the scientific method.

Basically, a scientific theory represents an attempt to explain and correlate natural laws in terms of fundamental postulates or assumptions. The postulates usually involve abstract concepts such as "gravity," "inertia," or "energy," which are themselves unexplained. Very often a theory is based on a ***model*** or representation of nature that provides a useful and convenient scheme for visualizing and describing natural phenomena. For example, the atomic theory is based on a model in which all matter is assumed to be composed of minute, distinct particles called atoms. Freud's theory of psychoanalysis contains a model in which the human mind is divided into three parts—the id, ego, and superego. In each of these theories the model is a basic assumption; thus we simply accept as given the existence of atoms (or egos) without attempting to prove their existence on the basis of observations.

Starting from the basic assumptions of a theory, the scientist then deduces conclusions using strict logical or mathematical reasoning. These conclusions are stated in the form of scientific laws. As we have learned, laws are descriptions of how nature behaves under certain conditions, and they can be tested by experimentation and observation. Obviously, if the laws that are derived from some theory are incorrect, the theory itself is incorrect, but if the laws predicted by a theory are found to be valid, the theory is a valid one. We see, therefore, that the validity of a theory depends on the validity of the scientific laws that are deduced from that theory.

In short, *a scientific theory is a conceptual scheme or model through which one can explain and deduce scientific laws.* To clarify this definition, let us consider the example of Newton's theory of gravitation. Newton proposed the concept of a universal force of gravity and deduced a mathematical formula for the gravitational force between any two bodies. Using basic assumptions about gravitation and the relationship between forces and motion, Newton was able to derive the law of falling bodies and the laws of planetary motion, and to account for ocean tides, the shape of the earth, and many other phenomena. Years later other scientists used Newton's principles to *predict* the existence and location of the planets Neptune and Pluto.

Thus Newton accounted for many scientific laws and phenomena in terms of the concept of gravitation. Yet Newton himself did not understand the cause of gravity. To the present day, physicists have no idea how or why bodies exert gravitational forces on one another, so the nature and cause of gravity is still not known. In this light, one may question whether the introduction of the concept of gravity has really increased our comprehension of nature. If a phenomenon is accounted for in terms of gravity when gravity itself is not understood, have we truly "explained" the phenomenon? In this sense the statement that "bodies fall because of gravity" really tells us nothing at all.

Nevertheless, the theory of gravitation is considered a highly successful and valuable theory because it relates a wide range of seemingly unconnected phenomena to a single concept. Moreover, many previously unknown and unexpected phenomena (such as the existence of Neptune and Pluto) have been predicted from the theory. Obviously, a certain level of understanding has been achieved, and even though the theory is incomplete, it must be regarded as a step in the right direction. To be sure, a *deeper* level of understanding would be attained if gravity itself could be explained in more fundamental terms. We might regard the lack of such an explanation as a basic shortcoming of science, but we should recognize that science is an unfinished business, which progresses as each generation of scientists builds on the work of previous generations. If we view science in this historical perspective, we come to realize that it is unfair to expect the scientists of today to possess all the answers. It is a task for future generations of scientists to improve on the present-day theory of gravitation by explaining why gravity exists.

On what grounds do we judge the merits of a scientific theory? As has already been noted, the validity of a scientific theory stands or falls on the validity of the theory's predictions. In fact, the main reason for the existence of a scientific theory is the prediction of scientific laws. Hence the ***predictive power*** of a theory is perhaps its most important feature. A theory's predictions (i.e., the scientific laws that can be rigorously deduced from the theory's basic assumptions) must, of course, be *testable*. That is, if a theory is used to predict the results of a certain experiment, that experiment ought to be one that can be performed fairly readily. A theory should not make predictions that cannot possibly be measured, for such predictions are scientifically meaningless. It is through a theory's predictions that its validity is measured; as more and more of its predictions are tested and found to be correct, the more solidly established the theory becomes. However, *a theory can never be considered true or completely verified* because there is always the possibility that the discovery of a new phenomenon or the performing of an experiment to greater precision will provide evidence that contradicts the predictions of the theory. In that case the theory is invalid, and it must be modified or even thrown out entirely.

A second criterion for judging a theory is its ***comprehensiveness.*** A comprehensive theory is one that can account for a wide range of phenomena; in other words, the more scientific laws that can be deduced from a theory, the more comprehensive the theory is. One of the basic goals of science is to construct more comprehensive theories and thereby reduce the total number of different or unrelated theories needed to account for all natural phenomena. For example, in the nineteenth century James Clerk Maxwell devised a theory from which he could show that electricity and magnetism are interrelated and that light is an electromagnetic phenomenon. Thus three separate theories for electricity, magnetism, and light—each with its own set of assumptions—were replaced by a single, comprehensive theory that accounted for all three phenomena in terms of a single set of assumptions. For many years Albert Einstein tried to find a way to incorporate gravitation and electromagnetism in a single theory. He was, unfortunately, not successful, but other theoreticians are attempting to continue his work. It is conceivable that scientists may eventually attain the ultimate theory, which would account for all known scientific laws in terms of the fewest possible assumptions and concepts.

Another measure of a scientific theory is its ***simplicity.*** We would like a theory to have a minimum number of basic assumptions, and these assumptions should be as uncomplicated as possible. The concepts and mathematical relationships of a scientific theory should not be so abstruse that only the most brilliant scientists and mathematicians can manipulate them. We should recognize, however, that the idea of simplicity is fundamentally a human aesthetic judgment. Scientists have adopted this criterion because of a philosophical belief that the universe ought to be governed by simple principles—simple, that is, in human terms. Unfortunately, the universe has not always proved to be so cooperative. On numerous occasions in the history of science, the experimental and observational evidence has necessitated the rejection of a simple theory in favor of one that is philosophically and mathematically more complex. Nevertheless, as a general rule, it is useful to adopt the principle that *if two theories are equal in all other respects, the simpler theory is adopted.*

Most scientists believe that these three criteria—***predictive power, comprehensiveness,*** and ***simplicity***—generally provide the best objective basis for judging the merits of a scientific theory. Yet sometimes a theory may fail to gain acceptance even though it meets all of these criteria. Scientists may reject a theory for nonscientific reasons—for instance, a theory may be incompatible with prevailing philosophical or religious beliefs. The theory of evolution and the idea that the earth is several billion years old are perhaps the most notable examples of scientific theories that were rejected at first (and are still not completely accepted today) because they conflicted with long-held religious tenets. In fact, very rarely has a radically new scientific theory won immediate acceptance. The weight of authority or tradition behind a long-established theory has often led scientists to ignore or discredit contrary experimental evidence and to reject out of hand a better theory, even though it meets all the objective criteria. Very often the acceptance of a new theory must await the arrival of a new generation of scientists, who are more willing to reject the old order. Like other members of the human species, scientists are not always objective, and they can sometimes display a remarkable reluctance to accept new ideas.

Let us now summarize the essential aspects of the scientific method. (See Figure 1-1.) There are three basic steps: (1) *the collection of scientific data* through observation and experimentation, (2) *the formulation of scientific laws,* which describe systematic (preferably mathematical) relationships between physical quantities and can be used for the prediction of future events, and (3) the *formulation of scientific theories,* which are conceptual models or systems from which one can explain or deduce scientific laws.

The discussion so far of the scientific method may seem somewhat abstract and academic. Certainly, without a background in science, it is not easy to comprehend

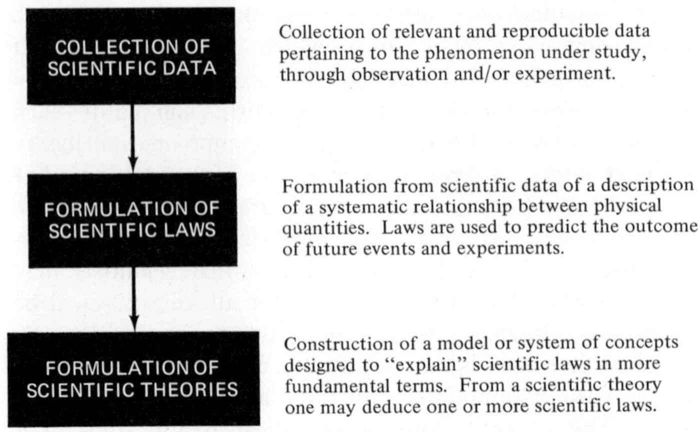

figure 1–1 **AN OUTLINE OF THE SCIENTIFIC METHOD**

fully this procedure. But the scientific method will be referred to often as we progress through the text, and specific historical examples illustrating the progression from data to laws to theories will be cited. If the reader studies these examples, he should eventually develop an understanding and appreciation of the scientific method and its applications.

1–3

THE SCIENTIST AS ARTIST

The description that has been given of the scientific method is incomplete and perhaps even misleading, because it tells us little about the way in which a scientist goes about discovering a law or formulating a theory. In fact, the methodology of scientific research cannot be reduced to a precise set of rules. How does the scientist know what assumptions to make in order to formulate a fruitful theory? Out of all the phenomena in nature, how does a scientist select one that is worth studying? How does he know which are the important quantities to measure? Since he cannot examine all of the possible relationships, he must rely on his accumulated knowledge and his intuition to help him find the best way to solve a particular problem. Inspiration and luck very often play a part; as we shall learn, some of the most important scientific advances have been the result of accidental discoveries. In short, scientific research is something of an art form, in which the scientist's creative talents play a significant role.

This image of the scientist as an artist is worth considering, for the scientist's goals and methods seem, at first glance, to be completely unrelated to those of the artist. The scientist's primary goal is to gain knowledge about nature and describe its workings in as simple a manner as possible. On the other hand, the artist's purpose is to get across a certain idea in a forceful manner, to evoke in others a certain emotional response, or simply to satisfy a need for self-expression. Unlike the scientist, the artist is not confined to the methodical description of the physical world. Yet the artist does operate under certain constraints. Traditionally, we have included among the arts such fields as music, dance, the visual arts (painting, sculpture, architecture, etc.), and various forms of literary expression such as prose, poetry, and drama. In each of these art forms a particular medium is stipulated (oil paint on canvas, the sounds made by specific musical instruments, the English language, etc.), and a set of rules is defined that determines how each medium is to be used. Thus we find, for example, that a classical symphony has a specific form and uses a particular type of musical scale or that only certain body movements are allowed in a classical ballet. The rules for each art form are flexible, of course; they vary from culture to culture and from century to century. Furthermore, these rules do not represent severe limitations on the use of the particular medium, so that the artist is allowed considerable freedom of choice in creating a work. The range of possibilities with each medium is so vast that the artist is free to make his choices in an intuitive and highly personal way. The notes of a musical scale can be arranged in an infinite number of different sequences. Out of all these possibilities, what made Beethoven choose the particular sequence and combination of notes that constitutes the Eroica Symphony? Paint can be placed on canvas in an infinite number of ways. Why and how did Picasso choose the particular set of images that he used to express his feeling toward war in "Guernica"? This is the mystery of the creative process, and it is the essence of art.

As we proceed with our study of the historical development of the great scientific theories, we will discover that this same creative process is decidedly at work in science. The scientist's medium—his paint on canvas, so to speak—is the scientific data that have been collected over the years through observation and experiment. The primary rule governing this art form is that the scientist must find some order in the available data. The scientist does not have quite the same freedom to create as does the artist, but there is nevertheless a certain element of choice in the way that the scientist proceeds toward his goal. As a consequence, each scientific theory reflects to some degree the philosophy, the aesthetic preferences, and even the personality of its author. The scientist does not merely discover that which is already inherent in nature; he creates, in a distinctly personal way, fundamental concepts that can be used to order our experiences and to simplify our description of the physical world. In this respect the formulation of a scientific theory is a creative act of the highest order. It is as incorrect to say that Einstein discovered the theory of relativity or that Freud discovered the subconscious as it is to say that Beethoven discovered the theme of the Eroica Symphony. Rather, each of these "discoveries" was a creation of a unique human intellect.

In certain ways the aims of the artist and those of the scientist are the same. Just as an artist may strive for beauty or harmony in his creative efforts, a scientist often takes into consideration such artistic criteria as symmetry, uniformity, and simplicity in the formulation of a theory. It is not unusual to hear a scientist speak of the beauty or elegance of a particular theory. Moreover, we should recognize that the need for self-expression exists for the scientist as well as for the artist. We tend to think of artistic endeavor as an emotional experience and of scientific research as a rational, objective, intellectual exercise. But imagine the joy and exhilaration that a scientist must feel when in a sudden flash of intuition he recognizes the key principle that will enable him to construct a successful theory. Albert Einstein once described his reaction when he realized that he had successfully completed the general theory of relativity: "I was beside myself with ecstasy for days." The solution of a scientific problem can be an immensely rewarding emotional experience. Thus despite significant differences in purpose and outlook, the artist and the scientist have much in common.

1–4

THE REALM OF THE SCIENTIST

As we have noted previously, the various fields of science are united in that they use the same methodology, even though they are concerned with different kinds of phenomena. Let us examine this idea more closely.

All of the various scientific disciplines can be classified into three main groups: the *physical sciences,* the *biological sciences,* and the *social sciences.* The physical sciences are concerned with the nature of matter. *Physics* is the most fundamental of all the sciences; the physicist attempts to understand the basic structure of all matter and to describe the various ways in which matter interacts with other matter. *Chemistry* deals specifically with the atom, but with a special viewpoint. The chemist is interested in the interactions between different atoms and in the properties of the substances that result from these interactions. (The study of the structure of the atom itself is within the province of the physicist and goes by the name of atomic physics.) *Geology* and *astronomy* are specialized physical sciences in which the basic principles of physics and chemistry are used to gain an understanding of the nature of the earth and of celestial bodies in general.

Traditionally, the biological sciences have been concerned with the nature of living things. Until fairly recently the cell was considered the fundamental biological unit, and scientists generally assumed that there was an essential difference between living and nonliving matter. Within the past few decades, however, there has been a dramatic breakdown of the boundary between the biological and the physical sciences. As a result of developments in the new fields of biochemistry and biophysics, the frontiers of biology have advanced beyond the cellular to the molecular level. Biologists have found that the DNA molecule seems to be the key to understanding the chemical basis for the transfer of genetic information in reproduction. Moreover, it seems likely that before long simple forms of life will be synthesized in the laboratory. It appears, then, that it is possible to understand living matter in terms of the principles of physics and chemistry.

The social sciences focus on man. *Psychology* studies the behavior of the individual as he responds to the stimuli of his environment. *Sociology* is concerned with the nature of group behavior and the development of social institutions. *Anthropology* deals with mankind as a whole; it compares the physical characteristics, geographical distribution, and cultural history of the various races of man.

It is a commonly held impression that the social sciences are "less scientific" than the biological and physical sciences. It is true that sociological theories do not have the mathematical rigor or universal applicability of atomic theory, but atoms are more amenable to measurement and experimentation than are human beings. The *method* of the sociologist is identical to that of the atomic physicist in that both are attempting to proceed from observations to laws to theories. The physicist is more successful in this venture because of the nature of the object of his study, but this does not make the sociologist "less scientific," so long as he is not misusing the scientific method. Thus in the application of the scientific method to a variety of subjects lies the unity of the sciences.

In our catalog of the various fields of science, we have thus far neglected to mention the study of *mathematics.* The reason for this omission is that mathematics

THE NATURE OF

PHYSICS

is not a science in the strictest sense. In particular, mathematics does not deal with observational data (i.e., data obtained through the senses) but with precisely defined abstractions such as points, straight lines, and circles. Conclusions about these abstractions are arrived at by rigorous deduction from certain basic definitions and axioms. Mathematical "laws" are different from scientific laws in that the former are considered to be rigorously true (if they follow logically from the basic definitions and axioms), whereas the latter are only approximations subject to considerations of accuracy and further experimentation. Thus mathematics is more properly classified as a form of logic, rather than as a field of science. Mathematics is, of course, the indispensable tool of the scientist, and the two fields have always been closely interrelated.

A distinction is commonly made between *science* and *technology*. Technology is concerned primarily with the practical applications of scientific laws and theories. Hence the electric motor, the transistor radio, color television, and so on are not scientific discoveries but rather technological advances, which were developed as useful consequences of the laws of electromagnetism. The basic principles on which such devices operate are deduced from well-established scientific laws, and the major problem in developing such devices is usually one of finding the proper materials and devising suitable design and construction techniques. For example, Newton predicted the possibility of orbiting space satellites in the late seventeenth century, but three hundred years elapsed before the first satellite was successfully launched. The scientific theory of space travel was available, but the technology was not; it was necessary to develop fuels that would supply sufficient energy, metal alloys that would withstand the high temperatures, and high-speed computers to perform the orbit calculations. Thus the feat of landing men on the moon is looked upon as a technological accomplishment rather than as a basic scientific advance.

1–5

THE STUDY OF SCIENCE: A PERSPECTIVE

We are now ready to embark on our study of the development of the modern-day picture of the physical universe. Beginning with ancient times and proceeding up to the present, we will trace the origin and evolution of the most significant physical laws and theories. In following this historical approach, we will be devoting some of our effort to the study of theories that are now considered false. What is the point of studying incorrect theories? The purpose is to emphasize the idea that present-day science is by no means complete or even correct. In every age the scientists of the time were convinced of the validity of theories that were later shown to be false—why should our age be any different? Viewing science from this historical perspective, we sense its organic nature. Science grows, then decays, then grows again, seemingly approaching, yet never attaining, perfect harmony. Authoritarianism and absolutism are the greatest deterrents to scientific progress; science grows best from the seeds of doubt and disbelief. It is hoped that the reader will be encouraged to develop a healthy skepticism toward authoritarian scientific (and nonscientific) pronouncements, and that he will recognize clearly that the modern-day body of scientific laws and theories represents a temporary stage of development subject to further correction and modification.

Through the historical approach one achieves an appreciation of science as a human intellectual endeavor. Because science has been developed by man, it is constricted by the limitations of human perception and intelligence. In learning about the lives and personalities of the great scientists, we realize that scientists are subject to all the human frailties; the false starts, the errors, the failures of vision of scientists throughout the ages are understandable in this perspective. Thus we will learn how science is "done." We will come to see it as a never-ending intellectual drama, rather than as a mere collection of formulas and problem-solving techniques.

QUESTIONS FOR REVIEW

1 Why do we say that a belief in the existence of a "real," or external, world (i.e., a world external to our senses) is a fundamental assumption of science?

2 (a) What is meant by the "orderliness of the universe"?
(b) How do we know that the universe will continue to be orderly in the future?

3 Outline the basic steps of the scientific method.

4 (a) What is a scientific law?
(b) What factors limit the validity of a scientific law?
(c) How is a scientific law overthrown?
(d) What is the value of a scientific law?

5 (a) What is a scientific theory?
(b) What is the purpose of a scientific theory?
(c) How is the validity of a scientific theory tested?
(d) Why is a scientific theory never considered "true," or finally proved?

6 (a) What are the basic criteria for judging the merits of a scientific theory?
(b) Which of these criteria is the most significant to the scientist?

7 Explain why the study of mathematics is not based on the scientific method.

8 What is the difference between science and technology?

EXERCISES

1 Look up the word "metaphysics" in a dictionary or encyclopedia and determine what it means. How does metaphysics differ from physics?

2 A solipsist is a person who believes that he alone exists and everything else in the world is a figment of his imagination. How would you prove to a solipsist that his belief is false?

3 (a) Rank the five human senses in order of their importance—that is, in terms of the value and amount of the sensory information they provide.
(b) Can you think of any way that a person can gain knowledge without using his five senses, directly or indirectly?

4 What did Albert Einstein mean when he said, "The eternal mystery of the world is its comprehensibility"?

5 How would you go about collecting a body of scientific data that would convince a skeptical scientist that some people possess extrasensory perception?

6 Why would most scientists reject the idea that "flying saucers" or UFOs exist and come from other planets?

7 Do you think animals are able to formulate scientific laws? Give specific examples.

8 (a) How does a scientific law differ from a judicial law?
(b) Can you think of any reasons why the same word ("law") is used to identify both of these entities?

9 What is the distinction between a hypothesis and a theory?

10 The presidents of the United States who were elected in the years 1840, 1860, 1880, 1900, 1920, 1940, and 1960 all died in office.
(a) Do you think that a *scientific* law can be formulated on the basis of these data?
(b) Is it reasonable to predict that whoever is elected president in 1980 will also die in office?

11 Indicate whether each of the following statements represents a law or a theory.
(a) The sun rises and sets every day.
(b) The universe began in a gigantic explosion about twenty billion years ago.
(c) Man is descended from the apes.
(d) When a quantity of gas is compressed, it becomes hotter.
(e) Vitamin C prevents colds.

12 The apparent brightness of a certain star, as seen from the earth, is observed over a period of time and is found to vary as shown in Figure 1–2.
(a) Formulate a scientific law based on these observations.
(b) Formulate one or more scientific theories designed to explain these observations.

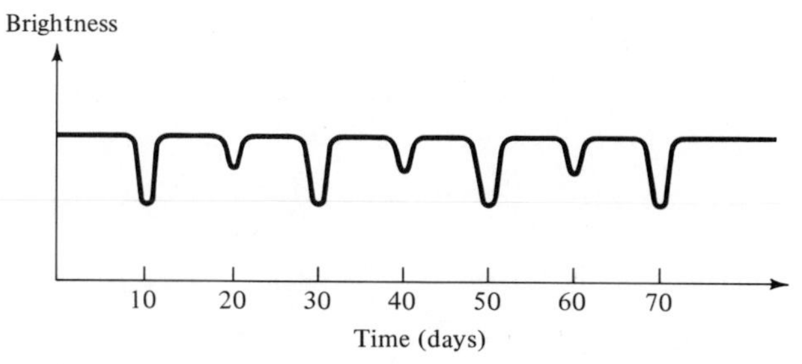

figure 1–2

13 Discuss whether or not the study of history can be considered a science. That is, does the historian use the scientific method? Is it possible to perform experiments in this field?

14 (a) Does the statement "1 + 1 = 2" represent an observation, a definition, a law, or a theory?
(b) Do you think it is possible to disprove this statement?

15 Einstein once said that "the present position of science can have no lasting significance."
(a) What did Einstein mean by this remark?
(b) Do you think it is completely justifiable?

One

FROM THE CAVE TO THE RENAISSANCE

COURTESY HALE OBSERVATORIES

The planet Saturn

The ancient astronomers identified Saturn as one of the five "wandering stars" or planets visible to the naked eye. The invention of the telescope in the seventeenth century led to the discovery of the magnificent and unique ring system around Saturn. Modern-day astronomers believe that the rings are composed of swarms of ice-covered rocks in orbit around the planet.

ASTRONOMY: THE FIRST SCIENCE

2

The heavens themselves, the planets and this centre,
Observe degree, priority and place,
Insisture, course, proportion, season, form,
Office and custom, in all line of order.

William Shakespeare
(*Troilus and Cressida*)

SCIENCE BEGAN with the first prehistoric men who took time out from their hunting and food gathering to wonder about the nature of the world around them. Probably, the first objects of human intellectual curiosity were the sun, moon, stars, and other celestial bodies, which proceeded in such regular and predictable patterns across the sky. The alternation of day and night, the changing of the seasons, the phases of the moon were events of great significance to prehistoric man, for the heavenly bodies determined whether he would have light or darkness, warmth or cold, and even whether he would have sufficient food to eat. Every aspect of human existence seemed to be controlled by the heavens. It is easy to understand why the celestial bodies were regarded as objects of religious worship in most of the early civilizations. It was of vital importance to record and study the motions of the celestial bodies so that their behavior could be predicted and understood.

Thus the science of astronomy was born. Over the many thousands of years of human history and prehistory, men have continued to study and speculate about the nature of the celestial realm in an attempt to understand the universe and how it relates to our own existence. Even to the present day, astronomy remains the area of science that holds the greatest interest and appeal for the nonscientist. As such, it provides a most excellent and appropriate starting point for our study of physical science.

2–1

BASIC ASTRONOMICAL OBSERVATIONS

Since the method of science is to proceed from data to laws and then to theories, we will begin our study of astronomy by describing the basic data pertaining to the motions of the prominent celestial bodies. Any person on the earth can observe these motions by carefully studying the sky over an extended period of time, using the naked eye and perhaps a few simple measuring instruments. The purpose here is simply to *describe* how the celestial bodies appear to move. There will be no attempt to *explain* these motions at first.

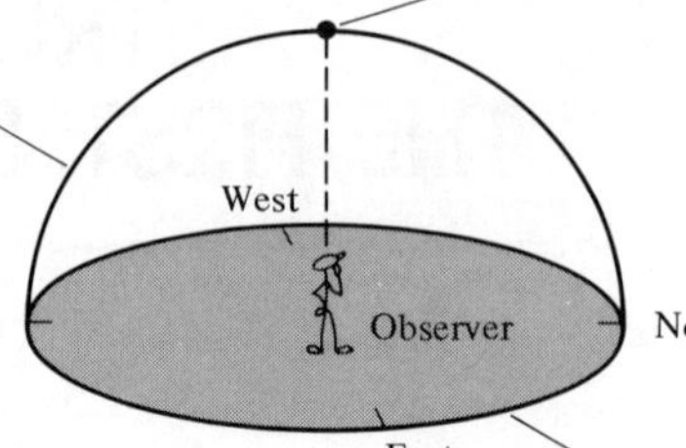

figure 2–1 **REFERENCE FRAME FOR OBSERVING THE SKY**

The observer sees the sky as a hemispherical dome covering a flat earth.

THE DIURNAL MOTION OF THE SKY

To an observer on the earth, the sky appears to be a great hemispherical bowl. The line where the sky meets the earth is called the ***horizon;*** the celestial bodies "rise" or "set" when they cross this line, either entering or leaving our field of view. We mark positions on the horizon by referring to the directions north, east, south, and west. The point in the sky that is directly overhead is called the ***zenith,*** and the imaginary line drawn from the north point on the horizon through the zenith to the south point is called the ***meridian.*** (See Figure 2–1.)

The first and most basic observation we can make is that the motions of the celestial bodies are cyclic. The most obvious cycle is, of course, the alternation of day and night which results from the motion of the sun. We observe that the day begins as the sun comes above the horizon in the east. The sun climbs higher in the sky, crosses the meridian, and begins to sink in the west. As the sun descends below the western horizon, the sky darkens and the stars become visible. During the night we see that the stars also appear to move from east to west across the sky. After a while the sky begins to brighten; the sun reappears on the eastern horizon; and the whole cycle begins again. This circling of the celestial bodies over a one-day period is referred to as a ***diurnal*** (daily) ***motion.***

When we examine the diurnal motions of the stars, we find that each star travels in a perfectly circular path across the sky (Figure 2–2). Some stars travel in large circles, some in small circles, and one star, Polaris, appears not to move at all. Polaris, which is sometimes called the ***pole star,*** is located at the center of all the circles in which the stars travel. Thus all the stars appear to trace out concentric circles about Polaris. (Actually, Polaris is about 1° away from the center. When seen in a telescope, Polaris itself traces out a small circle about the center. To the naked eye, however, Polaris appears stationary.)

The location of Polaris in the sky depends on the observer's position on the earth. A person standing at the north pole would see Polaris directly overhead at the zenith. At the north pole the circular paths that the stars trace out are parallel to the horizon (Figure 2–3a), so the stars never appear to rise or set.

A person looking at the sky from the latitude of New York would find that

Polaris is located at an angle of about 40° above the north point on the horizon. (In fact, the system of latitude used to locate points on the earth is based on the position of Polaris. The observer's latitude is equal to the angle that Polaris makes with the north point on the horizon.) The stars still appear to circle about Polaris, but most of them rise in the east, move southward until they cross the meridian, and then move back northward until they set in the west. (See Figure 2–3b.) A few stars (those within 40° of Polaris) never cross the horizon as they circle and remain in the sky continually. These are referred to as ***circumpolar stars.***

As we move farther south toward the equator, we find that Polaris appears lower

figure 2–2 **DIURNAL MOTION OF THE STARS**

The photograph shows a one-hour time exposure of the sky in the vicinity of Polaris. The trails of light show the circular motions of the stars during this time.

YERKES OBSERVATORY PHOTOGRAPH

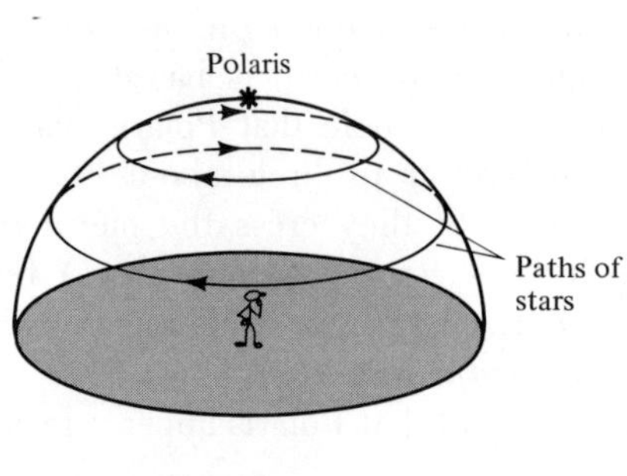

(a)

The north pole. Polaris is at the zenith. The stars circle around the sky, never rising or setting.

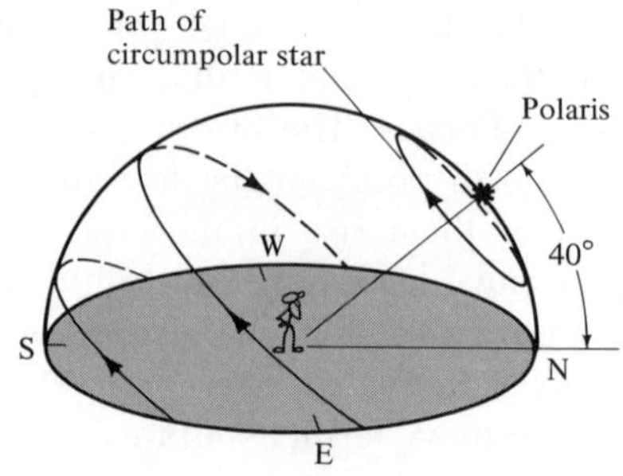

(b)

New York (latitude 40°N). Polaris is 40° above the north point on the horizon.

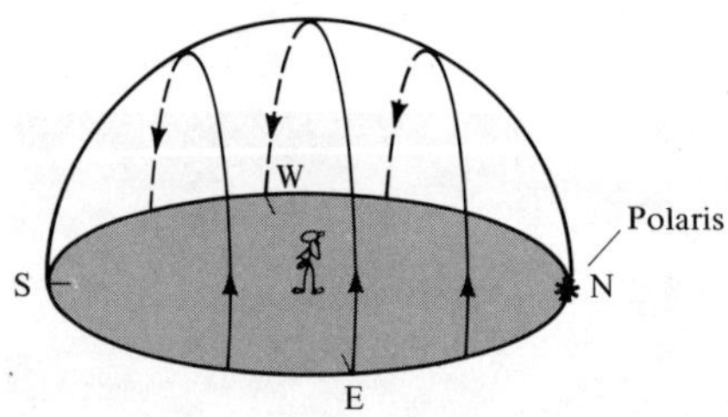

(c)

The equator. Polaris is on the northern horizon. All stars rise and set.

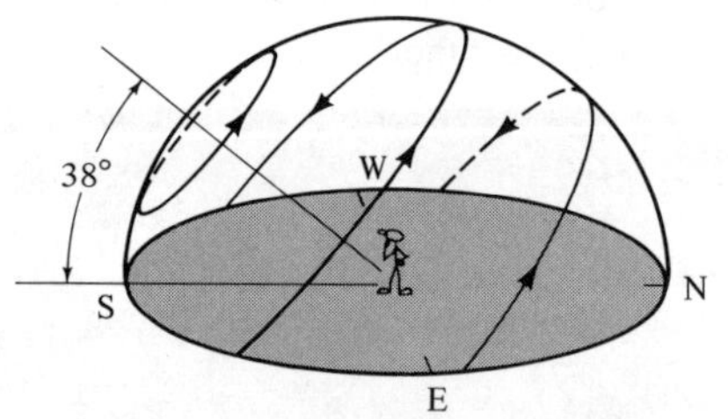

(d)

Melbourne, Australia (latitude 38°S). The stars appear to rotate about a point 38° above the southern horizon. Polaris is below the northern horizon and is not visible.

figure 2–3 **DIURNAL MOTION OF THE STARS AS SEEN FROM VARIOUS POINTS ON THE EARTH**

and lower in the sky. If we observed the sky from a point directly on the equator (latitude 0°), Polaris would be exactly on the horizon, and it would be difficult, if not impossible, to see (Figure 2–3c). At the equator every star appears to rise and set; the paths of the stars are vertical circles perpendicular to the horizon.

Polaris cannot be seen by anyone in the southern hemisphere. However, the stars appear to travel in concentric circles about a point in the sky directly over the *south* point on the horizon, as shown in Figure 2–3d. However, there are no bright stars located near the center of these circles, so there is no pole star for the southern hemisphere. An observer at the south pole finds a situation similar to that seen at the north pole; namely, the stars also appear to travel in concentric circles parallel to the horizon, never rising or setting. They appear, however, to rotate in a clockwise direction; at the north pole the stars rotate counterclockwise. Also, observers at the two poles of the earth see different stars, for the stars seen at the south pole are not visible at the north pole and vice versa.

One important aspect of the stars is that they are "fixed"; that is, they show no significant motion *relative to one another.* All stars exhibit diurnal motion, but the distances between the stars do not change as the stars move across the sky. Hence the arrangement of the stars in the sky is essentially the same today as it was five thousand years ago. Groups of the brighter stars form easily recognizable patterns called ***constellations.*** Even the city dweller, who must look at the stars through a smoky, hazy, and generally polluted atmosphere, can become familiar with the Big Dipper (the constellation Ursa Major), the prominent winter constellation Orion, or the W arrangement of the stars of Cassiopeia.

THE MOTION OF THE MOON

Another cyclic motion to be seen in the sky is that of the moon. The moon undergoes a periodic change in appearance. It is said to progress through a series of ***phases*** as it changes gradually from a crescent to a half to a full moon, shrinks back to a crescent, eventually disappears altogether, and then reappears as a crescent. The complete cycle of phases takes about $29\frac{1}{2}$ days and is the basis for the unit of time known as the ***month.*** During this cycle the moon also appears to move relative to the fixed stars. On a given night the moon will rise in the east, move across the sky, and set in the west, but the moon appears to move more slowly than the stars and gradually lags behind them. On the next night it will rise about 50 minutes (on the average) later than it did the night before, and it will have shifted about 13° relative to the background of the stars.

THE MOTION OF THE SUN

The sun also exhibits a motion relative to the fixed stars in addition to its diurnal motion. One can observe that with each successive sunrise there is a shift of slightly less than 1° in the position of the sun relative to the fixed stars. Like the moon, the sun appears to rotate around the sky at a slightly slower rate than the stars do, lagging behind the stars by nearly 1° each day. This shift in the sun's position is in a direction *opposite* to its diurnal motion. That is, the sun's diurnal motion is *westward* (from east to west), but the motion of the sun relative to the stars is *eastward* (from west to east). As a result, at each successive sunrise a given star appears 1° closer to the western horizon (Figure 2–4). In this manner the sun slowly drifts through the fixed stars. After $365\frac{1}{4}$ days the sun will have returned to its original position

figure 2–4 **THE EASTWARD MOTION OF THE SUN**

The sun drifts eastward 1° per day with respect to the fixed stars. Hence a given star will appear 1° farther west of the sun on each succeeding day.

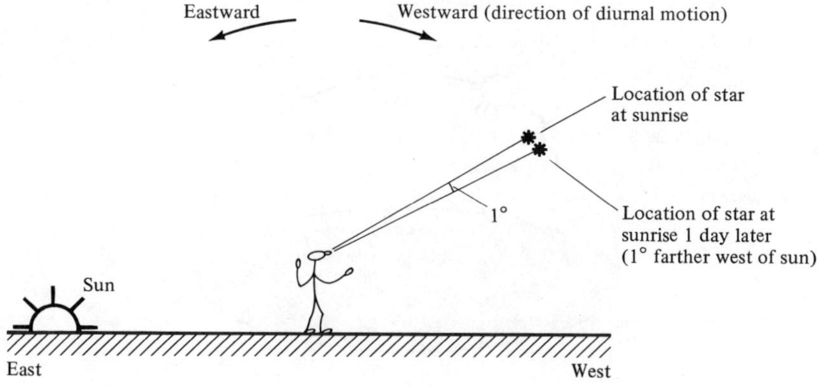

relative to the fixed stars. This cycle is the basis for the unit of time called the ***year,*** and is referred to as the ***annual motion*** of the sun.

Of course, the stars cannot be seen in the daytime, so the motion of the sun through the stars is not immediately obvious to the casual observer. Over a period of time, however, one notices the gradual shifting of the stars and constellations that appear in the night sky. A star that rises at 9 P.M. today will, after several months, rise at 7 P.M. A certain star may not be visible because it rises with the sun and is in the sky during the daytime. In six months, however, the star will be 180° away from the sun because of the sun's motion with respect to the stars. Instead of rising *with* the sun, the star will rise as the sun *sets* and will be visible throughout the night. Thus there are winter constellations and summer constellations; those constellations that are in the night sky during the winter are not visible in the summer because by that time the sun has moved into their part of the sky.

An additional, and considerably more important, aspect of the sun's annual motion is its drift in a *north–south* direction with respect to the stars. In addition to shifting eastward by 1° each day, the sun also shifts northward or southward by about $\frac{1}{4}$° (on the average) relative to the fixed stars. For half the year—from December to June—the sun moves steadily northward toward Polaris. On about June 22 (the exact date may vary by a day or two), the sun is closest to Polaris; it is 66.5° away from Polaris in the sky. At this point the sun reverses direction and begins to drift southward. From June to December the sun moves gradually farther away from Polaris. The sun is farthest from Polaris—113.5°—on about December 22. On this date the motion of the sun again reverses, and the sun begins to drift northward. Note that from June to December the sun shifts its north–south position by 47° (= 113.5° − 66.5°), which is slightly more than one-quarter of the distance from one horizon to the other. (See Figure 2–5.)

The effect of this aspect of the sun's annual motion is that at any given point

figure 2–5 **THE ANNUAL NORTH–SOUTH MOTION OF THE SUN**

The change in position of the noonday sun from June to December is shown. The sun is nearest to Polaris on June 22 and farthest from Polaris on Dec. 22. The position of Polaris in the sky depends upon the observer's latitude.

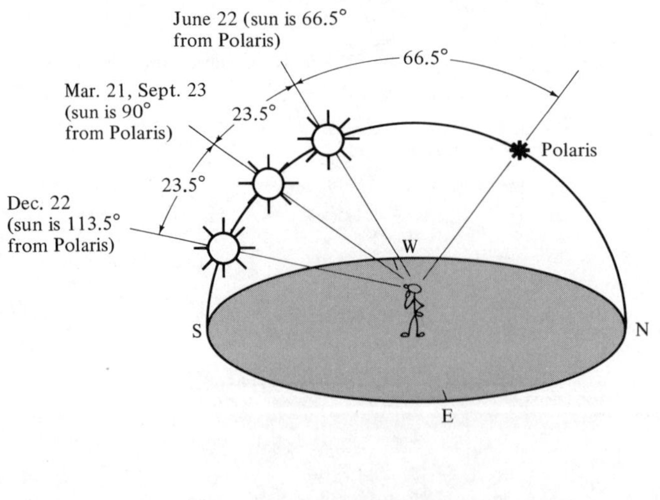

on the earth (1) the length of time that the sun is in the sky and (2) the altitude of the sun in the sky will vary over a period of $365\frac{1}{4}$ days. Since the sun is our source of heat and light, these changes in the sun's position are accompanied by changes in the amount of sunlight we receive, and our climate is affected accordingly. Thus the seasonal changes that we experience are related to the sun's annual motion.

Figure 2–6a shows how the sun's path across the sky varies over the year for an observer in New York. (Remember that Polaris is always 40° above the north point on the horizon for this observer.) On December 22 the sun is farthest (113.5°) from Polaris; at 12 Noon (when the sun is on the meridian and at its highest point in the sky), the sun is only 26.5° above the southern horizon. On this date the sun rises in the southeast, sets in the southwest, and is visible for only about 9 hours. As the sun spirals northward, each successive day finds the sun a little higher in the sky, and the number of daylight hours increases. On March 21 the sun is exactly 90° from Polaris; the sun rises due east, sets due west, and is in the sky for exactly 12 hours. On June 22, when the sun is closest to Polaris, the sun reaches its maximum altitude in the sky—73.5° above the southern horizon—and there are about 15 hours of daylight. From this day onward the sun spirals southward; its path across the sky gets lower and lower, and the amount of daylight decreases. On September 23 the sun is back to the same position it was in on March 21—namely, 90° away from Polaris. The sun continues to spiral lower and lower until about December 22, when it reaches its low point. The sun ends its southward spiral on this day, and the annual cycle begins anew.

The pattern of the sun's motion as seen by an observer at the equator is illustrated in Figure 2–6b. Here Polaris is located exactly on the northern horizon, so the sun is directly overhead when it is 90° from Polaris. This occurs at 12 Noon on March 21 and September 23. Note that the midday sun is never low in the sky at the equator; as a result, there are not the extremes of temperature that one experiences at higher latitudes, and there are no significant seasonal changes. There are 12 hours of daylight and 12 hours of night on each day of the year. Polaris cannot be seen from the southern hemisphere; it is always below the northern horizon. Consequently, the sun appears to get lower and lower in the sky from December to June as it drifts northward toward Polaris. The sun is lowest in the sky on June 22 and highest on December 22. Hence in the southern hemisphere the seasons are reversed; winter begins in June and summer begins in December. Figure 2–6c illustrates the sun's annual motion as seen in Melbourne, Australia (latitude 38°S).

At the poles there are six consecutive months of daylight followed by six consecutive months of darkness. At the north pole Polaris is seen directly overhead at the zenith; therefore during the time when the sun is more than 90° from Polaris (from September to March), it is below the horizon. The sun rises on March 21 and continues to spiral around the sky until it reaches its maximum altitude of 23.5° above the horizon on June 22. It then spirals back down to the horizon, setting on September 23. The sun is not seen again until the following March 21. At the south pole the situation is the same, except that the sun rises on September 23 and sets on March 21.

Thus every observer on the earth will see the sun spiral back and forth with respect to Polaris over the course of a year. The position of the sun with respect to the observer's horizon will depend, in turn, on the location of Polaris for that observer. (For an observer in the southern hemisphere Polaris will be below the horizon, so the sun's motion can be referred to the hypothetical southern pole star.)

figure 2–6 **ANNUAL CHANGES IN THE SUN'S DIURNAL MOTION AS SEEN FROM VARIOUS POINTS ON THE EARTH**

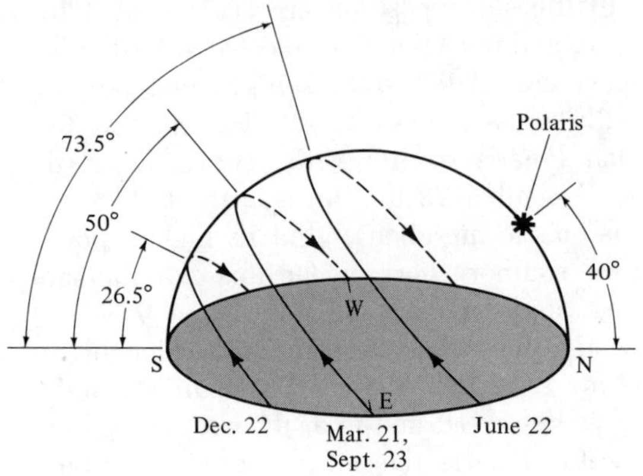

(a)

Path of the sun as seen from New York on the longest (June 22) and shortest (Dec. 22) days of the year and on the two days (Mar. 21 and Sept. 23) when the number of hours of day and night are equal.

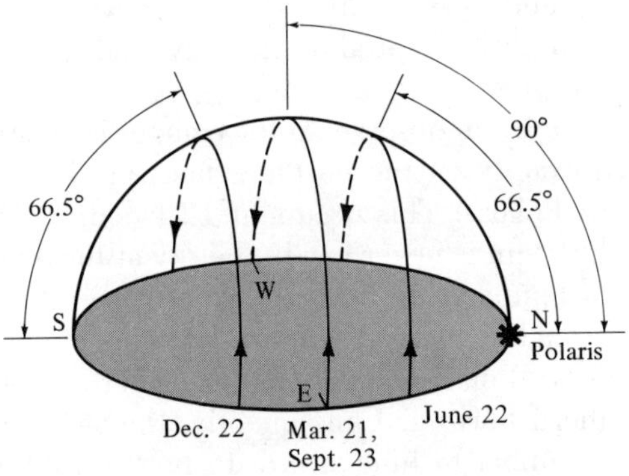

(b)

Path of the sun as seen from the equator. The sun passes through the zenith on Mar. 21 and Sept. 23. The sun is in the sky for 12 hours on every day of the year.

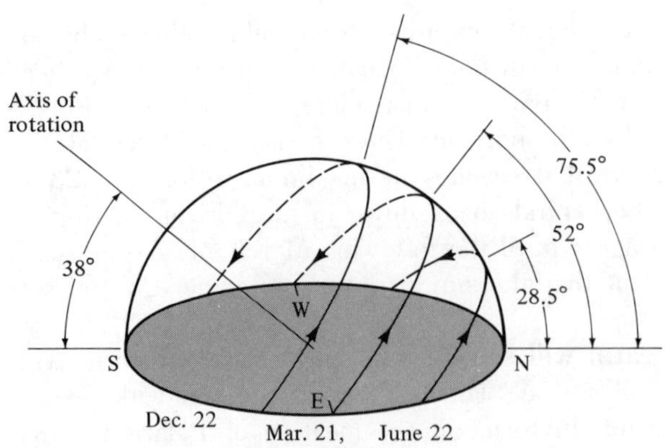

(c)

Path of the sun as seen from Melbourne, Australia (latitude 38° S). Note that the shortest day is June 22 and the longest day is Dec. 22.

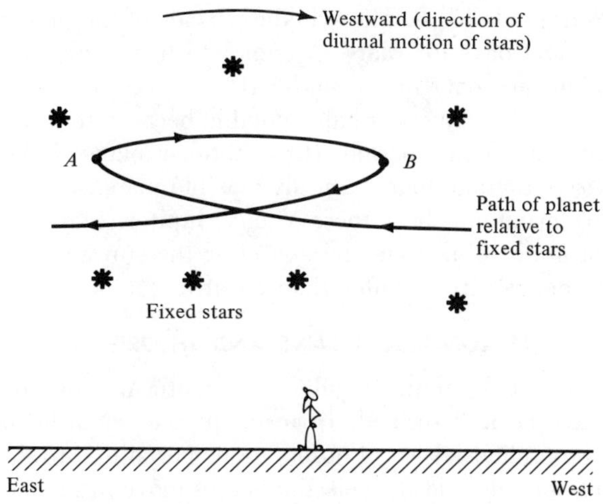

figure 2–7 **PATH OF A PLANET WITH RESPECT TO THE FIXED STARS**

At A the planet reverses its eastward motion and begins its retrograde loop. At B the planet reverts from westward to eastward motion. A planet will typically take several months to complete its retrograde motion.

THE MOTIONS OF THE PLANETS

It was previously stated that the stars show no motion with respect to one another. Actually, there are several objects in the sky that have the appearance of stars, yet change position relative to the fixed stars. The ancients referred to these objects as the "wandering stars," or ***planets*** (from the Greek word for "wanderer"). Five planets were known to the ancients: Mercury, Venus, Mars, Jupiter, and Saturn. Within the past two hundred years three more planets—Uranus, Neptune, and Pluto (none of which is visible to the naked eye)—have been discovered with the use of telescopes.* Like the sun and moon, the planets exhibit a slow change in position with respect to the stars in addition to their regular diurnal motion. However, whereas the sun and moon always tend to move eastward relative to the stars, the planets sometimes move eastward and sometimes move westward with respect to the stars. An illustration of the motion of a planet through the stars is shown in Figure 2–7. Note that generally the planet drifts from west to east through the stars, but at a certain point it reverses direction and moves from east to west for several months at a time. Eventually, the planet reverts to its eastward motion. The effect is that the planet slowly traces out a loop. This looping motion is referred to as ***retrograde motion.*** As a planet moves through the stars, it goes regularly into retrograde motion. That is, the time from the beginning of one loop to the beginning of the next is approximately constant. Each planet has a specific ***synodic period,*** which is the time between successive

* It is commonly believed that planets can be distinguished from stars because the stars "twinkle" and the planets do not. Actually, twinkling occurs because of disturbances in the atmosphere of the earth. Under certain conditions planets will also appear to twinkle. The planets are distinguished from the stars primarily by their motions relative to the stars.

retrograde loops. The synodic periods of the planets range from 116 days for Mercury to 780 days for Mars. A complete listing of synodic periods appears in the table of planetary data in Appendix B.

The planets are also notable because they vary rather considerably in brightness over a synodic period. (In general, a planet is brightest when it is in the middle of its retrograde loop.) The five visible planets are rather prominent objects in the sky. At maximum brightness Venus, Jupiter, Mars, and Mercury (in that order) are the brightest objects in the sky after the sun and moon, and only two stars (Sirius and Canopus) are brighter than Saturn.

ASTRONOMICAL LAWS AND THEORIES

Thus far the regular, systematic motions of the sun, moon, stars, and planets have been described. In terms of the scientific method, we have collected a rather comprehensive set of observations pertaining to the various celestial bodies. Moreover, in noticing that the celestial bodies move in regular, cyclic patterns, we have advanced to the stage in which we formulate *scientific laws*. That is, we have discovered a systematic relationship between two quantities: (1) the position in the sky of a celestial body as seen from a particular point on the earth, and (2) time. We can now use this relationship to *predict future events*. For example, the regular annual motion of the sun can be considered a basic law of astronomy and can be used to predict the position of the sun in the sky on any day of the year. An astronomer can predict the phase of the moon or the times of the rising and setting of Venus on any given date in the future. Yet in spite of our ability to make predictions, we are not completely satisfied. All we have is a set of disconnected laws; that is, we have one law for describing the motion of the sun, another law for the motion of the moon, and so on. Is there some underlying unity? Although we know *how* the celestial bodies move, we do not understand *why* they move the way they do. What we need is a *theory*—that is, a conceptual model of the universe that will enable us to explain all of the celestial motions. The historical development of two such theories—the ***geocentric*** (earth-centered) and ***heliocentric*** (sun-centered) models of the universe—will now be traced.

2–2

EARLY ASTRONOMY

The oldest existing astronomical records date back to about 2000 B.C. It seems evident, however, that these records were based on a long tradition of astronomical study. It is not unreasonable to suspect that such discoveries as the annual motion of the sun or the motions of the planets were made hundreds of thousands of years ago. In fact, it is possible that these discoveries were not made by modern man, *Homo sapiens,* but by one of his evolutionary ancestors—a clever Neanderthal, perhaps. However, this is mere speculation.

For the agriculturally based civilizations, a knowledge of the seasonal variations was vital. Thus the development of a calendar to keep track of the annual motion of the sun was one of the first great accomplishments of the ancient astronomers. Reliable records indicate that the Chinese accurately determined the length of the year well before 1300 B.C. The Chinese were rather careful observers of the sky. Besides recording the regular motions of the celestial bodies, they also noted the occurrence of many comets and other unusual events. The records of the Chinese have been of considerable help to modern astronomers, who have used them to fill in the gaps in the astronomical observations amassed by the western civilizations.

There is some evidence that Chinese astronomy was developed about 3000 B.C. It is written that two astronomers, Hsi and Ho, were beheaded in 2136 B.C. for failing to predict an eclipse. Although we now know that Hsi and Ho were legendary figures, the story indicates that the problem of "publish or perish" which besets the modern-day scholar may have been much more acute in the past.

STONEHENGE

One of the most remarkable testaments to the significance of astronomy in ancient times is Stonehenge, located near the town of Salisbury in southern England. Stonehenge (Figure 2–8) consists of a series of concentric circles of stones and holes. The main part of the monument is a ring of large, rectangular stones about 18 feet high. Within this ring are ten even larger stones arranged in a horseshoe pattern. Standing well outside the circle is another stone—the *heel stone.* It lies right on the main axis of Stonehenge, so that a person standing at the center of the ring finds that the top of the heel stone is exactly level with the horizon.

Stonehenge was erected in stages during the period from 1850 to 1500 B.C. A tremendous amount of human effort went into its construction. The largest stones weigh from 30 to 50 tons each; eighty of them were brought overland from a place

figure 2–8 **STONEHENGE**

Erected between 1850 and 1500 B.C., this impressive monument may have been an ancient astronomical observatory.

some 20 miles away and were then chipped down to the desired shape. Some of the smaller stones (weighing about 5 tons each) are of a type of rock found only in Wales. Eighty of these stones were brought to Stonehenge, a distance of some 200 miles.

Presumably, Stonehenge was the setting for religious or ritual ceremonies. But why were the stones arranged in that particular way? It was noticed some two hundred years ago that on the longest day of the year the sun first appears on the horizon at the tip of the heel stone. That is, the main axis of Stonehenge is aligned with the point of sunrise on June 22. If we look along this axis in the opposite direction, we find that it points to the position of sunset on the shortest day of the year. Although this alignment is certainly no coincidence, it still does not completely explain the existence of Stonehenge.

In 1963 a British astronomer, Gerald Hawkins, decided to check the possibility that the arrangement of Stonehenge had astronomical significance. An electronic computer was used to analyze all the possible alignments of the important stones. The computer found that almost every one of these alignments pointed to some significant position of the sun or moon on the horizon—the positions of sunrise and sunset on the longest and shortest days of the year, the positions of moonrise and moonset on these days, and so on. Hawkins also showed how Stonehenge could have been used to keep track of solar and lunar eclipses. Thus it appears that Stonehenge may have been used as a calendar or even as an astronomical observatory.

Who built Stonehenge? For many years the theory persisted that Stonehenge was a Druid temple. However, the Druids apparently did not arrive in Britain until about 500 B.C. Actually, almost nothing is known about the people who erected Stonehenge. They were the indigenous Stone Age inhabitants of Britain. Despite their obvious astronomical and engineering sophistication, they left no written records or other monuments. All that remains of them are some burial mounds, some pottery, ornaments, simple tools—and Stonehenge.

BABYLONIAN ASTRONOMY

The first great advances in the study of astronomy were made by the Babylonians, a Semitic people who lived in southern Mesopotamia (modern-day Iraq) in the valley of the Tigris and Euphrates rivers. The Babylonians made rather accurate measurements of the positions of the celestial bodies and kept extensive records of these positions. Star catalogs and planetary data have been found inscribed on clay tablets that date from 1800 B.C. The Babylonians are responsible for the division of the day into 24 hours, the hour into 60 minutes, and the minute into 60 seconds. (The Babylonians employed a complicated number system based on the number 60.) They were the first to measure positions in the sky in terms of angles; they divided the circle into $360°$ and devised the further subdivisions of the degree, namely, $1° = 60'$ and $1' = 60''$.* The Babylonians were predicting eclipses by 700 B.C., and by 400 B.C. they had developed mathematical procedures that enabled them to predict the future positions of the planets.

In the Babylonian religion each of the planets was identified with a god. The Babylonians believed that all human destiny was controlled by the sun, moon, stars, and planets. The course of a person's life was strongly influenced by the configuration

* $1'$ is read "one minute of arc" or "one arc-minute." Similarly, $1''$ is read "one second of arc" or "one arc-second."

of the various celestial bodies at the time of his birth. Moreover, if one could determine the future positions of the celestial bodies, one might be able to divine the future course of events on the earth. For these reasons the Babylonians kept very precise astronomical records and developed techniques for calculating future positions of the celestial objects. Thus Babylonian astronomy developed, not out of scientific curiosity, but out of the Babylonians' belief in astrology.

The Babylonian devotion to astrology is not difficult to understand. The Babylonians had no way to measure astronomical distances, and they considered all the celestial bodies to be equidistant and rather close to the earth. They could not determine the physical properties of the celestial bodies, nor could they give a natural explanation of celestial motions. In short, the heavens were a source of great mystery to the Babylonians. It is not surprising that they developed an extensive mythology around the celestial bodies and came to believe that these bodies exerted a powerful influence on human affairs. What is less comprehensible is that despite great advances in our knowledge of astronomy, biology, and psychology, a substantial segment of the human population still believes that one's personality and destiny are determined by the positions of the stars and planets at the instant of birth.

ANCIENT COSMOLOGY

The peoples of the various ancient civilizations held rather similar views as to the structure of the universe. They divided the world into two parts: the earth and the sky or "firmament"—the realm of the sun, moon, and other celestial bodies. The universe was not thought of as being particularly large, and it was believed to be completely enclosed by the bowl of the sky and the solid earth. The following picture is rather typical:

> The universe of the Egyptians was a rectangular box; the earth was its floor, the sky was either a cow whose feet rested on the four corners of the earth, or a woman supporting herself on her elbows and knees; later, a vaulted metal lid. Around the inner walls of the box, on a kind of elevated gallery, flowed a river on which the sun and moon gods sailed their barques, entering and vanishing through various stage doors. The fixed stars were lamps suspended from the vault, or carried by other gods. The planets sailed their own boats along canals originating in the Milky Way, the celestial twin of the Nile. Towards the fifteenth of each month, the moon god was attacked by a ferocious sow, and devoured in a fortnight of agony; then he was reborn again. Sometimes the sow swallowed him whole, causing a lunar eclipse; sometimes a serpent swallowed the sun, causing a solar eclipse. But these tragedies were, like those in a dream, both real and not; inside his box or womb, the dreamer felt fairly safe.[1 *]

At this stage in the development of astronomy, man did not conceive of the earth as a celestial body. Someone must have conjectured that if the sun sets in the west and then reappears the following morning in the east, it must have passed *under* the earth during the night, and therefore the earth must be floating in space. Yet such ideas did not catch on; perhaps in 1000 B.C. man was psychologically unready for such a concept of the world. It was, as we will see, the Greeks who eventually made the intellectual transformation from the firmament-and-earth view to the picture of a spherical earth suspended in the midst of the universe.

* References appear in Appendix D.

2–3

GREEK ASTRONOMY

The greatest of the ancient astronomers were the Greeks. They were relative latecomers to the field; we have no knowledge of any astronomy being done by the Greeks before 600 B.C. Nevertheless, within a period of a few hundred years they came to dominate astronomical thought. By A.D. 150 they had developed and perfected an astronomical system that was to be accepted until the seventeenth century.

Greek science and philosophy did not develop entirely in Athens; centers of learning were established in nearly all of the Greek settlements around the Mediterranean. One of the earliest centers arose in the cities of Ionia (a region along the west coast of modern Turkey, facing the Aegean Sea). The greatest of the Ionian philosophers was Thales of Miletus (640–545 B.C.). Thales and his followers set the course for Greek science by their rejection of mythological explanations of phenomena in favor of explanations based on purely natural causes. Thus, for example, the Ionians described the wind as being a current of air rather than the breath of the gods. They proposed that the moon shines by reflecting sunlight, and thus they were able to account for the phases of the moon in a natural way (Figure 2–9). The Ionian philosophers adopted the idea that all matter is composed of four elements—earth, air, fire, and water—and attempted to account for a whole range of natural phenomena in terms of the properties of these four elements and the interactions among them.

figure 2–9 **THE PHASES OF THE MOON**

The moon shines by reflecting sunlight. Since the moon is a sphere, only half of it is illuminated by the sun at any one time. The position of the moon in its path around the earth will determine how much of the illuminated side of the moon can be seen from the earth. As the moon moves from 1 to 5, we see progressively more of the illuminated side; from 5 back to 1, we see progressively less.

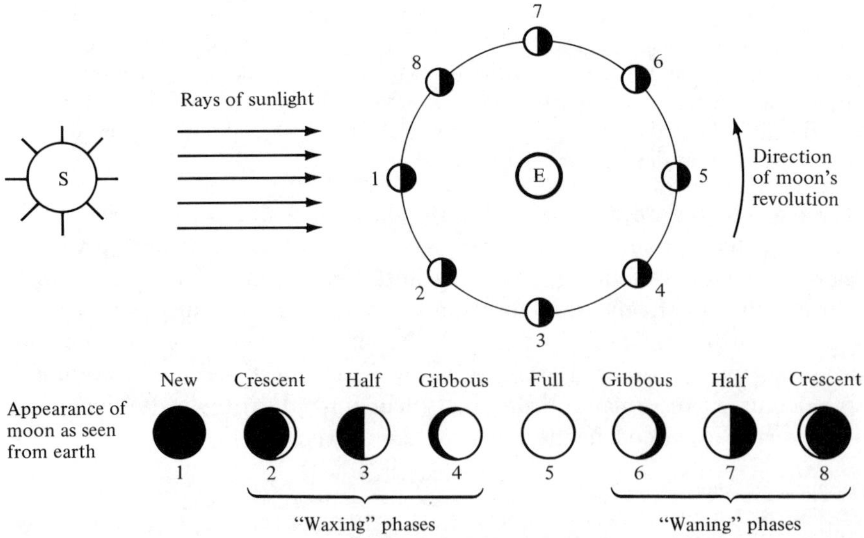

The cosmological models of the Ionians were about as primitive as those of the Babylonians and Egyptians, but they were significant in that they were totally free of any mythological explanations. Thales believed in a flat earth covered by a hemispherical heaven. There was much dispute among Thales and his followers as to whether the universe was surrounded by water or supported by air. Other members of the Ionian school believed in a spherical universe, with the earth suspended at the center. The sphere of the sky was supposed to be surrounded by invisible rings of fire; the sun, moon, and stars were merely holes in these rings, revealing the fire within.

THE PYTHAGOREANS

Around 500 B.C. another school of thought was founded under the leadership of Pythagoras. Little is known about the life of Pythagoras. He is reputed to have been born on the Aegean island of Samos in about 570 or 580 B.C., and he studied under one of the followers of Thales. He traveled extensively, finally settling in the town of Crotona in southern Italy in 530 B.C. Here a religious cult sprung up under Pythagoras. This Pythagorean Brotherhood, as it was called, was devoted to the purification of the soul through various forms of abstinence (such as not eating beans) and through the study of numbers. Because of their pure, abstract, and eternal nature, numbers were revered to the point of ecstasy;* through the study of numbers one could dissociate himself from all earthly distractions and thus cleanse his soul. As the center of this cult, Pythagoras was referred to as the Master and was raised to a semidivine status. With the death of Pythagoras in about 500 B.C., the Brotherhood began to crumble; by 450 B.C. political pressure had forced the Pythagoreans to disperse to other parts of the Greek world.

The Pythagoreans made many significant contributions to mathematics and geometry. Undoubtedly, their most important contribution was the introduction of mathematics into scientific inquiry. Their key discovery was that the pitch of a note produced by a stringed musical instrument depends on the length of the string. They then found that the lengths of the strings needed to produce the harmonic musical scale were related according to simple arithmetic ratios. This encouraged them to look for similar "harmonies" in nature. The belief that the underlying principles of nature were arithmetic relationships became the basis of the Pythagorean cosmology. For the first time the earth was considered to be a sphere (the sphere being the most perfect geometric form). The earth was at the center of a series of rotating concentric crystal spheres to which the heavenly bodies were attached. There were eight such spheres: one for the sun, one for the moon, one for each of the five planets, and one for the fixed stars. The sphere of the fixed stars rotated about an axis passing through Polaris and made one rotation each day. The spheres of the sun, moon, and planets each turned at slightly different speeds, and this difference in speeds accounted for the motions of these bodies relative to the fixed stars. The distances from the earth to these spheres were in proportion to the arithmetic ratios of the musical scale;

* The number 10 was especially revered because it is the sum of $1 + 2 + 3 + 4$. The Pythagoreans displayed this relationship in the *tetractys*, which became a magic symbol to them. Today it can still be found in bowling alleys throughout the world.

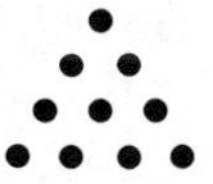

the Pythagoreans believed that as the spheres turned, each emitted its own particular musical note. Of all mortal men, only one could actually hear the "music of the spheres"; that man, of course, was Pythagoras.

Today, of course, no one believes in the music of the spheres—except as poetic imagery—but the idea that all nature is inherently mathematical has become one of the fundamental principles of science. In this respect Pythagoras must be regarded as a pivotal figure in the history of modern science.

PLATO AND ARISTOTLE

The geocentric Pythagorean cosmology, with its emphasis on the role of mathematics, soon entered the mainstream of Greek thought. In Athens the philosophical school led by Plato (428–348 B.C.) adopted the Pythagorean model and attempted to put it on a more solid philosophical foundation. Plato accepted the concept of a spherical universe centered on a spherical earth. The universe was thought to consist of two realms: the sublunary world (the earth) and the celestial world (the sun, moon, stars, and planets). Everything in the sublunary world was composed of earth, air, fire, and water and was subject to change, decay, and death. The celestial world, however, was constructed out of a mysterious fifth element, or "quintessence," and was eternal and unchanging. The natural motion of bodies in the sublunary world was a straight line, but in the celestial world the natural path was a circle. Plato thus established the principle that *all celestial motions are uniform circular motions.*

However, there was one glaring error in the original Pythagorean model; namely, it could not explain the retrograde motion of the planets. One simply could not expect the spheres to stop, reverse direction for a while, then stop and reverse again—this explanation would violate the principle of uniform circular motion. Accordingly, Plato posed the problem: What *combination* of uniform, concentric circular motions would account for the motions of the celestial bodies?

Plato's pupil Eudoxus (408–355 B.C.) solved the problem by assigning to each celestial body a combination of concentric spheres, each rotating at a different speed about a different axis. (See Figure 2–10.) The spheres were connected in such a way that the motion of the outermost sphere was transferred to the next inner sphere and so on. The outermost sphere rotated once a day, producing the observed diurnal motion, and the speeds of rotation and the inclinations of the axes of the inner spheres were adjusted so that the motions of the celestial bodies relative to the stars were accounted for reasonably well. Later, improvements were made to increase the accuracy of Eudoxus' system. Altogether, a total of fifty-five spheres was needed to account properly for the motions of the sun, moon, and planets.

Plato's most famous pupil was Aristotle (384–322 B.C.). Aristotle accepted the Pythagorean–Platonic model of the universe, but he tried to justify his belief on the basis of observations and natural causes rather than geometric and mathematical considerations. Aristotle believed in a spherical earth, but not because of the geometric perfection of the sphere. He pointed out that during a lunar eclipse the earth's shadow falls on the moon, and it is a curved shadow. Moreover, the change in the positions of the stars relative to the horizon that one observes when he travels north or south also indicates that the earth has a curved surface. The curvature of the earth explains why Polaris is seen directly overhead when viewed from the north pole, but is on the horizon when viewed from the equator. (See Figure 2–11.)

Aristotle considered the possibility that the earth moves and rejected the idea. He argued that if the earth moved, the stars would appear to shift positions relative

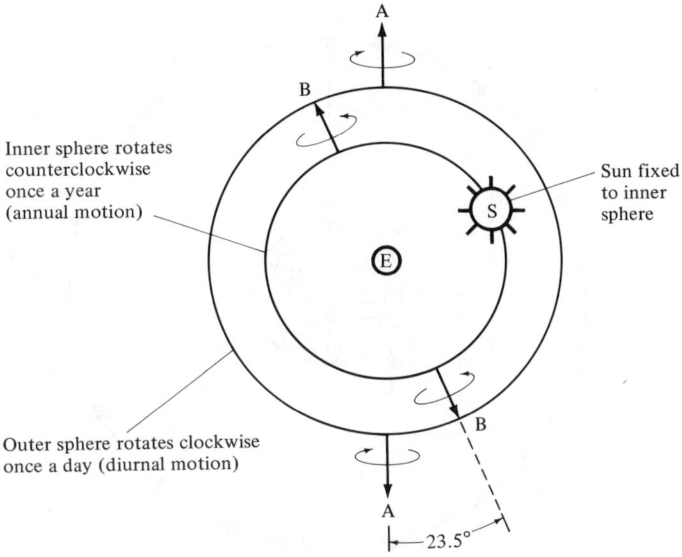

figure 2–10 **EUDOXUS' SYSTEM OF CONCENTRIC SPHERES FOR THE MOTION OF THE SUN**

The outermost sphere rotates clockwise about the axis A–A in one day. The inner sphere, to which the sun is attached, rotates counterclockwise once a year. The combined motions of the two spheres account for the diurnal and annual motions of the sun.

to one another as the position of the earth changed. As shown in Figure 2–12, the angle between two stars, as seen from the earth, would appear to change as the earth moved in its orbit. This effect is known as ***stellar parallax.*** However, the stars always appear fixed; stellar parallax is *not* observed; therefore, Aristotle reasoned, the earth is stationary.

THE EARTH IN MOTION

With Aristotle the geocentric model was firmly established. Nevertheless, theories involving the earth's motion were still being proposed. Heracleides (388–312 B.C.) believed that the diurnal motion of celestial bodies was due not to the daily turning of the celestial spheres, but to the rotation of the earth. His system, however, was still basically geocentric. The first truly heliocentric system was proposed by Aristarchus of Samos (310–230 B.C.). In his system the sun was placed at the center of the celestial universe, and the earth was moved to a new location as the third planet out from the center.

The heliocentric theory was unacceptable to the Greeks. It was totally contrary to the prevailing philosophical beliefs about the place of man in the universe; one critic went so far as to suggest that Aristarchus be censured for blasphemy. Moreover, there were strong scientific arguments against the heliocentric theory. The strongest of these was the absence of stellar parallax. Aristarchus' writings on the heliocentric theory are lost, so we do not know how he defended the theory. He probably suggested that the reason there was no stellar parallax was that the stars are so far away that

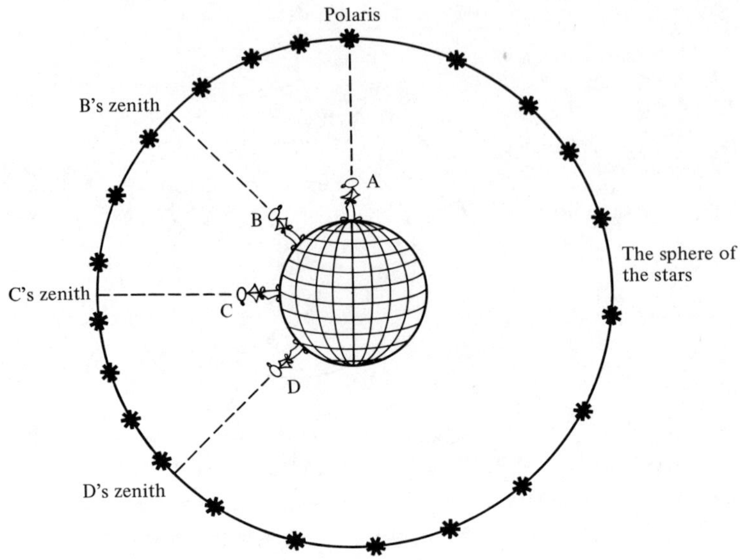

figure 2–11 **VIEWING THE SPHERE OF THE STARS FROM A SPHERICAL EARTH**

Observer A, located at the north pole, sees Polaris at the zenith. Observer B sees Polaris lower in the sky. For Observer C, at the equator, Polaris is 90° away from the zenith and lies on the horizon; Observer D, in the southern hemisphere, cannot see Polaris at all.

the orbital motion of the earth is negligible by comparison.* The Greeks simply could not fathom the idea of a universe as large as Aristarchus wanted to make it. If the stars are that far away, why are we able to see them? They would have to be truly immense in size. If the earth is both rotating on its axis and revolving about the sun, why don't we feel a strong wind? Why don't we have any sensation of the earth's motion? It seems quite understandable that the Greeks could not find any convincing reasons for rejecting the long-established geocentric model in favor of the heliocentric universe of Aristarchus.

THE ECCENTRIC AND THE EPICYCLE

Aristotle's most illustrious pupil was Alexander the Great. A brilliant general and leader, Alexander extended the boundaries of the Greek Empire into Egypt and as far east as India. In 330 B.C. Alexander captured the city of Babylon and with it the accumulated records of the Babylonian astronomers. The acquisition of several centuries' worth of precise astronomical data was of enormous value to Greek astronomers. By 300 B.C. the paths of the sun, moon, and planets were known to a degree of accuracy that clearly indicated that the celestial motions could no longer be accounted for in terms of the system of uniformly rotating concentric spheres proposed by Eudoxus.

There were a number of major difficulties. The first was that the planets vary

* This is, in fact, the correct explanation for the failure to observe stellar parallax. The first successful observation of stellar parallax was made in 1837—long after the argument between geocentrism and heliocentrism had been settled.

noticeably in brightness, and this variation indicates that the distances of the planets from the earth are not constant. Secondly, the sun and moon were both found to vary in apparent size, indicating that they too do not travel in perfectly circular orbits centered on the earth. Moreover, the sun, moon, and planets do not appear to move through the sky at constant speeds. The astronomers of the time recognized that a system of concentric spheres could not account for these variations in distance and speed and that the system would have to be scrapped. But they did not want to reject Plato's principle that all celestial motions are combinations of uniform circular motions.

Consequently, two new concepts were introduced into the geocentric system. One was the idea of the ***eccentric circle,*** or ***eccentric:*** a celestial body travels in a circular path at constant speed, but the center of the circle is *eccentric* (off-center) with respect to the earth. (See Figure 2–13.) Thus as the body moves in the circle, its distance, brightness, apparent speed, and (in the case of the sun and moon) its size appear to change when the body is viewed from the earth. In this manner uniform circular motion was preserved, but at the expense of displacing the earth as the center of celestial motion.

The second new concept to be introduced was that of ***epicyclic motion*** (Figure 2–14a). The planet moves in a small circle, the ***epicycle,*** and the center of the epicycle moves, in turn, in a larger circle, the ***deferent,*** with the earth at the center of the deferent. If the speeds of the planet along the epicycle and the center of the epicycle along the deferent were chosen properly, almost any path could be reproduced —including retrograde motion (Figure 2–14b). The epicycle-and-deferent system also preserved the concept of uniform circular motion, but more spheres had to be added to the system.

figure 2–12 **STELLAR PARALLAX**

As the earth moves from E_1 to E_2, the angle between the two stars A and B changes. Since angle 1 is greater than angle 2, the stars will appear farther apart in the sky when the earth is at E_1 than they will when the earth is at E_2. Hence the angular separation of the two stars should vary over the course of a year.

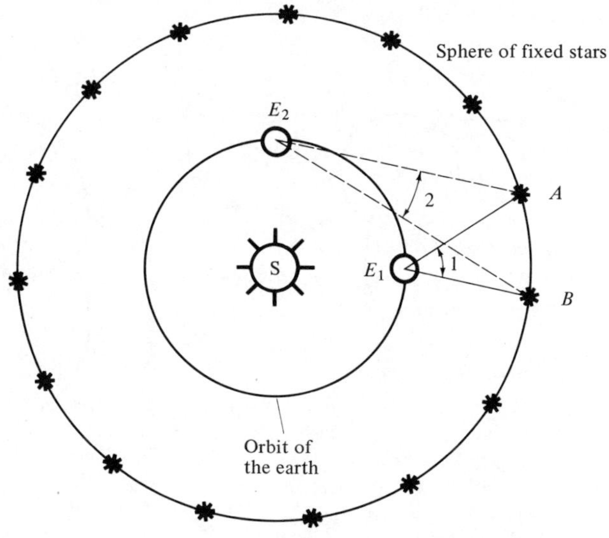

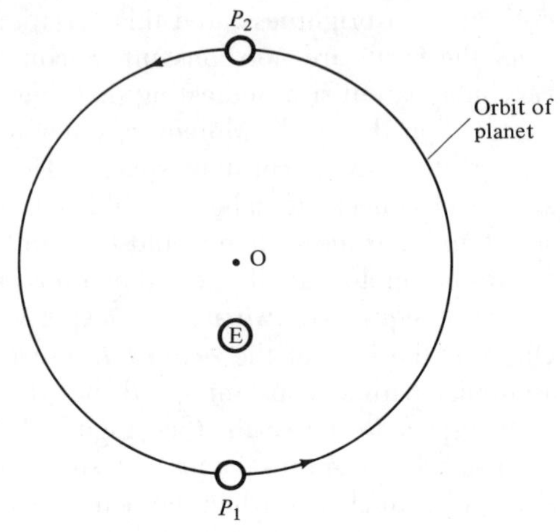

figure 2–13

THE ECCENTRIC CIRCLE

The planet travels with uniform speed on a circular orbit. The center of the circle at O does not coincide with the earth (E). When viewed from the earth, the planet will appear to be moving more rapidly at P_1 than at P_2.

The epicycle and the eccentric seem to have been first suggested by Apollonius of Perga in about 250 B.C., but he apparently did not attempt to calculate the eccentric and epicycle–deferent combinations that would fit the observed celestial motions. This arduous task was not begun until a century later, when the great astronomer and mathematician Hipparchus tackled the problem. By 150 B.C. the center of Greek culture had shifted to Alexandria in Egypt, where a university and the finest library in the world had been established. Here Hipparchus worked during the period from 160 B.C. to 130 B.C. Using the Babylonian data and his own very accurate observations, Hipparchus was able to use an eccentric circle to account for the observed variations

figure 2–14 **THE EPICYCLE**

(a) The planet moves with constant speed on the epicycle. At the same time, the center C of the epicycle is moving with constant speed on the deferent. The earth is at the center of the deferent. (b) As the planet revolves on the epicycle and the epicycle revolves on the deferent, the planet traces out loops that produce retrograde motion.

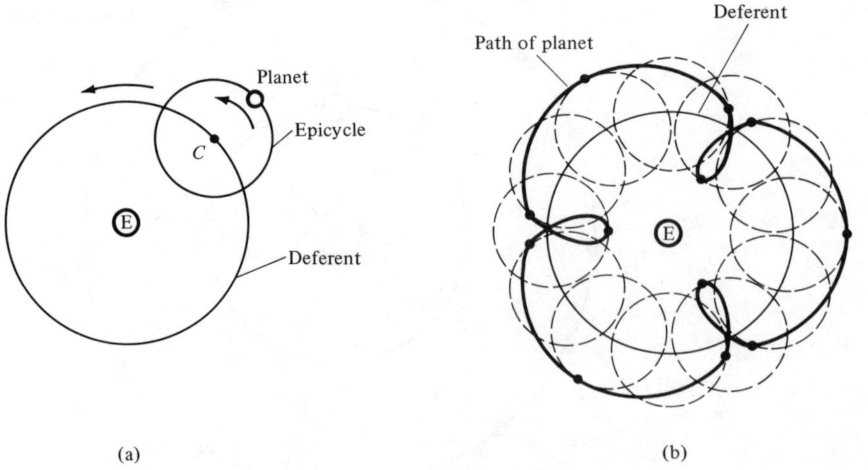

in the sun's motion. He also outlined a method for fitting epicycle–deferent combinations to the motions of the planets, but he did not complete these calculations, apparently because he did not have enough planetary data.

THE PTOLEMAIC SYSTEM

The work of Hipparchus remained unfinished for three centuries. Finally, in about A.D. 150 the Alexandrian astronomer Claudius Ptolemy perfected Hipparchus' methods and brought the geocentric model to its culmination. His great treatise, the *Almagest,** was a compilation of the accomplishments of Greek astronomy. Ptolemy's goal was to find geometric constructions that could be used to bring the calculated positions of the sun, moon, and planets into agreement with the most accurate observations available. To account for the observed motions of the planets, Ptolemy had to introduce moving eccentrics, combine eccentrics with epicycles, and even place epicycles on epicycles. Because of Ptolemy's desire to remain faithful to the Platonic dictum of uniform circular motion, the geocentric system became extraordinarily complicated. Altogether, about forty different spheres in uniform circular motion were needed to account for all the celestial motions.

In a sense the Ptolemaic system represented the *failure* of Greek astronomy. The Greeks had started out centuries before with the aim of finding natural causes for natural phenomena. The Ptolemaic system made no mention of causes, and it made no pretense of representing reality. Ptolemy intended it as a purely geometric representation for the avowed purpose of determining celestial motions. In this respect Ptolemy's outlook was no different from that of the early Babylonians.

In fact, the Babylonians were probably responsible for Ptolemy's attitudes. Babylonian astrology became enormously popular throughout the civilized world, which in Ptolemy's time was almost completely under Roman rule. The Romans were extraordinary engineers (some of the roads, bridges, and aqueducts that they constructed have been in continuous use for two thousand years), but they had no interest in pure science. Moreover, they were attracted to various forms of mysticism and cultism, and were strongly influenced by astrological considerations. Ptolemy himself wrote a book in which he defended astrology and described its methods. The decline of science had begun.

2–4

THE MIDDLE AGES

Astronomy reached a dead end with Ptolemy, and no significant advances were to be made for fourteen hundred years. During this period of time science did not just stand still; it fell into total collapse. From the time of Ptolemy until its capture by the barbarians in A.D. 476, Imperial Rome was in a state of accelerating intellectual decay. Mysticism and magic dominated virtually all thought. The rise of Christian theology during this period was another influential factor in the decline of science. Some theologians preached on the sinfulness of knowledge for knowledge's sake: Man should be concerned only with a knowledge of God and of the soul, so he would gain nothing from the study of nature. Since the Greeks were pagans, their writings

* "Almagest" is an Arabic corruption of the Greek words for "the great work." Ptolemy's original title for his work was *The Mathematical Treatise.*

were to be rejected. Before long theologians began to argue that the earth was flat, and cosmology reverted to the heaven-and-earth picture. By A.D. 500 science was back to the level of 600 B.C. With the fall of the Roman Empire, Europe was plunged into the Dark Ages—a state of political, social, and cultural disarray lasting nearly five centuries.

In the Middle East the Byzantine Empire provided some political stability, and science was at least able to hold its own. Greek manuscripts were copied and translated in Armenian and Syrian monasteries. However, the people who were most responsible for maintaining scientific knowledge during the Dark Ages were the Arabs. Following the rise of Islam and the Moslem conquests of the seventh century, the city of Baghdad became the cultural center of the world. Here Arab scholars translated into Arabic the works of Aristotle, Ptolemy, Euclid, and others, and the scientific traditions of Persia, India, and Greece commingled.* The Arabs made no significant scientific advances, but they did make some improvements in the Ptolemaic system on the basis of their own astronomical observations.

By the year 1000 Europe had begun to emerge from the Dark Ages. During the eleventh and twelfth centuries contacts with the Arabs increased, especially through the Arabic western outposts in Sicily and, in particular, Spain, which had become an important Moslem cultural center. Through these contacts the medieval European scholars learned of the long-lost writings of the ancient Greek philosophers, mathematicians, and astronomers. These rediscovered works were translated from Arabic into Latin and were studied with great interest. In particular, the writings of Aristotle had a profound effect on medieval philosophy. At first it was thought that Aristotle's scientific works conflicted with Christian beliefs, and for a while Aristotle's writings were banned. St. Thomas Aquinas (1225–1274) was mainly responsible for merging Aristotelian science with Christian theology, and before long Aristotle's doctrines were elevated to the status of religious dogma. The *Almagest* was also rediscovered, and further corrections were made in the Ptolemaic system. The Aristotelian universe of concentric spheres in uniform rotation about the earth was accepted as a tenet of religious faith, while at the same time Ptolemy's complicated system of epicycles and eccentrics was used to calculate the movements of the celestial bodies. Actually, the Ptolemaic system was looked on as a purely geometric construction that did not describe the "real" universe.

By the fifteenth century the revival of interest in knowledge that had begun with the rediscovery of the Greek classics had blossomed into the Renaissance. The constrictive medieval orthodoxy was replaced by the humanistic spirit; the obsessive contemplation of heaven and the afterlife gave way to free inquiry into the nature of man and his world. The growth of universities and the invention of the printing press led to wide dissemination of the new knowledge, and horizons were further widened by the discoveries of such explorers as Columbus and Vasco da Gama. The new freedom was perhaps most evident in the arts, where new approaches to painting, sculpture, architecture, and poetry were being advanced.

The sciences remained the last bastion of the old orthodoxy. Aristotelian ideas had been welded into the theological framework and were taught almost exclusively. In some of the universities the Pythagorean–Platonic doctrines were taught, but to teach these doctrines was to invite ecclesiastical disapproval. The conflict between

* Our system of "Arabic" numerals actually originated in India. It was adopted by the Arabs in about 800, and they transmitted it to Europe in the tenth century.

these two classical schools of thought was rather fundamental. In the Pythagorean–Platonic view the universe was designed according to mathematical relationships and could be understood through the study of geometry. To the Aristotelians the geometrical harmonies were of little importance; nature could be understood through logical reasoning from basic nonmathematical principles.

Such was the intellectual climate in 1496 when Nicolaus Copernicus came to Bologna to study law and astronomy.

2–5

COPERNICUS

Niklas Koppernigk (later Latinized to Nicolaus Copernicus) was born in the Polish city of Toruń in 1473. He began his college studies at the University of Cracow, then went to Italy to study at the universities of Bologna, Padua, and Ferrara, where he received a degree in canon law. In 1506 he returned to Poland to take up a position as a church administrator. He led a quiet, uneventful life; he never traveled, made few friends, and had little contact with other scholars or astronomers. He remained virtually isolated from the intellectual ferment of the time.

While a student Copernicus became interested in astronomy; at Bologna and Padua he was exposed to the Pythagorean–Platonic doctrines and also studied the works of Ptolemy. His dissatisfaction with the complexities of the Ptolemaic system led him to work out a comprehensive *heliocentric* theory. He outlined this system in the *Commentariolus,* a work that he never published. After his return to Poland, Copernicus set out to write a complete version of the heliocentric theory. By 1530 he had completed this treatise, which he titled *De Revolutionibus Orbium Coelestium* (*On the Revolution of the Celestial Spheres*). Once again, he refused to publish his work.

Nevertheless, Copernicus was not unknown. He allowed a few manuscripts of the *Commentariolus* to circulate among astronomers, and word of the new theory spread. Pope Clement VII attended a lecture on the Copernican system and was favorably impressed. Copernicus was encouraged to publish his full treatise, but he declined to do so. Then in 1539 Copernicus was visited by a young astronomer, Georg Joachim Rheticus, who became Copernicus' student and disciple. After much pleading, Rheticus managed to persuade Copernicus to publish his work. Rheticus spent the next few years checking the manuscript and supervising the printing.* Finally, in 1543 *De Revolutionibus* was published. Copernicus was given a copy of the book on the day he died.

Why was Copernicus so reluctant to see his work in print? Perhaps he was afraid of being accused of holding heretical views. It seems fairly certain that Copernicus really believed the earth moved and was not simply constructing a geometric system for calculating celestial motions. Nevertheless, Copernicus dedicated *De Revolutionibus* to Pope Paul III, and the preface to this work states that the heliocentric system is only a hypothesis and not meant to depict the truth. (However, the preface was not written by Copernicus, but by a Lutheran theologian, Andreas Osiander, and it

* Copernicus did not mention Rheticus' name anywhere in the book—presumably because Rheticus' Protestant affiliation might offend the Pope. Rheticus quite understandably lost all interest in Copernicus' work after this betrayal of his efforts.

is not certain whether or not Copernicus gave his approval.) Perhaps Copernicus may have felt that his work was a failure, for, as we shall see, the heliocentric theory that he devised was more complicated than the Ptolemaic system and did not offer any improvement in accuracy.

THE MOTIONS OF THE EARTH

At the beginning of the *Commentariolus*, Copernicus explained that he had rejected the Ptolemaic system because it was much too complicated. Copernicus believed that if the sun, and not the earth, was considered stationary at the center of the universe (Figure 2–15), all the celestial motions could be accounted for with fewer eccentrics and epicycles. It appears that Copernicus, rather than being the first modern astronomer, was actually trying to restore to astronomy the purity of the simple Platonic system based on the principle of circular motion at constant speed.

Copernicus explained that if the earth is moving, most of the observed motions of the sun, moon, planets, and stars are not real motions but *apparent* motions; that is, these bodies appear to move because the *observer* is moving rather than the bodies themselves. Thus all diurnal motions are apparent motions caused by the daily ***rotation*** of the earth on its axis; that is, celestial bodies appear to rise in the east and set in the west because the earth rotates from west to east. (An observer looking down on the north pole from above the earth would see the earth rotating counterclockwise.) The annual motion of the sun is also an apparent motion, which is a consequence of the yearly ***revolution*** of the earth around the sun (Figure 2–16). Since the earth completes a 360° revolution in $365\frac{1}{4}$ days, it moves in its own orbit through an angle of slightly less than 1° a day. Thus although the sun appears to shift its position relative to the stars by 1° a day, it is actually the earth that is moving 1° a day with respect to the stars.

The heliocentric theory must also account for the apparent annual north–south motion of the sun with respect to Polaris and the accompanying seasonal changes in the position of the sun in the sky shown in Figure 2–5. This effect can be explained

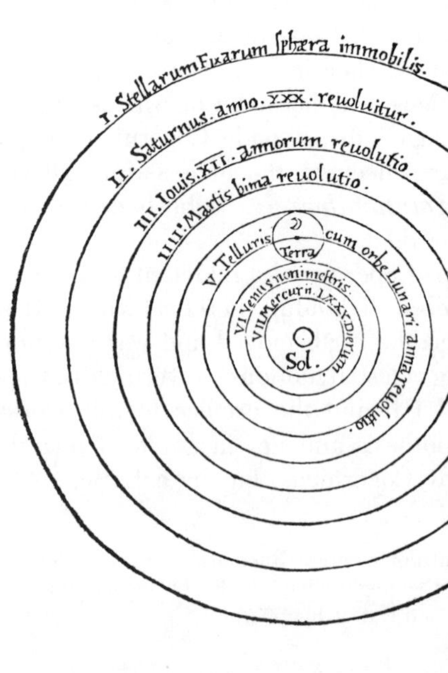

figure 2–15

THE HELIOCENTRIC SYSTEM

This diagram, drawn by Copernicus in *De Revolutionibus Orbium Coelestium*, belies the complexity of Copernicus' heliocentric system, since it does not show the many epicycles and eccentrics he used to maintain the concept of uniform circular motion. YERKES OBSERVATORY PHOTOGRAPH

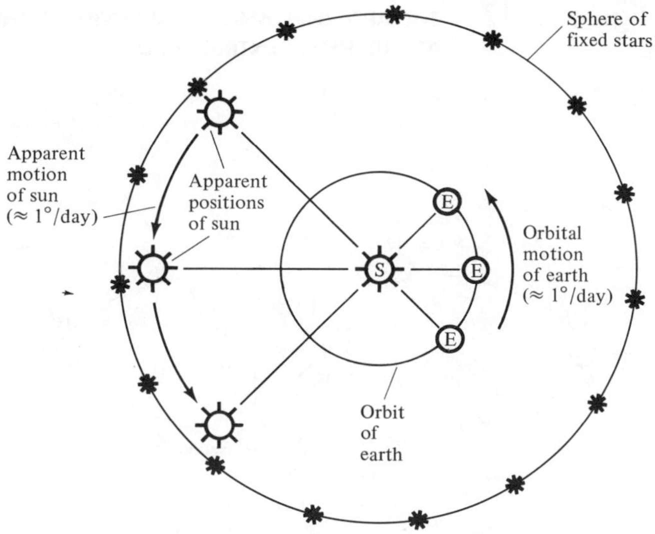

figure 2–16 **THE REVOLUTION OF THE EARTH**

The earth revolves in a nearly circular orbit around the sun. As seen from the moving earth, the sun will appear to move with respect to the fixed stars, making a complete revolution through the stars in the course of a year. The earth moves in its orbit at an average rate of slightly less than 1° per day; hence the sun will appear to shift position with respect to the stars at the same rate.

as follows: Let us assume that the orientation of the earth's axis is fixed in space; that is, the axis of rotation always points in the same direction. This is dictated by the observation that Polaris is always found directly over the north pole at all times of the year, from which it follows that the earth's axis must always point in the direction of Polaris regardless of the earth's position in its orbit (Figure 2–17a). The fact that Polaris does not change its apparent position in the sky over the course of a year (i.e., it exhibits no stellar parallax) attests to the great distance from the earth to Polaris, compared to which the diameter of the earth's orbit is negligibly small.

The annual north–south motion of the sun can be accounted for by assuming that the earth's axis is tilted at a 23.5° angle from the perpendicular to the plane of its orbit, as shown in Figure 2–17a. The earth maintains a constant angle of tilt as well as a constant orientation in the direction of Polaris over an entire revolution. However, the earth's axis does *not* maintain a constant orientation with respect to the *sun*. From March to September the north pole is tilted toward the sun, and the northern hemisphere receives the more direct rays of sunlight. The axis is most directly tilted toward the sun on June 22, when the angle Polaris–earth–sun is 66.5° (Figure 2–17b). From September to March the north pole is tilted away from the sun, which is now more directly overhead in the southern hemisphere. On December 22 the angle Polaris–earth–sun is 113.5° (Figure 2–17c). On March 21 and September 23 (Figure 2–17d) the tilt of the axis is neither toward nor away from the sun; the earth's axis now makes a 90° angle with the sun–earth line, and both hemispheres of the earth receive sunlight equally.

figure 2–17 **THE APPARENT ANNUAL MOTION OF THE SUN AS EXPLAINED BY THE HELIOCENTRIC THEORY**

The earth's axis is tilted 23.5° from the perpendicular to the orbital plane and always points in the direction of Polaris.

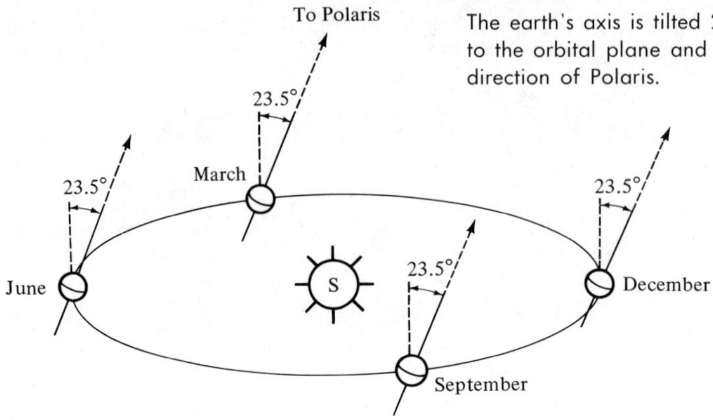

(a)

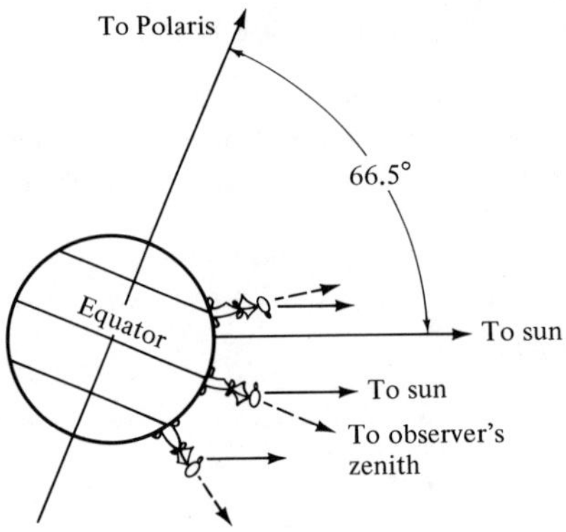

(b)

On June 22 the axis is tilted most directly at the sun. To an observer in the northern hemisphere the sun will appear higher in the sky than it will to an observer at the same latitude in the southern hemisphere.

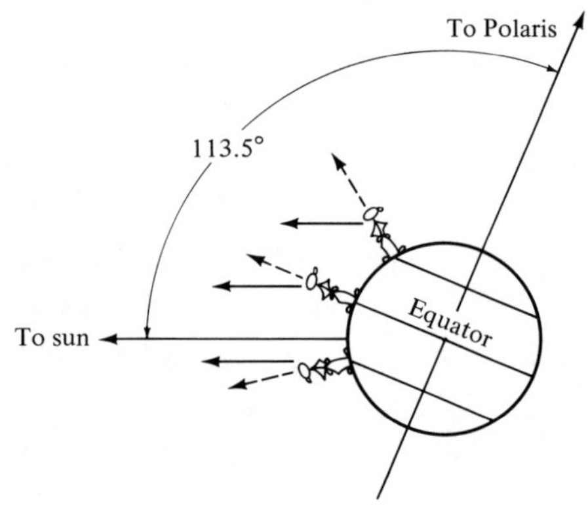

(c)

On Dec. 22 the axis is tilted most directly away from the sun, and the sun will be closer to the zenith in the southern hemisphere.

On Mar. 21 and Sept. 23 the tilt of the axis is not in the direction of the sun. (In the diagram the axis should be tilted out of the plane of the paper by 23.5°.) The sun is then overhead on the equator.

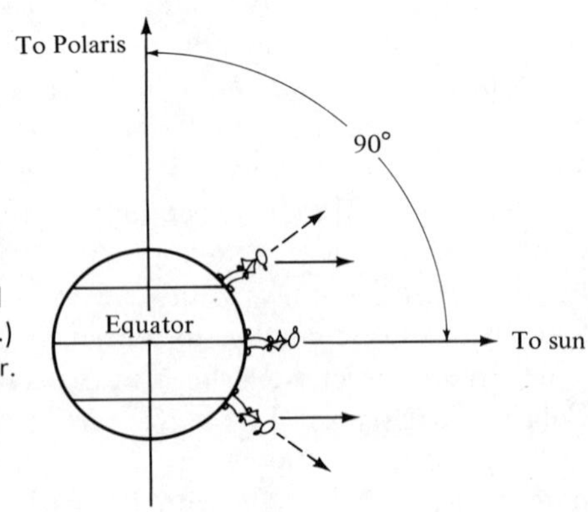

(d)

Hence in the heliocentric theory the annual motion of the sun and the associated seasonal changes experienced by an observer on the earth can be explained by assuming the following:

(1) The earth revolves in an orbit around the sun, taking $365\frac{1}{4}$ days to complete one revolution.
(2) The earth's axis is inclined at an angle of $23\frac{1}{2}°$ from the perpendicular to its orbital plane.
(3) The earth's axis always points in the same direction in space (i.e., in the direction of Polaris).

Nevertheless, the fact that the sun's annual motion can be explained through these assumptions does not *prove* that the earth moves, for these same motions can also be explained by the geocentric theory. And, of course, there were still the usual objections to the idea that the earth is in motion. Copernicus did attempt to answer these objections. We do not feel a violent wind as the earth rotates, he wrote, because the atmosphere rotates along with the earth at the same speed. To answer the thorny question of the absence of stellar parallax, Copernicus had to assume that the stars are very far away: "The distance from the earth to the sun is imperceptible in comparison with the height of the firmament." In addition, Copernicus was most careful to buttress with theological arguments his contention that the earth moves and to show that his hypothesis did not contradict Holy Scripture.

THE MOTIONS OF THE PLANETS

In the Copernican system the planets circle the sun and not the earth (which is now regarded as one of the planets). This arrangement offers several distinct advantages over the Ptolemaic system in explaining planetary motions.

In describing the motions and positions of the planets in the sky, astronomers have found it useful to refer to the ***elongation*** of a planet. The elongation measures the position of a planet with respect to the sun, as seen from the earth. It is defined as the angular distance between the sun and a planet in the sky. Since the observer on the earth is at the vertex of the angle, the elongation of a planet is simply the angle sun–earth–planet. (See Figure 2–18.) When the elongation of the planet is 180° (i.e., the planet is exactly opposite the sun in the sky), the planet is said to be in ***opposition.*** When the sun and planet lie along the same line (the elongation is 0°), the planet is said to be in ***conjunction.*** In general, a planet will be in either ***western elongation*** or ***eastern elongation,*** depending on whether the planet is seen to the west or to the east of the sun.

When we observe the planets, we find that Mars, Jupiter, and Saturn progress regularly from conjunction to opposition and back to conjunction. We note also that each of these planets is at its brightest when at opposition, and it gets fainter as it approaches conjunction. In the geocentric theory large epicycles had to be introduced to explain the retrograde motions and varying brightnesses of the planets. In the heliocentric theory these effects can be accounted for much more naturally. Copernicus recognized that the orbits of Mars, Jupiter, and Saturn are larger than that of the earth and referred to these three planets as the ***outer*** or ***superior planets.*** As the planets travel around the sun, all moving at different speeds (the earth, of course, is also in motion), their elongations change in a regular way. As shown in Figure 2–19, when the earth and an outer planet are on opposite sides of their respective orbits, the planet appears to be in line with the sun (in conjunction). When the earth

figure 2–18 THE ELONGATION OF A PLANET

The elongation is the angle that a planet makes with the sun in the sky, as seen by an observer on the earth.

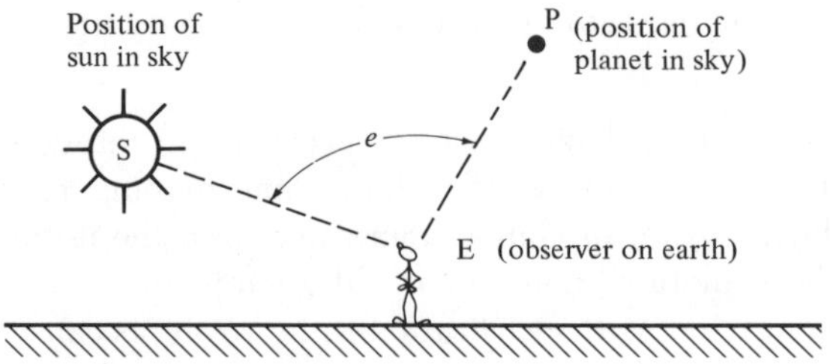

The elongation e of the planet is the angle sun–earth–planet ($\angle SEP$).

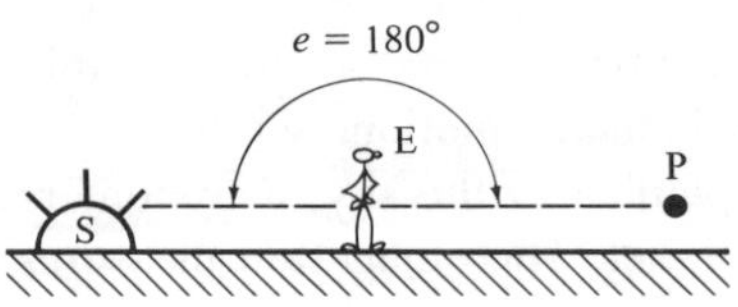

(a)

Opposition. The elongation of the planet ($\angle SEP$) is 180°. The planet is directly opposite the sun in the sky.

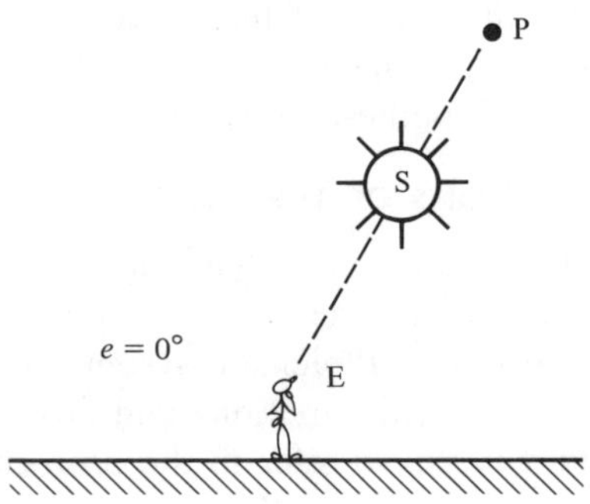

(b)

Conjunction. The elongation ($\angle SEP$) is 0°. The planet is in virtually a direct line with the sun and cannot be seen.

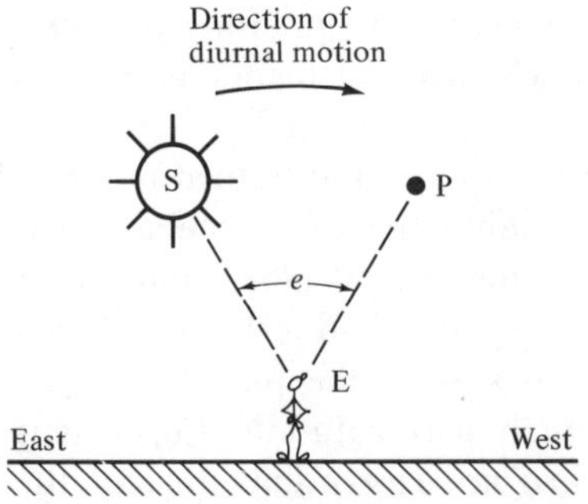

(c)

Western elongation. The planet is to the west of the sun. It rises before sunrise and sets before sunset.

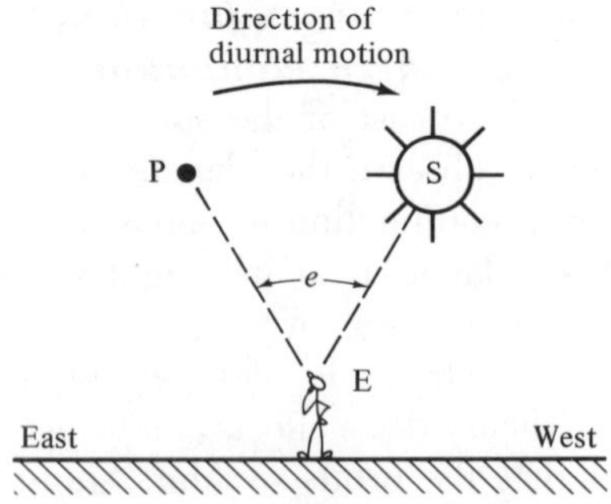

(d)

Eastern elongation. The planet is to the east of the sun. It rises after sunrise and sets after sunset.

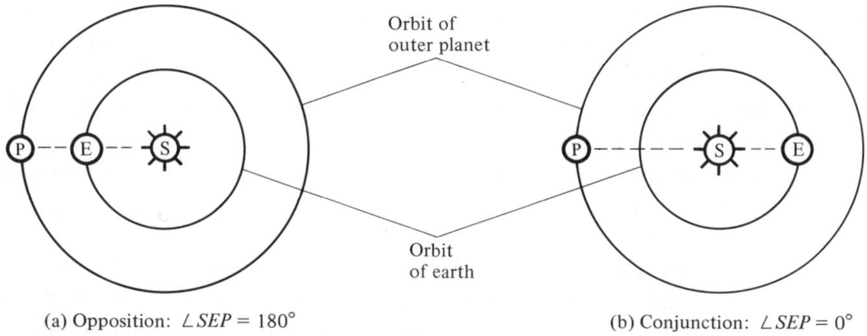

figure 2–19 **ELONGATIONS OF AN OUTER PLANET IN THE HELIOCENTRIC THEORY**

The orbits of the earth and an outer planet are arranged to show the planet in (a) opposition and (b) conjunction, as seen from the earth.

and the planet are on the same side of their orbits and directly in line with the sun, the planet appears to be in opposition (elongation = 180°). We would expect the planet to be brightest when it is in opposition, since at that point it is closest to the earth.

The apparent motions of Mercury and Venus are quite different from those of Mars, Jupiter, and Saturn. Mercury and Venus are always found near the sun; they are never seen in opposition. Venus is never more than 48° from the sun; that is, its ***maximum elongation*** is 48°. (See Figure 2–20.) Thus the position of Venus ranges from 48° west of the sun (maximum western elongation) to 48° east of the sun (maximum eastern elongation); the planet shuttles regularly from maximum western elongation to conjunction to maximum eastern elongation, then back to conjunction and to maximum western elongation. The same is true for Mercury, except that Mercury stays even closer to the sun than Venus does. The maximum elongation

figure 2–20 **OBSERVED MOTION OF VENUS**

Venus is never seen more than 48° away from the sun. Its elongation ranges from a maximum of 48° east of the sun to a maximum of 48° west of the sun.

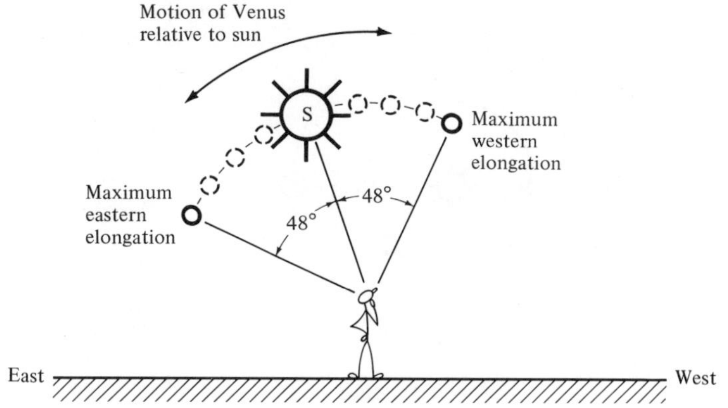

of Mercury is only 28°. (Actually, Mercury's maximum elongation varies from one synodic period to the next, ranging from 18° to 28°, because of the noncircular shape of Mercury's orbit.) Since Mercury and Venus are so close to the sun, they can be seen for only a few hours, either before sunrise or after sunset, depending on whether they are west or east of the sun.

In the geocentric theory the motions of Mercury and Venus were explained by assuming that these planets travel in rather large epicycles, such that the centers of the epicycles always remain in line with the sun. (See Figure 2–21a.) There was no explanation of why the epicycles of Mercury and Venus were coupled to the sun and the epicycles of Mars, Jupiter, and Saturn were not. The explanation that Copernicus offered using the heliocentric theory is considerably simpler. He proposed that Mercury and Venus are ***inner*** or ***inferior planets***—that is, they are closer to the sun than the earth is and their orbits are smaller than that of the earth. Since an inferior planet always remains inside the earth's orbit, it cannot possibly attain opposition (Figure 2–21b).

figure 2–21 **GEOCENTRIC AND HELIOCENTRIC EXPLANATIONS FOR THE MOTIONS OF VENUS AND MERCURY**

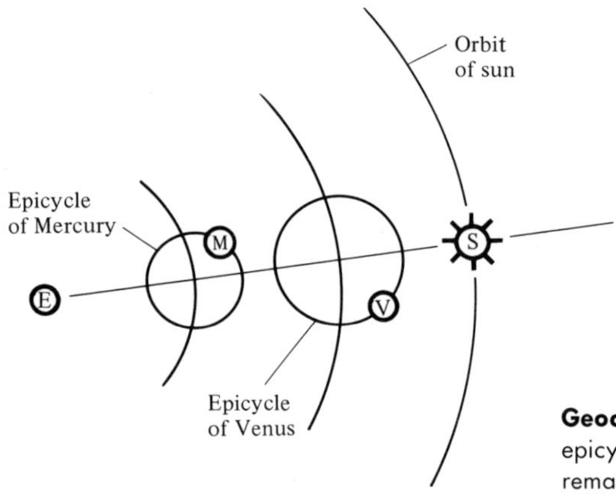

Geocentric system. The centers of the epicycles of Venus and Mercury always remain in direct line with the sun.

(a)

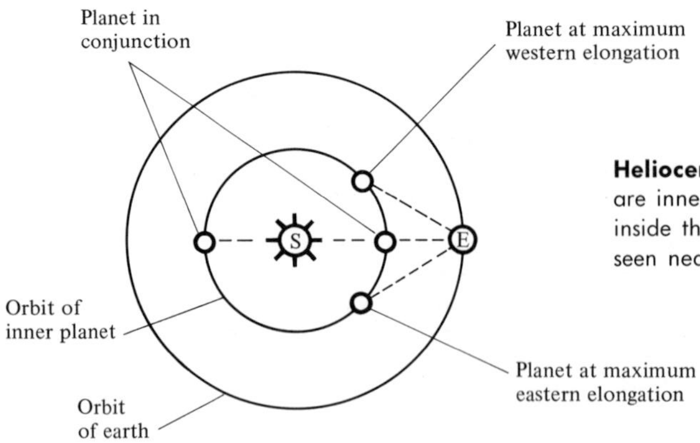

Heliocentric system. Venus and Mercury are inner planets. Since their orbits are inside the earth's orbit, they will always be seen near the sun.

(b)

RETROGRADE MOTION

To the ancient Greek astronomers, the most puzzling aspect of planetary motion was the retrograde motion regularly exhibited by the planets. The Greeks' devotion to Platonic principles required the use of the epicycle–deferent system to explain this phenomenon. Copernicus demonstrated that in the heliocentric theory retrograde motion occurs as an apparent motion caused by the motion of the earth. The planets do not travel with the same speeds; the closer a planet is to the sun, the faster it moves. Consequently, an inner planet periodically comes up from behind an outer planet, catches up with it, and passes it. When this happens, an observer on the inner planet sees the outer planet trace out a loop against the background of the fixed stars, as is shown in Figure 2–22. Note that when the outer planet is in opposition (position 6), it is in the middle of its retrograde motion. Similarly, an observer on the outer planet looking back at the inner planet also sees a retrograde loop being traced out.

figure 2–22 **RETROGRADE MOTION OF AN OUTER PLANET AS SEEN FROM AN INNER PLANET**

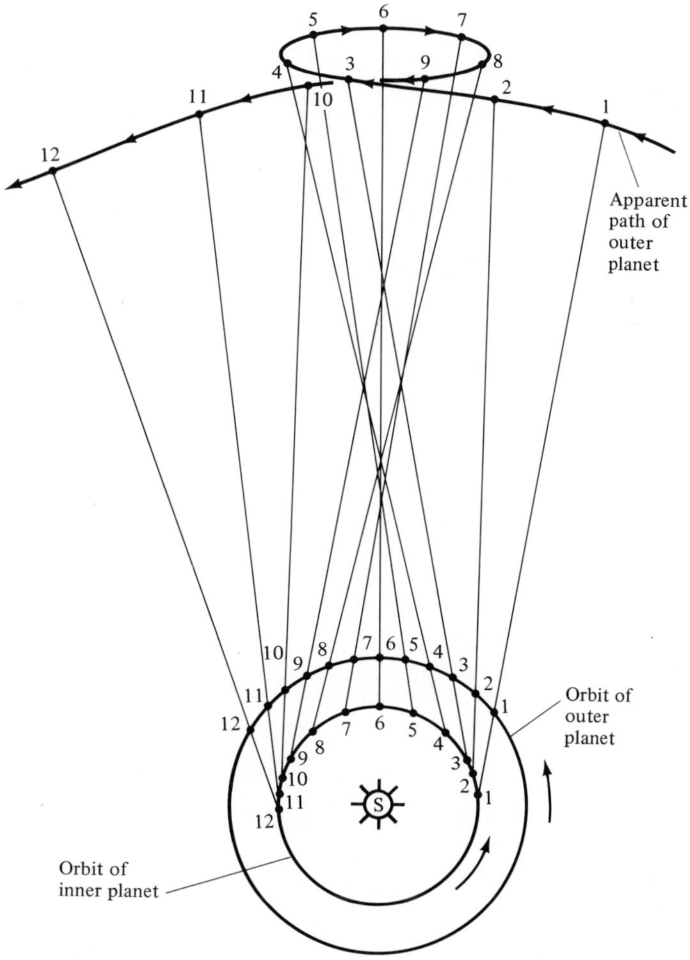

It follows that inner as well as outer planets show retrograde motion as seen from the earth; in fact, every planet exhibits retrograde motion as seen from every other planet.

We learned previously that the planets exhibit retrograde motion in a regular, periodic way. Each planet has a specific synodic period, which is the time from one retrograde loop to the next. In the heliocentric theory retrograde motion occurs whenever the planet goes into opposition; hence we can define the synodic period as *the time from one opposition to the next.* To illustrate this definition, let us refer to Figure 2–23. Suppose that the inner planet is initially at E_1 and the outer planet is at P_1, so that the outer planet will be in opposition, as seen from the inner planet. However, since the inner planet travels at a greater speed, it will move ahead of the outer planet. When the inner planet has reached E_2, the outer planet will have gone only as far as P_2 and will be in conjunction.

When the inner planet returns to E_1, completing one orbit around the sun, the outer planet will have reached P_1'. Finally, the inner planet catches up with the outer one at E_3. The outer planet (at P_3) is now in opposition again. Thus the synodic period (the time between successive oppositions) is the time required for the outer planet to go from P_1 to P_3. During this time the outer planet moves through the angle x ($= \angle P_1SP_3$). During the same interval of time, the inner planet moves more than one revolution through an angle $360° + x$. Therefore it follows that *one synodic period is the time required for an inner planet to gain 360°* (a full revolution) *on an outer planet.*

With this understanding of what happens during one synodic period, it becomes possible to determine how long it takes each planet to complete one orbit around the sun. In the heliocentric theory the time for one complete revolution is known as the ***sidereal period*** of the planet. Thus the sidereal period of the earth is one year or $365\frac{1}{4}$ days. The sidereal period of any other planet cannot be measured directly,

figure 2–23 **SYNODIC AND SIDEREAL PERIODS OF A PLANET**

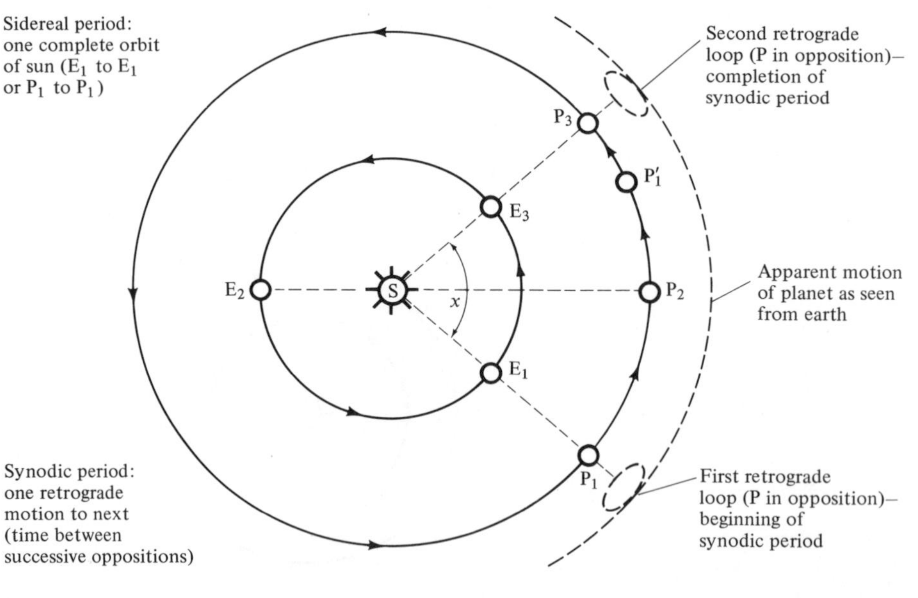

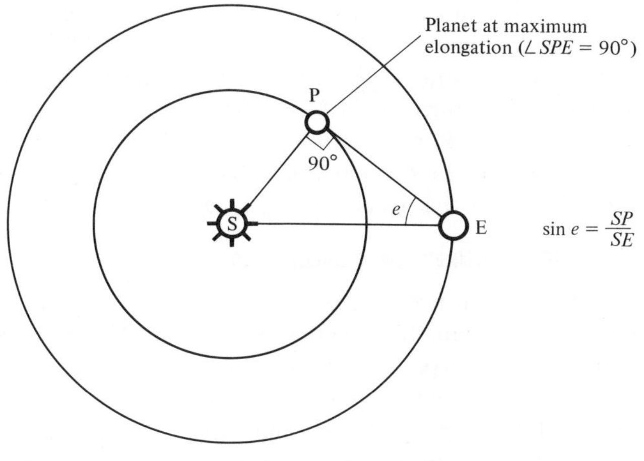

figure 2–24 **DETERMINATION OF THE ORBITAL RADIUS OF AN INNER PLANET**

At maximum elongation, the line from the earth to the planet is tangent to the orbit of the inner planet. Hence triangle *SEP* is a right triangle, and the sun-planet distance *SP* can be calculated by trigonometry.

but given the earth's sidereal period and the observed synodic period of a specific planet, it is a relatively simple matter to calculate its sidereal period. The sidereal periods of the major planets can be found in the Appendix (Section B–4).

ORDER AND DISTANCES OF THE PLANETS

In the geocentric theory the distances to the planets could not be calculated, and the order of the planets in the system was somewhat arbitrary. In the Copernican system, however, it is possible to calculate the radius of a planetary orbit, so that the order of the planets is completely determined. For an inner planet the orbital radius can be obtained rather easily from the observed value of the maximum elongation. As can be seen in Figure 2–24, when the planet is at maximum elongation, the line of sight from the earth (*EP*) is tangent to the orbit. Hence angle *SPE* is a right angle, and triangle *SPE* is a right triangle. Using trigonometry, we can find the sun–planet distance *SP* in terms of the sun–earth distance *SE*.

However, unless the sun–earth distance is accurately known (and it wasn't in Copernicus' time*), we can get only *relative* distances. Astronomers have found it convenient to define the average sun–earth distance as 1 ***astronomical unit*** (abbreviated to AU) and to measure all astronomical distances in AU. Thus from the observed maximum elongations of Venus and Mercury (angle *e* in Figure 2–24), it is possible to calculate that the sun–Venus distance is about 0.72 AU and the sun–Mercury distance is 0.39 AU. Hence we have established that the first three planets of the solar system are Mercury, Venus, and the earth, in that order.

The method for calculating the orbital radius for an outer planet is more complex, and it will not be discussed in any detail. Essentially, observations are made of the

* The value that Copernicus used, which came from an erroneous calculation made by Aristarchus, was 1200 times the earth's radius, or 4,800,000 miles. The presently accepted value is 92,956,000 miles.

changing elongation of the planet. From these data a triangle can be set up in which all angles are known, and one can use trigonometry to solve for the sun–planet distance in terms of the sun–earth distance. It has been found that the orbital radii of Mars, Jupiter, and Saturn are 1.52, 5.20, and 9.54 AU, respectively. Thus the relative distances of the known planets in the heliocentric system can be firmly established. Copernicus thereby demonstrated that the heliocentric system possesses an order that is not found in the geocentric system.

THE "FAILURE" OF COPERNICUS

We have seen that Copernicus was able to account for diurnal motion, the annual motion of the sun, the peculiar motions of Mercury and Venus, and the retrograde motions of the planets, and establish the order and sizes of the planetary orbits by using the heliocentric theory. This did not *disprove* the geocentric theory, however, for all of these phenomena could also be explained geocentrically, although in a somewhat more complicated manner. In this respect, we would have to say that Copernicus' heliocentric theory was a "better" theory than the geocentric theory because it was simpler.

Unfortunately, Copernicus was not able to maintain the simplicity of his system. The observational data indicated that the planets do not travel at uniform speeds around the sun and that the sun–planet distances are not constant. Therefore the planetary orbits could not be represented by simple circles centered on the sun. Copernicus firmly accepted the Platonic principle of uniform circular motion, so he began to add epicycles and eccentrics to the system in order to reproduce the motions of the planets with sufficient accuracy. He used nine circles to reproduce the motion of the earth alone and needed an epicycle on an epicycle to duplicate the moon's observed motions.

Hence the heliocentric system turned out to be much more complicated than Copernicus had expected. In his original rough calculations, described in the *Commentariolus,* Copernicus had predicted that only thirty-four circles would be needed, but when he worked the theory out in full, he found that the system contained forty-eight circles—eight *more* than in the Ptolemaic system! In addition, his predictions of planetary positions, using his own system, were no more accurate than those obtained with the Ptolemaic system. No doubt Copernicus was disappointed, and we can understand why he might have been so reluctant to publish his findings. His great work, *De Revolutionibus,* sold less than one thousand copies in the twenty years following its publication.

Why, then, is Copernicus listed among the greats of history? He was not the first person to suggest that the earth moves; in his writings he refers to the theories of Heracleides and Aristarchus. It is difficult to think of Copernicus as a man ahead of his times, for his system was firmly rooted in the Pythagorean–Platonic tradition of uniform circular motion. In this respect Copernicus was perhaps the last great ancient astronomer.

We may best think of Copernicus as a catalyst. Even though Copernicus' devotion to uniform circular motion led him to a scientific dead end, other astronomers could see the obvious advantages of the heliocentric theory and began to develop it along more fruitful lines. Thus by reviving the heliocentric theory, Copernicus triggered the overthrow of geocentrism. It is a curious fact, nonetheless, that one of the greatest intellectual revolutions of history was begun by a man who was much more of a conservative than a revolutionary.

2–6

KEPLER

During the first fifty years after his death, Copernicus' theory made only slight headway. Quite a few astronomers studied Copernicus' works, and some even taught the new theory in their classes. Nevertheless, not many people were willing to accept the idea that the huge, solid, heavy earth was in motion while the stars—tiny, ethereal points of light—were stationary. The first staunch defender of the Copernican theory was the German mathematician Johannes Kepler (1571–1630). As a student Kepler showed exceptional aptitude for mathematics and astronomy, although his main interest was in theology. When, upon graduation, he was recommended for a position as a teacher of mathematics, Kepler somewhat reluctantly accepted, giving up his hope of entering the ministry.

Philosophically, Kepler was a throwback to the Pythagoreans. He believed that mathematical harmonies were the basis of all natural phenomena and that the universe was built according to some geometrical scheme. He conceived of the Creator of the Universe as the Great Geometer in the Sky:

> Why waste words? Geometry existed before the Creation, is co-eternal with the mind of God, is God himself. . . . geometry provided God with a model for the Creation and was implanted into man, together with God's own likeness.[2]

Kepler adopted the Copernican theory as a student, but he was puzzled because the planetary distances did not seem to follow any simple mathematical pattern. Led on by his belief in the underlying mathematical harmony of the universe, Kepler devised a complex geometric model of the solar system in which the planetary spheres were either inscribed in or circumscribed about regular geometric solids. In this way, the distances of the planets could be given a geometric basis. Moreover, since there are only five regular geometric solids,* there could be only six planets in the scheme, which is the exact number of planets in the Copernican system.

However, Kepler found that the planetary distances predicted by his geometric model were only approximately equal to the observed distances. Therefore, he could not consider his hypothesis to be proved. He realized that he could go no further with the astronomical data available to him. Most of these observations dated back to Hipparchus, and they were full of inaccuracies. Fortunately for Kepler, the Danish astronomer Tycho Brahe, who was reputed to have amassed over twenty years' worth of the most accurate astronomical data ever known, was only a few hundred miles away and wanted Kepler to work with him.

TYCHO BRAHE

Born in 1546, Tycho Brahe began to study astronomy at the age of fourteen. Unlike Kepler (who struggled his entire life to achieve financial solvency), Tycho was a member of the nobility. His uncle had saved the life of King Frederick II, and out of gratitude the king in 1576 gave Tycho an island off the coast of Denmark

* A regular geometric solid is a completely symmetric solid whose faces are regular polygons. The ancient Greek geometers had proved that only five such solids exist: the tetrahedron, cube, octahedron, dodecahedron, and icosahedron.

and the funds to build an astronomical observatory there. Tycho supervised the construction of the largest and most accurate naked-eye observing instruments ever made. For the next two decades Tycho and his assistants carefully cataloged the positions of the stars and planets. Tycho jealously guarded this body of data; rather than release it for use by other astronomers, he planned to pass it on to his children as a legacy.

Two of Tycho's observations are worth mentioning. In 1572 a star suddenly flared up until it was brighter than Venus and could be seen in the daytime. (It was a star that violently exploded, producing a supernova. Kepler observed another such event in 1604.) Tycho's measurements of the star showed that it was a very distant object and not an atmospheric phenomenon. This discovery contradicted the Aristotelian notion that the stars were eternal and unchanging. Tycho made similar observations of the comet of 1577, showing that it was crossing the orbits of the planets. At that time comets were thought to be condensations of vapor in the atmosphere, and it was believed that the planetary spheres were impenetrable. Both discoveries did considerable damage to the philosophical structure of the geocentric theory.

Tycho also tried to observe stellar parallax, but he was unsuccessful. Unable to accept the idea that the stars could be so distant, he rejected the Copernican theory and proposed his own system, in which the earth is stationary and the sun and stars revolve about the earth, but the planets revolve around the sun. The Tychonic system gained much popularity among astronomers because it accounted for the observed planetary motions without requiring the earth to be in motion.

Tycho's high-handedness in dealing with both royalty and the peasants on his island eventually made him unwelcome in Denmark. In 1599 he went to Prague to become the imperial mathematician to Emperor Rudolph II. Tycho knew that he was not a good enough mathematician to find the proof for his model of the universe in his own data. He had heard of the brilliant young mathematician Kepler and invited Kepler to come to Prague as his assistant. Kepler, who was eager to get his hands on Tycho's data, accepted.

Their collaboration was a short and stormy one. Kepler apparently resented being treated as a mere assistant, and Tycho would not give him access to the data. Then in 1601 Tycho died when his bladder burst after he had overeaten at a banquet. Kepler then became the new imperial mathematician and set to work on Tycho's data.*

THE *ASTRONOMIA NOVA*

Kepler began by trying to determine the orbit of Mars, which was the most difficult of all the planets to fit. He succeeded in finding a circle that fit the new data rather well. At most, it deviated from the observed path of Mars by only 8 minutes of arc, a rather small angle. However, Tycho's data was accurate to 4′. The circle simply did not fit the facts, so Kepler rejected it.

Kepler realized that he would now have to compute from the observations the entire orbit of Mars, point by point, and then see what kind of a curve could be fitted to the orbit. Moreover, since the sun–Mars distance could only be determined relative to the sun–earth distance, Kepler would first have to compute the orbit of

* Some years later, Kepler explained to a friend how he managed to get the data: "I confess that when Tycho died, I quickly took advantage of the absence, or lack of circumspection, of the heirs, by taking the observations under my care, or perhaps usurping them. . . . I was in possession of the observations and refused to hand them over to the heirs."[3]

the earth. He spent several years making these exceedingly laborious calculations. In doing so, he found that the earth did not move around the sun with uniform speed, as Copernicus had insisted, but moved in such a way that its speed increased as its distance from the sun decreased. (This discovery, which became Kepler's second law, will be studied in more detail in the following section.) Having worked out the earth's orbit, Kepler returned to the orbit of Mars. After many more years of labor, he realized that the orbit of Mars was an *ellipse*. In his later works he was able to show that all the planets and the moon travel in elliptical orbits with varying speed. Thus Kepler became the first astronomer to reject openly the Platonic principle of uniform circular motion.

Kepler described his findings in his *Astronomia Nova*,* which was published in 1603. In this work he introduced an important new concept: the idea that there was a *force emanating from the sun that drove the planets around in their orbits.* Kepler thereby proposed that planetary motion has a *physical cause:* namely, a force of *gravitation* existing between the sun and the planets. But he could not find a way to relate mathematically the force of gravitation to the motions of the planets. This was achieved fifty years later by Isaac Newton, and it is he who is credited with the formulation of the theory of gravitation.

Even though Kepler's discoveries contradicted the ancient doctrine of circular motion, he never gave up his Pythagorean mysticism. In his next book, *The Harmony of the World,* Kepler purported to find all sorts of musical harmonies and geometric ratios in the orbits of the planets (although by now he had given up all hope of fitting the planetary distances to the geometric solids). Most of these "harmonies" were scientifically worthless, with one exception: a relationship between the periods and distances of the planets, which was to become Kepler's third law.

By this time Kepler was known as one of the leading astronomers of his day, but his existence was a miserable one. His main duty as the imperial mathematician was the casting of horoscopes, and his salary was invariably in arrears. During the last twelve years of Kepler's life, his homeland was ravaged by war. He was forced to wander from city to city to seek employment or to avoid religious persecution. At his death in 1630, he was unemployed and penniless.** It was almost fifty years after his death before the significance of Kepler's work came to be recognized.

KEPLER'S LAWS OF PLANETARY MOTION

Kepler's most important discoveries are his three laws of planetary motion. They are as follows:

(1) *The Law of the Ellipse.* Every planet travels in an elliptical orbit with the sun at one focus of the ellipse.
(2) *The Law of Equal Areas.* A planet moves such that the line from the sun to the planet sweeps out equal areas in equal times.
(3) *The Harmonic Law.* The ratio of the square of the period to the cube of the average sun–planet distance is the same for all planets.

* The full title of the work is *A New Astronomy Based on Causation or a Physics of the Sky Derived from Investigations of the Motions of the Star Mars Founded on Observations of the Noble Tycho Brahe.*

** Kepler's last work was the *Somnium* ("Dream"), a story about a voyage to the moon and the inhabitants thereof. This book, which was never completed, is generally regarded as the first work of science fiction ever written.

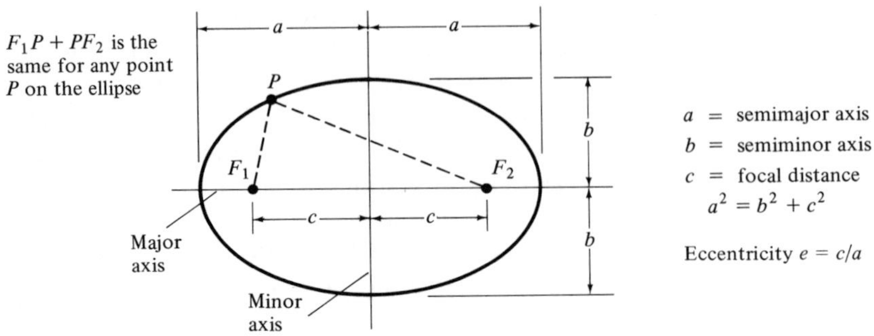

figure 2–25 **GEOMETRIC PROPERTIES OF THE ELLIPSE**

Let us now consider each of these in turn.

The Law of the Ellipse. The ellipse is a well-defined geometric figure whose properties had been carefully studied by the ancient Greeks. Given two fixed points F_1 and F_2 (see Figure 2–25), an ellipse has the property that for any point P on the ellipse, the sum of the two distances $F_1P + PF_2$ has the same value as for any other point on the curve. F_1 and F_2 are the ***foci*** of the ellipse. (Each one is called a ***focus.***) The distance c from the center of the ellipse to either F_1 or F_2 is called the ***focal distance.*** The distances a and b, measured from the center along the long and short axes of the ellipse, are called the ***semimajor axis*** and ***semiminor axis,*** respectively. It is possible to show that for any ellipse the three quantities a, b, and c are related by the equation

$$a^2 = b^2 + c^2 \tag{2–1}$$

The greater the disparity between a and b, the more flattened the ellipse will be. A useful measure of the "flatness" of an ellipse is the ***eccentricity,*** e, defined as follows:

$$e = \frac{c}{a} \tag{2–2}$$

An ellipse with an eccentricity of zero would have $c = 0$; the two foci would merge into one at the center. It follows from Eq. 2–1 that if $c = 0$, then $a = b$. The figure is actually a circle. Thus we may consider the circle as a special case of an ellipse; that is, a circle is an ellipse with zero eccentricity.

As the eccentricity of an ellipse increases, b becomes small in proportion to a, and c increases. (The foci move outward along the major axis.) In the extreme limit, b approaches 0; the ellipse approaches a straight line; and e approaches 1. Thus all ellipses have eccentricities between 0 and 1; the more flattened the ellipse, the closer e is to 1. A number of ellipses of different eccentricities are shown in Figure 2–26.

The law of the ellipse (Kepler's first law) states that a planet travels in an elliptical orbit with the sun at one focus. (See Figure 2–27.) Note that the sun is at one *focus* of the ellipse and *not at the center.* (The other focus is just an empty point in space.) From the geometry of the ellipse it is possible to prove the following:

(1) The planet is *closest* to the sun when it is on the long axis of the ellipse, on the same side of the center as the sun. This point is called the ***perihelion***

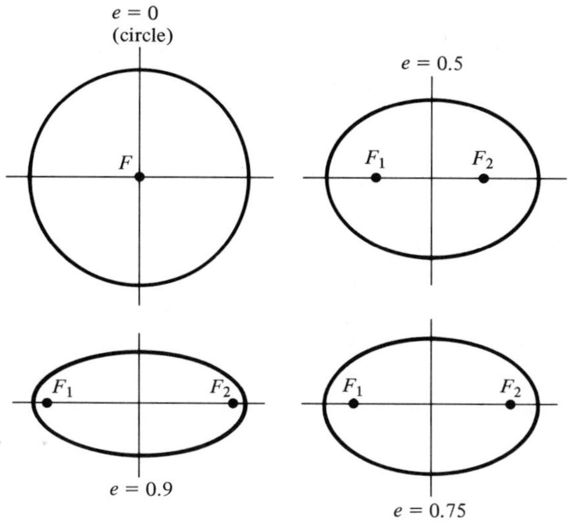

figure 2–26

ECCENTRICITY OF AN ELLIPSE

These four ellipses have different eccentricities, but the same major axis.

(from Greek *peri* = "near"; *helios* = "sun"). The perihelion distance (d_{peri}) is equal to $a - c$.

(2) The planet is *farthest* from the sun when it is on the long axis, on the opposite side of the center from the sun. This point is called the ***aphelion*** (from Greek *apo* = "away"). The aphelion distance (d_{ap}) is equal to $a + c$.

(3) The *average* sun–planet distance over one revolution is equal to a, the semimajor axis.

Given the aphelion and perihelion distances of a planet, it is a simple matter to determine the average distance, the focal distance, and the eccentricity of the orbit. Since

$$d_{\text{ap}} = a + c \tag{2–3a}$$
$$d_{\text{peri}} = a - c \tag{2–3b}$$

It is easy to show that

$$a = \tfrac{1}{2}(d_{\text{ap}} + d_{\text{peri}}) \tag{2–4a}$$

$$c = \tfrac{1}{2}(d_{\text{ap}} - d_{\text{peri}}) \tag{2–4b}$$

Hence

$$e = \frac{c}{a} = \frac{d_{\text{ap}} - d_{\text{peri}}}{d_{\text{ap}} + d_{\text{peri}}} \tag{2–5}$$

The orbits of the planets have rather small eccentricities; seven of the nine major planets have e less than 0.1. (See Appendix, Section B–4.) Consequently, the planetary orbits are not easily distinguishable from circles. Since the earth's orbit is not perfectly circular, the sun–earth distance is not constant. The *average* sun–earth distance (previously defined as 1 AU) is equal to the semimajor axis of the earth's orbit. In other words, $a = 1$ AU for the earth's orbit. Since the orbital eccentricity is 0.017, it follows that the sun–earth distance varies from 1.017 AU at aphelion to 0.983 AU at perihelion. The earth reaches perihelion on January 4 and aphelion on July 5. Hence the earth is actually *closest* to the sun in January (when it is winter in the northern hemisphere). However, the effects on climate of the variation in the sun–earth distance are relatively minor.

The Law of Equal Areas. Figure 2–28 shows that as the planet moves from P_1 to P_2, the line from the planet to the sun sweeps out an area. Kepler discovered that if the planet takes the *same amount of time* to go from P_3 to P_4 or from P_5 to P_6 as it does to go from P_1 to P_2, then the areas swept out will be the same. As the planet moves from P_1 to P_6, its distance from the sun is steadily decreasing. It follows that if areas I, II, and III are equal to one another, the distance P_5P_6 must be greater than P_3P_4, which in turn must be greater than P_1P_2. But the planet covers each of these distances in the same amount of time. Consequently, the planet must be moving at a *greater speed* when going from P_5 to P_6 than it is when going from P_3 to P_4, and its speed from P_3 to P_4 is greater than its speed from P_1 to P_2.

In essence, Kepler's second law shows that the planet moves around the sun with varying speed. The planet is at its minimum speed at aphelion, when it is farthest from the sun. As the sun–planet distance decreases, the planet's speed increases; at perihelion the speed is at a maximum.

The Harmonic Law. As a part of his effort to find mathematical harmonies in the universe, Kepler tried to find a mathematical relationship linking the sidereal period (P) and average orbital radius (a). Through a trial-and-error process, Kepler found that the quantity P^2/a^3 has the same numerical value for all the planets. (See Table 2–1.) We can put this result in the form of a simple equation:

$$\frac{P^2}{a^3} = K \qquad (2\text{–}6)$$

where K is a numerical constant that has the same value for all planets. The actual value of K depends on the units used to measure P and a. P can be measured in seconds, days, years, and so on; a can be measured in astronomical units, miles, feet,

figure 2–27 **ORBIT OF A PLANET ACCORDING TO KEPLER'S FIRST LAW**

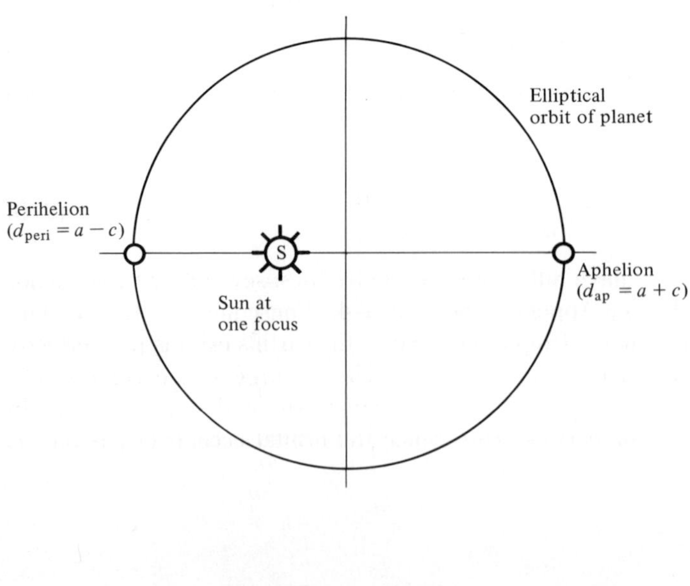

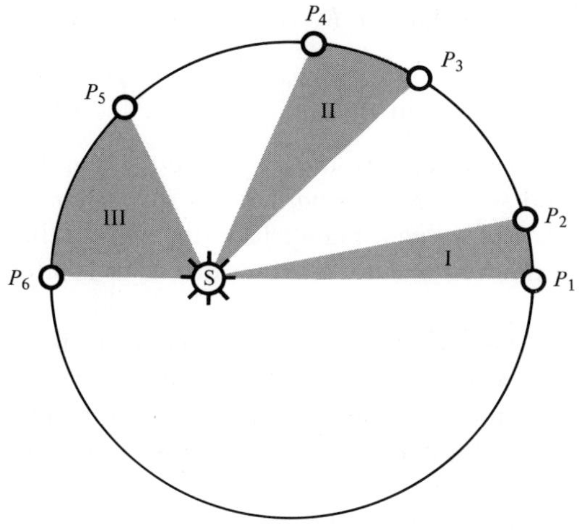

figure 2–28

THE LAW OF EQUAL AREAS (KEPLER'S SECOND LAW)

The areas I, II, and III are equal to one another. The planet takes the same amount of time to move from P_1 to P_2 as it does to move from P_3 to P_4 or from P_5 to P_6. Hence the speed of the planet increases as it moves from P_1 to P_6.

centimeters, and so on. K will have a different value for each combination of units chosen for P and a, but for any chosen set of units, K will have the same value for all planets.

The simplest choice of units are those used by Kepler himself—namely, years and AU. As can be seen in Table 2–1, if P is measured in years and a is measured in AU, then $P^2/a^3 = 1$ for all planets. In other words, $K = 1$ for all planets, *as long as* P *is measured in years and* a *in AU.* Suppose, for example, we choose seconds and centimeters as our units. Since 1 year $= 3.16 \times 10^7$ sec and 1 AU $=$ 92,956,000 miles $= 1.50 \times 10^{13}$ cm,*

$$K = \frac{P^2}{a^3} = \frac{(1 \text{ year})^2}{(1 \text{ AU})^3} = \frac{(3.16 \times 10^7 \text{ sec})^2}{(1.50 \times 10^{13} \text{ cm})^3} = 2.96 \times 10^{-25} \text{ sec}^2/\text{cm}^3$$

table 2–1 **EVIDENCE FOR KEPLER'S THIRD LAW**

Trial-and-error calculations using various combinations of multiples of P and a led to the discovery that $P^2/a^3 = 1$ for all planets orbiting the sun.

Planet	Sidereal period, P (yr)	Average orbital radius, a (AU)	P/a	P^2/a	P/a^2	P^2/a^3
Mercury	0.241	0.387	0.623	0.150	1.61	1.00
Venus	0.615	0.723	0.851	0.523	1.18	1.00
Earth	1.00	1.00	1.00	1.00	1.00	1.00
Mars	1.88	1.52	1.24	2.33	0.814	1.00
Jupiter	11.86	5.20	2.28	27.0	0.439	1.00
Saturn	29.46	9.54	3.09	91.0	0.324	1.00

* Students who are unfamiliar with the powers-of-ten arithmetic should consult Appendix A.

Obviously, years and AU give the simplest value for K. Kepler's third law can therefore be stated as follows:

$$P^2 = a^3 \qquad (P \text{ in years, } a \text{ in AU}) \tag{2–7}$$

In this form Kepler's third law is applicable to *any object in orbit about the sun,* including comets, artificial satellites, and so on. *K does not* equal 1 for the moon or any other planetary satellites. All satellites of a given planet have the *same* value of K, however. (Kepler himself showed that this was true for the four moons of Jupiter discovered by Galileo in 1610.)

Years later Newton was able to derive Kepler's third law mathematically from the theory of gravitation. He showed that the value of K for a given satellite system depends on the mass of the central body. K for the solar system, therefore, differs from the value of K for the satellite system of Jupiter, because the mass of the sun is different from the mass of Jupiter.

Kepler's laws of planetary motion allow us to make useful calculations pertaining to the orbits and periods of natural or artificial satellites. The following examples illustrate some simple problems that can be solved using the equations developed above.

EXAMPLE 2–1: Halley's comet has a perihelion distance of 0.6 AU and an aphelion distance of 35.2 AU. Calculate: (a) the average distance from the sun, (b) the eccentricity of the orbit, (c) the sidereal period.

SOLUTION: We use Kepler's first and third laws in conjunction. Since

$$d_{peri} = 0.6 \text{ AU} \qquad \text{and} \qquad d_{ap} = 35.2 \text{ AU}$$

we can calculate a, the average distance from the sun, using Eq. 2–4a:

$$a = \tfrac{1}{2}(d_{ap} + d_{peri}) = \tfrac{1}{2}(35.2 + 0.6) = \underline{17.9 \text{ AU}}$$

To determine the eccentricity we first calculate c using Eq. 2–4b:

$$c = \tfrac{1}{2}(d_{ap} - d_{peri}) = \tfrac{1}{2}(35.2 - 0.6) = 17.3 \text{ AU}$$

Then

$$e = \frac{c}{a} = \frac{17.3}{17.9} = \underline{0.966}$$

Finally, using Eq. 2–7, we obtain

$$P^2 = a^3 = (17.9)^3 = 5735$$

or

$$\underline{P = 76 \text{ years}}$$

EXAMPLE 2–2: An artificial satellite in orbit around the sun has a sidereal period of 8 years. Its closest approach to the sun is 1 AU. Calculate the eccentricity of the orbit.

SOLUTION: From the sidereal period $P = 8$ years, we can calculate the average radius a:

$$P^2 = a^3$$
$$8^2 = 64 = a^3$$
$$a = 4 \text{ AU}$$

The perihelion distance is 1 AU; thus

$$d_{\text{peri}} = a - c$$
$$1 = 4 - c$$
$$c = 3 \text{ AU}$$

The eccentricity is therefore

$$e = \frac{c}{a} = \frac{3}{4} = \underline{0.75}$$

THE SIGNIFICANCE OF KEPLER'S WORK

Kepler's contributions to the development of astronomy were of monumental importance. He rid astronomy of the doctrine of circular motion; in one stroke he did away with epicycles, eccentrics, and the like. He thereby reduced the forty-eight circles of the Copernican system to only seven ellipses (one for each of the planets and one for the moon) and greatly simplified the heliocentric theory. Kepler was the first to adduce a physical cause for planetary motion; his idea of a force emanating from the sun that causes the planets to move was a forerunner of the concept of gravitation.

Kepler's most substantial contribution was the three laws of planetary motion. Note that we refer to them as *laws* because they do nothing more than *describe* the way the planets move. We will learn in later chapters that Newton was able to *explain* Kepler's laws by deriving them mathematically from the theory of gravitation. Without Kepler's laws to guide him, Newton might not have been able to achieve his great success. Kepler's laws of planetary motion represent the starting point of modern astronomy.

Even though Kepler's heliocentric system was considerably less complicated and more accurate than the Copernican system, Kepler's discoveries did not win wide acceptance at first. The concept of uniform circular motion was nearly two thousand years old, and people were not willing to discard it. Moreover, Kepler had not disproved the geocentric theory; the old arguments, both theological and scientific, against the motion of the earth had still not been satisfactorily answered. Thus the Keplerian system could be looked on as just another mathematical hypothesis.

The biggest barrier against the acceptance of heliocentrism was that the combined authority of Aristotelian philosophy and Christian theology supported the geocentric theory. According to these views, the universe had been constructed by God for the edification of man, and clearly man belonged at the center. The universe was finite and rather small; outside of the sphere of fixed stars was heaven, the abode of God and His angels.

In the heliocentric universe man was no longer at the center; the earth was a mere planet. In this system the Aristotelian division of the universe into the sublunary world (made of the four elements) and the celestial world (made of the eternal, unchanging quintessence) did not make any sense. Furthermore, some astronomers were beginning to suggest that the universe is infinite and that the stars are spread through space rather than lying on a single sphere. In such a universe there could be no center. A few thinkers went so far as to suggest that there were other solar systems, other earths, and other civilizations. In the infinite universe man was not unique, and there was no place for a heaven.

Clearly, heliocentrism posed a serious threat to orthodoxy. Many of the newly formed Protestant sects adopted a fundamentalist view—that is, that the Bible was the only source of true knowledge, and it was to be interpreted literally. The official position of the Catholic Church was that the Holy Scriptures supported the geocentric view, but that they could be reinterpreted if there were convincing proof of the earth's motion. Astronomers were free to discuss heliocentrism, but only as a hypothesis. Early in the seventeenth century matters finally came to a head.

2–7

GALILEO

Galileo Galilei was born in 1564 in Pisa, Italy. He studied at the University of Pisa and was appointed to a professorship there at the age of twenty-five. Three years later he left to become a professor of mathematics at the University of Padua. A man of many contradictions, Galileo adopted the Copernican view when he was in his twenties, but did not espouse it openly until he was nearly fifty. In his lectures he taught the Ptolemaic system. He also corresponded with Kepler, but never accepted Kepler's discoveries. Galileo was a firm believer in circular motion and accepted the Copernican system, epicycles and all.

In 1609 Galileo learned of a new invention—an "optick tube" (later renamed the telescope) containing two lenses with which one could see distant objects magnified many times. He figured out the principle behind the device and constructed one for his own use. When Galileo pointed his telescope at the sky, the science of astronomy was revolutionized—astronomers, who had always been limited to naked-eye observations, now had a completely new view of the universe.

THE "MESSENGER FROM THE STARS"

Galileo constructed several telescopes (the most powerful of which had a magnification of 30×) and examined all of the important celestial bodies. In 1610 he published his discoveries in a short pamphlet, the *Sidereus Nuncius* ("Messenger from the Stars"). Galileo broke with scientific tradition by writing in colloquial Italian rather than in scholarly Latin. The work was widely circulated and created a sensation.

Galileo's discoveries contradicted the very foundations of Aristotelian physics and Ptolemaic astronomy. Galileo found that the sky was filled with an uncountable number of stars that were not visible to the naked eye. If God had put stars in the sky for the benefit of man, what was the point of making them invisible? Galileo observed that under magnification the planets looked like disks, but the stars remained points of light. The fact that the size of the stars could not be magnified showed that they were very far away, just as Aristarchus and Copernicus had predicted. When Galileo looked at the moon, he found that it had an irregular surface, with mountains and innumerable craters. Thus the moon had a surface much like the earth and was not perfectly smooth and spherical, as the Aristotelian view required.

Galileo discovered four small satellites traveling in circular paths around Jupiter (Figure 2–29). This observation showed conclusively that not all celestial bodies

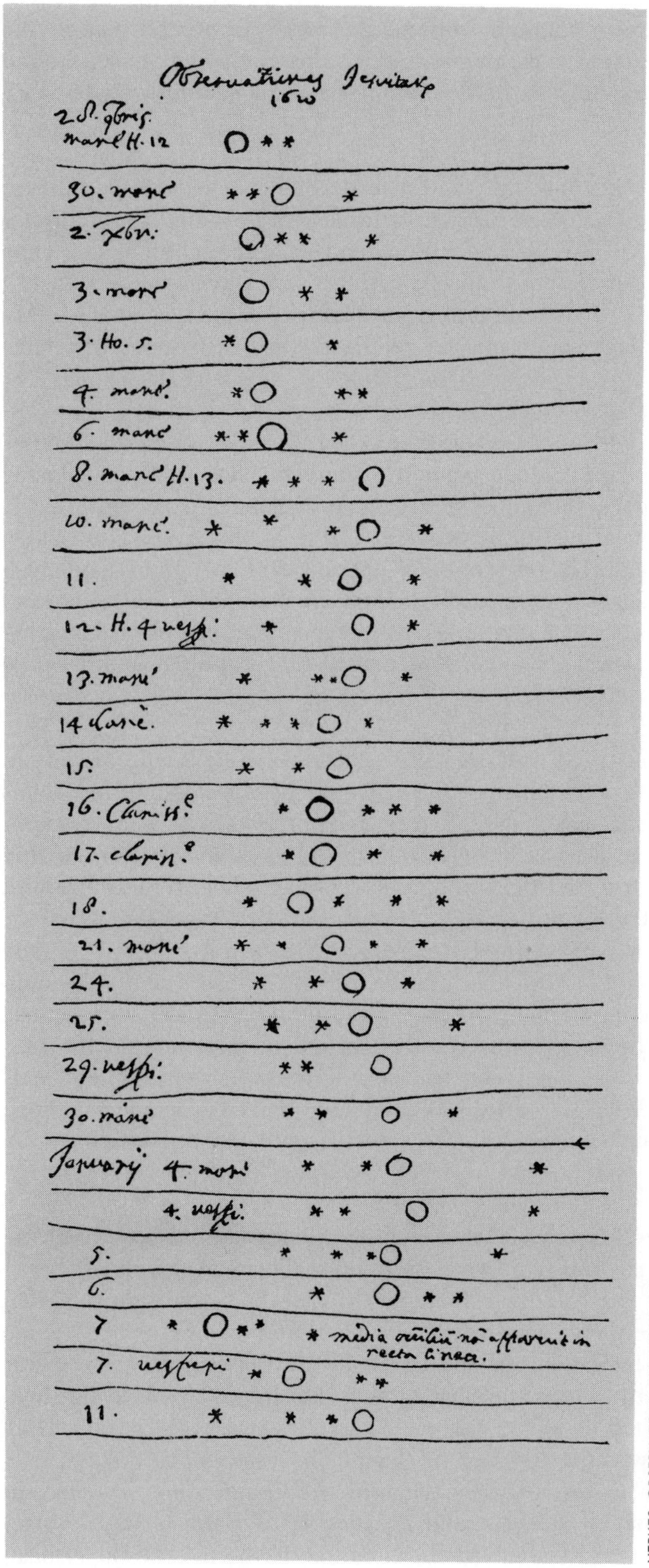

YERKES OBSERVATORY PHOTOGRAPH

figure 2–29

GALILEO'S OBSERVATIONS OF THE SATELLITES OF JUPITER

This page from Galileo's notebook shows the different positions of the satellites with respect to Jupiter as he observed them on successive nights.

revolve around the earth. It also countered one of the major arguments against the heliocentric system, for the earth was no longer unique in having a body going around it while it traveled around the sun. To Galileo, Jupiter and its satellites represented a miniature solar system.

At first, Galileo's discoveries were ridiculed. Some of the more fanatic Aristotelians refused to believe what they saw, and a few would not even look through a telescope. One astronomer, Francesco Sizzi, dispensed with the satellites of Jupiter by declaring "the satellites are invisible to the naked eye, and therefore can have no influence on the earth, and therefore would be useless, and therefore do not exist." In one of his letters to Kepler, Galileo described the reaction to his discoveries. "My dear Kepler," he wrote: "What would you say of the learned men here, who have steadfastly refused to cast a glance through the telescope? What shall we make of all this? Shall we laugh, or shall we cry?"

Before long, however, Galileo's discoveries were confirmed by Kepler and other well-known astronomers. In 1611 Galileo made a triumphant visit to Rome, during which he received many honors, including an audience with the Pope.

There were still more surprises in the heavens. Galileo's observations of Saturn showed that it was not spherical. (His telescope was not powerful enough to distinguish Saturn's rings.) He saw sunspots appear, grow, fade, and disappear, indicating that the sun was neither perfect nor unchanging. The discovery that was most damaging to the geocentric theory, however, was the fact that Venus exhibited a full set of phases, like the moon (Figure 2–30).

Venus appears as a large crescent, then gradually shrinks in size as it goes from crescent to half to three-quarter to full. In the Ptolemaic system Venus travels in an epicycle between the earth and the sun, and it would not be possible for Venus to exhibit a full or even a half phase. (See Figure 2–31a.) However, if Venus goes around the *sun*, a full set of phases could be seen, with the crescent phase occurring when Venus is close to the earth and the full phase occurring when Venus is farthest from the earth (Figure 2–31b).

Galileo's observations clearly showed that Venus circles the sun and not the earth. This discovery disproved the Ptolemaic system, which was based on the assumption that the earth is the center of all planetary motion. It did *not* prove the heliocentric system, however. Galileo's discovery showed only that Venus goes around the sun, not that the earth goes around the sun. Tycho's geocentric system, in which the planets circle the sun while the sun circles the earth, was still valid. Many astronomers adopted the Tychonic system as a compromise so that the earth could remain stationary.

THE TRIAL OF GALILEO

Galileo, however, studiously ignored the Tychonic system and asserted that the Copernican system had been rigorously proved. As additional evidence, Galileo offered a theory in which he showed that the ocean tides result from the combined revolution and rotation of the earth. The theory was totally erroneous and unconvincing. Galileo insisted that the Catholic Church should declare that the Copernican system was physically true and demanded a formal hearing by the Vatican. At this hearing (held in 1616), it was decided that belief in the earth's motion was contrary to Holy Scriptures and that Galileo was not to teach or defend the Copernican system.

Galileo remained silent for several years. Then in 1623 a new pope was elected. Pope Urban VIII was an amateur scientist and an admirer of Galileo. Shortly after his election, Pope Urban informed Galileo that he was free to discuss the Copernican

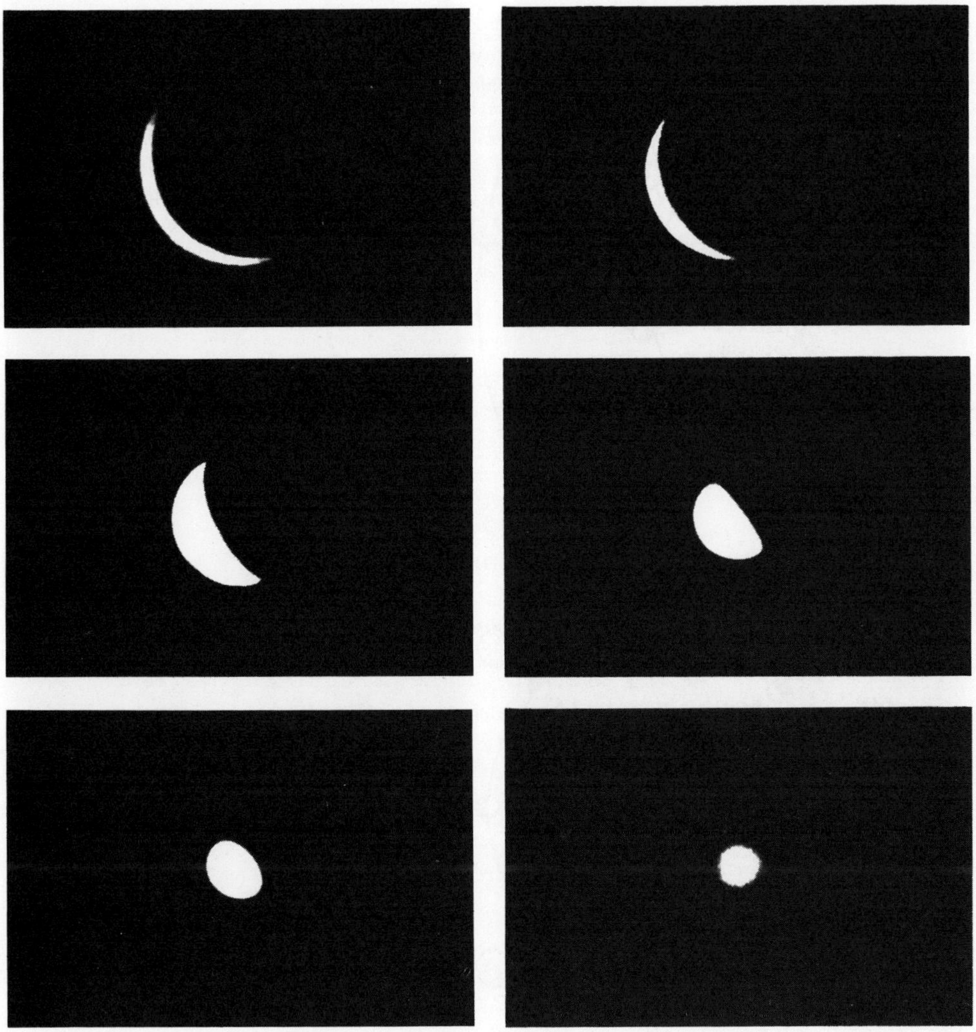

BROOKLYN COLLEGE OBSERVATORY PHOTOGRAPHS TAKEN BY ALLEN SELTZER

figure 2–30 **THE PHASES OF VENUS**

This sequence of photographs of the planet Venus (taken over a one-year period) shows that Venus exhibits a full set of phases. Note the decrease in the apparent size of Venus as it progresses from crescent to full.

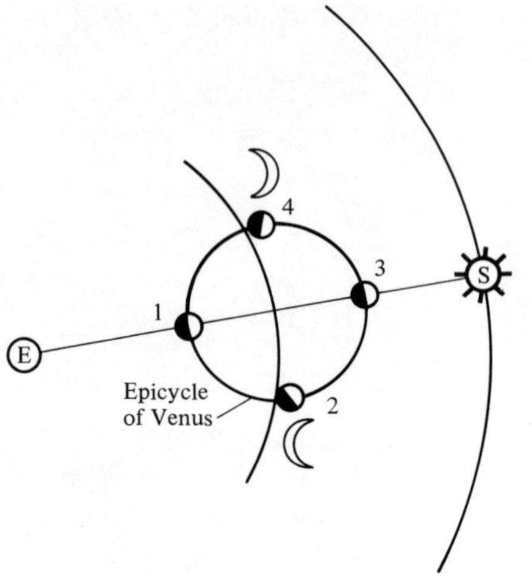

(a)

Geocentric system. The phases of Venus range from new (at 1 and 3) to crescent (at 2 and 4). No other phases are possible.

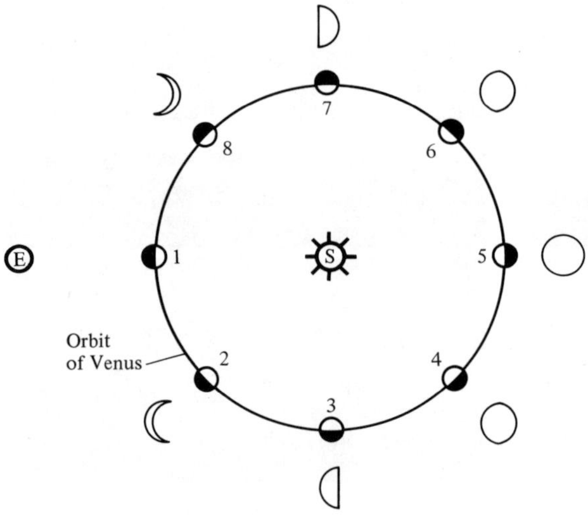

(b)

Heliocentric system. If Venus circles the sun, it exhibits a full set of phases. It is closest to the earth when new (at 1) and farthest from the earth when full (at 5).

figure 2–31 **THE PHASES OF VENUS ACCORDING TO THE GEOCENTRIC AND HELIOCENTRIC THEORIES**

system as long as he treated it as a hypothesis and not the truth. Galileo then set to work on the *Dialogue Concerning the Two Chief World Systems,* which he completed in 1630. This was a brilliantly written defense of the Copernican system. All of the old Aristotelian arguments against the earth's motion and in favor of geocentrism were demolished. Galileo also offered several "proofs" of the earth's motion, including his incorrect theory of the tides. He then managed to get the book published without its being carefully checked by the censors. The Pope, furious with what he considered to be Galileo's duplicity, stopped publication of the book and had Galileo brought before the Inquisition.

Galileo was accused, not of heresy (as is often thought), but of violating the injunction not to defend the Copernican system. At the time of his trial, Galileo was probably the most eminent scientist in Europe. He was also seventy years old and in poor health. Accordingly, he was handled with great care. At a time when most prisoners of the Inquisition were kept in windowless jail cells, Galileo was given a large apartment and a servant for the duration of his trial.

At first he insisted that he tried to *refute* the heliocentric theory in his book and had not done a good job of it. This convinced no one, so Galileo finally agreed to sign a statement in which he admitted his errors and swore that he no longer believed in the Copernican theory. He was sentenced to remain under house arrest for the rest of his life. During the years between his trial and his death in 1642, he completed his studies on motion; his greatest work, the *Dialogues Concerning Two New Sciences,* was smuggled out of Italy and published in Holland.

The conflict between Galileo and the Church was a sorry affair, which benefited neither side. Galileo provoked the battle, and the Church overreacted. As a result Galileo was humiliated, and the Church became a suppressor of free scientific inquiry. Most likely, if Galileo had not intervened, the geocentric theory would have fallen of its own weight anyway, and the heliocentric theory would have been accepted without a struggle. This was what happened in England, where Copernicanism was discussed freely, and it was to England that the center of scientific inquiry shifted following Galileo's trial.

QUESTIONS FOR REVIEW

1 What is a diurnal motion? Which celestial objects exhibit it?

2 What distinguishes Polaris from the other stars? How does its position in the sky depend on the observer's location on the earth? Where on the earth would Polaris appear at the zenith? On the horizon?

3 If the stars exhibit diurnal motion, why do we refer to them as "fixed"?

4 Describe the annual motion of the sun with regard to (a) its eastward motion relative to the stars and (b) its north–south motion relative to Polaris.

5 Why are some constellations referred to as winter (or summer) constellations? Where do the winter constellations go during the summer?

6 How did the ancient astronomers distinguish the planets from the stars?

7 Summarize the main contributions of (a) the Ionians and (b) the Pythagoreans to the development of scientific thought.

8 How did the Greek scientists explain the phases of the moon? What arrangement of the sun, moon, and earth will result in a half moon? A crescent moon? A full moon?

9 What principle did Plato establish with respect to the motions of celestial bodies? On what grounds did he justify this principle?

10 Compare the explanations offered by Plato with those offered by Aristotle as to why (a) the earth is round and (b) the earth is stationary.

11 What is stellar parallax? How does the existence or nonexistence of stellar parallax relate to the conflict between the geocentric and heliocentric theories?

12 Explain the concepts of (a) the eccentric circle and (b) epicyclic motion as they are used in the geocentric theory. Why was each of these concepts introduced?

13 Explain how each of the following motions is accounted for in the heliocentric theory.
(a) The diurnal motion of the stars around Polaris.
(b) The annual eastward drift of the sun through the constellations.
(c) The annual north–south motion of the sun.

14 What is meant by the elongation of a planet? List the minimum and maximum elongations of each of the five visible planets.

15 Describe the apparent motions of Mercury and Venus, as seen from the earth. How are these motions accounted for (a) in the geocentric theory and (b) in the heliocentric theory?

16 Describe the retrograde motion of a planet, as seen from the earth. How is retrograde motion explained (a) in the geocentric theory and (b) in the heliocentric theory?

17 Define and explain the distinction between the synodic period and the sidereal period of a planet.

18 What is an astronomical unit? Why were the planetary distances given in AU in Copernicus' time? Why is this unit still used today?

19 Outline the advantages and disadvantages of the Copernican system as compared to the Ptolemaic system. Explain why Copernicus did not disprove the geocentric theory. In what respect was Copernicus' work a failure?

20 What was the Tychonic system of the universe? What observations led Tycho to adopt this model?

21 State and explain the law of the ellipse (Kepler's first law).

22 State and explain the law of equal areas (Kepler's second law). Show that one can conclude from this law that the orbital speed of a planet is variable, with maximum speed at perihelion and minimum speed at aphelion.

23 State and explain the harmonic law (Kepler's third law). Under what circumstances will the constant K be equal to 1?

24 What was the significance of Galileo's discovery of Jupiter's satellites?

25 Discuss the importance of Galileo's discovery that Venus exhibits a full set of phases. Explain why this discovery *did not* prove that the earth is in motion.

EXERCISES

1 Drawing a diagram similar to those shown in Figure 2–6, show the position of Polaris and the paths of the sun on June 22, March 21, and December 22 for an observer at each of the following latitudes: (a) 60°N, (b) $23\frac{1}{2}$°N, and (c) 50°S. In each case calculate the maximum altitude of the noonday sun on the longest day of the year.

2 The Greek astronomers used the eccentric circle to account for the varying sizes and speeds of the sun and moon as they circle the earth. What other possible explanations could have been given? Why did the Greeks reject these alternatives?

3 In the heliocentric theory the rotation of the earth on its axis and the revolution of the earth around the sun are both assumed to be counterclockwise (looking down on the north pole). What astronomical observations indicate that both motions are in the same direction? In other words, how would celestial motions be affected if the revolution and rotation were in opposite directions?

4 What observational evidence tells us that in the heliocentric system (a) the earth's axis is inclined $23\frac{1}{2}$° from the vertical and (b) the earth's axis always points toward Polaris?

5 Show that (a) a planet that is in eastern elongation will set after the sun and will be seen as an "evening star" and (b) a planet in western elongation will rise before the sun and be seen as a "morning star."

6 When a planet has an elongation of 90°, it is said to be in ***quadrature.***
(a) Draw a diagram showing the relative positions of the sun, earth, and Mars when Mars is in quadrature, as seen from the earth.
(b) Is it possible for an inner planet to be in quadrature?

7 At what time will a planet rise and set (a) when it is in opposition? (b) when it is in western quadrature?

8 An outer planet will always appear brightest when it is in opposition. Explain why this is so, using (a) the heliocentric theory and (b) the geocentric theory.

9 A newly discovered planet has a maximum elongation of 60°, as seen from the earth. Where is its orbit located with respect to the other planets? What is the radius of its orbit (assumed circular)?

10 A planet is in a circular orbit around the sun with an orbital radius of 0.5 AU. What is its maximum elongation as seen from the earth?

11 Suppose that an astronomer is attempting to discover a new planet inside the orbit of Mercury. Where would he look for this planet? What difficulties would he encounter in his search?

12 Given any point P on an ellipse, the sum of its distances ($PF_1 + PF_2$) from the two foci is a constant.
(a) By placing P on the major axis, show that the sum $PF_1 + PF_2$ is equal to $2a$.
(b) By placing P on the minor axis, show that $a^2 = b^2 + c^2$.

13 Show that for an ellipse of eccentricity $e = \frac{1}{2}$, the length of the minor axis is 0.866 of the length of the major axis.

14 An artificial satellite is placed in an orbit around the sun with an average radius of 4 AU and an eccentricity of 0.1. What are the perihelion and aphelion distances?

15 A comet has an orbital eccentricity of 0.9. Its closest approach brings it to 2 AU from the sun. What is its maximum distance from the sun?

16 A satellite is in an orbit such that its aphelion distance is three times its perihelion distance. What is the eccentricity of the orbit?

17 A planet is found beyond Pluto with an average orbital radius of 100 AU. What is its sidereal period?

18 A comet has an orbit with a perihelion distance of 2 AU and an aphelion distance of 30 AU. Find its sidereal period.

19 Two planets are discovered going around the nearby star Alpha Centauri. Planet A has an orbit of radius 2 AU and a sidereal period of 4 years. Planet B has an orbital radius of 8 AU and a sidereal period of 32 years. Does this planetary system obey Kepler's third law? Explain your answer.

20 Show by means of a diagram that the outer planets do not exhibit a full set of phases, as seen from the earth.

21 Studies of the changing size of Venus as it goes through its cycle of phases indicate that it is six times farther from the earth when full than it is when in the new phase. Calculate the radius of the orbit of Venus from this information.

MOTION

3

I often wonder how it can be that these strict supporters of Aristotle's every word fail to perceive how great a hindrance to his credit and reputation they are, and how the more they desire to increase his authority, the more they actually detract from it.

Galileo Galilei

GALILEO'S *Dialogues Concerning Two New Sciences* represents an important milestone in the history of science. This treatise, summarizing Galileo's many years of research on motion, was published in 1638, but it is nevertheless strikingly modern in its language and approach. In it we find the first successful application of what we refer to today as the scientific method. Through mathematical analysis and experiment, Galileo successfully demonstrated that the natural motions of bodies could be represented exactly by simple mathematical equations. His discoveries provided one of the foundations on which Newton's tremendous achievements were based. Of even greater significance is the fact that Galileo's work heralded the downfall of the Aristotelian influence on scientific thought. It is for these reasons that Galileo is sometimes referred to as "the father of modern science."

We will begin with a brief summary of Aristotle's theory of motion. This will give us some feeling for the ideas that prevailed in Galileo's day and will enhance our appreciation of the originality and power of Galileo's achievements.

3–1

ARISTOTELIAN CONCEPTS OF MOTION

Aristotle and his followers analyzed motion in terms of what they believed to be the "natural" motions of bodies. In the Aristotelian system of motion, a body tends to move toward its natural resting place. Each of the four elements (earth, air, fire, and water) has its own natural motion, so that the natural motion of a body depends on its composition. Thus heavy bodies (made of earth and water) tend to move toward the earth (the natural resting place of these elements), and light bodies move naturally upward toward the sky (the home of air and fire). Consequently, the natural motion of any body in the terrestrial realm is in a straight line, either toward or away from the earth. Thus in the Aristotelian system a body does not fall toward the earth because of a gravitational force of attraction, but because it is the natural tendency of the body to seek the earth as its natural resting place.

Aristotle reasoned that a heavier object should have a greater tendency to seek its natural resting place than a lighter object would; therefore the heavier object should fall with the greater speed. He further assumed that an object falls at a constant speed, and that this speed depends on the resistance of the medium through which the object is falling. Specifically, the speed is inversely proportional to the resistance: the less resistance offered by the medium, the greater the speed of the fall. It follows from this assertion that in a vacuum—where no resistance would be offered—an object would fall with an infinite speed. This, Aristotle concluded, is an absurdity; therefore a vacuum cannot exist.

The celestial bodies (made of the mysterious fifth element) were thought to obey a different principle of motion: Their natural motion was assumed to be circular motion. That is, the celestial bodies have a natural tendency to move in circular paths, so that there is no need for any force to keep these bodies moving along in their orbits.

Any motion that is not a natural motion is, to Aristotle, a violent or *forced* motion and requires a continually applied force to maintain it. Thus a heavy object thrown vertically upward or a body moving horizontally is in forced motion and must be continually pushed to keep moving. This seems to be contrary to experience, for it would appear that when a ball is thrown, the impelling force is no longer being applied once the ball leaves the thrower's hand. Aristotle answered this by asserting that as the ball moves, the air that is pushed aside rushes into the space behind the ball and pushes the ball forward. This is hardly a convincing argument, however, for it follows that if a body is streamlined, less air will rush around to the back as it moves, and the body should come to a stop more quickly—yet our experience tells us that streamlining will increase the distance covered by a projectile.

The acceptance during the Middle Ages of Aristotle as the ultimate authority on all matters carried over into the realm of physics, and in spite of obvious inconsistencies in Aristotle's description of motion, his ideas were not seriously challenged until Galileo came on the scene.

3–2

GALILEO'S SCIENTIFIC METHOD

Galileo took a radically different approach to the problem of motion. Aristotle was mainly intent on explaining *why* objects moved (in terms of their tendencies to seek their natural places). Galileo, on the other hand, was interested only in a mathematical description of *how* objects moved. Like Kepler, Galileo believed that the universe ran according to mathematical laws (although he was not nearly as mystical in his belief as Kepler was). In Galileo's view, to understand a natural phenomenon meant to be able to describe the phenomenon in a precise mathematical relationship. Even more important was Galileo's emphasis on the role of observation and experimentation. Physical laws should not merely be designed to agree with preconceived notions as to how we *think* nature *ought* to operate; rather, they should describe phenomena *as they appear to our senses*. In Galileo's own words:

> Anyone may invent an arbitrary type of motion and discuss its properties; thus, for instance, some have imagined helices and conchoids as described by certain motions which are not met with in nature, and have very commendably established the properties which these curves possess in virtue of their definitions; but we have

> decided to consider the phenomenon of bodies falling with an acceleration such as actually occurs in nature and to make this definition of accelerated motion exhibit the essential features of observed accelerated motions. . . . In this belief we are confirmed mainly by the consideration that experimental results are seen to agree with and exactly correspond with those properties which have been, one after another, demonstrated by us.[1]

As an illustration of Galileo's method, let us consider his refutation of Aristotle's assumption that heavier bodies fall faster than lighter bodies. Galileo tested this hypothesis by dropping two different weights from the same height—legend has it that he used the Leaning Tower of Pisa*—and he found that they reached the ground at *nearly* the same time, with the heavier object hitting slightly ahead of the lighter object. The experiment clearly demonstrated, however, that a 100-lb object does not fall ten times faster than a 10-lb object, as the Aristotelians expected. Galileo explained that the slight difference was due to the effect of air resistance, and he proposed that in a vacuum all objects would fall with exactly the same speed. Further experimentation enabled Galileo to obtain a mathematical formula correctly descibing the motion of falling bodies.

If we are to understand Galileo's analysis of motion more fully, we must now proceed to study in detail the terminology and mathematical tools used in the description of motion.

3–3

MATHEMATICAL DESCRIPTION OF SPACE AND TIME

Our senses present us with a picture of an orderly universe, and we find it convenient to describe this order in terms of the basic concepts of space and time. First of all, we observe a universe made up of distinct objects, having specific sizes and shapes, and we notice that these objects appear separated from one another at specific positions in our field of vision. Our sense impressions thus lead us to conceive of objects as *localized in space.* Second, our perceptions are of a *changing* world; that is, objects often appear to change their positions in space. Besides an awareness of the present distribution of objects, we are also able to retain in our minds a memory of past distributions. Thus we perceive a *sequence* of states or events, leading to the concept of ***time.***

The concepts of space and time are rather difficult to comprehend, and many serious questions can be raised. How can we know, for example, whether all people perceive space and time in the same way? How can we be certain that our memories of past events are not always distorted? What do we mean when we speak of the "flow" of time? Questions such as these have been the source of philosophical speculation for thousands of years. For the scientist, however, the problem is not to understand the nature of space and time, but merely to establish some way of describing them in precise mathematical terms. For this purpose, we must make a distinction between *physical* space and time (i.e., the concepts of space and time used by physicists) and *personal* or *subjective* space and time (the way in which these notions are perceived by individual human beings).

* There is no reliable record that Galileo used the Leaning Tower for his experiment; however, it is not unlikely, since he was living in Pisa at the time, and the Leaning Tower was the tallest structure available.

MEASUREMENT OF TIME

We all realize that subjective time can sometimes pass slowly (e.g., when we are waiting for a bus on a cold night) and sometimes pass rapidly (as when we are reading a good book). Obviously, physical time must pass uniformly if it is to be a useful quantity. Physical time is therefore measured by referring to some uniformly recurring natural phenomenon, such as the diurnal motion of the sky or the motion of a pendulum. That is, when we say, "the sidereal period of Mercury is 88 days," we are comparing the time required for an orbital revolution of Mercury with a commonly accepted standard time interval—that is, the apparent uniform rotation of the stars about Polaris as seen from the earth. The ***day*** and the ***year*** are standard time units, precisely defined in terms of regular astronomical phenomena, and the ***hour, minute,*** and ***second*** are convenient subdivisions of these units.

We should recognize that the *units* by which physical time is measured are quite arbitrary. It is not necessary to use astronomical events to measure time. For example, we could make use of the fact that to reasonably good accuracy, the period of a pendulum of a given length is constant regardless of the angle through which the pendulum swings (a discovery made by the young Galileo). We could construct a standard pendulum of specific length and define a unit of time as the period of this standard pendulum. We could name the unit of time the "galileo" and make all time measurements in galileos. We might find, for example, that the time it takes for an object to fall from the top of the Leaning Tower of Pisa is 4 galileos. (That is, in the time that the object takes to fall, the standard pendulum makes four swings.) This would be a perfectly legitimate procedure for defining and measuring physical time. Eventually, however, we would find that a pendulum is not an accurate enough instrument for making time measurements of the precision needed in modern scientific experiments.

THE GEOMETRY OF SPACE

The measurement of space is a more subtle problem. Inherent in the procedure for measuring space is the use of geometry, which is a form of pure mathematics dealing with such abstract and essentially undefined concepts as points and straight lines. In geometry a set of axioms and postulates is stated, and various theorems are derived from them. Thus the statement "the sum of the angles of a triangle is 180°" is true only in the sense that it follows logically from the definitions, axioms, and postulates of Euclidean geometry. Since Euclidean geometry is an abstract mathematical system, there is no logical necessity for it to have any relationship to the physical world. However, our experience has been that the theorems of Euclidean geometry do seem to hold true in our world; for example, if we were to draw a triangle on a piece of paper and carefully measure its angles, we would find that the sum of the three angles is indeed 180°. Moreover, many natural objects assume geometric shapes: stars and soap bubbles are spheres; salt crystals are cubes; a stretched string will simulate a straight line; and optical experiments seem to indicate that light rays travel in straight lines in a vacuum. So it seems quite legitimate to assume that physical space is measured accurately using the principles of Euclidean geometry.

Although the Euclidean nature of space seems almost self-evident, this belief was called into question by Albert Einstein. In the general theory of relativity, Einstein demonstrated that the properties of physical space are *not* accurately described by Euclidean geometry. On the scale of the solar system, however, the deviation from Euclidean geometry is so small as to be virtually unnoticeable. Consequently, we will assume the use of Euclidean geometry for now and will not concern ourselves with the non-Euclidean nature of the universe until a later chapter.

Once we have decided to use Euclidean geometry to describe space, the question of the measurement of space becomes relatively simple. We describe the separation of points in space in terms of the ***distance*** between them. The distance is measured along the straight line joining the points in question. Here we are using the notion from Euclidean geometry that the straight line is the *shortest* distance between two points. The measurement of distance is made by comparison to some arbitrary standard. This standard can be set up simply by making two marks on a straight, rigid object (such as a steel bar) and defining the distance between the two marks as the basic measuring unit. The steel bar then becomes our ruler or measuring stick, and we use it to measure off any given distance.* Thus when we say that the distance between two points is 100 feet, we mean simply that the distance between the two points is 100 times the distance between the two marks that define the standard unit called the ***foot.***

UNITS OF DISTANCE MEASUREMENT

There have been almost as many different distance units as there have been civilizations and languages. The ancient Egyptians used the cubit (about 20 inches); the English formerly used the ell (approximately 4 feet), and the Chinese used to measure distances in ch'ih (= 14.1 inches). The unit most widely used throughout the modern-day world is the ***meter.*** The meter, with its subdivisions, is internationally recognized as the unit of distance for all scientific work. Until recently, 1 meter was defined as the distance between two marks on a platinum–iridium bar kept at specified temperature and pressure in a laboratory of the International Bureau of Weights and Measures at Sèvres, France. Because of uncertainties as to possible changes in the length of the bar with time, the meter was redefined in 1960 in terms of the wavelength of the light given off by the inert gas krypton. With this new definition distances can be measured to an accuracy of 1 part in 10^{10}. A meter is approximately one ten-millionth (10^{-7}) of the distance from the north pole to the equator, measuring along a meridian of longitude. (That is how the meter was initially defined.)

The meter is subdivided according to a decimal system. The most commonly used subunits and multiples of the meter are:

1 kilometer (km)	= 1000 meters (10^3 m)	or	1 meter = 10^{-3} km
1 centimeter (cm)	= 1/100 meter (10^{-2} m)	or	1 meter = 10^2 cm
1 millimeter (mm)	= 1/1000 meter (10^{-3} m)	or	1 meter = 10^3 mm
1 micron (μ)	= 1/1,000,000 meter (10^{-6} m)	or	1 meter = 10^6 μ

The choice of subunit to be used is usually based on convenience: The distance between cities is normally given in kilometers; laboratory measurements are more likely to be made in centimeters.

Most Americans are not well acquainted with the metric system. Instead, we are more familiar with the British system of measuring distances in inches, feet, or miles. The conversion factors between the units of the two systems are as follows:

* Another underlying assumption in the theory of the measurement of space is that the ruler does not change shape when moved from one location to another. One might imagine a universe in which the forces between atoms vary from place to place, so that a ruler would expand or contract as it was moved. Fortunately, our universe does not seem to behave that way.

1 inch = 2.54 cm	1 cm = 0.3937 inch
1 foot = 30.48 cm	1 m = 39.37 inches = 3.281 feet
1 mile = 1.609 km	1 km = 0.6214 mile

Because of the peculiar relationships between the British units (1 foot = 12 inches; 1 mile = 5280 feet), calculations done in this system tend to be unwieldy. The metric system is clearly more convenient to use. Unfortunately, we have become thoroughly accustomed to the British system, and a changeover to metric units would represent a major undertaking. Most of us have a subjective understanding of what a height of 6 feet 2 inches or a distance of 100 feet is like, but when we hear that an aspiring basketball player is 2.1 meters tall or that the vital statistics of the latest Italian starlet are 92-61-92 cm, we do not know how to visualize the measurements. Consequently, even though the British system is not commonly used by scientists, British and metric units will be used interchangeably in this text so that the reader can have some sort of feeling for the magnitudes of the quantities being discussed.

3–4
VECTORS

When we describe the separation of two points in space, it is not enough to give the distance between the two points; we must also specify a ***direction.*** This direction is usually given as an angle with respect to some commonly accepted reference system such as the x or y axis of a rectangular coordinate system (Figure 3–1) or the points of the compass. For example, an airline pilot who is flying from Washington, D.C., to New York must know not only the distance (205 miles) but also the fact that New York is located 3° east of north with respect to Washington (Figure 3–2).

figure 3–1 **DISTANCE AND DIRECTION**

The circle in (a) represents all the points that are a distance of 5 units from point O. In (b) the location of B relative to O is given by stating that B is 5 units from O *and* that the line joining these points makes a 30° angle with the x-axis. Thus the displacement of B from O is completely specified by giving a distance and a direction.

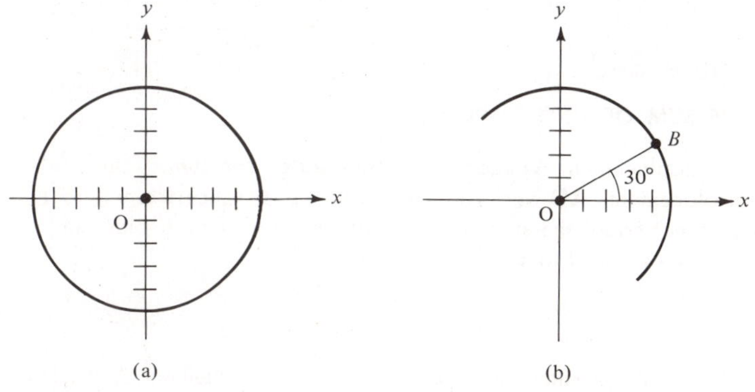

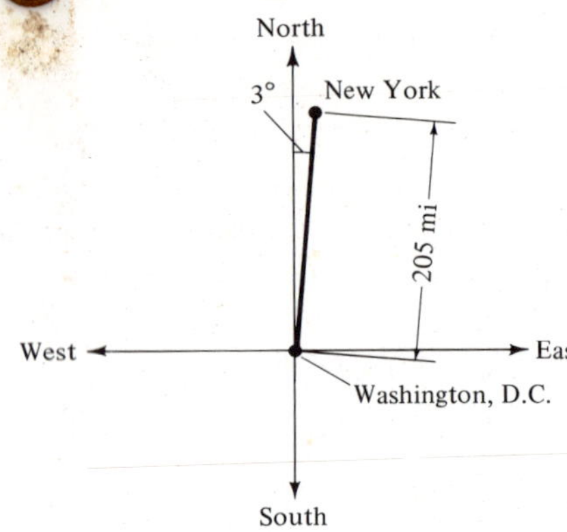

figure 3–2

DISTANCE AND DIRECTION

The geographical location of New York with respect to Washington, D.C., is specified by a distance and a direction with respect to geographical north.

In physics there are many quantities that must be described by both a ***magnitude*** (i.e., a numerical value) and a ***direction in space.*** Such quantities are called ***vectors.*** Quantities that are completely described by a magnitude alone are called ***scalars.*** Time, temperature, and volume are examples of physical quantities that have no spatial direction associated with them and are therefore scalar quantities. The separation of points in space is a vector quantity, which we call the ***displacement.*** The displacement vector consists of two parts: a distance and a direction. We see, then, that distance is the magnitude part of the displacement vector; by itself distance is a scalar quantity.

Scalar quantities can be added, subtracted, multiplied, and divided according to the usual rules of arithmetic. That is, scalars can be treated as pure numbers. Vectors, however, have their own rules of arithmetic. These rules are not arbitrarily defined; vectors were invented specifically as a mathematical tool for use in solving physical problems, so that the rules of vector arithmetic derive from the properties of those physical quantities that we have designated as vectors. To illustrate this point, let us consider the situation shown in Figure 3–3. The three points A, B, C are located at the corners of a right triangle. The displacement of B relative to A (vector **AB**) has a magnitude of 4 and points north.* Similarly, the displacement of C from B is represented by vector **BC** (magnitude 3, direction east). Now if the displacement **BC** is "added" to **AB,** we arrive at the point C, and we would like to consider the displacement **AC** as being the "sum" of the two displacements. But **AC** has a

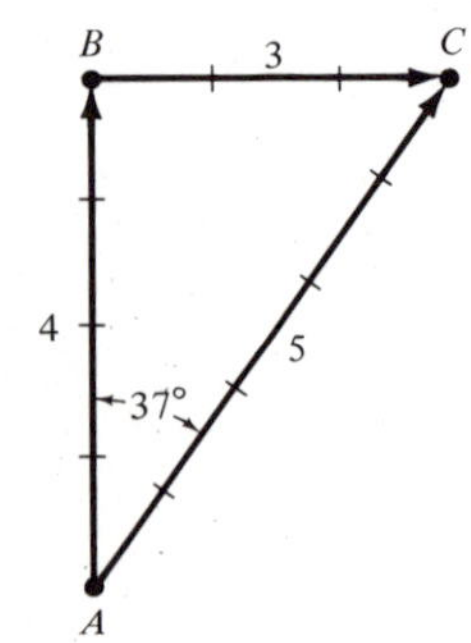

figure 3–3

THE SUM OF TWO DISPLACEMENTS

The displacement **AC** can be considered as the sum of the displacements **AB** and **BC**. However, the distance from A to C does *not* equal the sum of the distance from A to B plus the distance from B to C.

---

* The usual convention for representing vectors is as follows: **A** stands for the vector itself; A or $|\mathbf{A}|$ represents the *magnitude* of the vector only.

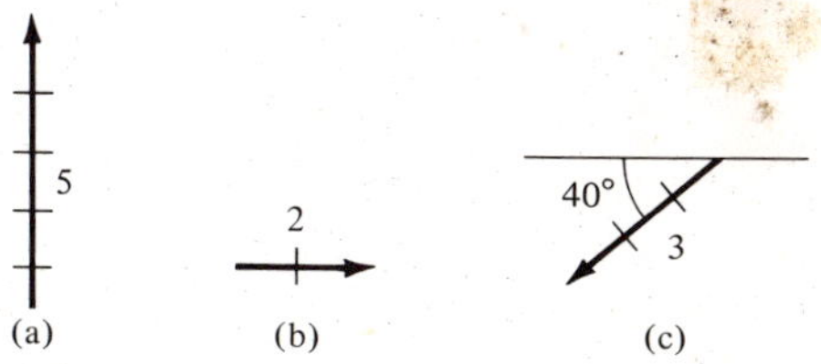

figure 3–4

GRAPHICAL REPRESENTATION OF VECTORS

(a) Magnitude 5, direction north; (b) magnitude 2, direction east; (c) magnitude 3, direction 40° south of west.

magnitude of 5 and a direction 37° northeast. Obviously, the magnitude of **AC** does not equal the sum of the magnitudes of **AB** and **BC,** so the rules of ordinary addition do not apply. If we are to define the operation of ***vector addition***

$$\mathbf{AB} + \mathbf{BC} = \mathbf{AC}$$

it must be done in a way that gives the proper magnitude and direction for the displacement **AC;** otherwise the definition will have no practical value for physicists.

GRAPHICAL METHOD OF VECTOR ADDITION

A rather simple procedure for properly adding vectors is the ***graphical method.*** We represent a vector by drawing an arrow, pointing in the specified direction, with a length proportional to the magnitude of the vector, as shown in Figure 3–4.

To add two vectors **A** and **B,** we draw them to scale and arrange them such that the *tail* of **B** coincides with the *tip* of **A,** as shown in Figure 3–5a. The *sum* or *resultant* **C** is the vector drawn from the *tail* of **A** to the *tip* of **B.** The magnitude and direction of **C** can then be measured directly from the diagram, or it can be calculated by using trigonometric formulas. The graphical method can be extended to determine the sum of any number of vectors (Figure 3–5b). Furthermore, it can

figure 3–5 **GRAPHICAL ADDITION OF VECTORS**

The vectors to be added are arranged tip-to-tail. The sum is the vector drawn from the tail of the first vector to the tip of the last vector. (a) Addition of two vectors. (b) Addition of more than two vectors. The resultant **R** will be the same regardless of the order in which the vectors are added.

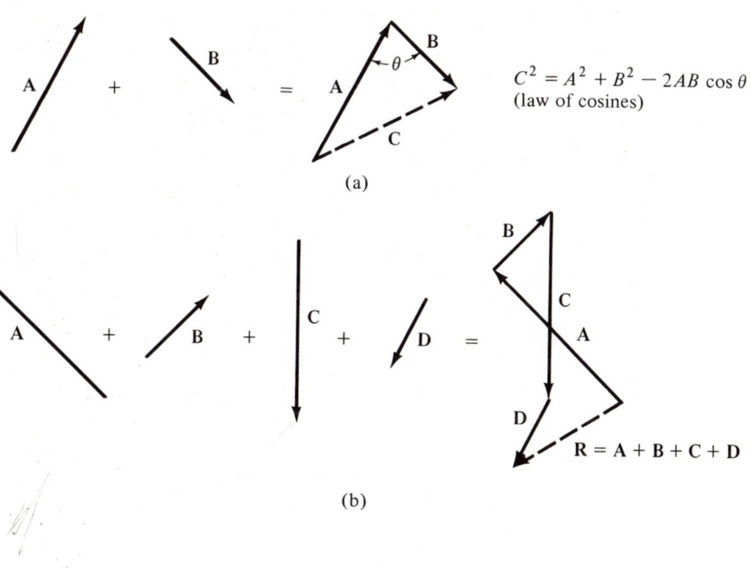

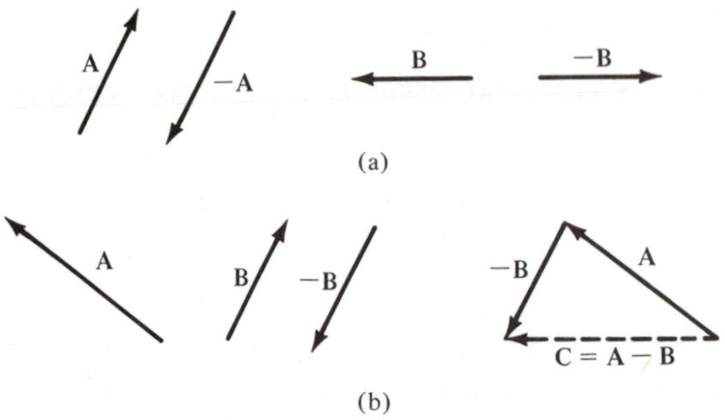

figure 3–6 **VECTOR SUBTRACTION**

(a) The negative of a given vector has the same magnitude but points in the opposite direction.
(b) Vector subtraction is performed by adding the negative of the vector to be subtracted:
A + (−**B**) = **A** − **B**.

be shown that the sum of two or more vectors will be the same regardless of the order in which the vectors are added. That is, vectors follow the ***commutative law:*** **A + B = B + A.**

Vectors can also be subtracted from one another. To understand how vector subtraction is performed, we must first define the *negative* of a vector, which is obtained by reversing the direction of the vector through 180° without changing its magnitude (Figure 3–6a). Thus **A** and −**A** have the same magnitude but point in opposite directions. Vector subtraction is performed by adding the negative of the vector to be subtracted (Figure 3–6b). That is, to obtain **A** − **B,** add the negative of **B** to **A,** since **A** + (−**B**) = **A** − **B.**

figure 3–7 **SPECIAL EXAMPLES OF VECTOR ADDITION**

(a) Collinear vectors; (b) perpendicular vectors.

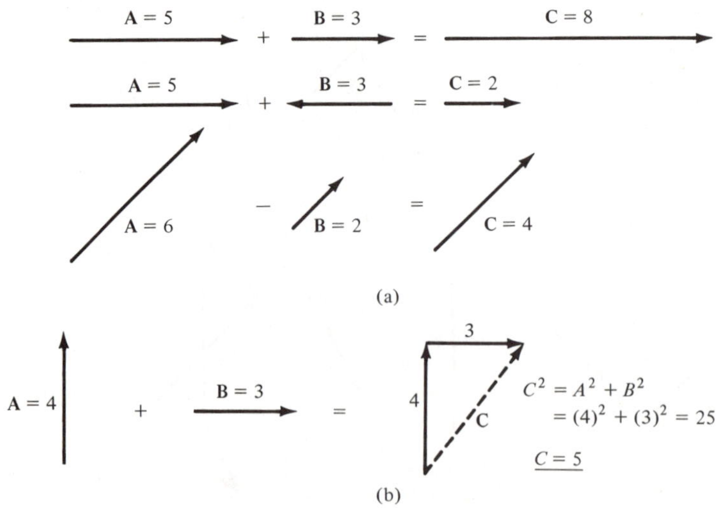

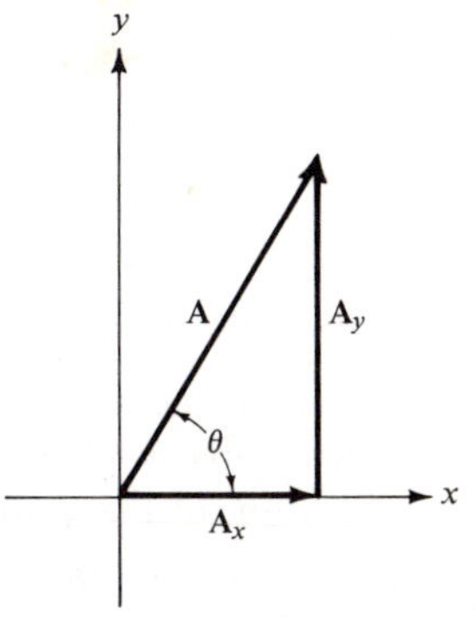

$$A^2 = A_x^2 + A_y^2$$

$$\cos\theta = \frac{A_x}{A} \qquad \therefore A_x = A\cos\theta$$

$$\sin\theta = \frac{A_y}{A} \qquad \therefore A_y = A\sin\theta$$

figure 3–8 **RESOLUTION OF A VECTOR**

The vector **A** can be decomposed into the sum of its x and y components.

There are two special cases of vector addition in which the graphical method can be avoided. First, if vectors are ***collinear*** (they lie along the same line), their vector sum or difference is the same as their numerical sum or difference, and they can be treated as if they were scalars (Figure 3–7a). Second, if two vectors are perpendicular to each other, they form a right triangle when arranged tip-to-tail (Figure 3–7b). Then their sum will be the hypotenuse of a right triangle, and the magnitude of the sum can be calculated by using the Pythagorean theorem. That is, if **A** and **B** are perpendicular to each other, the magnitude of their sum **C** is simply the square root of $C^2 = A^2 + B^2$.

VECTOR ADDITION BY RESOLUTION OF VECTORS

The addition of two perpendicular vectors by the Pythagorean theorem suggests the reverse procedure; that is, it is possible to decompose or ***resolve*** any vector into two components that are at right angles to one another. This operation can be performed as follows: Consider vector **A,** which makes an angle θ with the x axis, as shown in Figure 3–8. We can represent **A** as the sum of two vectors, $\mathbf{A}_x$ and $\mathbf{A}_y$, such that $\mathbf{A}_x$ points in the direction of the x axis and $\mathbf{A}_y$ points in the direction of the y axis. $\mathbf{A}_x$ and $\mathbf{A}_y$ are referred to as the ***x component*** and ***y component,*** respectively, of **A.** Since $\mathbf{A}_x$ and $\mathbf{A}_y$ are perpendicular to one another, they form a right triangle with **A,** so that

$$A_x^2 + A_y^2 = A^2 \tag{3–1}$$

From the rules of trigonometry, we find that*

$$A_x = A\cos\theta \qquad \text{and} \qquad A_y = A\sin\theta \tag{3–2}$$

This procedure for resolving any vector into components provides us with a general method for adding two or more vectors. The first step is to resolve all vectors into their x and y components. Then all the x components are added together. This step is simplified by the fact that all the x components lie along the same line and can therefore be added numerically. We do the same for the y components. These

* Students who are unfamiliar with these trigonometric functions should consult Appendix A.

two totals then represent the x component and y component of the sum vector, and they can be added together using the Pythagorean theorem.* The following example should clarify this procedure.

EXAMPLE 3–1: Given the vectors **A** (10, 45° east of north), **B** (12, 30° north of west), and **C** (8, south), find the sum **D** = **A** + **B** + **C**.

SOLUTION: The first step is to draw the three vectors on an x–y coordinate diagram, as shown in Figure 3–9a. We then resolve each vector into x and y components, using Eqs. 3–2. To facilitate the addition of the components, we adopt the convention that any component pointing *up* or to the *right* is *positive*, and any component pointing *down* or to the *left* is *negative*. This convention takes into account the fact that the components are added if they point in the same direction and are subtracted if they point in opposite directions. Thus A_x, A_y, and B_y are positive, and B_x and C_y are negative. (Note that **C** has no x component.) The components of the three vectors are tabulated as follows:

$$A_x = 10 \cos 45° = 10(0.707) = 7.07$$
$$B_x = -12 \cos 30° = -12(0.866) = -10.39$$
$$C_x = 0$$

$$A_y = 10 \sin 45° = 10(0.707) = 7.07$$
$$B_y = 12 \sin 30° = 12(0.500) = 6.00$$
$$C_y = -8$$

We now add the x components together. Note that $\mathbf{A}_x$ and $\mathbf{B}_x$ point in opposite directions and must actually be subtracted from one another—but the minus sign in front of B_x already takes its direction into account. Thus the x component of

figure 3–9

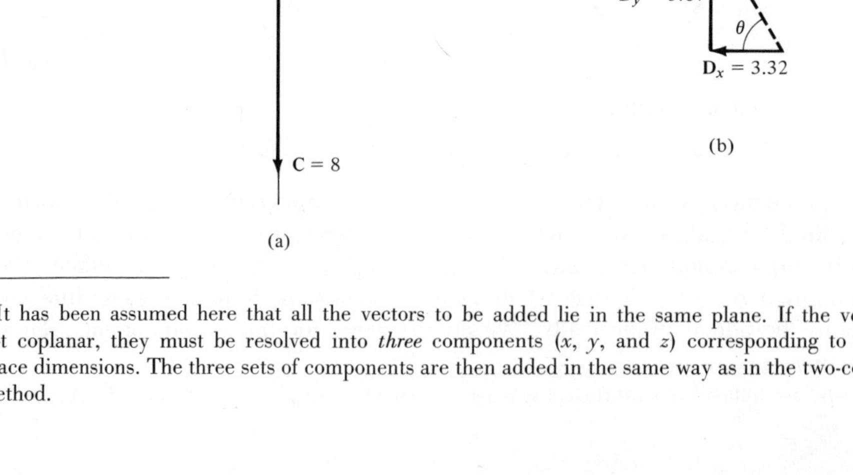

* It has been assumed here that all the vectors to be added lie in the same plane. If the vectors are not coplanar, they must be resolved into *three* components (x, y, and z) corresponding to the three space dimensions. The three sets of components are then added in the same way as in the two-component method.

the resultant vector **D** is

$$D_x = A_x + B_x + C_x = 7.07 - 10.39 + 0 = -3.32$$

The minus sign indicates that $\mathbf{D}_x$ points to the left. For the y component of the resultant, we have

$$D_y = A_y + B_y + C_y = 7.07 + 6.00 - 8 = 5.07$$

The sum is positive, indicating that $\mathbf{D}_y$ points upward. We now have the two components $\mathbf{D}_x$ and $\mathbf{D}_y$, which can be added to give the sum **D**, as shown in Figure 3–9b. Since $\mathbf{D}_x$ and $\mathbf{D}_y$ are perpendicular vectors, the Pythagorean theorem can be used to calculate the magnitude of **D**:

$$D = \sqrt{D_x^2 + D_y^2} = \sqrt{(3.32)^2 + (5.07)^2} = \sqrt{36.73} = \underline{6.06}$$

The angle θ that **D** makes with the x axis is determined as follows:

$$\tan\theta = \frac{D_y}{D_x} = \frac{5.07}{3.32} = 1.53 \qquad \text{so that} \qquad \underline{\theta = 56.8^\circ}$$

The direction of **D** is therefore 56.8° north of west.

3–5

KINEMATICS

The description of motion used in modern-day science is derived directly from the work of Galileo. The concepts of velocity and acceleration, which we are about to study, did not originate with Galileo, but in the *Dialogues Concerning Two New Sciences*, he gave the first precise definitions of these quantities and derived the basic equations of motion. Galileo was the founder of the science of ***kinematics***, which is concerned with the *mathematical description* of motion (as distinct from ***mechanics***, which deals with the *causes* of motion). His contributions to the study of motion served to establish the path along which physics has traveled for more than three hundred years. When Galileo's astronomical discoveries are taken along with his work on motion, we can see why he is considered one of the greatest scientific thinkers in human annals.

SPEED AND VELOCITY

The concepts of space and time are united in the study of motion. We observe that objects move through space, and that the amount of time that elapses when an object moves from one point to another point depends on the circumstances. Clearly, there is a difference in the ways a racing car and a snail traverse a distance of 100 yards. In everyday terminology we describe this difference by referring to the ***speed*** of a moving object. We learn, for example, that an automobile can travel (legally) at speeds up to 60 or 70 mi/hr, and a commercial jet plane travels at speeds around 500 mi/hr. Thus speed is a measure of the *rate* at which an object traverses a distance. In other words, the speed of an object is the ratio of the distance traveled to the time of travel. Speed is measured in units of distance/time—that is, mi/hr, ft/min, cm/sec, and so on.

Speed is a *relative* quantity in the sense that it is measured with respect to a specific frame of reference. At this moment the reader has a speed of 0 with respect to his chair, a speed of about 1000 mi/hr with respect to the center of the earth (as a result of the earth's rotation), a speed of about 18 mi/sec with respect to the sun (due to the earth's orbital motion), and a speed of some 200 mi/sec with respect to the center of the galaxy (arising from the sun's revolution about the galactic center). None of these can be considered one's "true" speed; they are all equally valid.

In physics it is often important to take the direction of motion into account. For this reason the quantity called ***velocity*** has been defined. Velocity is a vector quantity; the direction of the velocity vector is the same as the direction of motion. Thus velocity = speed + direction; in other words, speed is the magnitude part of the velocity vector. For example, an automobile with a velocity of 60 mi/hr north has a speed of 60 mi/hr. An automobile moving at 50 mi/hr north and one moving at 50 mi/hr south have the same speed but not the same velocity. Although the tendency is to use the words "speed" and "velocity" interchangeably in everyday language, the distinction between them is an important and necessary one for physicists.

AVERAGE VELOCITY

Let us examine the concept of velocity more carefully. Figure 3–10 depicts the motion of an automobile along a curved road. The automobile does not necessarily have a constant speed; it is likely that the driver will speed up or slow down as he turns the curve, and he may even have to stop for a traffic light. Moreover, the automobile clearly does not have a constant direction. It follows that the velocity is not constant, so we cannot describe the motion in terms of a single, specific velocity. However, we can calculate the *average* velocity $\mathbf{v}_{av}$, defined as the total displacement divided by the elapsed time. In equation form

$$\mathbf{v}_{av} = \frac{\mathbf{d}}{t} \tag{3–3}$$

where t is the amount of time that elapses during the motion, and $\mathbf{d}$, the displacement, is the vector joining the initial and final points of the path. The magnitude of this vector is the straight-line distance between the two points; for example, in Figure 3–10 the magnitude of the displacement is the length of the line AB, *not* the distance traveled along the curved road. The average velocity takes into account the different directions as well as the speeds associated with the motion. Hence it is different from the average speed, which is equal to the actual distance traveled divided by the time. For example, suppose that a person drives from New York to Boston, a road distance of about 230 miles, in 5 hours. His average speed is 230 mi/5 hrs = 46 mi/hr.

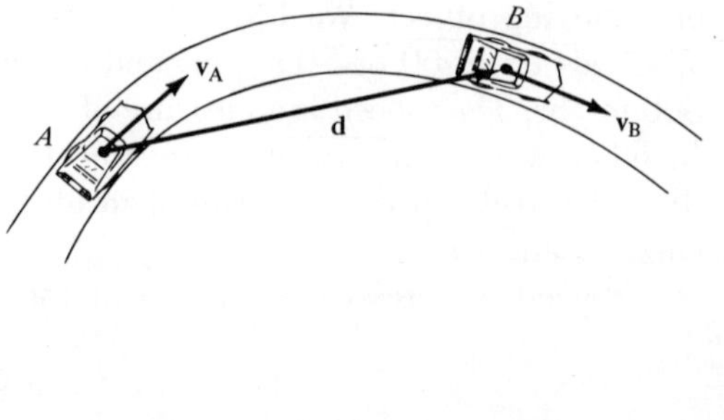

figure 3–10

AVERAGE VELOCITY

An automobile travels along a curved road from A to B. The velocity is not constant. The average velocity over the path from A to B is the straight-line displacement $\mathbf{d}$ divided by the time t required to travel from A to B.

However, his average velocity, which takes into account all the changes in direction along the road, depends on the straight-line displacement from New York to Boston, which is 188 miles northeast. Hence his average velocity is 188 mi/5 hrs = 37.6 mi/hr northeast. The distinction between speed and velocity becomes less important when an object is moving in a straight line. In this case the object's direction is constant, and the actual distance traveled is the same as the displacement. Hence in straight-line motion the average speed is equal to the magnitude of the average velocity.

INSTANTANEOUS VELOCITY

Generally, the average velocity tells us little about the actual motion of a body. To describe the motion in more detail, we should know the velocity at each point along the path for each instant of time. Such a quantity is known as the ***instantaneous velocity.*** It is defined as the average velocity taken over an infinitesimally small interval of time. To illustrate this concept, let us refer to Figure 3–11, in which the curve shown represents the path along which a body is moving. P is the position of the body at time t, and P' is some arbitrary point farther along on the path. The average velocity $\mathbf{v}_{av}$ between the points P and P' is the straight-line displacement $\mathbf{d}$ divided by t, the time required to move from P to P'. If, however, we move P' closer and closer to P, both the displacement and the time intervals will decrease, and $\mathbf{v}_{av}$ will change accordingly. When P' is moved very close to P, both $\mathbf{d}$ and t become very small. Mathematicians have been able to prove (using calculus) that when $\mathbf{d}$ and t simultaneously approach 0, $\mathbf{v}_{av}$ $(= \mathbf{d}/t)$ will approach a finite value. This limiting value of $\mathbf{v}_{av}$ is the instantaneous velocity $\mathbf{v}$. In other words, $\mathbf{v}$ is the limiting value of $\mathbf{v}_{av}$ as the average is taken over a smaller and smaller interval. The direction of $\mathbf{v}$ is along the tangent to the curve at P. When defined in this manner, the instantaneous velocity $\mathbf{v}$ can be thought of as the velocity of the body at a given instant of time.

UNIFORM MOTION

The simplest type of motion is constant-velocity or *uniform* motion. Uniform motion requires not only a constant speed but also a constant direction; that is, the object must be moving in a straight line. Motion along a straight line is called ***rectilinear*** motion, as opposed to ***curvilinear*** motion.

In constant-velocity motion the instantaneous velocity has the same value at all points along the path. In this case the average velocity has the same value as the instantaneous velocity. It follows from Eq. 3–3 that

$$\mathbf{v}_{av} = \mathbf{v} = \frac{\mathbf{d}}{t} \tag{3–4a}$$

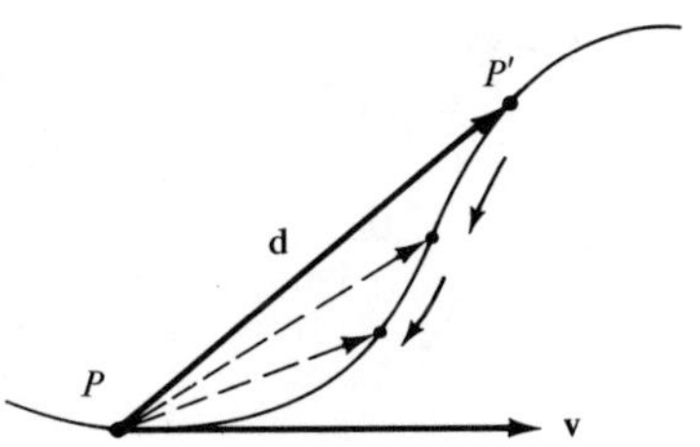

figure 3–11

INSTANTANEOUS VELOCITY

As P' is brought closer and closer to P, the average velocity $\mathbf{v}_{av}$ over the path PP' approaches the instantaneous velocity $\mathbf{v}$ of the body at P. Note that $\mathbf{v}$ is tangent to the curve at P.

Since the direction of motion is constant, d is the straight-line distance traveled by the body in a time t. Thus the magnitude of the instantaneous velocity is the same as the instantaneous speed, v:

$$v = \frac{d}{t} \tag{3–4b}$$

Very rarely in the physical world do we find objects moving with a constant velocity. Most moving bodies, from snails to satellites, change speed and direction rather frequently, except in the trivial case of when they are not moving at all ($\mathbf{v} = 0$ is, after all, a constant velocity). Consequently, Eq. 3–4b is used mainly for a few special cases of natural uniform motion or as an approximation for motion that is very nearly uniform.

ACCELERATED MOTION

To describe motion in which the velocity is not constant, we make use of a quantity called the ***acceleration,*** which measures the *rate of change of velocity with time.* Since the velocity of a body need not change at a uniform rate, we must first define the ***average acceleration,*** $\mathbf{a}_{av}$:

$$\mathbf{a}_{av} = \frac{\text{change in velocity}}{\text{change in time}} = \frac{\Delta \mathbf{v}}{\Delta t} \tag{3–5}$$

The symbol Δ (Greek delta) is standard mathematical notation for *the change in* a particular quantity. Thus if A represents some arbitrary quantity,

$$\text{the change in } A = \Delta A = (\text{final value of } A) - (\text{initial value of } A)$$

Let us suppose that an object is moving along a path as shown in Figure 3–12, and that we are interested in determining the average acceleration between points 1 and 2 on the path. We let point 1 be the initial point and point 2 the final point of the portion of motion we are studying. At time t_i, when the object is at 1, its

figure 3–12 AVERAGE ACCELERATION

An object moves from 1 to 2, along the path shown, in a time t. The change in $\mathbf{v}$ ($= \Delta \mathbf{v} = \mathbf{v}_f - \mathbf{v}_i$) must be found by drawing a vector diagram. The magnitude of $\mathbf{a}_{av}$ equals the magnitude of $\Delta \mathbf{v}$ divided by Δt. The direction of $\mathbf{a}_{av}$ is the same as the direction of $\Delta \mathbf{v}$.

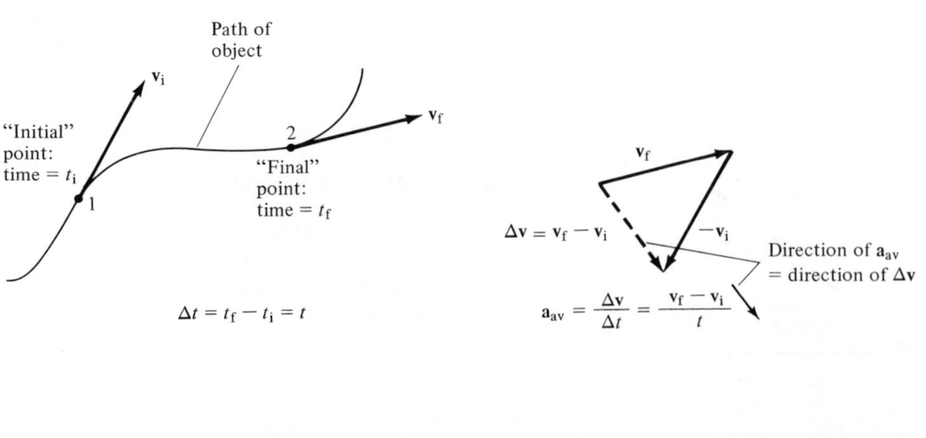

velocity is $\mathbf{v}_i$, and when the object reaches 2 at time t_f, its velocity is $\mathbf{v}_f$. Then

$$\Delta \mathbf{v} = \mathbf{v}_f - \mathbf{v}_i \qquad \text{and} \qquad \Delta t = t_f - t_i$$

The change in time, $\Delta t = t_f - t_i$, represents the *time interval* for the motion in question (i.e., the amount of time it takes for the object to move from point 1 to point 2). For simplicity, we will let $\Delta t = t_f - t_i = t$, where t is the time interval. Eq. 3–5 can now be written as follows:

$$\mathbf{a}_{av} = \frac{\mathbf{v}_f - \mathbf{v}_i}{t} \tag{3–6}$$

Thus $\mathbf{a}_{av}$ represents the average acceleration of the object between points 1 and 2. (The direction of $\mathbf{a}_{av}$ is the same as the direction of $\Delta \mathbf{v}$, as shown in Figure 3–12.) The ***instantaneous acceleration*** can be defined in the same way that the instantaneous velocity was defined previously—that is, by taking the limiting value of $\mathbf{a}_{av}$ as Δt is made smaller and smaller. The instantaneous acceleration thereby represents the value of the acceleration at a given instant of time.

From Eq. 3–5 it is readily seen that the units in which acceleration is measured are the units of velocity/time = (distance/time)/time = distance/(time)^2. Thus ft/sec^2, cm/sec^2, and km/hr^2 are legitimate units of acceleration. A mixed unit such as (mi/hr)/sec is also legitimate, but often inconvenient to use. A unit such as ft/sec^2, although convenient to write, may conceal the meaning of the term "acceleration." For example, the equation $a = 20$ ft/sec^2 is actually an abbreviation of $a = 20$ (ft/sec)/sec, which means that *the velocity is increasing by 20 ft/sec every second.* A negative acceleration, or ***deceleration,*** means that the object is slowing down. Thus $a = -20$ ft/sec^2 means that the velocity is *decreasing* by 20 ft/sec every second. An object moving with an acceleration of zero must have a constant velocity; remember that with a constant velocity *both* the magnitude *and* the direction of the velocity are unchanging. Thus an object traveling with a constant speed on a curved path is accelerating because its direction is not constant.

MOTION WITH CONSTANT ACCELERATION

Constant acceleration means, of course, that the acceleration is constant both in magnitude and in direction. If we consider only straight-line motion, then the velocity and acceleration vectors will all lie on the same line, and we can treat **v** and **a** as scalars. Furthermore, if the acceleration is constant in magnitude, the average and instantaneous accelerations will be the same. If an object accelerates uniformly from an initial speed of v_i to a final speed of v_f in a time interval t, the instantaneous acceleration will be

$$a = \frac{v_f - v_i}{t}$$

or

$$v_f = v_i + at \tag{3–7}$$

This equation is valid for straight-line motion with constant acceleration. By constant acceleration we mean a motion in which the speed is changing at a uniform rate. As an illustration of such a situation, the instantaneous speed v at regular intervals of time has been tabulated for a hypothetical straight-line motion, and the acceleration over each interval has been calculated:

t (sec)	v (ft/sec)	a (ft/sec^2)
0	10	
		30
1	40	
		30
2	70	
		30
3	100	
		30
4	130	
		30
5	160	

Note that the speed increases at a steady rate of 30 ft/sec every second. In other words, the acceleration has a constant value of 30 ft/sec^2. Now consider the following data:

t (sec)	v (ft/sec)	a (ft/sec^2)
0	10	
		10
1	20	
		20
2	40	
		40
3	80	
		80
4	160	

In this situation the speed is *not* increasing at a steady rate, and the acceleration is *not* a constant quantity. Equation 3–7 would not be applicable for such motion.

When the acceleration is constant and the speed is changing at a uniform rate, the average speed over a given time interval will be the speed that is midway between the initial and final values. (For example, if a car accelerates uniformly from 20 mi/hr to 60 mi/hr, the average speed over this time will be 40 mi/hr.) The average speed can be obtained by averaging the initial and final values; that is,

$$v_{av} = \tfrac{1}{2}(v_i + v_f) \tag{3–8}$$

Let us write Eq. 3–3 in the form $d = v_{av}t$ and substitute Eqs. 3–7 and 3–8. We obtain an equation for the distance traveled during a uniformly accelerated motion:

$$d = v_{av}t = \tfrac{1}{2}(v_i + v_f)t = \tfrac{1}{2}(v_i + v_i + at)t$$

Combining terms gives us

$$d = v_i t + \tfrac{1}{2}at^2 \tag{3–9}$$

If the velocity is constant, $a = 0$; then Eq. 3–9 reduces to $d = vt$ (Eq. 3–4b), which is the equation for constant-velocity motion. (When the velocity is constant, $v_i = v_f = v$.) Thus Eq. 3–9 also covers the case of constant-velocity motion.

The two equations (3–7 and 3–9)

$$v_f = v_i + at$$

$$d = v_i t + \tfrac{1}{2}at^2$$

represent the basic kinematic equations for straight-line, constant-acceleration motion and can be used for the solution of various motion problems. For certain types of problems, the arithmetic can be reduced considerably by using the equivalent set of equations (3–3 and 3–8):

$$d = v_{av}t$$

$$v_{av} = \tfrac{1}{2}(v_i + v_f)$$

Another convenient equation can be obtained by solving Eqs. 3–7 and 3–9 simultaneously to eliminate t:

$$v_f^2 = v_i^2 + 2ad \tag{3–10}$$

This equation is often useful in solving problems in which the time does not appear explicitly. The following examples will serve to illustrate how the various kinematic equations are used to solve specific motion problems.

EXAMPLE 3–2: A driver accelerates his car uniformly from 0 to 80 ft/sec in 10 sec. What is his acceleration? What distance does he cover during this time?

SOLUTION: We are given the following information:

$$v_i = 0 \qquad v_f = 80\ \text{ft/sec} \qquad t = 10\ \text{sec}$$

To find the acceleration, we use Eq. 3–7:

$$\begin{aligned} v_f &= v_i + at \\ 80\ \text{ft/sec} &= 0 + a(10\ \text{sec}) \\ a &= (80\ \text{ft/sec})/(10\ \text{sec}) = \underline{8\ \text{ft/sec}^2} \end{aligned}$$

The distance can then be calculated using Eq. 3–9:

$$\begin{aligned} d &= v_i t + \tfrac{1}{2}at^2 \\ &= (0)(10\ \text{sec}) + \tfrac{1}{2}(8\ \text{ft/sec}^2)(10\ \text{sec})^2 \\ &= \underline{400\ \text{ft}} \end{aligned}$$

An alternative method for determining the distance is to calculate the average speed;

$$v_{av} = \tfrac{1}{2}(v_i + v_f) = \tfrac{1}{2}(0 + 80\ \text{ft/sec}) = 40\ \text{ft/sec}$$

and then substitute this into Eq. 3–3:

$$d = v_{av}t = (40\ \text{ft/sec})(10\ \text{sec}) = \underline{400\ \text{ft}}$$

EXAMPLE 3–3: A driver traveling at a speed of 100 ft/sec sees a traffic light 500 feet ahead. What uniform acceleration must he provide by stepping on his brakes in order to bring the car to a stop at the light?

SOLUTION: We are given the information that d = 500 feet, v_i = 100 ft/sec, and

$v_f = 0$. In order to find the acceleration, we must first find out the time required to stop the car. Eqs. 3–3 and 3–8 can be used as follows:

$$\begin{aligned} v_{av} &= \tfrac{1}{2}(v_i + v_f) \\ &= \tfrac{1}{2}(100 \text{ ft/sec} + 0) = 50 \text{ ft/sec} \\ d &= v_{av}t \\ 500 \text{ ft} &= (50 \text{ ft/sec})t \\ t &= 10 \text{ sec} \end{aligned}$$

Then we use Eq. 3–7 to obtain a:

$$\begin{aligned} v_f &= v_i + at \\ 0 &= (100 \text{ ft/sec}) + a(10 \text{ sec}) \\ a &= -(100 \text{ ft/sec})/(10 \text{ sec}) = \underline{-10 \text{ ft/sec}^2} \end{aligned}$$

The minus sign indicates that the car is decelerating; that is, its speed *decreases* by 10 ft/sec every second. As an alternative procedure, we could have calculated the acceleration directly from Eq. 3–10:

$$\begin{aligned} v_f^2 &= v_i^2 + 2ad \\ 0^2 &= (100 \text{ ft/sec})^2 + 2a(500 \text{ ft}) \\ a &= -(100)^2/(2)(500) = \underline{-10 \text{ ft/sec}^2} \end{aligned}$$

CIRCULAR MOTION: AN EXAMPLE OF NONUNIFORM ACCELERATION

There will be no attempt here to deal with motion in which the acceleration is changing, since in most cases the mathematics required to solve such problems is beyond the scope of this book. However, there is one such situation that yields a rather simple solution and is worth mentioning. This is the case of ***uniform circular motion,*** in which an object is moving in a circular path with constant speed (Figure 3–13). This is an *accelerated* motion, for although the magnitude of the velocity is constant, the direction is constantly changing. (In circular motion the direction of the instantaneous velocity is always tangent to the circle.) Through the use of vector calculus, a simple equation for the magnitude of the acceleration can be obtained. This equation (first derived by Christian Huygens in 1669) is as follows:

$$a = \frac{v^2}{r} \tag{3–11}$$

where v is the speed and r is the radius of the circle. Furthermore, it can be shown that the direction of the acceleration will always be toward the center of the circle. (That is, the acceleration vector lies along the radius.) In circular motion the acceleration is called the ***centripetal acceleration.***

The fact that the acceleration always points toward the center should not be taken to mean that the acceleration has a constant direction. The direction is not *constant in space*—that is, at the top of the circle **a** points south; at the bottom it points north, and so on. (See Figure 3–13.) Thus in uniform circular motion both **v** and **a** are constant in magnitude but changing in direction, so that neither one is a constant vector.

Consequently, a centripetal acceleration of magnitude $a = v^2/r$ is required to keep a body moving with constant speed in a circular path. If the acceleration is

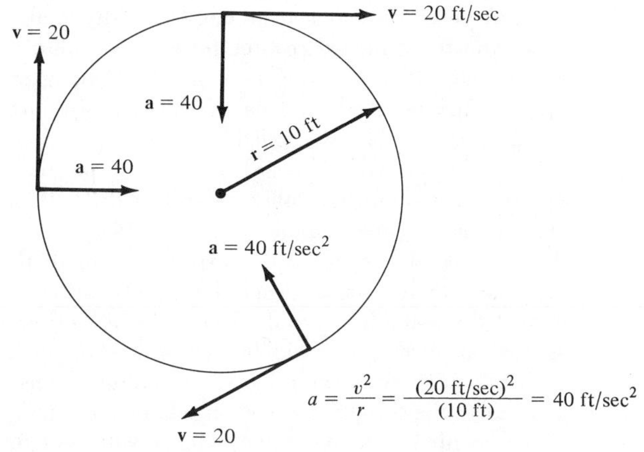

figure 3–13 **UNIFORM CIRCULAR MOTION**

A body is moving on a circular path of radius 10 ft with a uniform speed of 20 ft/sec. The acceleration has a magnitude of 40 ft/sec^2 and always points toward the center of the circle.

greater or less than this value, the speed will not be constant, and the body will deviate from a circular path. The concept of uniform circular motion has many physical applications. We will return to it in the next chapter.

3–6

NATURAL MOTION

Thus far we have derived several equations for describing the motions of bodies. It should be noted that these kinematic equations, as they are called, are *not* physical laws. They are not obtained by experimentation, but are derived mathematically from the basic *definitions* of velocity and acceleration. With this in mind, we return to the question posed by Galileo: Can we give a precise mathematical description of the natural motions of bodies in the physical world? In other words, can the kinematic equations, derived from basic definitions, be employed to describe the *observed* motions of bodies? Through careful experimentation and ingenious deductive reasoning, Galileo showed that natural motions could in fact be represented in simple mathematical terms. In this manner, he was able to refute convincingly the Aristotelian theories of motion.

FREE FALL

Perhaps Galileo's most significant discovery was the law of falling bodies. Galileo arrived at his law through his experimental studies of motion on an inclined plane. He rolled a ball down planes of various inclinations and observed that in every case the distance covered was proportional to the time squared. Equation 3–9 shows that for an object initially at rest ($v_i = 0$), $d = \frac{1}{2}at^2$. Galileo's observation that the ratio of d/t^2 is a constant quantity showed that the ball moved with a *constant acceleration.* In the *Dialogues Concerning Two New Sciences,* Galileo tells us how he performed this experiment:

> A piece of wooden moulding or scantling, about 12 cubits long, half a cubit wide, and three finger-breadths thick, was taken; on its edge was cut a channel a little more than one finger in breadth; having made this groove very straight, smooth, and polished, and having lined it with parchment, also as smooth and polished as possible, we rolled along it a hard, smooth, and very round bronze ball. Having placed this board in a sloping position, by lifting one end some one or two cubits above the other, we rolled the ball, as I was just saying, along the channel, noting, in a manner presently to be described, the time required to make the descent. We repeated this experiment more than once in order to measure the time with an accuracy such that the deviation between two observations never exceeded one-tenth of a pulse-beat. Having performed this operation and having assured ourselves of its reliability, we now rolled the ball only one-quarter the length of the channel; and having measured the time of its descent, we found it precisely one-half of the former. Next we tried other distances, comparing the time for the whole length with that for the half, or with that for two-thirds, or three-fourths, or indeed for any fraction; in such experiments, repeated a full hundred times, we always found that the spaces traversed were to each other as the squares of the times, and this was true for all inclinations of the plane, i.e., of the channel, along which we rolled the ball. . . .
>
> For the measurement of time, we employed a large vessel of water placed in an elevated position; to the bottom of this vessel was soldered a pipe of small diameter giving a thin jet of water, which we collected in a small glass during the time of each descent, whether for the whole length of the channel or for a part of its length; the water thus collected was weighed, after each descent, on a very accurate balance; the differences and ratios of these weights gave us the differences and ratios of the times, and this with such accuracy that although the operation was repeated many times, there was no appreciable discrepancy in the results.[2]

Noting that the ball moved with constant acceleration regardless of the angle of inclination (although the actual value of the acceleration depends on the steepness of the incline), Galileo reasoned that if the inclination were increased to 90° the ball would still fall with constant acceleration, but it would now be in free fall. (See Figure 3–14.) He therefore concluded that freely falling bodies move with a constant *acceleration,* rather than with constant speed, as Aristotle had thought.

Now it is quite apparent that a light object such as a feather or a balloon or a raindrop does *not* fall with constant acceleration. If a raindrop were to fall from an altitude of 10,000 feet with constant acceleration, it would hit the ground with a speed of over 500 mi/hr! Galileo correctly reasoned that the motion of such bodies was retarded by the resistance of the medium in which they were falling. He assumed that the resistance would increase in proportion to the speed of the falling body, in which case the body would attain a ***terminal speed.*** That is, a body falling in a medium accelerates up to its terminal speed and falls at constant speed thereafter. The value of the terminal speed depends on the nature of the body and the density of the medium. Galileo proposed that in the absence of a medium, all bodies would fall with constant acceleration; furthermore, the value of this acceleration would *not* depend on the size or weight of the body.* Consequently, Galileo formulated the law

* Galileo's prediction has been dramatically verified by an experiment performed on the moon. During the Apollo 15 mission in August 1971, one of the astronauts dropped a hammer and a feather simultaneously from the same height. In the airless environment both bodies fell with exactly the same motion, as predicted.

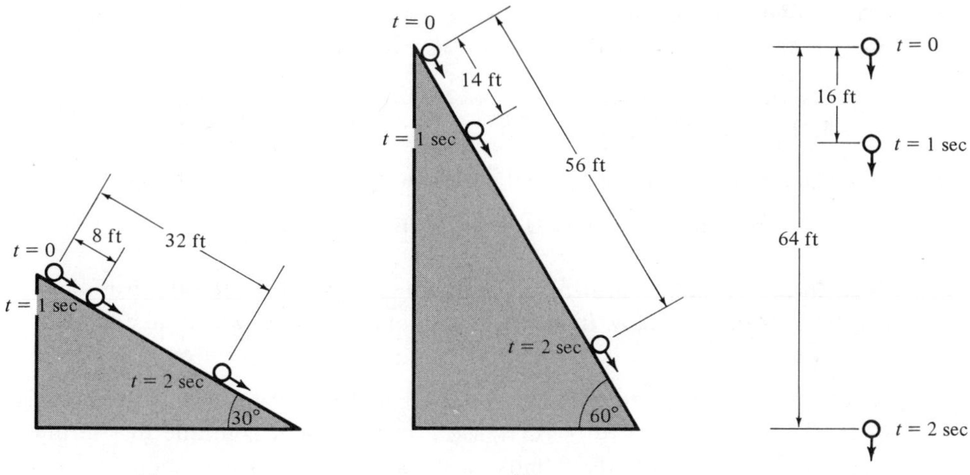

(a) Inclination = 30°: $a = 16$ ft/sec^2

(b) Inclination = 60°: $a = 28$ ft/sec^2

(c) Inclination = 90° (free fall): $a = 32$ ft/sec^2

figure 3–14 **GALILEO'S INCLINED PLANE EXPERIMENT**

Galileo found that the ball moved such that $d \propto t^2$ regardless of the angle of inclination. He assumed that this would also be the case for an inclination of 90°, that is, for free fall. Motion in which d is proportional to t^2 corresponds to motion with constant acceleration.

of falling bodies, which can be stated in the following manner:

All objects allowed to fall freely in a vacuum will fall with the same constant acceleration.

The numerical value of this acceleration, which we call the ***acceleration of gravity*** and represent by the letter g, can be experimentally measured rather easily. It is found that g varies slightly over the surface of the earth, ranging from 32.26 ft/sec^2 at the poles to 32.09 ft/sec^2 at the equator. For our purposes, however, we will let

$$g = 32 \text{ ft/sec}^2$$

In metric units

$$g = 9.8 \text{ m/sec}^2 = 980 \text{ cm/sec}^2$$

(since 32 feet = 9.8 meters).

We are now able to give a simple mathematical description of the motions of objects in free fall. Unless otherwise stated, we will assume that air resistance is negligible. Since bodies fall with constant acceleration in the absence of resistance, the equations derived previously (Eqs. 3–7 to 3–10) are applicable, with the restriction that $a = +g$ or $-g$ depending on whether the body is being accelerated or decelerated by gravity.

EXAMPLE 3–4: An object is released from the top of a 400-foot building. How long does it take for the object to reach the ground, and with what speed does it hit?

SOLUTION: We are given the following data: $v_i = 0$, $d = 400$ feet, and $a = g = 32\ \text{ft/sec}^2$. To find the time, we use Eq. 3–9:

$$d = v_i t + \tfrac{1}{2}at^2$$
$$400 = (0)t + \tfrac{1}{2}(32)t^2 = 16t^2 \qquad \text{so that} \qquad \underline{t = 5\ \text{sec}}$$

To find the final speed (i.e., the speed just before impact), we use Eq. 3–7:

$$v_f = v_i + at = 0 + (32)(5) = \underline{160\ \text{ft/sec}}$$

EXAMPLE 3–5: A ball is thrown vertically upward. It leaves the thrower's hand with a speed of 96 ft/sec. How high will it rise, and how long will it take to return to the thrower's hand?

SOLUTION: As the ball rises, it is being decelerated by gravity. Thus during this part of its motion, $a = -g = -32\ \text{ft/sec}^2$. The ball will continue to rise until its speed is 0. The ball then falls with a positive acceleration of $a = +g = 32\ \text{ft/sec}^2$.

It is convenient to break up the motion into two parts. On the upward part $v_i = 96\ \text{ft/sec}$, $v_f = 0$, and $a = -32\ \text{ft/sec}^2$. Then, using Eq. 3–10, we obtain

$$v_f^2 = v_i^2 + 2ad$$
$$0^2 = (96)^2 + 2(-32)d \qquad \text{or} \qquad \underline{d = 144\ \text{feet}}$$

The time required to reach this height is found by Eq. 3–7:

$$v_f = v_i + at$$
$$0 = 96 - 32t \qquad \text{or} \qquad \underline{t = 3\ \text{sec}}$$

For the downward motion $v_i = 0$, $a = +32\ \text{ft/sec}^2$, and $d = 144$ feet. Using

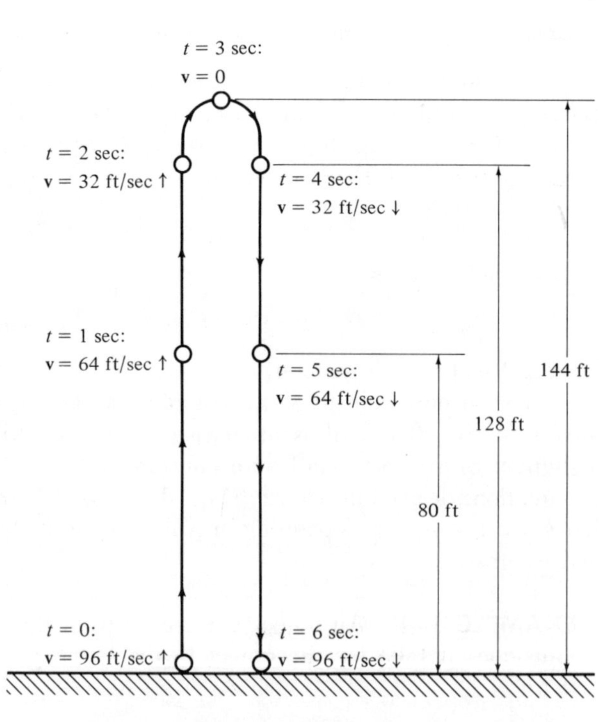

figure 3–15

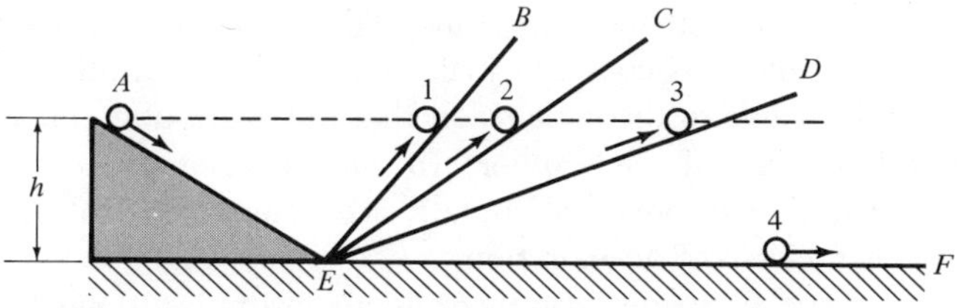

figure 3–16 **MOTION UP AN INCLINED PLANE**

A ball released from rest at A and allowed to roll up the various inclines EB, EC, and ED will rise to points 1, 2, and 3, respectively. That is, the ball will rise as far as the original vertical height h, regardless of the angle of inclination. Galileo postulated that a ball rolling on a horizontal plane (EF) would continue to move indefinitely with constant velocity.

Eq. 3–9, we find

$$d = v_i t + \tfrac{1}{2}at^2$$
$$144 = (0)t + \tfrac{1}{2}(32)t^2 \qquad \text{or} \qquad \underline{t = 3 \text{ sec}}$$

The motion of the ball is completely symmetrical; it rises for 3 sec, then falls for 3 sec. At each point on its upward path, its speed will be the same as its speed at the corresponding point on the downward path. (See Figure 3–15.) The ball will return to the thrower's hand at the same speed (96 ft/sec) with which it left.

HORIZONTAL MOTION

Galileo's experiments with inclined planes led him to speculate on the nature of horizontal motion. Galileo allowed a ball to roll down one inclined plane and up another, and found that if friction was held to a minimum, the ball would return to the same height from which it was initially released (Figure 3–16). This was the case regardless of the inclination of the second plane. In other words, the ball "seeks its own height." Galileo then reasoned that if the second plane were horizontal, the ball would continue to roll indefinitely and would do so at a constant velocity, since it would not be retarded or accelerated by gravity as it moved. In Galileo's own words: "Any velocity once imparted to a moving body will be rigidly maintained as long as the external causes of acceleration and retardation are removed, a condition which is found only on horizontal planes."[3]

Once again Galileo is in complete opposition to Aristotle. Whereas Aristotle believed that for a body to move with constant velocity a force must be continually applied, Galileo states that a body will continue to move with constant velocity *unless* a force is applied. Thus Galileo presents us with the first inkling of the concept of ***inertia***—the natural tendency of bodies to maintain their state of motion—which will be discussed more fully in the next chapter.

PROJECTILE MOTION

In the *Dialogues Concerning Two New Sciences,* Galileo also dealt with the motion of projectiles. The motion of a projectile is twofold; it is moving horizontally and vertically at the same time. The problem is to determine how the horizontal and vertical motions combine with and affect one another. Through deductive reasoning, Galileo came to the conclusion that the horizontal and vertical motions are *completely independent* of each other. In the absence of any forces (other than gravity), a projectile will move horizontally with constant velocity while at the same time falling with a constant acceleration.

The following argument is an example of the kind of reasoning Galileo used in asserting the independence of horizontal and vertical motion. If we allow a ball to fall from a height of 4 feet, we find that the ball takes $\frac{1}{2}$ sec to reach the ground. Now suppose we are inside a train that is moving with a constant velocity. If we perform the same experiment, we find that the ball appears to fall in a vertical straight line and hits the floor after $\frac{1}{2}$ sec. To an observer outside the train, however, the ball is also moving horizontally (its velocity is the same as that of the train) and it appears to fall along a curved path (Figure 3–17). However, the time of fall is still $\frac{1}{2}$ sec. Clearly, the horizontal motion has no effect on the vertical motion—whether the ball is moving horizontally or not, it still takes $\frac{1}{2}$ sec to traverse a vertical distance of 4 feet. Furthermore, the ball's vertical motion does not affect its horizontal motion; while falling, the ball still travels horizontally with the same speed as the train. Thus it follows that the horizontal and vertical motions of the ball are independent of one another.

Let us examine projectile motion in more detail. Suppose a ball is thrown horizontally and allowed to fall. If air resistance is negligible, the ball will move with a constant velocity in the horizontal direction. That is, it will cover equal horizontal distances in equal intervals of time ($d_{\text{horiz}} = vt$). However, the ball is also falling toward the earth with a constant acceleration of 32 ft/sec^2. Since at $t = 0$ the velocity of the ball is entirely horizontal, *the initial velocity in the vertical direction is 0.* Because the horizontal and vertical motions are independent, the ball falls as if it were being released from rest. Therefore since $v_{\text{i}} = 0$, the vertical motion is described by the equation $d_{\text{vert}} = \frac{1}{2}at^2$, where $a = g$. The motion of the ball is thus a combination of horizontal motion at constant velocity and vertical motion at constant

figure 3–17 MEASURING THE TIME OF FALL IN A MOVING TRAIN

(a) An observer inside the train sees the ball fall vertically, with no apparent horizontal motion. (b) An outside observer sees the ball travel on a parabolic path, since the ball appears to have the same horizontal motion as the train. Both observers measure the same time of fall.

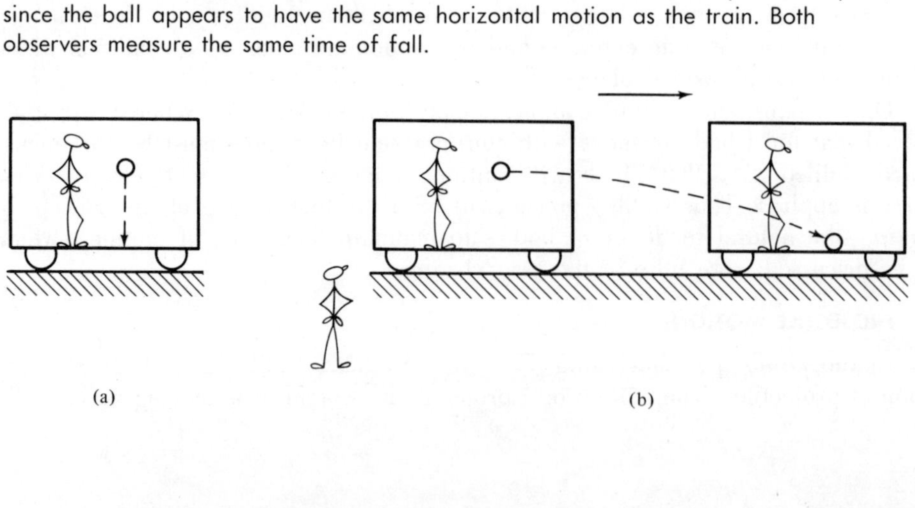

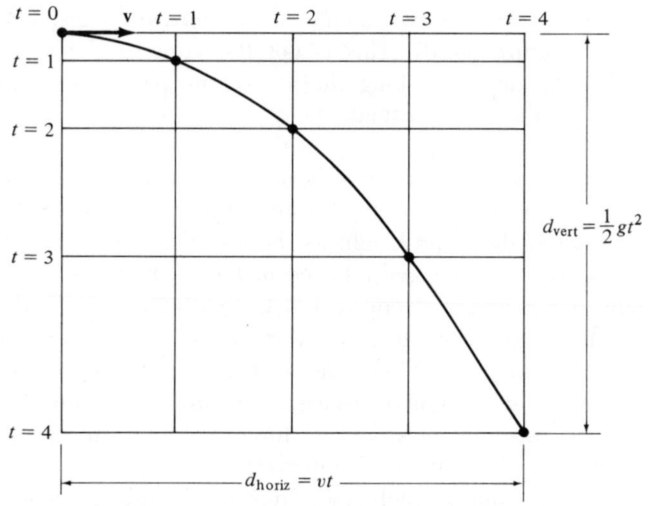

figure 3–18 **THE PATH OF A HORIZONTALLY PROJECTED BODY**

The motion is a combination of constant velocity in the horizontal direction and constant acceleration in the vertical direction.

acceleration. The combined motion is shown in Figure 3–18. (The path of the ball is the geometric curve known as the ***parabola.***)

EXAMPLE 3–6: A bullet is fired horizontally with an initial velocity of 3000 ft/sec from a gun held 4 feet above the ground. What horizontal distance will the bullet travel, if there is no air resistance?

SOLUTION: The bullet will travel for as long as it takes to fall a distance of 4 feet. At $t = 0$ the bullet has a horizontal velocity of 3000 ft/sec and a vertical velocity of 0. We construct the following table:

Vertical	Horizontal
$d = 4$ feet	$d = ?$
$v_i = 0$	$v = 3000$ ft/sec
$a = 32$ ft/sec^2	$a = 0$
$t = ?$	$t = ?$

First we find from the vertical motion how long it takes the bullet to reach the ground:

$$d_{vert} = v_i t + \tfrac{1}{2}at^2$$
$$4 = (0)t + \tfrac{1}{2}(32)t^2 \qquad \text{or} \qquad \underline{t = \tfrac{1}{2}\text{ sec}}$$

After $\frac{1}{2}$ sec the motion will stop, since at that time the bullet will hit the ground. Therefore the bullet will travel horizontally for $\frac{1}{2}$ sec. Thus

$$d_{horiz} = vt = (3000\text{ ft/sec})(\tfrac{1}{2}\text{ sec}) = \underline{1500\text{ ft}}$$

Note that we have calculated the *horizontal* distance traveled by the bullet (d_{horiz} in Figure 3–18). This is not the same as the length of the bullet's path, which is the distance along the arc of the parabola (a quantity that can be computed with the use of calculus).

The independence of horizontal and vertical motion leads to some consequences that seem to be contrary to our intuition. For example, let us suppose that a bullet is held at the same height as the gun described in Example 3–6, and that the bullet is simply released and allowed to fall at the same instant that the gun is fired. Since the bullet that is dropped and the bullet that is fired are both in free fall from the same height with an initial vertical velocity of zero, *they will both hit the ground at the same time* if the effects of air resistance are negligible.

In the foregoing discussion of projectile motion, it was assumed that there was no air resistance or any other forces acting on the projectile besides gravity. In many common circumstances, however, air resistance is not a negligible factor, and the deviation from parabolic motion is considerable. The game of baseball depends heavily on such deviations. In the absence of air resistance, a baseball will always travel in a parabolic path. However, the effect of air resistance on a spinning baseball can cause it to curve, sink, or hop. Baseball without air resistance would be a hitter's paradise.

3–7

THE OVERTHROW OF ARISTOTELIAN PHYSICS

Galileo's refutation of Aristotle's theories of motion gave a clear indication that the long reign of Aristotelian physics was about to come to an end. The idea that circular motion is the natural motion of celestial bodies had already been challenged by Kepler. After showing that the planets travel in elliptical rather than circular paths, Kepler had proposed that the planets are driven by a force emanating from the sun. Therefore planetary motion is a forced motion, rather than a natural motion.

In the terrestrial realm, Aristotle's theory of falling bodies was disproved by Galileo's experiments. In addition, Galileo proposed that horizontal motion (a forced motion in the Aristotelian system) is a natural motion, and that an unhindered body would continue to move at constant velocity. This idea was carried one step further by René Descartes, the great French philosopher–mathematician, who proposed that constant-velocity motion is the natural motion of all bodies throughout the universe.

The revolutionary ideas proposed by Kepler, Galileo, and Descartes did not gain immediate acceptance. Aristotelian physics was but one part of a whole system of thought that had held sway for nearly two thousand years. The theories advanced by Kepler, Galileo, and Descartes were simply not of a large enough scope to overthrow completely the Aristotelian view. Yet by the end of the seventeenth century, the rejection of Aristotelian physics was virtually complete. This came about through the efforts of a single man, Isaac Newton. Building on the work of his predecessors, Newton constructed a completely new system of physics, out of which emerged a radically new picture of the universe. The Newtonian system and the world view that it inspired will be the subject of the next part of this text.

QUESTIONS FOR REVIEW

1 Summarize Aristotle's analysis of motion. What distinction did the Aristotelians make between (a) celestial and terrestrial motion and between (b) natural and forced motion?

2 According to Aristotle, why do bodies fall? What factors influence the motion of a falling body?

3 Discuss the various astronomical and nonastronomical methods for the measurement of time.

4 How is the distance between two points measured? What assumptions are inherent in this procedure?

5 What is the difference between a vector and a scalar?

6 Describe the graphical method for the addition (and subtraction) of vectors.

7 Define and discuss the distinction between (a) displacement and distance, (b) speed and velocity, and (c) average velocity and instantaneous velocity.

8 What is meant by the term "acceleration"? Explain in words the meaning of the following equations:
(a) $a = 15\ \text{ft/sec}^2$
(b) $a = -10\ \text{m/sec}^2$

9 Give examples of (a) motion with constant acceleration, (b) motion with changing acceleration, and (c) motion with zero acceleration.

10 State the kinematic equations that describe the motion of a body with constant acceleration.

11 Explain why uniform circular motion is accelerated motion.

12 What is centripetal acceleration? What is the equation for the acceleration required to keep a body moving in a circle at constant speed?

13 State the law of falling bodies. Under what conditions is it valid? How did Galileo arrive at this law?

14 What is meant by the acceleration of gravity? What is the numerical value of this quantity?

15 Compare Galileo's and Aristotle's analyses of (a) free fall and (b) horizontal motion.

16 When a body is thrown so that it is moving in two directions (i.e., horizontally and vertically) at the same time, how do the horizontal and vertical motions affect each other? What kinematic equations describe the motion of the body?

EXERCISES

1 Convert the following quantities to their equivalents:
(a) 20 cm to inches.
(b) 6 feet 2 inches to meters.
(c) 100 meters to yards.
(d) 93 million miles to kilometers.

2 Show that (a) 100 mi/hr = 44.7 m/sec and (b) 15 mi/hr = 22 ft/sec.

3 Calculate the magnitude and indicate graphically the direction of the sum of the following vectors:
(a) **A** = 3 west; **B** = 4 west
(b) **A** = 3 north; **B** = 4 south
(c) **A** = 6 north; **B** = 8 west

(d) $\mathbf{A} = 8$ east; $\mathbf{B} = 3$ west; $\mathbf{C} = 12$ north
(e) $\mathbf{A} = 1$ pointing 30° east of north; $\mathbf{B} = 1$ pointing 30° east of south

4 Given the vectors $\mathbf{A} = 5$ north, $\mathbf{B} = 8$ south, $\mathbf{C} = 12$ west, and $\mathbf{D} = 8$ east, find the magnitude and direction of
(a) $\mathbf{A} - \mathbf{B}$
(b) $\mathbf{C} - \mathbf{A}$
(c) $\mathbf{A} + \mathbf{B} - \mathbf{C}$
(d) $\mathbf{A} - \mathbf{B} + \mathbf{C} - \mathbf{D}$

5 Find the magnitude and direction (using the method of resolution of vectors) of the vector $\mathbf{D} = \mathbf{A} + \mathbf{B} + \mathbf{C}$ in each of the following. (Note: $\sin 53° = 0.8$; $\cos 53° = 0.6$. For other angles, see table of sines and cosines in Appendix A.)

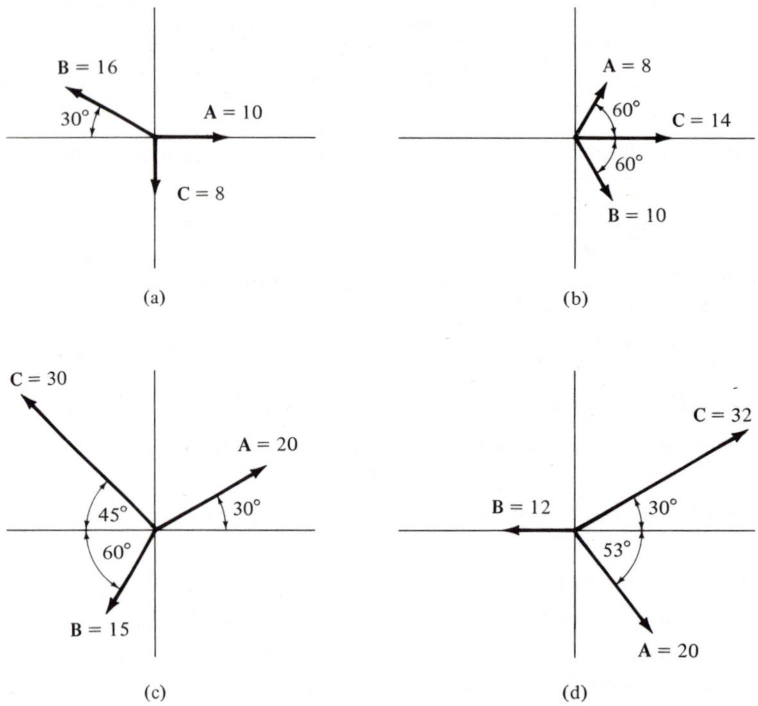

6 A car is driven at a speed of 30 mi/hr for the first 2 hours and at 50 mi/hr for the next 3 hours. What is the average speed for the entire trip?

7 A bus driver has to cover a total distance of 400 miles in 8 hours. During the first 3 hours he covers 120 miles. With what average speed must he travel for the remainder of the trip?

8 A car is driven at uniform speed on a circular track 400 feet in diameter. The car completes one full lap around the track in 31.4 sec. Calculate (a) the average speed for one lap, (b) the average velocity for one lap, (c) the average speed for half a lap, and (d) the average velocity for half a lap.

9 Light travels through space with a constant speed of 186,300 mi/sec (3×10^{10} cm/sec). What distance will a light beam travel in 1 year? (Note: This distance is the astronomer's unit of length known as the ***light year.***)

10 A car starting from rest is accelerated uniformly for 8 sec. In this time a distance of 240 feet is covered. What is the speed of the car after 8 sec?

11 A moving body passes point A at a speed of 30 ft/sec. At this point the body begins to accelerate uniformly. Ten seconds later it passes point B at a speed of 50 ft/sec. How far apart are points A and B?

12 A car is traveling at a speed of 60 ft/sec when the driver applies the brakes. The car is brought uniformly to a stop in 15 sec. Calculate (a) the acceleration, (b) the average speed, and (c) the distance covered.

13 Derive Eq. 3–10 from Eqs. 3–7 and 3–9.

14 A car is driven on a circular track at a constant speed of 80 ft/sec. The track has a radius of 400 feet. What is the magnitude and direction of the acceleration of the car?

15 A feather and a rock are released simultaneously from the same height above the ground. Which hits the ground first? How would Galileo and Aristotle each have explained this result?

16 A ball is released from rest from the top of a tall building. It hits the ground after 10 sec.
(a) How high is the building?
(b) What is the speed of the ball at the moment of impact?

17 A ball starting from rest rolls down an incline and covers a distance of 12 feet in 1 sec. Determine (a) the acceleration of the ball, (b) the distance covered and the speed of the ball after 2 sec, and (c) the distance and speed after 3 sec.

18 A boy wants to throw a ball onto the roof of a 100-foot building. What is the minimum initial velocity he must give the ball so that it will just reach the roof?

19 A ball is thrown vertically upward with an initial velocity of 128 ft/sec.
(a) What is the maximum height to which the ball will rise?
(b) What is the position, velocity, and acceleration of the ball after 5 sec?

20 The acceleration of gravity at the surface of the moon is one-sixth of g at the surface of the earth. If a ball is thrown vertically upward on the moon with an initial speed of 128 ft/sec, to what height will it rise? How does this answer compare with the answer to 19(a)?

21 A bullet is fired horizontally from a height of 144 feet and strikes the ground after having traveled a horizontal distance of 8400 feet. With what speed did the bullet leave the gun?

22 A gun is known to have a muzzle velocity of 2000 ft/sec. (The muzzle velocity is the velocity with which the bullet leaves the gun.) A marksman sets up a target, paces off 1000 feet, holds his gun horizontally 6 feet above the ground, and fires. He hits the bull's-eye. How high is the center of the target above the ground?

Two

THE NEWTONIAN UNIVERSE

THE BETTMANN ARCHIVE

Isaac Newton

One of the greatest geniuses in human history, Newton created a system of scientific principles that completely revolutionized man's picture of the universe. This portrait of Newton at the age of forty-seven is by Godfrey Kneller.

THE PRINCIPLES OF MOTION

4

Then ye who now on heavenly nectar fare,
Come celebrate with me in song the name
Of Newton, to the Muses dear; for he
Unlocked the hidden treasures of Truth:
So richly through the mind had Phoebus cast
The radiance of his own divinity.
Nearer the gods no mortal may approach.

Edmund Halley
(from *The Ode Dedicated to Newton*)

IN SCANNING the history of human endeavor, we find certain works of creative genius that are so monumental in their scope it is difficult to believe they were encompassed by the vision of a single man. Among these works we might include the plays of Shakespeare, Dante Alighieri's *Divina Commedia,* or the symphonies of Beethoven. We must certainly place in this category the *Principia* of Isaac Newton, for in this astounding book Newton achieved the unification of virtually all of physics and astronomy on the basis of a few simple principles. The magnitude of Newton's accomplishment was such that the influence of the Newtonian world view spread beyond science into the realm of philosophical and political thought.

A brief summary of the contents of *Philosophiae Naturalis Principia Mathematica* (*Mathematical Principles of Natural Philosophy*) will serve to illustrate why Newton is ranked among the greatest creative geniuses. The *Principia,* as it is commonly called, begins with some basic definitions—mass, momentum, inertia, force—and states the three principles of motion. In Book I, *The Motion of Bodies,* Newton derives Kepler's laws from basic principles and shows that motion in elliptical, parabolic, and hyperbolic orbits can be accounted for by a centripetal force that varies inversely as the square of the distance. The motion of two bodies about their center of mass is also discussed. In Book II, *The Motion of Bodies in Resisting Media,* the effects of resistance on motion are derived, and the motions of fluids are analyzed. Finally, in Book III, *The System of the World,* Newton states the law of gravitation and then proceeds to account for the detailed motions of the earth, the moon, and the planets and their satellites. A thorough explanation of ocean tides is also provided, and the orbits of comets are discussed in detail.

In itself the *Principia* is an incredible achievement; yet Newton's contributions do not end here, for Newton found that many of the problems he dealt with could not be solved with the existing mathematical tools, and he had to develop his own mathematical techniques. In so doing, he invented differential and integral calculus. In the next few chapters we will take a closer look at some of the discoveries in the *Principia.* Before doing so, let us digress long enough to take a look at the facts of Newton's life in the hope that we may gain some insight into the nature of his genius.

4–1

ISAAC NEWTON

Newton was born on Christmas day, 1642, in the small village of Woolsthorpe in the English midlands. His father died before Newton was born, and when Newton was three years old, his mother remarried and left him in the care of his grandmother. Newton was a brilliant student (although not a child prodigy) and was admitted to Cambridge University in 1661. There Newton came under the tutelage of a famous mathematician, Isaac Barrow. In 1669 Barrow stepped down in favor of his pupil, and at the age of twenty-seven Isaac Newton became the Lucasian Professor of Mathematics.

In 1665 England was swept by the plague. When Cambridge closed down, Newton returned to Woolsthorpe to live with his mother. During the next two years Newton, working in almost complete isolation, made important discoveries in optics, invented calculus, and sketched out much of the *Principia*. Unfortunately, Newton's notebooks from this time are lost, so we have no idea how he managed to achieve such a prodigious amount in such a short period of time. All we have from Newton himself is this masterpiece of understatement: "All this was in the two plague years of 1665 and 1666, for in those days I was in the prime of my age for invention, and minded Mathematics and Philosophy more than at any time since."[1]

At this stage in his life, Newton did not like the idea of having to defend his theories against criticism and wanted to avoid being thrust into public view. There were also certain mathematical difficulties that he had not yet overcome, so he made no attempt to publish his work.

Newton was by no means the only one concerned with the problem of accounting for planetary motion. Many scientists, notably Robert Hooke, Edmund Halley, and Christopher Wren, were trying vainly to determine the force required to make a planet move in an elliptical orbit. In 1684 Newton casually admitted to Halley that he had solved the problem many years before. At Halley's insistence Newton put the finishing touches on his calculations, and the *Principia* was presented to the world in 1687.

With the publication of the *Principia*, Newton's interest in physics came to an end. He became preoccupied with the study of alchemy and spent almost thirty years working on this subject without making a single notable discovery. He also wrote voluminously on historical chronology (i.e., the dating of historical events of ancient times) and the interpretation of biblical prophecy. Having successfully systematized the physical world, Newton was apparently intent on systematizing the theological world as well.

In 1693 Newton suffered an apparent mental breakdown from which he recovered fairly rapidly. Following this episode, Newton decided to leave Cambridge and move to London, where he was accepted as a member of the English aristocracy. In 1696 Newton was appointed Warden of the Mint, a post to which he devoted great energy. He personally interrogated and prosecuted many counterfeiters and was greatly responsible for maintaining the stability of the English coinage.

In 1703 Newton was elected president of the Royal Society, an organization of professional and amateur scientists. Under Newton's administration the influence of the Royal Society grew tremendously, and before long Newton had become the virtual dictator of the British scientific establishment, with great influence over the filling of vacant professorships and administrative posts. He was also involved in several personal feuds—with Hooke, who claimed that he had discovered the law of gravitation; with John Flamsteed, the astronomer royal, who could not supply Newton with astronomical data at a fast enough rate; and with the great German philosopher-mathematician Gottfried Leibniz, who had independently invented calculus and whom Newton accused of plagiarism. The feud with Hooke was a particularly bitter one, and to avoid further controversy Newton withheld the publication of his researches on optics—a field in which Hooke had also worked—until Hooke's death. Thus in 1704 the *Opticks* was published, describing significant discoveries that Newton had made some forty years before.

In 1705 Newton was knighted—the first scientist to be so honored. He never married, and although he managed to amass a large personal fortune, he lived rather simply. At his death in 1727 Sir Isaac was given a state funeral and was buried in Westminster Abbey.

The details of Newton's life tell us that he was a man with tremendous inner energy and a single-mindedness that enabled him to pursue scientific problems and counterfeiters with the same relentless force. We can also identify the psychological determinants that may have shaped his complex personality; certainly, the fact that Newton's greatest period of productivity came when he was reunited with his mother during the plague years is grist for the psychoanalytical mill. Yet we cannot begin to understand how Newton was able to unlock the secrets of the universe with such seeming ease. Unable as we are to explain the riddle of Newton's genius, we must proceed to the simpler problem of understanding his contributions to physics.

THE NEWTONIAN SYSTEM OF MECHANICS

Mechanics is the branch of physics that is concerned with the nature and causes of the motions of bodies. In the first part of the *Principia,* Newton developed a complete theory of mechanics, based in part on the work begun by Galileo, Kepler, and Descartes. Newton proposed a system of concepts and principles through which one could establish a mathematical relationship between the *cause* of a motion and the motion itself. The relationship is embodied in three fundamental principles of motion, which are stated at the beginning of the *Principia.*

It is common practice to refer to these three principles as "Newton's laws of motion." However, they are not laws in the sense that we have defined a law, which is as an observed relationship between measured quantities. We shall learn that the Newtonian system of mechanics is *not* a collection of experimental observations, but a *system of concepts* useful for analyzing motion. The greatness of Newton's achievement lies in the simplicity, comprehensiveness, and predictive power of this system.

In the following, Newton's principles will be stated in modern terminology rather than in their original form, so as to make their meaning clearer to us.

4–2

NEWTON'S FIRST PRINCIPLE: THE CONCEPT OF INERTIA

In the previous chapter we learned that Aristotle classified the motions of bodies as being either *natural* (vertical terrestrial motion and the circular motion of celestial bodies) or *forced* (horizontal terrestrial motion). Newton similarly chose to distinguish between natural and forced motion, but he based his distinction between the two types on an entirely different principle. He postulated that *it is the natural tendency of bodies to move in a straight line at constant speed.* In other words, constant-velocity motion is assumed to be the natural motion of a body. Thus any motion in which the velocity is not constant (i.e., any accelerated motion) is an unnatural or *forced* motion. Newton's hypothesis is embodied in the ***first principle of motion,*** which can be stated* as follows:

A body will remain in a state of rest, or in uniform motion in a straight line, unless acted on by an unbalanced force.

Implicit in the first principle are the following ideas:

(1) No force is required to keep a body moving at a constant velocity.
(2) If a body is accelerating, a force must be acting on it.
(3) If an unbalanced force (one not canceled out by any counterforce) is acting on a body, the body must accelerate.
(4) A body will continue to accelerate for as long as an unbalanced force is continually applied. Once the force is removed or canceled, the body will revert to constant-velocity motion.

Newton emphasized the fact that when a body is at rest, its velocity is constant ($\mathbf{v} = 0$). Since the state of rest is a state of constant-velocity motion, the natural tendency of a body at rest is to remain at rest. Thus in the Newtonian system stationary bodies and bodies moving with constant velocity behave according to the same principle: Their natural tendency is to maintain their state of motion.

Unlike Aristotle, Newton did not make any distinction between terrestrial and celestial motions. He proposed that the celestial bodies also have a natural tendency to move in a straight line at constant speed. Consequently, the circular or elliptical orbits of the planets indicate that the planets' orbital motions are *forced* motions, as Kepler had previously suggested. Newton later identified the force that causes the planets to deviate from their natural motion as the force of gravitation.

Newton considered the tendency to maintain a constant-velocity motion to be an innate property of all material bodies. This property has been given the name ***inertia.*** In the *Principia* Newton states: "Inertia . . . is a power of resisting by which every body . . . continues in its present state, whether it be of rest or moving uniformly forwards in a straight line."[2]

Thus the statement "All bodies possess inertia" is simply an abbreviated way of saying, "All material bodies have a natural tendency to maintain constant-velocity motion." Similarly, constant-velocity motion is sometimes referred to as ***inertial motion.***

Why is constant-velocity motion the natural motion of a body? Newton made no attempt to justify this hypothesis. The concept of inertia represents nothing more than Newton's *belief* as to how matter behaves in the universe and is philosophically

* Another, more up-to-date statement of this principle has been kindly provided by Professor Edward Green of Brooklyn College:

The first law of Newton I sing;
But my voice has a relevant ring:
"An object left free
Of hassles will be
Engrossed in just doing its thing."

no different from Aristotle's assertion that circular motion is the natural motion of celestial bodies. The eminent success of the Newtonian system of mechanics is the strongest justification that can be given for accepting Newton's concepts over those of Aristotle.

Nevertheless, it is not unreasonable to go one step further and to pose the question: "Why do bodies possess inertia?" At the present stage of the development of physics, the answer to this question remains a deep mystery. Theoretical physicists have made attempts to explain the inertia of a body in terms of its interactions with other bodies, but these attempts have not been very successful. For the present we must simply accept inertia as a fundamental, inherent, unexplained property of matter.

THE CONCEPT OF MASS

We have learned that in the Newtonian system of mechanics constant-velocity motion is regarded as natural motion, and accelerated motion is regarded as forced motion. A force is required to produce and maintain the acceleration of a body; once the impressed force ceases to act, the body reverts to its natural inertial motion.

Through observation and experimentation, certain facts emerge with respect to the relationship between a given force and the acceleration produced by that force. Primarily, we observe that a given force does not necessarily produce the same acceleration when applied to different bodies. We can imagine an experiment in which two identically shaped footballs are set up on kicking tees. One football is filled with air, and the other is filled with concrete. If someone attempts to kick each ball with identical swings of the foot (thereby applying the same force in each case), it will be quite evident to him that he has not imparted the same acceleration to both footballs.

The conclusion that one inevitably comes to after performing experiments of this type is that the resistance that bodies offer to a change in their state of motion is different for different bodies. In other words, *some bodies possess more inertia than others.* Therefore we must look on the inertia of a body as a *quantity;* that is, we say that a body possesses a certain *amount* of inertia.

To describe this aspect of inertia, Newton introduced the concept of ***mass.*** The mass of a body is a *numerical measure of the amount of inertia that the body possesses.* If body A has a greater mass than body B, then body A offers greater resistance to a change in its motion than does body B. Hence the greater the mass of a body, the greater the force required to change its motion by a given amount. Also, if the same force is applied to two bodies having different masses, the body with the greater mass will be accelerated less by the force. In other words, the larger the mass of an object, the smaller the acceleration that will be produced by a given force.

By doing experiments, we can determine the mathematical relationship between the mass of a body and the acceleration produced by a given force. As an example, let us consider the following experiment. Two lightweight carts are connected together by a rope, and a compressed spring is placed between them, as shown in Figure 4–1. A number of identical steel cubes serve as test bodies. On cart A we place one cube, and on cart B we place two cubes, so that the mass of cart B is twice the mass of cart A. (We assume that the masses of the carts themselves are very small compared to the masses of the cubes.) When the rope is cut, the spring expands, pushing the carts apart with equal force. The expansion of the compressed spring causes the two carts to accelerate in opposite directions, and the accelerations of the two carts can be easily measured.

When we perform this experiment, we find that the acceleration of cart A is

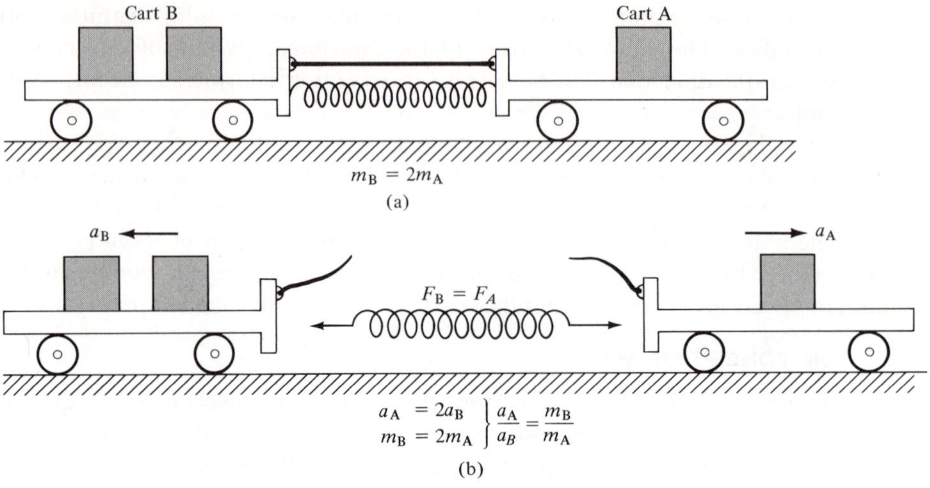

figure 4–1 **THE RELATIONSHIP BETWEEN MASS AND ACCELERATION**

(a) Two carts, A and B, are held together by a rope. Between them is a compressed spring. Cart B holds two identical steel cubes; cart A has one such cube. Hence B has twice the mass of A. (b) When the rope is cut, the carts are pushed apart by the spring, which exerts an equal force on each cart. Measurement of the accelerations of the two carts shows that cart A has twice the acceleration of cart B, in inverse proportion to their masses.

twice the acceleration of cart B. But cart B has twice the mass of cart A. Thus it appears that there is a simple inverse relationship between the mass and the acceleration. If we change the experiment by placing *three* steel cubes in cart B, we find that cart B has *one-third* the acceleration of cart A. Continuing the experiment with various combinations of cubes, we find in every case that the ratio of the accelerations is inverse to the ratio of the masses. In mathematical form this relationship is written

$$\frac{a_A}{a_B} = \frac{m_B}{m_A} \tag{4–1}$$

where m_A and m_B represent the masses and a_A and a_B represent the accelerations of carts A and B. The cart-and-spring technique is just one of several experimental methods by which the relationship between mass and acceleration can be tested. The results are always the same, provided the bodies being compared are subject to the same force.

On the basis of these results, a procedure can be set up for measuring the mass of any object. We first select some arbitrary object and define it as the standard mass unit. In the metric system the unit of mass is the ***kilogram;*** the standard kilogram is a platinum–iridium cylinder kept at the International Bureau of Weights and Measures. Thus just as the meter has been arbitrarily defined as the unit by which distances are measured, the kilogram has been arbitrarily defined as the unit by which the mass of a body is measured. Since mass is the measure of the amount of inertia that a body possesses, the standard kilogram can be said to possess one unit of inertia.

The mass of any body can be determined by comparing it with the standard kilogram or its equivalent. The comparison can be made by subjecting the given body and the standard kilogram to the same force and measuring the accelerations produced. The mass of the given body is then calculated using Eq. 4–1. For example, if the given body is observed to have one-fourth the acceleration of the standard kilogram, the body must have four times the mass. Hence its mass is 4 kilograms (4 kg).

There is no inherent reason why the mass of a given body should be a constant quantity. We might expect that a body's resistance to a force would depend on the nature or the magnitude of the force. It is not unreasonable to expect that the inertia of a body is greater when the body is at rest than when it is already in motion. However, all of the experimental evidence available to Newton showed that the mass of a body is the same under all conditions. Thus a basic assumption in the Newtonian system is that *the mass of a body is a constant quantity.*

Newton's assumption that a body has a constant mass under all conditions was consistent with his belief that mass is a measure of the "quantity of matter" in a given body. But it is now recognized that there is no consistent way to define or measure the amount of matter in a body. Surely the size or volume of an object is not the only determining factor; a cubic foot of lead clearly has a greater mass than a cubic foot of air. Newton himself rejected the idea that the *weight* of a body is a measure of its quantity of matter, for experiments showed that the weight of a body varies from place to place on the earth. (The distinction between weight and mass will be discussed more fully later in the chapter.)

Nor is it possible to define the amount of matter in terms of the number of atoms in a given body, for experiments in atomic physics have shown that the individual atoms of different substances have different masses. For instance, the mass of a single lead atom is about thirteen times the mass of a single oxygen atom. Hence *two* oxygen atoms must represent a *smaller* amount of matter than *one* lead atom. Research into the structure of the atom has demonstrated that the atoms of the various elements differ in the number of particles (protons and neutrons) contained in the nucleus—but further research has revealed that the mass of a given atomic nucleus does *not* equal the sum of the masses of the protons and neutrons in the nucleus.* Consequently, it is not possible to define the amount of matter in terms of the number of particles contained in a given body.

Hence physicists have come to reject Newton's definition of mass as a measure of the quantity of matter on the grounds that the term is ambiguous and cannot be rigorously defined. The modern viewpoint is to regard mass as nothing more than a measure of the inertia of a body. Since inertia is presently regarded as a fundamental and unexplained property of matter, it follows that mass must also be considered a fundamental property of bodies that cannot be explained or understood in terms of other concepts.

The Newtonian assumption that the mass of a body is a constant was eventually disproved by experiments performed in the early 1900s. These experimental findings were in agreement with the predictions made by Einstein in the theory of relativity that the mass of a body increases with its speed. The mass increase turns out to be noticeable only at extremely high speeds (of the order of 20,000 mi/sec or more), which explains why the mass increase was not evident in the experimental data

*The explanation for this effect is that protons and neutrons tend to lose their independent identities inside the nucleus; a portion of their mass is converted to the energy required to bind the nucleus together. The details will be discussed in Chapter 18.

available to Newton. We may say, at least, that Newton's assumption of a constant mass is a very good approximation at common speeds.

In view of the evidence that mass depends on speed, it becomes rather difficult to understand how mass can be considered a measure of the amount of matter in a body. (How can the amount of matter in a body change as the body moves faster or slower?) However, if we regard mass as simply the measure of the inertia of a body, the dependence of mass on speed becomes less mysterious. This dependence indicates that *the inertia of a body increases as the body's speed increases.* The faster a body is moving, the more it resists a change in its motion, or in other words, it is easier to accelerate a body when it is moving slowly than when it is moving rapidly.

We will consider the mass–speed relationship in more detail when we study the theory of relativity in Chapter 12. In our examination of the Newtonian system, however, we will maintain the assumption that a body has a constant mass, keeping in mind that this assumption is a very good approximation at all except extremely high speeds.

4–3

NEWTON'S SECOND PRINCIPLE: THE CONCEPT OF FORCE

We have discussed the effects of forces on bodies, but we have not as yet defined precisely what we mean by the term "force." In particular, we should like to know how to identify a force, how to measure its strength, and how to determine numerically its effect on the motion of a body. It would seem at first glance that a force is a rather self-evident thing; we can feel through our muscular and nervous systems what it is like to exert a force or to have a force exerted on us. In fact, Aristotle defined a force simply as "a push or pull," and Newton considered force to be an intuitive notion and did not provide a precise definition of it in the *Principia*. However, a number of nineteenth-century physicists (most notably Ernst Mach and Heinrich Hertz of Germany) stated that these descriptions of force are much too vague, and they provided a more exact definition of force.

The most preferable definition of force is the following: *A force is any action that can change the state of motion of a body.* In any situation or circumstance in which we observe the acceleration of a body, we assume that a force must be acting on the body. Thus "force" is the general term that is used to describe any cause of accelerated motion. Depending on the circumstances under which accelerated motion is observed, a force may be classified as gravitational, electric, magnetic, frictional, and so forth. To illustrate the sense of this definition of force, let us take the example of a body in free fall toward the earth. Our observations indicate that the body is in accelerated motion. (It is seen to fall with a constant acceleration of 32 ft/sec^2.) In the Newtonian system this is a motion that requires a force; that is, since the body is continually accelerating, a force must be continually acting on it. We identify this force as the force of gravity. Note, however, that *we do not observe the force of gravity itself; we observe only the acceleration of the falling body.* The existence of a force of gravity is *inferred* from the observed acceleration.

The magnitude of a given force can be determined from Newton's second principle, which defines precisely the mathematical relationship between force, mass, and acceleration. This relationship is based on experiments showing that when the same force is applied to two different bodies, the resulting accelerations are inversely proportional to the masses of the bodies. This relationship, given previously in Eq. 4–1, can be written as follows:

$$m_A a_A = m_B a_B \qquad (4\text{–}1a)$$

In this form the equation states that the product of mass and acceleration is the same for all bodies subjected to the same force. It is therefore reasonable to define force by the equation

$$\mathbf{F} = m\mathbf{a} \qquad (4\text{–}2)$$

This is the modern-day form of Newton's second principle. It *defines* the magnitude of a force acting on a body as the product of the mass of the body and the acceleration produced by the force. Note that Eq. 4–2 is a *vector* equation, in keeping with the observation that a body always accelerates in the same direction as the applied force—that is, the force and acceleration vectors always point in the same direction.

Further evidence for the vector nature of forces is obtained when one observes a situation in which two or more separate forces act on a body (Figure 4–2). Experiments reveal that the acceleration of the body depends on the *vector sum* of all the forces acting on it. We refer to the vector sum of the forces as the ***net force*** on the body. Thus when two or more forces are acting on a single body, it is the net force that is substituted into Eq. 4–2 to determine the acceleration of the body. If the net force is zero (the forces cancel each other out), the body will not accelerate. In this situation the body is said to be in ***equilibrium,*** and it will maintain its inertial motion.

The simplicity of the second principle belies the great power it bestows on the

figure 4–2 **FORCE AS A VECTOR QUANTITY**

If two equal forces are applied in opposite directions to the same body, as shown in (a), they will counteract each other so that the net force is zero. If two forces are applied in the same direction, as in (b), the body will experience a net force of 2**F** and will move to the right. If the forces are not collinear, as in (c), the body will move in the direction of the resultant. In each case the net force is the vector sum of the individual forces.

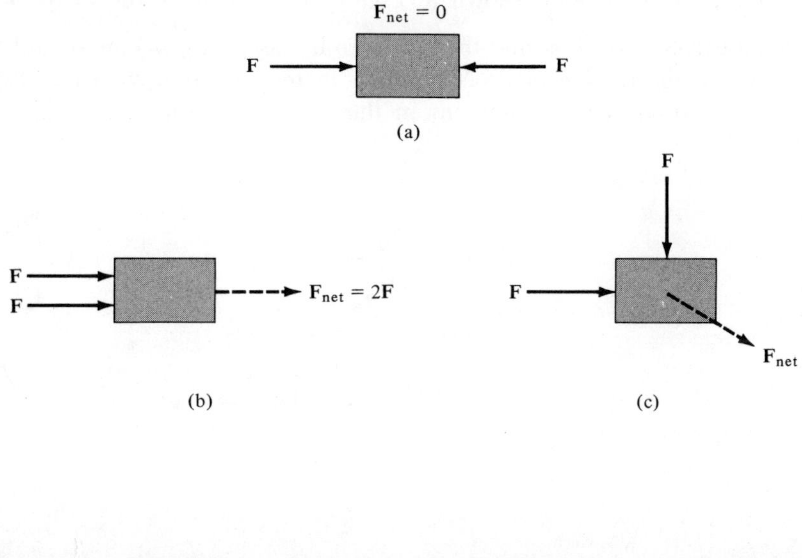

physicist as a means of understanding and predicting events in the universe. In the Newtonian view the universe consists entirely of masses being acted on by forces, and if one can identify all the forces acting on a body, one can completely determine the future motions of that body.

UNITS OF FORCE: SYSTEMS OF UNITS

According to Eq. 4–2, force should have units of mass × acceleration. Therefore if mass is measured in grams and acceleration in cm/sec^2, force will have the units of g-cm/sec^2. To avoid the use of this rather cumbersome unit, the word ***dyne*** has been coined to stand for the unit of force:

$$1 \text{ dyne} = 1 \text{ g-cm/sec}^2$$

One dyne is the force required to give a mass of 1 gram an acceleration of 1 cm/sec^2.

At this point let us digress briefly to discuss the main systems of units used in modern-day science. In a given system the units assigned to more complex physical quantities are defined in terms of the units chosen for the fundamental quantities of distance, mass, and time. There are two systems of metric units: the CGS (centimeter-gram-second) system and the MKS (meter-kilogram-second) system. It is easy to see that the dyne is the CGS unit of force, since it is defined in terms of the gram, centimeter, and second. In the MKS system force is measured in kg-m/sec^2. The tradition has been to name MKS units after famous scientists, and the MKS unit of force has been most appropriately named:

$$1 \text{ newton (N)} = 1 \text{ kg-m/sec}^2$$

Since 1 kg = 10^3 grams and 1 meter = 10^2 cm, it follows that 1 N = 10^5 dynes. In the British system the unit of mass is called the ***slug,*** and the unit of force is the ***pound:***

$$1 \text{ pound (lb)} = 1 \text{ slug-ft/sec}^2$$

The following conversion factors apply:

mass units: 1 slug = 14.59 kg = 1.459×10^4 grams
force units: 1 lb = 4.448 N = 4.448×10^5 dynes

EXAMPLE 4–1: A body having a mass of 20 kg is acted on by a 15-N force and a 20-N force, as shown in Figure 4–3. Calculate the body's acceleration.

SOLUTION: We first find the net force by taking the vector sum of the two forces. We obtain a value of 25 N pointing in the direction shown in Figure 4–3. The acceleration vector will point in the same direction. Using $F_{net} = ma$, we find

figure 4–3

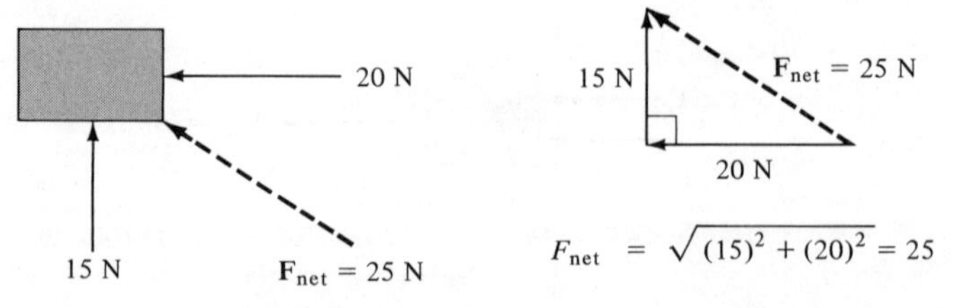

that the magnitude of the acceleration is

$$a = \frac{F_{net}}{m} = \frac{25\ \text{N}}{20\ \text{kg}} = \underline{1.25\ \text{m/sec}^2}$$

WEIGHT VERSUS MASS

Newton was the first physicist to make a distinction between weight and mass, and it is important that we have a clear understanding of this distinction. Mass, as has been noted, is a measure of the inertia of a body. *Weight is the force of gravity exerted on a body.* Whereas mass is a fundamental quantity, weight is a particular kind of force; namely, it is the gravitational force exerted by the earth, or by any large astronomical body, on a given object at or near the body's surface. The weight of an object varies from place to place; one weighs less at the equator than at the poles and less on a mountaintop than at sea level. An astronaut who weighs 180 lb on the earth weighs 30 lb on the moon because the moon exerts one-sixth the gravitational force of the earth. On the other hand, in the Newtonian system the mass of a body is a constant quantity and has the same value wherever the body is found. According to Newton, inertia is an inherent property of a body and should not depend on the presence of other bodies, but the weight of a body is the result of an interaction with another object and therefore should depend on the location of the body.

It is a simple matter to derive a formula relating weight and mass from the equation $F = ma$. Weight is the force of gravity; therefore we set $F = w$. The acceleration associated with this force is g, the acceleration of gravity; hence $a = g$. We thereby obtain

$$w = mg \tag{4–3}$$

On the surface of the earth, where $g = 32\ \text{ft/sec}^2 = 9.8\ \text{m/sec}^2 = 980\ \text{cm/sec}^2$, we find that a *mass* of 1 kg has a *weight*

$$w = mg = (1\ \text{kg})(9.8\ \text{m/sec}^2) = 9.8\ \text{N}$$

Similarly, 1 gram weighs 980 dynes, and 1 slug weighs 32 lb. On the moon, where $g = 5.3\ \text{ft/sec}^2$, 1 slug weighs only 5.3 lb.

Much confusion is caused by the way in which mass and weight are dealt with in everyday use. In countries using the metric system, people describe quantities by their mass instead of their weight; meat, vegetables, and so on are sold by the "kilo" (short for kilogram). A European would say, "I weigh 80 kilos." This is common usage but incorrect terminology. He means that his *mass* is 80 kilos. (His weight is 784 N.) In countries where the British system is used, one gives his weight in pounds, which is proper usage.

One sometimes sees the conversion factor 1 kg = 2.2 lb. As it stands, this is an incorrect comparison because it equates mass with weight. What it means is that a mass of 1 kg has a weight of 2.2 lb *on the surface of the earth* (where $g = 32\ \text{ft/sec}^2$). On the moon 1 kg weighs only 0.36 lb.

The equation $w = mg$ provides us with another method for determining the mass of a body; namely, we measure the body's weight and then divide by the local value of the acceleration of gravity. Weight can be measured by a spring scale, a device that operates on the principle that the amount by which a spring stretches is proportional to the stretching force. (This effect was discovered by Robert Hooke

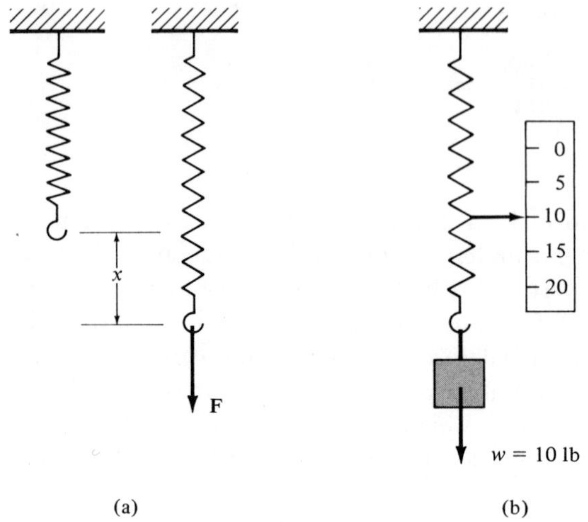

figure 4–4 **PRINCIPLE OF A SPRING SCALE**

(a) The amount x by which the spring elongates is proportional to the magnitude of the applied force **F**. (b) The scale is calibrated by marking off the elongations produced by known weights.

(Newton's adversary) and is known as Hooke's law.) A spring scale can be calibrated by measuring the elongation produced by known weights (Figure 4–4).

A procedure for determining mass that does not depend directly on the local value of g utilizes the equal-arm balance, shown in Figure 4–5. The pointer is at zero when the weights in the two pans are equal. According to Eq. 4–3, if the weights are equal, the masses are also equal, for whatever the local value of g happens to be, it will be the same for both weights. Thus one determines the mass of an unknown object by balancing it with a known mass.

More will be said about the distinction between weight and mass in the next chapter, after we have learned about the principle of gravitation.

EXAMPLE 4–2: An astronaut weighs 160 lb on the earth. How much will he weigh on Mars, where the acceleration of gravity is 12.3 ft/sec^2?

SOLUTION: The weight of the astronaut on Mars can be calculated from the equation $w = mg$, where $g = 12.3\ \text{ft/sec}^2$. However, we have to know the mass of the astronaut, which can be calculated from the information that the astronaut weighs 160 lb on earth, where $g = 32\ \text{ft/sec}^2$. Thus

$$m = \frac{w}{g} = \frac{160\ \text{lb}}{32\ \text{ft/sec}^2} = 5\ \text{slugs}$$

Since mass is a constant quantity, the astronaut will have a mass of 5 slugs on Mars also. Therefore his weight on Mars will be

$$w = mg = (5\ \text{slugs})(12.3\ \text{ft/sec}^2) = \underline{61.5\ \text{lb}}$$

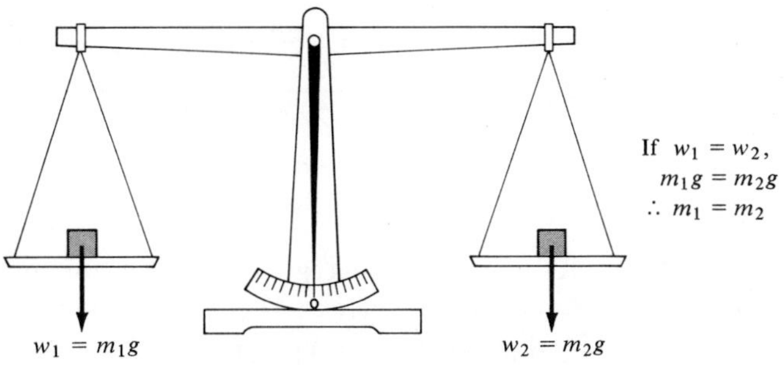

figure 4–5 **THE EQUAL-ARM BALANCE**

The device is balanced (pointer at zero) when the weights in the two pans are exactly equal. If $w_1 = w_2$, then $m_1 = m_2$, independently of the local value of g.

FRICTION

In applying Newton's principles of motion, we must deal with a particular class of forces known as ***resistance forces.*** These are forces that are generated in response to some motion or force and that vary in strength according to the magnitude of the initial motion or force. One force of this type is ***air resistance,*** which is the force that air exerts against any body moving through it. The magnitude of this force is, in general, dependent on the speed and shape of the moving body. Of course, air is not the only substance that exerts such a force; a resistance of this kind is exerted against bodies moving through any gaseous or liquid medium.

Perhaps the most familiar resistance force is ***friction,*** which arises whenever there is motion between two surfaces that are in contact. Although the exact nature of friction is not precisely understood, friction apparently results from irregularities in the two contacting surfaces and ultimately derives from electrical attractions between the atoms at the surfaces. Friction can be reduced by polishing the surfaces or by using a lubricant. The invention of the wheel represented a tremendous advance for transportation because (everything else being equal) rolling friction is always less than sliding friction. The force of friction depends, of course, on the nature of the contact surfaces and on the weight of the body to be moved. The friction force also varies according to the force being applied to produce the motion.

Although we tend to think of friction as a retarding (and therefore inefficient) force, we must realize that most forms of locomotion would be impossible without it. Anyone who tries to walk or drive an automobile over a sheet of ice will recognize the value of frictional forces. It is friction that enables us to stop our motions once we have started them. The automobile brake, for example, uses the friction generated when the brake shoes are forced against the brake linings to stop the car wheels from turning. The sliding (rather than rolling) friction between the tires and the road brings the automobile to a safe and controlled stop.

EXAMPLE 4–3: A 40-gram block is pushed along a horizontal surface by a force of 800 dynes. The measured acceleration is 16 cm/sec^2. What is the frictional force between the block and the surface?

figure 4–6

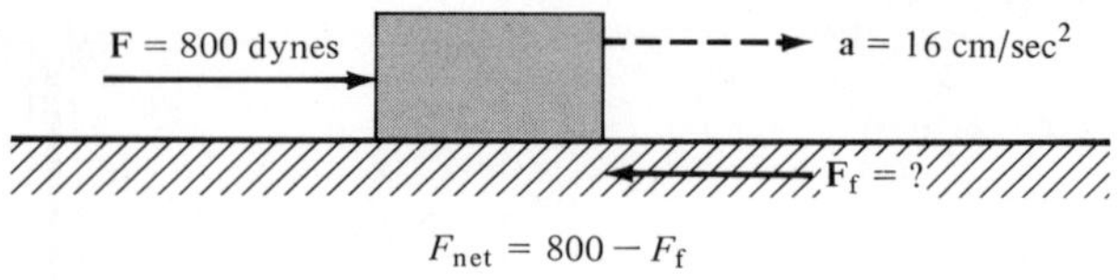

SOLUTION: As diagramed in Figure 4–6, the unknown frictional force $\mathbf{F}_f$ acts as a retarding force on the block. The acceleration of the block results from the combined 800-dyne force and $\mathbf{F}_f$. Since $\mathbf{F}_{net} = m\mathbf{a}$, we have

$$F_{net} = 800 - F_f = ma = (40 \text{ grams})(16 \text{ cm/sec}^2) = 640$$

so that

$$F_f = 800 - 640 = \underline{160 \text{ dynes}}$$

4–4

NEWTON'S THIRD PRINCIPLE: ACTION AND REACTION

The Newtonian system of mechanics is completed by the ***third principle of motion,*** which, more than any of the other principles, is Newton's own unique creation. The third principle can be stated as follows:

If one body exerts a force on another body, the second body will exert an equal and opposite force on the first body.

Let us consider body A, which is exerting a force on body B. The nature of this force is entirely arbitrary—it may be gravitational, electrical, frictional, and so on. According to the third principle, body B will exert an equal and opposite force on body A—that is, the force will be equal in magnitude but opposite in direction to the force exerted by A on B (Figure 4–7). Newton's third principle is sometimes referred to as the ***principle of action and reaction,*** and it can be stated rather succinctly:

For every action there is an equal and opposite reaction.

figure 4–7 **NEWTON'S THIRD PRINCIPLE: ACTION AND REACTION**

If body A exerts a force $\mathbf{F}_{AB}$ on body B, then body B exerts a force $\mathbf{F}_{BA}$ on body A that is equal in magnitude but opposite in direction to $\mathbf{F}_{AB}$. This principle holds regardless of the kind of force involved.

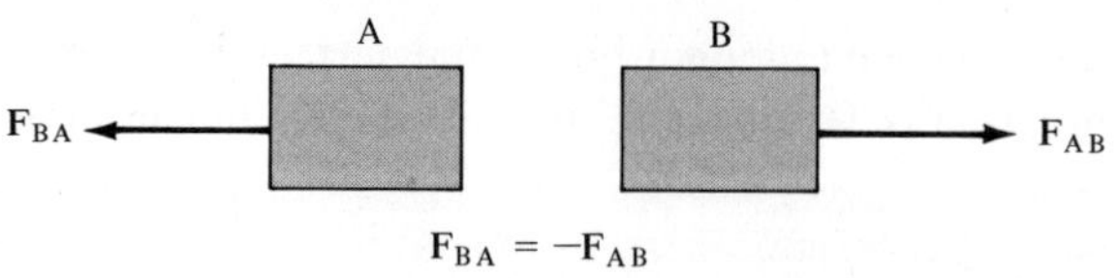

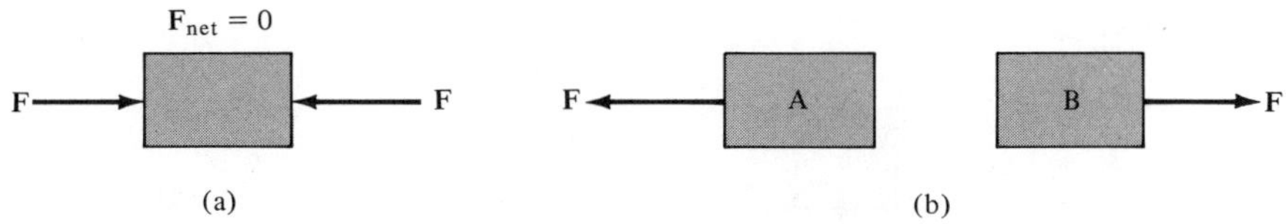

figure 4–8 **NONCANCELLATION OF ACTION–REACTION FORCES**

In (a) two equal and opposite forces act on the same body. These forces are the result of two different interactions; hence they are not an action–reaction pair. The net force is zero. In (b) the two forces are members of an action–reaction pair (caused by the same interaction); they act on separate bodies and therefore do not cancel each other. The net force on A is **F**; the net force on B is also **F**. Both bodies will be accelerated.

Essentially, the third principle is a postulate about the nature of forces—namely, that *forces always exist in pairs*. In Newton's view force is an *interaction* between bodies. Therefore one cannot reasonably think of a single, isolated force acting on one body without another body being affected in some way. In every interaction the two paired forces are equal in strength and exist simultaneously. Hence there is no proper way to identify one force as an "action" and the other as the "reaction"—they are entirely equivalent.

A common misconception is that since the action and reaction forces are equal and opposite, they cancel each other. Equal and opposite forces cancel each other *only if they are acting on the same body*. (See Figure 4–8a.) Since the forces in an action–reaction pair act on different bodies, they can never cancel each other. Note that in Figure 4–8b body B has only one force acting on it and should be accelerated to the right; similarly, body A should be accelerated to the left by the single force acting on it.

In Figure 4–9 the example of a 20-lb block resting on a rigid surface is analyzed. There are two pairs of action–reaction forces at play. The block has two distinct forces acting on it: its weight (the force of gravity) and the upward force exerted by the surface. These two forces do not constitute an action–reaction pair, since they are the result of two separate interactions. Since weight is the force of the earth's attraction on the block, the reaction to the weight is an upward pull exerted *by* the block *on* the earth. The upward force exerted by the surface on the block has as its reaction the downward force of the block on the surface. Thus the two forces on the block are not an action–reaction pair but the result of two different interactions, and they can be added together to give a net force of zero on the block.

One somewhat puzzling aspect of this discussion concerns the interaction between the block and the earth. According to the third principle, the block exerts the same force on the earth that the earth exerts on the block. Intuitively, one might expect the earth to exert a larger force, since it is the larger body. But this is not the case: According to Newton, the two forces are of equal strength. We must think of these two forces not as separate and independent forces, but as twin manifestations of a mutual interaction. Though it may seem contrary to our intuition, the validity of this principle has been thoroughly borne out by experimentation and observation. For instance, the third principle tells us that the gravitational force exerted by the earth on the moon is the same as the gravitational force exerted by the moon on the earth.

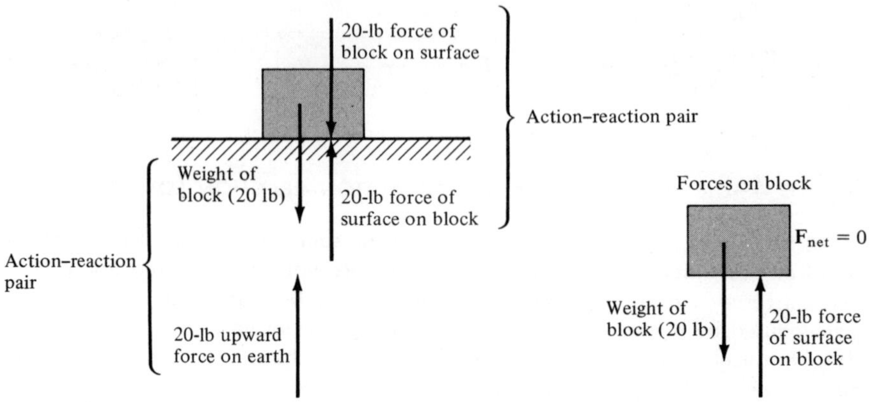

figure 4–9 **ANALYSIS OF ACTION–REACTION FORCES**

In the case of a 20-lb block at rest on a rigid surface, two pairs of action–reaction forces are involved.

By accepting this principle, we can predict with considerable accuracy the motions of the earth and the moon. Any other assumption leads to incorrect predictions.

It is important to recognize that although both members of an action–reaction pair of forces are equal in magnitude, their *effects* may be quite different. Specifically, the accelerations produced by the two equal forces will be inversely proportional to the masses of the bodies on which they are acting, in accordance with Newton's second principle. Let us suppose that the 20-lb block is removed from the surface and is allowed to fall freely. The third principle states that both the block and the earth experience the same unbalanced force, and they will both accelerate toward one another. However, the mass of the earth is enormously large (about 10^{24} times greater) compared to the mass of the 20-lb block; hence the acceleration of the earth will be extremely small compared to that of the block. Consequently, although the forces on the block and on the earth are equal, the accelerations produced by these forces are quite different. The block will fall toward the earth with an acceleration of 32 ft/sec^2, and the earth will be given an upward acceleration of about 10^{-23} ft/sec^2. The latter is so small that it is completely undetectable.

SOME APPLICATIONS OF NEWTON'S THIRD PRINCIPLE

The principle of action and reaction enables us to account in very simple terms for a great many of the phenomena we encounter in our everyday lives. To begin with, let us analyze the process of walking. When we take a step, we press our foot on the ground, pushing backward. Because of friction, we exert a backward force on the ground. The reaction force is the equal and opposite force exerted by the ground on our foot. It is this force, which acts in the forward direction, that propels us in the direction we want to go (Figure 4–10a).

The same principle applies to the motion of wheeled vehicles. In an automobile or train the engine is linked mechanically through the drive shaft and axle in such a way as to cause the wheels to rotate. Because of friction between the wheels and the ground (or tracks), the wheels exert a backward force on the ground, and the

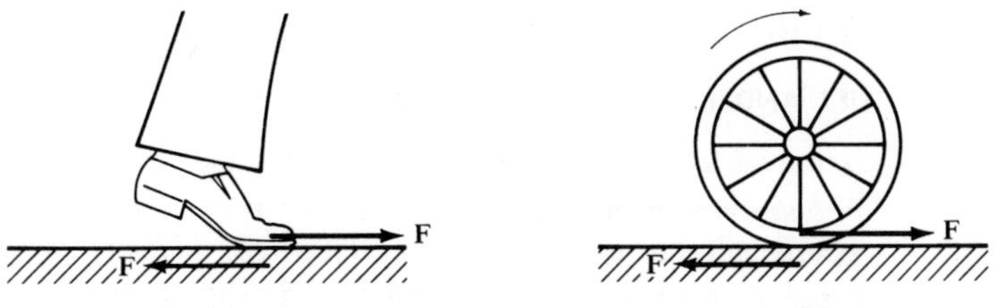

figure 4–10 **EXAMPLES OF THE ACTION–REACTION PRINCIPLE**

(a) We are able to walk because of the action–reaction pair of friction forces generated by the contact of our shoes with the ground. (b) A wheel is made to rotate mechanically. The friction force acting at the point where the rotating wheel makes contact with the ground results in a pair of forces, one of which pushes the wheel forward. In the absence of friction the wheel would spin without moving forward.

ground exerts a forward force on the wheels. This latter force drives the vehicle forward (Figure 4–10b). In the absence of friction the wheels would simply spin in place without any forward motion.

Another common phenomenon that can be readily explained by the third principle is the recoil of a gun. When a gun is fired, it exerts a large force on the bullet, accelerating the bullet out of the gun barrel. The bullet exerts a reaction force on the gun in the opposite direction, causing the gun to "kick," or recoil. The forces on the bullet and the gun are equal in magnitude; however, the mass of the gun is considerably larger than the mass of the bullet. Accordingly, the forward acceleration of the bullet is much greater than the backward acceleration of the gun.

Newton's third principle is fundamental to the operation of jet and rocket engines. In a jet engine air is drawn in at the front, compressed and heated, and then forced out the rear at very high speed. Since the engine is pushing backward on the expelled air, the air must push forward on the engine, thus providing forward thrust to the airplane. A rocket engine works on the same principle, using the reaction force of the expelled gases on the rocket to push the rocket forward. The difference between a rocket and a jet is that a jet uses the surrounding air, and a rocket makes its own gases. (See Figure 4–11.) Usually, a rocket contains a liquid fuel that turns to gas upon combustion and is forced through a nozzle in the rear of the rocket at very high speed. Since it produces its own gases, a rocket can operate in a vacuum, whereas a jet engine can operate only in the atmosphere.

A common misconception is that a rocket operates because the exhaust gases push off on the ground or the surrounding air. As we have seen, the rocket does not need any surrounding air; it works because the exhaust gases push back on the *rocket*. In fact, a rocket operates most efficiently in a vacuum, since neither the rocket nor the exhaust gases are retarded by air resistance. In any case, the student who has held this misconception need not feel too bad, for he is in good company. An editorial in The New York *Times* of January 13, 1920, commenting on the work of rocket pioneer Robert H. Goddard, made the following pronouncement:

figure 4–11

JET AND ROCKET PROPULSION

In the jet engine (a) the surrounding air enters through the intake and is forced out of the exhaust; in the rocket (b) gas produced by combustion of fuel is forced out of the exhaust. In each case the reaction to the force exerted on the expelled gases is a forward thrust exerted by the gases on the engine or rocket.

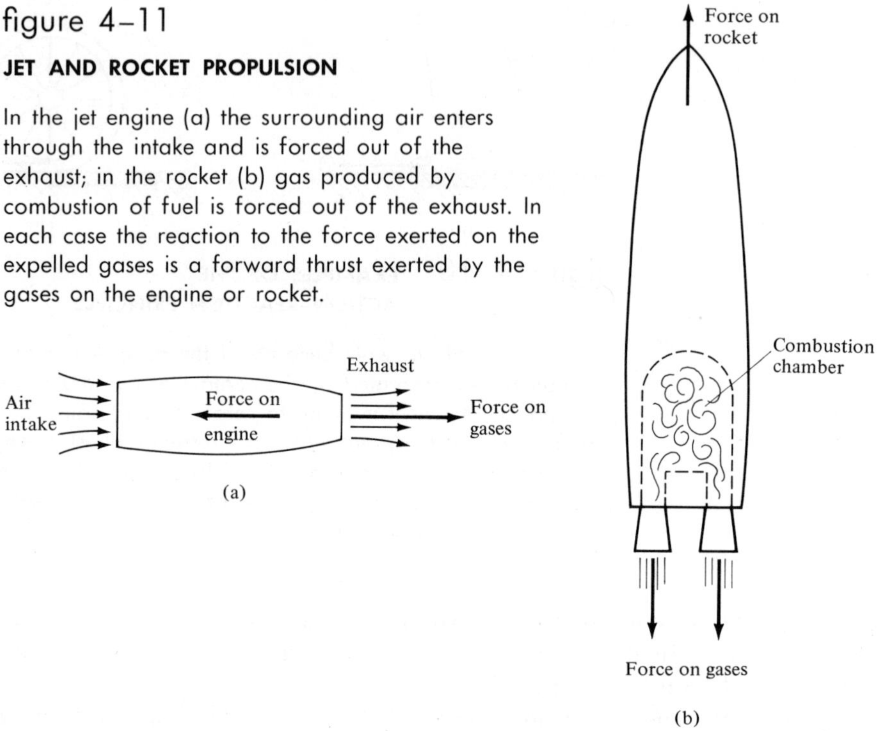

> That Professor Goddard . . . does not know the relation of action to reaction, and of the need to have something better than a vacuum against which to react—to say that would be absurd. Of course, he only seems to lack the knowledge ladled out daily in high schools.

On July 17, 1969 (the day that Apollo 11 took off for the first manned landing on the moon), the *Times* acknowledged its mistake:

> Further investigation and experimentation have confirmed the findings of Isaac Newton in the 17th century, and it is now definitely established that a rocket can function in a vacuum as well as in an atmosphere. The *Times* regrets the error.

4–5

CIRCULAR MOTION REVISITED

According to Newtonian principles, uniform circular motion is a forced motion. That is, a continually applied force is necessary to maintain the motion of a body in a circular path. In the previous chapter we learned that the acceleration required to maintain motion in a circle at constant speed must have a magnitude

$$a = \frac{v^2}{r}$$

Applying Newton's second principle ($F = ma$), we find that the force needed to maintain this acceleration must have the value

$$F = \frac{mv^2}{r} \tag{4–4}$$

where m is the mass of the body traveling in the circular path.

In uniform circular motion the acceleration vector always points toward the center of the circle. The force that causes this acceleration must act in the same direction, that is, toward the center. Therefore to maintain a body in uniform circular motion a force of magnitude mv^2/r, directed toward the center of the circle, must be continually applied to the body. Such a force is called a ***centripetal force.***

The centripetal force must be provided by some interaction with the moving body. For example, when a ball is whirled around on the end of a string, the centripetal force is supplied by the cohesive force or ***tension*** in the string, which prevents the string from stretching. That is, the tension in the string acting on the *ball* provides the centripetal force that keeps the ball moving in a circle. When a racing car is driven on a circular track, the centripetal force on the car is provided by the friction between the wheels and the track. For an artificial satellite traveling in a circular orbit about the earth, the centripetal force is provided by the force of gravity exerted by the earth on the satellite. Note that the satellite is *not* acted on by gravity *and* a centripetal force, but by gravity *alone*—that is, gravity *is* the centripetal force in this case.

At any point on the circular path, the direction of the centripetal force (and the centripetal acceleration) is always perpendicular to the direction of the instantaneous velocity. If, at any instant, the centripetal force were suddenly to disappear, the body would revert to inertial motion and fly off on a straight line tangent to the circle (Figure 4–12). Thus the effect of the centripetal force is to deviate the body inward from its straight-line inertial motion onto a circular path.

This description of circular motion in terms of an inward-directed centripetal force is sometimes confusing because it seems contrary to our experience of curved motion. When we are in a train that is rounding a curve or in an automobile that is making a sharp turn or riding on a merry-go-round, we get the sensation of being *pushed outward* away from the center. It seems more reasonable, therefore, to think that in circular motion a body is acted on by an outward-directed or ***centrifugal*** force.

However, when we examine the situation more closely in terms of Newtonian principles, we recognize that the centrifugal force is an illusion. Suppose a man is

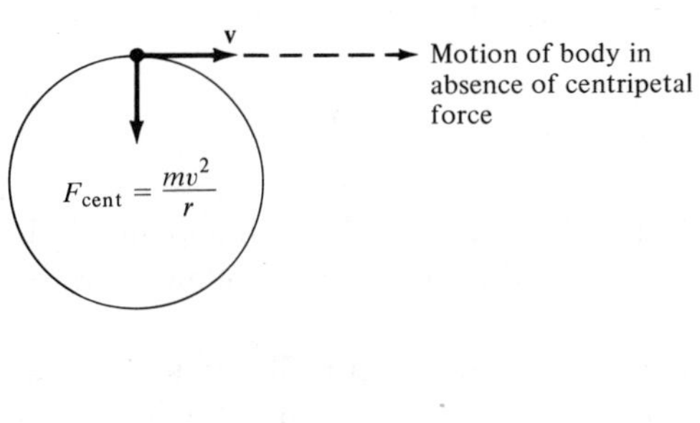

figure 4–12

CIRCULAR MOTION UNDER A CENTRIPETAL FORCE

To maintain uniform circular motion, a body must be acted on by a force directed toward the center of the circle. If F_{cent} were suddenly set to zero, the body would move off on a straight line.

standing in a train that is going around a sharp curve (Figure 4–13). The tendency of his body is to maintain its inertial motion, that is, to continue moving in a straight line. But his feet, held in place on the floor of the train by friction, turn the curve with the train. Thus his body is not being pushed outward by a centrifugal force; rather, his feet are being pulled *inward* by a centripetal force provided by friction while the rest of his body is trying to maintain its inertial motion. What he is experiencing is not a centrifugal force, but *his own inertia*. Similarly, on a merry-go-round we find ourselves leaning outward because the upper half of our body is trying to move in a straight line while the lower half, constrained to move with the horse, is being pulled inward.

Another example of this phenomenon is provided by the spin drying of clothing in a washing machine. As the drum starts to spin rapidly, the clothing and water begin to fly out in a straight line. Their inertial tendency is to continue to move outward in a straight line, but the clothing is prevented from doing so by the wall of the drum. That is, the *inward* force exerted by the wall on the clothing supplies the centripetal force that keeps the clothing moving in a circular path. The water, however, is able to escape because small holes have been punched in the wall; the water passes out through these holes in a straight-line path because there is no centripetal force acting to deflect the water from its inertial motion. The clothing cannot pass through these holes, so that after a few minutes of rapid spinning nearly all the water has left, and the clothing remains.

To maintain a body in uniform circular motion, then, the force on the body must be directed *inward*. The outward "centrifugal force" that one experiences is an imagined or "fictitious" force that is actually the sensation of one's own inertial tendency.

4–6

PRESSURE

In many situations the magnitude of a force is not the most suitable measure of its effectiveness. It is obvious, for example, that a force of 5 lb applied to the skin will be more painful when applied by a needle than when applied by the blunt end of a pencil. This is so because the force is being concentrated on a smaller area when the needle is used. As a general rule, a force is more intense when concentrated over a smaller area. It is useful, therefore, to have a physical quantity that measures the *intensity* of a force. For this purpose the concept of ***pressure*** has been defined as the *force per unit area*. More precisely, the pressure associated with a force is equal to the magnitude of the component of the force perpendicular to the surface in question, divided by the magnitude of the area itself. That is,

$$P = \frac{F_{\perp}}{A} \tag{4-5}$$

The perpendicular component is used because the parallel component of the force ($\mathbf{F}_{\parallel}$) is tangential to the surface and tends to slide the surface sideways rather than compress it. (See Figure 4–14.)

Pressure is a scalar quantity. It has units of force/area, that is, lb/ft^2 (or lb/in.^2) in the British system and dynes/cm^2 or newtons/m^2 in the CGS and MKS systems.

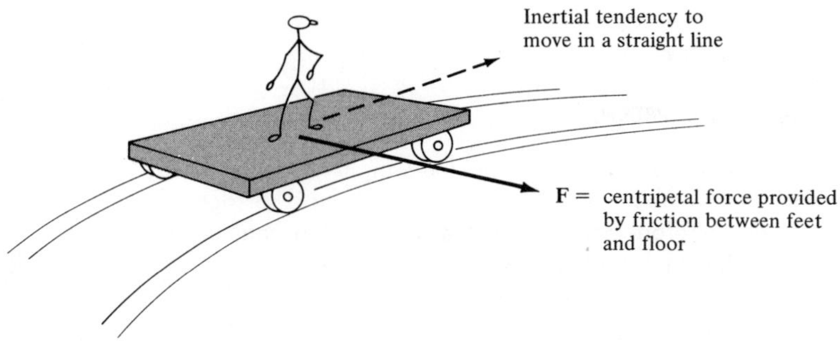

figure 4–13 **THE EXPERIENCE OF "CENTRIFUGAL FORCE"**

The natural tendency of a body traveling on a curved path is to move in a straight line, but a centripetal force (often provided by friction) pulls the body onto the curve. A person, finding himself leaning outward, will mistake his inertial tendency for an outward (centrifugal) force.

HYDROSTATIC PRESSURE

The concept of pressure is particularly useful for describing the properties of liquids and gases. It is well known, for example, that when a body is submerged in a liquid, the liquid exerts a compressive force on the body. This force becomes progressively greater as the depth increases; at great depths the compressive force can become great enough to crush the hull of a submarine. The reason this compressive force exists in a liquid lies in the nature of liquids themselves. A liquid does not have a fixed shape and lacks a rigid internal structure. As a consequence, the weight of the upper layers of liquid is supported totally by the deeper layers. Therefore the compressive force at a given depth is the result of the weight of the overlying liquid. Consequently, any object submerged in a liquid must support the weight of the liquid directly above it (Figure 4–15).

With this in mind we can derive an equation for calculating the pressure at a given depth in the liquid. Consider a surface area A submerged to a depth h in the liquid. The compressive force on this surface is equal to the weight w of the liquid contained in the column directly above the surface. This column of liquid has a height h and cross-sectional area A, as shown in Figure 4–15. The volume V of the column is equal to the column's area times its height; that is,

$$V = Ah$$

The weight of the column of liquid is related to its volume through a physical

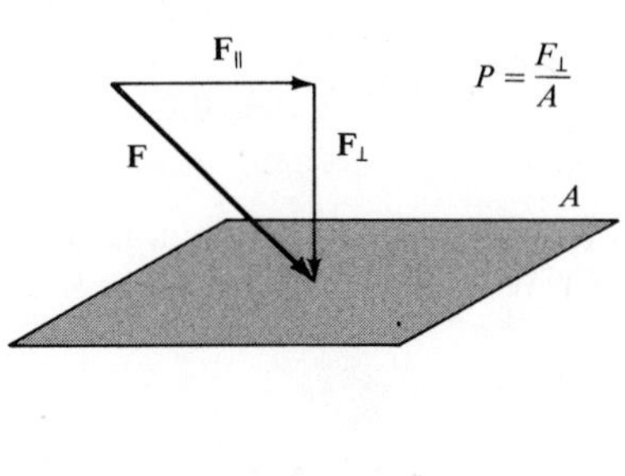

figure 4–14

DEFINITION OF PRESSURE

A force $\mathbf{F}$ acts uniformly over a surface area A. The pressure is defined as the magnitude of the force component perpendicular to the surface ($F_{\perp}$) divided by the area.

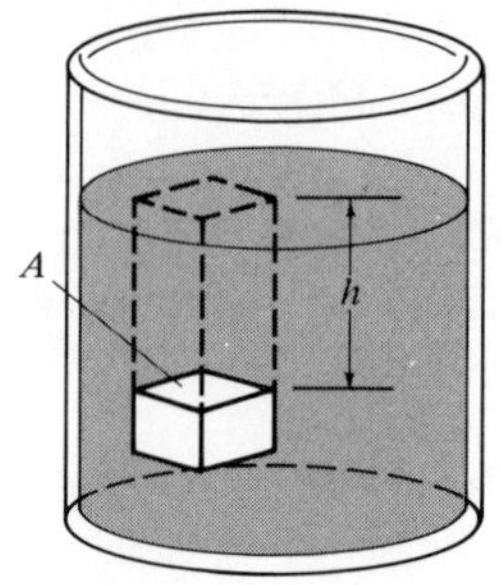

figure 4–15

HYDROSTATIC PRESSURE

A submerged body must support the weight of a column of liquid directly above it. The pressure on the top surface of the body therefore depends on the weight of a column of liquid of height h and cross-sectional area A.

quantity called ***density.*** The density ρ (rho) of a body is defined as the ratio of its mass to its volume:*

$$\rho = \frac{m}{V} \tag{4–6}$$

Density is a particularly useful quantity, since it is an intrinsic property of substances. That is, under similar conditions different samples of a given substance will generally have the same density, regardless of size or shape. The units of density are g/cm^3 (CGS), kg/m^3 (MKS), and slugs/ft^3 (British). One can convert from metric to British units by using the factor 1 g/cm^3 = 1.94 slugs/ft^3. The densities of some common substances are listed in Table 4–1.

Let us rewrite Eq. 4–6 in the form $m = \rho V$. It follows that the mass of the column of liquid is equal to its density times its volume. From Eqs. 4–3 and 4–6 we can obtain the weight of the column:

$$w = mg = \rho V g$$

Now let us calculate the pressure, using Eq. 4–5. The compressive force $F_\perp$ will be equal to the weight of the column of liquid. Hence

$$P = \frac{F_\perp}{A} = \frac{w}{A} = \frac{\rho V g}{A}$$

table 4–1 **DENSITIES OF SOME COMMON SUBSTANCES**

(Measured at 0°C and standard atmospheric pressure)

To convert these densities to British units (slugs/ft^3), multiply the given value by 1.94.

Substance	Density (g/cm^3)	Substance	Density (g/cm^3)
Aluminum	2.7	Mercury	13.6
Copper	8.96	Alcohol	0.80
Iron	7.9	Water	1.00
Gold	19.3	Air	0.001293
Iridium	22.5	Carbon dioxide	0.001977
Ice	0.92	Hydrogen	0.00009

* Sometimes density is defined as the ratio of the *weight* to the volume. Thus we distinguish between the *mass density* ($\rho = m/V$) and the *weight density* ($d = w/V$). It is easy to see that $d = \rho g$. In this text we will use the mass density unless otherwise specified.

Substituting for the volume $V = Ah$, we have

$$P = \frac{\rho(Ah)g}{A}$$

The area A cancels out, leaving

$$P = \rho gh \tag{4–7}$$

According to this equation, the pressure generated by the weight of the liquid is directly proportional to the depth below the surface. It also follows that the pressure is the same along any horizontal layer in the liquid (that is, at all points the same vertical distance from the surface).*

EXAMPLE 4–4: Calculate the hydrostatic pressure in water at a depth of 100 feet.

SOLUTION: The density of water according to Table 4–1 is $1\ \text{g/cm}^3 = 1.94$ slugs/ft^3. At a depth of 100 feet the hydrostatic pressure is

$$P = \rho gh = (1.94\ \text{slugs/ft}^3)(32\ \text{ft/sec}^2)(100\ \text{ft})$$

or

$$P = \underline{6210\ \text{lb/ft}^2} = \underline{43.1\ \text{lb/in.}^2}$$

GAS PRESSURE

The fact that a gas will exert forces against any surface it comes in contact with is seen in many familiar phenomena. A balloon or an automobile tire can be inflated well beyond its normal shape by filling it with air; the air must exert some force, since the balloon or tire is prevented from returning to its natural shape (Figure 4–16a). Similarly, when a gas is contained in a piston-and-cylinder arrangement (Figure 4–16b), the piston is prevented from falling under its own weight by the upward force exerted by the contained gas. Furthermore, if the gas is heated rapidly, the force exerted by the gas increases, and the piston is driven *upward*. This principle

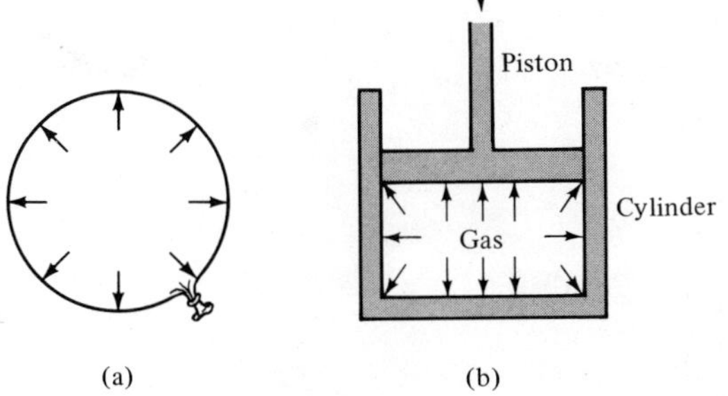

figure 4–16

GAS PRESSURE

In (a) the gas in a balloon exerts an outward force on the inside surface of the balloon. This force prevents the balloon from contracting to its natural size. In (b) the weight of the piston is supported by the upward force exerted by the gas in the cylinder.

---

*Credit for this discovery and the discovery of many other phenomena related to pressure in gases and liquids belongs to Blaise Pascal (1623–1662). An intensely spiritual man, Pascal became distressed by the fame and ego gratification he received from his scientific and mathematical discoveries. He eventually gave up his scientific researches to devote himself to religious meditation.

is basic to the operation of the automobile engine, in which the upward motion of the piston is transmitted by mechanical linkage to the wheels of the automobile. The analysis of such phenomena shows that a contained gas exerts force uniformly against the walls of its container; consequently, it is more useful to speak of the *pressure* exerted by the gas rather than the total force.

The first important step toward an understanding of gas pressure came about in the 1640s and 1650s, when a series of experiments by several scientists led to the conclusion that *the atmosphere exerts pressure.* It had been known for thousands of years that if one end of a tube is immersed in a liquid and the air is removed from the other end, the liquid will rise in the tube (Figure 4–17a). This is the basic principle of the operation of the drinking straw and the siphon. The ancient Greeks explained this phenomenon by referring to the axiom "Nature abhors a vacuum"; that is, the liquid rushes up the tube to occupy the space from which the air has been removed. This belief was commonly held into the seventeenth century. The first complication arose when very long suction tubes were constructed and it was found that water could not be made to rise any higher than 34 feet in a tube. When this was brought to Galileo's attention, he questioned why nature's abhorrence of a vacuum should stop at a height of 34 feet.

Galileo himself was not able to offer a full explanation of this phenomenon. It was Evangelista Torricelli, Galileo's close associate during the last few months of Galileo's life, who showed that the water was being forced up the tube by the pressure of the atmosphere rather than being drawn up by the presence of a vacuum. Torricelli concluded that the 34-foot limit represented the maximum height of water that the atmosphere could support; that is, the force exerted by the atmosphere is equal to the weight of a 34-foot column of water. As a test of this hypothesis, Torricelli substituted mercury for water in the tube and found that the mercury rose to a height

figure 4–17 **ATMOSPHERIC PRESSURE**

(a) One end of a tube is inserted in a liquid that is open to the atmosphere. When air is removed from the other end, atmospheric pressure on the open surface pushes the liquid up the tube. (b) In the barometer one end of the tube is sealed. The space above the liquid surface at the closed end is effectively a vacuum, so that the pressure at A is zero. The weight of the column of liquid is supported by atmospheric pressure.

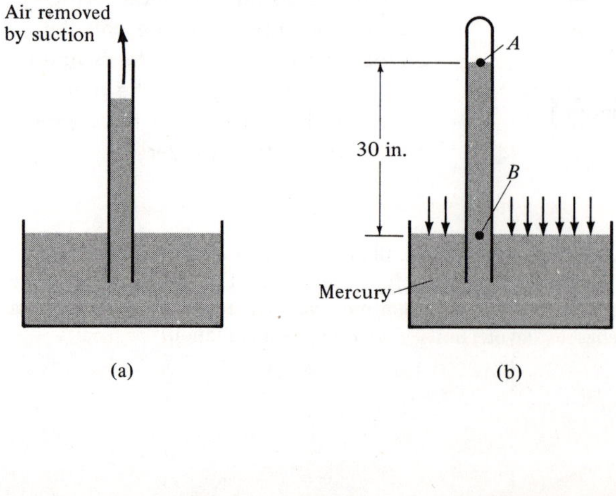

of only 30 inches. Since mercury is 13.6 times denser than water, it follows that the height of the mercury supported by the atmosphere would be 1/13.6 the height of the water. If we divide 34 feet by 13.6, we obtain 2.5 feet = 30 inches, in agreement with the experimental result.

To demonstrate his discovery, Torricelli took a tube sealed at one end and filled it with mercury. He then inverted the tube into an open dish of mercury (Figure 4–17b). Some of the mercury left the tube until the height of the column was 30 inches (the maximum height that the atmosphere could support). Since no air could get into the empty space at the closed end of the tube, it was evident that this space was not occupied by matter—thus Torricelli demonstrated that a vacuum could exist in nature. Torricelli's device (first constructed in 1643) was the prototype of the modern ***barometer,*** which is used to measure atmospheric pressure.

An important test of Torricelli's theory was performed by Blaise Pascal in 1646. Pascal proposed that if the mercury column is supported by the pressure of the atmosphere, the height of the column should be less at higher altitudes, where the atmosphere is thinner. Pascal's brother-in-law was persuaded to take a barometer to the top of a mountain. As he ascended, he found that the mercury level gradually dropped; at the top, the height of the mercury column was only 27 inches.

It is a simple matter to calculate the magnitude of the pressure exerted by the atmosphere. Referring to Figure 4–17b, we note that the pressure at point A is zero (since the space above A is evacuated). The pressure at B is equal to the hydrostatic pressure at a depth of 30 inches in mercury. The density of mercury is 13.6 times that of water—in British units (13.6)(1.94) = 26.4 slugs/ft^3. Using Eq. 4–7, we find

$$P_B = \rho g h = (26.4\ \text{slugs/ft}^3)(32\ \text{ft/sec}^2)(30/12\ \text{ft}) = 2110\ \text{lb/ft}^2$$
$$(2110\ \text{lb/ft}^2)(1\ \text{ft}^2/144\ \text{in.}^2) = \underline{14.7\ \text{lb/in.}^2}$$

Since the weight of the mercury column is supported entirely by the force exerted by the atmosphere, P_B must equal the atmospheric pressure. Thus at sea level (where the height of the mercury column is, on the average, 30 inches)

$$P_{\text{atm}} = 14.7\ \text{lb/in.}^2$$

When measured in metric units, the height of a mercury column at sea level is 76 cm (30 inches $\approx$ 76 cm). Thus in the CGS and MKS systems

$$P_{\text{atm}} = 1.01 \times 10^6\ \text{dynes/cm}^2 = 1.01 \times 10^5\ \text{N/m}^2$$

A unit of pressure commonly used in physics and chemistry is the ***atmosphere*** (abbreviated ***atm***), which is defined as being equal to standard sea level pressure. Thus 1 atm = 14.7 lb/in.^{2}; 2 atm = 29.4 lb/in.^{2}, and so on. It is also common usage to give the pressure directly in terms of the height of the mercury column in a barometer, that is, in units of inches or centimeters of mercury. (The chemical symbol of mercury is Hg.) Thus

$$1\ \text{atm} = 14.7\ \text{lb/in.}^2 = 1.01 \times 10^6\ \text{dynes/cm}^2 = 76\ \text{cm of Hg} \approx 30\ \text{in. of Hg}$$

Sea level atmospheric pressure does not remain constant, but fluctuates in time. Typically, the atmospheric pressure at a given location ranges from about 29 to 31 inches of Hg, with 30 inches being the average value. These pressure changes are the result of the motions of large air masses through the atmosphere, and it is these motions that are responsible for changes in our weather. The "lows" and "highs" that one finds on a weather map represent regions of the atmosphere that are at relatively low or high pressure; temperature and humidity conditions within these regions are mainly responsible for the kind of weather they bring. Thus the barometer is an essential tool to the weather forecaster. A typical weather report invariably contains a statement such as "The barometer is 29.9 inches and falling." Although this may conjure up an image of an expensive barometer tumbling down from its peg on the wall, what the statement really means, of course, is that the height of the mercury column in the barometer is 29.9 inches, and the height of the column is slowly decreasing as the atmospheric pressure drops.

QUESTIONS FOR REVIEW

1. State and explain the first principle of motion.
2. What is meant by the term "inertia"? What is inertial motion? Can we explain why bodies possess inertia?
3. What is meant by the term "mass"? Discuss the methods by which the mass of a body can be measured.
4. Why has Newton's concept of mass as a measure of the quantity of matter in a body been rejected?
5. Why did Newton assume that the mass of a body is a constant quantity? What is the modern-day viewpoint with regard to the validity of this assumption?
6. What is the definition of a force? State the principle that defines how the magnitude and direction of a force are measured.
7. In what units is force measured in the CGS, MKS, and British systems?
8. Explain the distinction between mass and weight.
9. State and explain the principle of action and reaction. Give examples of motions that can be explained in terms of this principle.
10. Explain why the two members of an action–reaction pair of forces can never cancel each other out, even though they are always equal and opposite.
11. Explain in basic terms how a rocket is propelled. Why is it possible for a rocket to work in a vacuum?
12. Explain the difference between centripetal force and centrifugal force. Why is the latter considered a "fictitious" force?
13. State the definitions of (a) pressure and (b) density.
14. Explain why the hydrostatic pressure in a liquid depends on (a) the depth below the surface and (b) the density of the liquid.
15. How did Torricelli and Pascal prove that the atmosphere exerts pressure?
16. Why is a barometer useful in weather forecasting?

EXERCISES

1. A body is observed to be moving with a constant velocity. Can we infer that *no* forces are acting on it? Explain your answer.
2. In the *Principia* Newton defines "quantity of matter" (mass) as the product of density times volume. Criticize this definition.
3. In the Newtonian system how can the motion of a body with a nonconstant acceleration be explained?
4. Two bodies, A (mass = 20 kg) and B (mass = 50 kg), repel each other through a mutual

interaction of an unspecified nature. No other forces act on them. Body A is given an acceleration of 10 m/sec^2 as a result of the interaction. What is the acceleration of body B?

5 What force is required to give a 10-kg mass an acceleration of 20 cm/sec^2?

6 A 30-kg object is acted on by three forces: 10 newtons north, 4 newtons south, and 8 newtons west. What is the magnitude and direction of the acceleration of the object when acted on by these forces?

7 A body with a mass of 100 grams is acted on by two forces: a force of 5000 dynes east and an unknown force, **F**. The body has an acceleration of 30 cm/sec^2 west. What is the magnitude and direction of **F?**

8 Calculate (a) the mass in slugs of a 160-lb man, (b) the weight in dynes of a body of mass 300 grams, and (c) the weight in newtons of a body of mass 60 kg.

9 A 3200-lb car is decelerated uniformly from 60 ft/sec to 30 ft/sec in 4 sec. What is the magnitude of the decelerating force?

10 A 16-lb weight is lifted upward by a 50-lb force. What is the acceleration of the weight?

11 A 64-lb block is initially at rest on a rough horizontal surface. A force of 30 lb applied for 4 sec is required to accelerate the block to a speed of 20 ft/sec. What is the magnitude of the frictional force between the block and the surface?

12 Using the Newtonian principles of motion, explain:
(a) how a bird flies.
(b) how a fish swims.
(c) how a saltshaker works.
(d) how a helicopter hovers above the ground.
(e) why catsup can be made to come out of the bottle by hitting the back of the bottle.
(f) why seat belts save lives.
(g) why an inflated (but unsealed) balloon flies around the room when it is released.
(h) why the passengers in a train are thrown backward when the train starts suddenly.

13 In each of the following situations, identify the centripetal force:
(a) A ball is whirled around on a string.
(b) A satellite travels in a circular orbit around the earth.
(c) A car is driven on a flat circular road.
(d) A toy train goes around on a circular track.
(e) An electron travels in a circular orbit around the nucleus of the atom.

14 A 10-gram ball is whirled around on the end of a 20-cm string. If the tension in the string is 50 dynes, what is the speed of rotation?

15 A 3200-lb automobile travels at a constant speed of 60 ft/sec on a circular track 400 feet in diameter. What is the magnitude of the centripetal force required to keep the car on the track?

16 Calculate the weight in pounds of a cubic foot of (a) water, (b) air, and (c) gold.

17 What is the hydrostatic pressure at a depth of 10,000 feet in the ocean? (Assume that seawater has a constant density of 1.025 g/cm^3.)

18 Determine the height of the liquid column in a barometer at 1 atm pressure if the liquid in the barometer is (a) alcohol, (b) bismuth ($\rho = 10$ g/cm^3).

19 What would the height of the mercury column be in a barometer at the surface of
(a) Venus (surface pressure = 90 atm)?
(b) Mars (surface pressure = 0.005 atm)?

20 Discuss at least two methods for using a barometer to determine the height of a building.

THE SYSTEM OF THE WORLD

5

I began to think of gravity extending to the orb of the Moon. . . . I deduced that the forces which keep the Planets in their Orbs must be reciprocally as the squares of their distances from the centers about which they revolve: and thereby compared the force requisite to keep the Moon in her Orb with the force of gravity at the surface of the Earth, and found them answer pretty nearly.

Isaac Newton

IF THE PRINCIPLES OF MOTION had been Newton's only contribution to physics, his place among the greatest scientists would have been assured. Yet the formulation of the principle of gravitation was his crowning achievement. In *The System of the World* (Book III of the *Principia*), Newton—armed with the mathematical tools that derived from his invention of calculus—successfully applied the results of his studies of motion and gravitation to explain virtually all of the celestial motions known to the astronomers of his time. Of equal significance, Newton once and for all destroyed the Aristotelian notion that the terrestrial and celestial realms are made of different substances and governed by different physical laws: The fall of an apple, the motion of the moon, and the comings and goings of comets were all shown to be explainable under the same simple principle.

5–1

THE PRINCIPLE OF UNIVERSAL GRAVITATION

Actually, the concept of gravitation did not originate entirely with Newton. Nearly a half century before, Kepler had suggested that the planets were driven by a force originating in the sun. Galileo and Descartes had both speculated that the natural motion of bodies is in a straight line; this principle, combined with the concept of centripetal force (first formulated by the Dutch scientist Christian Huygens) required that the planets be continually acted on by a force of attraction originating in the sun. Finally, Huygens and Hooke had both shown that for circular orbits the force exerted by the sun must vary inversely with the square of the distance in order to account for Kepler's third law.

This result can be obtained fairly simply. The centripetal force required to keep a planet in a circular orbit of radius R is (see Eq. 4–4)

$$F = \frac{Mv^2}{R}$$

where M is the mass of the planet, and v is its orbital speed. The orbital speed of a planet in a circular orbit can be evaluated in the following way. Since the planet is moving at a constant speed, the value of the speed is equal to the distance traveled in one revolution (which is just the circumference of the orbit) divided by the time for one revolution (which is, by definition, the sidereal period). That is,

$$v = \frac{d}{t} = \frac{\text{circumference}}{\text{period}} = \frac{2\pi R}{P}$$

Thus

$$v^2 = \frac{4\pi^2 R^2}{P^2}$$

Substituting this term into the centripetal force equation yields

$$F = \frac{M(4\pi^2 R^2/P^2)}{R} = \frac{4\pi^2 MR}{P^2}$$

According to Kepler's third law (Eq. 2–6), for a circular orbit $P^2 = KR^3$, where R is the radius of the circle. Substituting for P^2 in the denominator of the above expression gives

$$F = \frac{4\pi^2 MR}{KR^3}$$

or

$$F = \left(\frac{4\pi^2}{K}\right)\frac{M}{R^2}$$

The quantity $4\pi^2/K$ is a constant, which has the same value for all planets. Thus the centripetal force (presumed to emanate from the sun) required for a circular orbit is directly proportional to the mass of the planet and inversely proportional to the square of the distance from the sun to the planet (the radius of the orbit).

However, this calculation is legitimate only for circular orbits, and neither Huygens nor Hooke nor any of the many other scientists who tackled this problem were able to derive a formula for the force that would make a planet move in an *elliptical* orbit. With the use of calculus, Newton was able to show that the inverse-square relationship was applicable to elliptical orbits as well. In particular, Newton was able to derive Kepler's three laws by assuming that a planet is acted on by a force of attraction exerted by the sun that varies inversely with the square of the distance from the sun to the planet.

It might seem that history has failed to give proper credit to Hooke and Huygens, who really established the fundamental inverse-square relationship while Newton added a mathematical refinement. But Newton's insight carried him far beyond the work of his two contemporaries, for he did not attribute the power of gravitation just to the sun and planets; *he postulated that gravitation is a property of all bodies in the universe.* Thus we can justly attribute to Newton and Newton alone the ***principle of universal gravitation:***

Every body in the universe attracts every other body with a mutual force that is directly proportional to the product of their masses and inversely proportional to the square of the distance between their centers.

We can restate this principle in mathematical form in the following way: given two bodies, A and B, with masses m_A and m_B, respectively, each will be attracted toward the other with a force

$$F_{\text{grav}} = \frac{G m_A m_B}{d^2} \tag{5-1}$$

where d is the distance between the two bodies. (See Figure 5–1.)

Equation 5–1 is often referred to as the "law of gravitation." It is a law in the sense that Newton derived the mathematical formula from observations; he arrived at the dependence on d^2 by working backward from Kepler's laws. However, it is a basic principle of Newton's *theory* of gravitation that the inverse-square relationship applies to *all* bodies in the universe, not just the planets of the solar system.

The factor G in Eq. 5–1 is called the ***universal constant of gravitation.*** (The student must be careful not to confuse G with g, the acceleration of gravity of bodies near the surface of the earth.) G is a numerical constant of proportionality (much like the K in Kepler's third law) and must be determined experimentally. Newton himself did not attempt to establish its numerical value; the first successful measurement of G was made by Henry Cavendish in 1798. (The experiment is described in Figure 5–2.) The modern-day value of G is:

$$\begin{aligned} G &= 6.67 \times 10^{-8}\ \text{cm}^3/\text{g-sec}^2 \\ &= 6.67 \times 10^{-11}\ \text{m}^3/\text{kg-sec}^2 \\ &= 3.43 \times 10^{-8}\ \text{ft}^3/\text{slug-sec}^2 \end{aligned}$$

The small numerical value of G indicates that gravity is an *extremely weak force.* Using Eq. 5–1, we find that the gravitational force between two 160-lb bodies ($m = 5$ slugs) placed 1 foot apart is 0.00000086 lb, which is negligibly small and easily overcome by air resistance, friction, and so on. This explains why all the objects in a room do not fly together as a result of their mutual attraction. It is only on an astronomical scale (where we are dealing with very massive objects such as stars and planets) that gravitation becomes a significant force. In fact, it was Cavendish's experiment to measure G that first demonstrated the existence of a gravitational force between nonastronomical objects.

The force of gravitation is a *mutual* force in that the forces on the two bodies are equally strong. Given two bodies, one of large mass and one of small mass, the

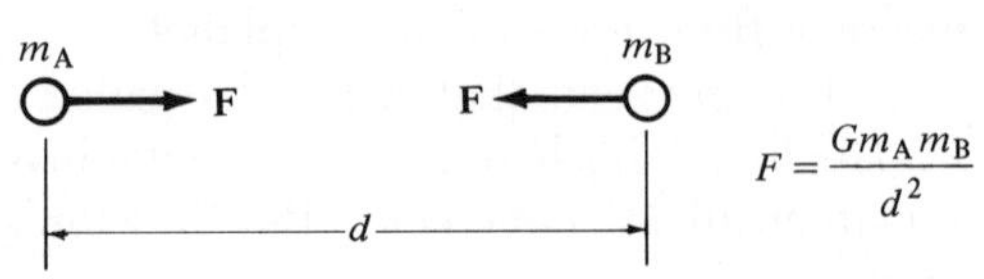

figure 5–1

THE FORCE OF GRAVITATION

The mutual forces on the two bodies are equal in magnitude and opposite in direction, as required by the principle of action and reaction.

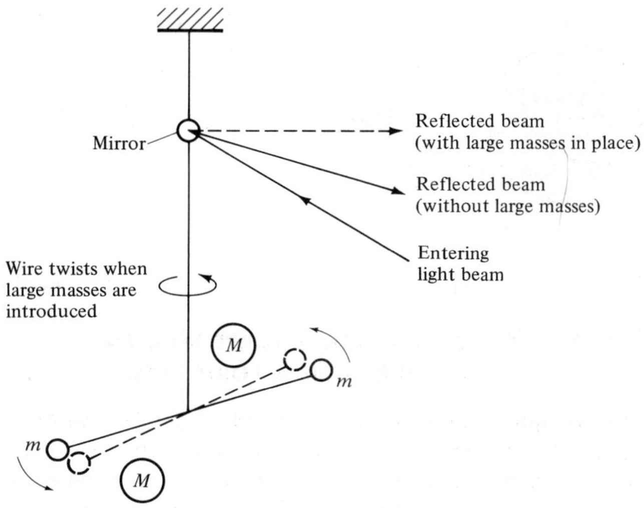

figure 5–2 **THE CAVENDISH EXPERIMENT FOR MEASURING *G***

Two small bodies, each of mass m, are attached to a rigid rod, which is held at the center by a vertical wire. When two large bodies, each of mass M, are placed as shown, the gravitational attraction will cause the rod to rotate, twisting the wire. The angle of twist can be determined by measuring the deflection of a beam of light reflected off the mirror. If the properties of the wire are known, it is possible to measure the force of gravitational attraction between the large and small bodies. The numerical value of G can then be calculated by using the law of gravitation.

larger mass does *not* exert the greater force or have the greater force exerted on it. The gravitational forces on the two bodies, in fact, represent an action–reaction pair; gravitation, like any other force, follows the principle of action and reaction.

As has already been noted, the strength of the mutual gravitational force between two bodies varies in inverse proportion to the square of the distance between them. Thus if the distance between two bodies is doubled, the gravitational force on each body will be one-fourth as strong—that is, if d is *increased* by a factor of 2, F will *decrease* by a factor of $2^2 = 4$. Similarly, if d is tripled, F will be one-ninth as strong and so on.

THE GRAVITATIONAL ATTRACTION OF A SPHERICAL BODY

The principle of universal gravitation assumes that the bodies in question are point masses, which means that the bodies are dimensionless points (or at least that their dimensions are negligible in comparison to the distance between them). How, then, can Eq. 5–1 be applied to extended bodies, such as the earth? When Newton first sketched out the theory of gravitation in 1665–1666, his intuition told him that the inverse-square law could be applied to large bodies by letting d equal the distance from the center of one body to the center of the other. However, Newton was unable at first to obtain a rigorous mathematical proof of this assumption.

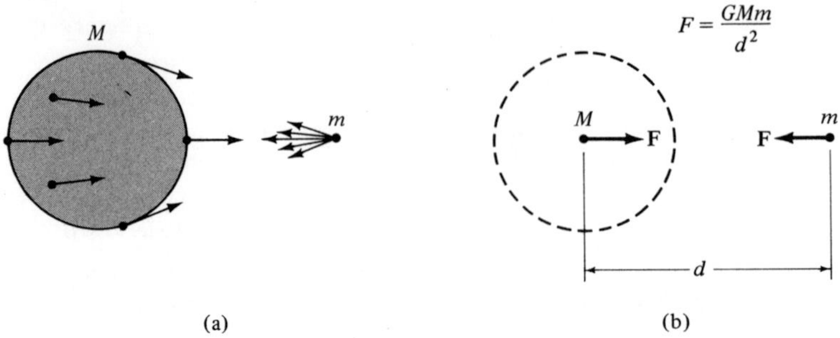

figure 5–3 **GRAVITATIONAL FORCE BETWEEN A SPHERE AND A POINT MASS**

(a) If the sphere (mass M) is considered as a collection of point masses, then the force on the point mass m is the sum of all the forces exerted by each of the point masses in the sphere. By means of calculus, it can be shown that the force on m is the same as if the entire mass of the sphere were concentrated at the center, as shown in (b).

The difficulty that confronted Newton is illustrated in Figure 5–3. The problem is to calculate the gravitational force between the large sphere M and the point mass m. We consider the large sphere to be an aggregate of many point masses, each of which exerts a force on m according to Eq. 5–1. How does one compute the vector sum of all the forces acting on m? To solve this problem, Newton had to develop the mathematical techniques of integral calculus.* Finally, in 1685 Newton was able to prove that *any spherically symmetrical distribution of mass acts as if all the mass were concentrated at the center of the sphere.* That is, for computational purposes it can be assumed that a sphere behaves as if it were a point mass located at its center.

This conclusion has a significant restriction: *it holds only when the body of mass m is completely outside of the sphere.* If the body were inside, there would be some cancellation of forces acting in opposite directions, so that the net gravitational force on the body would be less than it would be if the body were on the surface of the sphere (Figure 5–4). If the body is at the center of the sphere, there is equal force on it from all directions, so that the net force is zero.

GRAVITY ON THE EARTH

Newton's proof that a spherical body behaves like a point mass located at its center enabled him to deduce many significant consequences of the theory of gravitation. The inverse-square formula could now be applied to calculate the gravitational force that the earth exerts on bodies at or above its surface. (Remember that Newton's proof does not apply to bodies *below* the surface.) The earth is very nearly spherical in shape, so it can be treated as if it were a point mass. Hence to calculate the gravitational force that the earth exerts on a body, we substitute for d in Eq. 5–1

* Newton first invented differential calculus, which he called "the method of fluxions." Then he developed "the method of inverse fluxions," which we now call integral calculus. The terminology and notation used in present-day calculus come from the work of Leibniz.

the distance from the body to the center of the earth. If the body is located at the surface, that distance is the radius of the earth (approximately 4000 miles).

Suppose a body of mass m is located a distance R from the center of the earth. (R must be greater than or equal to the radius of the earth.) Let M_E represent the mass of the earth. According to Eq. 5–1,

$$F_{grav} = \frac{GM_E m}{R^2}$$

The acceleration produced by this force can be determined from Newton's second principle ($F = ma$), which applies to *all* forces:

$$F_{grav} = ma = \frac{GM_E m}{R^2}$$

The mass of the body appears on both sides of the equation and cancels out. Thus

$$a = \frac{GM_E}{R^2} \tag{5–2}$$

Note that the acceleration produced by the gravitational force on a given body is independent of the mass of the body. This result tells us that all bodies, regardless of their masses, fall toward the earth with the same acceleration. This is none other than Galileo's law of falling bodies, which has thus been shown to be a direct consequence of the theory of gravitation. Moreover, at the surface of the earth, where $R = R_E$ (the radius of the earth), the acceleration of a falling body is the quantity we call the acceleration of gravity, g_E. That is,

$$g_E = \frac{GM_E}{R_E^2} \tag{5–3}$$

figure 5–4 **GRAVITATIONAL FORCE ACTING ON A POINT MASS INSIDE A SPHERE**

In (a) the gravitational force exerted by the part of the sphere to the right of the point mass partially cancels the force exerted by the matter to the left, so that the net force *decreases* as the point mass moves from the surface to the center. (b) When the point mass is exactly at the center, the forces acting on it are symmetrically distributed, and the net force is zero.

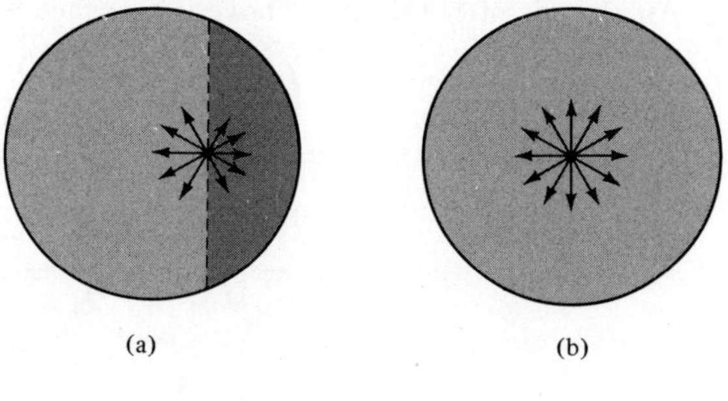

(a) (b)

Before Newton, the value of g_E could be obtained only by experiment, and the question "Why is the acceleration of gravity equal to 32 ft/sec^2 and not some other number?" could not be answered. Now we see that g_E is related to more fundamental physical constants—namely, the mass and radius of the earth and the gravitational constant.

Moreover, the theory of gravitation reveals a significant limitation of the law of falling bodies. Galileo had proposed that all bodies fall toward the earth with a *constant* acceleration. But Eq. 5–2 indicates that the acceleration of a falling body depends on its distance from the center of the earth. The acceleration of gravity is 32 ft/sec^2 *at the earth's surface,* but it has a smaller value above the earth's surface. Thus as a body falls toward the earth, the body's acceleration gradually *increases* as it approaches the surface.

Galileo's assumption of a constant acceleration for free fall is therefore not strictly correct, but it stands as a good approximation unless the falling body is a considerable distance above the earth's surface. The acceleration of gravity decreases in inverse proportion to the square of the distance to the center of the earth, so "a considerable distance" means a distance that is significant compared to the radius of the earth. At an altitude of 100 miles above the earth's surface (4100 miles from the center), the value of g has decreased by a factor of $(4000)^2/(4100)^2 = 0.952$; that is, at this altitude $g = (0.952)(32) = 30.5$ ft/sec^2. In other words, in the first 100 miles above the earth's surface, the acceleration of gravity decreases only by about 5 percent. Obviously, g changes very little over distances of a few hundred feet or so and can be safely considered to be a constant.

Since weight is related to the acceleration of gravity through the equation $w = mg$, it follows that the weight of a body increases or decreases in direct proportion to the value of g. Consequently, the weight of a body *decreases* the farther the body is from the surface of the earth. A body weighing 100 lb on the earth's surface weighs 95.2 lb at an altitude of 100 miles above the surface (since g decreases by a factor of 0.952 at this altitude).

EXAMPLE 5–1: Calculate the acceleration of a falling body at an altitude of 8000 miles above the earth's surface. What is the weight at this altitude of a spaceship that weighs 180 tons on the earth's surface?

SOLUTION: The acceleration of a falling body decreases in inverse proportion to the square of the distance from the center of the earth. This principle can be written in mathematical form by dividing Eq. 5–2 by Eq. 5–3:

$$\frac{a}{g_E} = \frac{GM_E/R^2}{GM_E/R_E{}^2} = \frac{R_E{}^2}{R^2} = \left(\frac{R_E}{R}\right)^2$$

An altitude of 8000 miles above the surface is 12,000 miles from the center. Thus

$$\frac{a}{g_E} = \left(\frac{R_E}{R}\right)^2 = \left(\frac{4000}{12{,}000}\right)^2 = \left(\frac{1}{3}\right)^2 = \frac{1}{9}$$

Therefore

$$a = \frac{1}{9}g_E = \frac{32}{9} = \underline{3.56 \text{ ft/sec}^2}$$

Since the acceleration of gravity is one-ninth as great as at the surface, the weight of a body will also be one-ninth as great at an altitude of 8000 miles above the surface. Thus a 180-ton spaceship will weigh $\frac{180}{9} = \underline{20 \text{ tons}}$ at this altitude.

In addition, Eq. 5–3 can be used to predict the acceleration of gravity at the surface of any other celestial body. For example, on the moon's surface the acceleration of gravity g_M depends in the same way on the mass (M_M) and radius (R_M) of the moon:

$$g_M = \frac{GM_M}{R_M^2} \tag{5–4}$$

To get a numerical value of g_M, we must have information about the mass and radius of the moon. Astronomical observations show that the mass of the earth is 81.3 times the mass of the moon, and the earth's radius is 3.67 times that of the moon. Let us divide Eq. 5–4 by Eq. 5–3:

$$\frac{g_M}{g_E} = \frac{GM_M/R_M^2}{GM_E/R_E^2} = \frac{M_M/M_E}{(R_M/R_E)^2}$$

According to the given information,

$$\frac{M_M}{M_E} = \frac{1}{81.3} \quad \text{and} \quad \frac{R_M}{R_E} = \frac{1}{3.67}$$

Thus

$$\frac{g_M}{g_E} = \frac{(1/81.3)}{(1/3.67)^2} = \frac{(3.67)^2}{81.3} = \frac{1}{6.04} = 0.166$$

so that

$$g_M = 0.166 g_E = (0.166)(32 \text{ ft/sec}^2) = 5.33 \text{ ft/sec}^2$$

The prediction that the acceleration of gravity on the moon is approximately one-sixth of that on the earth has been amply verified by the astronauts who have walked on the moon. Thanks to the theory of gravitation, scientists knew in advance what the acceleration of gravity would be on the moon and could design the astronauts' equipment accordingly.

EXAMPLE 5–2: An astronaut exploring a distant solar system lands on the planet Vulcan. He finds that he weighs twice as much as he does at the surface of the earth. The radius of Vulcan is known to be half that of the earth. What is the mass of Vulcan?

SOLUTION: According to the equation $w = mg$, the weight of a body is directly proportional to the acceleration of gravity. Therefore if the astronaut weighs twice as much on Vulcan as he does on earth, g on Vulcan must be twice as large as g on the earth. That is,

$$\frac{w_V}{w_E} = \frac{mg_V}{mg_E} = \frac{g_V}{g_E} = 2$$

The mass of Vulcan can be found from the equation

$$\frac{g_V}{g_E} = \frac{GM_V/R_V{}^2}{GM_E/R_E{}^2} = \frac{M_V/M_E}{(R_V/R_E)^2}$$

The radius of Vulcan is half the radius of the earth; thus $R_V = \frac{1}{2}R_E$. Therefore

$$\frac{g_V}{g_E} = 2 = \frac{M_V/M_E}{(\frac{1}{2})^2}$$

Hence

$$2(\tfrac{1}{2})^2 = M_V/M_E$$

so that

$$\underline{M_V = \tfrac{1}{2}M_E}$$

That is, the mass of Vulcan is half the mass of the earth.

THE APPLE AND THE MOON

We recall that Newton derived the inverse-square force law (Eq. 5–1) as a means of accounting for Kepler's laws of planetary motion. In Newton's time gravity was thought of only as the force that causes bodies to fall to the earth, and it was assumed that a force of a different nature was responsible for the motions of celestial bodies. What led Newton to believe that terrestrial gravity and the cause of celestial motions were governed by the same principle?

Over the years a certain amount of mythology has developed about the great scientists and their discoveries. One of the most popular "myths" is that Newton got the idea for the theory of gravitation when he saw an apple fall from a tree. Remarkably, the story appears to be true. Newton's friend and biographer, William Stukeley, tells of a conversation that he had with Sir Isaac in 1726:

> After dinner, the weather being warm, we went into the garden and drank tea, under the shade of some apple trees, only he and myself. Amidst other discourse, he told me he was just in the same situation as when formerly the notion of gravitation came into his mind. It was occasioned by the fall of an apple, as he sat in a contemplative mood.[1]

Apparently, Newton had been speculating on the nature of the force that keeps the moon in its orbit, and when the apple fell, he realized that perhaps the apple and the moon were being acted on by the same force. Could the earth's gravity extend as far as the moon?

Newton decided to test this idea by making a simple calculation. He began by determining the centripetal acceleration of the moon from observational data. For simplicity, he assumed that the moon travels in a circular orbit at constant speed. The moon's orbital speed is then

$$v = \frac{\text{circumference}}{\text{period}} = \frac{2\pi R}{P}$$

Here R is the radius of the moon's orbit, which is nearly 239,000 miles = 1.26×10^9 feet. P is the sidereal period of the moon, which is 27.3 days =

2.36×10^6 sec. Substituting in the above equation gives

$$v = \frac{2\pi(1.26 \times 10^9 \text{ feet})}{(2.36 \times 10^6 \text{ sec})} = 3350 \text{ ft/sec}$$

The centripetal acceleration of the moon is therefore

$$a = \frac{v^2}{R} = \frac{(3350 \text{ ft/sec})^2}{(1.26 \times 10^9 \text{ feet})} = 0.0089 \text{ ft/sec}^2$$

This is the observed value of the centripetal acceleration produced by the force that keeps the moon moving in its (nearly) circular orbit.

Now Newton considered the hypothesis that this force is the gravitational attraction of the earth for the moon, as described by Eq. 5–1. It is well known that the gravitational force of the earth causes all bodies at the surface to fall with an acceleration of 32 ft/sec^2, but (as we learned in the previous section) this acceleration should diminish above the surface of the earth according to the inverse-square relationship. What would be the acceleration of gravity of an apple if it were moved to the distance of the moon?

The distance from the center of the earth to the center of the moon is approximately 240,000 miles, so that when the apple is moved from the surface of the earth to the distance of the moon's orbit, its distance from the center of the earth is increased by a factor of $(240{,}000)/(4000) = 60$. The gravitational acceleration should therefore *decrease* by a factor of $(60)^2$. Hence the acceleration of gravity at the moon's distance from the earth should be

$$a = \frac{32 \text{ ft/sec}^2}{(60)^2} = 0.0089 \text{ ft/sec}^2$$

Thus when Newton calculated the acceleration of a falling body at the moon's distance from the earth, *on the assumption that the earth's gravity extends to the moon and follows an inverse-square law,* he obtained the same numerical value (0.0089 ft/sec^2) as the observed centripetal acceleration of the moon in its orbit.* This agreement between observation and theory convinced Newton that the centripetal force that keeps the moon in its orbit is provided by the gravitational attraction of the earth. (See Figure 5–5.)

In effect, Newton saw the moon as a body undergoing projectile motion (Figure 5–6). The moon is moving in two directions at once; it has an acceleration toward the earth of 0.0089 ft/sec^2 and at the same time is moving sideways with an orbital speed of 3350 ft/sec. If the moon had no orbital speed, it would fall toward the earth in a straight line, just as an apple does (that is, with an acceleration that would be 0.0089 ft/sec^2 initially but would increase as the moon got closer to the earth). The orbital speed deviates the moon sideways as it falls toward the earth, and in fact the speed is sufficiently large to deflect the moon far enough so that it never hits the earth as it falls (path d in Figure 5–6).

How is the orbital speed of the moon maintained? The moon's present-day orbital

*This is the acceleration of the entire moon toward the earth. It should not be confused with g_M (= 5.33 ft/sec^2), which is the acceleration of a falling body at the surface of the moon caused by the moon's gravitational attraction.

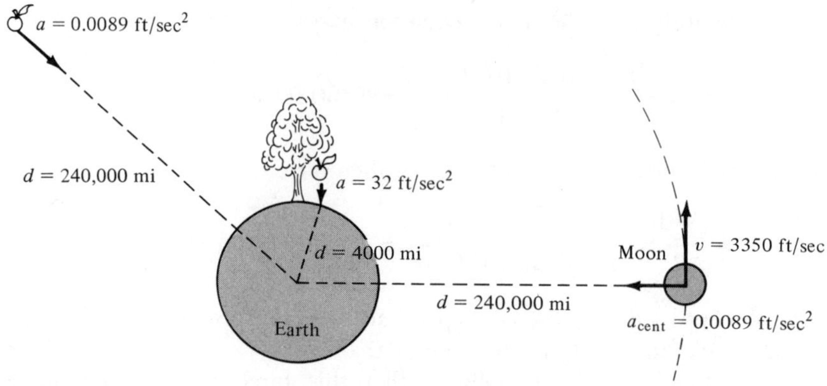

figure 5–5 **THE APPLE AND THE MOON**

At the surface of the earth, an apple falls with an acceleration of 32 ft/sec^2. If removed to a distance of 240,000 mi (the radius of the moon's orbit), the apple would have an acceleration of $32/(60)^2 = 0.0089$ ft/sec^2. This is the same as the centripetal acceleration of the moon in its orbit. We can conclude that the centripetal acceleration of the moon is caused by the gravitational attraction of the earth.

speed is essentially the same as it was when it became a satellite of the earth.* According to the principle of inertia, *no* force is required to keep a body moving at a constant velocity. That is, the moon's inertial tendency is to travel in a straight line at constant speed (path *e* in Figure 5–6). The effect of the earth's gravitational pull is to supply the centripetal force that keeps the moon on a closed curved path. The moon's orbital motion is therefore a consequence of a balance between the moon's inertial motion (which requires no force to maintain) and the gravitational attraction of the earth. Thus the earth's gravitational pull is the *only* force needed to keep the moon in its orbit.

In demonstrating that the earth's gravity extends to the moon, Newton took a step of enormous scientific and philosophical significance. For the first time, it was shown that terrestrial and celestial phenomena were governed by the same physical principle. Gravity, which causes all bodies to fall to the earth, became Universal Gravitation, responsible for all celestial motions.

KEPLER'S THIRD LAW: THE MASSES OF CELESTIAL BODIES

As part of his solution of the problem of planetary motion, Newton was able to derive Kepler's third law from basic principles. As we saw in Chapter 2, Kepler obtained this law by trial and error, using the known periods and distances of the major planets with respect to the sun. Kepler had found that (Eq. 2–6)

$$\frac{P^2}{a^3} = K$$

* Just how the moon became a satellite of the earth is still a matter of conjecture. The two main theories are that (a) the moon formed in place at the same time that the earth formed and (b) the moon was captured by the earth. Further exploration of the moon may eventually settle this question.

where K has the same numerical value for all planets. Newton derived the following equation from the theory of gravitation:

$$\frac{P^2}{a^3} = \frac{4\pi^2}{G(M + m)} \tag{5–5}$$

In this equation a, P, and m represent, respectively, the average radius, the sidereal period, and the mass of the orbiting body, and M is the mass of the central body. If we apply this equation to a planet in the solar system, M becomes the mass of the sun (M_s), and m becomes the mass of the planet (m_p). Comparing Newton's theoretical result (Eq. 5–5) with Kepler's law, we have

$$K = \frac{4\pi^2}{G(M_s + m_p)}$$

This result tells us that the constant K in Kepler's third law depends on the masses of the sun and the planet. It therefore appears that K should be *different* for each of the planets, since the value of K depends on the mass m_p of the particular planet! However, the masses of the planets are very small compared to the mass of the sun; so, to a good approximation, $M_s + m_p \approx M_s$. This approximation introduces at most a 0.1 percent error (for Jupiter, which has a mass of about 0.001 of the sun's mass). With the accuracy of modern-day astronomical data, K has been found to vary very slightly from planet to planet, as predicted by Newton. In the data available to Kepler in 1600, these variations were not apparent. Thus we find that Kepler's third law is only approximate.

Newton's version of Kepler's third law can be used to calculate the mass of the sun. If the mass of the planet is neglected, the equation for K becomes

$$K = \frac{4\pi^2}{GM_s}$$

In Chapter 2 we learned that if P is measured in years and a in AU, K will have a numerical value of 1 yr^2/AU^3. However, if we are to determine the mass of the sun in grams, K must have proper CGS units. That is, P must be in seconds

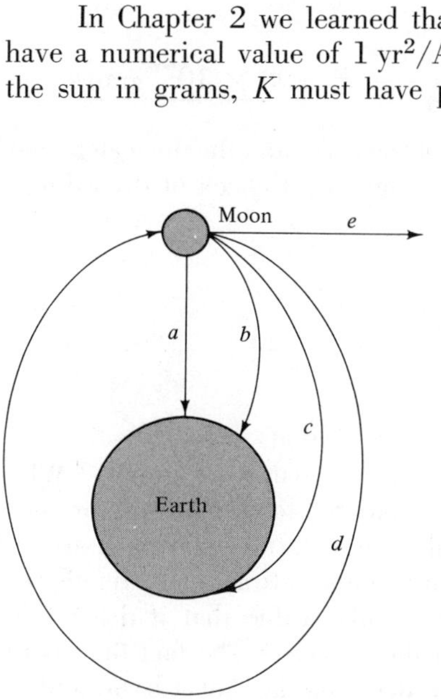

figure 5–6

THE MOON AS A PROJECTILE

In the absence of any orbital (sideways) speed, the moon would fall to the earth along path a. Given an orbital speed, the moon is deflected as it falls. For paths b and c the orbital speed does not deflect the moon sufficiently and it hits the earth. On path d the orbital speed is large enough to prevent the moon from hitting the earth, so the moon remains in an orbit about the earth. Thus the moon is continually falling toward the earth. Path e represents the path that would be taken by the moon if the gravitational force suddenly disappeared—that is, the moon would move on a straight line at constant speed.

and a in centimeters. With these units (see page 61) K has a value of 2.96×10^{-25} sec^2/cm^3. Solving the above equation for M_s and substituting numerical values yields

$$M_s = \frac{4\pi^2}{GK} = \frac{4\pi^2}{(6.67 \times 10^{-8}\ \text{cm}^3/\text{g-sec}^2)(2.96 \times 10^{-25}\ \text{sec}^2/\text{cm}^3)}$$

$$= \underline{1.99 \times 10^{33}\ \text{grams}}$$

Equation 5–5 can be used in similar fashion to determine the mass of any body that is the center of an orbital motion. For example, the mass of a planet can be determined from the orbital data of any of the planet's satellites. In particular, the mass of the earth can be calculated by measuring the average orbital radius, a, and sidereal period, P, of the moon (or any orbiting artificial satellite) and substituting them into Eq. 5–5. When this calculation is performed, one obtains

$$M_E = 5.98 \times 10^{27}\ \text{grams}$$

for the mass of the earth.

In comparing the mass of the earth with the mass of the sun (1.99×10^{33} grams), we discover that the mass of the sun is 330,000 times greater than that of the earth. Hence the theory of gravitation has given us our first indication of the enormous size of the sun. Only a few centuries before men had argued that the earth was too large a body to be in motion, and that the sun—a small, ethereal object—revolved around the earth once each day.

The mass of the earth can also be calculated by an entirely different procedure based on the dependence of the acceleration of gravity at the surface of the earth, g_E, on the mass of the earth ($g_E = GM_E/R_E^2$). Since g_E, G, and R_E can all be measured experimentally, the mass of the earth can be computed. Solving for M_E and substituting $g_E = 977$ cm/sec^2 and $R_E = 6.378 \times 10^8$ cm (the values at the equator in CGS units) yields

$$M_E = \frac{g_E R_E^2}{G} = \frac{(977\ \text{cm/sec}^2)(6.378 \times 10^8\ \text{cm})^2}{(6.67 \times 10^{-8}\ \text{cm}^3/\text{g-sec}^2)} = \underline{5.98 \times 10^{27}\ \text{grams}}$$

The agreement between this value of the mass of the earth and the one calculated from Newton's form of Kepler's third law serves as a significant check of the validity of the theory of universal gravitation.

5–2

WHAT IS GRAVITY?

In spite of the success of Newton's theory of gravitation, many questions of a fundamental nature remain unanswered. Just what is the nature of gravity? Why is it an innate property of all matter? Why is the gravitational force inversely proportional to d^2 and not to d or d^3? Why is there only a gravitational force of attraction—unlike electric and magnetic forces, which can be either attractive or repulsive? Why does the gravitational constant G have the particular value that it does? Can we be certain that G has the same value throughout the universe? The fact that there are no clear-cut answers to any of these questions reflects the lack of a basic understanding of the nature of gravity in modern-day science.

Newton himself was loath to speculate on the nature of gravity. In a letter written to his friend Richard Bentley in 1692, Newton said: "You sometimes speak of gravity as essential and inherent in matter. Pray do not ascribe that notion to me, for the cause of gravity is what I do not pretend to know."[2] Newton believed that it was sufficient for the purposes of science to demonstrate the mathematical form of the principle of gravitation, and that an understanding of the nature of gravity was not necessary. In later years, however, Newton did allow himself to speculate on the cause of gravity, as we shall see shortly.

ACTION AT A DISTANCE

The one feature of the principle of gravitation that is perhaps most difficult to accept is the notion that bodies can exert gravitational forces on one another, even though they are not in contact. This is the concept of ***action at a distance.**** How is the gravitational force of the sun transmitted to the earth, and vice versa, when the two bodies are separated by 93,000,000 miles of nearly empty space? In fact, the nature of the intervening medium has no effect whatsoever on the strength of the gravitational force. For that matter, if two bodies are placed in a vacuum, the mutual gravitational attraction is undiminished, even though there is no physical connection or material substance whatsoever between them.

Newton's opinion of the action-at-a-distance principle was quite clearly stated in another letter to Bentley:

> That gravity should be innate, inherent, and essential to matter, so that one body may act upon another at a distance through a vacuum, without the mediation of anything else by and through which their action and force may be conveyed from one to another, is to me so great an absurdity that I believe that no man who has in philosophical matters a competent faculty of thinking can ever fall into it.[3]

Newton was also beset by the problem of explaining how light traveled from place to place, and so he came to accept the existence of the ***ether,*** an invisible, weightless, and highly rarefied substance that pervades all of space. To Newton, the ether was highly suitable as a medium for the transmission of light and gravity, as well as for electric and magnetic forces. In his earlier writings Newton had attacked the propounding of hypotheses that could not be directly verified by experiment, but in later years Newton detailed a highly speculative theory in which he proposed that gravitation is the result of fluctuations in the density of the ether.

By the latter half of the nineteenth century, the belief was widely held that gravitation and electromagnetic forces are transmitted from one body to another through the ether. The concept of a ***field*** was introduced to provide a more detailed picture of how these forces are transmitted. A gravitational (or electromagnetic) field was conceived of as a state of disturbance or strain of the ether created by the presence of masses (or electric charges). These disturbances in turn affect other bodies, forcing them to move in a particular direction. Hence it could be said that the earth and moon do not interact directly; the earth disturbs the ether, and the ether acts on the moon. In this manner the problem of action at a distance was seemingly solved.

Unfortunately, the success of this approach was rather short-lived. Around the beginning of the twentieth century, a number of experiments were performed that

*The action-at-a-distance effect is not a property of gravitation alone; electric and magnetic forces also appear to be action-at-a-distance forces that vary as the inverse square of the distance.

cast serious doubt as to the existence of the ether. (The rise and fall of the ether hypothesis will be discussed in considerable detail later in the text.) These experiments led inescapably to the conclusion that if the ether exists at all, it certainly could not be the medium for the transmission of gravitational or electromagnetic forces. With the demise of the ether hypothesis, physicists once again had to assume that these forces could be transmitted through empty space. The concept of a gravitational or electromagnetic field remains, but a field now has to be viewed as an unexplained entity that can exist in a vacuum. As such, the field concept no longer provides us with a better understanding of the nature of gravity; it is merely a convenient way of describing gravitational phenomena. Thus the ultimate nature of gravity is as much of a mystery to the modern-day physicist as it was to Isaac Newton.

INERTIA AND GRAVITATION

One aspect of gravitation that may provide a clue as to its ultimate nature is the peculiar relationship between gravitation and inertia. This relationship is most pointedly illustrated when we re-examine in detail the derivation from Newtonian principles of the law of falling bodies. Newton derived the law by substituting the gravitational force between the earth and a falling body of mass m into the equation $F = ma$:

$$F_{\text{grav}} = \frac{GM_{\text{E}}m}{R_{\text{E}}^2} = ma$$

The next step is to cancel the m on both sides of the equation, with the result that the acceleration $a = GM_{\text{E}}/R_{\text{E}}^2$ is independent of the mass of the falling body. Herein lies the mystery.

The m that appears in the equation $F = ma$ is a measure of the *inertia* of the body. It measures the resistance of a body to *any* force. Let us call this the ***inertial mass.*** The m that appears in the equation $F = GMm/d^2$ determines the strength of a *particular* interaction involving the body—namely, the gravitational force. Let us call this the ***gravitational mass.*** The inertial and gravitational masses must be exactly equal; otherwise we could not cancel them in the above equation. If they *did not* cancel, we would arrive at the conclusion that the acceleration of a falling body *does* depend on its mass (either inertial or gravitational, or both), which is contrary to experience. The theory of gravitation would be incorrect.

Therefore, we see that Newton did not really derive the law of falling bodies; rather, the law of falling bodies (an experimental result) *forced him to assume that the inertial and gravitational masses must be equal.*

The equality of the inertial and gravitational masses is extremely puzzling to physicists. Why should gravitation, of all the many forces in nature, be the only one to be so intimately related to inertia, which is supposed to be an inherent property of bodies that is independent of the nature of the force being exerted? This connection between inertia and gravity—two completely different and otherwise unrelated properties of bodies—has thus far eluded an explanation. In fact, Einstein looked on the equality of inertial and gravitational mass as a fundamental principle of nature and used it as a starting point for the general theory of relativity. This theory provided yet another conceptualization of gravity. Einstein's theory of gravitation is far too complex to summarize in a few sentences, so we will defer a discussion of it until a later chapter. In any case, the equality of the inertial and gravitational masses is clearly an important clue as to the fundamental nature of gravitation (and inertia as well). This equality must be explained by future physicists if a deeper comprehension of gravitation and inertia is to be achieved.

In view of the lack of a fundamental understanding of the nature of gravity, it seems remarkable that Newton's theory of gravitation could have brought about such an enormous advance in our understanding of the universe. But this is characteristic of the way in which science progresses. It is not necessary to understand gravity fully in order to have a successful theory of gravitation. Simply by accepting the concept of gravitation as an axiom, we can account for a wide variety of seemingly unrelated phenomena in relatively simple terms. Thus in view of its simplicity, comprehensiveness, and predictive power, the theory of gravitation must be considered an outstanding theory. The attempts on the part of some physicists to "understand" gravity represent the next stage in man's attempt to probe ever more deeply until the ultimate answers are found—if, in fact, such answers exist.

In the remainder of this chapter we will examine some of the consequences of Newton's discoveries. We will begin by considering how Newtonian principles can be applied to the solution of some familiar astronomical problems.

5–3

ORBITAL MOTION

Historically, the first problem to which Newton applied his theories of motion and gravitation was that of orbital motion. Newton wanted to prove that a body in motion under the influence of a gravitational force could travel in an elliptical orbit around the center of that force—that is, it was Newton's intention to show that Kepler's laws of planetary motion could be derived mathematically from the basic principles of the theories of motion and gravitation.

The essentials of the problem are shown in Figure 5–7. A small body of mass m (a planet) is acted on by a force $\mathbf{F}$ as a result of the mutual gravitational interaction between the small body and a larger body of mass M (the sun). This force $\mathbf{F}$ is directed along the line joining the two bodies. According to Newton's second principle, the acceleration of the smaller body points in the same direction. In addition, the smaller body has a velocity $\mathbf{v}$, which points in some arbitrary direction. (Obviously, if $\mathbf{v} = 0$ or if $\mathbf{v}$ points along the line joining the two bodies, the smaller body will fall directly in a straight line onto the larger body, and the problem becomes a trivial one.) It will be assumed that the larger body is considerably more massive than the smaller one, so that the larger body's acceleration is negligible. That is, we will consider the larger body to be stationary.

If the two bodies are a distance d apart, the mutual gravitational force of

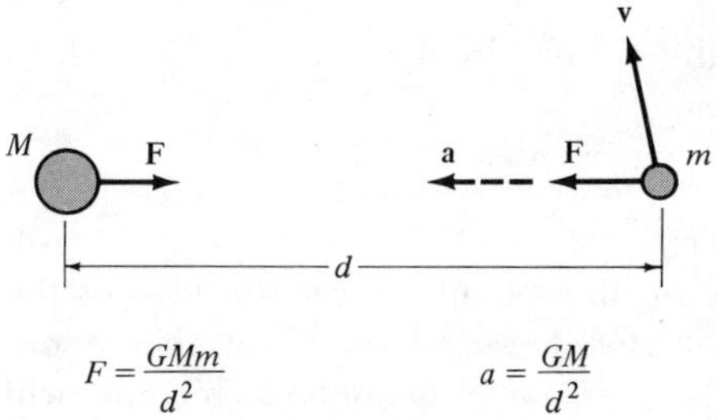

figure 5–7

BASIC DIAGRAM FOR THE PROBLEM OF ORBITAL MOTION

The small body of mass m is moving under the influence of its mutual gravitational interaction with the larger body of mass M.

attraction is $F_{grav} = GMm/d^2$. Setting $F_{grav} = ma$, we obtain

$$a = \frac{GM}{d^2} \tag{5-6}$$

where a represents the acceleration of the smaller body. The absence of the mass m in this equation reflects the fact that a planet also obeys the law of falling bodies; its orbital motion is independent of its mass.

If the initial position and velocity of the smaller body are known, it should be possible, in principle, to use Eq. 5–6 to determine the body's position and velocity at any other time, thus allowing us to trace out the path or ***orbit*** of the body. However, the solution of this problem becomes rather involved. For noncircular orbits d is variable, and therefore the acceleration does not have a constant magnitude. The equations of motion and methods of solution developed in Chapter 3 do not apply in this case, and the problem cannot be solved by algebraic techniques. The methods of differential and integral calculus must be used—many of which were developed by Newton for the specific purpose of solving the orbital motion problem.

The solution of this problem leads to some interesting and often rather surprising results. We are aware, of course, that Newton was able to prove that a planet will travel in an elliptical orbit about the sun at one focus (Kepler's first law). In addition, Newton showed that under certain conditions the orbit can be a circle, a parabola, or a hyperbola. The latter two orbits are of particular interest because *they are not closed orbits.* This means that a body acted on by the gravitational force of the sun is not necessarily bound to the sun, but can escape from the effective influence of the sun's attraction.

The methods of Newtonian mechanics enable us to identify the particular factors that determine the type and shape of an orbit under specific conditions. If the velocity (both magnitude and direction) of the orbiting body and its distance from the center of force are known at *any one point on its path,* the nature of the *entire* orbit can be determined. In particular, it is possible to learn whether the orbit is bounded (elliptical or circular) or unbounded (parabolic or hyperbolic). Equations have been developed that enable one to calculate the specific shape (average radius, eccentricity, etc.) of the entire orbit, given the velocity and distance at one point on the orbit.

Of particular interest is the rather simple equation by which one can determine whether a body is in a bounded or an unbounded orbit. Suppose that at a particular instant an orbiting body has a speed v when it is a distance R from the center of a large body of mass M. Let us define the ***escape speed,*** v_{esc}, by the equation

$$v_{esc} = \sqrt{\frac{2GM}{R}} \tag{5-7}$$

It can be shown that:

(1) If v is less than v_{esc}, the orbit is elliptical or circular.
(2) If v is equal to v_{esc}, the orbit is parabolic.
(3) If v is greater than v_{esc}, the orbit is hyperbolic.

(This result will be derived in the next chapter.)

Thus to find the kind of orbit a body is in, one compares the speed v of the body at a given point in its orbit with the calculated escape speed for that point. If v is greater than or equal to v_{esc}, the body is in an unbounded orbit and will escape from the effective influence of the gravitational force. If v is less than v_{esc}, the orbit will be an ellipse or a circle. In order for a body to remain in a circular orbit of radius R, it must have an orbital speed

$$v_c = \sqrt{\frac{GM}{R}} \tag{5–8}$$

A speed slightly greater or smaller than v_c will result in an elliptical orbit. The relationship between the various kinds of orbits is illustrated in Figure 5–8.

If a body in orbital motion is acted on only by the single gravitational force of attraction of the central mass, the body's orbit will be stable and unchanging. An orbit can be altered from one form to another by the application of an additional force, which has the effect of changing the magnitude and/or the direction of the orbital velocity of the body.

CENTER-OF-MASS MOTION

To this point, we have discussed orbital motion on the assumption that the mass of one body is very large relative to the mass of the other body ($M \gg m$). The acceleration of the larger body was assumed to be negligible and that body was

figure 5–8 **ORBITAL MOTION**

The orbiting body is at point P, a distance R from the central mass M. At this point the body has a velocity $\mathbf{v}$ perpendicular in direction to R. The condition for a circular orbit (orbit 2) is $v_c = \sqrt{GM/R}$. If $v < v_c$ (orbit 1), the orbit will be elliptical, with P as the aphelion point. If $v > v_c$, the orbit will also be elliptical, with P as the perihelion point (orbit 3). If $v = v_{esc} = \sqrt{2GM/R}$, the orbit will be parabolic (orbit 4); and if $v > v_{esc}$, the orbit will be hyperbolic (orbit 5).

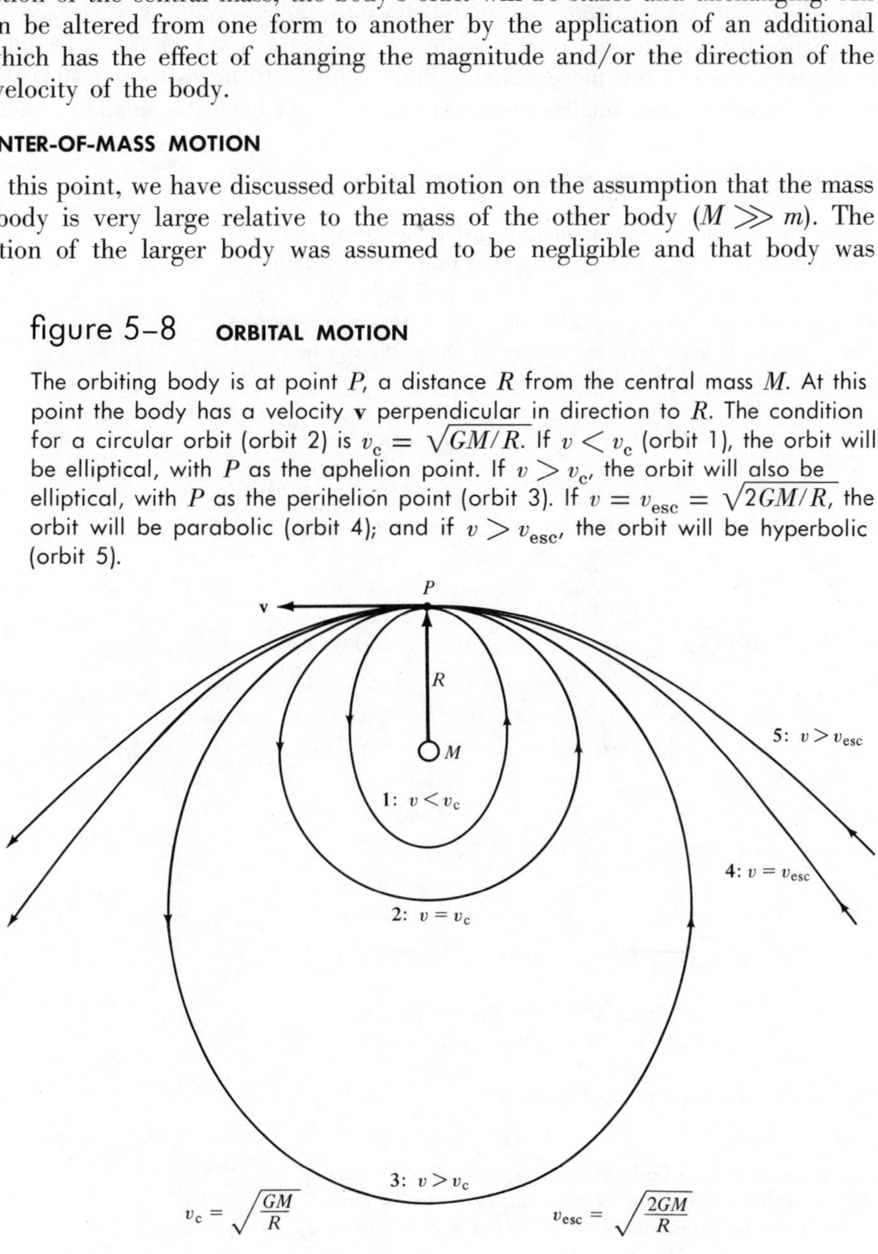

considered to be stationary; the smaller body was assumed to move in an orbit with the larger body fixed at the focus of the orbit. We will now eliminate this simplifying assumption and consider what kind of motion arises when the masses of the two bodies are more comparable.

Under the influence of their mutual gravitational attraction, the two bodies will accelerate toward one another with accelerations that are inversely proportional to their masses. As a consequence, the larger body as well as the smaller one must have an orbital motion. Newton was able to prove that in a two-body system both bodies travel in orbits about a common point, called the ***center of mass,*** which lies on the line joining them. The location of the center of mass depends on the relative masses; the distances from the two bodies to their center of mass are inversely proportional to the masses of the bodies. (See Figure 5–9.) Thus the center of mass is always closer to the more massive body. This is to be expected: since the more massive body has the smaller acceleration, it should have the smaller orbit.

figure 5–9 CENTER-OF-MASS MOTION

Under the influence of their mutual gravitational attraction, two bodies revolve around a common point known as the center of mass. The distances from the bodies to the center of mass are inversely proportional to their respective masses. Their motion is such that the center of mass always lies on the line joining the two bodies.

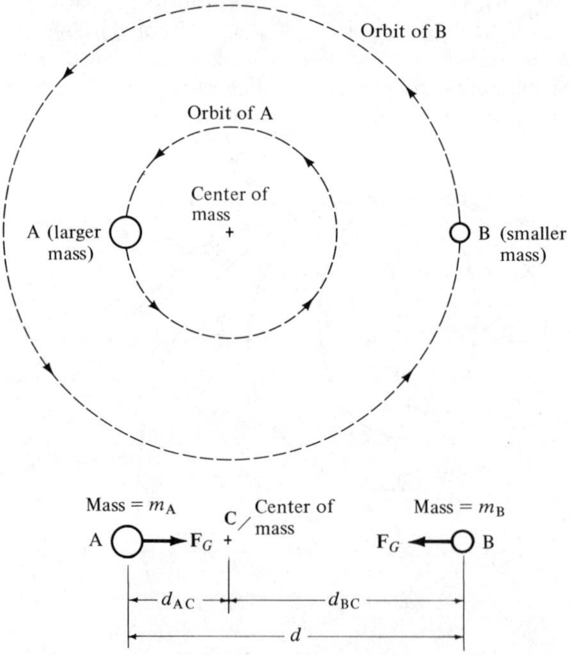

Equation for locating center of mass:

$\underline{m_A d_{AC} = m_B d_{BC}}$

d_{AC} = distance from body of mass m_A to center of mass C
d_{BC} = distance from body of mass m_B to center of mass C
d = distance between centers of bodies A and B ($d = d_{AC} + d_{BC}$)

The earth and moon constitute a two-body system, so they both revolve around a common center of mass. From precise measurements of the motions of the earth and moon, it can be determined that the center of mass is located about 2900 miles from the center of the earth. Given that the distance between the center of the earth and the center of the moon is approximately 239,000 miles, one can calculate that the mass of the earth is 81.3 times the mass of the moon. It is the center of mass of the earth–moon system, rather than the center of the earth itself, that travels in an elliptical orbit about the sun.

ARTIFICIAL SATELLITES

It is in the past two decades that Newton's orbital equations have been put to their most impressive use. In October 1957 an artificial satellite—Sputnik I—was placed in a stable orbit around the earth. Within a few years after this event, artificial satellites were placed in orbit about the moon and were sent to the vicinity of Venus and Mars. With the successful launching of Apollo 11 in July 1969, man for the first time was able to set foot on another planetary body. Yet although the exploration of space by unmanned and manned satellites is a rather recent human development, we must remember that the theory on which these accomplishments are based was produced nearly three hundred years ago! Although space travel has been theoretically feasible since the seventeenth century, it could not take place until several *technological* problems had been solved—such as the development of suitable rocket fuels and of the computers and other electronic equipment needed to communicate with and control space flight.

For orbital motion about the earth, the center of the earth is at the focus of all elliptical, parabolic, and hyperbolic orbits. Equations 5–7 and 5–8 are also applicable to earth orbits, except that M now represents the mass of the earth and R is measured from the center of the earth. Thus the escape speed for a satellite that is launched directly from the surface of the earth becomes

$$v_{\mathrm{esc}} = \sqrt{\frac{2GM_{\mathrm{E}}}{R_{\mathrm{E}}}}$$

When the appropriate numerical values are substituted, $v_{\mathrm{esc}} = 6.94\ \mathrm{mi/sec} \approx 25{,}000\ \mathrm{mi/hr}$.

The development of the technological capability for orbiting artificial satellites has proved to be of great benefit to mankind. Photographs and measurements made of the earth from altitudes of hundreds of miles have brought about improvements in weather forecasting, and communications satellites have made possible intercontinental live television broadcasting. Moreover, the opportunity to make astronomical measurements of planets, stars, and galaxies from above the surface and atmosphere of the earth has resulted in the collection of vast amounts of data not formerly available. This information explosion has ushered in a whole new era in astronomy, comparable to the one that occurred when Galileo first turned his telescope to the heavens.

WEIGHTLESSNESS

In an orbiting spacecraft astronauts find that common objects—pencils, tools, bits of food—remain suspended in midair, and they themselves are able to float effortlessly around their spacecraft with no sense of up or down. The conditions resemble those that one would find if placed in a completely gravity-free environment. The astronauts are experiencing what is commonly called ***weightlessness.***

Obviously, the spacecraft is *not* in a gravity-free location. The spacecraft is kept in orbit by the earth's gravitational attraction; if there were no force acting on the spacecraft, it would move away in a straight line. An orbiting spacecraft travels at an altitude of a few hundred miles above the earth, where the gravitational force is only slightly less than the force that would be exerted at the earth's surface. (Previously, we calculated that at an altitude of 100 miles above the earth, the gravitational force is 0.952 of the force at the surface.) It is quite apparent, then, that the experience of weightlessness is not the result of the absence or substantial weakening of gravitational force.

In fact, one does not have to be in an orbiting spacecraft to experience weightlessness. Suppose a person holding a book is in an elevator and the elevator cable snaps. The person, the book, and the elevator are now all in free fall with the same acceleration. If the person lets go of the book, the book would appear to float in midair. This is because the person, book, and elevator are all falling at exactly the same rate and therefore maintain the same positions relative to one another. If the person gave a slight upward push to the book, it would appear to float to the ceiling. (The upward force would temporarily decrease the downward acceleration of the book, causing the book to fall at a slower speed than the elevator.) All the events in the falling elevator (except the final crash) would give the appearance of a weightless state, and the person inside (if he did not know he was falling) could easily believe that he was in a gravity-free location.

Conditions inside the falling elevator and in an orbiting spacecraft are identical—and for a good reason: the spacecraft and everything in it are all falling toward the earth with the same acceleration. (See Figure 5–10.) Like the moon, the spacecraft

figure 5–10 WEIGHTLESSNESS AND FREE FALL

(a) In a freely falling elevator, all objects fall with the same acceleration. To a person inside the elevator, any object inside would appear to be suspended in the air. (b) Similarly, when in orbit, the spacecraft, the astronaut, and any other objects inside the spacecraft are all falling toward the earth with the same acceleration. Both the elevator rider and the astronaut are experiencing "weightlessness," in that it appears to them that there is no gravity.

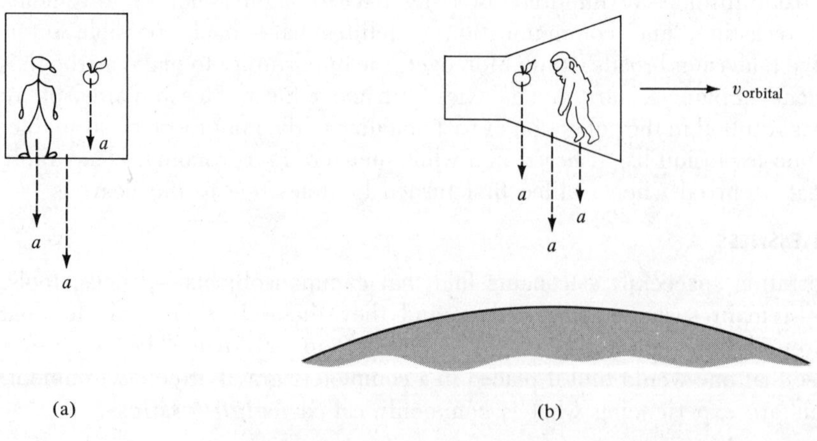

(a) (b)

is acted on only by the earth's gravitational force, which provides the centripetal force needed to keep the spacecraft in orbit. The spacecraft is continually accelerated toward the earth but is prevented by its sideways (orbital) motion from hitting the earth.

We see, therefore, that the weightlessness that prevails in an orbiting spacecraft is not true weightlessness caused by the absence of gravity. Rather, it is an *apparent* weightlessness resulting from the fact that all objects inside the spacecraft, as well as the spacecraft itself, are in free fall with the same acceleration.

COMETARY ORBITS

The sudden appearance of a comet is one of the more spectacular astronomical events. Appearing first as a hazy starlike object in the night sky, a comet will gradually grow in size and brightness and develop an extremely long tail. It is visible for several months and then slowly fades away. In pre-Newtonian times the appearance of a comet was a totally unpredictable event, and comets were regarded with superstitious awe, particularly as omens or warnings of impending disasters.

Newton, however, considered the possibility that comets were regular members of the solar system, moving under the influence of the sun's gravitation. His theory was verified by Edmund Halley, who showed that the observed motions of several comets could be accounted for by assuming that these bodies were traveling in hyperbolic or highly eccentric elliptical orbits around the sun. In calculating these orbits, Halley found that the comets of 1531, 1607, and 1682 had rather similar orbits. He proposed that they were the *same* comet on an elliptical orbit with a period of approximately 76 years. When the comet reappeared in 1758 as Halley had predicted, it was named "Halley's comet." Unfortunately, Halley was not able to appreciate this honor, since he had died in 1742. (Further examination of old astronomical records indicates appearances of Halley's comet dating back to Roman times. It reappeared in 1835 and 1910, and is expected to be visible again in 1986.)

To date, the orbits of nearly six hundred comets have been calculated. Most of them have long, elliptical orbits with sidereal periods of more than 200 years. Of these, about a hundred have nearly parabolic orbits, with eccentricities between 0.95 and 0.999. There is also a group of about a hundred "short-period" comets (sidereal period less than 200 years) with considerably less eccentric orbits.

One important feature of cometary orbits is that they can be altered considerably by the gravitational attraction of the planet Jupiter, which is the largest planet in the solar system. (Saturn, the second largest planet, may also affect cometary orbits to a lesser extent.) A comet moving under the gravitational influence of the sun alone should have an unchanging orbit. However, if the comet comes within the vicinity of Jupiter, its velocity can be increased or decreased considerably by the additional gravitational force, and rather severe changes in the shape of the orbit can result. It is possible, for example, for a comet in a hyperbolic or highly eccentric elliptical orbit to be deviated into a considerably less eccentric orbit, and it is generally believed that the short-period comets were produced in this manner. One case in point is Brook's comet, which came within 0.01 AU of Jupiter in 1886 and had its period changed from 29 years to 7.1 years as a consequence.

It is also possible for a comet in a highly eccentric orbit to gain enough speed in such an encounter for its orbit to change from elliptical to parabolic or hyperbolic. In fact, a few comets have disappeared after having been observed for several orbits; presumably, they were deviated into hyperbolic orbits by Jupiter and escaped from the solar system.

One may ask why an encounter with a comet does not produce a similar distortion in the orbit of Jupiter or of any of the other planets. The answer is that the masses of comets are relatively quite small (at most, about one-millionth of the earth's mass). Thus although the comet and Jupiter exert equal forces on one another, the acceleration experienced by Jupiter will be negligible compared to that of the comet, and Jupiter's orbit will be virtually unaffected.

THE DISCOVERY OF NEPTUNE AND PLUTO

We shall now learn about what is undoubtedly the most impressive example of the predictive power of the theory of gravitation: the prediction, on theoretical grounds, of the existence and location of additional planets in the solar system.

Into the eighteenth century only six planets were known to be in the solar system—the five known to the ancient astronomers and the earth. Then in 1781 the planet Uranus was accidentally discovered by William Herschel, a professional musician and amateur astronomer. Herschel was looking at the stars through a telescope when he noticed an object that appeared as a small disk rather than as a point of light. In subsequent observations Herschel noticed that the object moved with respect to the stars. Eventually, the orbit was calculated, and Uranus was recognized as a major planet of the solar system. (For nearly sixty years after its discovery, the new planet was called Herschel, until finally the mythological name was universally agreed on. In Roman mythology Uranus was the father of Saturn and grandfather of Jupiter.)

As it turned out, Uranus had been observed and plotted on star charts as far back as 1690 without its being recognized as a planet. (One astronomer, Pierre Lemonnier, observed it on twelve separate occasions between 1750 and 1769 without noticing its motion relative to the stars!) With these additional observations it became possible to determine the orbit of Uranus rather precisely. The orbit was found to deviate somewhat from an ellipse, but it was assumed that the deviations were due to the gravitational attractions of Jupiter and Saturn. Yet when these effects were taken into account, small deviations still remained. By 1840 the discrepancy between the computed and observed positions of Uranus was 2′, an intolerably large error.

The possibility that Newton's theory of gravitation was incorrect was entertained by a few astronomers, but the more likely explanation was that the discrepancy was caused by the gravitational attraction of another planet beyond the orbit of Uranus. By this time scientists had developed mathematical techniques that made it possible to determine the orbit and position of the hypothetical planet from the observed discrepancies. These extremely laborious calculations were first completed in 1845 by the young English mathematician John C. Adams, and independently by the French astronomer Urbain Leverrier in 1846. Adams communicated the results of his computations to several English astronomers, but none of them would take the trouble to make a serious search for the new planet. Meanwhile, Leverrier was having similar difficulties in France. Finally, Leverrier wrote to Johann Galle, an assistant at the Berlin Observatory, asking him to look for the planet. On the night of September 23, 1846 (the same day that he received Leverrier's letter), Galle discovered the planet less than 1° away from the position predicted by Leverrier.

The name Neptune was suggested for the newly discovered planet (although for a time Leverrier had asked that it be called Leverrier). When the existence of Adams' calculations became known, there was considerable controversy as to whether Adams or Leverrier should get credit for the discovery. Today they are generally granted equal recognition. In any case, the prediction of the existence and nearly exact location of a planet of the solar system invisible to the naked eye was considered a major triumph for Newton's theory of gravitation.

Before long the search for a planet beyond Neptune had begun. Small discrepancies still remained in the orbit of Uranus, and these were used by the American astronomer Percival Lowell to calculate the location of the trans-Neptunian planet. Lowell searched unsuccessfully for the planet until his death in 1915. The search continued until 1930, when Clyde Tombaugh at the Lowell Observatory in Arizona discovered the planet, which was subsequently named Pluto. It was found some 6° away from Lowell's last predicted position.

The calculation of the positions of Neptune and Pluto from deviations in the orbit of Uranus was made possible by a mathematical procedure known as ***perturbation theory.*** Any given planet in the solar system is being acted on by the gravitational forces of the eight other planets and the sun. Because the other planets are all moving at different speeds, the net force on the given planet changes in a very complicated way, making an exact mathematical analysis of its motion impossible to obtain.

Naturally, the sun exerts the greatest force on the given planet, and by comparison the forces exerted by the other planets are many orders of magnitude smaller. This fact permits the use of the approximation techniques of perturbation theory. The method of perturbation theory is first to calculate the orbit of the planet, neglecting entirely the other planets—that is, by assuming a two-body system consisting of the sun and the planet. The small deviations, or ***perturbations,*** produced by each of the other planets can then be computed (using equations specially derived for the purpose) and added on, one by one. The procedure can be done in reverse; the perturbations caused by the known planets can be subtracted, one by one, from the *observed* orbit, and any remaining deviations from a purely elliptical orbit can then be attributed to the effects of an unknown planet or planets.

Newton was bothered by the possibility that the perturbations in a planet's orbit might build up with time, causing the planet to deviate considerably from an elliptical orbit and possibly escape from the solar system. He therefore proposed that God had to intervene personally from time to time and reduce the deviations. Newton thus viewed the universe as an imperfect machine that had to be readjusted occasionally by its Creator. As it turned out, Newton's fears were unjustified. At the beginning of the nineteenth century, the French mathematician Pierre Simon de Laplace published his great treatise *Celestial Mechanics,* in which the *Principia* was brought up to date, with particular emphasis on the application of the theory of gravitation to the motions of the planets. Laplace was able to prove that the perturbations of planetary orbits would never build up substantially, so that the orbits would always remain stable.*

THE FLATTENING OF THE EARTH

Careful surveying of the earth has shown that its shape is not exactly spherical. Actually, the earth has a greater diameter across the equator than from pole to pole. Hence the earth appears to be flattened at the poles; it has an elliptical shape in cross section and more closely resembles the geometric solid known as the ***oblate***

*Laplace was once presented to Napoleon, who asked him how he had managed to write a book about the system of the universe without making a single reference to its Creator. Laplace replied, much to Napoleon's amusement, "I had no need of that hypothesis."

spheroid. The difference between the equatorial and polar diameters is small—7926.4 miles as compared to 7899.8 miles—so that the earth is flattened by 1 part in 298. The degree of flattening or ***oblateness*** is defined as follows:

$$\text{Oblateness} = \frac{\text{equatorial diameter} - \text{polar diameter}}{\text{equatorial diameter}}$$

Hence the oblateness of the earth is 1/298.

In the *Principia* Newton was able to show that the flattening of the earth is a consequence of the earth's rotation. Each point on the earth's surface completes one rotation in 24 hours. During this time a point on the equator travels a distance of nearly 25,000 miles (the circumference of the earth at the equator), whereas a point at the latitude of New York travels in a circle of only 19,000 miles circumference. At the poles the distance traveled is 0 miles. It follows that different points of the surface of the earth have different speeds of rotation, ranging from more than 1000 mi/hr at the equator to 0 mi/hr at the poles.

We have learned that to maintain a body in circular motion, a centripetal force must act on it to overcome its inertial tendency to move in a straight line. Thus a body on the earth's surface must be acted on by a centripetal force large enough to keep it on the surface. Since the speed of rotation is greatest at the equator, a body at the equator will have the greatest tendency to fly off in a straight line. Hence the centripetal force needed to keep a body on the earth's surface as it rotates is greatest at the equator and decreases to zero at the poles. The agent that provides the required centripetal force is the gravitational attraction of the earth for the body.

We know that bodies are bound to the earth's surface by the gravitational attraction of the earth. In fact, the earth itself is held together mainly by the mutual gravitational attraction of its constituent parts. The theory of gravitation shows that the earth would be spherical in shape if not for its rotation. But on a rotating earth the inertial tendency of the constituent parts to fly off in a straight line must be counteracted. A part of the gravitational attraction must provide the centripetal force. Since the required centripetal force is greatest at the equator, it follows that a body at the surface of the earth will be less tightly bound to the earth at the equator than it would be at the poles. The result is that the solid body of the earth cannot maintain a spherical shape. The earth tends to bulge out at the equator and thus takes on the shape of an oblate spheroid.

The oblateness of the earth is not especially large because the earth does not rotate that rapidly. Conceivably, the earth's rotation could be so great that the gravitational attraction would not be sufficient to provide the required centripetal force. The earth would become unstable, and material would fly off at the equator.

Newton derived the result that for a rotating body held together by gravitational forces, the oblateness is equal to the ratio of the centripetal acceleration (v^2/R) at the equator to the gravitational acceleration ($g = GM/R^2$). At the earth's equator $v^2/R = 0.111$ ft/sec^2 and $g = 32.1$ ft/sec^2. Hence the predicted oblateness is

$$\frac{\text{equatorial centripetal acceleration}}{\text{equatorial gravitational acceleration}} = \frac{0.111\ \text{ft/sec}^2}{32.1\ \text{ft/sec}^2} = \frac{1}{289}$$

This is in rather close agreement with the observed oblateness of 1/298. The difference can be attributed to the fact that the earth is held together partly by chemical forces and not entirely by gravitational forces. Thus Newton was able to use his theories of motion and gravitation to predict rather accurately the shape of the earth. As additional evidence, Newton referred to the planet Jupiter, which is observed to be eighteen times more oblate than the earth. He was able to attribute the shape of Jupiter to its rapid rotation (its period of rotation is less than 10 hours) and large size (its diameter is 11.2 times larger than that of the earth). The speed of rotation at the equator of Jupiter is calculated to be about 28,000 mi/hr! On the other hand, the sun and the planet Venus, which rotate very slowly, appear to be perfectly spherical in shape.

5–4

THE TRIUMPH OF HELIOCENTRISM

From the death of Copernicus in 1543 to the publication of the *Principia* in 1687, the central issue of astronomy was the question of whether the universe is heliocentric or geocentric. With the immense success of Newton's theories of motion and gravitation (which are based on the assumption that the solar system is heliocentric), it seemed that the controversy had been finally settled in favor of the heliocentric theory. Additional astronomical observations made during the century following Newton's death seemed to confirm this view.

From the time of Aristotle the absence of any observable stellar parallax had served as a major barrier against the acceptance of the heliocentric theory. The argument that the stars were too far away to exhibit parallax was hardly a convincing one; it was inconceivable that the universe could be so large. The most astute naked-eye observers from Hipparchus to Tycho Brahe were unable to find a single star that exhibited a measurable parallax, and even the invention of the telescope did not improve the situation. Eventually, however, stellar parallax was observed. The first unequivocal measurements were reported in rapid succession by Friedrich Struve in 1837, Friedrich Bessel in 1838, and Thomas Henderson in 1839. By that time there were not many supporters of the geocentric theory, so the discovery was somewhat anticlimactic.

STELLAR ABERRATION

Actually, the most convincing piece of evidence for the revolution of the earth was discovered in 1729, only two years after Newton's death, and more than a hundred years before stellar parallax was finally observed. In 1727 two English astronomers, Samuel Molyneux and James Bradley, made a series of observations of the position of the star Eltanin, in an attempt to measure the parallax of this star. They found that the star did indeed appear to move in a small circle with respect to the more distant stars over a period of a year. They realized, however, that the motion was *not* stellar parallax, since their observations showed that the star had its maximum shift in position at the time of year when it should exhibit no parallax at all. Observations of several other stars showed similar results. The two men found that all of these stars moved in circles 20.5″ in radius, regardless of their distances.

In 1728 Molyneux died, and Bradley was left to figure out the puzzle alone. A year later he announced his solution: the apparent shift in the position of a star is due to the motion of the earth with respect to the light coming from that star. Bradley called this effect ***stellar aberration.***

Stellar aberration is very similar to the common phenomenon that we observe when driving a car in a rainstorm. When the car is stationary, the rain appears to be falling vertically, but when the car is moving, the rain appears to be falling toward the car at an angle. The rain is, of course, not really falling at an angle; it only appears that way because the car is moving toward the raindrops as they are falling. Thus there is an *apparent* change in the direction from which the rain is seen to come because of the motion of the observer.

We can illustrate this effect in a somewhat different way. Consider a long tube that is used to collect rainwater, as shown in Figure 5–11. If the tube is stationary and held vertically, rain will fall through it without hitting the sides. However, if the tube is moving horizontally, the falling raindrops will collide with the sides of the tube. This can be avoided by tilting the tube in the direction of motion at the proper angle. To the person holding the tube, it *appears* as if the rain were coming from the direction in which the tube is tilted.

The same analysis applies to light traveling down a telescope tube. If the telescope is on a moving earth, the telescope must be tilted in the direction of motion in order to receive the light from a star that is overhead. Consequently, the starlight appears to be coming from a direction that is not the true direction of the star (Figure 5–12a). As the earth travels in its orbit, the direction of its motion is constantly changing; consequently, the position in the sky of a given star appears to change. As a result of stellar aberration, each star appears to move in a small elliptical orbit over the course of a year. (See Figure 5–12b.) The maximum displacement of the star from its true position occurs when the telescope is moving perpendicular to the direction of the starlight. It can be shown that the telescope must be tilted in the direction

figure 5–11 **RAIN FALLING THROUGH A MOVING TUBE**

In (a) the tube is stationary, and raindrops fall through without hitting the sides of the tube. If the tube were moving sideways while being held vertical, the raindrops would collide with the sides of the tube as they fell. In (b) the tube is tilted at an angle so that raindrops entering the top at A and B will fall through the moving tube and leave at A' and B', without hitting the sides of the tube.

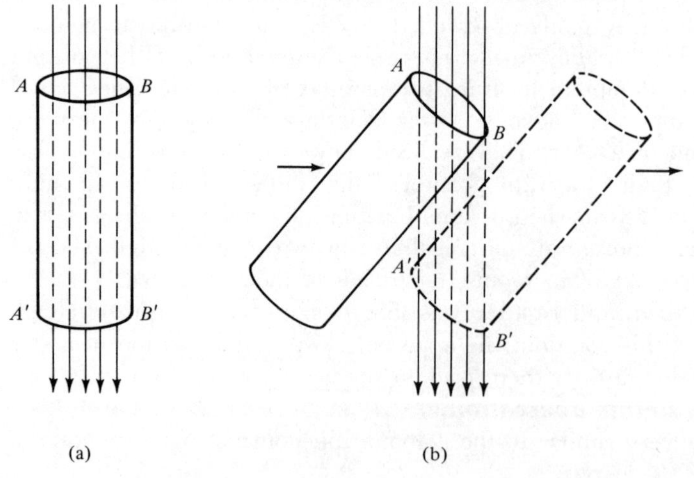

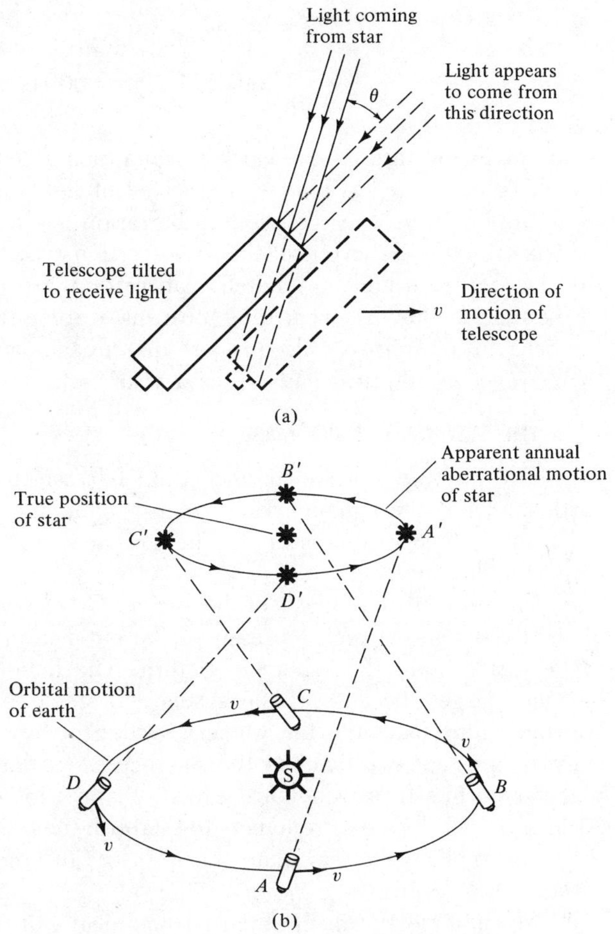

figure 5–12 **STELLAR ABERRATION**

(a) If the telescope is moving relative to the light from a star, the telescope must be tilted in order to receive the light. The angle of aberration θ represents the angle through which the telescope must be tilted from the true direction. (b) When viewed through a telescope attached to the earth, the apparent position of a star will appear to change as a consequence of the changing direction of motion of the earth. Over the course of one year, the star will appear to travel in an orbit.

of motion through an angle θ such that

$$\tan \theta = \frac{v}{c} \tag{5–9}$$

where v is the speed of the telescope, and c is the speed of light. The amount of stellar aberration seen from the earth therefore depends only on the ratio of the orbital speed of the earth to the speed of light and is independent of the distance to the star. The maximum deflection observed is 20.5″. Substituting this value into

Eq. 5–9 gives

$$\tan 20.5'' = 0.0001 = \frac{v}{c}$$

which gives the ratio of the earth's orbital speed to the speed of light. Hence if the earth's orbital speed is known, the speed of light can be calculated and vice versa.

Bradley's discovery of stellar aberration has to be considered extremely strong evidence for the heliocentric theory—for if the earth did not move with respect to the stars, there would be no stellar aberration. If one is to deny the earth's motion in the light of this discovery, one must make some rather peculiar assumptions about the behavior of light—or else propose that all stars are acted on by a force that makes them travel in elliptical paths once a year!

THE SIZE OF THE UNIVERSE

The discovery of stellar aberration left astronomers with little doubt that the earth, and not the sun, moves with respect to the fixed stars. The triumph of the heliocentric theory was assured. The sun now replaced the earth as the stationary center of the universe.

The geocentric universe of the ancient Greeks was a rather small one. The sphere of the fixed stars—which represented the outer limit of the universe—was presumed to lie just beyond the sphere of Saturn. The heliocentric universe was thought to be much larger (because of the absence of stellar parallax), but the astronomers of Newton's time had only the vaguest inkling of how vast the universe really is. We learned in Chapter 2 that the Renaissance astronomers could determine the sizes of planetary orbits in astronomical units—that is, in terms of the sun–earth distance. Although this allowed astronomers to establish the *relative* sizes of the planetary orbits, they were not able to calculate the *absolute* distances in miles until the sun–earth distance was known.

The first measurement of the astronomical unit was made in 1672 by Jean Richer and Giovanni Cassini. Their method was to make two observations, from two widely separated points on the earth, of the position of Mars against the background of the stars (Figure 5–13). From the amount of parallax observed and the known distance in miles between the two observation points on the earth, the distance to Mars in miles could be calculated trigonometrically. The distance to Mars in miles could then be compared with the known distance to Mars in astronomical units at the time of observation, and the length of the astronomical unit in miles could thereby be calculated. Richer and Cassini obtained a value of 86 million miles for the length of the astronomical unit by this method.

By 1900 the accepted value had been established as 93 million miles on the basis of more precise measurements. Further improvements in accuracy were obtained in the 1960s, when the astronomical unit was measured by an entirely new method. A large radio transmitter–receiver was used to bounce a radar beam off the planet Venus. Since the speed of radio waves is accurately known, the travel time from transmission to reception of the echo could be used to compute the distance to Venus in miles. This value could then be compared with the known distance to Venus in astronomical units. With this procedure the uncertainty in the length of the astronomical unit has been reduced to a few hundred miles. The modern-day value of the astronomical unit has been established at

$$1\ \text{AU} = 92{,}956{,}200 \pm 300\ \text{miles} = 149{,}598{,}500 \pm 500\ \text{km}$$

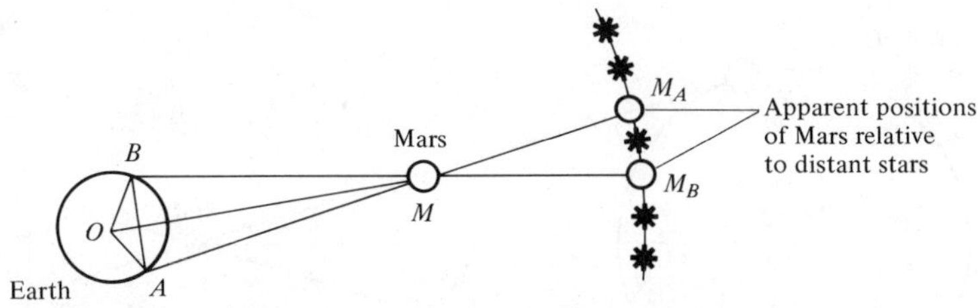

figure 5–13 **A METHOD FOR MEASURING THE LENGTH OF THE ASTRONOMICAL UNIT**

Two observations of the position of a nearby body (Mars in this example) relative to the distant stars are made from two points A and B on the earth. From the observed parallax of Mars, the angles BMO and OMA can be determined. Since the distances OA, OB, and AB are known in miles, the distance OM can be calculated in miles using trigonometry. This value can then be compared with the known value of OM in AU to determine the length of the astronomical unit in miles.

Knowing the length of the astronomical unit, we are now able to calculate the actual distances and sizes of the bodies of the solar system. Mercury, the planet nearest to the sun, has an orbit of average radius 0.387 AU. Thus its average distance from the sun in miles is (0.387 AU)(92,956,200 mi/AU) = 36,000,000 miles. Pluto, the most distant (a = 39.4 AU) of the planets, is on the average 3.66 billion miles from the sun. From the apparent size of a celestial body as it appears in the sky and its known distance, we can calculate the body's actual size. For example, the sun subtends an angle of 32′ (about $\frac{1}{2}°$) in the sky at a distance of approximately 93 million miles. Using simple trigonometry (see Figure 5–14), we find that the sun's actual radius is about 430,000 miles. The sun is therefore more than 100,000 times larger than the earth.

The distances to the nearby stars can be determined through the measurement of stellar parallax. All stars appear to travel in small circular or elliptical orbits over

figure 5–14 **MEASURING THE SIZE OF THE SUN**

(a) The sun subtends an angle of 32′ ($\approx \frac{1}{2}°$) as seen by an observer on the earth. (b) A right triangle is constructed with the sun–earth distance as the hypotenuse. The radius of the sun can be calculated by trigonometry.

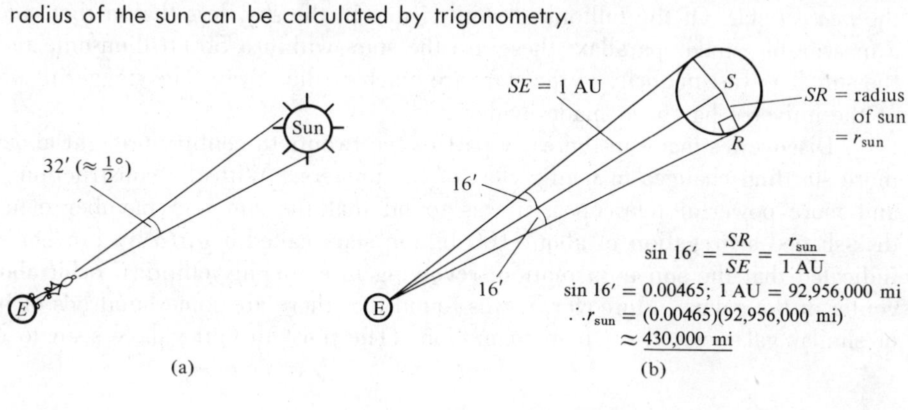

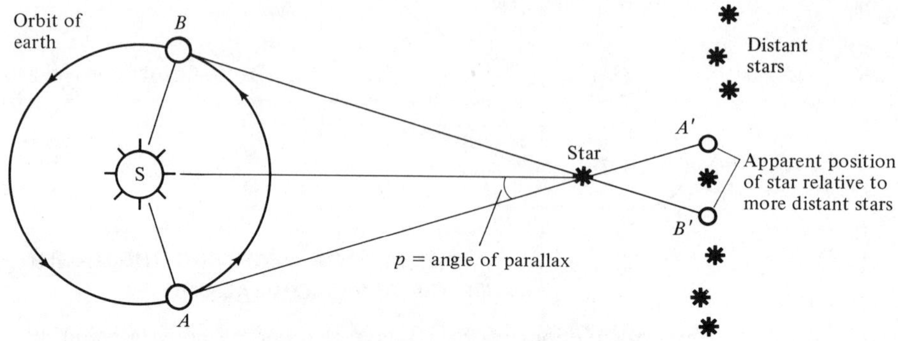

figure 5–15 **MEASUREMENT OF STELLAR DISTANCES BY STELLAR PARALLAX**

As the earth moves on its orbit from A to B, the star will appear to shift position from A' to B' with respect to the background of the more distant stars. From the measured angle of parallax p and the known sun–earth distance, the distance from the sun to the star may be calculated.

the course of one year. For most stars this motion is caused by stellar aberration. However, when the effect of stellar aberration is subtracted off, some stars still exhibit a much smaller elliptical orbit, which is due to stellar parallax. The method for measuring stellar distances from their parallax is shown in Figure 5–15. We see that as the earth moves in its orbit, a relatively nearby star will appear to change its position relative to the more distant "fixed" stars. Obviously, the more distant the star is from the earth, the smaller the angle through which it appears to move. It turns out that the stars are so distant from the earth that the angle of parallax (angle p in Figure 5–15) for the *nearest* star, known as Proxima Centauri, is only 0.75″, or 1/4800 of 1°. The reason it is so difficult to observe stellar parallax now becomes apparent. It takes an extremely careful observer and a rather precise measuring instrument to discern the shift in position of a star through such a small angle!

Once the angle of parallax is known, the distance to the star can be readily determined by trigonometry. For Proxima Centauri the calculated distance is 25,600,000,000,000 (25.6 trillion) miles. When one tries to contemplate such a distance, one can begin to understand why the ancient Greeks or even the Renaissance thinkers were unwilling to accept the heliocentrists' explanation for the absence of stellar parallax. The size of the solar system is negligible by comparison; Proxima Centauri is 7000 times farther from the sun than Pluto. Yet this is the distance to the *nearest* star! Of the billions of stars visible in the sky, less than a thousand have a measurable stellar parallax; these are the stars within a 500-trillion-mile radius of the sun. The vast majority of the stars are much farther away. The staggering vastness of the universe has been made evident.

Discoveries made in the early part of the twentieth century brought about even more startling changes in man's view of the universe. With the construction of new and more powerful telescopes, it was found that the sun is a member of a huge, disk-shaped aggregation of about 100 billion stars called a ***galaxy.*** Further studies indicated that the sun is in motion, revolving in a roughly elliptical orbit about the center of the galaxy. Moreover, it was found that there are some hundreds of billions of similar galaxies, all of them in motion. (The most distant galaxy seen to date is estimated to be about 5×10^{22}, or 50 billion trillion, miles away.) As a consequence, neither the earth, the sun, nor the center of our galaxy (nor of any other galaxy) can be considered certifiably to be stationary and located at the center of the universe; in fact, it may be physically impossible to locate this point.

In the light of these discoveries, we are forced to admit that *the universe is neither heliocentric nor geocentric.* As a consequence, man's view of his own place in the universe has been radically altered. In the geocentric universe man stood at the center; the universe was his home and had been designed for his benefit. Man's place in the universe was downgraded somewhat by the heliocentric theory. Yet although the earth itself was no longer at the exact center, the heliocentric universe was still centered on man, for among all the stars in the heavens, the sun, with its system of planets (including man's home, the earth) occupied the most eminent position. But in the modern-day model of the universe, the sun is a star of no particular distinction and occupies no position of special importance. Nor is the solar system especially unique; it is generally believed that perhaps as many as half the stars in the sky have planetary systems. It is not unreasonable to estimate that there are millions of planets in our galaxy alone capable of supporting life as we know it.

Thus in the space of a few centuries, the cozy little universe of the ancients has given way to a universe of mind-boggling size, in which the sun, the earth, and its inhabitants are of seemingly no significance. To dwell on the philosophical, psychological, and religious implications of this state of affairs is, unfortunately, outside of the scope of this book.

THE LIMITATIONS OF THE NEWTONIAN SYSTEM

In the face of the modern-day picture of the universe, we must look at the Newtonian system of mechanics in a new light. Newton's theories of motion and gravitation were based on the assumption that the universe is heliocentric, but now we realize that *both* the sun and the earth are in motion with respect to the stars. The Newtonian system need not be rejected for this reason, however; it is still valid to refer all motion in the *solar system* to the sun, and then consider separately the motion of the sun with respect to the other stars.

Nevertheless, one may raise the argument that for theoretical purposes it is just as logical to consider any one point as the center of all motion as it is to consider any other point. Thus it is conceivable to assume that the earth is stationary and nonrotating and to devise a mathematical theory that will refer all motion to a stationary earth. However, although such a theory can be constructed in principle, it would be enormously complicated and virtually unusable because of its mathematical complexity. If we set out to devise a theory in which all motion in the solar system is referred to a stationary *sun,* however, we arrive at a much simpler result—namely, Newtonian mechanics. If one assumed the earth, or Mars, or the moon were the center of the system, one would arrive at a different set of formulas for the laws of motion and gravitation. (For example, a theory of motion based on the geocentric system would require the introduction of new forces in addition to gravitation to account for stellar aberration, the flattening of the earth, retrograde motion, etc.)

Thus the present-day attitude is to adopt the heliocentric model of the solar system—and along with it, Newton's theories of motion and gravitation—*not because it is strictly true, but because it is simpler.* We have, in effect, invoked the basic principle (stated in Chapter 1) that *if two theories are equal in all other respects, we adopt the simpler theory.*

Finally, we must respect the fact that the Newtonian theoretical system does have its limitations. In our century the precision of astronomical observations has been greatly increased, and some Newtonian predictions that appeared to be exactly verified by older observations are now in disagreement with newer, more accurate observations. In order to gain theoretical agreement with these observations, physicists have turned to Einstein's theory of relativity, and Newtonian mechanics has been relegated to the status of an approximate theory. Nevertheless, for most applications Newtonian theory is quite adequate; the differences between Newtonian and relativistic predictions are rather minute, and most astronomical observations are not accurate enough to distinguish between the two theories. Furthermore, although relativity is a more accurate theory, it is considerably more complex mathematically. For all practical purposes, then, it is legitimate to use Newtonian mechanics to solve most problems and to turn to relativity theory only in the more extreme cases.

QUESTIONS FOR REVIEW

1 State the principle of universal gravitation. How is this principle distinct from the law that describes the force that moves the planets around the sun?

2 What is the universal constant of gravitation? How is its numerical value determined?

3 What assumption can be made about the gravitational force exerted by a spherical body? Under what conditions will this assumption be *invalid?*

4 Explain why the weight of a body decreases (a) as it moves away from the surface of the earth and (b) as it moves inside the earth toward the earth's center.

5 Give the equation for the acceleration of gravity g at the surface of a spherical body of mass M and radius R. How is this equation derived?

6 How did Newton explain Galileo's law of falling bodies? Explain why this law is only approximately correct.

7 Explain how Newton demonstrated that the earth's gravitational force extends to the moon and obeys an inverse-square law.

8 Show how orbital motion can be explained in terms of the concepts of inertia and gravitation. Identify the force or forces that act on an orbiting body.

9 How did Newton demonstrate that Kepler's third law is only approximately correct? According to Newton, what factors determine the value of the constant K in Kepler's third law?

10 Discuss the difficulties physicists encounter in attempting to explain the cause and nature of gravity. What hypothesis did Newton propose to solve the action-at-a-distance problem?

11 Distinguish between the inertial mass and the gravitational mass of a body. Why do we assume that they are always equal?

12 Discuss the different kinds of orbital paths that a body can have when moving under the influence of a gravitational force. What factors determine the type of orbit that a body will move in?

13 State and explain the equation for the escape speed. What is the significance of this quantity?

14 Explain why the members of a two-body system must both orbit around a point called the center of mass. What factors determine the location of this point?

15 Explain why astronauts in an orbiting spaceship experience "weightlessness." In particular, explain why weightlessness in orbit cannot be due to the absence of gravitational force.

16 Summarize the method by which the existence and location of a new planet can be predicted.

17 How did Newton account for the oblate shape of the earth? What factors determine the oblateness of a rotating body?

18 What is stellar aberration? What was the significance of its discovery?

19 Describe the methods used to determine the length of the astronomical unit.

20 Explain how the diameter of a planet in miles can be calculated.

21 Can we consider the heliocentric theory to be proved? If not, why do we accept it? Explain your answer.

EXERCISES

1 Two bodies a distance D apart exert a mutual gravitational force Q on each other. What is the magnitude of this force when the distance is changed to $2D$? To $4D$? To $D/2$? To $D/5$?

2 Calculate the magnitude of the gravitational force (a) between two 1600-lb bodies placed 10 feet apart and (b) between two 1000-kg bodies placed 10 meters apart.

3 Suppose a hole is drilled through the earth passing through its center and a ball is dropped into the hole. Describe the motion of the ball, assuming friction and air resistance are negligible. In particular, explain why the ball does *not* stop at the center of the earth.

4 Calculate the acceleration of gravity at an altitude of (a) 400 miles above the surface of the earth, (b) 4000 miles above the surface of the earth, and (c) 40,000 miles above the surface of the earth.

5 A body weighs 320 lb on the surface of the earth. To what altitude must the body be raised in order for its weight (i.e., the gravitational force acting on it) to be 20 lb?

6 A hypothetical planet has a mass three times that of the earth, and its radius is twice as large.
(a) What is the acceleration of gravity at the surface of this planet?
(b) If a person weighs 160 lb on earth, what will he weigh on this planet?

7 The mass of Jupiter is 318 times the mass of the earth, and Jupiter's radius is 11.2 times that of the earth. What is g at the surface of Jupiter?

8 An astronaut landing on an unknown planet finds that the acceleration of gravity is $8\ \text{ft/sec}^2$ at the surface. The mass of the planet is nine times that of the earth. What is the radius of the planet?

9 An astronaut who weighs 150 lb on earth finds that he weighs 200 lb on the planet Vulcan. The radius of Vulcan is 1000 miles. How does the mass of Vulcan compare with the mass of the earth?

10 A spaceship is about to land on the planet Krypton. At an altitude of 9000 miles above the surface, the acceleration of gravity is $3\ \text{ft/sec}^2$. At an altitude of 3000 miles above the surface, $g = 12\ \text{ft/sec}^2$. From these data determine (a) the radius of Krypton and (b) the acceleration of gravity at the surface.

11 Suppose that the radius of the moon's orbit were 120,000 miles instead of 240,000 miles. What would the centripetal acceleration of the moon be?

12 What information would be needed in order to calculate the mass of Jupiter?

13 Suggest methods that could be used to determine (a) the mass of a planet that has no satellites and (b) the mass of a star.

14 Criticize and correct the following statement: The fact that all bodies fall to the earth with the same acceleration implies that the earth exerts the same force on all bodies.

15 Does the statement "Bodies fall because of gravity" really explain why bodies fall? Explain your answer.

16 Derive the equation $v_c = \sqrt{GM/R}$ for a body in a circular orbit. (Hint: set the centripetal force on a body in an orbit of radius R equal to the gravitational force on that body.)

17 Calculate the escape speed for a body (a) at the surface of the moon ($M_M = M_E/81.3$; $R_M = R_E/3.67$), (b) at the surface of Jupiter ($M_J = 318\ M_E$; $R_J = 11.2\ R_E$), and (c) at an altitude of 8000 miles above the surface of the earth.

18 In a double-star system one star has a mass five times greater than the other. The two stars are 24 AU apart. Where is the center of mass of the system located?

19 By setting the centripetal acceleration at the equator equal to the gravitational acceleration, calculate the rotational speed at which the earth would become unstable at the equator. What would be the length of the day (the period of rotation of the earth) at this speed?

20 Can the measurement of stellar aberration be used to determine the distance to a star? Explain your answer.

21 (a) Using the known values of the average radius and sidereal period of the earth's orbit, calculate the average orbital speed of the earth in miles per second.
(b) Using the answer to (a) and the stellar aberration data given in the chapter, calculate the speed of light in miles per second.

22 Explain why the angle of parallax of a star decreases as the distance to the star increases.

MOMENTUM AND ENERGY

6

I do not know what I may appear to the world; but to myself I seem to have been only like a boy, playing on the sea shore, and diverting myself, in now and then finding a smoother pebble or a prettier shell than ordinary, whilst the great ocean of truth lay all undiscovered before me.

Isaac Newton

THE ENORMOUS SUCCESS of the Newtonian synthesis had much the same effect on human thought as did the rediscovery of the works of Aristotle in the medieval era. The famous couplet written by Alexander Pope

Nature and Nature's laws lay hid in night,
God said: "Let Newton be!", and all was light.

reflected the commonly held opinion that Sir Isaac Newton had been chosen as the one man to whom the secrets of the universe were truly and finally revealed. Scientists adopted the basic principle that all physical phenomena were explainable in terms of Newtonian mechanics; indeed, this principle was applied for nearly two hundred years with great success and without serious conflicts.

The Newtonian view was extended even beyond science into the realm of political thought. By analogy to the Newtonian picture of the universe as a system of distinct particles that move according to certain physical laws, political theorists sought to view human society as a group of distinct individuals acting in accordance with certain political and economic laws. Thus such concepts as the law of supply and demand or Gresham's law came to be looked on by some thinkers as "natural laws," which governed human society in the same way that Newton's laws governed the motions of bodies.

During the eighteenth and nineteenth centuries, physics was concerned mainly with the assimilation of the results of the *Principia* and the application of the Newtonian method in explaining additional natural phenomena. The outstanding mathematicians and physicists of this era devoted their energies to developing new mathematical techniques with which more complex problems could be tackled.

Theoreticians attempted to discover more fundamental principles of mechanics from which Newton's principles themselves could be derived. Of particular value was the development of the concept of a ***conservation principle.*** In general, a conservation principle states that under certain specified conditions a given physical property will "be conserved" (will remain constant) during an interaction or motion of some kind. An alternative way of expressing a conservation principle is to state that the specific property (mass, for example) "can be neither created nor destroyed." This is perhaps a more dramatic way of signifying the constancy of the given property.

The use of conservation principles has won great favor among physicists; in fact, much of present-day physical research is aimed at finding additional physical quantities that are conserved (particularly in nuclear and subnuclear reactions). Moreover, conservation principles—because of their simplicity and universality—have great aesthetic appeal for physicists, and it is held that conservation principles are the most fundamental of physical principles. The fact that certain conservation principles were derived from Newton's three principles of motion is an historical accident, it is argued; actually, Newton's principles should be looked on as consequences of the conservation principles, and not the other way around. In any case, the conservation principles of mechanics are not new and independent physical principles but simply Newton's principles cast in a different form. From a practical point of view, therefore, it is immaterial whether one considers the conservation principles or Newton's principles as being more fundamental, since they lead to the same conclusions.

This chapter will deal with three physical properties that obey conservation principles: ***linear momentum, angular momentum,*** and ***energy.*** We will find that these conservation principles often provide a simple and extremely useful means of describing and predicting natural phenomena.

6–1

LINEAR MOMENTUM

In the *Principia,* Newton defined the "quantity of motion" of a body as the product of its mass and its velocity. Today we refer to this quantity as the ***linear momentum,* p:**

$$\mathbf{p} = m\mathbf{v} \tag{6–1}$$

Linear momentum is a vector quantity; its direction is the same as that of the velocity vector. The units of linear momentum in the CGS, MKS, and British systems are g-cm/sec, kg-m/sec, and slug-ft/sec, respectively.

The linear momentum of a body is, in a certain respect, a better measure of the body's motion than just the mass or the velocity alone. Let us compare a high-speed bullet with an automobile moving at 5 mi/hr. The bullet has a small mass and a high velocity, whereas the automobile has a large mass and a small velocity. Yet a great deal of force is required to stop each of them. Even though the masses and velocities are quite dissimilar, the *product* of mass and velocity—that is, the linear momentum—is large in both cases.

The relationship between force and momentum can be demonstrated using Newton's equation of motion:

$$\mathbf{F} = m\mathbf{a} = m\frac{(\mathbf{v}_\mathrm{f} - \mathbf{v}_\mathrm{i})}{\Delta t} = \frac{m\mathbf{v}_\mathrm{f} - m\mathbf{v}_\mathrm{i}}{\Delta t}$$

According to Eq. 6–1, $m\mathbf{v}_\mathrm{f} = \mathbf{p}_\mathrm{f}$ = the final momentum, and $m\mathbf{v}_\mathrm{i} = \mathbf{p}_\mathrm{i}$ = the initial momentum. Thus we have

$$\mathbf{F} = \frac{\mathbf{p}_\mathrm{f} - \mathbf{p}_\mathrm{i}}{\Delta t} = \frac{\Delta\mathbf{p}}{\Delta t} = \frac{\text{change in linear momentum}}{\text{change in time}} \tag{6–2}$$

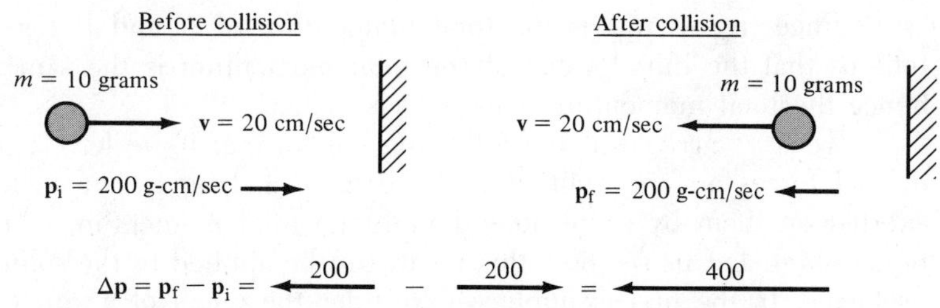

figure 6–1 **CHANGE IN LINEAR MOMENTUM**

The ball rebounds from the wall without a change in speed, but the linear momentum changes from 200 pointing east to 200 pointing west. The change in the linear momentum $\Delta\mathbf{p}$ is therefore 400 west.

This equation tells us that the effect of an applied force is to change the linear momentum of a body. More precisely, according to Eq. 6–2, *the force is equal to the rate of change of linear momentum.* (This is, in fact, the way that Newton originally stated his second principle in the *Principia.*) We see that if the net force is 0, the linear momentum of a body remains constant.

Because linear momentum is a vector, a change in the linear momentum means a change in either its magnitude or its direction. Suppose, for example, a ball with a mass of 10 grams is thrown at a wall with a speed of 20 cm/sec. It hits the wall and rebounds with the same speed. Before and after the ball hits the wall, the magnitude of the linear momentum is 200 g-cm/sec, but there has been a change in direction. Thus the *change* in the linear momentum is 400 g-cm/sec. (See Figure 6–1.) This change has been brought about by the force that the wall exerted on the ball.

CONSERVATION OF LINEAR MOMENTUM

Thus far we have seen that if a single body has no net force acting on it, its linear momentum is constant; that is, its linear momentum is conserved. Now suppose we have *two* bodies, A and B, exerting mutual forces on one another. Body A exerts a force $\mathbf{F}_{AB}$ on B, and this force causes a change in the linear momentum of B. At the same time B exerts a force $\mathbf{F}_{BA}$ on A, causing a change in the linear momentum of A. Since $\mathbf{F}_{AB}$ and $\mathbf{F}_{BA}$ are (by the third principle) equal and opposite, the changes in the linear momenta of A and B will also be equal and opposite. That is, if A's momentum increases by a certain amount, B's momentum will *decrease* by the same amount. It follows that the *total* linear momentum of the two bodies will remain constant.

Let us derive this result mathematically. According to Newton's third principle,

$$\mathbf{F}_{AB} = -\mathbf{F}_{BA}$$

Substituting from Eq. 6–2 gives

$$\frac{\mathbf{p}_{B_f} - \mathbf{p}_{B_i}}{\Delta t} = -\frac{\mathbf{p}_{A_f} - \mathbf{p}_{A_i}}{\Delta t}$$

Canceling and rearranging terms, we have

$$(\mathbf{p}_A + \mathbf{p}_B)_i = (\mathbf{p}_A + \mathbf{p}_B)_f$$

Since $(\mathbf{p}_A + \mathbf{p}_B)$ is the total momentum of A and B together, this equation tells us that the initial value of the total momentum is the same as the final value; hence the total momentum does not change.

We have arrived at the following conclusion: if we have two bodies that exert mutual forces on one another and are not acted on by any external forces (forces exerted on them by some other body), the total momentum of the two bodies will be constant. Let us see how this result can be applied to the solution of some simple problems. In the first example we consider the recoil of a gun when it is fired.

EXAMPLE 6–1: A 0.01-kg bullet is fired from a gun with a speed of 600 m/sec. The gun has a mass of 5 kg. With what speed will the gun recoil?

SOLUTION: Before the gun is fired, both the bullet and gun are at rest. Therefore each has a velocity and momentum of 0. The total linear momentum of the gun and bullet combined will be 0. But the total linear momentum is a constant quantity, so it should be 0 *after* the gun is fired. It may seem paradoxical for the total linear momentum to be 0 even though the gun and bullet are both in motion, but we must remember that linear momentum is a *vector* quantity. Thus if the gun and bullet go off in *opposite* directions with linear momenta of the same magnitude, the vector sum of their linear momenta will be 0. Thus

$$(m\mathbf{v})_{\text{gun}} + (m\mathbf{v})_{\text{bullet}} = 0$$

Equating the *magnitudes* of the two linear momenta and substituting, we obtain

$$(5\text{ kg})v_{\text{gun}} = (0.01\text{ kg})(600\text{ m/sec}) = 6\text{ kg-m/sec}$$
$$v_{\text{gun}} = 6/5 = \underline{1.2\text{ m/sec}}$$

This gives the magnitude of the recoil velocity of the gun; the gun's direction will be the opposite of that of the bullet.

The next example involves a collision between two bodies. In the collision the bodies exert equal and opposite forces on each other, so that their total linear momentum is the same before and after the collision. Once again, it is important to keep in mind the vector nature of linear momentum.

EXAMPLE 6–2: A metal ball of mass 10 kg is traveling to the right with a speed of 20 m/sec. It collides with a second ball of mass 5 kg, traveling to the left with a speed of 30 m/sec. As a result of the collision, the first ball rebounds to the left with a speed of 12 m/sec. (See Figure 6–2.) What will the velocity of the second ball be after the collision?

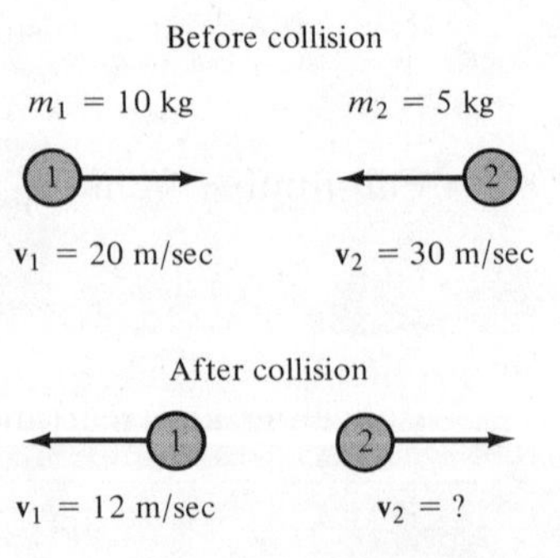

figure 6–2

SOLUTION: We calculate the total linear momentum of the two balls before and after the collision, then equate the two results. To simplify the vector arithmetic, let us assume that a body moving to the right has a positive momentum and a body moving to the left has a negative momentum.

Before the collision:

$$\begin{array}{ll} \text{Ball 1: } \mathbf{p}_i = m_1\mathbf{v}_1 = (10\text{ kg})(20\text{ m/sec} \longrightarrow) = 200\text{ kg-m/sec} \longrightarrow & = +200 \\ \text{Ball 2: } \mathbf{p}_i = m_2\mathbf{v}_2 = (5\text{ kg})(30\text{ m/sec} \longleftarrow) = 150\text{ kg-m/sec} \longleftarrow & = -150 \\ \hline \text{Total linear momentum} = (\mathbf{p}_{tot})_i = \quad 50\text{ kg-m/sec} \longrightarrow & = +\ 50 \end{array}$$

Since the two balls are moving in opposite directions, their total momentum will be the *difference* between the individual values.

After the collision:

$$\begin{array}{ll} \text{Ball 1: } \mathbf{p}_f = m_1\mathbf{v}_1 = (10\text{ kg})(12\text{ m/sec} \longleftarrow) = 120\text{ kg-m/sec} \longleftarrow & = -120 \\ \text{Ball 2: } \mathbf{p}_f = m_2\mathbf{v}_2 = (5\text{ kg})(v_2\text{ m/sec} \longrightarrow) & = +\ 5v_2 \\ \hline \text{Total linear momentum} = (\mathbf{p}_{tot})_f = & -120 + 5v_2 \end{array}$$

The velocity $\mathbf{v}_2$ is, of course, the unknown quantity in this problem. The direction of $\mathbf{v}_2$ is unknown, but we will assume that it is moving to the right. (If v_2 turns out to be negative, then our guess is wrong, and the ball is moving to the left.) To find v_2, we make use of the principle that the total linear momentum will be the same before and after the collision. Thus

$$\begin{aligned} (\mathbf{p}_{tot})_i &= (\mathbf{p}_{tot})_f \\ +50 &= -120 + 5v_2 \\ v_2 &= 170/5 = \underline{34\text{ m/sec}} \end{aligned}$$

Since v_2 turns out to be a positive number, it follows that the second ball rebounds to the right after the collision with a speed of 34 m/sec. Hence

$$\underline{\mathbf{v}_2 = 34\text{ m/sec} \longrightarrow}$$

These two examples demonstrate that the momentum concept can be useful in solving certain types of problems. But the major significance of this concept becomes most evident when we extend it from two interacting bodies to a large number of bodies. Let us consider a system of an arbitrary number of bodies, as illustrated in Figure 6–3. We divide the forces acting on these bodies into two groups: the ***external forces*** (those exerted by bodies outside the system) and the ***internal forces*** (those exerted by bodies within the system). It follows from what was shown previously that the internal forces cannot change the total momentum of the system, because every change in the momentum of a body within the system caused by an internal force will be matched by an equal and opposite change in the momentum of some other body within the system (the body on which the reaction force acts). An external force will change the total momentum, however, since the reaction force acts on a body that is outside the system. Thus we have the ***principle of conservation of linear momentum:***

In the absence of external forces, the total linear momentum of a system of bodies remains constant.

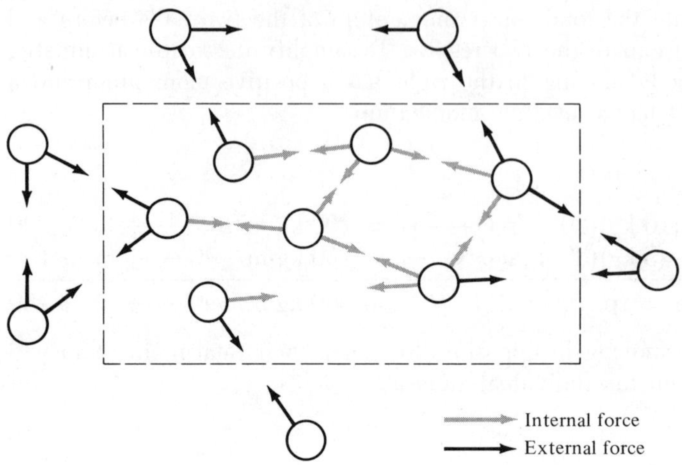

figure 6–3 **INTERNAL AND EXTERNAL FORCES**

The system of bodies under consideration is enclosed by the dashed lines. Internal forces are those exerted between bodies within the system; they do not change the total linear momentum of the system. External forces do change the total linear momentum of the system.

This is a rather significant and far-reaching principle. Suppose we consider the entire universe as a system of bodies on which no external forces are acting. By the very definition of the universe, there are no objects external to it; hence there can be no external forces. We therefore conclude that *the total linear momentum of the universe is constant!* Linear momentum can be exchanged between bodies within the universe, but it cannot be created or destroyed and cannot be transferred out of the universe. The total linear momentum of the universe is the same now as it was billions of years ago and as it will be billions of years in the future.

The importance of the conservation of linear momentum is underscored by the fact that this principle has been shown experimentally to hold true on an atomic and nuclear level as well as on an astronomical scale. It is one of the most firmly rooted principles of physics.

6–2

ANGULAR MOMENTUM

The principle of conservation of linear momentum applies specifically to bodies in straight-line motion. It is not entirely surprising to find that a quantity such as linear momentum is conserved in a body moving with constant velocity in the absence of external forces. A more interesting result is that there is another quantity that is conserved under certain circumstances in *rotational* motion, even though a continually applied force is required for this type of motion. This quantity is called ***angular momentum.*** In this section we will see that the principle of conservation of angular momentum, which follows as a direct consequence of Newton's second principle of motion, can be an extremely effective tool for analyzing certain seemingly complex physical phenomena.

TORQUE

Up to this point we have dealt mainly with the effects of a force when the force is used to push or pull a body along a straight line. Now let us consider what happens when a force is used to twist or rotate a body. We apply forces in this manner in a variety of everyday situations—to open or close a bottle cap, wind a watch, turn a doorknob, change a light bulb, or tighten a screw or nut. Often we find that our hands and wrists are not strong enough to produce the desired effect, so we make use of a tool such as a wrench. We find that the effectiveness of a wrench depends not only on the force applied, but also on the *length* of the wrench; the longer the wrench, the easier it is to tighten or untighten something.

We can translate these experiences into useful mathematical and physical concepts. The twisting effect of a force depends on (1) the magnitude of the applied force, (2) the distance from the point of application of the force to the axis of rotation (i.e., the length of the wrench), and (3) the angle at which the force is applied—obviously, if the force is applied *along* the wrench, it will not produce any rotation; the best effect is obtained when the force is applied *perpendicular* to the arm of the wrench (Figure 6–4a).

Taking all of these factors into account, physicists have defined a quantity called

figure 6–4 **DEFINITION OF TORQUE**

(a) The torque on a nut is produced by using a wrench to rotate the nut. (b) The torque is the product of the perpendicular component of the applied force and the length of the wrench.

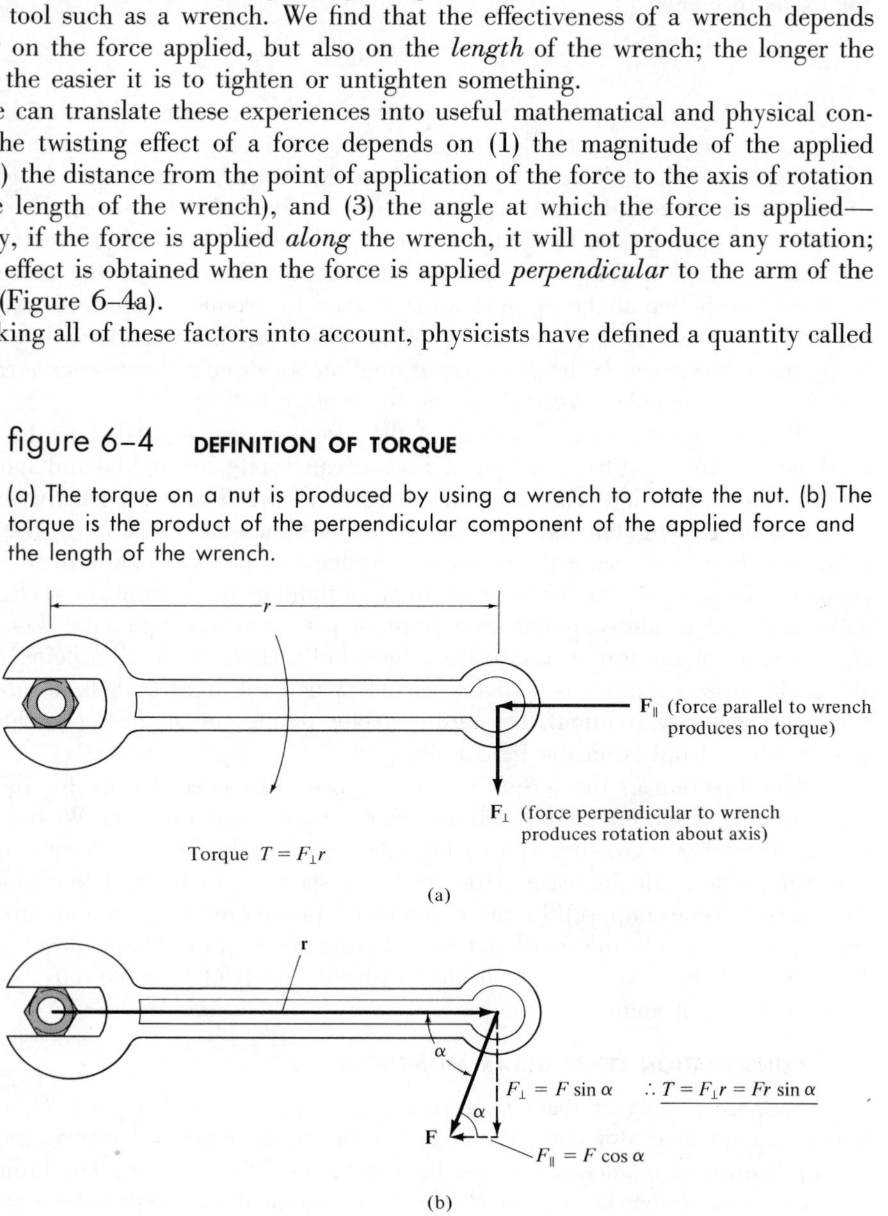

torque, which measures the twisting effect of a force. Let r be the distance from the axis of rotation to the point of application of the force, and $F_{\perp}$ the component of the force perpendicular to r. The magnitude of the torque, T, can be defined as the product of these two quantities:

$$T = F_{\perp} r \tag{6–3a}$$

This equation can be modified if we take the vector nature of the quantities involved into account. Let α be the angle between the **F** and **r** vectors (Figure 6–4b). Then the component of **F** perpendicular to **r** will be $F_{\perp} = F \sin \alpha$. Hence the equation for torque becomes

$$T = Fr \sin \alpha \tag{6–3b}$$

The effect of an applied torque depends not only on the magnitude of the torque, but also on its direction. Hence torque must be defined as a *vector* quantity. It has been found useful to define the direction of the torque as pointing *along* the axis of rotation, perpendicular to the plane formed by the **r** and **F** vectors. For example, if a wrench is being used to tighten a nut, the direction of the torque vector is along the axis of the bolt (Figure 6–5a). Specifically, the direction of the torque vector is defined such that if the applied force is seen to produce a *clockwise* rotation of the body on which it acts, then the torque vector points along the axis of rotation *away* from the viewer. If the direction of rotation is viewed as being *counterclockwise,* the torque vector points *toward* the viewer (Figure 6–5b).

This definition of the direction of the torque vector is related to the fact that most people are right-handed, and a person can rotate his right hand more easily in a clockwise direction. Thus nuts and bolts are threaded so that a clockwise rotation of a wrench will tighten the nut; screws are designed so that a clockwise rotation of a screwdriver will cause the screw to advance into the wood (Figure 6–5c). Light bulbs are designed so that a clockwise rotation tightens the bulb in the socket. Hence a clockwise torque always points away from the person who is applying it. For instance, the direction of the torque vector on a light bulb when the bulb is being tightened (by a clockwise rotation) is *into* the socket. When the light bulb is unscrewed (by a counterclockwise rotation), the torque vector points out of the socket toward the person whose hand is on the light bulb.

The direction of the torque vector has been discussed mainly for the sake of completeness; it will not be of much importance to us in what follows. We have learned that a torque has a twisting or rotating effect on a body; hence a torque applied to a rotating object will increase or decrease the speed of rotation of that object. (See Figure 6–6 for examples.) In more advanced physics texts, equations are derived relating the torque to the rotational velocity and acceleration that a torque produces. For our purposes, however, we can deal with the effects of a torque only as it relates to the concept of angular momentum.

CONSERVATION OF ANGULAR MOMENTUM

Suppose a body of mass m is rotating on a path about a point O, and when it is at a point P on this path, the body has an instantaneous velocity **v.** (See Figure 6–7a.) The linear momentum of the body at point P is $\mathbf{p} = m\mathbf{v}$. The radius vector **r** is the vector drawn from O to P. The linear momentum **p** will have a component

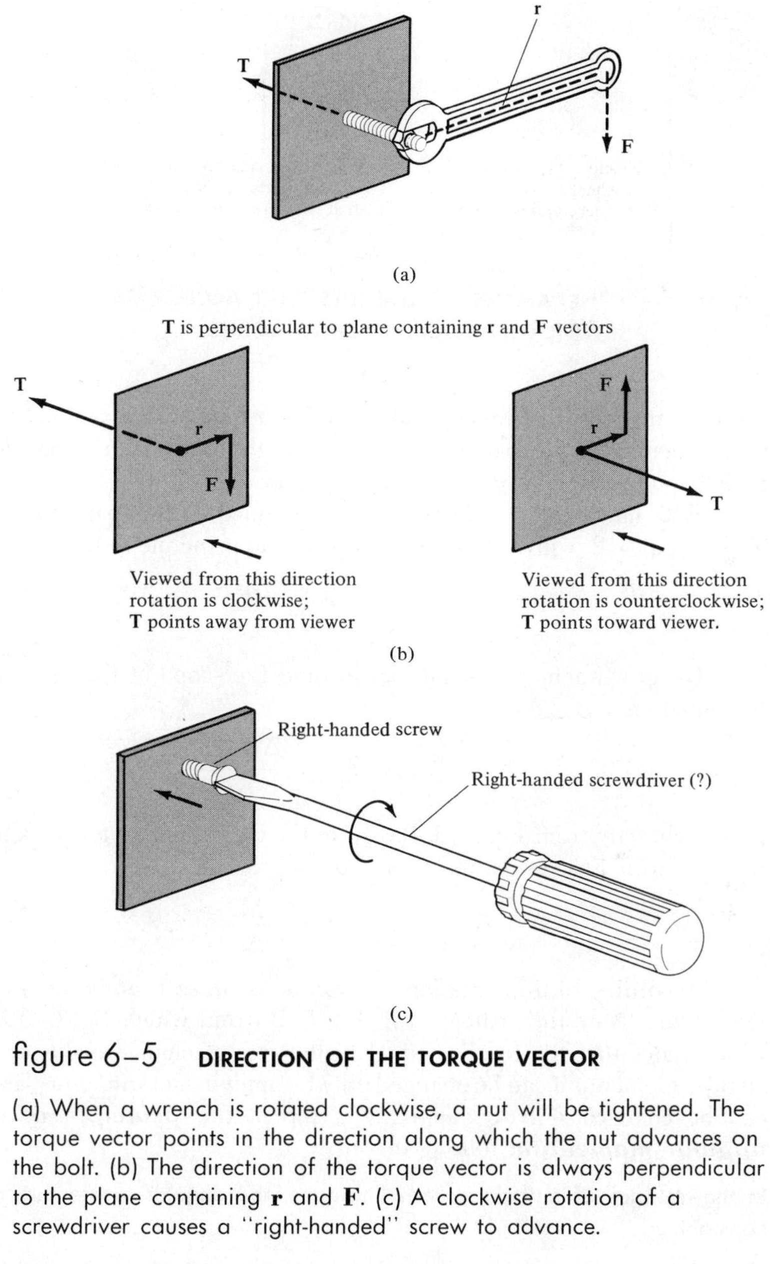

figure 6–5 **DIRECTION OF THE TORQUE VECTOR**

(a) When a wrench is rotated clockwise, a nut will be tightened. The torque vector points in the direction along which the nut advances on the bolt. (b) The direction of the torque vector is always perpendicular to the plane containing **r** and **F**. (c) A clockwise rotation of a screwdriver causes a "right-handed" screw to advance.

$\mathbf{p}_\perp$ that is perpendicular to **r.** Let us define the magnitude of the ***angular momentum*** **L** relative to point O by the equation:

$$L = p_\perp r = mv_\perp r = mvr \sin\theta \qquad (6\text{–}4)$$

where θ is the angle between the **p** and **r** vectors.

Note that the angular momentum has been defined as the product of **r** and **p** in the same way that the torque has been defined as the product of **r** and **F.** In

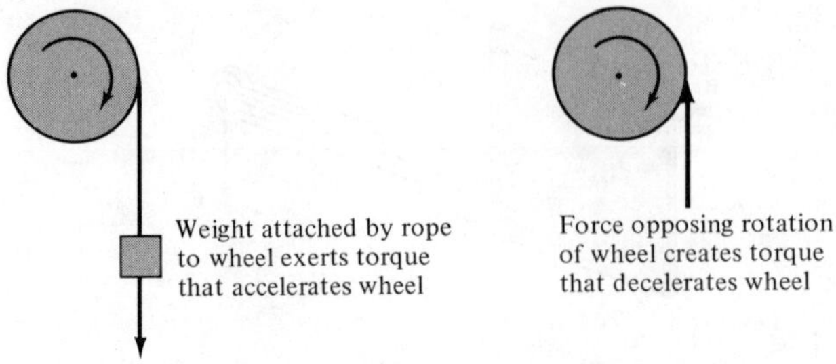

figure 6–6 **EXAMPLES OF TORQUES THAT ACCELERATE OR DECELERATE ROTATING BODIES**

the same manner, the direction of the angular momentum vector $\mathbf{L}$ is defined as being perpendicular to the plane of $\mathbf{r}$ and $\mathbf{p}$; if the body is moving clockwise about O, then $\mathbf{L}$ points away from the viewer (Figure 6–7b).

Why has angular momentum been defined in this particular way? Let us refer back to Eq. 6–2, which states Newton's second principle in terms of linear momentum:

$$\mathbf{F} = \frac{\Delta \mathbf{p}}{\Delta t}$$

Using mathematics somewhat beyond the scope of this text, one can show that the equation

$$F_{\perp} r = \frac{\Delta(p_{\perp} r)}{\Delta t} \tag{6–5}$$

follows directly from Eq. 6–2. But from the definitions of torque and angular momentum given in Eqs. 6–3a and 6–4, Eq. 6–5 becomes

$$\mathbf{T} = \frac{\Delta \mathbf{L}}{\Delta t} \tag{6–6}$$

According to this equation, *the torque is equal to the rate of change of angular momentum.* Note the analogy with Eq. 6–2 (from which Eq. 6–6 has been derived), which states that the force is equal to the rate of change of linear momentum. Thus angular momentum can be changed only by applying a torque, just as linear momentum can be changed only by applying a force. The ***principle of conservation of angular momentum*** follows directly:

In the absence of any net external torques, the angular momentum of a body remains constant.

Thus as a result of some seemingly mysterious mathematical definitions and manipulations, we have found another conservation principle hidden in Newton's principles.

To familiarize ourselves with the concept of angular momentum, let us take the example of a body moving in a circle with constant speed, and calculate $\mathbf{L}$ with respect to the center of the circle. In this case $\mathbf{r}$ corresponds to the radius of the circle; also for uniform circular motion $\mathbf{v}$ is always tangent to the circle and perpendicular to $\mathbf{r}$. Therefore $\theta = 90°$; $\sin \theta = 1$; and

$$L = mvr \sin \theta = mvr \tag{6–7}$$

Since m, v, and r are constant, **L** will be constant in magnitude. The direction of **L** is perpendicular to the plane of the circle and is also constant. Thus in the case of uniform circular motion, the angular momentum relative to the center remains constant, even though circular motion is accelerated motion requiring a continually applied force. However, the existence of a force does not necessarily imply the existence of a torque, for if the **F** and **r** vectors lie along the same line, the torque is 0. (If $\alpha = 0°$, $\sin \alpha = 0$; then $Fr \sin \alpha = 0$). The accelerating force for uniform circular motion is the centripetal force, which always points toward the center of the circle. Since the centripetal force acts along the radius vector, it produces no torque, so the angular momentum remains constant. The name ***radial force*** is usually applied to any force collinear with the radius vector. In general, a radial force will not change the angular momentum of a body with respect to the center of rotation.

Suppose a solid body is set in rotation about its center. The torques produced by internal forces (i.e., forces between the individual particles of the body) will cancel out because of the action–reaction principle; consequently, in the absence of any net external torques, the angular momentum of the body with respect to its center remains constant in magnitude and direction. If there is friction, the friction force will act in a direction opposite to the rotation and will create a torque, causing the angular

figure 6–7 **DEFINITION OF ANGULAR MOMENTUM**

(a) The angular momentum $\mathbf{L} = \mathbf{p}_\perp \mathbf{r}$, where $\mathbf{p}_\perp$ is the component of the linear momentum that is perpendicular to the radius vector **r**. (b) The direction of the angular momentum vector **L** is always perpendicular to the plane of **r** and **v**.

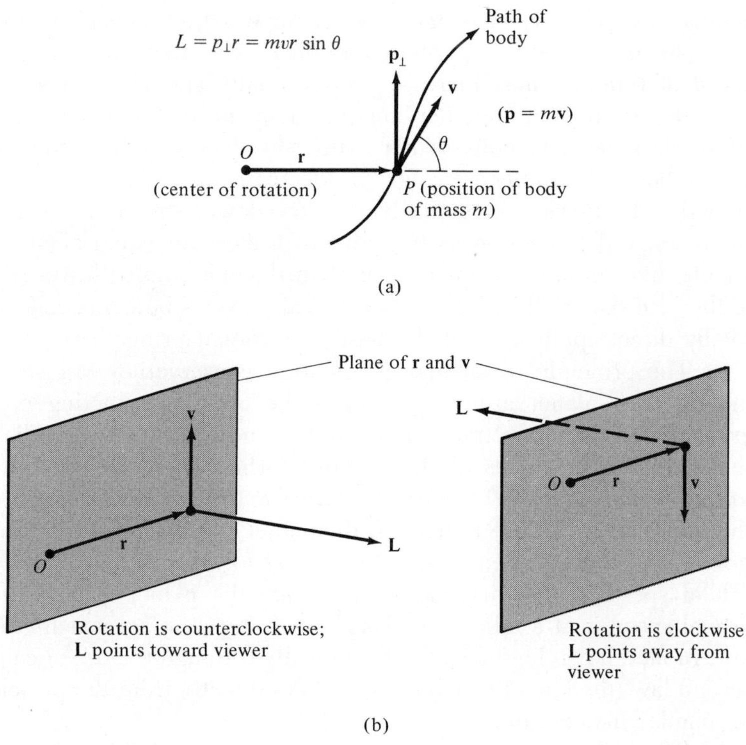

momentum (and thereby the speed of rotation) to decrease. In the absence of friction forces, however, we would expect a wheel, once set into motion, to continue to rotate indefinitely at a constant speed. Thus the wheel appears to have some sort of "rotational inertia"; that is, it remains in constant rotation unless acted on by a torque. Yet no part of the wheel is moving in a straight line, and circular motion, as we know, is accelerated motion. How can accelerated motion be continued indefinitely? To say that the wheel continues in constant rotation because its angular momentum is conserved is perhaps to beg the question; it seems to conjure up a picture of angular momentum as some metaphysical entity that rotating bodies possess in fixed amounts.

In fact, rotational motion can be explained *without* resorting to the concept of angular momentum. Each portion of the wheel has an inertial tendency to fly off in a straight line, but is prevented from doing so by the cohesive forces that maintain the wheel as a solid body. Thus "rotational inertia" arises as a balance between straight-line inertia and the internal forces that preserve the shape of the rotating body. The concept of angular momentum provides a mathematical and verbal shorthand description of this balance and allows us to explain rotational phenomena in much briefer terms. We should keep in mind, however, that the principle of conservation of angular momentum arises from an artful mathematical manipulation of Newton's principles of motion.

Consider the following common situations: an ice skater goes into a spin and, by pulling her arms closer to her body, makes herself spin faster with seemingly little effort. High divers or gymnasts turning somersaults increase their rate of spin by tucking in their arms and legs. These phenomena can be understood rather simply in terms of the concept of angular momentum. If a body is rotating such that its angular momentum is conserved, the quantity $L = mvr$ must remain constant in magnitude. Consequently, if any one of the three factors—mass, speed, or radius—changes in any way, there must be a compensating change in the other factors. For example, if the radius of rotation is decreased, then (assuming m is constant) the speed of rotation must increase proportionally and vice versa. Once the ice skater has set herself spinning, her angular momentum is constant (assuming negligible friction). When she pulls in her arms, she decreases the radius of rotation of part of her body. To compensate for the decrease in r, v must increase and her rate of spin thereby increases. Similarly, to slow down, she simply extends her arms; as r increases, v decreases so as to maintain a constant value of mvr. The power of the angular momentum concept as an analytical tool is amply illustrated by this explanation of the "ice skater effect," for it is not at all obvious how this effect could be accounted for by direct application of the basic Newtonian principles.

The principle of angular momentum conservation can also be applied to the motion of a planet around the sun. The only force acting on the planet is the gravitational force of attraction of the sun, and this force is a radial force with respect to the sun at the focus of the elliptical orbit. Therefore we can conclude that *the angular momentum of a planet relative to the sun is constant.* It follows that as the sun–planet distance r changes, the planet's orbital speed v must change in inverse proportion. Hence when a planet is closest to the sun (r has its minimum value) the orbital speed is at a maximum, and when the planet is farthest from the sun the orbital speed is at a minimum. This conclusion is in agreement with Kepler's second law. In fact, it can be shown mathematically (through the use of calculus) that Kepler's second law (the law of equal areas) follows directly from the principle of conservation of angular momentum.

The conservation of angular momentum is as solidly established and fundamental as the conservation of linear momentum. It is an extremely useful principle in all branches of physics, and we shall invoke it many times in the remainder of the text.

6–3

WORK AND ENERGY

We are about to consider yet another transformation of Newton's laws that leads to the definition of one of the most important physical quantities: ***energy.*** The usefulness of the concept of energy and its associated conservation law is underscored by the extent to which this term has been taken over in everyday language. We speak of being "full of energy"; certain foods are described as "energy foods"; we may characterize someone as having an "energetic personality." In common usage, energy is thought of as a kind of driving power that somehow controls our level of activity. In the following sections we shall study the physicist's definition of energy, and we will attempt to understand why the term has gained such wide acceptance both within and outside of science.

WORK

We begin by defining a quantity called ***work,*** which measures the motion-producing effects of a force. Work represents a combination of the strength of a particular force and the motion produced by that force. Specifically, work is defined as the product of the magnitude of the force on an object and the distance through which the object moves as a result of the force.

Figure 6–8 shows a body being acted on by a constant force **F** as the body moves to the right through a displacement **d.** Let us decompose **F** into two components, $F_\parallel$ and $F_\perp$, which are, respectively, parallel and perpendicular to the motion (i.e., to the displacement vector **d**). The perpendicular component is unrelated to the motion; acceleration and deceleration in the direction of the motion will be produced by the parallel component only. Thus the work depends on the product of the parallel component $F_\parallel$ and the distance d. The work W is

$$W = F_\parallel d = (F \cos \theta)d = Fd \cos \theta \tag{6–8}$$

Work is defined as a scalar quantity; there is no direction attached to it. Work is measured in units of force × distance. In the three systems of units the standard work units are:

British: ***ft-lb***
CGS: dyne-cm = $\text{g-cm}^2/\text{sec}^2$ = ***erg***
MKS: newton-m = $\text{kg-m}^2/\text{sec}^2$ = ***joule***

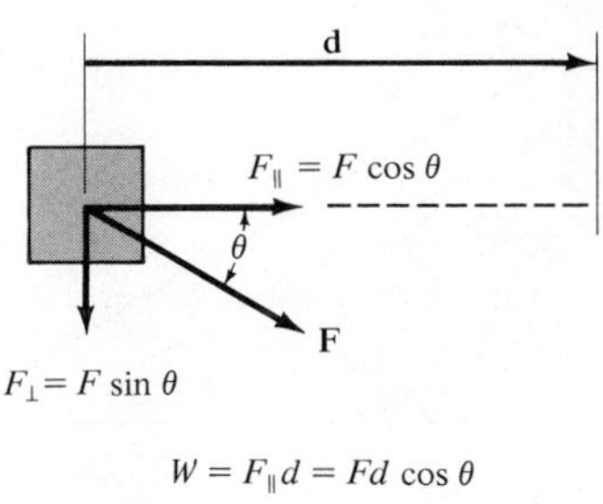

figure 6–8

DEFINITION OF WORK

The body has a displacement **d** under the influence of an applied force **F**. The angle θ is the angle between the **F** and **d** vectors. The work W done by the force is defined as the product of the displacement and the component of the force parallel to the displacement.

Using the standard conversion factors, one can calculate that 1 ft-lb = 1.356 joules and 1 joule = 10^7 ergs.

Remember that work is done only when there is an applied force *and* motion associated with that force. A weight lifter holding 400 lb over his head is doing no work, since there is no motion—he did work only while he was lifting the weight. Also the net work done on a body that is moving at constant velocity is zero because in order for the velocity to be constant, the net force must be zero.

Let us examine the effect of the $\cos\theta$ term in Eq. 6–8. When the force and displacement are in the same direction, $\theta = 0°$, $\cos\theta = 1$, and $W = Fd$. When **F** and **d** are perpendicular to one another, $\theta = 90°$, $\cos\theta = 0$, and $W = 0$. Thus no work is done by a force that is perpendicular to the displacement. For example, no work is done on a body in circular motion, since the centripetal force is always perpendicular to the circular path. If **F** and **d** are in opposite directions, $\theta = 180°$, $\cos\theta = -1$, and $W = -Fd$. Hence negative work is done when a force is in *opposition* to the motion (thereby causing deceleration). Friction and other resistance forces always do negative work on a body.

Let us now look at a few examples to clarify how work is calculated.

EXAMPLE 6–3: A 50-lb object is accelerated horizontally a distance of 20 feet by a horizontal force of 40 lb. How much work is done?

SOLUTION: The weight of the object is represented by a force pointing downward (Figure 6–9). Since the weight is perpendicular to the displacement, the weight force does no work. The work done by the 40-lb force is

$$W = Fd\cos\theta = (40\text{ lb})(20\text{ feet})(\cos 0°) = (40)(20)(1) = \underline{800\text{ ft-lb}}$$

figure 6–9

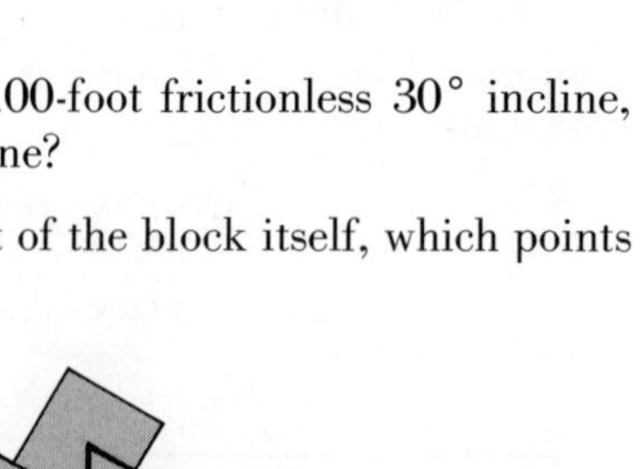

EXAMPLE 6–4: A 30-lb weight slides down a 100-foot frictionless 30° incline, as shown in Figure 6–10. How much work is done?

SOLUTION: The force doing the work is the weight of the block itself, which points

figure 6–10

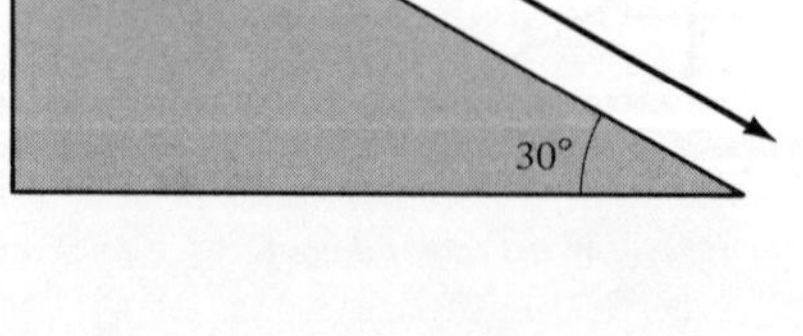

vertically downward. The angle between the force and displacement vectors is 60°. Therefore the work done by gravity is

$$W = Fd\cos\theta = (30\ \text{lb})(100\ \text{feet})(\cos 60°) = (30)(100)(\tfrac{1}{2}) = \underline{1500\ \text{ft-lb}}$$

EXAMPLE 6–5: A 2-kg block is pushed at constant velocity along a rough level surface by a horizontal force of 10 newtons. How much work is done when the block is moved a distance of 20 meters?

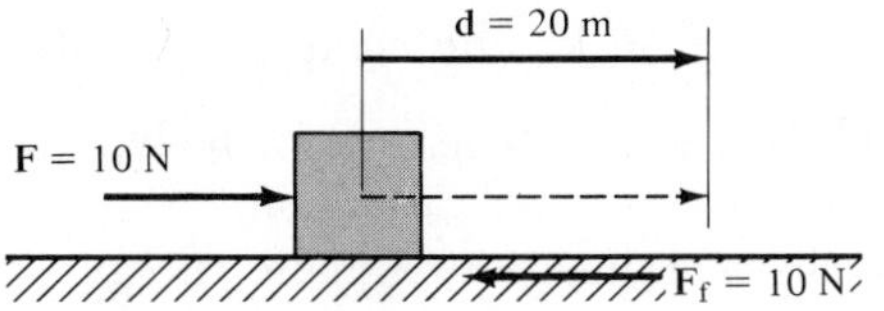

figure 6–11

SOLUTION: If the block is moving at constant velocity, the net force on it must be 0. Therefore the 10-newton applied force is being balanced exactly by a 10-newton friction force. (See Figure 6–11.) The applied force acts in the same direction as the motion, whereas the friction force is opposite to the motion. The work done by the 10-newton applied force is

$$W = Fd\cos\theta = (10\ \text{N})(20\ \text{m})(\cos 0°) = (10)(20)(1) = 200\ \text{joules}$$

The work done by the friction force is

$$W = Fd\cos\theta = (10\ \text{N})(20\ \text{m})(\cos 180°) = (10)(20)(-1) = -200\ \text{joules}$$

Thus the work done by the two forces is equal in magnitude but opposite in sign. Positive work is done by the applied force, and an equal amount of negative work is done by the friction force. Hence the *net* work done on the block is *zero*.

KINETIC ENERGY

At this point the purpose of calculating the work done by a force is not at all obvious. The significance of this quantity will now be demonstrated. Suppose a force of some kind is applied to a body of mass m. The work done by the force is, according to Eq. 6–8,

$$W = F_{\parallel}d$$

The effect of the force will be to accelerate the body. Substituting $F = ma$ gives

$$W = (ma)d$$

where a is the acceleration along the path of motion produced by the force. From the equations of motion, we have

$$v_f^2 = v_i^2 + 2ad$$

where v_i and v_f represent the initial and final speed of the body when the force is applied. Solving for d and substituting gives

$$W = (ma)\left(\frac{v_f^2 - v_i^2}{2a}\right)$$

or
$$W = \tfrac{1}{2}mv_f^2 - \tfrac{1}{2}mv_i^2 \tag{6–9}$$

This equation tells us that if work is done on a body by a force, the quantity $\frac{1}{2}mv^2$ will change. This quantity has been given the name ***kinetic energy.*** Hence the kinetic energy KE of a body of mass m and speed v is defined as

$$KE = \tfrac{1}{2}mv^2 \tag{6–10}$$

Kinetic energy, like work, is a scalar quantity, and it has the same units as work (ft-lb, ergs, or joules). Eq. 6–9 can now be written as follows:

$$W = KE_f - KE_i = \Delta KE = \text{change in kinetic energy} \tag{6–11}$$

where KE_i and KE_f represent the initial and final kinetic energy of a body, respectively.

In this manner, we have defined a new physical quantity—kinetic energy—that is a measure of the motion of a body. The relationship of work to kinetic energy is given by Eq. 6–11: *the work done on a body by a force is equal to the change in the kinetic energy of that body*. Note that if the work done is *positive,* the kinetic energy *increases;* if the work done is *negative,* the kinetic energy *decreases;* and if no work is done, the kinetic energy remains constant.

This definition permits yet another mathematical description of motion, in which work and energy are used as the basic quantities instead of force and momentum. The definition of kinetic energy originates with the work of Leibniz, who called kinetic energy the *vis viva* (living force). For many years a controversy raged between the followers of Leibniz and those of Newton as to whether kinetic energy or momentum was the more fundamental measure of motion. The modern physicist considers the two quantities to be equally fundamental, and the two formulations (work–energy and force–momentum) are equivalent. This equivalence can be demonstrated by the following example, which we will solve two ways.

EXAMPLE 6–6: A racing car (weight 1440 lb) is moving at a speed of 200 ft/sec. It is brought to a stop by opening a parachute, which exerts a constant retarding force on the car. The car travels a distance of 1500 feet before coming to a stop. What is the retarding force exerted by the parachute on the car?

SOLUTION 1 (FORCE–MOMENTUM METHOD): According to Newton's second principle, the force is equal to the rate of change of momentum; that is, $F = (mv_f - mv_i)/t$. The first step is to calculate the time required to stop the car. We know that $v_i = 200$ ft/sec $v_f = 0$, and $d = 1500$ feet. The average speed is

$$v_{av} = \frac{v_i + v_f}{2} = \frac{200\text{ ft/sec} + 0}{2} = 100\text{ ft/sec}$$

Hence the time is

$$t = \frac{d}{v_{av}} = \frac{1500\text{ feet}}{100\text{ ft/sec}} = 15\text{ sec}$$

The mass of the car is

$$m = \frac{w}{g} = \frac{1440\text{ lb}}{32\text{ ft/sec}^2} = 45\text{ slugs}$$

The force is therefore

$$F = \frac{mv_f - mv_i}{t} = \frac{(45)(0) - (45)(200)}{15} = \underline{-600 \text{ lb}}$$

The minus sign indicates that the force decelerates the car.

SOLUTION 2 (WORK–ENERGY METHOD): We find the work done by the retarding force by using Eq. 6–11. Since $m = 45$ slugs, $v_i = 200$ ft/sec, and $v_f = 0$,

$$W = \tfrac{1}{2}mv_f^2 - \tfrac{1}{2}mv_i^2 = \tfrac{1}{2}(45)(0) - \tfrac{1}{2}(45)(200)^2 = -900{,}000 \text{ ft-lb}$$

From the definition of work we have $W = Fd\cos\theta$. We are given that $d = 1500$ feet. Also, since the force opposes the motion of the car, $\theta = 180°$. Hence

$$W = Fd\cos\theta$$
$$-900{,}000 = F(1500)(\cos 180°) = -1500F$$

or

$$F = \frac{-900{,}000}{-1500} = \underline{600 \text{ lb}}$$

That is, the magnitude of the retarding force is 600 lb. Although the two methods are equivalent, certain problems are analyzed and solved much more easily by the work–energy method than by the force–momentum method and vice versa. This is the major advantage of having two separate formulations of Newtonian mechanics.*

GRAVITATIONAL POTENTIAL ENERGY

In the previous section it was shown that the work done on a body by any arbitrary force changes the kinetic energy of the body. Now we will deal with the work done by specific types of forces. We will begin by analyzing the effects of the gravitational force exerted by the earth on a body in terms of work and energy; then we will see whether the results of this analysis can be applied to other types of forces as well.

Suppose a body of mass m is held at a height h_i above the ground, as shown in Figure 6–12a. When the body is released, it begins to fall with an accelerated motion. As it falls, its velocity increases, and consequently its kinetic energy increases. The source of this increase in kinetic energy is the work done by the gravitational force that the earth exerts on the body. The greater the distance that the body falls, the greater the kinetic energy will be. It becomes very tempting to take the point of view that the body, by virtue of its position in a gravitational field, possesses an amount of "stored" or "potential" energy that is converted into kinetic energy of motion when the body is allowed to fall. Is it possible to put this point of view into mathematical form?

Let us calculate the work done by the gravitational force exerted on the body. The displacement is equal to the distance the body falls, so that if the body falls

* In particular, time does not appear in the work–energy method. Thus problems in which the time is given explicitly or is required as part of the solution must be solved by the force–momentum method.

figure 6–12 GRAVITATIONAL POTENTIAL ENERGY

PE_G ($= mgh$) is defined as the gravitational potential energy relative to the earth's surface of a body whose height above the surface of the earth is h. (a) A body falls from point 1 to point 2. A decrease in gravitational potential energy is associated with the positive work done by gravity. (b) A body rising from 2 to 1 experiences an increase in gravitational potential energy associated with the negative work done by gravity.

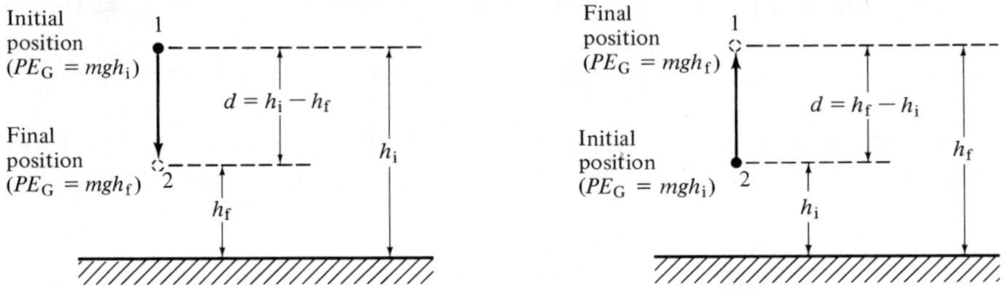

(a)

$$\begin{aligned} W &= Fd\cos\theta \\ &= mg(h_i - h_f)(1) \\ &= mgh_i - mgh_f \\ h_i > h_f&;\ \therefore\ \underline{W > 0} \end{aligned}$$

(b)

$$\begin{aligned} W &= Fd\cos\theta \\ &= mg(h_f - h_i)(-1) \\ &= mgh_i - mgh_f \\ h_f > h_i&;\ \therefore\ \underline{W < 0} \end{aligned}$$

from point 1 (at a height h_i above the ground) to point 2 (at a height h_f above the ground), the displacement $d = h_i - h_f$. The force acting on the body is equal to the weight of the body; thus $F = w = mg$. It is assumed here that the weight of the body does not change as the body falls, which means that g must have a constant value over the distance the body falls. This in turn requires that the displacement d is small compared to the radius of the earth. (A more precise derivation for the case in which g is not constant will be given later in the chapter.)

As the body falls, the force **F** and the displacement **d** are in the same direction; hence $\theta = 0°$ and $\cos\theta = 1$. The work done by gravity is therefore

$$W = Fd\cos\theta = (mg)(h_i - h_f)(1) = mgh_i - mgh_f \qquad (6\text{–}12)$$

We now define a quantity called the ***gravitational potential energy***, PE_G:

$$PE_G = mgh \qquad (6\text{–}13)$$

where h is the height of the body (of mass m) above the ground. Note that the higher the body is above the ground, the greater its potential energy is. Eq. 6–12 now becomes

$$W = PE_{G_i} - PE_{G_f} = -(PE_{G_f} - PE_{G_i}) = -\Delta PE_G \qquad (6\text{–}14)$$

As the body falls, positive work is done by gravity, and the potential energy of the body decreases. Thus gravitational potential energy has been defined such that the work done by gravity is equal to the *decrease* in the potential energy. Similarly, if a body *rises* in a gravitational field (Figure 6–12b), its gravitational potential energy *increases*, but the work done by gravity is *negative*. (**F** and **d** point in opposite directions.) Thus the work done by gravity is equal to *minus* the change in the gravitational potential energy.

In the above discussion gravitational potential energy is measured relative to the surface of the earth; that is, for a body at the surface of the earth, $PE_G = 0$ (since $h = 0$). It is permissible, however, to measure the gravitational potential energy with respect to any convenient surface. One may choose the top of a table or the bottom of a 10-foot hole as the level of zero gravitational potential energy and measure h from this reference level. This can be done because it is the *change* in gravitational potential energy that is important, and the change depends only on the distance between the initial and final points ($d = h_i - h_f$). This distance, of course, is the same regardless of the choice of reference level. Thus gravitational potential energy is a *relative* quantity that can be measured with respect to any arbitrary reference level.

6–4

CONSERVATION OF MECHANICAL ENERGY

Now let us see why the concepts of work, kinetic energy, and gravitational potential energy are useful. We have learned that the work done by gravity is equal to minus the change in the gravitational potential energy. But previously we found that the work done by *any* force is equal to the change in the kinetic energy. Hence by combining Eqs. 6–11 and 6–14, we obtain

$$PE_{G_i} - PE_{G_f} = KE_f - KE_i$$

This equation states that if a body is falling in a gravitational field, the *decrease* in the gravitational potential energy is equal to the *increase* in the kinetic energy. On the other hand, if the body is rising instead of falling—as when a ball is thrown vertically upward—the body decelerates as it rises, and therefore its kinetic energy decreases. At the same time, however, the gravitational potential energy is *increasing* (since h is increasing). In this case, the *increase* in the gravitational potential energy would be equal to the *decrease* in the kinetic energy. The fact that a decrease in one quantity is matched by a compensating increase in another quantity is a clear indication that a conservation principle is at work. We simply rearrange the terms of the above equation to obtain

$$KE_i + PE_{G_i} = KE_f + PE_{G_f} \tag{6–15}$$

We see that as the body moves from one point to another under the influence of a gravitational force, the *sum* of its kinetic and gravitational potential energies remains constant. We now propose that kinetic energy and gravitational potential energy are two forms of the same quantity, which we shall call the ***total mechanical energy*** E. Thus

$$E = KE + PE_G \tag{6–16}$$

According to Eq. 6–15, the total mechanical energy of a body is a constant quantity; kinetic energy can be transformed into gravitational potential energy or vice versa, but their sum remains constant. We have arrived at a new conservation principle—***the principle of conservation of mechanical energy:***

If the only external forces acting on a body are gravitational forces, the total mechanical energy (kinetic energy plus gravitational potential energy) of the body remains constant.

Looking back, we see that we have specifically defined several new physical quantities (namely work, kinetic energy, and gravitational potential energy) in such a way that another conservation principle can be generated. It may seem that these concepts have been artfully contrived for the sole purpose of devising a new conservation principle. Moreover, this principle is *not* a new and independent physical law, but simply a reformulation of the Newtonian principles of motion and gravitation. Kinetic energy and potential energy are, in effect, little more than inertia and gravitation in disguise. Yet these manipulations are justifiable in view of the usefulness, simplicity, generality, and aesthetic appeal of the principle of conservation of mechanical energy.

The following examples demonstrate how this conservation principle can be used to solve specific problems.

EXAMPLE 6–7: A ball is thrown vertically upward. It leaves the thrower's hand with a speed of 96 ft/sec. How high will it rise?

SOLUTION: Let point 1 be the thrower's hand and point 2 the top of the ball's trajectory. (See Figure 6–13.) From Eq. 6–15

$$KE_1 + PE_{G_1} = KE_2 + PE_{G_2}$$

At point 1, $h_1 = 0$, $v_1 = 96$ ft/sec. At point 2, $v_2 = 0$, $h_2 = H$. Substituting $KE = \frac{1}{2}mv^2$ and $PE_G = mgh$ gives

$$\tfrac{1}{2}m(96)^2 + 0 = 0 + mgH$$

We cancel m on both sides and solve for H:

$$H = \frac{(96)^2}{2g} = \frac{(96)(96)}{2(32)} = \underline{144 \text{ feet}}$$

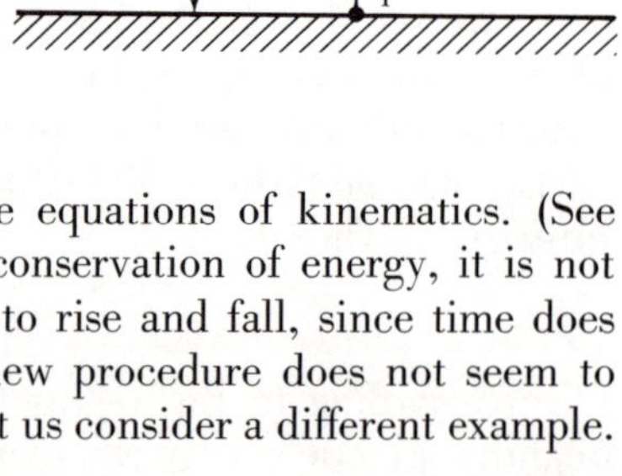

figure 6–13

This problem can be solved directly using the equations of kinematics. (See Example 3–5, page 94.) Moreover, when one uses conservation of energy, it is not possible to determine the time required for the ball to rise and fall, since time does not appear explicitly in the energy principle. The new procedure does not seem to offer any great advantage over the old one. But now let us consider a different example.

EXAMPLE 6–8: A block starts from rest and slides down the inclined plane shown in Figure 6–14. The surface of the plane is frictionless. What is the speed of the block when it reaches the bottom of the incline?

SOLUTION: We compute the kinetic energy and the gravitational potential energy at points 1 and 2, then substitute them into Eq. 6–15.

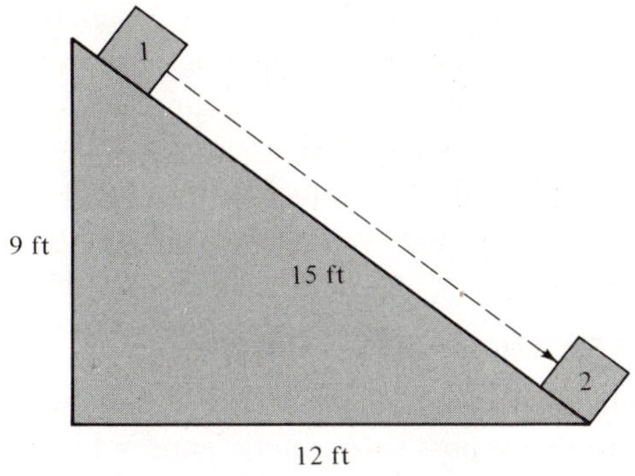

figure 6–14

At point 1:

$$v_1 = 0 \qquad \therefore KE_1 = 0$$
$$h_1 = 9 \text{ feet} \quad \therefore PE_{G_1} = m(32 \text{ ft/sec}^2)(9 \text{ feet})$$

(Remember that h is the *vertical* height above the ground.)
At point 2:

$$h_2 = 0 \quad \therefore PE_{G_2} = 0$$
$$KE_2 = \tfrac{1}{2}mv_2{}^2$$

(v_2 is the unknown quantity in the problem.)

We now substitute:

$$KE_1 + PE_{G_1} = KE_2 + PE_{G_2}$$
$$0 + m(32)(9) = \tfrac{1}{2}mv_2{}^2 + 0$$
$$2(32)(9) = v_2{}^2 = (64)(9)$$
$$v_2 = (8)(3) = \underline{24 \text{ ft/sec}}$$

This problem would be considerably more difficult to solve by any other method. We would have to compute the acceleration of the block as it slides down the plane. (The acceleration is *not* g, since the block is not in free fall.) Moreover, if the surface of the incline were curved instead of perfectly plane, the acceleration of the block would not even be constant, and we would not be able to solve the problem at all by any of the methods learned previously. None of these considerations enter into the solution of this problem by the energy principle. The final speed is determined only by the initial speed and the initial vertical height of the block. If the initial height is 9 feet and the block starts from rest, the same final speed (24 ft/sec) will be attained regardless of the steepness or shape of the incline. Obviously, the steepness of the incline must have some effect on the motion; it determines the *time*. The steeper the incline, the less time required for the block to reach the bottom. Thus the energy principle is most useful for solving problems in which we are concerned only with the initial and final position and speed of the body, and not with the time of motion or the specific path that the body travels.

For example, the principle of conservation of mechanical energy helps us to understand one of the results of Galileo's experiments with inclined planes. Namely, if a body is allowed to roll down one plane and up another it will return to the same vertical height, regardless of the steepness of the incline. (See Figure 3–16.) Initially, when the ball is at the top of the first incline, its energy is entirely gravitational potential energy. As the ball rolls down the incline, the gravitational potential energy is converted into kinetic energy. As the ball rolls up the second incline, its kinetic energy is converted back into gravitational potential energy. Conservation of energy tells us that the final gravitational potential energy must be equal to the initial value (since the initial and final kinetic energies are both zero). It follows that the ball must rise to the same vertical height, regardless of the steepness of the incline.

CONSERVATION OF MECHANICAL ENERGY AND ORBITAL MOTION

The principle of conservation of mechanical energy is especially suitable for an analysis of orbital motion. A body in orbital motion possesses kinetic energy by virtue of its motion, and gravitational potential energy because it is acted on by a gravitational force. No other forces of any consequence act on the body. Hence the total mechanical energy of an orbiting body will be conserved. That is,

$$E = KE + PE_G = \text{constant} \tag{6–17}$$

The kinetic energy will, of course, be equal to $\frac{1}{2}mv^2$. However, we *cannot* use $PE_G = mgh$, for this formula was originally derived on the assumption that the body is acted on by a constant gravitational force (i.e., that g is constant). Obviously, this condition is not met in the case of orbital motion, so that at this point we must derive the more general formula for gravitational potential energy (of which $PE_G = mgh$ is a special case).

As before, the procedure is to determine the work done by the gravitational force on a body that is rising or falling in a gravitational field. Let us calculate the work done as a body of mass m falls a distance D from point 1 to point 2, as shown in Figure 6–15. Normally, we would calculate the work by substituting into the equation $W = FD\cos\theta$. In this situation, however, F is not constant over the entire distance D. Since the strength of the gravitational force depends on the distance, F will steadily increase as the body moves from 1 to 2, and work will not be done

figure 6–15 **GRAVITATIONAL POTENTIAL ENERGY**

As the body of mass m moves from 1 to 2, the strength of the gravitational force increases, so that work is not done at a constant rate. The gravitational potential energy is a negative quantity, which decreases as the body moves from 1 to 2.

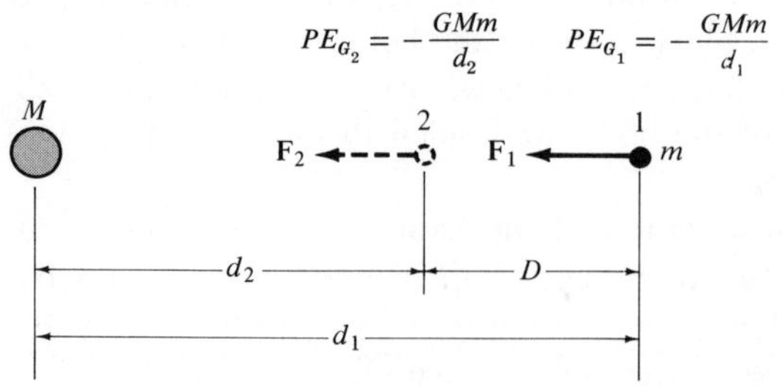

at a uniform rate. A more complicated mathematical procedure (involving the use of calculus) must be used to obtain an equation for the work done. The result turns out to be

$$W = \frac{GMm}{d_2} - \frac{GMm}{d_1} \qquad (6\text{–}18)$$

Since d_1 is greater than d_2, GMm/d_2 will be greater than GMm/d_1. Hence the work done by the mutual gravitational force on the body of mass m will be a positive quantity as the body falls.

We now define the gravitational potential energy as follows:

$$PE_G = -\frac{GMm}{d} \qquad (6\text{–}19)$$

In this equation d is the distance between the centers of the bodies of masses m and M. Thus $PE_{G_1} = -GMm/d_1$ is the gravitational potential energy of the body at point 1, and $PE_{G_2} = -GMm/d_2$ is the gravitational potential energy at point 2. Eq. 6–18 can then be written as follows:

$$W = PE_{G_1} - PE_{G_2}$$

That is, the work done is equal to the decrease in the potential energy. According to Eq. 6–19, the gravitational potential energy is always a *negative* quantity. When the body is an infinite distance away from the center of mass M ($d = \infty$), PE_G takes on its *maximum* value, which is zero. As the body falls toward the center of mass, its gravitational potential energy decreases (becomes more negative).* Similarly, as the body rises in the gravitational field, its potential energy increases. (As d increases, PE_G becomes less negative.)

Sometimes the idea of a negative energy is confusing, but the reader should realize that it is the *change* in the potential energy that is of interest, rather than the actual value of the potential energy at any point. The choice of the level of zero gravitational potential energy is relatively arbitrary. There are distinct advantages to letting $PE_G = 0$ at $d = \infty$: (1) it gives Eq. 6–19 the simplest form mathematically, and (2) it makes sense physically—for as d becomes infinite, the gravitational force approaches 0; thus a body an infinite distance away will have no force acting on it and will not fall in toward the center of mass. That is, when the body is an infinite distance away, its potential for motion in the gravitational field is zero.

The equation $PE_G = -GMm/d$ is the exact expression for the gravitational potential energy associated with the gravitational force between two bodies. The equation $PE_G = mgh$ (Eq. 6–13) is an approximate formula, which is valid only when g remains virtually constant as the height h changes. This would be the case for small changes in height near the surface of a large body such as the earth. The mathematical proof that $PE_G = mgh$ is a valid approximation to $PE_G = -GMm/d$ under such conditions is given in Appendix C.

We can now return to our analysis of orbital motion using the principle of conservation of energy. Substituting for KE and PE_G in Eq. 6–17 gives

$$E = \tfrac{1}{2}mv^2 - \frac{GMm}{R} = \text{constant} \qquad (6\text{–}20)$$

* Remember that negative numbers increase as they approach 0 (as their absolute magnitudes decrease). Thus -5 is *smaller* than -3; -3 is smaller than -2, and so on.

That is, a body in orbital motion moves so that its total energy E has a fixed numerical value. As a simple application of this principle, let us consider the motion of a planet in an elliptical orbit around the sun. As the planet moves from aphelion to perihelion, its distance R from the sun decreases; consequently, its potential energy decreases (becoming more negative). Since the total energy is a constant quantity, this decrease in potential energy must be compensated for by an increase in the kinetic energy. At perihelion the potential energy has its smallest value, so that the kinetic energy has its maximum value. Therefore the orbital speed will be greatest at perihelion and smallest at aphelion, in agreement with Kepler's second law.

The solution of the problem of orbital motion is considerably simplified when cast in terms of the principles of conservation of mechanical energy and conservation of angular momentum. As the orbiting body moves, its total mechanical energy E and its angular momentum L are both conserved, so both E and L are constant quantities. The properties of the orbit (size, shape, eccentricity, etc.) can be determined in terms of E and L. In particular, it is possible to show that the type of orbit is determined by the total energy. That is

(1) If $E < 0$, the orbit is an ellipse.
(2) If $E = 0$, the orbit is a parabola.
(3) If $E > 0$, the orbit is a hyperbola.

For an elliptical orbit, the total energy is a negative quantity, which means that the kinetic energy (always a positive quantity) is smaller in absolute value than the potential energy (always a negative quantity). If the kinetic energy is numerically larger than the potential energy, the total energy E will be positive, and the orbit will be a hyperbola. The boundary between unbounded and bounded orbits occurs when the total energy is zero. Thus Eq. 6–20 becomes

$$0 = \tfrac{1}{2}mv^2 - \frac{GMm}{R}$$

Solving for v gives

$$v = \sqrt{\frac{2GM}{R}}$$

which is the equation for the escape speed, v_{esc}. (See Eq. 5–7.) Thus the equation for the escape speed follows directly from the principle of conservation of energy. It is easy to see that if

$$\begin{array}{llll} v > v_{esc} & \text{then} & E > 0 & \text{(hyperbolic orbit)} \\ v = v_{esc} & \text{then} & E = 0 & \text{(parabolic orbit)} \\ v < v_{esc} & \text{then} & E < 0 & \text{(elliptical orbit)} \end{array}$$

in agreement with the conditions stated on page 148.

CONSERVATION OF ENERGY AS A GENERAL PRINCIPLE

Because the work done by a gravitational force could be expressed in terms of a change in gravitational potential energy, it was possible to derive the principle of conservation of mechanical energy. However, this principle is rather limited in that it applies only to bodies acted on by gravitational forces. Our next task is to see whether the principle of conservation of energy can be expanded to include the effects of other types of forces besides gravity.

A great many of the forces in nature can be included in the principle of conservation of energy. The force between two electric charges, the force between two magnets, the force between the particles of the nucleus of an atom, and the force required to stretch a spring are examples of forces for which the work done by the force can be expressed in terms of a change in potential energy. That is, when a body is moved from point 1 to point 2 as a result of any of these forces, the work done by the force can be written in the form

$$W = PE_1 - PE_2$$

where the specific formula for PE depends on the particular nature of the force. Thus one can define the electric potential energy PE_{E}, the magnetic potential energy PE_{M}, the nuclear potential energy PE_{N}, and so on, associated with each of these forces. A force that can be treated in this manner is known as a ***conservative force.***

A change in the potential energy associated with any conservative force is compensated for by a change in the kinetic energy, as was the case for the gravitational force (which is, of course, a conservative force). Thus we can expand Eq. 6–15 to include all conservative forces:

$$KE_{\mathrm{i}} + PE_{G_{\mathrm{i}}} + PE_{\mathrm{E_i}} + PE_{\mathrm{M_i}} + PE_{\mathrm{N_i}} + \cdots = KE_{\mathrm{f}} + PE_{G_{\mathrm{f}}} + PE_{\mathrm{E_f}} + PE_{\mathrm{M_f}} + PE_{\mathrm{N_f}} + \ldots$$

We can now define the ***total energy*** E of a body as the sum of its kinetic and various potential energies:

$$E = KE + PE_G + PE_{\mathrm{E}} + PE_{\mathrm{M}} + PE_{\mathrm{N}} + \ldots \tag{6–21}$$

The principle of conservation of energy can now be stated in more general terms:

If the only forces acting on a body are conservative forces, the total energy of the body remains constant.

There are, however, several rather common forces that *cannot* be described in terms of a change in potential energy. Friction and air resistance are examples. Such forces are referred to as ***nonconservative forces.*** One basic distinction between conservative and nonconservative forces is that the work done by a conservative force is recoverable, whereas the work done by a nonconservative force is not. For example, a body at a given height above the earth's surface will have a specific amount of potential energy. When the body is moved to another height, its potential energy is changed, but when it is returned to its original height, it will reacquire its initial potential energy. In other words, any change in the potential energy of a body can be nullified simply by returning the body to its initial position.

This cannot be done, however, when the body is acted on by a friction force. Friction and other resistance forces always act in opposition to motion. As we noted previously, such forces can only do negative work; thus any changes in energy brought about by friction are nonrecoverable. For example, suppose an automobile is traveling on a level road at 60 mi/hr. The brakes are applied, and frictional forces in the brake system and between the tires and the road bring the automobile to a stop. The kinetic energy of the automobile has been reduced to zero, and there has been no change in the gravitational potential energy (since the vertical height h has not changed). There is no physical process by which friction could be used to return the automobile to a speed of 60 mi/hr. Moreover, it appears that energy has not been conserved, since the decrease in kinetic energy does not seem to have been compensated for in any way.

We have come to a serious difficulty, because a rather severe limitation has been placed on the principle of conservation of energy. It would appear that energy is not conserved when nonconservative forces are acting. Since nonconservative forces (particularly friction) are rather prevalent in nature, the usefulness of the conservation of energy principle seems rather doubtful. One may without undue irreverence paraphrase the principle as follows: *Energy is conserved, except when it isn't.*

Yet more careful observations indicate that there are additional factors to be considered. Friction between bodies is always accompanied by heating of the bodies—as when we rub our hands together to warm them or when we rub two sticks together to start a fire. An automobile tire will be heated by the friction between the tire and the road, and the brakes of an automobile will heat up when they are applied. A space capsule will glow red-hot as a result of air resistance when it enters the atmosphere.

In general, the action of nonconservative forces is associated with the generation of heat, and this is clearly an important clue. It would seem that the principle of conservation of energy could be extended to include nonconservative forces if heat could be related to energy in some way. As it turned out, it was nearly 150 years after the death of Newton when this problem was finally solved, and until that time the principle of conservation of energy received relatively little notice. The story of the events leading to the discovery of the relationship between heat and energy and the subsequent revival of the principle of conservation of energy will be the subject of the next chapter.

QUESTIONS FOR REVIEW

1 Explain, in general terms, what a conservation principle is.

2 Define linear momentum. In what units is it measured? How is the direction of the linear momentum vector determined?

3 How can Newton's second principle be stated in terms of linear momentum?

4 (a) State the principle of conservation of linear momentum.
(b) On which of the three Newtonian principles of motion is it based?
(c) Under what conditions will the linear momentum of a *single body* be conserved?
(d) Under what conditions will the linear momentum of a *system of bodies* be conserved?

5 Define (in words and also by an equation) the quantity called torque. In what units is it measured? How is the direction of the torque vector determined?

6 What effect does an applied torque have on a body? Under what conditions will an applied force *not* produce an applied torque?

7 Define angular momentum. In what units is it measured? How is the direction of the angular momentum vector determined?

8 What is the relationship between torque and angular momentum?

9 What change in the angular momentum of a body is produced by an applied torque when the $\mathbf{T}$ and $\mathbf{L}$ vectors point in the same direction? In opposite directions?

10 State the principle of conservation of angular momentum. Under what conditions will the angular momentum of a body remain constant even though a force is acting on the body?

11 State the definition of work. In what units is it measured?

12 Under what conditions will an applied force *not* do work on a body? When does a force do negative work on a body?

13 State the definition of kinetic energy. In what units is it measured?

14 What is the relationship between the work done by an accelerating force on a body and the kinetic energy of the body? What happens to the kinetic energy when positive work is done? Negative work? No work?

15 Give the equation for the gravitational potential energy of a body near the surface of the earth. In what units is the gravitational potential energy measured? What is the *exact* equation for gravitational potential energy?

16 What is the relationship between the work done by gravity on a body and the gravitational potential energy of the body?

17 State and explain the principle of conservation of mechanical energy. How is the total mechanical energy of a body defined? Under what conditions is the total mechanical energy *not* conserved?

18 (a) Give the equation for the total energy of a body in orbital motion.
(b) Why is the gravitational potential energy of an orbiting body always a negative quantity?
(c) How is the type of orbit related to the total energy?

19 What is the difference between a conservative force and a nonconservative force? Give examples of each. What is the effect of a conservative or a nonconservative force on the total energy of a body?

EXERCISES

1 For each of the following, calculate the *change* in the linear momentum:
(a) $m = 10$ kg; $\mathbf{v}$ changes from 20 m/sec east to 30 m/sec east.
(b) $m = 20$ grams; $\mathbf{v}$ changes from 30 cm/sec east to 20 cm/sec west.
(c) $m = 100$ grams; $\mathbf{v}$ changes from 30 cm/sec east to 40 cm/sec north.

2 Using the data from the previous question, determine the magnitude and direction of the applied force that produces the change in linear momentum, assuming that in each case the change takes place over a time interval of 10 sec.

3 Body A (mass = 5 kg), moving to the right at a speed of 30 m/sec, strikes body B (mass = 10 kg), which is stationary. As a result of the collision, A is brought to a stop. What is the velocity of B after the collision?

4 A body of mass 20 grams, traveling at a speed of 10 cm/sec, is hit from behind by a second body of mass 30 grams, traveling in the same direction at a speed of 30 cm/sec. After the collision the first body continues to move in the same direction with a speed of 20 cm/sec. What is the speed and direction of the second body after the collision?

5 Two boxcars, A and B, have masses of 5000 kg and 3500 kg, respectively. Boxcar A, moving at 10 m/sec, collides with B, which is stationary. The two cars couple together as a result of the collision. With what speed will they travel after they have coupled together?

6 In Figure 6–6 the wheel has a radius of 20 feet, and the body at the end of the rope weighs 50 lb. What is the torque (magnitude and direction) produced by the weight about the center of the wheel?

7 What is the direction of the torque vector used to turn the doorknob on your front door? Check to see if all the doorknobs in your house or school turn in the same direction. Is there any reason why doorknobs are designed to turn in one particular direction?

8 A car ($m = 1000$ kg) is driven around a circular track with a 200-meter radius at a constant speed of 20 m/sec. Calculate the angular momentum of the car relative to the center of the track.

9 A ball of mass 50 grams is swung in a circle at a constant speed of 40 cm/sec on the end of a 30-cm-long string. Calculate (a) the angular momentum of the ball, (b) the centripetal force on the ball, and (c) the torque produced by this force.

10 A rotating wheel with a radius of 2 meters has an initial angular momentum of 6000 kg-m^2/sec. A friction force of 15 newtons is applied tangentially to the edge of the wheel so as to bring the wheel to a stop.
(a) What is the torque produced by the friction force?
(b) For how long must this force be applied before the wheel will come to a complete stop?

11 A well-thrown football always spins or spirals as it flies through the air; hence the football has angular momentum about its own axis.
(a) What is the direction of the angular momentum vector of the football, as seen by the passer? As seen by the receiver?
(b) Does the answer to (a) depend on whether the passer is right-handed or left-handed?

12 Explain why angular momentum is conserved for (a) a body in uniform circular motion, (b) a body acted on by a centripetal force, (c) a planet in orbital motion, (d) a freely spinning wheel, and (e) a freely rotating sphere.

13 The earth possesses an *orbital* angular momentum by virtue of its revolution around the sun and a *spin* (rotational) angular momentum by virtue of its rotation on its axis.
(a) In which directions do the two angular momentum vectors point?
(b) Which of the two is greater in magnitude? Explain your answer.

14 If there are no significant torques acting on the earth as it rotates, what can be said about the magnitude and direction of the earth's spin angular momentum? Can you now explain why the earth's axis always points at Polaris as the earth orbits the sun?

15 Using the principle of conservation of angular momentum, show that for a planet in an elliptical orbit about the sun, the ratio of the perihelion speed to the aphelion speed is inversely proportional to the ratio of the perihelion distance to the aphelion distance.

16 A 9000-lb truck crosses a 200-foot bridge at a constant speed of 20 mi/hr. Assume that friction is negligible. How much work is done on the truck?

17 A 12-lb block sits at the top of a 100-foot frictionless slide that is inclined at a 45° angle. The block is allowed to slide down. How much work is done by gravity?

18 A 15-kg mass falls a distance of 4 meters. How much work is done by gravity?

19 A 20-lb block is lifted vertically a distance of 40 feet by a 30-lb force. Calculate (a) the work done by the 30-lb lifting force, (b) the work done by gravity, and (c) the *net* work done on the body.

20 A 20-gram mass is at rest on a horizontal frictionless surface. A 40-dyne horizontal force is then applied until the body has moved a distance of 100 cm. (a) What is the total work done on the body? (b) What is the final kinetic energy of the body? (c) What is the body's final speed?

21 Using the data and solution to Example 6–2, calculate (a) the total kinetic energy of the two balls before the collision, and (b) the total kinetic energy of the two balls after the collision. Is kinetic energy conserved in the collision? Can you account for the missing energy? (The collision takes place on a level surface.)

22 A body with a mass of 30 grams is lowered from a height of 200 cm above the earth to 80 cm above the earth. Calculate (a) the initial potential energy of the body, (b) the final potential energy, and (c) the change in the potential energy.

23 A ball is thrown vertically upward and rises to a maximum height of 196 feet. Using the principle of conservation of mechanical energy, calculate (a) the initial speed, and (b) the speed when the ball is 132 feet above the ground.

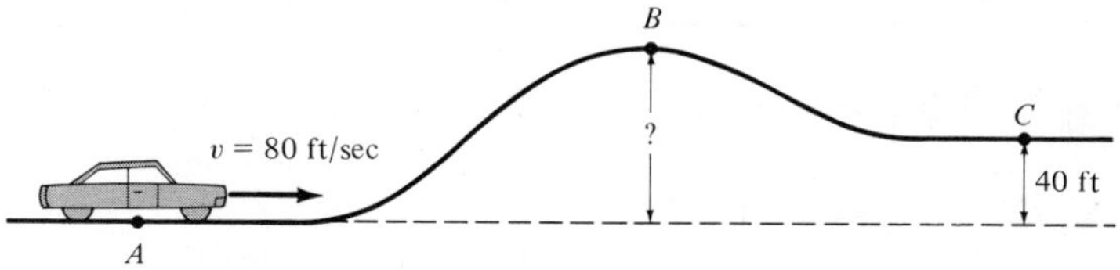

figure 6–16

24 A car is traveling along level ground with a speed of 80 ft/sec (Figure 6–16). It coasts uphill to point *B*, where its speed is 40 ft/sec.
(a) What is the height of the hill?
(b) What is the speed of the car when it reaches point *C*?

25 A satellite of mass m_s is in a circular orbit of radius R about a planet of mass m_p. Show the following:
(a) The total energy of the satellite is

$$E = -\frac{Gm_s m_p}{2R}$$

(Hint: Use Eq. 5–8 for the orbital speed.)
(b) The kinetic energy of the satellite is half of its gravitational potential energy; that is, $KE = -\frac{1}{2}PE_G$.

26 A 10-kg block starts from rest (point 1 in Figure 6–17) and slides down the inclined plane, the surface of which is frictionless. Upon reaching the bottom (point 2), the block continues to slide along the rough horizontal surface for a distance of 50 meters, until it comes to a stop (point 3).
(a) Calculate the speed of the block at the bottom of the incline.
(b) How much work is done on the block by the friction force between points 2 and 3?
(c) What is the magnitude of the friction force?

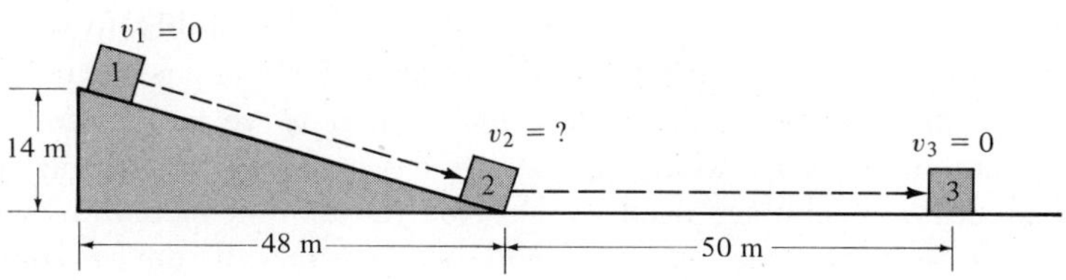

figure 6–17

HEAT AND THE NATURE OF MATTER

7

To engage in experiments on heat was always one of my most agreeable employments. . . . Indeed, I was often prevented by other matters from devoting my attention to it, but whenever I could snatch a moment I returned to it anew, and always with increased interest.

Benjamin Thompson (Count Rumford)

UP TO THIS POINT we have been concerned exclusively with the description and explanation of the motions of bodies in space and time. Reduced to essentials, the equations of kinematics and the principles of mechanics provide an abstract representation of the information we obtain about the world through our visual and temporal senses. Now we are going to deal with the information transmitted by another human sense—the sense of touch. We will be especially interested in a particular aspect of the sense of touch—namely, the sensations that we have learned from childhood to classify as "hot" or "cold." The ability to sense hotness and coldness, which we shall refer to as the ***thermal sense,*** is clearly a primary human sense; we are no more able to describe what it is we feel when we feel "hot" or "cold" than to describe what we see when we see the color red. It seems clear that the scientific vocabulary used to describe the motion of bodies will be inadequate to describe heat phenomena, so it will be necessary to devise some new physical quantities.

The most elementary observations indicate that the hotness or coldness of a body is primarily related to some internal property of the body rather than to the motion of the body as a whole. But thus far we have avoided any description of the *internal* properties of bodies; we have been dealing with the motions of bodies and objects without any reference to their internal structure. In order to gain a theoretical understanding of the nature of heat, we must examine the nature of matter itself.

The physicists of the eighteenth and nineteenth centuries believed that all physical phenomena could be encompassed within the framework of the Newtonian system of mechanics. Thus one of their goals was to construct a theory explaining heat phenomena in terms of Newtonian principles. The successful development of such a theory was one of the great triumphs of physics. By the middle of the nineteenth century a remarkable synthesis of the theories of heat, motion, and the structure of matter had been attained, resulting in a further vindication of the Newtonian mechanical view of the universe.

7–1

THE CONCEPT OF TEMPERATURE

Since the scientific vocabulary established up to this point in the text is inadequate to describe heat phenomena, the first step in the study of heat is necessarily the defining of suitable new physical quantities. Since all heat phenomena are ultimately related to the human thermal sense, the most fundamental quantity required is a scientific measure of the hotness or coldness of bodies. This quantity is called ***temperature.*** Of all the quantities defined in the sciences, temperature is one of the most well known and widely used by nonscientists. It is employed to determine such diverse things as one's state of health, what sort of clothing one should wear when going out of doors, and whether one's food is sufficiently cooked. Yet few people have a clear understanding of what temperature really is. Before we can explain how temperature is defined and measured, we must first describe a few important heat-related phenomena.

SOME THERMAL PROPERTIES OF MATTER

The thermal sensations that we receive vary in intensity; that is, we are able to sense gradations in hotness and coldness along a scale ranging from "very hot" to "very cold." In particular, we can detect changes in the degree of hotness or coldness of a given body, and we can make an obvious, but nevertheless important, observation that the hotness or coldness of a body is not an immutable property of that body. That is, the degree of hotness or coldness of a body changes rather readily (usually on contact with other bodies); in fact, some effort is required to maintain a body at a constant degree of hotness or coldness.

Examining the situation more carefully, we discover a rather significant phenomenon: *when a hot body comes in contact with a cold body, the hot body gets colder, and the cold body gets hotter.* This effect is one of the most fundamental physical laws, and no exceptions to it have ever been found. There is no known circumstance under which a cold body brought into contact with a hot body can become colder (or vice versa). Furthermore, if contact between the two bodies is maintained, the cooling of the hot body and the warming of the cold body will continue until both bodies reach the same state of hotness or coldness. At this point the bodies are said to be in ***thermal equilibrium.***

As is well known, material substances can be classified into three distinct forms or ***states:*** *solid, liquid,* or *gas.* Under certain conditions a substance brought in contact with another substance may change its state instead of (or in addition to) changing its degree of hotness or coldness. For example, an ice cube placed in a glass of warm soda will melt (i.e., change from solid to liquid); water placed in contact with a very cold surface can solidify into ice; a liquid in contact with a hot surface can boil or evaporate (change from liquid to gas). One of the important thermal properties of matter, then, is that a change in the degree of hotness or coldness can bring about a change of state.

Another significant thermal property of bodies is that virtually all substances expand in size as they become hotter and shrink as they become cooler. (The most notable exception to this rule is water, which expands slightly as it cools when it is near its freezing point.) The propensity of bodies to undergo ***thermal expansion*** is, for most practical purposes, a considerable nuisance. In the design and construction of roads, bridges, buildings, and so on, care must be taken to account for the fact that the construction materials will shrink and expand (at different rates, depending on the material) with changes in the weather. However, thermal expansion is of great value to the scientist, for it provides him with a basis for defining and measuring temperature.

THE THERMOMETER

Physicists have defined the quantity called temperature in order to provide a quantitative measure of the degree of hotness or coldness of a body. Briefly, temperature is measured as follows: the measuring device, or ***thermometer,**** contains some suitable thermometric substance, which undergoes a measurable amount of thermal expansion. The temperature of this substance is defined in terms of the substance's thermal expansion (relative to some reference level) at a given degree of hotness or coldness. To measure the temperature of another body, we make use of the tendency of two bodies in contact to attain thermal equilibrium. Bodies in thermal equilibrium have the same degree of hotness and coldness; hence *they have the same temperature.* Thus if we bring the thermometer in contact with another body and allow them to come to thermal equilibrium, the temperature of the second body will be the same as the temperature of the thermometer (as measured by the thermal expansion of the thermometric substance).

The simplest and most common form of thermometer consists of a sealed glass tube containing a liquid (usually mercury or alcohol) as the thermometric substance. Thermal expansion results in a change in the level of the liquid in the tube. The thermometer is calibrated by making equally spaced markings on the tube and numbering them. In this way a number (measured in ***degrees***) is associated with the level of the liquid, thereby defining the numerical value of the temperature.

Thus when we say, "The temperature is 65°," we mean that the substance in a thermometer that has been allowed to come into thermal equilibrium with the surrounding air has, through thermal expansion, risen to the level of the 65° marking on the thermometer. Of course, the numerical value of the temperature is somewhat arbitrary in that it depends on how we choose to calibrate the thermometer. Yet even though there can be many different temperature scales, each of them can be used to establish a quantitative measure of the relative hotness and coldness of bodies. By defining temperature in this way, we have a scientifically measurable quantity that is directly related to the human thermal sense. As such, temperature must be regarded as a fundamental physical quantity on the same order as mass, distance, and time. Just as mass is a measure of the inertia of a body, temperature is a measure of its hotness or coldness.

TEMPERATURE SCALES

Since the thermometer is a scientific measuring instrument, it must be constructed and calibrated such that scientists in different laboratories around the world are measuring temperatures in exactly the same way. It is most practical, then, to calibrate a thermometer in terms of some easily reproducible physical condition. Commonly, the freezing and boiling points of water are used for this purpose.

* The first thermometer was constructed by Galileo in 1593. It is remarkable that in the light of Galileo's other work, the invention of the thermometer must be considered one of his minor achievements.

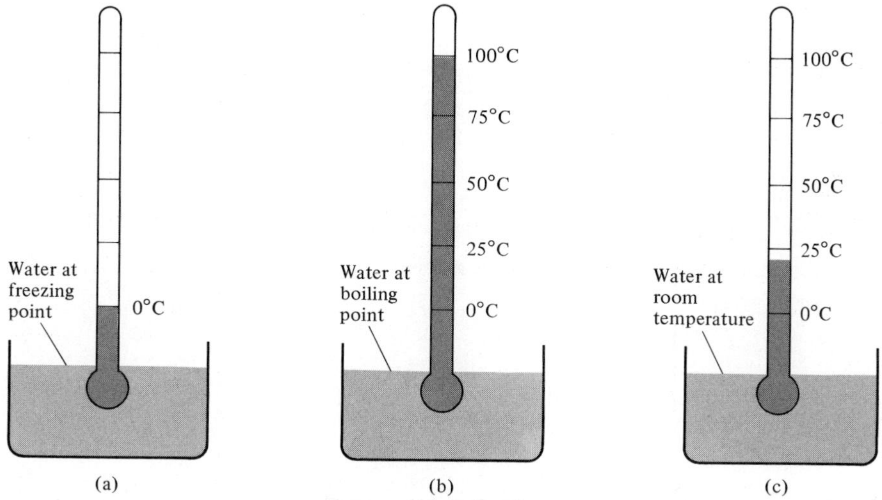

figure 7–1 **CALIBRATION OF A THERMOMETER (CELSIUS SCALE)**

(a) The thermometer is placed in contact with water at its freezing point and allowed to come to equilibrium. The level of the thermometric substance is marked as 0°C. (b) The thermometer is now placed in water at its boiling point. The new level is marked as 100°C. Then the interval between the 0° and 100° levels is subdivided, and the thermometer is calibrated. (c) When the thermometer is placed in water at room temperature, the level rises to the 20°C mark. The temperature of the water is therefore 20°C.

First, the thermometer is placed in water at its freezing point (where water and ice coexist without any ice melting or any water freezing). The level of the thermometric substance is marked on the thermometer. Then the thermometer is placed in water at its boiling point. The thermometric substance will expand and rise to a new level, which is also marked. These two fixed points calibrate the thermometer. The distance between them is then subdivided and the markings numbered. Clearly, there is no limit to the number of ways in which this last step can be done, and in fact there have been many different temperature scales in use at one time or another. Today, the two most commonly used scales are the ***Fahrenheit*** and the ***Celsius*** or ***centigrade*** scales.*

On the Celsius scale the distance between the two fixed points is divided into 100 equal parts or ***degrees.*** That is, 1/100 of the distance between the two fixed points is defined as ***1 Celsius degree*** (C°). The freezing point is marked as 0°C; the boiling point thus becomes 100°C. (See Figure 7–1.) The Celsius scale is used in countries that have adopted the metric system and is commonly employed in all scientific work.

The Fahrenheit temperature scale was invented by the German scientist Gabriel Fahrenheit in 1714. Fahrenheit's original temperature scale used the human body temperature as one fixed point and the freezing point of a salt–water mixture as the

*In the sciences the internationally recommended name is Celsius, in honor of the Swedish scientist Andreas Celsius, who introduced this scale in 1742. Thus the abbreviation "°C" stands for both "degrees Celsius" and "degrees centigrade" (the latter is still widely used).

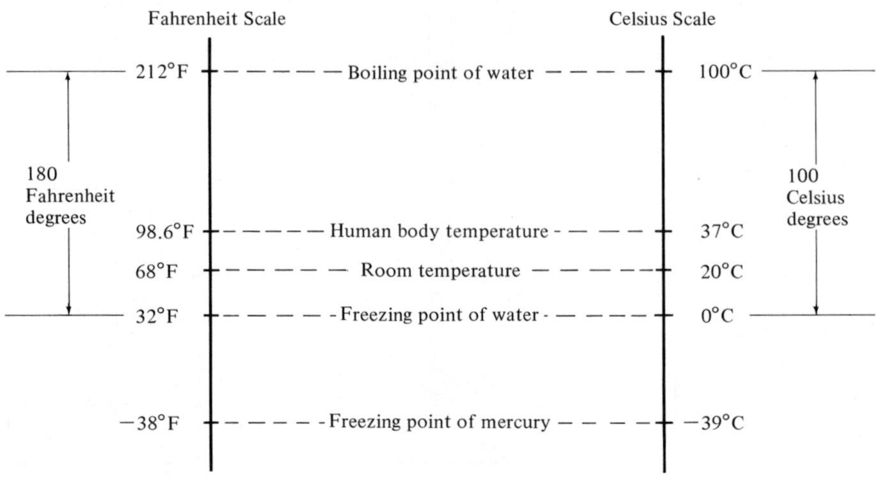

figure 7–2 **COMPARISON OF THE FAHRENHEIT AND CELSIUS TEMPERATURE SCALES**

other. (Since the latter was the lowest temperature Fahrenheit could produce in his laboratory, he defined it as 0°.) On this scale the freezing point of water is at 32°F and the boiling point is at 212°F. These two temperatures define the modern-day Fahrenheit scale, which is used in countries adopting the British system of units.*

Temperatures can be converted from Celsius to Fahrenheit and vice versa rather easily. Both the modern Fahrenheit and Celsius temperature scales use the freezing and boiling points of water as their fixed points. However, on the Fahrenheit scale the interval is divided into 180 parts rather than into 100, as in the Celsius scale (Figure 7–2). Therefore a Fahrenheit degree is $100/180 = 5/9$ of a Celsius degree. Fahrenheit temperatures can be converted to Celsius by using the equation

$$T_C = \frac{5}{9}(T_F - 32) \tag{7–1a}$$

To convert from Celsius to Fahrenheit, we use Eq. 7–1a in the form

$$T_F = \frac{9}{5}T_C + 32 \tag{7–1b}$$

The measurement of temperature by a thermometer should not depend on the choice of thermometric substance. One might suppose that since different substances have different rates of thermal expansion, thermometers using different thermometric substances would give different temperature readings for the same degree of hotness or coldness. However, this difficulty is circumvented by the very way that temperature is defined. Different thermometric substances will rise to different levels in the glass tube when placed in boiling water, but *all of these levels are defined as 100°C (or 212°F)*. Therefore, different thermometers assign the same temperature to a given degree of hotness or coldness, even though different amounts of expansion are produced.

Generally, the temperature range over which the thermometer is to be used

*Americans who travel in Europe are sometimes surprised to discover that in continental countries, weather reports give temperatures in centigrade. In Great Britain both Fahrenheit and centigrade readings are reported.

determines the choice of thermometric substance. A main consideration is that the thermometric substance should not change state within the given temperature range. For most common uses mercury has proved to be the most suitable substance; it remains in the liquid state from −39°C to 357°C.

ABSOLUTE TEMPERATURE SCALES

Both the Fahrenheit and Celsius scales have certain notable shortcomings. One problem is the existence of negative temperatures. On these scales negative temperatures have no special physical significance; they are simply temperatures that are lower than the arbitrarily defined zero levels. In fact, the arbitrary way in which the Fahrenheit and Celsius scales are defined limits their physical significance; there could be any number of equally valid scales. Is it possible to find one temperature scale that is somehow more fundamental than all the others?

This question was partially answered by the discovery that there is a lower limit to temperature; that is, there is a minimum temperature, commonly referred to as ***absolute zero,*** that represents the coldest possible state of material objects. The existence of absolute zero was first indicated by the results of experiments performed on gases by the French scientists Jacques Charles and Joseph Gay-Lussac at the turn of the nineteenth century.

All gases exhibit thermal expansion; in fact, their rate of thermal expansion is considerably greater than that of solids and liquids. Charles measured the changes in the volume of a gas caused by changes in temperature when the gas was maintained at a constant pressure. The results of such an experiment are shown in Figure 7–3. The measured values of the volume V and the temperature T fall on a straight line, indicating that the change in volume is directly proportional to the change in temperature. In particular, when the temperature is increased to 273°C, the volume is twice as much as it is at 0°C. By extension, if we *reduce* the temperature to −273°C, the volume would be zero. Furthermore, the numerical value of the temperature of the zero-volume point turns out to be independent of the kind of gas used, for Charles and Gay-Lussac's experiments showed that all gases expand and contract at virtually the same rate. These results suggest that −273°C is the lowest possible natural

figure 7–3 **THERMAL EXPANSION OF A GAS**

The dots represent the measured volumes of a fixed weight of gas at constant pressure for different values of the temperature. These points fall on a straight line. When the line is extended in the direction of lower temperatures, the zero-volume point is reached at −273°C.

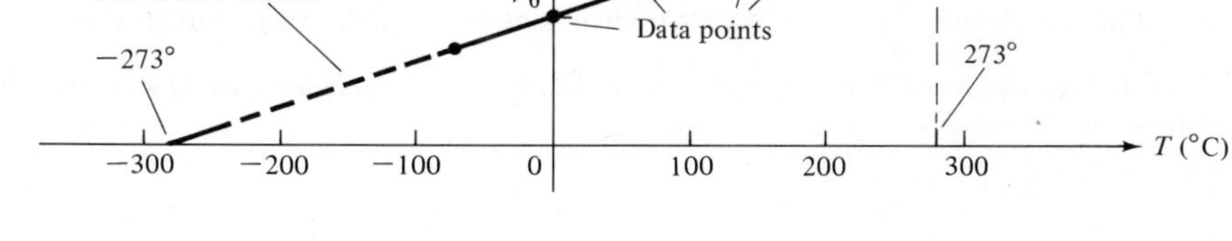

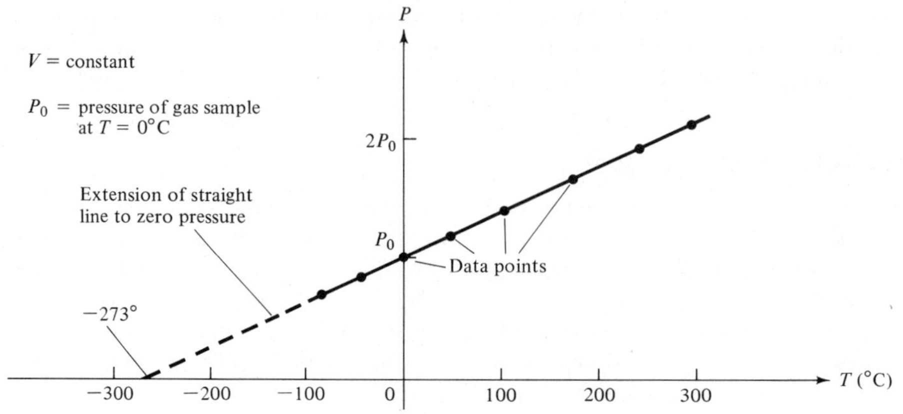

figure 7–4 **RELATION OF PRESSURE TO TEMPERATURE**

Experiments show that for a fixed volume of gas the pressure is directly proportional to the temperature. The limit of zero pressure would be reached at a temperature of −273°C.

temperature. (Precise measurements give a value of −273.15°C, which equals −459.67°F, for this temperature.) Further corroboration is provided by the experimental measurement of the change in *pressure* of a fixed volume of gas when the temperature is changed. As shown in Figure 7–4, a gas cooled to −273°C would presumably exert zero pressure.

Clearly, a temperature scale becomes less arbitrary if its zero point corresponds to the lowest possible natural temperature. In other words, on such a scale 0° should be equivalent to −273°C. A scale defined in this manner is called an ***absolute temperature scale.*** On an absolute temperature scale, a negative temperature has no physical meaning.

The Celsius temperature scale can be readily converted to an absolute scale simply by adding 273° to the Celsius temperature. This defines the ***Kelvin temperature scale*** (named after Lord Kelvin,* who made important contributions to the theory of heat). The Celsius and Kelvin scales are related by the equation

$$T_K = T_C + 273° \qquad (7\text{–}2)$$

Thus on the Kelvin scale the freezing and boiling points of water are 273°K and 373°K, respectively. (See Figure 7–5.) An absolute Fahrenheit scale can be generated in a similar manner by adding 460° to the Fahrenheit temperature. This defines the ***Rankine*** scale, which is not frequently used in scientific work.

THE IDEAL GAS

Implicit in the method for determining the numerical value of absolute zero is an assumption about the properties of gases that merits careful examination. The experimental data on the thermal expansion of a gas shown in Figure 7–3 indicate a straight-line relationship between volume and temperature at high temperatures only, and *it is assumed that this relationship holds for low temperatures as well.* This

*Lord Kelvin was born William Thomson in 1824. He was knighted by Queen Victoria for his contributions to the development of telegraphy.

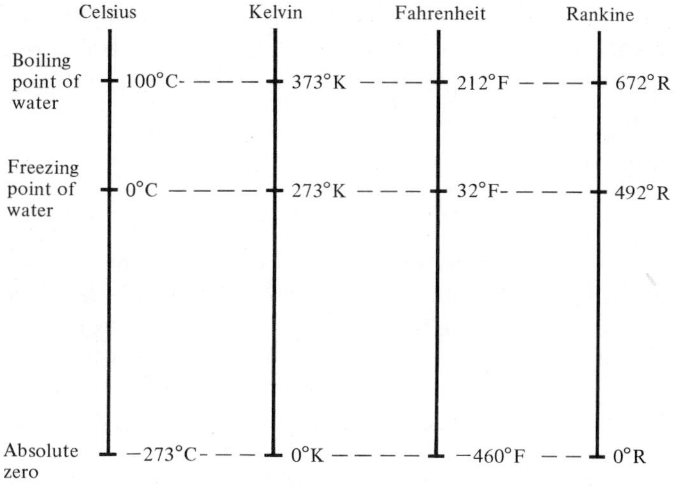

figure 7–5 **COMPARISON OF THE MAJOR TEMPERATURE SCALES**

The Kelvin and Rankine scales are absolute temperature scales derived from the Celsius and Fahrenheit scales, respectively.

assumption is one that ought to be verifiable; with modern techniques we can attain temperatures much lower that those available to Charles and Gay-Lussac, and thus we should be able to extend their measurements. When this is done, we find that the above assumption *does not hold strictly true for any known gas*. The explanation for this is quite simple: at very cold temperatures all gases liquefy! As liquids, their thermal expansion properties deviate considerably from the assumed straight-line behavior shown by the dotted line in Figure 7–3. This result would seem to invalidate totally the conclusion that −273°C is the lowest possible temperature.

To get around this difficulty, physicists proposed the concept of an ***ideal gas*** that has exactly the properties implicit in the assumption—namely, that it remains a gas down to absolute zero and maintains the straight-line relationship between volume and temperature. The ideal gas is an extremely useful concept in physics, even though no real gases exhibit such behavior. For the purpose of predicting the properties of gases at given temperatures and pressures, the ideal gas model has value because at high temperatures and/or low pressures real gases *approach* ideal gas behavior. In other words, the ideal gas model can be used to approximate (with rather good accuracy) the properties of real gases under these conditions.

Of even greater importance, the absolute temperature scale can be defined in terms of the theoretical properties of an ideal gas. That is, −273°C represents the *lowest possible temperature of an ideal gas* and is therefore taken as the absolute zero of the ideal gas temperature scale. Strong justification for using the ideal gas as the basis of an absolute temperature scale was provided by Lord Kelvin, who was able to show on the basis of a rather complex theoretical argument that the absolute zero of the ideal gas scale is, in fact, the lowest possible natural temperature consistent with the laws of physics.

The properties of the ideal gas—on which the absolute temperature scale is based—can be put into specific mathematical form. It has been assumed that an ideal gas follows the straight-line relationship between volume and temperature shown in Figure 7–3 down to absolute zero. In Figure 7–6 this relationship has been redrawn with temperatures plotted in °K instead of in °C. The volume is now directly proportional to the absolute temperature. That is,

$$\frac{V}{T} = k_1 \qquad (7\text{–}3)$$

where k_1 is a constant whose value depends on the pressure and the mass of the gas sample. Equation 7–3 is known as ***Charles' law.*** This equation provides a means for measuring the absolute temperature. Suppose a fixed weight of gas maintained at constant pressure has a volume V_1 at a known temperature T_1. (The volume could be measured, for example, at the freezing point of water, for which $T_1 = 273°\text{K}$.) When this gas is brought into contact with a body at a different temperature, the volume will change to a new value V_2. The new temperature T_2 is given by the following equation (derived from the fact that V/T is constant):

$$\frac{V_1}{T_1} = \frac{V_2}{T_2} \quad \text{(at constant pressure)} \qquad (7\text{–}4)$$

Thus the absolute temperature can be determined by measuring the change in volume of an ideal gas. But if no ideal gases exist, how can this measurement be made? Although it is true that no real gases exhibit ideal gas behavior at all temperatures and pressures, there are some gases that deviate only slightly from ideal gas behavior down to rather low temperatures. Helium, for instance, remains in the gaseous state down to 3°K, and its deviations from ideal gas properties are relatively small and easily corrected for. A gas thermometer using helium as a thermometric substance can be used to measure temperatures almost to absolute zero with an accuracy of about 0.001°.*

Another relationship between the properties of an ideal gas follows from the experimental measurements made by Robert Boyle, a contemporary and close friend of Sir Isaac Newton. Boyle found that when the pressure and volume of a fixed mass of gas are varied while the gas is being maintained at a constant *temperature*, the following equation holds:

$$PV = k_2 \qquad (7\text{–}5)$$

where k_2 depends on the temperature and mass of the gas. This equation is known as ***Boyle's law.*** As in the case of Charles' law, Eq. 7–5 only *approximates* the behavior of real gases; it is more accurate at high temperatures than at low ones. Real gases deviate considerably from Boyle's law for temperatures near their melting points. Thus Boyle's law exactly describes the behavior of an ideal gas.

Boyle's law and Charles' law can be combined to give a single ***ideal gas law:***

$$\frac{PV}{T} = K = \text{constant} \qquad (7\text{–}6)$$

*It has been shown on theoretical grounds that it is impossible to reduce the temperature of any object to absolute zero. Temperatures extremely close to absolute zero have been attained. As of 1970 the lowest temperature that had been reached in the laboratory was 0.00085°K.

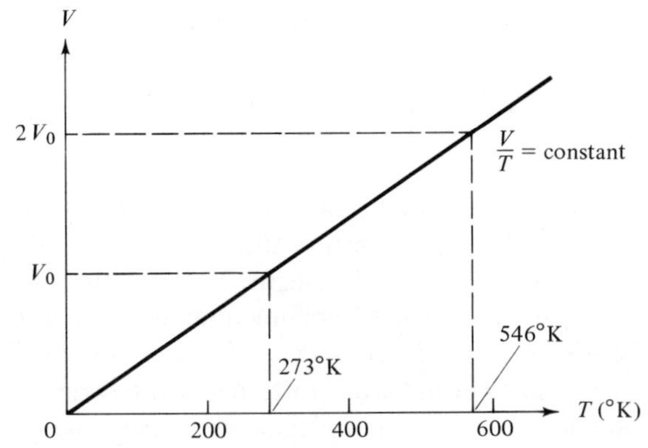

figure 7–6

CHARLES' LAW

Figure 7–3 has been redrawn with the temperature measured on the Kelvin scale. The straight line represents the behavior of an ideal gas: at constant pressure the volume of the gas is directly proportional to the absolute temperature.

The numerical value of K depends on the type of gas and the mass of the gas sample. Equation 7–6 *defines* an ideal gas; that is, an ideal gas is one that obeys this equation. The ideal gas equation can be used to approximate the properties of real gases. If the pressure, volume, and temperature, P_1, V_1, T_1, of a fixed mass of gas are known, and any of these quantities is changed, the new pressure, volume, and temperature, P_2, V_2, T_2, are related by the equation

$$\frac{P_1V_1}{T_1} = \frac{P_2V_2}{T_2} \tag{7–7}$$

The ideal gas law is important not only because it allows us to predict the properties of gases, but also because it enables us to learn something about the nature (specifically the internal structure) of gases. We will return to the ideal gas concept later on in this chapter.

EXAMPLE 7–1: A sample of air at a temperature of 27°C and 1 atm pressure occupies a volume of 6 liters. If the temperature is decreased to 0°C and the pressure is doubled, what volume will the gas occupy?

SOLUTION: The initial conditions are given as follows: $P_1 = 1$ atm, $V_1 = 6$ liters, $T_1 = 27°\text{C} = 300°\text{K}$. We also know that $P_2 = 2$ atm and $T_2 = 0°\text{C} = 273°\text{K}$. We can substitute directly into Eq. 7–7:

$$\frac{P_1V_1}{T_1} = \frac{P_2V_2}{T_2}$$

$$\frac{(1\text{ atm})(6\text{ liters})}{300°\text{K}} = \frac{(2\text{ atm})V_2}{273°\text{K}}$$

$$V_2 = \frac{(273)(6)}{(300)(2)} = \underline{2.73\text{ liters}}$$

A word about units: any convenient units can be used for P and V, as long as the same unit is used on both sides of the equation. It is essential, however, that T be converted to an absolute temperature.

7–2

THE NATURE OF HEAT

We have learned that temperature is a quantity that is a measure of the degree of hotness or coldness of a body. Now let us turn our attention to those phenomena that bring about changes in the temperature of a body. To begin, let us return to what is usually considered the most fundamental thermal phenomenon: *when a hot body comes into contact with a cold body, the hot body gets colder and the cold body gets hotter, until both bodies reach thermal equilibrium.* In terms of the concept of temperature, we can describe this phenomenon by saying that when two bodies at different temperatures are placed in contact, they interact in such a way that their temperatures are equalized. But what is involved in this interaction? Obviously, *something*—possibly a material substance—is being transferred from one body to the other. It seems reasonable to assume that the transfer is taking place *from* the hot body *to* the cold body, for it appears that the hot body is losing something and the cold body is gaining something in the interaction (although it could well be the other way around). Now let us introduce the concept of ***heat.*** Without making any attempt to determine what heat is, we shall define heat as *that which is transferred from a hot body to a cold body when the bodies are placed in contact.* A clear distinction between *heat* and *temperature* is being made here; temperature is a measure of the degree of hotness or coldness of a body, and heat is something that changes the temperature and/or state of a body when added to or removed from that body.

Heat is, of course, one of the most familiar scientific concepts, and attempts to understand the nature of heat date back to the ancient Greeks. Two main lines of thought developed. The earliest theory held that heat was a material substance contained in bodies, perhaps in the form of a fluid. (Even today we speak of heat as "flowing" from one body to another.) By the seventeenth century a theory that heat is a form of motion had been proposed; such luminaries as Newton, Descartes, Boyle, and Hooke were supporters of this explanation. Both theories had their strengths and weaknesses, and a consistent theory of heat was not developed and accepted until the 1840s.

MEASUREMENT OF HEAT

Before examining theories of the nature of heat, we must learn how heat is measured. Fortunately, the techniques of heat measurement or ***calorimetry*** do not depend on any detailed assumptions as to the nature of heat. Therefore the mathematical relationships to be developed can be applied with equal validity to either of the two theories of heat.

The person most responsible for the development of calorimetry is the Scottish physician-scientist Joseph Black. Black, who was successively a professor of anatomy, medicine, and finally chemistry at the University of Edinburgh,* formulated and tested his theories of heat measurement in the 1760s, but published virtually none of his results. It was not until several years after his death in 1799, when his lecture notes

* One of Black's students at the University of Edinburgh was James Watt, whose invention and perfection of the steam engine, based in part on the principles Watt learned from Black, was one of the most significant sources of the Industrial Revolution.

were edited and published, that the extent of Black's researches on heat became generally known.

The mathematics of heat measurement, as developed by Black and refined by later generations of scientists, are based on some rather simple observations. Let us suppose that water is being heated over a fire. The flame, which is really a flow of hot gas, can be considered a continual source of heat. Two observations are evident: (1) the longer the water is heated, the greater the temperature increase, and (2) the greater the amount of water, the longer the water has to be heated before it is brought to a given temperature. It follows that the amount of heat added is proportional both to the mass of material being heated and to the increase in temperature produced. In mathematical terms we have

$$Q \propto m\,\Delta T \tag{7–8}$$

where Q is the amount of heat added, m is the mass, and ΔT $(= T_{\mathrm{f}} - T_{\mathrm{i}})$ is the change in temperature.

Now suppose we take two equal masses of water, one at 80°C and the other at 40°C, and mix them together in a thermally insulated container. We find, not surprisingly, that the measured equilibrium temperature of the mixture is 60°C—exactly halfway between the two initial temperatures. However, when we mix 2 kg of water at 80°C with 1 kg of water at 40°C, the equilibrium temperature is 66.7°C. These two results can be easily understood if we assume that the heat lost by the warm water is equal to the heat gained by the cold water. This assumption may be looked on as a statement of a principle of *conservation of heat*—that is, heat cannot be created or destroyed, only transferred from one body to another. Let us apply this principle to the example of the two equal samples of water. We assume that the heat lost by the sample of water initially at 80°C is equal to the heat gained by the sample of water at 40°C. From Eq. 7–8 it follows that

$$\begin{aligned}&(\text{mass of water at } 80^\circ\text{C})(\text{decrease in temperature})\\ &\qquad = (\text{mass of water at } 40^\circ\text{C})(\text{increase in temperature})\end{aligned}$$

The transfer of heat ends when both samples of water are at the same equilibrium temperature T. Hence the warm water decreases in temperature from 80°C to T, while the cold water temperature rises from 40°C to T. Substituting into the above equation (remember that both masses are equal) gives us

$$m(80 - T) = m(T - 40)$$

Solving for T, we obtain

$$\underline{T = 60^\circ\text{C}}$$

In the second example there are 2 kg of water at 80°C and 1 kg of water at 40°C. Equating the heat loss to the heat gain gives us an equation in terms of the final equilibrium temperature T:

$$(2\text{ kg})(80 - T) = (1\text{ kg})(T - 40)$$

Solving this equation gives

$$\underline{T = 66.7^\circ\text{C}}$$

which is in agreement with what is observed.

One of Black's important discoveries was that different substances have different capacities for absorbing or releasing heat. This is evident when equal masses of two different substances are brought into contact. For example, if 1 kg of iron at 80°C is immersed in 1 kg of water at 40°C, the equilibrium temperature is found to be 43.7°C. In the transfer of heat from the iron to the water, the same amount of heat has resulted in a much greater temperature change in the iron than in the water. We interpret this result as indicating that water has a greater *capacity for heat* than iron does.

To put this result in mathematical form, we define a quantity called the ***specific heat*** s. The specific heats of all substances are measured in comparison to water, which is defined as having a specific heat of 1. The specific heat is not truly constant for any given substance, but at ordinary temperatures and for relatively small temperature changes, the variations in s are rather small.

The relationship between heat and temperature can now be defined by the following equation:

$$Q = ms\,\Delta T \tag{7–9}$$

where Q is the amount of heat added or removed, m is the mass of the sample, and ΔT is the change in temperature. Using Eq. 7–9, we can determine the specific heat of iron from the given data in the following way:

$$\text{heat lost by iron} = \text{heat gained by water}$$
$$m_{\text{iron}}s_{\text{iron}}(80^\circ - 43.7^\circ) = m_{\text{water}}s_{\text{water}}(43.7^\circ - 40^\circ)$$

The masses of iron and water both equal 1 kg; also by definition $s_{\text{water}} = 1$. Thus we have

$$(1)(s_{\text{iron}})(36.3) = (1)(1)(3.7)$$

Solving for s_{iron} we obtain

$$\underline{s_{\text{iron}} = 0.102}$$

Hence the specific heat of iron is about one-tenth that of water.

We have not as yet defined a unit for the measurement of heat. In the metric (CGS) system the unit of heat is the ***calorie.*** One calorie (1 cal) is defined as the amount of heat required to raise the temperature of 1 gram of water by 1 C° (specifically from 14.5°C to 15.5°C, since the specific heat of water is not exactly constant from 0° to 100°C: it varies from about 0.997 to 1.01). The word "calorie" is familiar to weight watchers. Actually, the unit used by dieticians is the Calorie (with a capital C), which is equal to 1 kilocalorie (1 kcal = 1000 cal). The caloric value of a given type of food is determined by actually burning a sample in pure oxygen and measuring the amount of heat released. Thus a piece of chocolate cake, for example, will release 200,000 cal of heat when burned.

In the British system of units the heat unit is the Btu (short for British thermal unit). One Btu is defined as the amount of heat required to raise the temperature of 1 lb (= 1/32 slug) of water by 1 F° (from 58.5°F to 59.5°F). It can be shown that

$$1 \text{ Btu} = 252 \text{ cal}$$

In both metric and British systems the specific heat of water is taken to be 1. As a consequence, the specific heats of other substances also have the same numerical value in both systems. The specific heats of some common substances are listed in the table of thermal properties in Appendix B.

EXAMPLE 7–2: A 6-kg block of aluminum (specific heat $s = 0.21$ cal/g-C°) is heated to 450°C and then immersed in an insulated container holding 20 kg of water initially at a temperature of 30°C. What will the final temperature of the water be?

SOLUTION: Let T be the final temperature of the water. Since the water and aluminum are in equilibrium, T is also the final temperature of the aluminum. That is, the temperature of the aluminum falls from 450°C to T, while the temperature of the water rises from 30°C to T. Equating the heat lost by the aluminum to the heat gained by the water, we have (using Eq. 7–9)

$$\begin{aligned} m_{\text{Al}}s_{\text{Al}}\,\Delta T_{\text{Al}} &= m_{\text{water}}s_{\text{water}}\,\Delta T_{\text{water}} \\ (6)(0.21)(450^\circ - T) &= (20)(1)(T - 30^\circ) \\ 567 - 1.26T &= 20T - 600 \\ 21.26T &= 1167 \\ \underline{T} &\underline{= 54.9^\circ\text{C}} \end{aligned}$$

CHANGE OF STATE: LATENT HEAT

It is well known that if heat is continually added to a block of ice, the temperature of the ice will rise to 0°C, at which point the ice will melt. Further addition of heat will bring the temperature of the liquid water up to 100°C, where the water will boil and change to steam. Black made careful studies of the changes of state that occur when water (or any other substance) is heated. He made the important discovery that *a change of state is a constant-temperature process.* That is, when a block of ice is melting, the temperature of the ice and of the melted water *remains at 0°C until the ice has completely melted.* Similarly, the temperature of a pot of boiling water will remain at 100°C until all the water has boiled off. Thus the melting and boiling of a substance represent situations in which the addition of heat *does not produce a change of temperature.* Changes of state caused by removal of heat follow the same rule: during the condensation of a gas to a liquid or the freezing of a liquid into a solid, the temperature remains constant.

Black gave the name of ***latent heat*** to the amount of heat required to bring about a change of state. The heat transferred in the solid–liquid transformation is known as the ***latent heat of fusion;*** it is 80 cal/g (144 Btu/lb) for water. The latent heat of fusion is the same for both freezing and melting. That is, 80 calories must be added to melt 1 gram of ice at 0°C, and 80 calories must be removed to freeze 1 gram of water at 0°C. For the liquid–gas transformation, 539 cal/g (970 Btu/lb) must be added or removed; this is referred to as the ***latent heat of vaporization.*** The numerical values of the latent heats of fusion and vaporization (as well as the temperatures of the melting and boiling points) change according to the pressure exerted on the substance. Latent heats at atmospheric pressure of a few common substances are listed in the table of thermal properties in Appendix B. Latent heats are normally given in calories per gram, so that to determine the quantity of heat required to produce a change of state in a given sample, one must multiply the latent heat by the mass of the sample.* Thus for a change of state the amount of heat added or removed is

$$Q = mL \tag{7–10}$$

where L is the latent heat of fusion or vaporization, as the case may be. Thus Black's conclusions can be summarized as follows: the addition or removal of heat under normal circumstances will result in either a change of temperature (governed by Eq. 7–9) or (if the substance is at its melting or boiling point) a change of state (Eq. 7–10).

EXAMPLE 7–3: Calculate the amount of heat required to change 10 grams of solid mercury at its melting point to gaseous mercury at its boiling point.

SOLUTION: The process occurs in three steps: (1) change of state from solid mercury to liquid mercury, (2) raising the temperature of liquid mercury from melting point to boiling point, and (3) change of state from liquid to gas. The heat required for each step must be calculated separately using Eqs. 7–9 and 7–10. The relevant data are taken from the table of thermal properties in Appendix B.

(1) *Change of state:* latent heat of fusion = 2.7 cal/g.

$$Q_1 = mL = (10\text{ grams})(2.7\text{ cal/g}) = \underline{27\text{ cal}}$$

(2) *Change of temperature:* melting temperature = −39°C; boiling temperature = 357°C; specific heat = 0.03 cal/g-C°.

$$Q_2 = ms\,\Delta T = (10\text{ grams})(0.03\text{ cal/g-C}°)(357 - (-39)) = \underline{118.8\text{ cal}}$$

(3) *Change of state:* latent heat of vaporization = 71 cal/g.

$$Q_3 = mL = 10(71) = \underline{710\text{ cal}}$$

$$\text{total } Q = Q_1 + Q_2 + Q_3 = \underline{855.8\text{ cal}}$$

THE CALORIC THEORY

In 1800 the prevailing theory was that of heat as a material substance. Even though Newton himself had suggested that heat was a form of motion, this idea did not catch on, mainly because a detailed theory that accounted for the common thermal phenomena in terms of heat as motion had not been worked out. On the other hand, Black (using the results of his calorimetric experiments) and others advanced a rather effective theory of heat as a material substance.

According to this theory, heat consists of an invisible substance that attaches itself to the fundamental particles of matter. The French chemist Antoine Lavoisier† gave the name of ***caloric*** to this substance, listing it as one of the chemical elements,

*In the British system the latent heat is given in Btu/lb, so that it must be multiplied by the *weight* of the sample in pounds.

†Antoine Lavoisier (1743–1794) was one of the founders of modern chemistry. His career was ended abruptly by the guillotine during the Reign of Terror. Even though he was recognized as one of the greatest living French scientists, he was condemned as a member of the nobility and a part-time tax collector for the French monarchy. He was also accused by Marat of stopping the circulation of air in Paris with his experiments. His death was justified with the statement "The Republic has no need of savants."

and the theory soon become known as the ***caloric theory*** of heat. According to this theory, when a body is heated, caloric flows into it and surrounds each atom of matter in the body. The temperature is a measure of the amount of caloric surrounding each individual particle. Caloric was assumed to be weightless, since careful measurements showed that the weight of a body does not change when it is heated or cooled. It was also assumed that caloric repels itself. This assumption, as we will see, is a very useful one for explaining the thermal properties of matter. As a material substance, caloric was supposed to obey a conservation law; that is, it cannot be created or destroyed, only transferred from one body to another.

On the basis of these few simple assumptions, the proponents of the caloric theory were able to explain virtually all thermal phenomena. As caloric is added to a body and surrounds the atoms, the repulsive force between the caloric counteracts the attractive forces between the atoms. As more caloric is added, the increased repulsion causes the atoms to move farther apart—thus explaining the thermal expansion of bodies. Further addition of caloric causes the arrangement of atoms to become disordered, with the result that the body changes from a solid to a liquid. Eventually, the repulsion becomes so great that the attractive forces are completely overcome, and the atoms separate from one another—resulting in a change of state from liquid to gas. The observation that different substances have different specific heats, melting and boiling points, rates of thermal expansion, and so on, could be explained in terms of the relative strengths of the attractions that the atoms of different substances have for caloric or for one another. The fundamental observation that heat always flows from a hot body to a cold body could be accounted for rather easily by the caloric theory: a hot body contains more caloric than a cold body; since caloric is self-repulsive, it is forced out of the hot body into the cold body, where there is less caloric and therefore less repulsion.

Black had to make further assumptions as to the properties of caloric in order to explain his discovery that there is no temperature change during a change of state. He proposed that the caloric enters into a sort of chemical combination with the atoms of matter during a change of state (i.e., ice + caloric = water). Thus the caloric is rendered *latent* (that is, it does not produce a temperature change). When a liquid freezes, this latent caloric is released and becomes *sensible.* In other words, it is detectable by the human thermal sense as heat.

The observation that friction causes heating posed some difficulty for the caloric theory, for when two bodies are rubbed against one another, their temperatures increase, even though there clearly is no caloric being added to them from any outside source. Proponents of the caloric theory hypothesized that the friction causes the caloric in the bodies to be forced out and made sensible. Specifically, they proposed that friction reduces the specific heats of the bodies, so that caloric escapes because of the reduced capacity of the bodies to retain heat.

Thus supporters of the caloric theory were able to account for the common thermal phenomena in a relatively simple and straightforward manner. Some found it difficult to accept the idea of a material substance being weightless, but this objection was really more philosophical than scientific. The comprehensiveness of the caloric theory was the overriding factor. Moreover, the concept of heat as a fluid was in concord with the then-prevalent notion that electricity and magnetism were carried by electric and magnetic fluids. It is not at all surprising that scientists at the beginning of the nineteenth century considered the caloric theory a highly successful theory.

THE FAILURE OF THE CALORIC THEORY

The overthrow of the caloric theory began with one of the most extraordinary characters to be found in the pages of history. Benjamin Thompson was born in 1754 in the town of Woburn, in the British colony of Massachusetts. His formal education was rather limited; poverty forced him to leave school at the age of thirteen and take a job as a storekeeper's apprentice. When he was seventeen years old, Thompson moved to Boston to work in another store. Here he took the opportunity to sit in on science lectures at Harvard College whenever possible. At the age of nineteen Thompson married a rich, thirty-two-year-old widow, and his fortunes began to change.

He was soon able to buy a commission as a major in the Colonial army. However, his sympathies were more with the British than with the Americans. When suspected (with good reason) of collaborating with the British, Thompson fled to England, abandoning his wife and daughter. In England he became involved in the war effort against the Colonies and amassed a large fortune by accepting bribes from manufacturers of military equipment. At this time he was apparently involved in selling British military secrets to the French. In 1783 Thompson went to Bavaria, where he rose quickly in the government hierarchy (even though he was acting as a British secret agent for the first few years of his stay). In 1791 he was made a count of the Holy Roman Empire; he took the name of Count Rumford, by which he is more commonly known. In 1795 Rumford found it necessary to leave Bavaria. He returned briefly to England, then moved on to Paris, where he became an adviser and confidant of Napoleon. Rumford died in 1814, an honored, widely respected, and very wealthy man.

What is perhaps most remarkable about Rumford is that he could find time in his incredibly active life for scientific inquiry. Yet he made many important contributions in the theoretical and applied sciences, and his fame as a scientist is well deserved. Rumford's life has been summarized rather accurately and succinctly as one of "unswerving honesty in science and unconscionable duplicity in politics."[1]

Rumford's interest in the theory of heat was stimulated by the observations he made when, as minister of war of Bavaria, he was in charge of the production of cannon. At that time the hole in the cannon was cut by friction between the cannon block and the boring tool (which was kept turning by horsepower). Rumford was struck by the considerable amount of heat generated in the process; it seemed to him that the quantity of heat released was far greater than the heat content of the boring tool and cannon combined. Subsequently, Rumford devised an experiment (shown in Figure 7–7) to show that a continual supply of heat could be produced by friction—enough to bring a large quantity of water to a boil! Rumford also collected the chips of metal cut out by the boring process and found that the specific heat of the metal chips had not changed, contrary to the caloric theory's explanation of friction.

According to the caloric theory, Rumford noted, heat is a material substance, and the cannon and boring tool can contain only a finite amount of caloric. Yet Rumford's experiment showed that through friction these bodies could be made to generate an inexhaustible supply of heat! Rumford concluded, therefore, that *heat could not be a material substance.* He decided he could explain his results only "by taking refuge in the very old doctrine which rests upon the supposition that heat is nothing but a vibratory motion taking place among the particles of the body."[2] Rumford believed the mechanical rotating motion of the boring tool was being

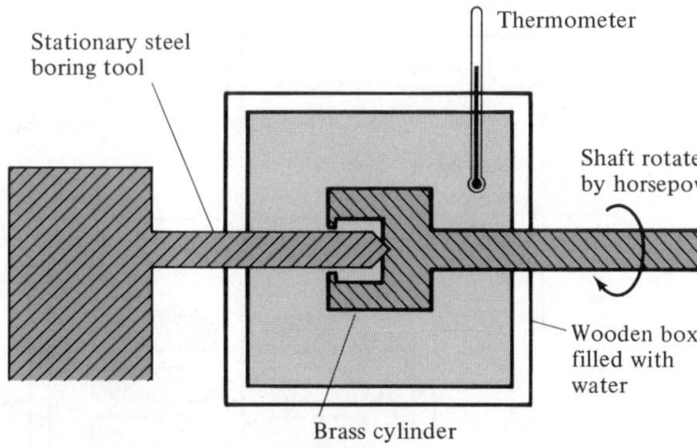

figure 7–7 **RUMFORD'S EXPERIMENT**

Friction between the steel boring tool (kept stationary) and the rotating brass cylinder generates heat, which is transmitted to the surrounding water. Continual turning of the cylinder causes the temperature of the water to rise at a steady rate. In one of Rumford's experiments with this apparatus, approximately 25 lb of water was made to boil after $2\frac{1}{2}$ hours of turning. The experiment showed that an inexhaustible supply of heat can be generated through friction.

transmitted to the individual particles of the boring tool and cannon. The resultant increase in the vibratory motion of the particles was then responsible for the sensation of heat.

The caloric theory was not immediately rejected on the basis of Rumford's findings, however. The theory was too firmly entrenched in the minds of scientists. Moreover, Rumford did not make any quantitative measurements on the amount of heat generated in relation to the mechanical work done, and in particular there was still no detailed theory of thermal phenomena in terms of heat as atomic motion. The caloric theory was not discarded until nearly fifty years after Rumford's energetic bones had been laid to rest.

By the 1840s scientists generally agreed that if heat is related to motion, a given amount of motion should produce a specific amount of heat. Hence there should be a direct mathematical relationship between the amount of motion and the quantity of heat generated by that motion. This belief was given a firm experimental basis by James Prescott Joule (1818–1889), an English brewery owner and amateur scientist. Joule believed that heat was a form of energy and therefore chose to measure the "amount of motion" in terms of the quantity of *work* done. Through a series of ingenious experiments, Joule was able to show that the amount of heat generated was in direct proportion to the mechanical work expended.

Figure 7–8 depicts one of Joule's experiments. A quantity of water in an insulated container was heated by the motion of paddle wheels turning through the water. The paddle wheels were driven mechanically by a slowly falling weight. The amount of work done was measured by the decrease in potential energy of the falling weight, and the heat generated was determined from the increase in temperature of the water. Joule's experiments showed that the application of 778 ft-lb of mechanical

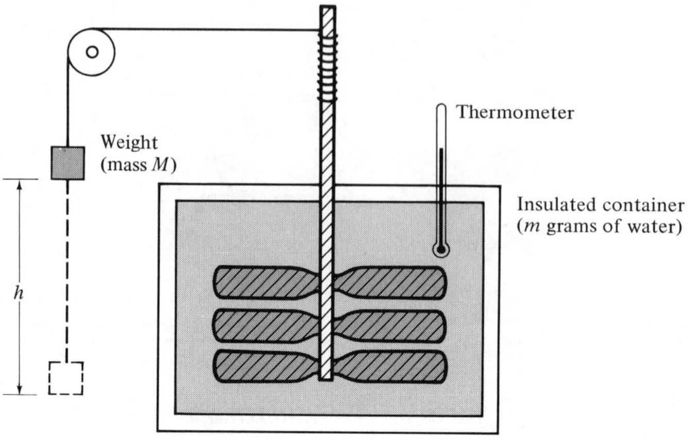

Mechanical work done $= Mgh$ Heat added $= ms\Delta T$

figure 7–8 **JOULE'S PADDLE WHEEL EXPERIMENT**

The weight is allowed to fall, causing the paddle wheel to rotate. The action of the paddles against the water causes the temperature of the water to increase. The amount of mechanical work done is determined by calculating the decrease in potential energy as the weight falls. If the mass of the water is known, the amount of heat added to the water by the action of the paddles can be calculated. The mechanical equivalent of heat is determined by equating the mechanical work done to the heat added.

work, if completely converted to heat, would raise the temperature of 1 lb of water by 1 F°; that is, 778 ft-lb of work is convertible to 1 Btu of heat. This ratio is known as the ***mechanical equivalent of heat:***

$$778 \text{ ft-lb} = 1 \text{ Btu}$$

In metric units

$$4.185 \times 10^7 \text{ ergs} = 4.185 \text{ joules} = 1 \text{ calorie}$$

These equations establish the rate at which mechanical work can be converted to heat. They can also be interpreted in another way: if heat is to be considered a form of energy, then the calorie and Btu are actually *energy units.* In this respect, the mechanical equivalent of heat represents a conversion factor between different units for the same quantity, in the same way that 2.54 cm = 1 inch.

Joule's experiments helped to confirm the thoughts of many of his contemporaries as to the nature of heat. The interchangeability of heat and work had been made evident not only by Joule's experiments in which work was converted to heat, but also by such devices as the steam engine (Figure 7–9) in which mechanical work could be extracted *from* heat. Thus by 1850 the caloric theory could no longer be taken seriously. Rumford had shown experimentally that heat could not be a material substance and had asserted that heat was a form of motion. This belief was firmly buttressed by Joule's demonstration of a direct relationship between heat and work, which, as we have learned, is in turn related to the kinetic energy of motion. The real significance of these developments, however, is not so much that they brought about the downfall of the caloric theory, but that by relating heat to work (and thereby to energy), they enabled scientists to extend the applicability of the principle of conservation of energy to cover all transformations in nature.

7-3

HEAT AS A FORM OF ENERGY

The middle of the nineteenth century saw the rapid development of the science of ***thermodynamics,*** which is the study of the transformations that occur between heat and mechanical energy. The growth of this new science was stimulated in part by Rumford's and Joule's discoveries and in part by the Industrial Revolution. The invention of the steam engine had had an enormous impact on society, and scientists and engineers were eager to increase the efficiency of this device.

The first important step in the development of thermodynamics was the extension of the principle of conservation of energy to include heat phenomena. In 1842 a

figure 7-9 **THE CONVERSION OF HEAT TO WORK**

Heat is used to boil water to make steam. Like any other gas, steam will expand when heated, and the pressure generated by this expansion can be used to produce motion. (a) The steam is forced through a nozzle and directed at a *turbine* (a series of blades attached to a wheel), setting it in rotation. (b) In a standard steam engine, the steam expands against a piston, setting it in motion. Through mechanical linkage the motion of the piston can be used to turn a wheel.

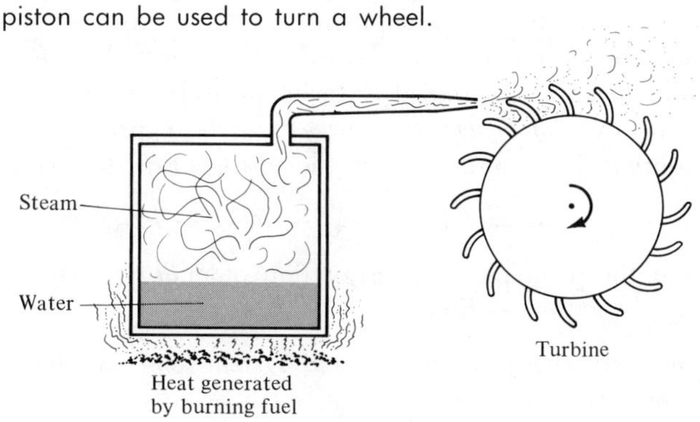

(a)

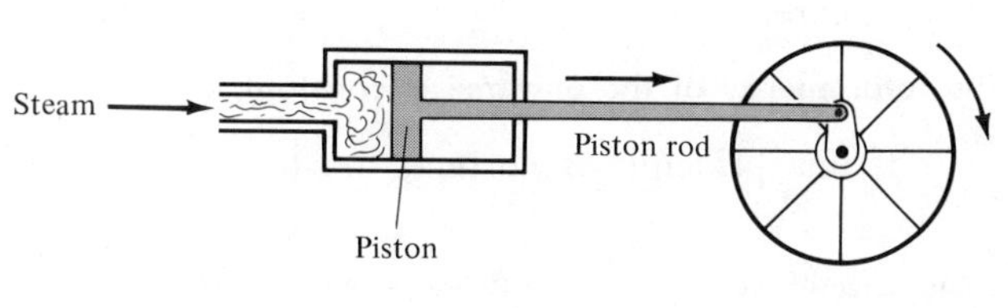

(b)

German physician, Julius Robert Mayer, published a paper in which he proposed that heat is a form of energy, equivalent to kinetic and potential energy, and that energy cannot be created or destroyed. Unfortunately, Mayer's work was largely ignored. Despondent over this lack of recognition and beset by personal problems, Mayer attempted suicide in 1849 and never regained full mental health afterward. In 1847 Joule presented the results of his experiments and asserted that energy is conserved; his work was enthusiastically supported by Thomson (the future Lord Kelvin), who was one of the founders of thermodynamics. In that same year the principle of conservation of energy was independently proposed by a brilliant young German scientist, Hermann von Helmholtz.* Working from these beginnings, a host of other scientists developed the basic principles and mathematical theory of thermodynamics.

THE FIRST LAW OF THERMODYNAMICS

In thermodynamic theory the principle of conservation of energy is known as the ***first law of thermodynamics.*** In pre-nineteenth-century physics only two kinds of energy were recognized—kinetic energy and gravitational potential energy—and energy was thought to be conserved only in those systems in which all the forces are conservative. Now several new forms of energy were introduced. Heat was recognized as a form of energy possessed by a body by virtue of the motions of the individual particles that make up the body. In this respect heat was classified as ***internal energy,*** and a distinction was made between the heat *transferred* from one body to another and the heat (or internal energy) *contained in* a body. Electric, magnetic, and chemical effects were also encompassed in the principle. For example, Joule did experiments to show that the heat–energy relationship is also obeyed when heat is generated in an electric circuit; that is, the work done by electric forces is converted to heat at a rate of 778 ft-lb per Btu. Helmholtz made a particular effort to show that living matter derives its motive power from the energy released in the chemical reactions governing the digestion of food.

With these developments in mind, we can now revise the principle of conservation of energy as described in Chapter 6. As before, the work done by conservative forces can be expressed in terms of a change in potential energy, but now the work done by *nonconservative* forces can be expressed in terms of a change in the heat content (internal energy) of the system. Equation 6–21 can now be expanded to read

$$E = KE + PE_{\mathrm{G}} + PE_{\mathrm{E}} + PE_{\mathrm{M}} + \cdots + \text{internal energy}$$

and the principle of conservation of energy, alias the first law of thermodynamics, can be stated as follows:

The total energy of an isolated system (one on which no outside forces are acting) is constant.

The universe is an isolated system (by definition, there is nothing outside of it); therefore:

The total energy of the universe is constant.

If this principle is correct, the total energy of the universe does not change

*The range of Helmholtz' interests was staggering. He made significant contributions to the physiology of the eye, the physiology of hearing, the theory of color vision, thermodynamics, and electromagnetic theory.

in time; all the processes that occur in the universe—the formation of galaxies, the birth and death of stars, the evolution of life forms, the very motions of our bodies—proceed without increasing or decreasing the energy content of the universe. The energy of any individual body or group of bodies can be increased (or decreased), but not without a corresponding decrease (or increase) in some other part of the universe. Thus any event, any change that takes place in the universe, can be (at least in principle) explained in terms of a transformation from one form of energy to another.

The principle of conservation of energy is so highly valued that it has proved more fruitful to dissolve apparent contradictions by postulating new forms of energy and new transformation processes than to give up the principle. (This was the case when Einstein proposed as a consequence of the theory of relativity that mass could be converted into energy, and vice versa, according to the equation $E = mc^2$.) On this basis, conservation of energy has managed to survive several profound revolutions in physics (notably relativity and quantum theory) and has risen to a position where it is revered as the most fundamental of all scientific principles.

THE SECOND LAW OF THERMODYNAMICS

Consider an automobile, traveling at high speed, which is brought to a complete stop when the driver applies the brakes. Where does the kinetic energy of the automobile go? We now recognize that the kinetic energy of the motion of the automobile has been transformed into an equivalent amount of heat, raising the temperature of the brake system, the tires, and the road. Yet there is no process by which the heat residing in these bodies could be collected together and transformed into work so as to return the automobile to its original speed. Similarly, there is no way by which the heat transmitted to the water by the paddle wheel in Joule's experiment could be returned to the paddle wheel, causing the paddle wheel to begin turning and raising the weight. Although processes of this sort would not violate the conservation of energy, they are never observed in the natural world! Here we see a clear limitation of the first law of thermodynamics. It does tell us that heat and work are equivalent, but it does *not* tell us whether it is possible to convert one completely to the other. Another principle is required to complete the picture.

We can think of a great many physical processes in which a quantity of work is converted entirely to heat—Rumford's and Joule's experiments, for example; in fact, any process in which motion is brought to a stop by friction represents complete conversion of work to heat. However, we know of *no physical process in which the net effect is the conversion of a quantity of heat entirely to work.* This restriction is not based on any theoretical argument; the complete conversion of heat to work is certainly permissible according to the first law of thermodynamics. It is simply a statement of how nature operates, according to our observations. Therefore, we must regard it as a basic law of nature—the ***second law of thermodynamics:****

It is impossible to devise a process that has as its only net effect the conversion of a given quantity of heat entirely into mechanical work.

Notice that the second law stipulates that the complete conversion of heat to work cannot be the *only net effect* of the process. It is possible to devise a process in which a given quantity of heat is supplied and an equivalent amount of work is

*There are, as we will see, many different but entirely equivalent ways to state this law.

done, but this will not occur without some ultimate change in the temperature or heat content of the working substance (i.e., the substance that absorbs the heat and does the work) or the surroundings. Thus in such processes some "hidden" source always supplies additional heat energy that is not converted into work.

A machine devised for the purpose of converting heat to work is known as a ***heat engine.*** Included in this category are steam engines, automobile and airplane engines, diesel engines, and so on. To illustrate how these machines operate, let us take the example of the steam engine. The essential steps in the operation of a steam engine are as follows (see Figure 7–10):

(1) A quantity of water is heated in a boiler (the heat is produced by the combustion of coal or some other fuel) and evaporates into steam.
(2) The steam enters a cylinder and, by expanding, drives a piston to produce mechanical motion.
(3) The spent steam is cooled and condensed back into water, which is returned to the boiler to be reheated.

This is an example of a ***cyclic process,*** because a fixed amount of working substance (the water) is returned at the end of the cycle to the same state that it was in at the beginning of the cycle. Thus there is no *net* change in the properties of the working substance; the water simply acts as the *carrier* of heat energy. In step 1 a given quantity of heat (Q_{in}) is introduced and transferred to the water. Some of this heat is transformed into work (W_{out}) in step 2, where the pressure of the steam sets the piston into motion. In step 3 the remainder of the heat energy originally added to the steam is removed (Q_{out}) in order to condense the steam back to water at the same temperature as in the beginning of the cycle. Experience has shown that there is no way to extract all of the heat added in step 1 and convert it completely to work in step 2; some of the added heat always remains to be removed in step 3. The impossibility of converting heat entirely to work in the steam engine is described in general terms by the second law of thermodynamics.

The formulation of the second law was stimulated by the work of Sadi Carnot,

figure 7–10 **THERMODYNAMICS OF A STEAM ENGINE**

Energy is introduced in the form of heat (Q_{in}). Some of this energy is converted to mechanical work (W_{out}). A portion of the energy remains as unconverted heat to be removed from the system (Q_{out}).

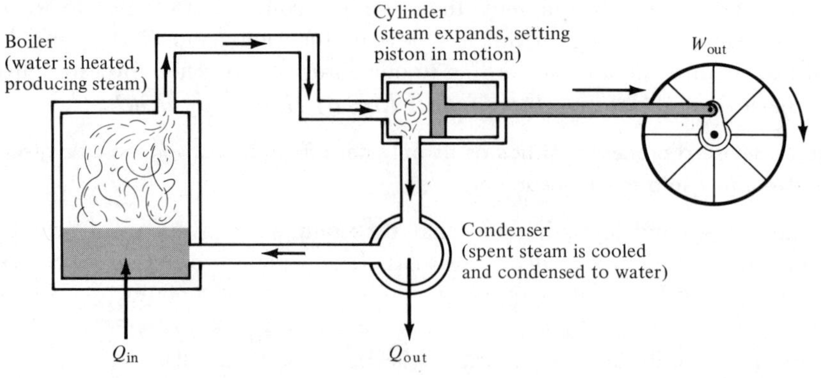

a French engineer. In 1824 Carnot published a paper entitled *Reflections on the Motive Power of Fire,* which was a theoretical analysis of the extraction of work from heat in steam engines. Carnot was able to show that heat cannot be converted entirely to work in a cyclic process, *not even in an ideal heat engine*—one that is perfectly insulated and frictionless and uses an ideal gas as its working substance.

Carnot's work was largely ignored at first, partly because he formulated his theory in terms of the caloric theory and did not clearly understand the connection between heat and energy. However, the work eventually came to the attention of such scientists as Lord Kelvin, who recognized its significance. Kelvin and the German physicist Rudolph Clausius were primarily responsible for improving and expanding Carnot's arguments.

Kelvin preferred to state the second law in the following manner:

It is impossible in a cyclic process to remove heat from a hot body and convert this heat to work without transferring some heat to a colder body.

This statement reflects the fact that in a cyclic heat engine (such as the steam engine) some of the heat absorbed from the burning fuel (the hot body) is unavoidably transferred to colder bodies (the pipes, cylinder, piston, condenser, etc.) and cannot be converted into mechanical work. Thus one necessary by-product of the operation of a heat engine is the transfer of a quantity of heat from a hot body to a cold body.

Clausius, however, chose to state the second law in a different, though entirely equivalent, form:

It is impossible for a cyclic process to have, as its sole effect, the transfer of heat from a cold body to a hot body.

Clausius' statement of the second law says, in effect, that heat cannot *by itself* flow from a cold body to a hot body. In this way we are reminded that the second law is ultimately related to the fundamental observation that heat flows spontaneously from a hot body to a cold body.

Clausius' form of the second law can be applied to the operation of refrigerators and air conditioners, since the purpose of these devices is to transfer heat from a cold body to a hot body (e.g., from the air inside a refrigerator to the air in the room, or in the case of the air conditioner, from the air inside a room or building to the air outside). Of course, refrigeration can be achieved simply by putting a large block of ice in the room, for ice will absorb a large quantity of heat (144 Btu/lb) in the process of melting. However, this is not a *cyclic* process; refrigeration will terminate when the ice has melted. A standard refrigerator or air conditioner operates in a cycle such that the properties of the working substance (called the ***coolant*** or ***refrigerant***) remain unchanged.

The basic refrigeration cycle is as follows (see Figure 7–11):

(1) The liquid coolant comes into contact with the air to be refrigerated. The coolant is initially at a lower temperature than the air; therefore heat flows *from* the inside air *to* the coolant, thus cooling the air. The heat transferred to the coolant (Q_{in}) causes it to evaporate.

(2) The evaporated coolant is compressed; this process elevates the coolant temperature so that it is higher than that of the outside air.

(3) The coolant is now brought into contact with the outside air, and heat flows *from* the coolant *to* the outside air. The loss of heat (Q_{out}) causes the coolant to condense back to a liquid (but at high pressure).

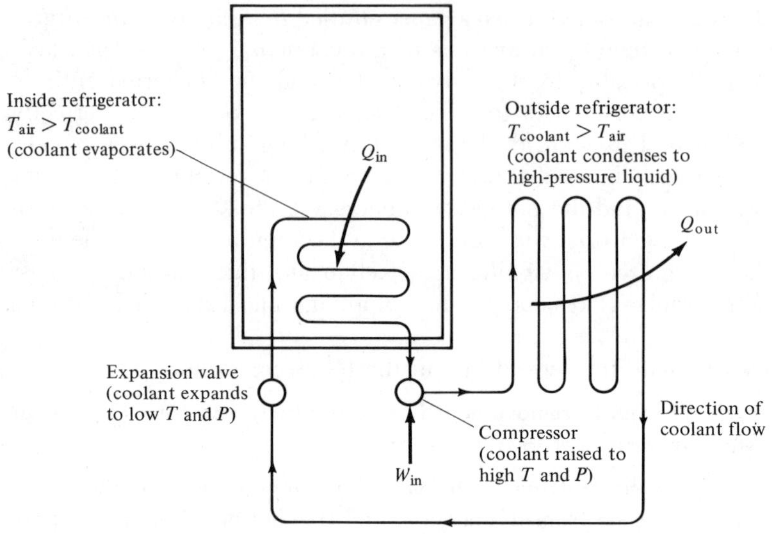

figure 7–11 **THERMODYNAMICS OF A REFRIGERATOR**

Heat (Q_{in}) is transferred from the air inside the refrigerator to the coolant. Additional energy (W_{in}) is supplied to the coolant to raise its temperature and pressure. Heat (Q_{out}) is then transferred from the coolant to the air outside the refrigerator.

(4) The high-pressure liquid coolant is allowed to expand; this process lowers the temperature of the coolant until it is below that of the inside air. The cycle is now complete.

We see, then, that the net effect is the transfer of heat from the colder inside air to the hotter outside air, without any *net* change in the coolant. However, this *cannot* occur spontaneously, according to Clausius' form of the second law. In fact, work (W_{in}) must be done on the coolant in order to compress it (step 2). Thus for a refrigerator to transfer heat from a cold body to a hot body energy must be supplied to the system. In common refrigerators and air conditioners this energy is supplied in the form of the electric energy needed to drive the compressor.

The first and second laws of thermodynamics place fundamental restrictions on the operation of machines in general. The first law, which is the principle of conservation of energy, tells us that we cannot get more energy out of any machine than we put into it—in effect ruling out the possibility of a perpetual motion machine.* The second law adds a further restriction by telling us that it is impossible to convert heat energy entirely to mechanical work—that is, that a heat engine with 100 percent efficiency cannot be built. With these results in mind, a physicist once proposed the following statements of the laws of thermodynamics:

First Law: You can't win.

Second Law: You can't break even.

* By a perpetual motion machine we do *not* mean a device in which friction has been completely eliminated so that there is no loss of energy to heat. *All* natural processes entail some sort of frictional heat loss, so that a perpetual motion machine cannot be constructed in this way. What is needed is a device that produces more energy than it is supplied with, so that excess energy is available to make up for heat losses.

As we will discover when we discuss the kinetic theory, there is more than just a passing resemblance between the laws of thermodynamics and the operating principles of a Las Vegas gambling casino.

7–4

THE KINETIC THEORY

By relating heat to energy, the laws of thermodynamics give a firm indication of a connection between the theory of heat and the theory of motion. This connecting link was of great significance to nineteenth-century physicists, for it suggested the possibility of a theory of heat in which all heat phenomena could ultimately be related to Newton's laws of motion. A theory of this sort was considered highly desirable, for it would represent yet another triumph for the already tremendously successful Newtonian view in which all phenomena in nature are ultimately explainable in terms of forces and motion. Since a fundamental assumption of such a theory is that heat is related to the kinetic energy of motion of the particles of a body, a picture of the internal structure is essential, for there must be some knowledge of what the particles of matter are like and how they are arranged in various bodies. At this point, then, we will turn our attention to the development of the atomic theory of matter.

THE ATOMIC THEORY

The origins of the atomic theory go back to the ancient Greeks. Tradition credits Leucippus (about 500 B.C.) as the founder of the theory, but Democritus (460–370 B.C.) was the greatest ancient exponent of atomism. In his view all material objects were made up of solid, impenetrable, and indivisible particles called ***atoms*** (from the Greek word *atomos*, meaning "indivisible"). Atoms were assumed to be the fundamental units of the universe; the world consisted entirely of atoms and the empty space in which they moved, and all else was illusion. In the words of Democritus: "By convention sweet is sweet, by convention bitter is bitter, by convention hot is hot, by convention cold is cold, by convention color is color. But in reality there are atoms and the void."[3]

However, Democritus defended the atomic theory on philosophical rather than scientific grounds. When Aristotle (who could not accept the existence of a vacuum and consequently believed that matter could be subdivided infinitely) eventually rejected the atomic theory, it was discarded. Nearly two thousand years elapsed before the theory was to be resurrected.

The concept of the atom first reappeared in modern times in the more speculative writings of such thinkers as Descartes, Boyle, and Newton. However, it was the English schoolteacher John Dalton (1766–1844) who revived atomism and placed it on a firm scientific foundation. Dalton was concerned with the problem of explaining the laws of chemical combination. Paramount among these was the ***law of definite proportions,*** which states that elements combine to form compounds according to fixed weight ratios. For example, when any quantity of water is decomposed into oxygen and hydrogen, it is always found by direct measurement that the weight of the oxygen in the sample is eight times the weight of the hydrogen. That is, there is a fixed 8:1 weight ratio of oxygen to hydrogen in any sample of water. Similarly, the measured weight ratio of carbon to oxygen in any given amount of carbon monoxide is always 3:4. By 1800 it had been well established through extensive experimentation that there is always a fixed proportion between the weights of the constituents in any chemical compound.

Dalton was able to provide a simple explanation of the law of definite proportions by making the following assumptions:

(1) Elements consist of tiny, indivisible particles called atoms, which are unchanged in chemical reactions.
(2) All atoms of the same element are identical; however, atoms of different elements have different properties and, in particular, have different weights.
(3) Chemical compounds are formed when the atoms of one element combine with the atoms of other elements to form ***molecules*** of the compound.

Thus elements combine to form compounds in a fixed weight ratio because the atoms of the individual elements themselves have fixed weights. If, for instance, we consider a carbon monoxide molecule to consist of one carbon atom and one oxygen atom, then in any given quantity of carbon monoxide there will be an equal number of carbon and oxygen atoms. If we find by experiment that there is a 3:4 ratio between the weights of carbon and oxygen in any sample of carbon monoxide, it means that the ratio of the weight of a single carbon *atom* to the weight of a single oxygen *atom* must be 3:4. In other words, an atom of carbon has three-fourths the weight of an atom of oxygen.

Dalton's findings, published in 1808, were well received, for the merits of his atomic hypothesis were quite evident. However, the experimental determination of accurate values for the relative weights of the atoms of different elements was filled with complications. (Dalton himself was a rather poor experimenter; his own experimental results for relative weights served, if anything, as a hindrance to the work of later experimenters.) It would be well outside the scope of this chapter to discuss any of these difficulties in detail; suffice it to say that it took chemists over fifty years to resolve the problems that arose and to establish a consistent table of atomic weights.

By 1860 the atomic theory had been developed to the point where it was fully consistent with the laws of chemical combination. In addition, a coherent picture of the structure of matter had begun to evolve. The key feature of this picture was that it was based on a *kinetic* theory of matter; that is, the basic particles (atoms or molecules)* were considered to be in *constant random motion.* The characteristic properties of the three states of matter—solid, liquid, and gas—could be generally accounted for in terms of the balance between the motions of the individual molecules and the restraining forces of attraction between the molecules. The concept of heat as a form of energy—specifically, the kinetic energy of the individual molecules—implied that all heat phenomena could ultimately be explained in terms of random molecular motion. Thus the atomic theory emerged as the unifying principle by which the laws of chemical combination, the properties of matter, and the nature of heat could all be encompassed within the realm of physics.

THE KINETIC THEORY OF HEAT

The first test of the kinetic theory was that it had to be able to explain thermal phenomena even more successfully than did the caloric theory. To begin with, the theory had to explain how heat is transferred from hot to cold bodies. If heat is not

*There may be some confusion as to the distinction between atoms and molecules. Normally, atoms are considered to be the basic units of a chemical *element,* and they cannot be changed by chemical means. A molecule is the basic unit of any chemical *compound* and consists of two or more atoms. In the text we will use the word "molecule" to refer to the fundamental unit of the substance.

a material substance, then what is it that is being transferred? How, for that matter, does the kinetic theory account for the varying degrees of hotness and coldness that can be attained by bodies?

To answer these questions, we must proceed from the basic assumption that heat is a manifestation of the random motion of the molecules of a body. Thus the distinction between a hot body and a cold one is that the molecules of the hotter body are in more rapid motion (and therefore possess more kinetic energy) than the molecules of the colder body. When two bodies are placed in contact, heat is transferred as a result of collisions between the molecules of the two bodies. In general, when a fast-moving body collides with a slower-moving one, the faster body will slow down, and the slower one will speed up; that is, the faster body will lose energy to the slower one. (This is necessary because both momentum and energy must be conserved in the collision.) Consequently, the molecules of a hot body will be slowed down by collisions with the molecules of a colder body, and there will be a net transfer of energy from the hotter body to the colder one. Thus in this context heat "flow" is viewed as the transfer of kinetic energy from the faster-moving molecules of the hot body to the slower-moving molecules of the colder body. We can now understand why heat always travels from hot bodies to cold bodies. In collisions it is not possible for a faster-moving body to gain energy at the expense of a slower-moving one—therefore a hot body cannot get hotter when placed in contact with a colder body.

It is *not* assumed in the kinetic theory that *all* molecules in a body have the same speeds or that all the molecules in a hot body are moving faster than all the molecules in a colder body. Rather, it is believed that the molecules tend to distribute themselves over a range of speeds, so that at any given instant there are some fast-moving molecules and some slow-moving ones in the same body. A molecule will not maintain any given speed for very long, for its speed will be changed frequently as a result of collisions with other molecules. It is obviously a practical impossibility to attempt to follow the motion and changes in speed of every molecule in a body, so that in the kinetic theory we are forced to deal in statistical averages. We will postpone until later a discussion as to how such averages are obtained, but let us assume at this point that it is possible to determine the *average speed* and *average kinetic energy* of the molecules of a given body. Then the distinction between a hot body and a cold body is that the average kinetic energy of all the molecules in the hot body is greater than the average for the cold body. It follows, moreover, that the attainment of thermal equilibrium between two bodies placed in contact means that as a result of many collisions the molecules of the two bodies have the same average kinetic energy.

It now becomes possible to take the significant step of interpreting the concept of temperature in terms of the kinetic theory. Since two bodies in thermal equilibrium are said to have the same temperature, it follows that *the temperature of a body is related to the average kinetic energy of the molecules of that body.* (Some mathematical justification for this assertion will be provided in the next section.) We can now also make a clear-cut distinction between heat and temperature. The *temperature* measures the *average* kinetic energy of the molecules, whereas the *heat* contained in a body is the *total* kinetic energy of all the molecules. Thus a cup of hot coffee has a higher *temperature,* but less *heat,* than the Arctic Ocean. The molecules in the coffee have a higher average kinetic energy than those in the Arctic Ocean, but because there are so many more molecules in the Arctic Ocean, its total heat content is much greater than that of the cup of coffee.

The phenomenon of heating by friction, which could not be handled successfully by the caloric theory, can be easily understood in terms of the kinetic theory. In any situation where there is heating by friction, there is always some outside source supplying mechanical energy (such as a boy scout rubbing two sticks together or a horse turning the boring tool in Rumford's experiment). This mechanical energy is transmitted to the individual molecules of the two bodies and adds to their random motion. The average molecular speed is thereby increased, and as a consequence the temperatures of both bodies rise.

Next we turn to the problem of the relationship between heat, temperature, and the states of matter. In the kinetic theory the internal structure of matter is determined by the balance between the intermolecular forces of attraction (which tend to bring the molecules into a highly regular structure) and the thermal motion of the molecules (which tends to disorder any regularities). In the solid state the intermolecular forces predominate; these forces (which are actually the electric forces of attraction between molecules) tend to keep the molecules in relatively fixed positions. The thermal motions of the molecules consist of small vibrations about these equilibrium positions. In some solids the molecules line up in a very regular geometric array; the structure of such solids can be visualized as a sort of lattice with the molecules located at the intersections. (See Figure 7–12a.) The molecules are rather closely packed, being no more than a few diameters apart. When the solid state of matter is depicted in this way—as a highly ordered arrangement of molecules held rather tightly in place by intermolecular forces—it is easy to understand why solid matter is hard, not easily deformed, and able to withstand the application of considerable force without breaking.

When a solid is heated, the thermal motions of the molecules are increased; that is, the molecules tend to vibrate more rapidly and with greater amplitude. This has the effect of making the molecules increase their distances from one another, thereby weakening the force of attraction between neighboring molecules. The increased separation of molecules from one another suggests that the body should expand in size. Thus the kinetic theory is able to account for the common phenomenon of thermal expansion. In addition, the weakening of the intermolecular forces upon expansion accounts for the fact that solids tend to become more malleable and less able to withstand stress as they are heated.

figure 7–12 **KINETIC THEORY MODEL OF THE STATES OF MATTER**

(a) In the solid state the molecules are bound together into a relatively rigid, orderly structure. (b) In the liquid state the overall structure breaks down; the molecules are bound together in clumps of a few molecules each. (c) In the gaseous state the molecules are completely independent of one another and are in random motion.

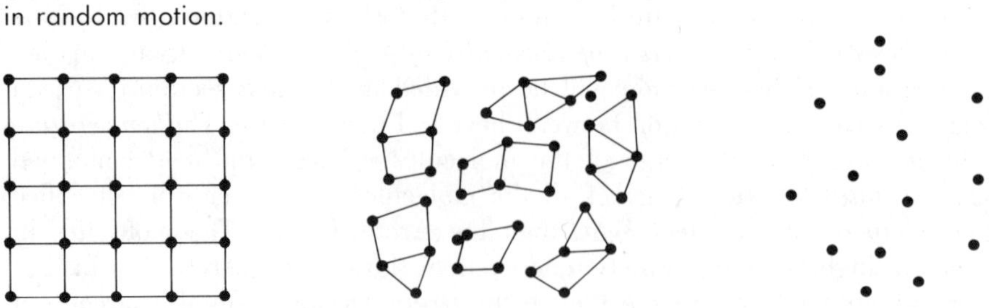

(a) Solid (b) Liquid (c) Gas

With further addition of heat, the vibratory motion of the molecules becomes so great that a rigid structure can no longer be maintained. The orderly, overall array of molecules that characterizes the solid state breaks down into small clumps of perhaps five to ten molecules bound together (Figure 7–12b). This breakdown marks the transition from the solid to the liquid state. With the loss of the regular solid-state structure, matter can no longer maintain a fixed shape; this is why liquids always assume the shape of their container. The lack of strong cohesion between the small clumps of molecules is what allows us to "pour" liquids, to mix them with one another, and to move objects through them. Nevertheless, in the liquid state the molecules are still packed together rather closely, so that liquids are almost as difficult to compress as are solids.

When heat is added to a substance at its melting point, the heat goes to changing the state rather than to raising the temperature—this is the phenomenon of *latent heat* first discovered by Black. This effect is quite understandable in terms of the kinetic theory, for at the melting point the heat energy being added (the latent heat of fusion) is used to overcome the intermolecular forces rather than to increase the kinetic energy of motion of the molecules themselves. Furthermore, we can account for the fact that different substances melt at different temperatures in terms of the strength of the intermolecular forces: the stronger these forces are, the more energy is required to overcome them and the higher the temperature of the melting point.

As compared to the solid state, the liquid state is one of relative disorder. The molecules are in rather rapid thermal motion; they continually break away from one clump of molecules and reattach themselves to another. Nevertheless, the intermolecular forces are still strong enough to keep the molecules packed closely together. With the further addition of heat, however, individual molecules may gain sufficient kinetic energy to overcome these forces and escape from the liquid surface. These escaped molecules are now in completely free random motion, virtually independent of any intermolecular forces. Their only interactions with other molecules occur when they directly collide with them. Under these conditions matter is in the form of a gas (Figure 7–12c).

The process of vaporization marks the transformation from the liquid to the gaseous state. At the boiling point of a substance, heat transmitted to its molecules provides them with the energy to overcome completely the attractive forces of the other molecules; that is, the energy supplied serves to counteract the potential energy associated with the attractive forces rather than to increase the kinetic energy of the molecules. Thus no temperature change is seen during the vaporization process.

Because the molecules of a gas are unaffected by any cohesive forces and are moving at high speeds, a gas will tend to expand to fill its container. The low density of gases under normal temperature and pressure conditions and the ease with which gases can be compressed are evidence for the relatively large distances between the molecules in a gas. As an example of the expansion that occurs when matter changes from liquid to gas, consider the fact that a quantity of liquid water at 100°C and 1 atm pressure will expand to 1600 times its original volume when completely evaporated!

Thus on the basis of a few simple assumptions, we have been able to use the kinetic theory to account for the major thermal properties of matter. Yet what we have discussed represents a rather superficial sampling of the wide range of phenomena that can be accounted for in terms of the kinetic theory. When developed in full mathematical detail, the kinetic theory is able to make rather accurate predictions of the values of such properties as the specific heat, viscosity, and electrical conductivity of a given substance. The mathematical theory of the properties of solid matter—more commonly known as ***solid state physics***—has been advanced to a rather high degree of complexity, and as a result many important technological advances have been made, ranging from the development of new metal alloys to the invention of the transistor. Few scientific theories are as comprehensive as the kinetic theory, and in this respect it must be considered one of the most successful theories in the realm of science.

THE KINETIC THEORY OF GASES

As an introduction to some of the mathematical aspects of the kinetic theory, we will now examine in more detail the application of the theory to the nature of the gaseous state. The kinetic theory of gases makes the following assumptions about the properties of the molecules of a gas:

(1) The molecules are very far apart. (That is, the distances between the molecules are large compared to the size of the molecules themselves.)
(2) Intermolecular forces are negligible (except in collisions between the molecules).
(3) The molecules are in constant, random motion.
(4) In collisions between molecules, momentum and kinetic energy are conserved.

In Chapter 4 we noted that a gas is able to exert *pressure;* that is, it exerts an outward force on the walls of its container. According to the kinetic theory, the pressure that a gas exerts is a result of the forces exerted by molecules as they collide with the walls of the container. (This explanation was, in fact, first proposed by Daniel Bernoulli in 1738.) Proceeding from this assumption, we can, through a relatively simple calculation, arrive at some rather interesting results.

Figure 7–13a depicts a cube, length l on a side, which contains a gas. It is assumed that the cube contains N molecules of the gas; each molecule has a mass m and is moving with an average speed v. We propose to determine the pressure of the contained gas by computing the forces exerted by the molecules as they collide with the walls of the cube. Because of the symmetry of the container and the fact that we consider the molecules to be in random motion, the force exerted on any one face of the cubical container will be the same as the force exerted on any other face. Let us then compute the force on side $ABCD$, which is drawn perpendicular to the x axis.

To determine the force exerted by a single molecule as it collides with the wall, we will use Newton's second law in the form *force equals the rate of change of linear momentum* (Eq. 6–2):

$$F = \frac{\Delta p}{\Delta t} = \frac{\Delta(mv)}{\Delta t}$$

Since the molecules are in random motion, they will not, on the average, collide head on with the wall. The velocity vector $\mathbf{v}$ will generally have three components: $\mathbf{v}_x$, $\mathbf{v}_y$, and $\mathbf{v}_z$ corresponding to the three space directions (Figure 7–13b). In a collision with side $ABCD$, only the component $\mathbf{v}_x$ will be changed, since $\mathbf{v}_y$ and $\mathbf{v}_z$ are both parallel to this side. Thus as a result of collisions with side $ABCD$, $\mathbf{v}_y$ and $\mathbf{v}_z$ remain constant, while $\mathbf{v}_x$ is changed in direction (Figure 7–13c) through an angle

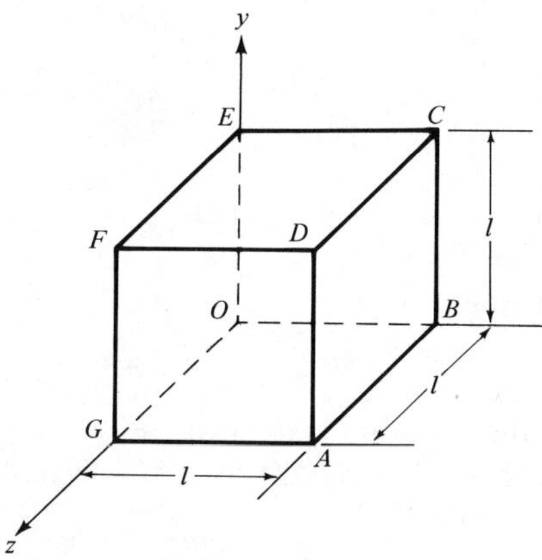

(a)

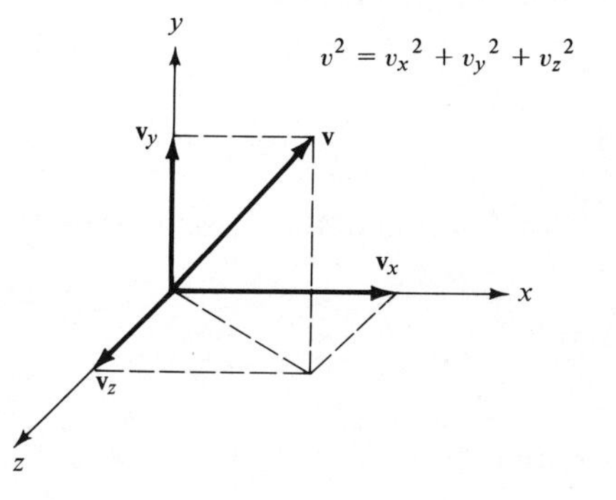

(b)

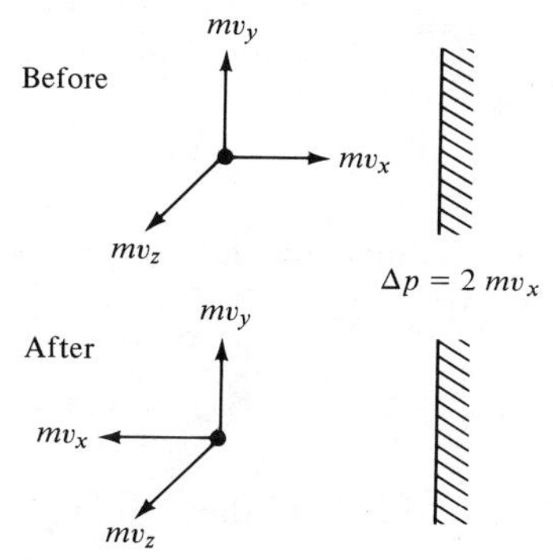

(c)

figure 7–13 **DIAGRAM FOR DERIVATION OF IDEAL GAS LAW**

of 180°. There is no change, however, in the *magnitude* of $\mathbf{v}_x$, since the molecule cannot lose kinetic energy as a result of the collision. But linear momentum is a vector quantity. Hence the change in direction implies a change in the linear momentum. The magnitude of the change is

$$\Delta p = mv_x - (-mv_x) = 2mv_x$$

Therefore

$$F = \frac{2mv_x}{t}$$

In this equation t represents the average collision time. To estimate t, we assume that the *next* collision with the side *ABCD* will occur after the molecule has traveled the width of the cube, bounded off side *EFGO*, and traveled back to *ABCD*. Thus between collisions the molecule will travel a distance of $2l$ in the x direction with a speed v_x. The time between successive collisions with *ABCD* is

$$t = \frac{\text{distance}}{\text{speed}} = \frac{2l}{v_x}$$

Substituting into the previous equation gives

$$F = \frac{2mv_x}{2l/v_x} = \frac{mv_x^2}{l}$$

On the average the motion of a molecule should not show a preference for any one direction; that is, we expect $v_x = v_y = v_z$. The total speed v is related to its components by the equation*

$$v^2 = v_x^2 + v_y^2 + v_z^2$$

If $v_x = v_y = v_z$, the equation reduces to $v^2 = 3v_x^2$; hence $v_x^2 = v^2/3$. Thus the average force exerted on side $ABCD$ by a single molecule can be written as

$$F = \frac{mv^2}{3l}$$

There are N molecules in the container; therefore the total force exerted on $ABCD$ by all the molecules is simply

$$F_{tot} = \frac{Nmv^2}{3l}$$

Since by definition the pressure $P = F/A$, and the area of $ABCD$ is l^2, we have

$$F_{tot} = PA = Pl^2 = \frac{Nmv^2}{3l} \qquad \text{or} \qquad Pl^3 = \tfrac{1}{3}Nmv^2$$

But l^3 is the volume V of the box; thus

$$PV = \tfrac{1}{3}Nmv^2$$

Rearranging the terms somewhat, we have

$$PV = (\tfrac{2}{3}N)(\tfrac{1}{2}mv^2)$$

We can identify $\frac{1}{2}mv^2$ as the ***average molecular kinetic energy*** KE_{av}. Then the equation becomes

$$PV = (\tfrac{2}{3}N)(KE_{av}) \tag{7–11}$$

Now let us compare this result with what is already known about the properties of gases. Earlier in the chapter we learned about the ideal gas law, which was based in part on the experimental findings of Boyle and Charles. The ideal gas law (Eq. 7–6) states that $PV/T = K$, or

$$PV = KT$$

Experiments have shown that the constant K depends on the mass of the gas sample. The mass in turn depends on N, the number of molecules of the gas in the sample. Let us therefore set $K = Nk$, where k is a new constant. The ideal gas law becomes

$$PV = NkT \tag{7–12}$$

*Remember that for a two-dimensional vector $A^2 = A_x^2 + A_y^2$. It is easy to show by using the Pythagorean theorem that in three dimensions $A^2 = A_x^2 + A_y^2 + A_z^2$.

If the kinetic theory is to be a useful theory, its predictions must concur with the results of experiments. In particular, we would expect Eq. 7–11 (which follows from the premises of the kinetic theory) to be in agreement with Eq. 7–12 (which is based on experimental measurements of the properties of gases). Comparing Eqs. 7–11 and 7–12, we have

$$PV = (\tfrac{2}{3}N)(KE_{\text{av}}) = NkT$$

so that

$$KE_{\text{av}} = \tfrac{3}{2}kT \tag{7–13}$$

Thus we have a mathematical basis for the statement (made in the previous section) that *the temperature of a gas is directly proportional to the average molecular kinetic energy.*

Note that the temperature being considered here is the *absolute* (Kelvin) temperature. Equation 7–13 thus provides us with an explanation for the existence of a lower limit to temperature (i.e., absolute zero). Note that if $T = 0$, $KE_{\text{av}} = 0$. In order for the average kinetic energy to be 0, all the molecules must have zero speed. Thus according to the kinetic theory picture, *at absolute zero, molecular motion stops.*

The constant k in Eq. 7–13 can be evaluated from experimental data on gases. It is found that k has the same value for all gases. This number has come to be known as the ***Boltzmann constant,*** and it is considered a fundamental physical constant. In CGS units the Boltzmann constant has the value

$$k = 1.380 \times 10^{-16}\ \text{erg}/^\circ\text{K}$$

The equations that have been derived from the assumptions of the kinetic theory can now be used to calculate some properties of molecules. For example, we can use Eq. 7–12 to calculate the number of molecules in a given volume of gas. For instance, in 1 cm^3 of any gas at $T = 0^\circ\text{C}$ (273°K) and $P = 1$ atm pressure (1.013×10^6 dynes/cm^2), the number of molecules is

$$N = \frac{PV}{kT} = \frac{(1.013 \times 10^6\ \text{dynes/cm}^2)(1\ \text{cm}^3)}{(1.38 \times 10^{-16}\ \text{erg}/^\circ\text{K})(273^\circ\text{K})} = 2.69 \times 10^{19}\ \text{molecules}$$

This number (known to scientists as the ***Loschmidt number***) gives us our first inkling as to the minuteness of the molecule. We have found that 1 cm^3 of a gas (which is matter in a relatively *rarefied* form) contains nearly 27,000,000,000,000,-000,000—read as 27 quintillion—molecules.

Next, it is possible to determine the mass of a single atom. By direct weighing, one finds that the mass of 1 cm^3 of hydrogen gas (H_2) at 0°C and 1 atm pressure is equal to 8.99×10^{-5} gram. Since this mass of hydrogen is known to contain 2.69×10^{19} molecules, it follows that the mass of *one* hydrogen molecule is

$$\frac{\text{mass of 1 cm}^3}{\text{no. of molecules in 1 cm}^3} = \frac{8.99 \times 10^{-5}\ \text{gram}}{2.69 \times 10^{19}\ \text{molecules}}$$
$$= 3.34 \times 10^{-24}\ \text{gram/molecule}$$

Since a hydrogen molecule is known to consist of two hydrogen atoms, the weight of a single hydrogen atom is half of 3.34×10^{-24} gram, or

$$m_{\text{H atom}} = 1.67 \times 10^{-24} \text{ gram}$$

Certainly, the fundamental particle of matter must be quite tiny indeed.

Finally, we are able to estimate the average speed of the molecules in a gas. From Eq. 7–13,

$$KE_{\text{av}} = (\tfrac{1}{2}mv^2)_{\text{av}} = \tfrac{3}{2}kT$$

so that*

$$v_{\text{av}} = \sqrt{\frac{3kT}{m}} \qquad (7\text{–}14)$$

For hydrogen molecules at $0°\text{C} = 273°\text{K}$,

$$v_{\text{av}} = \sqrt{\frac{3(1.38 \times 10^{-16} \text{ erg}/°\text{K})(273°\text{K})}{3.34 \times 10^{-24} \text{ gram}}} = 1.84 \times 10^{5} \text{ cm/sec}$$

which is slightly more than 1 mi/sec! Thus through the kinetic theory we have obtained our first picture of the nature of matter. A thimbleful of gas turns out to contain a fantastically large number of extremely tiny bodies, which are in perpetual random motion at enormous speeds.

7–5

STATISTICAL MECHANICS

The first attempts at a mathematical description of molecular motion in gases (made primarily by Joule and Clausius) were, as we have seen, relatively fruitful. However, the inherent weakness of this approach lay in its oversimplification of the situation, particularly in the assumption that (for calculational purposes) all molecules are moving with the same average speed. A more rigorous kinetic theory was needed, one by which scientists could determine the ***velocity distribution*** of the molecules—that is, an equation giving the range of velocities of the molecules and the number of molecules moving at any given velocity under specified conditions. In principle, this problem could be solved by the use of the classical methods of Newtonian mechanics, for if the positions and velocities of the molecules are known at any given instant, their future paths can be calculated. However, since there are on the order of 10^{19} molecules in each cubic centimeter to be considered, it is clearly highly impractical to adopt such an approach—for the solution of 10^{19} simultaneous equations is far beyond the capacity of even the most sophisticated modern-day computer.

In the face of the impossibility of gaining *exact* knowledge about the velocities of individual molecules, physicists had no recourse but to settle for *statistical* knowledge. With such information they would be able to predict the *overall* behavior of a large aggregate of molecules, although nothing could be said about the behavior of any single molecule. A similar situation arises when we attempt to predict whether a coin flipped into the air will come down heads or tails. We cannot predict with

* Actually, Eq. 7–14 does not give the average speed, but the square root of the average squared speed ($\sqrt{v^2{}_{\text{av}}}$) which is usually called the root mean square (rms) speed. A more precise mathematical treatment gives $v_{\text{av}} = \sqrt{8kT/\pi m}$.

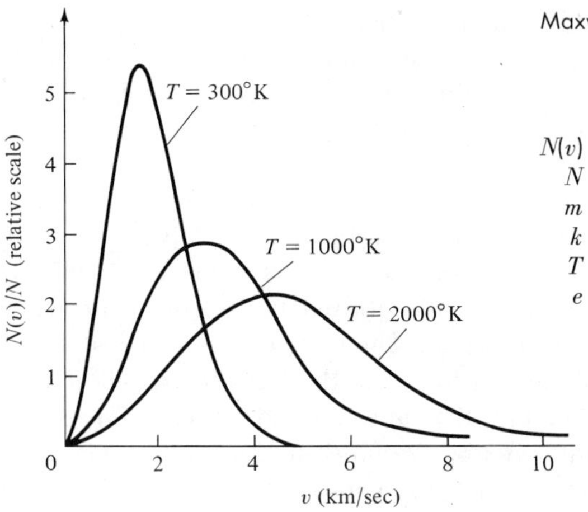

Maxwell–Boltzmann distribution equation:

$$\frac{N(v)}{N} = 4\pi\left(\frac{m}{2\pi kT}\right)^{3/2} v^2 e^{-mv^2/2kT}$$

$N(v)$ = number of molecules having speed v
N = total number of molecules
m = mass of molecule
k = Boltzmann constant (1.38×10^{-16} erg/°K)
T = absolute temperature (°K)
e = 2.718 = base of natural logarithms

figure 7–14 **THE MAXWELL–BOLTZMANN DISTRIBUTION OF MOLECULAR SPEEDS**

Note that at higher temperatures the molecules tend to be distributed over a wider range of speeds. The distributions shown are for hydrogen gas (H_2).

certainty what the result of a single flip will be; however, we *can* predict that if the coin is flipped one thousand times, it will come down heads very nearly five hundred times. Similarly, given a large population of human beings, we can predict what percentage of them will live past the age of seventy, even though we cannot predict when any *given* individual from that population will die. When dealing with individuals, the best that can be done is to state a ***probability.*** This is a number that represents the likelihood of a given event (on a scale in which 0 represents impossibility and 1 represents certainty) based on past observations. Thus we may say that the probability of a coin coming up heads is $\frac{1}{2}$ (meaning that it is equally likely to come up heads or tails), or that the probability of a sixty-year-old male living to age seventy is 0.30 (based on the observation that of the sixty-year-old males in similar circumstances for which data have been obtained, 30 percent have lived to age seventy). To be sure, statistical theories lack the certainty and precision that we have come to expect from science, but they are not without value.

The first successful application of the principles of probability and statistics to the problem of determining the distribution of molecular speeds was made by James Clerk Maxwell, who in 1860 derived a formula for the distribution of molecular speeds in gases. In 1885 Ludwig Boltzmann expanded upon Maxwell's work and placed Maxwell's results on a firmer theoretical basis. The equation for the distribution of molecular speeds has come to be known as the ***Maxwell–Boltzmann distribution;*** it is shown graphically in Figure 7–14. Boltzmann showed mathematically that when a collection of molecules has a constant total energy E that is to be divided among the molecules, collisions will cause the energy to be apportioned among the molecules so as to lead to the Maxwell–Boltzmann distribution of speeds as the equilibrium state. Moreover, it was shown that the Maxwell–Boltzmann distribution is the most *probable* distribution (i.e., the one with the greatest probability of occurrence). We might imagine a situation in which the total energy E is concentrated in one particle while the others are at rest, or one in which all molecules have exactly the same energy.

Such distributions are possible, but highly unlikely—that is, their probability is low. Of all the possible ways of distributing energy among the molecules, the Maxwell–Boltzmann distribution has the highest probability of occurrence. Thus Boltzmann proposed that the molecules of a gas tend toward the most probable distribution as the equilibrium state.

Underlying Boltzmann's proposal is the principle that an ordered arrangement of the molecules is less likely to occur than a disordered arrangement. A highly ordered arrangement of a system of molecules, such as one in which all molecules have the same speed or one in which all the energy is concentrated in one molecule while the others are at rest, would have a low probability. The system of molecules always tends to the Maxwell–Boltzmann distribution, which has the highest probability.

It follows by extension of this principle that any isolated system of particles will tend to attain its most probable state. A few examples will make this principle clear. Consider a brand-new deck of cards. When the package is opened, it is found that the cards are arranged in a highly ordered manner—in numerical sequence by suit. Once the deck is shuffled, however, the order is destroyed, and after repeated shuffling, there is no semblance of an orderly sequence to the cards. Each successive shuffle produces a different, more disordered, arrangement. What is the likelihood that the original sequence will turn up as a result of one of these shuffles? Experience tells us that the probability that the original sequence will reappear is exceedingly small. Obviously, there are many more ways to arrange fifty-two cards in a disordered sequence than in an ordered sequence. Thus the effect of shuffling is to change the system from an ordered (and less probable) arrangement to a disordered (and more probable) one.

Let us consider another example. Salt is carefully poured into a jar that is already half filled with pepper. We have a relatively ordered system—all the pepper grains are on the bottom, and all the salt grains are on top. We now begin to shake the jar. After sufficient shaking the salt and pepper grains should be completely mixed. By shaking the jar, we have destroyed the original order. The probability is exceedingly small that by continued shaking we might eventually separate the mixture back into a layer of salt and a layer of pepper. The salt–pepper system has attained a state of maximum disorder.

This statistical approach provides us with a theoretical interpretation for the second law of thermodynamics, which deals with the impossibility of converting heat entirely into work. Such a process would involve the transformation of disordered motion (the random motion of molecules) entirely into ordered motion (the mechanical motion of a body). For example, the complete conversion of heat to work in a steam engine would require that the molecules bounce off the piston in such a way that they transfer *all* their kinetic energy to the piston (causing it to move)—that is, all the molecules would come to a stop after colliding with the piston. Although such an event is not impossible, it is highly improbable—so improbable that it has never been observed. The statistical kinetic theory thus provides us with an added insight as to the meaning of the second law—namely, that the complete conversion of heat into work is not strictly impossible. It could conceivably occur as the result of a chance statistical fluctuation, although the likelihood of such an event is exceedingly small.

ENTROPY: THE HEAT DEATH OF THE UNIVERSE

One of Rudolph Clausius' important contributions to thermodynamics was the invention of the concept of ***entropy.*** Although the exact definition of this quantity is a bit beyond the scope of this chapter, we can say that entropy is a physical property of a body that changes whenever heat is added or removed. Bodies can be considered to possess a quantity of entropy just as they possess a certain amount of energy.

Clausius was able to prove that *in an isolated system in which real physical processes take place, the total entropy of the system increases.* The universe is an isolated system; therefore the entropy of the universe is steadily increasing. It is possible for real processes to occur in which the entropy of a body decreases, but these are always accompanied by a larger increase in entropy in some other part of the universe. Thus there is always a *net* increase in the total entropy of the universe. This increase in entropy can be considered a mathematical measure of the "unavailability" of heat energy, which, by the second law, cannot be converted entirely back into mechanical work. Clausius was thus able to express the first and second laws of thermodynamics in the following way: "The energy of the universe is constant; the entropy of the universe tends always toward a maximum."

In the kinetic theory the meaning of entropy becomes more apparent: *entropy is a measure of the degree of disorder of a system.* Consequently, the entropy increases in any process in which ordered motion is changed into disordered motion. The conversion of work to heat by friction is an example of such a process; the ordered, mechanical motion of a body as a whole becomes converted into the disordered, random motion of its molecules. Accordingly, any process involving friction always involves an increase in entropy. The flow of heat from a hot body to a cold body is also a process in which a relative order (the faster-moving molecules in one body, the slower-moving molecules in the other) is destroyed.

The concept of entropy has considerable physical and even philosophical meaning. Processes in which mechanical energy is converted into heat, processes involving friction, processes in which heat flows from a hot body to a cold body, all result in an increase in entropy. Thus the statement "the entropy of the universe tends always to a maximum" contains the message that the trend of these processes cannot be reversed, that all sources of energy will eventually be degraded into heat, that heat will continue to flow from hot bodies to cold bodies until all objects in the universe are at the same temperature. In this ultimate state—often referred to as the *heat death of the universe*—no energy transfer is possible; the entropy of the universe will then be at a maximum. All the matter in the universe will have reached a state of maximum disorder in which *all* motion is random molecular motion. This inexorable march of the universe toward its heat death provides us with the knowledge that the lifetime of the universe is finite, for if the universe has always been—if it has existed for an infinite time—it would surely have reached its heat death by now.

On the other hand, if we consider the present status of the universe, we find that there is a great deal of usable energy still available. Moreover, *all* sources of energy must be dissipated before the heat death of the universe is reached—which means that all atoms must disintegrate to release the energy stored within their nuclei. The heat death of the universe is certainly a long way off—tens of billions of years, at least (although there is no way to calculate how long it will take).

Finally, we must remember that the laws of thermodynamics follow from observations made here on earth, and we must recognize the dangers of generalizing to the universe as a whole. We are not at all certain that the physical principles we have devised are applicable uniformly throughout the universe, nor are we confident that we have identified all possible sources of energy. Thus the prospect of the heat death of the universe, although it leads to interesting philosophical speculations, is nothing more than a fanciful notion to the practicing scientist.

QUESTIONS FOR REVIEW

1 State the law that describes what happens when two bodies of different degrees of hotness and coldness come into contact. What is meant by the term "thermal equilibrium"?

2 Give a definition of the term "temperature." Specifically, what property of a body does the term describe?

3 Explain how a thermometer works. Identify the basic physical principles involved in its use as a device for measuring the temperature of a body.

4 Explain why the temperature measured by a thermometer is independent of the choice of thermometric substance.

5 How are the Fahrenheit and Celsius temperature scales defined? What are the shortcomings of these temperature scales?

6 Describe the experiments of Charles and Gay-Lussac on the thermal expansion of gases. How did the results of their experiments point to the existence of a minimum temperature?

7 Explain how the numerical value of the absolute zero temperature is determined.

8 How is the Kelvin temperature scale defined? What is the relationship between the Kelvin and Celsius temperature scales?

9 Describe the properties of an ideal gas. If no such gases exist in nature, what is the value of this concept?

10 State and explain (a) Charles' law, (b) Boyle's law, and (c) the ideal gas law.

11 State and explain the equation relating the heat added to, or removed from, a body to the change in temperature produced (assuming no change in state takes place). In what units is heat measured?

12 What does the specific heat of a substance measure? How is specific heat defined?

13 Describe the changes that take place in the temperature and state of a substance at its melting or boiling point when heat is added.

14 What is meant by the latent heat of fusion of a substance? The latent heat of vaporization?

15 Outline the basic assumptions of the caloric theory. Show how the caloric theory can be used to explain (a) the flow of heat from a hot body to a cold body, (b) thermal expansion and contraction, (c) change of state and latent heat, and (d) heating by friction.

16 Describe Rumford's "cannon-boring" experiment. What conclusions did Rumford come to as a result of this experiment?

17 Describe Joule's experiment demonstrating a direct relationship between heat and work. What is meant by the mechanical equivalent of heat?

18 State and discuss the first law of thermodynamics. How does it differ from the principle of conservation of mechanical energy discussed in Chapter 6?

19 State and explain the second law of thermodynamics.

20 Outline the basic stages in the operation of the steam engine. In which of the stages is heat added? In which stage is work done? In which stages is heat removed? How does the second law of thermodynamics limit the efficiency of the steam engine?

21 What is Clausius' form of the second law of thermodynamics? How does it relate to the operation of a refrigerator or air conditioner?

22 State the basic assumptions of the kinetic theory of heat. Show how the kinetic theory is used to explain the thermal phenomena listed in question 15.

23 How is the distinction between heat and temperature explained in the kinetic theory? What is the significance of absolute zero in the kinetic theory?

24 According to the kinetic theory, what causes the pressure that a gas exerts on the walls of its container? On what properties of the gas molecules does the pressure depend?

25 What is the Maxwell–Boltzmann distribution? How does the concept of probability enter into the derivation of the Maxwell–Boltzmann distribution?

26 How is the concept of entropy explained in the kinetic theory?

27 What is meant by the heat death of the universe? How is it related to the concept of entropy?

EXERCISES

1 Convert the following Fahrenheit temperatures to their Celsius equivalents: 0°F, 50°F, 98.6°F, 180°F.

2 Convert the following Celsius temperatures to their Fahrenheit equivalents: −10°C, 20°C, 35°C, −40°C.

3 A temperature scale in common use in eighteenth-century Europe was the *Reaumur scale.* On this scale the freezing and boiling points of water were 0°R and 80°R, respectively. Construct a formula relating (a) the Reaumur and Celsius temperature scales, and (b) the Reaumur and Fahrenheit scales.

4 Investigate the operation of the common device known as the thermostat. On what basic thermal properties of matter is its operation based?

5 Discuss whether the ideal gas law would be suitable to describe the properties of steam (water in gaseous form) at 1 atm pressure when the steam is at a temperature of (a) 900°F, and (b) 150°F.

6 Using the ideal gas laws, explain why (a) an automobile tire becomes overinflated after a long drive, (b) the spray coming out of an aerosol can is cold to the touch, and (c) a sealed, air-filled balloon deflates when placed in a refrigerator.

7 A sample of gas occupies a volume of 10 liters at a temperature of 27°C. The temperature is increased to 127°C, and the pressure is kept constant. Assuming ideal gas behavior, what will be the volume of the gas at the new temperature?

8 A rigid steel container is filled with oxygen at a temperature of 77°C and 1 atm pressure. The container is heated until the pressure is doubled. What will the new temperature be?

9 One cubic foot of air is at a pressure of 76 cm of mercury and a temperature of −73°C. Compute the temperature of the air when the volume is reduced to 0.5 ft^3 and the pressure is changed to 38 cm of mercury.

10 Twenty grams of alcohol ($s = 0.6$ cal/g-C°), initially at a temperature of 50°C, is mixed with 80 grams of water at a temperature of 10°C in an insulated container. What is the equilibrium temperature of the mixture?

11 A quantity of copper ($s = 0.09$ cal/g-C°) heated to 300°C is placed in an insulated container filled with 50 grams of water at 20°C. When equilibrium is reached, the water temperature is found to be 50°C. What is the mass of the copper?

12 How many calories of heat must be added to change 30 grams of ice ($s = 0.5$ cal/g-C°) at a temperature of −10°C to water at 40°C?

13 An ice cube at 0°C is dropped into a glass containing 100 grams of water at 50°C. After the ice has melted and equilibrium has been reached, the temperature of the water is 20°C. What was the mass of the ice cube? (Hint: Don't forget the water produced by the melting of the ice cube.)

14 A 120-lb person walks up ten flights of stairs (a vertical distance of 100 feet). How much work does he do on himself? What is the heat equivalent in calories? (Note: 1 Btu = 252 cal.)

15 The *horsepower* (hp) is a unit measuring the rate of energy production or consumption. It is defined as follows: 1 hp = 550 ft-lb/sec.

(a) How much work is done by a 100-hp engine in 1 hour?

(b) If this work were completely converted to heat energy, how many Btu of heat would be generated?

16 A 1600-lb automobile traveling at a speed of 50 ft/sec is brought to a stop. How much heat is generated in the brakes, tires, and road?

17 For any given substance the heat of fusion is the same for both the melting and the freezing process. That is, the heat absorbed in the change from solid to liquid is the same as the heat removed in the change from liquid to solid. Show that this is required by the first law of thermodynamics; that is, if the heat of fusion were *not* the same for both changes, energy could be created out of nothing.

18 In a single cycle of a heat engine, 5000 Btu of heat is absorbed in the boiler, and 2000 Btu is released to the condenser and surroundings. What is the work output of the engine in ft-lb?

19 One ton of ice, melting in air, absorbs 200 Btu/min from the air. A "1-ton" air conditioner cools a room at an equivalent rate; that is, it removes 200 Btu/min from the surrounding air. Given a room 20 feet × 20 feet × 10 feet filled with air (weight density 0.08 lb/ft^3, s = 0.24 Btu/lb-F°), how long will it take for a 1-ton air conditioner to cool the room from 80°F to 50°F?

20 Prove that equal volumes of any gas at the same temperature and pressure contain the same number of molecules. (Hint: Use Eq. 7–12.) This is known as *Avogadro's hypothesis* (after Amedeo Avogadro, who proposed it in 1811 on the basis of the observed properties of gases, many years before the kinetic theory was proposed).

21 The average molecular speed in hydrogen gas at 0°C has been calculated to be 1.84×10^5 cm/sec.

(a) What is the average molecular speed for oxygen gas at 0°C? (Note: The mass of an oxygen molecule is sixteen times the mass of a hydrogen molecule.)

(b) What is the average molecular speed for hydrogen gas at 273°C?

22 When the heat death of the universe is attained, the entire universe will be at the same temperature. Explain why this temperature *cannot* be absolute zero.

Three

LIGHT, ELECTROMAGNETISM, AND THE ETHER HYPOTHESIS

NATIONAL RADIO ASTRONOMY OBSERVATORY

A radio telescope

This huge dish, 300 feet in diameter, collects and focuses the radio waves emitted by planets, stars, and distant galaxies. In the latter half of the nineteenth century, it was discovered through a combination of theory and experiment that light is an electromagnetic phenomenon. The development of the radio telescope was one of the many practical applications of this discovery.

LIGHT AS A WAVE

8

Much as I venerate the name of Newton, I am not therefore obliged to believe he was infallible.

Thomas Young

CERTAINLY, the most important of the human senses is the sense of sight. From a philosophical point of view, we may say that the visual sense represents the strongest link between our conscious minds and the world of external objects. More than any other, the visual sense provides us with a steady and intense stream of information through which we are able to gain some understanding of the nature of the external world.

In analyzing the phenomenon of sight, scientists have separated it into two aspects. First, there is some sort of transmission from the object being viewed to the retina of the human eye. Second, the impulse received by the retina is transmitted to the human brain, and a visual image impinges on our consciousness. The mechanism by which the eye and brain conspire to produce the visual sense is only dimly understood and remains a difficult problem for physiology. The first aspect falls in the realm of physics, where it is a central concern. The "thing" that is transmitted from the object to the eye is what we call ***light.***

Understanding the nature of light, as well as the processes by which it is generated, transmitted, and absorbed, has always been one of the most important goals of physics. In fact, theories of light and vision date back to antiquity. In the Book of Genesis we find that light was brought into being on the very first day of the Creation. Thus light seems to have been thought of as having an existence independent of its sources—for the sun, moon, and stars were not created until the fourth day. One of the early Greek schools of thought held that light proceeded *from* the eye *to* the object being viewed, but this theory did not gain much support. The focusing properties of lenses and mirrors were well known to the ancients; legend has it that Archimedes defended the city of Syracuse against attack by using mirrors to focus the sun's rays on the sails of the enemy ships, setting the sails afire. The laws of reflection and refraction were studied by the Greeks; among these early investigators was the astronomer Ptolemy, who made measurements of the refraction of light in water. During the Middle Ages the Arabs made important contributions to optics, and their translated works were an important source for the scientists of the Renaissance.

Since the time of Galileo the problem of the nature of light has been the subject of intense study; it has been the concern of such giants of physics as Newton, Maxwell, and Einstein. Yet, as we shall discover, our understanding of the nature of light remains clouded even at the present time, despite the attention that has been devoted to it.

In keeping with the procedures of the scientific method, we will begin with a descriptive survey of the basic light phenomena and some of the mathematical laws governing the behavior of light. Then we will proceed in this and the following chapters to study the various theories of the nature of light that have been developed to explain these phenomena.

8–1

THE PROPERTIES OF LIGHT

Let us first consider the most fundamental and obvious attributes of light. From the observation that objects cast distinct shadows when held up to a light source, it follows that light travels in straight lines. We will refer to this property as the ***rectilinear propagation*** of light. We also know that light can travel over great distances, for we can see stars in the sky that are many trillions of miles away.

THE SPEED OF LIGHT

The simplest observations tend to show that light travels at an extremely high speed; indeed, the earliest theories held that light was transmitted instantaneously from the source to the observer. In the *Dialogue Concerning Two New Sciences* Galileo reported on his attempts to measure the speed of light. He and an assistant stationed themselves with lanterns on two mountains about a mile apart. Galileo uncovered his lantern; his assistant, upon seeing the lantern, was to uncover his own lantern. Galileo hoped to measure the time delay between the uncovering of his own lantern and the instant he saw his assistant's lantern. Needless to say, the experiment was not one of Galileo's most successful. He could conclude only that the speed of light was very high.

After Galileo's failure it was realized that the travel time of a light ray could be measured only if the distance between the light source and the observer was extremely large. Accordingly, several attempts were made to determine the speed of light on the basis of astronomical observations.* One example is the measurement of stellar aberration (discussed in Chapter 5). It will be recalled that the magnitude of stellar aberration depends on the speed of light—there would be no stellar aberration if light had an infinite speed. However, to determine the speed of light from stellar aberration the orbital speed of the earth must be known (see Eq. 5–9), and this in turn requires an accurate value for the radius of the earth's orbit. In general, all astronomical measurements of the speed of light are based on the astronomical unit. This presented scientists of the eighteenth and early nineteenth centuries with a serious dilemma—because of the extremely high speed of light, the travel time of light could apparently be measured only over astronomical distances; but the values of all astronomical distances are ultimately dependent on the value of the astronomical unit, which was not accurately known. Finally, in the latter half of the nineteenth century

*The first successful measurement of the speed of light was made by the Danish astronomer Olaus Roemer in 1676. Roemer's method was based on observations of the satellites of Jupiter.

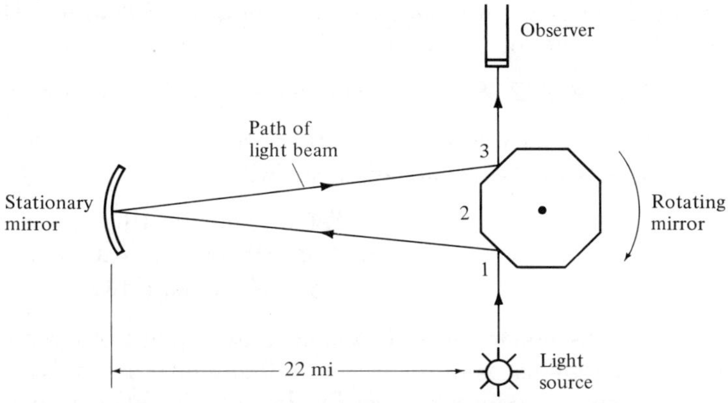

figure 8–1 **MICHELSON'S APPARATUS FOR MEASURING THE SPEED OF LIGHT**

The rotating mirror must make one eighth of a rotation in the time the light beam travels to the stationary mirror and back in order for the source to be seen by the observer.

successful nonastronomical methods were devised for measuring the speed of light to a high degree of accuracy.

The man most closely associated with the accurate determination of the speed of light is the American physicist Albert Michelson (1852–1931), who devoted the greater part of his life to this single measurement. Michelson's method involved a modification of the techniques first used by A. H. L. Fizeau in 1849 and Léon Foucault in 1862 to measure the speed of light. Michelson's apparatus is shown in Figure 8–1. A beam of light is focused on one face (face 1) of a small, eight-sided mirror that can be rotated. The light beam is reflected to a much larger stationary mirror and then back to another face (face 3) of the eight-sided mirror. The beam is then focused by a small lens, so that the observer can see a sharp image of the source.

When the mirror is set in rotation, the image will no longer be seen because face 3 is not at the proper angle to reflect the light beam to the observer. The image will be seen, however, if the mirror is rotating fast enough so that face 2 moves to the position of face 3 (i.e., the mirror makes one-eighth of a rotation) in the time that it takes for the light beam to travel from the rotating mirror to the stationary mirror and back. Because of the high speed of light, the stationary mirror must be placed as far away as possible. In his most successful measurement Michelson placed his rotating mirror on Mount Wilson in California with the stationary mirror on Mount San Antonio, 22 miles away.* The mirror made one-eighth of a rotation in the time the light beam traveled 44 miles. Michelson carefully measured the rate of rotation of the mirror (a rate of over 30,000 rotations per minute was needed), and from this he was able to calculate the speed of light.

Michelson repeated and refined his measurements over a period of fifty years, from 1879 until the time of his death. He arrived at a value for the speed of light

* In an earlier version of this experiment, Michelson placed the stationary mirror on top of the Washington Monument, locating his rotating mirror in a laboratory some 2 miles away. At the time Michelson was working under a $5000 grant voted to him by the United States Congress.

of 186,285 mi/sec, with an uncertainty of about 30 mi/sec. The most recent (November 1972) value of the speed of light is

$$c = 186{,}282.3960 \pm 0.0007 \text{ mi/sec} = 299{,}792.4562 \pm 0.0011 \text{ km/sec}$$

The symbol c has traditionally been used to represent this physical constant. For the purposes of computation, we will use

$$\begin{aligned} c &= 186{,}000 \text{ mi/sec (British)} \\ &= 3 \times 10^{10} \text{ cm/sec (CGS)} \\ &= 3 \times 10^{8} \text{ m/sec (MKS)} \end{aligned}$$

This discussion should be qualified by noting that c represents the speed of light in *free space*—that is, in the absence of any material medium. To earlier generations of scientists, c signified the speed of light in the ether; in modern terms c is referred to as the speed of light in a vacuum. In material media light travels with a speed different from c.

THE REFLECTION OF LIGHT

To continue our survey of the properties of light, let us now consider how light interacts with matter. All substances will reflect, transmit, or absorb light to varying degrees. Most natural materials are ***opaque;*** that is, they do not transmit any light, but reflect a portion and absorb the remainder of the light incident upon them. A perfectly black object is one that absorbs 100 percent of the incident light, but this represents an ideal situation, which is not attained by any natural substance. ***Transparent*** materials such as glass or water transmit most of the light incident upon them, but also reflect and absorb light to a lesser extent. It is readily observed that the absorption of light is a heating process, from which we must conclude that light is associated with, carries, or is itself a form of energy.

The reflection of light from smooth surfaces occurs in a rather specific way, as shown in Figure 8–2. We define the ***angle of incidence*** i as the angle between the incident light beam and the perpendicular or ***normal*** to the surface. Similarly, the ***angle of reflection*** r is the angle between the reflected beam and the normal. Observations show that *the angle of incidence equals the angle of reflection* ($i = r$). This is the ***law of reflection.***

THE REFRACTION OF LIGHT

When we look into a pool of shallow water, the bottom seems closer than it really is. When a spoon is placed in a glass of water, it appears foreshortened and

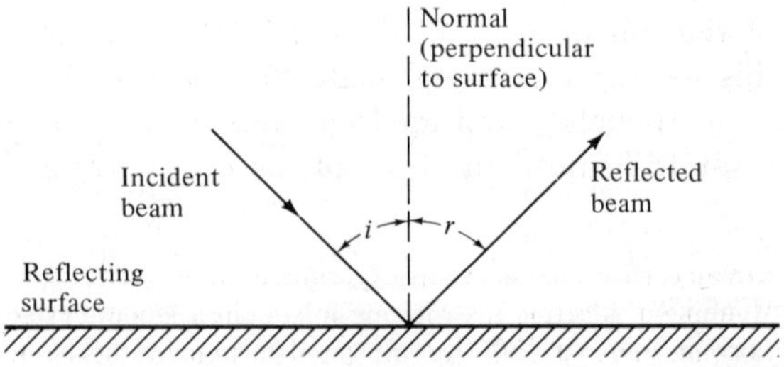

figure 8–2

THE LAW OF REFLECTION

The angle of incidence i is equal to the angle of reflection r.

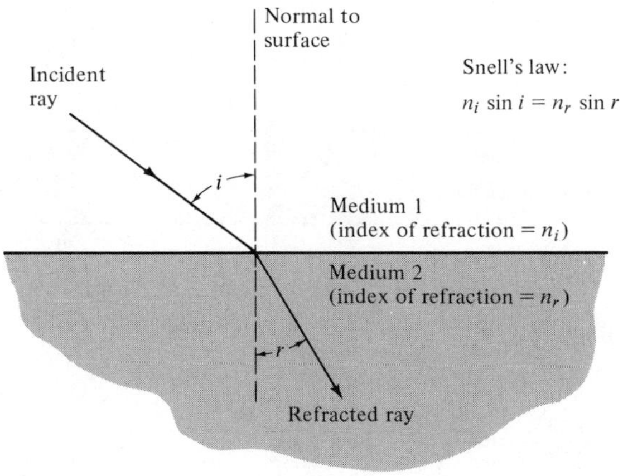

figure 8–3

THE REFRACTION OF LIGHT

When light travels from one medium to another, it changes direction. A light beam bends toward the normal when it passes from the medium with the lower index of refraction to the medium with the higher index. The mathematical relationship between the angles of incidence and refraction is given by Snell's law.

bent. These two phenomena are consequences of the ***refraction*** of light, which is the bending or change in direction of a light beam when the beam passes from one medium into another.

The amount of bending that occurs is described by the ***law of refraction,*** also known as ***Snell's law*** (after Willebrord Snell, who deduced the law from experiment in 1621). Snell found that the sines of the angles of incidence and refraction (see Figure 8–3) are in a fixed ratio for light going from one medium to another; that is,

$$\frac{\sin i}{\sin r} = \text{constant}$$

It is preferable to define the constant as the ratio of two quantities n_i and n_r such that

$$\frac{\sin i}{\sin r} = \frac{n_r}{n_i} \qquad (8\text{–}1a)$$

so that Snell's law can be written as

$$n_i \sin i = n_r \sin r \qquad (8\text{–}1b)$$

Each of the two constants n_i and n_r is known as the ***index of refraction*** of the medium; n_i is the index of refraction of the medium in which the incident light beam is traveling, and n_r is the index of refraction of the medium in which the refracted ray travels. By definition $n = 1$ for a vacuum. Experiments show that in any material medium, n is always greater than 1. For example, $n = 1.33$ for water and 1.50 for plate glass. For air $n = 1.00029$; however, we can assume $n_{air} = 1$ in simple calculations.

EXAMPLE 8–1: A ray of light is incident at an angle of 60° to the vertical on a piece of parallel-sided plate glass. At what angle does the ray emerge from the glass?

SOLUTION: Referring to Figure 8–4, we first find the angle of refraction r_A at the first surface, using Snell's law:

figure 8–4

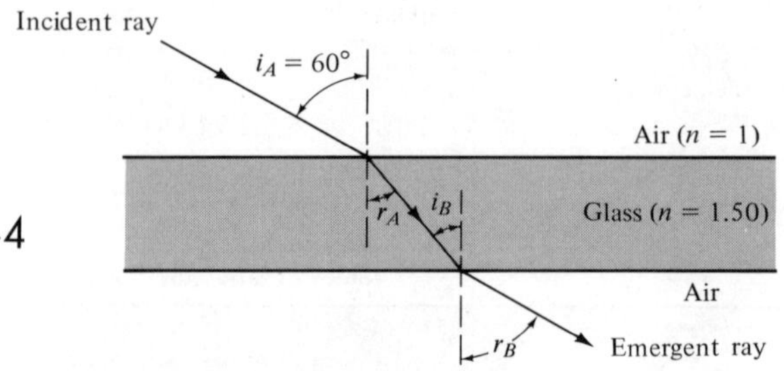

$$n_{air} \sin i_A = n_{glass} \sin r_A$$
$$(1) \sin 60° = (1.50) \sin r_A$$
$$\sin r_A = \frac{\sin 60°}{1.5} = \frac{0.866}{1.5} = 0.577$$

Using a table of trigonometric functions, we find

$$r_A = 35.3°$$

Now let us consider what happens at the second surface. Because the sides are parallel, $r_A = i_B$; that is, the angle of refraction at the first surface equals the angle of incidence at the second surface. Substituting into Snell's law a second time shows that the angle of refraction at the second surface, r_B, is equal to the original angle of incidence (i.e., $r_B = 60°$). Thus the emergent ray will be parallel to the incident ray, but displaced sideways from it.

In Example 8–1 we found that a parallel-sided piece of glass will displace a ray of light without changing the direction of the light ray. However, if the sides of the glass are not parallel, the light ray will be deflected from its original path. (See Figure 8–5a.) This effect is responsible for the properties of optical lenses. A lens having two concave surfaces will cause parallel beams of light to diverge (Figure 8–5b), whereas a lens with two convex surfaces will bring parallel light rays to a focus (Figure 8–5c). The ***focal length*** of a lens (defined as the distance from the center of the lens to the focus) is determined by the curvatures of the two surfaces, as well as by the index of refraction of the glass.

The human eye itself contains a convex lens, which serves the purpose of bringing light from an object to a sharp focus on the retina. The lens is flexible enough to be stretched or compressed by the supporting muscles; these changes in the shape change the focal length of the lens so that near and distant objects can be seen sharply. ***Nearsightedness*** (myopia) occurs when the lens cannot bring the light from distant objects to a focus on the retina (Figure 8–6a), and only nearby objects can be seen sharply. This defect is corrected by placing a diverging lens in front of the eye. ***Farsightedness*** (hyperopia) occurs when the focus of the relaxed lens is behind the eye (Figure 8–6b); then nearby objects cannot be focused sharply. A converging lens will correct this defect.

It is of interest that the optical properties of lenses can be deduced without recourse to a theory of light. That is, we do not need to know anything about the nature of light except that it travels in straight lines and obeys the laws of reflection and refraction. In fact, most of the properties of lenses were worked out by trial and error. Both the telescope and the microscope were invented in 1609, twelve years *before* Snell's law was discovered!

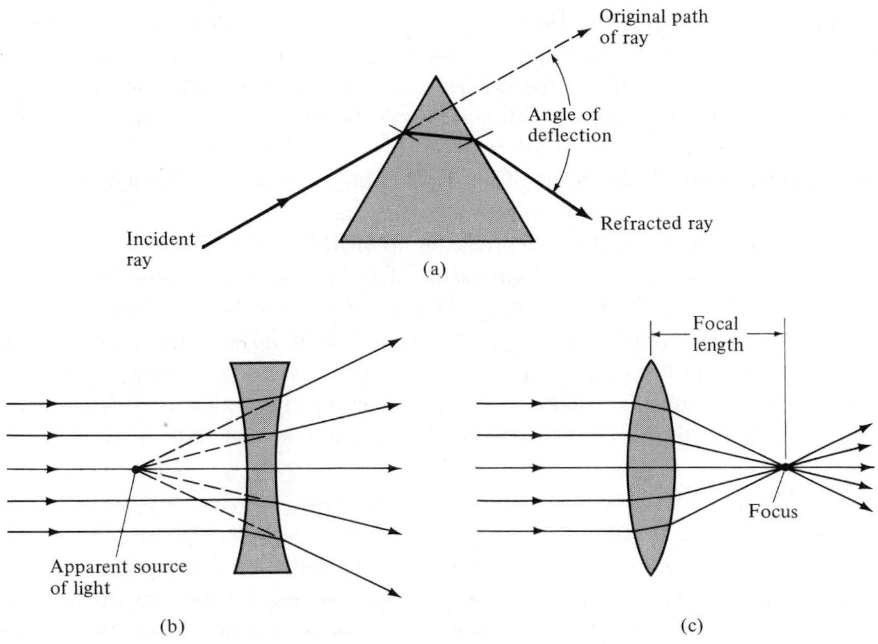

figure 8–5 **REFRACTIVE PROPERTIES OF PRISMS AND LENSES**

(a) A triangular prism deflects a light ray from its original path. (b) A double concave lens diverges parallel beams of light, making the light appear to come from a point close to the lens. (c) A double convex lens converges parallel beams of light to a focus.

COLOR

One of the most significant aspects of light is that we perceive objects as having ***color.*** The questions of how the eye and brain produce the sensation of color, why we see the colors that we do, and the very nature of color perception pose great difficulty for physicists, psychologists, and physiologists alike. We are taught in early childhood to identify colors by name, but the sensation defies definition. How can one describe verbally what he sees when he sees the color red? Man derives much aesthetic pleasure from the colors he sees in nature as well as those he has learned to produce artificially. For the scientist this pleasure is often enhanced by his attempts to understand how and why colors are produced.

The first worthwhile theory of color was developed when Isaac Newton, at the age of twenty-three, began to experiment with a prism. At that time it was well known that a prism would change a beam of white light into a spectrum of colors in the order red, orange, yellow, green, blue, violet, similar to the order seen in a rainbow (Figure 8–7a). However, the prevailing view was that the colors were somehow produced by the prism itself. Through a series of simple experiments, Newton was able to prove otherwise.

Newton allowed a narrow beam of sunlight to pass through a prism, producing a spectrum. He then took a second prism, inverted with respect to the first, and placed it in the beam of colored light (Figure 8–7b). He found that the second prism could be made to combine the spectrum back into a beam of white light. Newton concluded from this result that white light is actually *composed* of all the colors of the spectrum. Thus what we see as white light is simply the way in which our visual sense responds when a combination of all the colors (in approximately equal intensities) strikes the retina. As a further test, Newton allowed the spectrum to fall on a screen with a slit that allowed a single color to pass through (Figure 8–7c). He found that when ***monochromatic*** light (light of a single color) was passed through a prism, it could not be broken up into any additional components.

Newton reasoned that the breaking up of white light into its component colors—a phenomenon known as the ***dispersion*** of light—occurs because the different colors are refracted by different amounts. That is, the index of refraction for light in glass depends on the color of the light. Since red light is bent the least and violet light the most, the index of refraction must be greatest for violet light and smallest for red light. For most substances the variation in the index of refraction across the spectrum is relatively small; for example, in water $n = 1.329$ for red light and 1.338 for violet.

Having identified the nature of white light, Newton was able to account for what he referred to as "the permanent colors of natural bodies." Those bodies that do not themselves emit light are seen by the reflected light given off by luminous bodies such as the sun, gas flames, and light bulbs. Depending on its composition, a given substance reflects certain colors and absorbs others. Therefore, roses are red because when white light shines on them, they reflect the red light and absorb all other colors. Grass appears green because it contains a substance that strongly reflects the green part of the spectrum while absorbing the red and violet ends.

Light that is absorbed by bodies is converted into heat. If two cloths, one white and one black, are left in the sun, the black one will feel much hotter to the touch. The white cloth appears white because it reflects nearly all the incident sunlight; the black cloth, on the other hand, absorbs a major fraction of the sunlight, and the absorbed sunlight is converted to heat. This is why dark clothing is uncomfortable to wear in the summertime.

This concludes our summary of some of the better known optical phenomena. We will now consider the theories of light that have been proposed to account for these phenomena.

figure 8–6 CORRECTION OF COMMON VISUAL DEFECTS

(a) A diverging lens corrects for nearsightedness, in which light from a distant object comes to a focus in front of the retina. (b) Farsightedness occurs when the focal point of the eye is behind the retina. This defect is corrected by a converging lens.

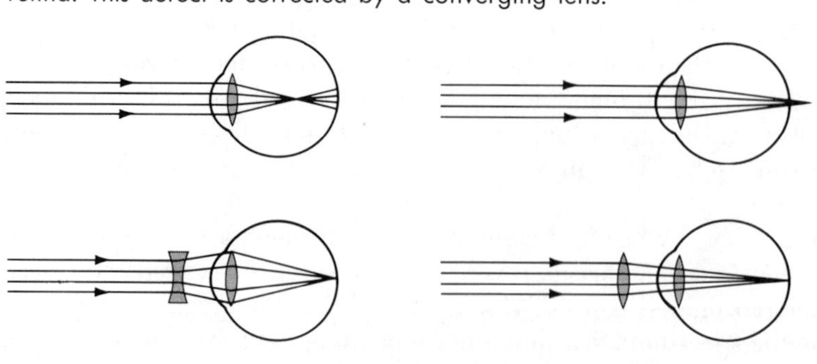

(a) (b)

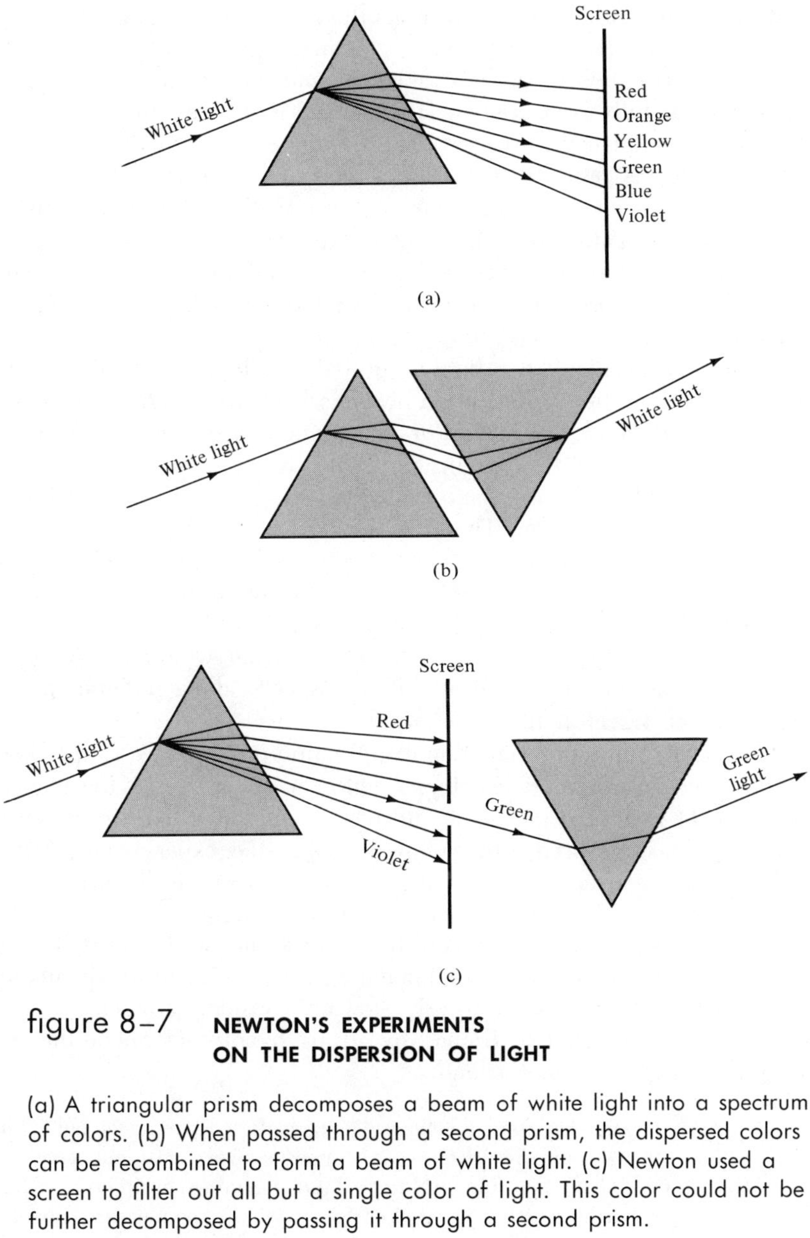

figure 8–7 **NEWTON'S EXPERIMENTS ON THE DISPERSION OF LIGHT**

(a) A triangular prism decomposes a beam of white light into a spectrum of colors. (b) When passed through a second prism, the dispersed colors can be recombined to form a beam of white light. (c) Newton used a screen to filter out all but a single color of light. This color could not be further decomposed by passing it through a second prism.

8–2

NEWTON'S PARTICLE THEORY OF LIGHT

In the history of physics there have been two main lines of thought as to the nature of light. The ***particle*** or ***corpuscular*** theories have assumed that light consists of a stream of tiny particles; the ***wave*** theories propose that light is a disturbance that propagates through a medium, much like waves traveling on the surface of the ocean or sound traveling through air. Both theories were first proposed in detail in the middle of the seventeenth century, and during this period the conflict between the two theories was sharply delineated.

The wave theory was first sketched out by Robert Hooke in 1665 and was evolved more fully by Christian Huygens in 1678. Essentially, the theory proposed that light is a wave disturbance that propagates through the ether—the mysterious substance that is assumed to fill up all of space. (See page 145.) However, Newton was loath to accept the wave theory because it did not account satisfactorily for the rectilinear propagation of light. He believed that if light is a wave, it should bend into shadows and around corners. Whereas sound and water waves will bend around obstacles, light does not; objects held in the path of light appear to cast distinct, sharp shadows.

Newton speculated that light is emitted from luminous bodies as a rapidly moving stream of tiny particles. Reflection occurs when these particles cannot penetrate into a medium and simply bounce off or are repelled by the surface. The law of reflection can then be explained by applying the principles of conservation of energy and linear momentum. Refraction is explained by assuming that the approaching light particle is attracted to the medium. The amount of refraction depends on the strength of the force of attraction between the particle and the medium; in general, the denser the medium, the greater the force of attraction. To explain the dispersion of light, Newton supposed that each color is made up of a different kind of light particle, having weaker or stronger attractions to a given medium. When white light enters a prism, the different colors of light travel at different speeds along different paths, but in free space they all travel at the same speed.

According to the explanation of refraction provided by the particle theory, light should speed up whenever it enters a denser medium. Hence the theory predicts that the speed of light will be *greater* in glass or in water than in air or free space. In Newton's time this prediction could not be tested experimentally, but as we shall find out later, it eventually proved to be of considerable importance to scientists in deciding between the two conflicting theories of light.

Newton's particle theory seemed to work rather well; it also fit into the mechanical view of nature in which all phenomena are ultimately explainable in terms of forces between particles. However, Newton considered much of his theory to be speculative and refused to frame any strong hypotheses about the nature of light. As Newton explained to a friend:

> It is true that from my theory I argue the corporeity of light, but I do it without any absolute positiveness . . . and make at most but a very plausible consequence of the doctrine, and not a fundamental supposition. . . . But I knew that the properties which I declared of light were in some measure capable of being explicated not only by that, but by many other mechanical hypotheses, and therefore I chose to decline them all and speak of light in general terms, considering it abstractedly as something or other propagated every way in straight lines from luminous bodies, without determining what that thing is.[1]

In spite of such disclaimers, Newton's theory was strongly attacked by Hooke, and the debate was sharpened by the animosity the two men had developed for one another. When Newton finally allowed the *Opticks* to be published, however, Hooke was dead and Newton had achieved a reputation as the greatest scientific genius who ever lived. Simply on the strength of Newton's reputation, the wave theory was abandoned. For nearly a hundred years no reputable scientist would defend it—for how could one fail to trust the intuition of the man who had figured out the secrets of the celestial universe?

But as we have seen, Newton recognized the shortcomings of the particle theory and certainly did not have as much faith in it as his adulating followers did. Eventually, a number of experiments made in the early 1800s turned the tables in favor of the wave theory of light. Before we can take up the study of these events, however, we must learn more about the properties of waves.

8–3

WAVE MOTION

For many people the word "wave" is reminiscent of a pleasant day at an ocean beach or possibly of an unfortunate shipboard experience with *mal de mer.* Very often the word is used in common speech to describe some phenomenon or disturbance that moves from place to place, as in the expressions "cold wave" and "crime wave." In physics a number of diverse and seemingly unconnected phenomena come under the category of waves. In addition to the aforementioned ocean or water waves, there are sound waves, seismic (earthquake) waves, and the waves that can be generated by shaking one end of a taut string (hereafter to be referred to as string waves).

The connecting link that entitles all these effects to be classified as waves is that they are all *disturbances that propagate through a medium.* When we shake one end of a string, the "bump" that is produced moves along the string with a finite speed and eventually arrives at the other end. Although the wave travels from one end of the string to the other end, there is no transfer of *matter* from one end to the other; each section of string simply bobs up and down about its normal position as the wave passes. In the same way, when a rock is thrown into a small lake, a series of small waves travel on the surface of the water until they reach the opposite shore. It is the wave disturbance alone that crosses the lake; there is no transfer of water from one side to the other. In short, when a wave propagates through a medium, each point in the medium simply oscillates about a fixed equilibrium position, and there is no *net* motion of the medium itself.

Yet something is transferred from one part of a medium to another when a wave passes. The surface of a lake is perfectly still; as the wave passes, the surface begins to bob up and down. Thus each portion of the surface, initially at rest, is set into motion; hence the surface now possesses *energy* and *momentum* that it did not have before. Thus a wave effects the transfer of energy and momentum from one point to another in a medium, without the transfer of matter. We can therefore characterize a wave as *a disturbance in a medium by which energy and momentum are transmitted through the medium.*

THE PROPAGATION OF TRANSVERSE WAVES

Let us examine more closely the process by which a disturbance moves through a medium. In general, in order for a wave to propagate there must be some restoring force inherent in the internal structure of the medium, so that when a portion of the medium is displaced away from its normal or equilibrium position, the restoring force pulls or pushes it back to equilibrium. In a solid this restoring force is usually provided by the intermolecular forces that maintain the rigid shape of a solid body.

In a string these intermolecular forces manifest themselves in the tension force by which a string resists being stretched.

Figure 8–8 shows how the tensional restoring force brings about the propagation of a wave along a string. Assume that the string is distorted into the shape shown at step A. The tension forces exerted by the portions of the string immediately adjacent to point 1 combine to give a net *downward* force on this point; that is, they act to restore point 1 to its equilibrium position. However, the net force acting at point 2 is *upward,* because of the distorted shape of the string. (Point 2 is being pulled toward point 1.) Acted on by these forces, point 1 moves toward its equilibrium position, but point 2 moves *away* from equilibrium. A short time later the string will have the shape shown at step B. Now the upward force on point 1 acts to decelerate it, and it is brought to a stop. (Actually, because of inertia, it will overshoot somewhat.)

figure 8–8 PROPAGATION OF A TRANSVERSE WAVE ON A STRING

The motion of a single wave pulse along a string is shown. The sequence of steps depicts the shape of the string at equal time intervals, starting with the initial displacement of the string (Step A). The net force at each of several points on the string is indicated by an arrow.

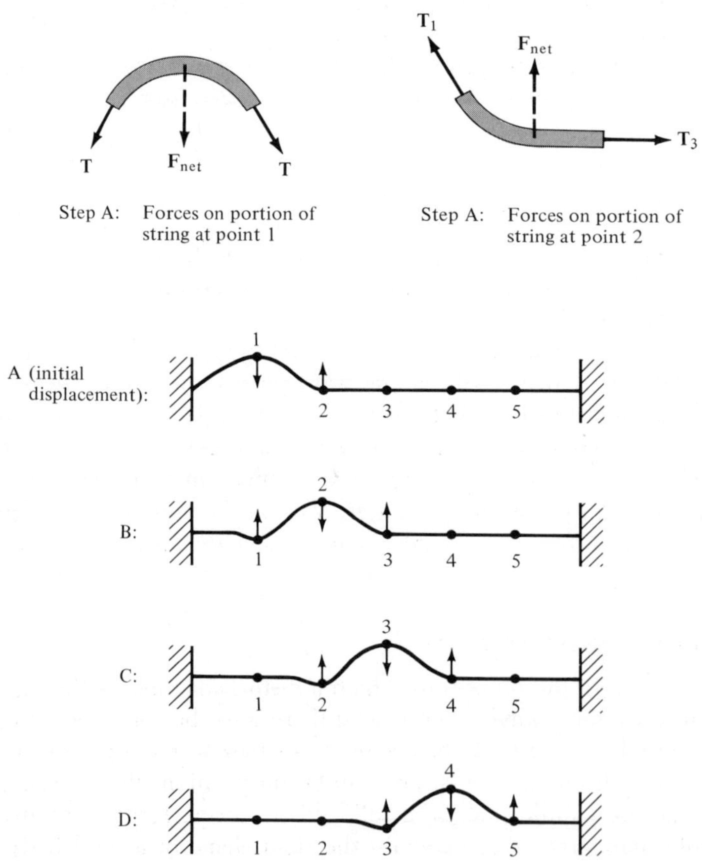

By the same process, the disturbance will proceed to points 3, 4, and 5 in turn; the wave pulse is thus seen to travel along the string from left to right. Moreover, energy (and momentum) are initially supplied to the string by the outside force that produces the initial displacement; through the tension force, each section of string does work on the adjacent section, causing the transfer of energy along the string as the wave progresses.

Note that the motion of each portion of the string is at right angles to the direction in which the wave travels. A wave of this type, in which *the direction of motion of the medium is perpendicular to the direction of propagation of the wave,* is known as a ***transverse wave.***

Two factors determine the speed at which a wave propagates along a string. The greater the *tension* in the string, the greater the restoring force and the quicker the string will snap back to its equilibrium position. Working against this factor is the *inertia* of the string; that is, a more massive string will respond less quickly to the restoring force. Theoretical analysis shows that the speed of a wave on a string is given by the equation

$$v = \sqrt{\frac{T}{m/L}} \tag{8–2}$$

where m is the mass and L the length of the string, and T is the tension force (the force applied to stretch the string).

The ripples that are produced by a disturbance in a small body of water are propagated in much the same way. These waves travel on the surface of the water; the restoring force is provided by the surface tension that exists at the boundary of any liquid.* Deep-sea ocean waves represent a much more complex phenomenon. Small-scale water waves such as pond ripples are basically transverse waves, but in an ocean wave each portion of water travels in a circle as the wave passes. (See Figure 8–10c). Ocean waves are generated and driven by winds and travel with speeds that are roughly proportional to the wind speeds. As an ocean wave approaches land, its characteristics change, and the speed becomes proportional to the water depth. Thus as an ocean wave moves in toward the shore, it slows down. If the ocean bottom is not smooth, the waves will bend, since they move faster over the deeper portions and slower over the shallower regions. When the wave is close enough to shore so that the height of the wave is about equal to the water depth, the water particles cannot complete their circular motions. The top of the wave then moves faster than the bottom, and finally the wave breaks.

THE PROPAGATION OF SOUND: COMPRESSIONAL WAVES

The question of how sounds are transmitted from place to place has naturally been of considerable interest to scientists for thousands of years. Once again, we are dealing with a primary human sense—the sense of hearing or audition—and, as in the case of vision, we shall divide the phenomenon into two parts: (1) the transmission of an impulse from the source to the ear and (2) the conversion of that impulse by

*Surface tension is the result of the intermolecular forces in a liquid. The molecules at the surface experience a net inward force, the overall effect of which is to reduce the amount of surface area to a minimum. This is why liquids form spherical drops, for a sphere has the minimum surface-to-volume ratio of any geometric shape.

the ear and brain into an auditory sensation. We will be concerned here only with the first of these.

Let us mention a few effects connected with the propagation of sound. One can readily observe that sounds can be heard around corners and can penetrate through apartment-house walls and other presumably solid materials. It is also evident that sounds are associated with energy, for extremely loud noises can break windows or puncture eardrums—not to mention the legendary opera singers who could shatter a glass with their voices. Atmospheric conditions seem to affect our ability to hear sounds; a sound can be heard more readily when one is downwind from the source than when one is upwind. All of these phenomena indicate that sound is a wave disturbance that propagates through air and other media. A simple test of this hypothesis is provided by an experiment in which a bell or siren is placed in a container from which the air can be pumped out. As the air is removed, the sound gets weaker and weaker, and eventually cannot be heard at all. Thus a medium is necessary for the propagation of sound.

The mechanism by which a sound wave travels through a medium will now be described. The initial disturbance is generated by some vibrating body such as a bell, a string, or a human vocal cord. The vibration of this body causes the medium in contact with it to be compressed, as shown in Figure 8–9b. This compressed region of the medium will attempt to expand, and in doing so, it compresses the previously undisturbed region adjacent to it (Figure 8–9c). In fact, the initially compressed region will tend to overexpand until it becomes rarefied; that is, the pressure in this region is now *lower* than it was before the disturbance began. By the repetition of the process, a pulse of compression and rarefaction will travel through the medium. If the source continually vibrates such pulses are produced at a regular rate, and the medium will contain a series of regions of alternate compression and rarefaction. When these pulses come into contact with a human eardrum, the pressure differences set the eardrum into vibration. These vibrations are transformed by the ear and brain into an auditory sensation.

Not all waves of this type will evoke a conscious auditory response, however. If the rate of vibration is too high or too low, there will be no sensation of sound. (Generally, the human brain will respond to waves with vibration rates from about 20 to 20,000 vibrations/sec.) We will give the name of ***compressional waves*** to the phenomenon described, keeping in mind that compressional waves must have the proper vibration rate (and energy) in order to evoke hearing. By this definition, a sound wave is simply a compressional wave of the right vibration rate, and a sound is the physiological response produced when such waves fall on the ear. (These definitions, incidentally, are not universally accepted; many scientists refer to *any* compressional wave as a sound wave, regardless of the vibration rate or energy.)

As we have seen, a compressional wave consists of a pressure disturbance that travels through a medium. No particles of the medium are transferred from the source to the ear; the individual particles simply oscillate back and forth as the pressure changes. Note that the particles oscillate in a direction *parallel* to the motion of the wave. This type of wave is therefore known as a ***longitudinal*** wave. (See Figure 8–10 for a comparison of transverse and longitudinal waves.)

The nature of compressional waves is such that these waves can be transmitted by virtually any medium, whether gaseous, liquid, or solid. The speed at which a compressional wave travels through a medium generally depends on the rigidity of the medium—that is, the tendency of the particles to snap back to their equilibrium

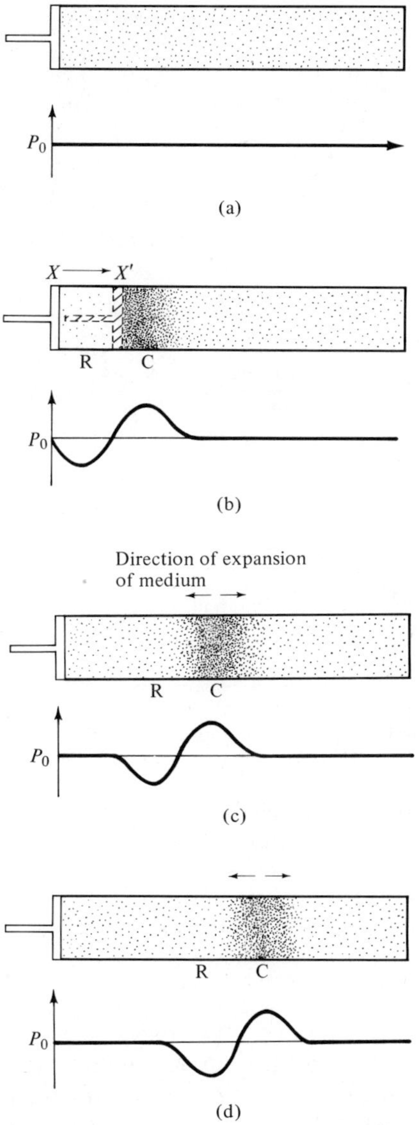

figure 8–9

PROPAGATION OF A COMPRESSIONAL WAVE THROUGH A MEDIUM

(a) The distribution of particles in an undisturbed medium is shown. (b) The pulse is generated when the piston is moved into the medium, from X to X'. Regions of compression (C) and rarefaction (R) are produced in the medium. (c) and (d) The disturbance propagates through the medium as a result of the expansion of the compressed portion of the medium. The graphs show the distribution of pressure in the medium at each state; P_0 represents the equilibrium pressure in the undisturbed medium.

positions after having been displaced by the pressure wave. Inertia also enters as a retarding factor; the more massive the particles, the slower they will respond. Thus the speed of compressional waves in media is governed by an equation of the type $v \sim \sqrt{(\text{rigidity})/(\text{density})}$, in analogy with Eq. 8–2. Further application of this equation to gases yields the result that for a given gas the speed depends on the temperature. (Specifically, $v = C\sqrt{T}$, where C is a constant that is different for each gas, and T is the absolute temperature.) The speeds of compressional waves in some representative media are listed in Table 8–1. For rough calculations a value of 1100 ft/sec (= 750 mi/hr) can be used for the speed of sound in air at normal temperatures.

The earth, being composed of fairly rigid solid material, is a good medium for the transmission of both compressional and transverse waves. These ***seismic waves***

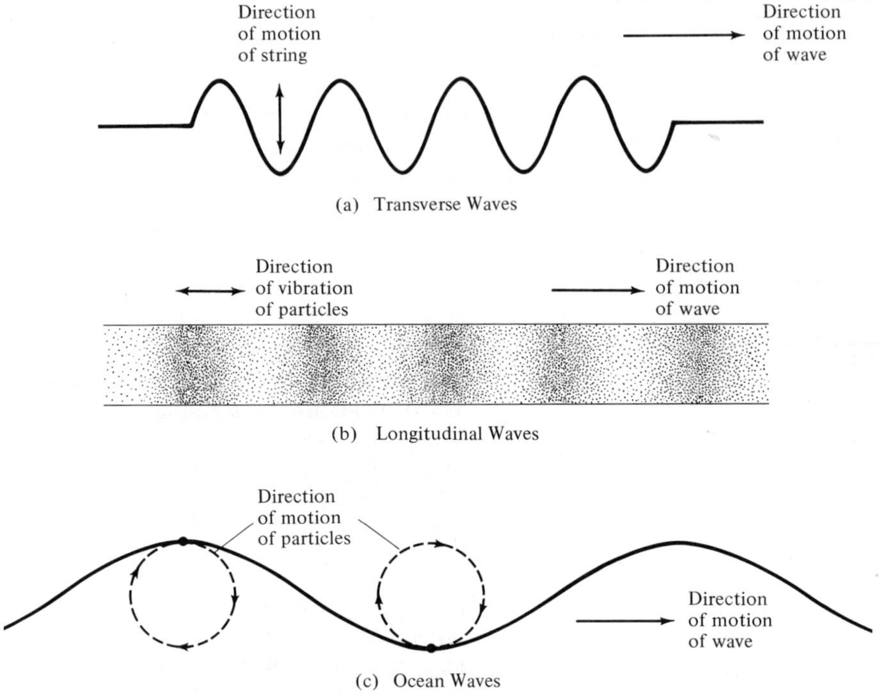

figure 8–10 **TRANSVERSE AND LONGITUDINAL WAVE PROPAGATION**

(a) A series of waves are traveling on a string: the direction of vibration of each portion of the string is perpendicular to the path of the wave. This is an example of a *transverse* wave. (b) The compressions and rarefactions produced by compressional waves are shown: the particles of the medium vibrate in a direction parallel to the path of the wave. This is an example of a *longitudinal* wave. (c) In an ocean wave, the particles travel in circular orbits; thus the motion is both transverse and longitudinal.

are generated by any large-scale disturbance inside or on the surface of the earth, such as an earthquake, a volcanic eruption, or a nuclear explosion. They will travel in all directions through the interior and on the surface of the earth.* By measuring the paths and travel times of these waves, geologists can gain a great deal of information about the structure and composition of the earth's interior. The data can be used to calculate the wave speeds, from which the density at various points in the earth can be determined. Of particular interest is the observation that transverse waves are not transmitted through a deep interior region of the earth. It is known that a liquid will transmit longitudinal (compressional) waves, but not transverse waves. (Transverse waves only travel on the *surface* of a liquid.) On this basis, it is concluded that a part of the earth's interior is in the liquid state; it is presumably molten metal or rock.

*Such disturbances also generate large ocean waves called *tsunamis* (more commonly known as tidal waves—which is a misnomer, since they have nothing to do with tides). Tsunamis have been known to travel across thousands of miles of open ocean at speeds of up to 450 mi/hr. Upon reaching shore, they may rise to heights of 100 feet and can be extremely destructive.

table 8–1 **SPEEDS OF COMPRESSIONAL WAVES IN VARIOUS MEDIA**

Material	Speed (ft/sec)	Material	Speed (ft/sec)
Air (0°C)	1087	Aluminum	16,750
Hydrogen (0°C)	4150	Steel	17,100
Mercury	4610	Glass	18,050
Water	4790	Diamond	60,400

These techniques are now being applied to the study of the moon. Using the seismometers (devices that measure the intensity of seismic waves) placed on the moon by the Apollo astronauts, scientists are now able to obtain data pertaining to moonquakes, meteorite impacts, and so on. These data are presently being used to determine the interior structure of the moon. Such information will be of great value to astronomers who are interested in formulating theories of the origin of the moon and of the solar system as a whole.

PROPERTIES OF WAVE TRAINS

In the previous section the mechanical process by which waves propagate through a medium was illustrated by studying the motion of a single wave pulse, produced by a single disturbance of the medium. It is of greater interest to study the properties of a continuous sequence or ***train*** of identical wave pulses generated in the medium by a source that vibrates in a continuous and regular manner.

For such a wave train the ***frequency*** f is defined as the rate at which waves are generated; that is, the number of waves produced per unit time. This is, of course, the same as the vibration rate of the source of waves. Traditionally, frequency has been measured in such units as waves/sec, vibrations/sec, and cycles/sec (cps). A few years ago, however, scientists decided to name the frequency unit the ***hertz*** (1 Hz = 1 vibration/sec) in honor of Heinrich Hertz, whose contributions to physics will be considered in another chapter. In the case of sound waves the frequency determines the ***pitch*** of the sound. Thus the "highness" or "lowness" of a sound is the way in which our auditory sense makes us aware of differences in frequency. As mentioned previously, the frequency range of human hearing is from about 20 to 20,000 Hz (although the upper limit tends to decrease with age); middle A on a musical instrument corresponds to a frequency of 440 Hz.

The ***period*** of a wave is the time required for a single wave pulse to move past any given point in the medium. An alternative definition of the period is the time required for a particle of the medium to make one complete oscillation about its equilibrium position. For a wave train the period is inversely proportional to the frequency; if, for example, waves are being generated at a rate of 200 waves/sec, each wave will be generated in 1/200 of a second. Thus the period P is related to the frequency f by the equation

$$P = \frac{1}{f} \tag{8–3}$$

The ***wavelength*** is the distance between two successive points in the medium that are in the same state of disturbance. Thus for sound waves the wavelength is the distance between two successive compressions or two successive rarefactions. For transverse waves on a string or on the surface of water, the wavelength is the distance between two successive "crests" or "troughs," as shown in Figure 8–11. The Greek letter λ (lambda) is the symbol commonly used to represent wavelength.

From the definitions of wavelength and period, it follows that a wave pulse will travel a distance equal to its wavelength in a time equal to its period. Since waves move with a constant speed in a uniform medium, we can relate these quantities by using the distance equation $d = vt$. Substituting $d = \lambda$ and $t = P$ yields

$$\lambda = vP$$

Substituting $P = 1/f$ (Eq. 8–3) and rearranging terms, we obtain

$$v = f\lambda \qquad (8\text{–}4)$$

This useful equation relates three important properties of wave trains. The speed v is a property of the medium; the frequency f is determined by the wave source. The wavelength λ therefore is dependent on the other two quantities. Note that in a given medium (constant v) the wavelength and frequency are *inversely proportional.* That is, low-frequency waves have large wavelengths, and as the frequency increases, the wavelength decreases. Eq. 8–4 is applicable to both transverse and longitudinal waves, but not to ocean waves, whose speeds depend on wind speed, ocean depth, and so on.

EXAMPLE 8–2: Determine the wavelength of the sound wave produced by the middle A key on a piano.

SOLUTION: Middle A has a frequency of 440 Hz. The speed of sound in air is approximately 1100 ft/sec. We substitute into Eq. 8–4:

$$v = f\lambda$$

therefore

$$\lambda = \frac{v}{f} = \frac{1100\ \text{ft/sec}}{440\ \text{Hz}} = \underline{2.5\ \text{feet}}$$

Thus when this piano key is struck, a vibrating string sets the air into motion, producing compressions 2.5 feet apart.

figure 8–11 **WAVELENGTH AND AMPLITUDE**

The wavelength λ is the distance between successive crests or troughs. The amplitude is the maximum displacement of the medium from equilibrium.

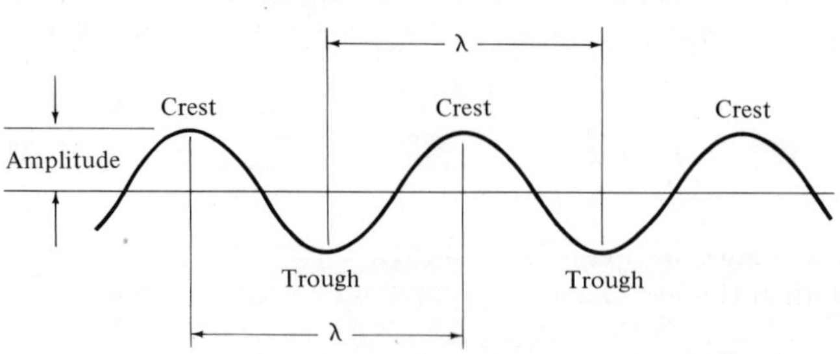

The ***amplitude*** of a wave is the maximum displacement of a particle of the medium from its undisturbed position. For transverse waves the amplitude is the "height" of the wave. (See Figure 8–11.) In longitudinal waves the amplitude is the maximum side-to-side displacement of the particles of the medium. In an ideal medium a wave traveling in a straight line should be transmitted with a constant amplitude. In a real medium, however, there is dissipation of the wave due to friction, heat generation, and so on. Thus as a single wave pulse travels through a medium, its amplitude gradually diminishes.

When a portion of a medium is set into motion by an oncoming wave, that portion gains energy. The amount of energy carried along by the wave is measured by the motion of the medium, which in turn depends on both the amplitude and the frequency of the wave. In general, the energy transfer through the medium is proportional to A^2f^2. For sound waves the amplitude of the wave determines the ***loudness*** of a sound. Loudness can be related to the pressure excess (the difference between the maximum pressure developed in a compression and the normal undisturbed air pressure). The human ear is a highly sensitive instrument; the threshold for hearing corresponds to an excess pressure of only 4×10^{-9} lb/in.^{2}. At this pressure excess, energy is being transferred to the eardrum at a rate of approximately 10^{-16} watt. (1 watt = 1 joule/sec. The watt is a measure of ***power,*** which is defined as the *rate* at which work is done or energy is expended—that is, power = work/time.) It has been said that if the hearing threshold were a few orders of magnitude lower, we would be able to hear the random motion of air molecules striking our eardrums. At the other extreme, an energy input of about 10^{-4} watt represents a sound so loud as to cause pain. Thus the human ear can respond comfortably to sound waves ranging in energy over a factor of a trillion.

So far we have been considering waves as if they travel in one direction only. This is the case when the medium is essentially one-dimensional; in this category we can include transverse waves on a string or longitudinal compressional waves in a long narrow rod. (Such waves can be produced simply by hitting one end of a steel rod with a hammer, for example.) In a two-dimensional medium, such as the surface of a lake, a steel plate, or the head of a drum, waves proceed in all directions. Figure 8–12 shows the pattern of waves produced in such a medium by a point source.

figure 8–12 **WAVE PATTERN PRODUCED BY A POINT SOURCE IN A TWO-DIMENSIONAL MEDIUM**

The lines represent the wave crests (the points of maximum displacement). At a great distance from the source, the wave crests appear as nearly parallel plane waves.

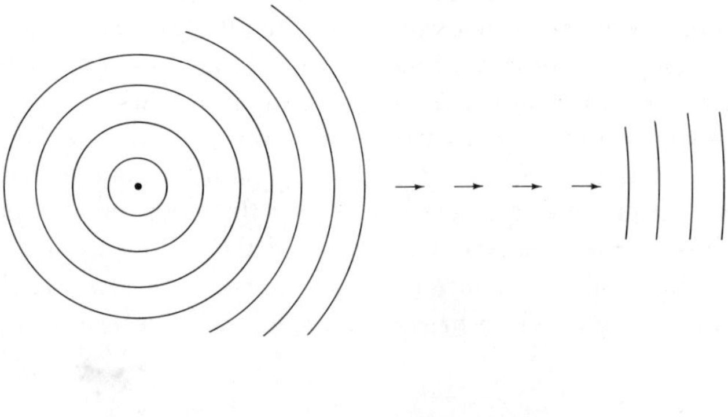

The successive crests of the waves in the wave train are evenly spaced concentric circles. At a great distance from the source, the curvature of these waves becomes so small that to an observer the wave crests appear as parallel lines. In this state the waves are referred to as ***plane waves.*** In three dimensions the wave crests produced by a point source are concentric spheres. (This is the pattern for sound waves in air.)

In two- and three-dimensional wave trains, each wave spreads out over a larger area as it progresses outward into the medium. Conservation of energy requires that the energy in each wave become less concentrated as the wave expands. As a consequence, the amplitude of the wave decreases as the wave's distance from the source increases, even in an ideal medium. This is why the loudness of a sound diminishes as one moves farther from the source.

WAVES FROM A MOVING SOURCE: THE DOPPLER EFFECT

The discussion thus far has been based on the assumption that the wave source is stationary with respect to the medium, so that all the waves produced by a source emanate from the same point. Now let us consider what happens when the wave source moves through the medium as it generates waves. What is the pattern of the water waves produced by a moving boat or the sound waves generated by a jet plane or by the siren of a fire engine?

In the wave pattern shown in Figure 8–12, successive wave fronts comprise a series of concentric circles emanating from the source at the center. This pattern will be changed if the source is in motion, since successive waves will emanate from different points in space. However, each wave propagates through the medium by radiating outward from its point of origin, *independent of the motion of the source or of other waves.* In other words, once a wave is emitted it is completely independent of its source, and its motion depends entirely on the properties of the medium in which is is traveling. In particular, the wave speed is not affected by the motion of the source.

Figure 8–13a shows the relationship between two successive waves emitted by a moving source, when the source is moving with a speed less than the wave speed for the given medium. The first wave was emitted when the source was at point 1, so that the wave front is a circle with point 1 as its center. By the time the second wave was emitted, the source had moved to point 2, so the second wave is not concentric with the first. As the first and second waves continue to radiate out from their respective points of origin, a third wave will begin to spread out from point 3, where the source is now located. The resulting wave pattern will have the appearance shown in Figure 8–13b.

When we examine this wave pattern, we find that the wavelength (the distance between successive wave fronts) is smaller in the direction of motion of the source than it is in any other direction. Consequently, an observer located in front of the source will measure a *shorter* wavelength than he would if the source were stationary, and an observer behind the source will measure a *longer* wavelength. Since wavelength and frequency are inversely related, it follows that if the source is *approaching* an observer, the observer will receive waves with a *higher* frequency than those emitted by a stationary source ($f_{obs} > f_{source}$); if the source is moving *away* from an observer, the observed frequency will be *less* than the source frequency ($f_{obs} < f_{source}$). This shift in the observed frequency of the waves from a moving source (as compared to the frequency emitted by the source) is known as the ***Doppler effect,*** after the Austrian physicist Christian Doppler, who described the phenomenon in 1842.

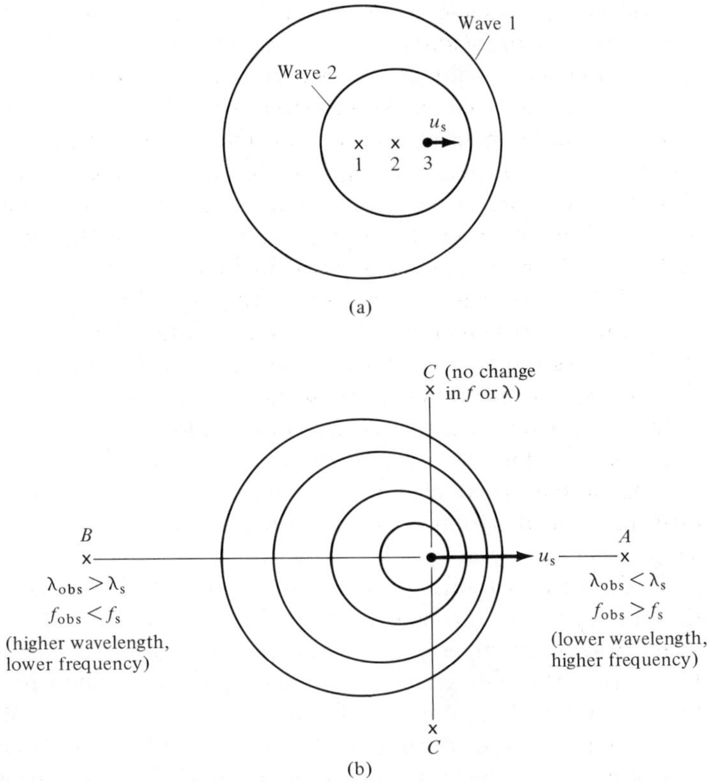

figure 8–13 **THE DOPPLER EFFECT FOR A MOVING SOURCE**

The wave source is in motion with a speed u_s (which is less than the speed of the waves themselves). (a) The waves emitted by the source do not emanate from the same point, so a pattern of nonconcentric waves is generated. (b) An observer at A, who is receiving waves from an approaching source, observes a frequency higher than the source frequency. An observer at B, who is receiving waves from a receding source, observes a lower frequency. An observer at C observes no change in frequency.

For sound waves the frequency determines the pitch of the sound heard by the ear. Thus if a moving source of sound waves (such as a fire engine siren) is approaching an observer, the sound that the observer hears will have a higher pitch than it would if the source were stationary. If the source is receding from the observer, the pitch will be lower. A familiar example of this effect is the sudden drop in pitch that is heard when a speeding fire engine passes by. The drop in pitch occurs as the sound source changes from approaching to receding from the observer.

The size of the frequency shift increases as the speed of the source increases. The size of the shift is also a function of the observer's position with respect to the source. The observed frequency is largest for an observer directly in the line of motion ahead of the source (at A in Figure 8–13b) and is lowest for an observer directly behind the source (at B). At all other points the frequency is somewhere between these extremes; if the observer is 90° to the line of motion (at C), he will not experience any Doppler shift at all. Note that *the size of the frequency shift does not depend on the distance to the observer.*

A frequency shift is also observed if the *observer* is moving toward or away from a stationary source. An observer who is moving toward a group of oncoming waves will meet them at a faster rate than he would if he were stationary; thus he observes a higher frequency, just as if the wave source were moving toward him. Hence the Doppler effect results from any *relative motion* between the source of the waves and the observer. The equation governing the Doppler effect when both source and observer are in motion is given in Figure 8–14.

If the wave source is moving with a speed equal to or greater than the wave speed for the particular medium, a rather different wave pattern will result. Figure 8–15a shows the wave pattern produced when the speed of the source is equal to the wave speed. Since the source is moving just as fast as its own waves, at the front of the wave pattern each successive wave will be superimposed on all previous waves. Consequently, the wave amplitude at this point will be extremely large. An object moving at the speed of sound produces a "sonic boom," in which the pressure disturbance at the front of the wave pattern can be large enough to burst eardrums, shatter windows, crack walls, and so forth.

The pattern shown in Figure 8–15b results if the source is moving faster than the wave speed. In this case a "shock wave," consisting of a region of high amplitude where waves are superposed, is created and travels out at an angle to the path of the source. This shock wave is similar to the bow waves produced by boats moving through water. The sonic shock waves generated by jet planes traveling at speeds greater than the speed of sound can also cause considerable damage. For this reason planes are generally not allowed to fly over land at supersonic speeds except at relatively high altitudes.

The Doppler effect has numerous technological applications. In general, it can be used to determine the speed of a moving object, even if the distance to the object

figure 8–14

EQUATION FOR THE DOPPLER SHIFT

It is assumed that the source and observer are moving on the same straight line at constant speeds.

$$\frac{f_{obs}}{f_s} = \frac{v - u_{obs}}{v - u_s}$$

v = wave speed in medium
u_{obs} = speed of observer
u_s = speed of source
f_{obs} = frequency measured by observer
f_s = frequency emitted by source

Note: The formula is correct for the diagram as shown. If either the source or the observer is moving in a direction opposite to that shown, the sign of u_s or u_{obs} in the equation should be reversed, as follows:

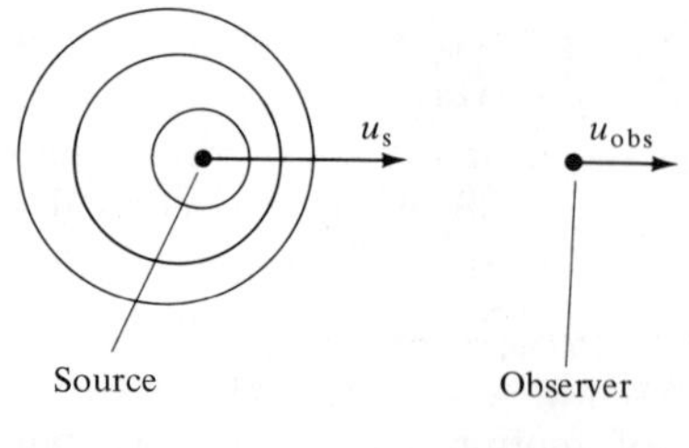

Direction of		*Sign of*	
Source	*Observer*	u_s	u_{obs}
→	→	−	−
→	←	−	+
←	→	+	−
←	←	+	+

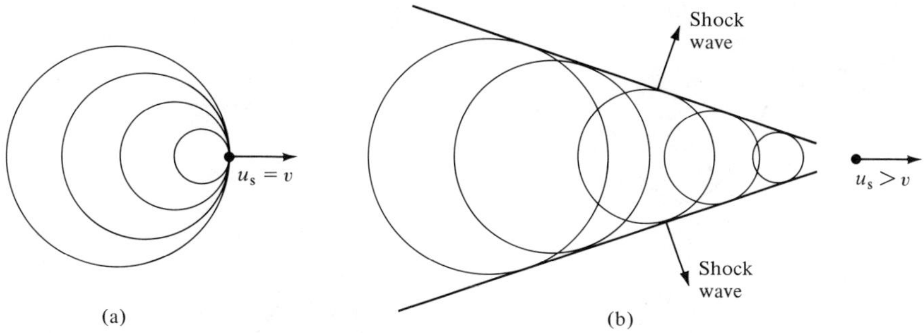

figure 8–15 **WAVES FROM A MOVING SOURCE**

The diagrams illustrate the wave patterns produced by sources moving with speeds (a) equal to the wave speed and (b) greater than the wave speed. In either case a region of very high amplitude is created where the waves are superimposed. If the waves are sound waves in air, a sonic boom is produced.

is not known. For example, the police use the Doppler effect to detect highway speeders. An electronic device designed for this purpose transmits radio waves of a known frequency and receives the waves reflected back by the car. Since the reflected waves will be shifted in frequency, a comparison of the transmitted and reflected frequencies will give the speed of the car.

MATHEMATICAL REPRESENTATION OF WAVES

In order to analyze further the properties of waves, we have to represent them in mathematical form. Because of their periodic nature, the trigonometric sine and cosine functions are eminently suitable for this purpose. In Figure 8–16 the trigonometric equation $y = K \sin px$ has been plotted for several different values of K and p. It is evident from these graphs that wave trains of different amplitudes and wavelengths can be depicted by selecting appropriate values for K and p, respectively. K directly determines the amplitude A of the wave, and p is related to the wavelength λ by the equation $p = 2\pi/\lambda$. Thus a train of waves can be represented by the equation*

$$\psi = A \sin \frac{2\pi x}{\lambda} \tag{8–5}$$

where A is the amplitude and λ the wavelength of the wave. The symbol ψ (the Greek letter psi) is used to designate the ***wave function*** of a wave. The wave function describes in mathematical form the state of the medium as the wave disturbance passes through it. Specifically, ψ represents the displacement from equilibrium of a particle of the medium that is at a distance x away from the source of the disturbance. Thus ψ can be used to represent the transverse displacement of a vibrating string or the longitudinal displacement of the medium in a compressional wave—to cite two examples.

*The equation $\psi = A \cos (2\pi x/\lambda)$ would be equally suitable, since the sine and cosine functions have the same shape. The sine function sets $\psi = 0$ at $x = 0$ (zero displacement at the source), and the cosine function depicts the wave when there is maximum displacement at the source. (See Fig. 8–17b.)

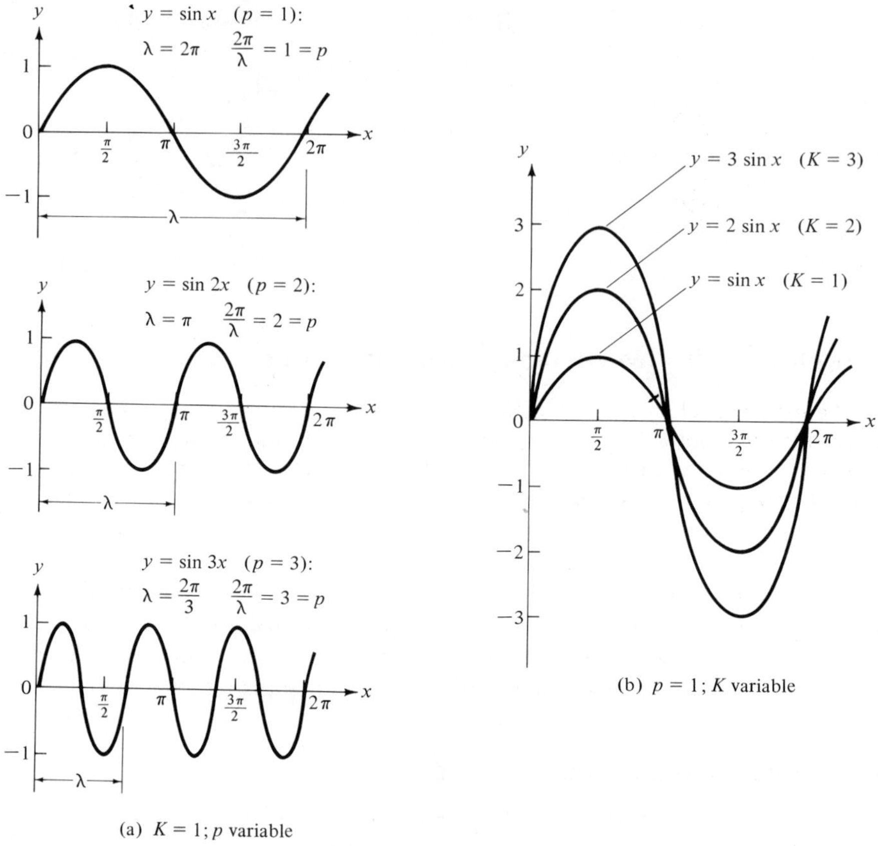

figure 8–16 MATHEMATICAL REPRESENTATION OF WAVES

A wave train can be described by the function $y = K \sin px$, where K and p are numbers whose values can be varied. (a) As p changes, the wavelength changes. Note that in each case $p = 2\pi/\lambda$. (b) As K changes, the amplitude changes. In all graphs x (the distance from the wave source) is measured in radians (π radians $= 180°$).

The wave function given by Eq. 8–5 is shown in Figure 8–17a. This equation has the limitation that it depicts the wave *at a given instant of time;* that is, it does not describe the *motion* of the wave through the medium as time progresses. In order to show the motion of the wave, the wave function must include the wave speed v and must depend on time as well as distance. It can be shown that the wave function

$$\psi = A \sin \frac{2\pi}{\lambda}(x - vt)$$

meets these criteria. (See Figure 8–17c.) In the form represented by this equation the wave will move to the right (along the positive x axis) as time progresses. If $x + vt$ is used instead of $x - vt$, the wave will progress to the left.

Sine and cosine wave functions are of great value in the analytical study of wave motion. Of particular significance is the fact that *virtually any arbitrary wave shape can be decomposed into a sum of sine and cosine wave functions.* Such a sum

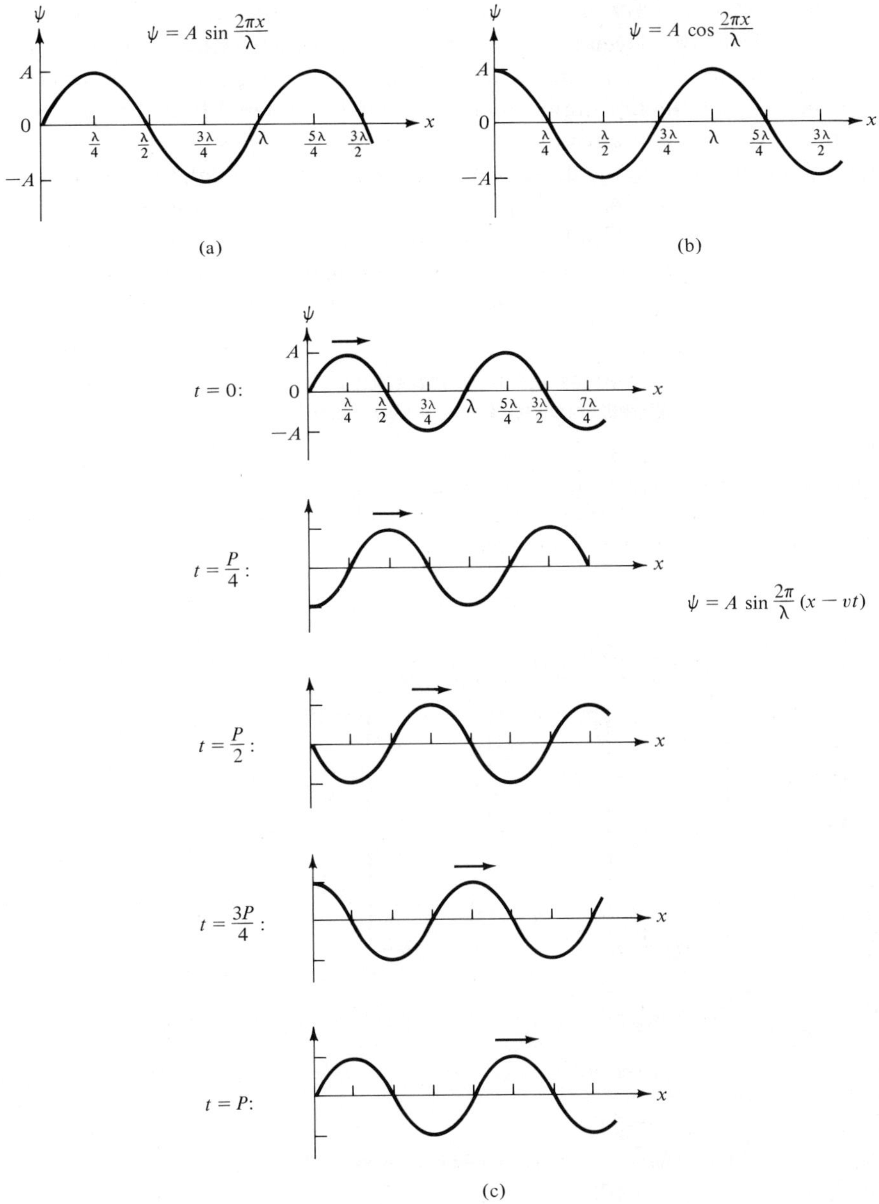

figure 8–17 **THE WAVE FUNCTION OF A WAVE**

The wave function ψ describes mathematically the displacement of the medium produced by a wave. (a) ψ at the instant when the displacement at the source is zero. (b) ψ at the instant when the displacement at the source is maximum. (c) The motion of a wave through a medium is depicted by a time-dependent wave function. The wave is seen to move to the right as time progresses. Note that the wave has moved one wavelength to the right after a time equal to its period P has elapsed.

is known as a ***Fourier series,*** after the French mathematical physicist Jean Fourier (1768–1830), who developed this mathematical theory for his study of heat flow. Thus any wave motion can be mathematically analyzed in terms of the properties of the sine and cosine functions into which it can be decomposed (the "Fourier components" of the wave). Some interesting examples of Fourier series are shown in Figure 8–18.

One further point about the mathematical representation of waves is worth noting. A wave function of the form $\psi = A \sin 2\pi x/\lambda$ represents an *infinitely long* wave train (Figure 8–19a). In order to represent a *finite* wave train consisting of a fixed number of wave pulses traveling through the medium (Figure 8–19b), one

figure 8–18 **EXAMPLES OF FOURIER SERIES REPRESENTATION OF WAVE SHAPES**

The square wave (a) and triangular wave (b) can each be represented mathematically as a sum of an infinite series of sine and cosine wave functions of different amplitudes and wavelengths. The sum of a finite number of terms approximates the given wave shapes.

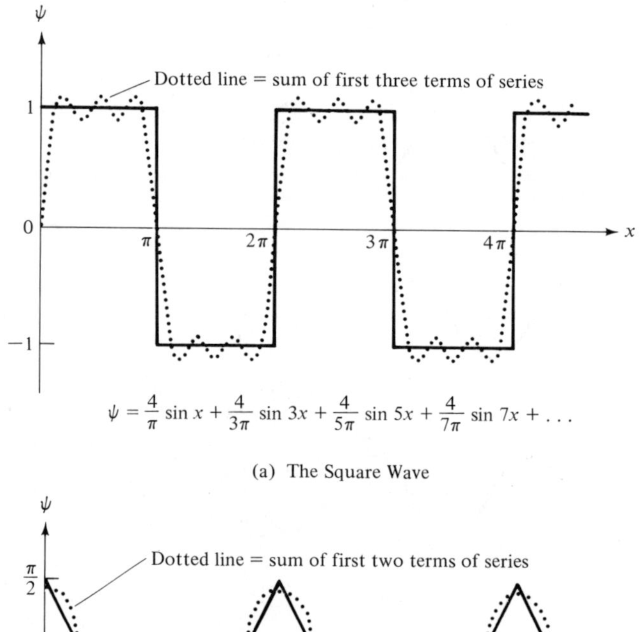

$$\psi = \frac{4}{\pi}\sin x + \frac{4}{3\pi}\sin 3x + \frac{4}{5\pi}\sin 5x + \frac{4}{7\pi}\sin 7x + \ldots$$

(a) The Square Wave

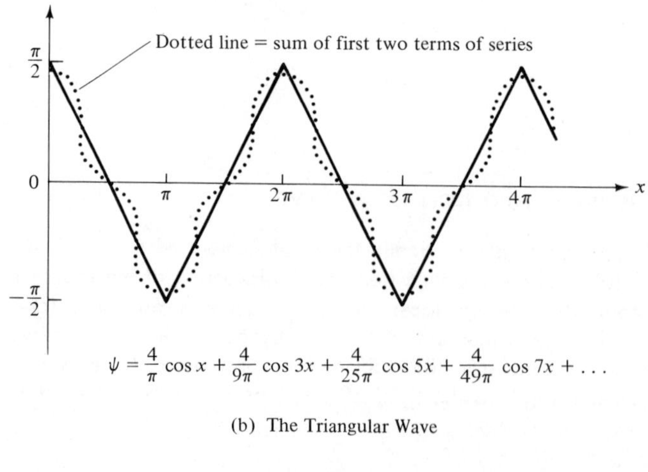

$$\psi = \frac{4}{\pi}\cos x + \frac{4}{9\pi}\cos 3x + \frac{4}{25\pi}\cos 5x + \frac{4}{49\pi}\cos 7x + \ldots$$

(b) The Triangular Wave

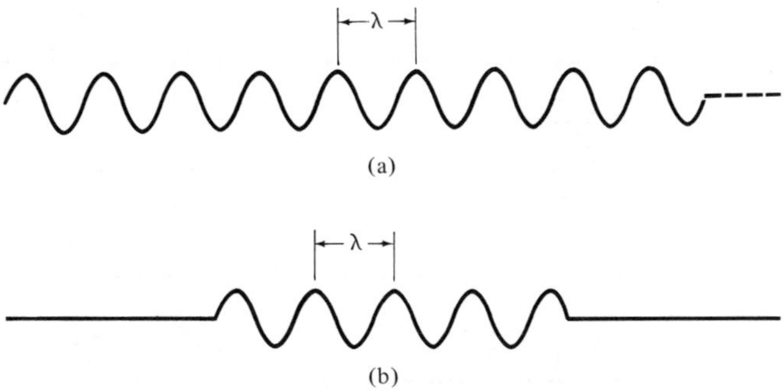

figure 8–19 **INFINITE AND FINITE WAVE TRAINS**

The infinite wave train (a) is represented by a single sine wave function ($\psi = A \sin 2\pi x/\lambda$). The finite wave train (b) must be represented as a Fourier sum of sine wave functions with wavelengths differing slightly from λ.

must use a Fourier series. That is, a finite wave train cannot be precisely represented by a single sine wave of a given wavelength, but must be represented as a sum of a number of sine waves of slightly different wavelengths. In general, the shorter the wave train, the greater the range of wavelengths needed to represent it.

WAVE INTERFERENCE

What happens when two different wave trains that are traveling in different directions meet at some point in a medium? Application of basic Newtonian principles dictates that *the displacement of the medium at that point will be equal to the sum of the separate displacements caused by the individual waves.* This is sometimes known as the ***superposition principle,*** since it predicts that the two waves will simply be superimposed on one another. Once past the point of intersection, the waves will continue to move as before, just as if there had been no interaction. Thus two wave patterns do not "collide" or reflect off one another; they merely interpenetrate. This principle is amply borne out by observation and experiment. One simple observation can be made by watching the surface of a body of water when it is raining; the wave patterns produced by the raindrops striking the surface will be seen to move through each other.

To further illustrate the superposition principle, let us imagine two identical wave pulses traveling toward one another from opposite ends of a stretched string (Figure 8–20). At the point where the waves meet, the displacement of the string is equal to the arithmetic sum of the two waves. Afterward the wave pulses continue to move along the string without any change in shape or amplitude. In Figure 8–20a the displacements are in the same direction, so that when the two waves coincide, the displacement of the string is increased; this is known as ***constructive interference,*** or ***reinforcement.*** In Figure 8–20b the displacements are opposite, so that at the moment of coincidence the two pulses tend to cancel each other out; this is known as ***destructive interference,*** or ***cancellation.*** If the wave pulses are identical, at that instant the string will have no displacement anywhere along its length. However, the two wave pulses have not destroyed each other; although the displace-

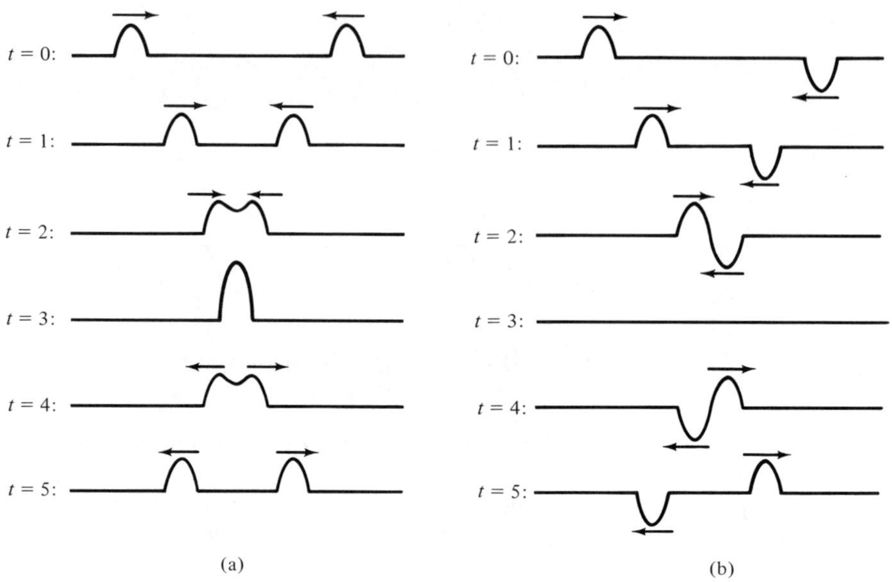

figure 8–20 **INTERFERENCE BETWEEN TWO WAVE PULSES ON A STRING**

Constructive interference (a) and destructive interference (b) occur when the two traveling wave pulses meet at $t = 3$.

ment of the string may be zero at that instant, the string is still in motion. A moment later the pulses reappear and continue on their separate paths.

Now let us suppose that two wave trains, traveling along different paths, are going to meet at some point in a medium. Each individual wave will cause the particles of the medium at that point to vibrate in a certain way. According to the superposition principle, the net vibration produced by the two waves will be the sum of the individually produced vibrations; in other words, there will be interference at the point of intersection. However, the kind of interference that occurs will depend on what part of their wave pattern the waves are in when they arrive. If the two waves arrive so that they produce the same displacement and the same direction of motion (as when crest meets crest or trough meets trough), reinforcement occurs (Figure 8–21a). The two waves are then said to be ***in phase.*** The amplitude of the displacement at the point of intersection will be twice that produced by a single wave.

If the two waves arrive so that the displacements and motions produced are in opposite directions (crest meets trough), cancellation results (Figure 8–21b). The two waves are shifted by a half wave (π radians, or 180°) with respect to one another and are said to be ***180° out of phase.*** In this case there is no net displacement *at any time* at the point of intersection. If the two waves arrive out of phase by other than 180°, a displacement of intermediate amplitude results, as shown in Figure 8–21c.

STANDING WAVES

One notable example of wave interference is the production of standing waves on a string. Suppose a string is fixed at one end, and a wave pulse is sent along the string. When the pulse reaches the fixed end, it will be reflected back in the opposite direction, but with its displacement reversed, as shown in Figure 8–22. If instead a wave train is generated on a string with fixed ends, the waves at the beginning of the train will be reflected back, and there will be interference between these inverted, reflected waves and the oncoming waves of the train that have not yet reached the end of the string. In general, this interference will produce a combined wave pattern, which will travel back and forth along the string at a rate determined by the frequency of the original wave train. There are certain frequencies, however, that produce stationary wave patterns in which waves no longer appear to travel along the string. Each point on the string vibrates with a constant amplitude. Certain points on the string will be motionless at all times; these points are called ***nodes.*** Other points will vibrate with maximum amplitude; these are the ***antinodes.*** Intermediate points will vibrate with smaller amplitudes. Such a pattern is called a ***standing wave.*** Some examples of standing wave patterns are shown in Figure 8–23.

It is a simple matter to calculate the frequencies that will produce standing waves. The string in Figure 8–23 has a length L and is fixed at both ends; thus both ends

figure 8–21 **INTERFERENCE BETWEEN WAVES**

Two sine waves of equal amplitude, moving along different paths, meet at the same point. The displacement as time progresses produced by each wave and by the combined waves is shown. (a) Two waves in phase. (b) Two waves 180° (half a wave) out of phase. (c) Two waves 90° (one fourth of a wave) out of phase.

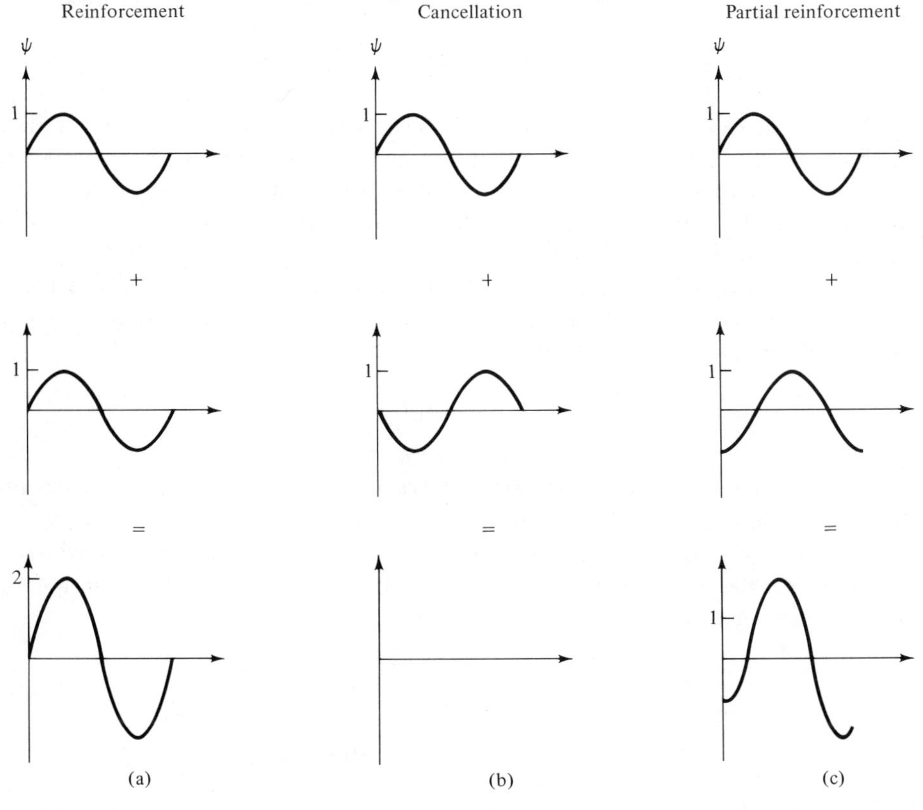

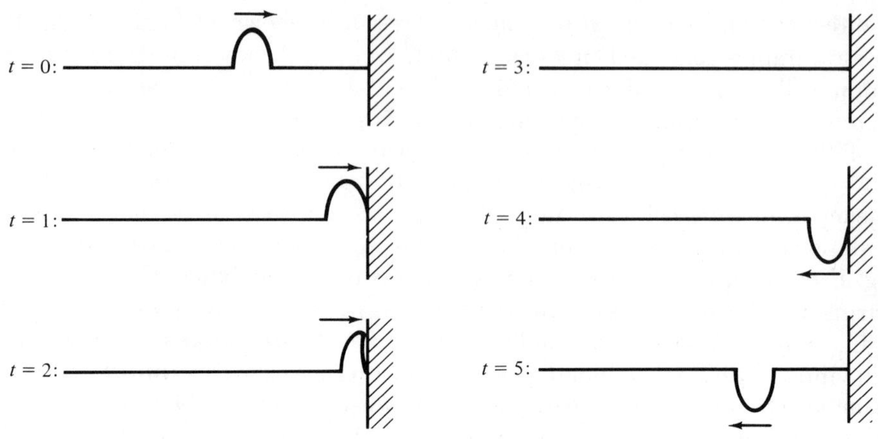

figure 8–22 **REFLECTION OF A WAVE PULSE**

At the fixed end of a string, the displacement must always be zero. Note that the wave pulse is inverted by the reflection.

must be nodes. The longest standing wave that can fit on this string is necessarily one in which a half wave falls between the ends, because a longer wave could not have nodes at both ends. This wave is shown in Figure 8–23a. It follows that the wavelength of this wave is twice the length of the string; that is, $\lambda_1 = 2L$. To calculate the frequency f_1 of this wave we use the equation $v = f\lambda$. Thus we have

$$\lambda_1 = 2L \qquad \text{and} \qquad f_1 = \frac{v}{2L}$$

We will call f_1 the ***fundamental frequency*** of the string. The fundamental frequency is the lowest frequency that can generate standing waves when applied to a string of length L that is fixed at both ends. Note that f_1 also depends on the wave speed v, which in turn is related to the tension and weight of the string. (See Eq. 8–2.)

The next possible standing wave pattern has a node in the middle of the string, as shown in Figure 8–23b. Here the wavelength of the standing wave is equal to the length of the string itself. Thus

$$\lambda_2 = L \qquad \text{and} \qquad f_2 = \frac{v}{L}$$

It is customary to call f_2 the ***first overtone.*** Note that $f_2 = 2f_1$; any frequencies between f_1 and f_2 will not generate standing wave patterns.

Successively higher overtones correspond to those frequencies for which 3 half waves fit on the string, 4 half waves fit on the string, and so on. The appropriate wavelengths and frequencies are as follows:

$$\lambda_3 = \frac{2L}{3} \qquad f_3 = \frac{3v}{2L} \qquad \text{(second overtone)}$$

$$\lambda_4 = \frac{L}{2} \qquad f_4 = \frac{2v}{L} \qquad \text{(third overtone)}$$

$$\lambda_5 = \frac{2L}{5} \qquad f_5 = \frac{5v}{2L} \qquad \text{(fourth overtone)}$$

and so on.

We can combine these results into a single formula. The wavelengths and frequencies that give standing wave patterns are:

$$\lambda_n = \frac{2L}{n} \qquad (n = 1, 2, 3, 4, 5, \ldots) \tag{8–6a}$$

$$f_n = \frac{nv}{2L} \qquad (n = 1, 2, 3, 4, 5, \ldots) \tag{8–6b}$$

This analysis of standing waves is useful for understanding the workings of musical instruments. When a stretched string (as on a guitar or violin) is plucked, it will vibrate mainly as a standing wave of its fundamental frequency. The vibration of the string sets the surrounding air (as well as the body of the instrument to which it is attached) into vibration at the same frequency, producing a traveling sound wave that proceeds outward. When the sound wave falls on an eardrum, the eardrum will also vibrate at this frequency, and thus one hears the sound produced by the string. Different frequencies can be generated by changing the string tension or weight, thereby changing the wave speed. Also, one can temporarily shorten the string by placing one's finger on it: decreasing the length of the string increases the fundamental frequency of the string, resulting in a sound of higher pitch.

figure 8–23 **STANDING WAVE PATTERNS**

At certain vibration frequencies, standing waves are generated on a string fixed at both ends. The range of vibration at each point of the string is shown. The antinodes (A) vibrate with maximum amplitude; the nodes (N) are always at rest. For each standing wave pattern, the relationship between the wavelength λ and the string length L is given.

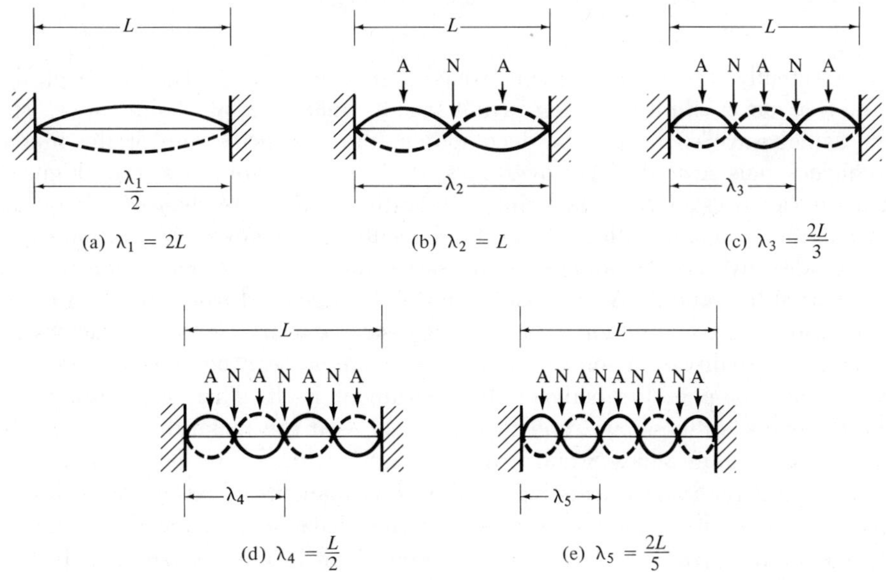

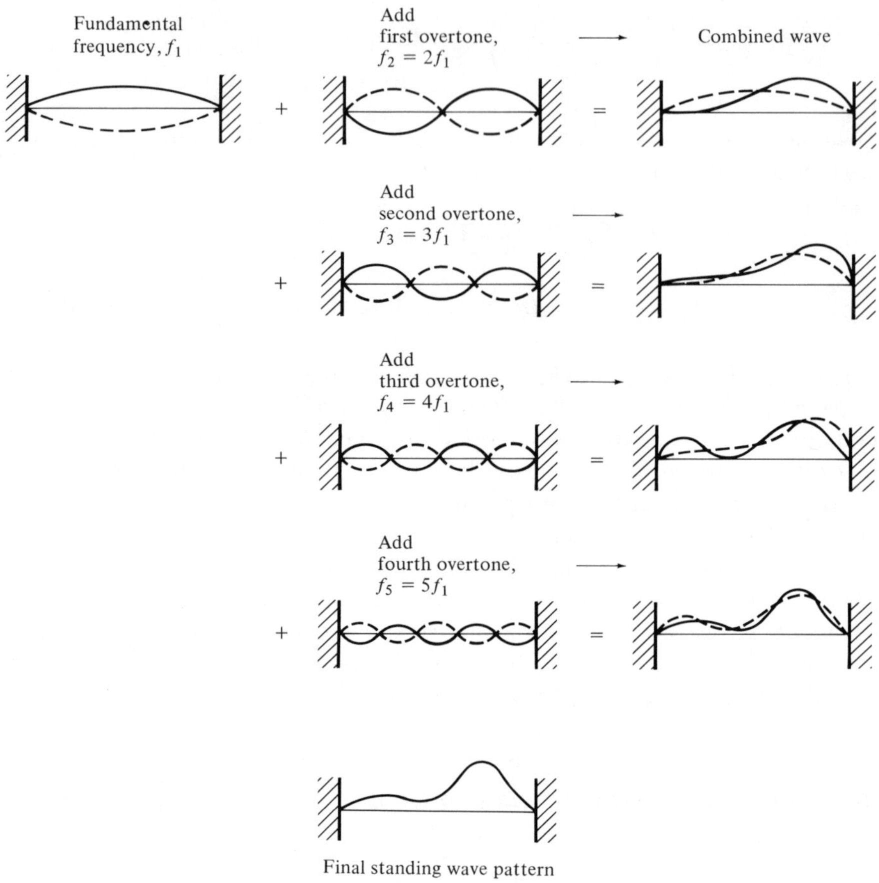

figure 8–24 **OVERTONES**

A complex standing wave pattern is produced by combining the fundamental frequency and several overtones of different amplitudes.

Actually, the picture is somewhat more complicated. The shape of the wave produced on a vibrating string is not usually that of a pure sine wave. However, the actual wave shape can be decomposed into a Fourier sum of the fundamental frequency plus several of the overtones with smaller amplitudes. (See Figure 8–24.) Thus it can be said that the string is vibrating with more than one frequency at a time. Different combinations of the fundamental frequency with overtones at various amplitudes give rise to sounds of the same pitch but of different *quality;* it is for this reason that middle A on a piano can be distinguished from middle A on a violin. The shape and construction of a musical instrument are significant factors in determining the quality of a sound, for certain overtones can be amplified as a result of vibrations set up in the body of the instrument itself. Thus such subtle factors as the thickness, stiffness, age, and type of wood used can make the difference between an ordinary violin and a Stradivarius.

One of the more interesting musical developments of recent years has been the growing popularity of electronic music. Devices have been invented that can generate sounds of any given frequency and combine them in any chosen way. It is possible to combine a given frequency with any number of other frequencies (not only the given frequency's overtones), since one is no longer restricted to producing sounds by standing waves. Thus sounds can be produced electronically with qualities that cannot be duplicated by any of the traditional instruments. To some traditional music lovers, the development of electronic music is another example of the pernicious influence of technology on modern-day life.

HUYGENS' PRINCIPLE: WAVE DIFFRACTION

In his discussions of wave motion, Huygens proposed a geometric principle that can be used to predict how a wave front will move through a medium. Given the position and shape of a wave front at some instant of time, *each point on this wave front can be considered as a point source of secondary wavelets.* The shape of the wave front at a later time can then be constructed from the superposition of the secondary wavelets. This procedure, known as ***Huygens' principle,*** can be illustrated as follows. Consider the circular wave AB shown in Figure 8–25a. We consider each point on AB as a new source of waves and draw small circles to indicate the spreading of these secondary wavelets. Each secondary wavelet radiates outward as a circular wave front from its point of origin. Each wavelet travels at the wave speed v of the particular medium; hence in a given time t the wavelet will travel a distance vt. The superposition of these wavelets will produce a line of reinforcement along $A'B'$, which can be obtained by drawing the envelope enclosing the wavelets. (Note that only the forward motion of the wavelets is considered.) Thus $A'B'$ represents the new wave front after an interval of time t has elapsed; this wave front is a section of a circle concentric with AB, in agreement with observations of wave motion. In Figure 8–25b the same procedure is used to show the advancement of a plane wave.

Huygens' principle is especially useful in explaining the ***diffraction*** of waves, which is the spreading and bending of waves that occurs when they encounter obstacles. Some examples of the wave patterns produced when plane waves encounter obstacles of various kinds are shown in Figure 8–26. In Figure 8–26a plane waves are shown striking a barrier with a small opening. The waves can advance only at the opening, which acts essentially as a point source of circular waves. When plane waves encounter a barrier with a wider opening (Figure 8–26b), the amount of

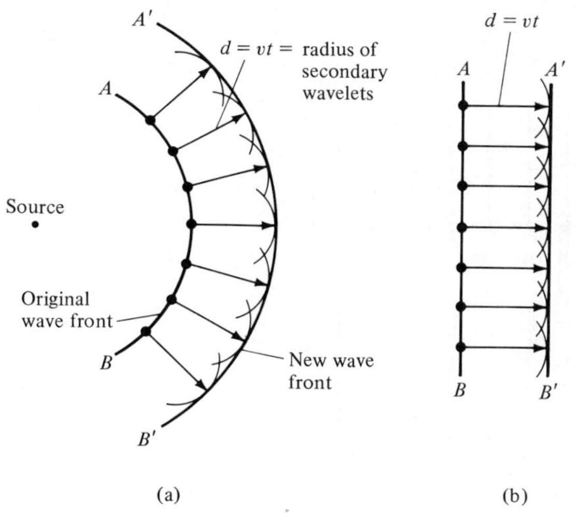

figure 8–25

HUYGENS' PRINCIPLE

Each point on the wave front AB is considered a source of secondary wavelets. The line $A'B'$ enclosing these wavelets represents the new wave front after a time interval t has elapsed. The motions of (a) circular waves and (b) plane waves are shown.

diffraction that occurs is lessened. In general, the factor that determines the amount of diffraction caused by an opening in a barrier is the ratio λ/d of the wavelength λ of the incoming waves to the width d of the opening.

Wave diffraction is rather common in nature; it is because of this phenomenon that sounds can be heard around corners. (Sound waves have relatively large wavelengths and therefore diffract considerably.) The diffraction of water waves by obstacles is also readily apparent. Figure 8–26c, for example, shows the wave pattern produced by a breakwater placed near a beach to weaken incoming waves.

The use of Huygens' principle to explain diffraction is illustrated in Figure 8–27. When the wavelength is large compared to the size of the opening, there are fewer secondary wavelets emanating from the opening; hence the emerging wave front will be nearly circular. For a larger opening the interference between the emerging secondary wavelets will tend to produce a less circular wave front, and the diffraction will be less pronounced.

figure 8–26 **DIFFRACTION**

Plane waves encountering a barrier experience a change in direction. Shown are the wave patterns produced by (a) a barrier with a small opening, (b) a barrier with a wider opening, and (c) an obstructing barrier.

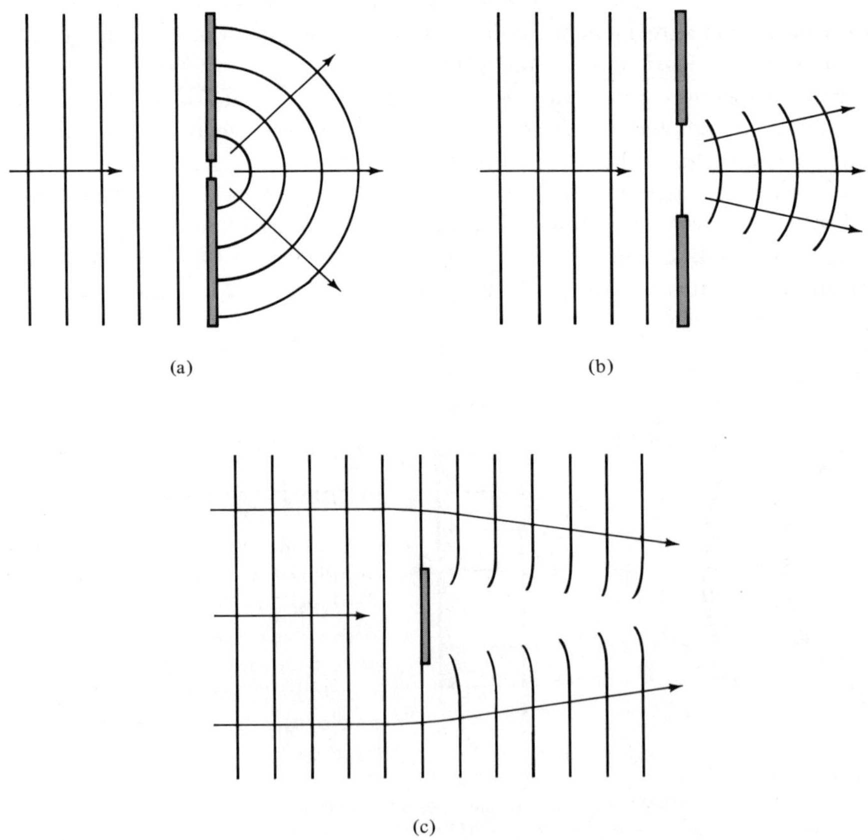

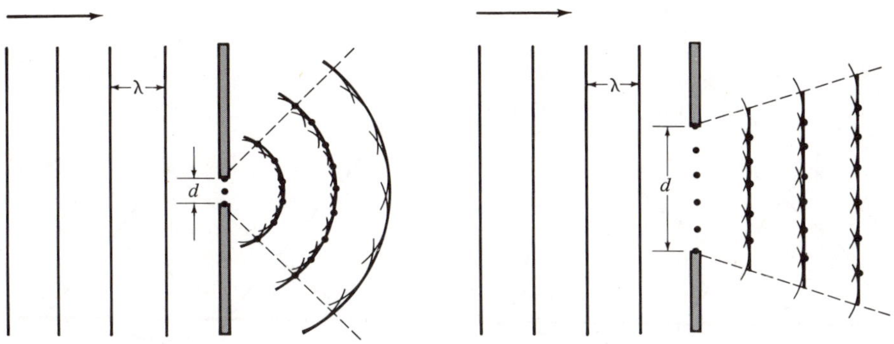

figure 8–27 **EXPLANATIONS OF DIFFRACTION IN TERMS OF HUYGENS' PRINCIPLE**

The constructions using secondary wavelets show that there is less spreading when the wavelength λ is small compared to d, the width of the opening. Wavelets will spread into the regions outside the dashed lines, but there is little or no reinforcement of these wavelets in these regions. Consequently, the wave amplitudes there are small compared to those along the main wave front.

8–4

THE WAVE THEORY OF LIGHT

Although Huygens' principle is valid for all wave motion, Huygens specifically intended it for the purpose of showing that light propagates as a wave. Huygens proposed his wave theory of light in his *Treatise on Light,* which was presented to the Paris Academy in 1678. In this work Huygens proposed that light is a longitudinal wave disturbance that propagates as a series of compressions and rarefactions set up in the ether.

Of particular interest to us is the use of Huygens' principle to show that waves obey the law of refraction. Figure 8–28a shows a series of plane waves advancing along a path that makes an angle i with the perpendicular to the interface between two media. We will assume that the waves travel with speed v_i in medium 1 and with speed v_r in medium 2. We will also assume that v_r is *less* than v_i; that is, the wave *slows down* upon entering the second medium. Now let us apply Huygens' principle to the wave front AB, which has one end in contact with the boundary between the two media. Let t be the time required for the wavelet generated at B to reach the boundary. Since this wavelet moves with a speed v_i, the distance BB' is equal to $v_i t$. However, the wavelet generated at A moves entirely in medium 2; therefore in the same time interval this wavelet will travel a distance $AA' = v_r t$. Note that AA' is smaller than BB' because v_r is less than v_i. Wavelets generated at points between A and B will travel for part of the time at speed v_i and part at speed v_r. In accordance with Huygens' principle, the new wave front will be the line $A'B'$. Now the wave front is entirely in medium 2 and will continue to move parallel to itself. Note, however, that as a consequence of the change in wave speed, the path of the wave fronts has changed direction; the wave fronts now make an angle r with

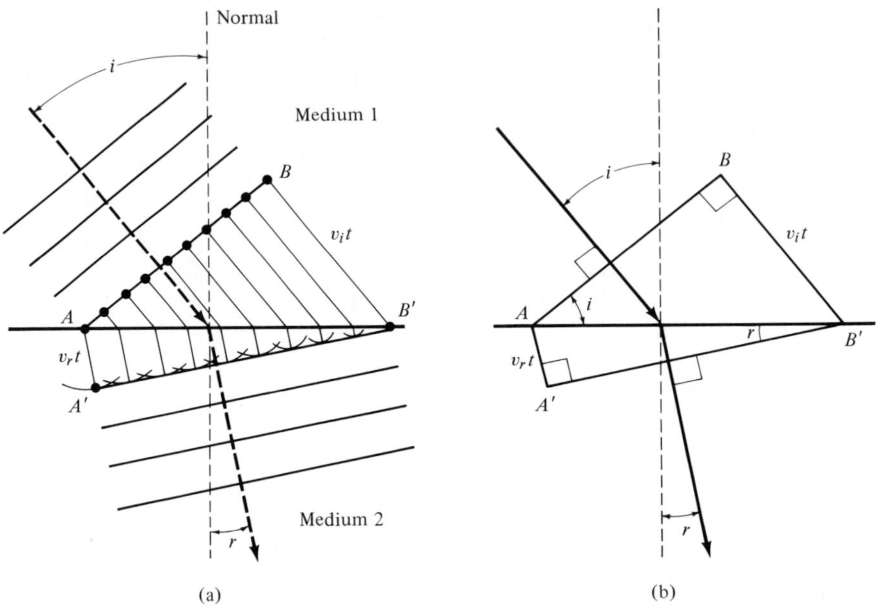

figure 8–28 **DERIVATION OF SNELL'S LAW USING HUYGENS' PRINCIPLE**

The secondary wavelets of wave front AB change speed as they enter a different medium. Hence the direction of the entire wave front is changed (see text).

the normal. From the geometry of the situation (Figure 8–28b) it can be seen that $\angle BAB' = i$ and $\angle AB'A' = r$. Thus

$$\sin i = \frac{BB'}{AB'} = \frac{v_i t}{AB'} \qquad \text{and} \qquad \sin r = \frac{AA'}{AB'} = \frac{v_r t}{AB'}$$

Combining these two equations, we obtain

$$\frac{\sin i}{\sin r} = \frac{v_i}{v_r} = \text{constant} \tag{8–7}$$

Thus Snell's law of refraction can be derived from Huygens' assertion that light is a wave. But we learned previously that refraction can also be explained by the *particle* theory of light. However, there is an important difference in the assumptions made in the two theories. Suppose the angle of incidence is greater than the angle of refraction (as is the case when light goes from air to glass). The particle theory proposes that light *speeds up* when it goes from air to glass, whereas the wave theory proposes that light *slows down.* Clearly, the contradiction between the particle and wave theories could have been resolved immediately by comparing the speeds of light in air and glass. Unfortunately, in 1678 there were no reliable methods for determining the speed of light in any medium, so this crucial experiment could not be performed.

Huygens' presentation of the wave theory of light had some notable shortcomings. For one thing, he did not offer any explanation of the phenomenon of colors. Of greater importance to Newton was Huygens' inability to satisfactorily account for the rectilinear propagation of light. In fact, the diffraction of light had already been demonstrated by Francesco Grimaldi (1618–1663), a professor of mathematics at Bologna, who observed a very slight bending of a narrow beam of light around obstacles placed in its path. Both Huygens and Newton knew of this discovery. But Huygens could not explain why the diffraction of light was so much smaller than the diffraction of sound or water waves; he had not worked the wave theory out in enough detail to show the dependence of diffraction on the wavelength. Newton, on the other hand, had proposed that light particles passing close to the edge of a body experience a slight attractive force, which causes small deviations from the normal straight-line path. Here the matter rested, with Newton's particle theory in the ascendancy, until the beginning of the nineteenth century.

THE EXPERIMENTS OF YOUNG AND FRESNEL

The revival of the wave theory of light was brought about by the discoveries of two men—Thomas Young of England and Augustin Fresnel of France. Young was a man of prodigious accomplishments. We are told that Young could read at the age of two; by the age of four he had read the Bible through twice. (Fresnel, on the other hand, was barely able to read as an eight year old.) By the age of fourteen Young was familiar with seven languages; he maintained his interest in linguistics throughout his lifetime, making important contributions to the translation of Egyptian hieroglyphics. Young studied medicine and was a practicing physician in London for many years. He became interested in the physiology of vision and proposed a theory of color vision that remains the foundation of the modern-day theory. This interest led naturally to his experiments on the nature of light.

In the years from about 1800 to 1804, Young devised and perfected an experiment that clearly demonstrated that light exhibits wave properties. Young's experiment is illustrated in Figure 8–29. Light from a source falls on an opaque screen in which there is a narrow slit (S_1). The light that passes through this slit falls on a second screen in which there are two closely spaced slits, S_2 and S_3, a distance d apart (d is typically no more than a few millimeters). The pattern produced by the light that passes through these two slits falls on a third screen, where it can be observed. According to the particle theory of light, one would expect to see two bright lines—the images of the two slits—on the screen, and nothing more. Quite to the contrary, Young observed a pattern of fifteen to twenty alternating bright and dark lines on the screen.

To explain this phenomenon, Young introduced the principle of wave interference. He proposed that the dark lines are caused by destructive interference between light waves that are 180° out of phase with one another, and the bright lines fall where light waves, being in phase, interfere constructively. To show how the interference pattern is produced, Young applied Huygens' principle, as shown in Figure 8–30. Slit S_1 acts as a point source of waves. (In three dimensions the actual wave fronts are cylindrical, concentric about the slit.) The wave fronts emanating from S_1 fall simultaneously on S_2 and S_3; each of these, in turn, acts as a point source of waves. However, because the waves emanating from S_2 and S_3 are produced by the same wave front (from S_1), *they will always start out in phase with one another.* Reinforcement (producing bright lines) will then occur wherever the crests of one wave front intersect the crests of the other wave front, and cancellation (producing dark lines) will occur wherever crests and troughs intersect.

The importance of the first slit (S_1) should be stressed, for without it there would be no interference pattern. The light given off by the source is emitted randomly from all parts of the source, and the light given off from one part is not necessarily in phase with the light from any other part. If the light from the source were allowed to fall directly on the double slit (S_2 and S_3), there would be no guarantee that the waves emanating from one slit would start out in phase with the waves from the other slit. In that case reinforcements and cancellations would not always occur at the same points, and there would be no fixed interference pattern. The first slit ensures that waves always leave S_2 and S_3 in phase.

A simple calculation can be used to determine the points where reinforcement and cancellation should occur. Referring to Figure 8–31a, we find that there will be a bright line, B_0, in the center of the screen. Since this point is equidistant from S_2 and S_3, waves leaving the two slits in phase will travel the same path length to B_0 and will arrive in phase. The next set of bright lines are at B_1 and B_1'. If reinforcement occurs at B_1, it means that the waves from S_2 and S_3 arrive in phase; yet B_1 is closer to S_2 than to S_3. It follows that the path lengths S_2B_1 and S_3B_1 must differ by one full wavelength if the waves are to meet in phase. That is,

$$S_3B_1 - S_2B_1 = \lambda \qquad \text{and} \qquad S_2B_1' - S_3B_1' = \lambda$$

By the same argument, it can be seen that the next pair of bright lines (B_2 and B_2') fall where the distances to the two slits differ by *two* full wavelengths. In general, then, bright lines will occur wherever the path lengths differ by an exact number of wavelengths. At points in between successive bright lines, various degrees of cancellation occur because the distances to the two slits do not differ by an exact

figure 8–29 **YOUNG'S EXPERIMENT**

Light enters through a narrow slit S_1, then passes through two closely spaced slits, S_2 and S_3. A pattern of alternating bright and dark lines appears on the screen.

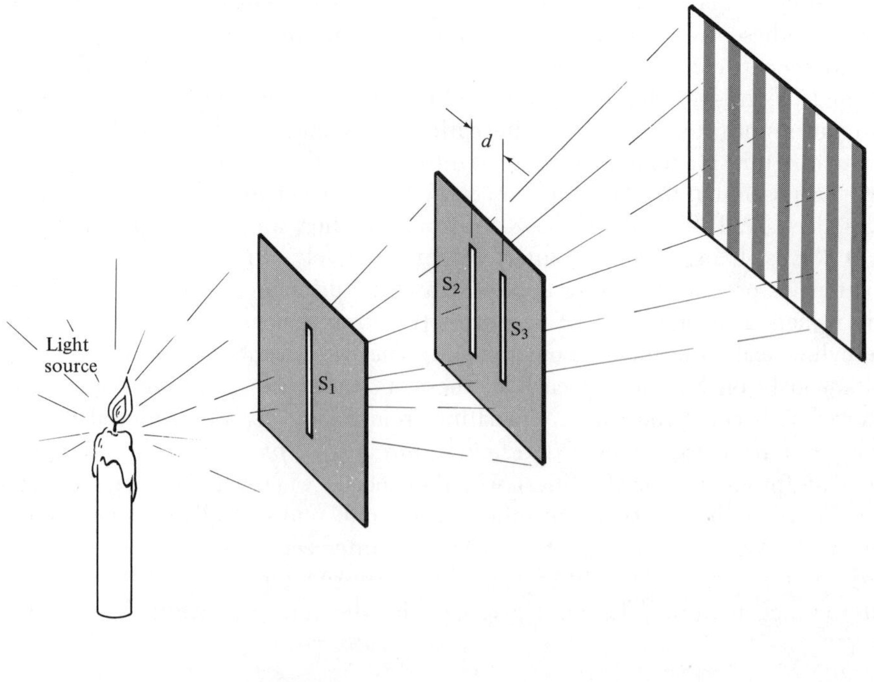

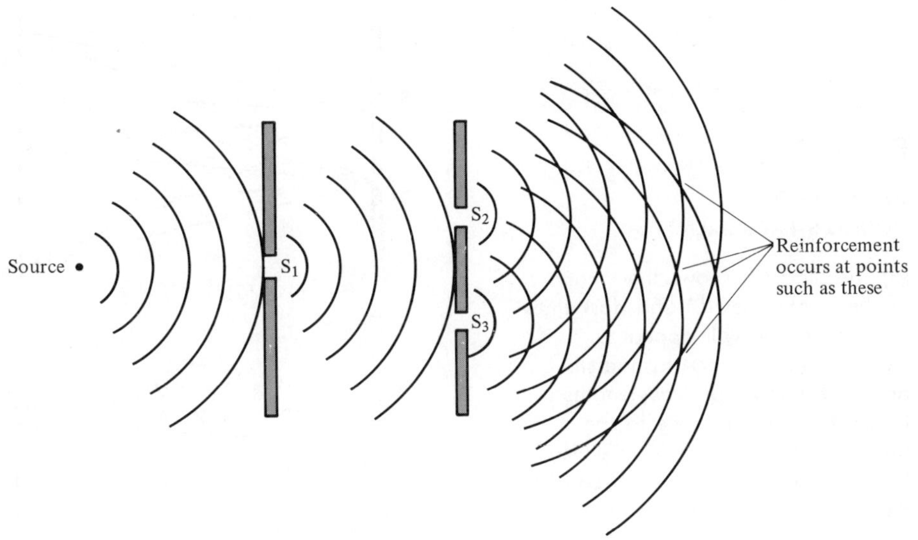

figure 8–30 **INTERFERENCE PATTERN PRODUCED BY YOUNG'S EXPERIMENT**

Waves emanating from S_2 and S_3 are in phase with each other. Bright lines (reinforcement) will occur wherever the crests of the two wave fronts intersect.

number of wavelengths, and the waves arrive out of phase. At points D_1, D_1', D_2, D_2', and so on, complete cancellation occurs because the waves always arrive 180° out of phase.

From the geometry of his experiment, Young was able to calculate the wavelength of light. The path lengths S_2B_1 and S_3B_1, which differ by a full wavelength, are drawn in Figure 8–31b. The line S_2Q has been drawn so that $S_2B_1 = QB_1$ (forming the isosceles triangle S_2B_1Q). Therefore the line segment S_3Q, which represents the difference in length between S_2B_1 and S_3B_1, is equal to the wavelength λ. Through a geometric proof (which is a bit too lengthy to be shown here), one can relate λ geometrically to the quantities d (the distance between the slits), l (the distance from the slits to the screen), and x_{B_1} (the spacing between the central line B_0 and the first bright line B_1). In general, x_{B_1} is very small compared to l, and as a result a simple approximate equation for λ can be obtained.* The equation is as follows:

$$\lambda = \frac{x_{B_1}d}{l}$$

Since the nth bright line (B_n) occurs where the path lengths differ by n full wavelengths, we can generalize this equation as follows:

$$n\lambda = \frac{x_{B_n}d}{l} \qquad (n = 1, 2, 3, 4, \ldots) \tag{8–8}$$

where x_{B_n} is the distance from the central line to the nth bright line. Since d, x_{B_n}, and l can be readily measured, the wavelength λ of the light from the source can

*The student who is proficient in geometry is encouraged to try to derive this relationship.

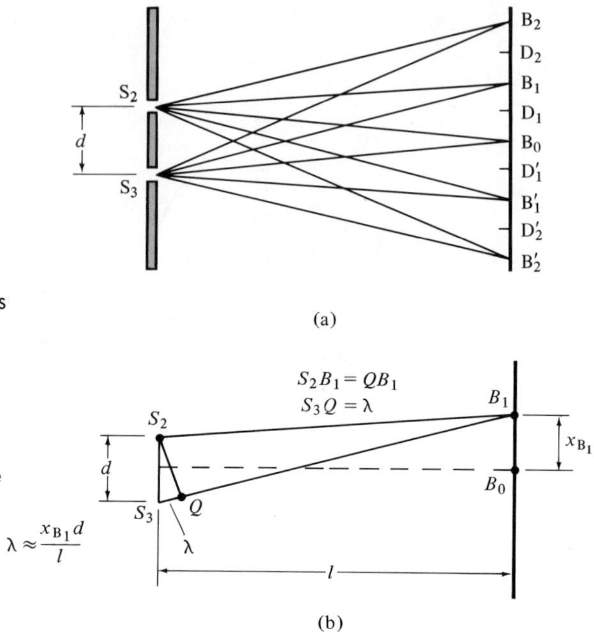

figure 8–31

ANALYSIS OF YOUNG'S EXPERIMENT

(a) The diagram shows the location of the bright lines (B) and dark lines (D). Bright lines will appear wherever the distance to one slit is an exact number of wavelengths longer than the distance to the other slit. (b) Diagram from which the equation for wavelength can be derived geometrically (see text).

be calculated. At first, Young used white light, and the lines were smeared out and fringed with color. He then worked with monochromatic light (which he could produce with prisms and filters), and the lines became sharp and distinctly separated. Young's measurements then showed that *each color has its own characteristic wavelength.*

In a typical setup of Young's experiment, the two slits were 0.05 cm apart, and the screen on which the interference pattern appeared was placed 100 cm from the slits. At this distance the spacing between the central and first bright line was found to be 0.13 cm when red light was used. Substituting $d = 0.05$ cm, $l = 100$ cm, $x_{B_1} = 0.13$ cm, and $n = 1$ in Eq. 8–8 gives

$$(1)\lambda = \frac{x_{B_1} d}{l} = \frac{(0.13)(0.05)}{100} = 0.000065$$

so that one obtains

$$\lambda = 6.5 \times 10^{-5} \text{ cm}$$

for the wavelength of red light.

When violet light is used, the bright lines are found to be even closer together; the corresponding wavelength turns out to be about 4×10^{-5} cm. Hence Young discovered a direct correlation between wavelength and position in the spectrum: *the wavelength decreases regularly from the red end to the violet end.* Young also noted that the minuteness of the wavelength of light explains why the diffraction of light is barely visible. (Sound waves, for example, have wavelengths of the order of a few feet.)

Thus, with a single experiment, Young made three important discoveries:

(1) The production of light interference illustrates that light has wave properties.
(2) The color of light is dependent on the wavelength.

(3) The wavelength of light is extremely short, being on the order of 5×10^{-5} cm (about 1/50,000 of an inch).

Unfortunately for Young, the specter of Newton was still stalking through the halls of the Royal Society, so that when Young presented his findings, he was strongly attacked. In disgust Young abandoned his researches on light and turned to his other interests.

Many years later Fresnel took up the study of light, not knowing of Young's previous discoveries. In 1815 he independently demonstrated light interference by a method similar to Young's. In France the worship of Newton had not quite reached the heights that had been attained in England, so that when Fresnel made his findings known, he received some encouragement from several influential French scientists. Among these was François Arago, who had the unhappy task of informing Fresnel that his discoveries had been made by Young fourteen years earlier. Fresnel and Young then began a correspondence that fully revived Young's interest in the subject.

Fresnel demonstrated the diffraction of light in a number of different experiments. He worked out the theory of interference and diffraction in great detail, proving mathematically the dependence of diffraction on wavelength. Fresnel was able to show convincingly that the rectilinear propagation of light was not in disagreement with the wave theory, owing to the very short wavelength of light. This removed Newton's main objection to the wave theory. In spite of these advances, however, most physicists remained unconvinced of the wave nature of light.

THE POLARIZATION OF LIGHT

One particularly thorny problem prevented the wave theory from gaining greater acceptance. This was the phenomenon of the ***polarization*** of light. There are several naturally occurring crystalline solids that have the property of transmitting light more efficiently in one direction than in any other. One such material is ***tourmaline.*** A single crystal of tourmaline is transparent to light, but the crystal reduces the intensity of a beam of light by about half. If a second tourmaline crystal is held with its axis parallel to that of the first, *all* the light that passes through the first crystal will pass through the second. As the second crystal is rotated, however, it transmits less and less light; when its axis is at a right angle to that of the first crystal, it transmits *no light at all.* (See Figure 8–32a.)

Both Huygens and Newton studied this peculiar phenomenon. Huygens could offer no reasonable explanation; to Newton it was an additional reason for rejecting the wave theory. To account for the effect, Newton conjectured that the particles of light were not symmetrically shaped. He imagined that light particles are ellipsoidal, rather than spherical; thus light will only pass through tourmaline if the long axis of the ellipsoid is aligned with the crystal axis—otherwise the light particles will not fit. Thus the particles that are aligned to fit through the first tourmaline crystal will not be able to pass through the second if the axis of the second crystal is rotated 90° with respect to the first.

The proponents of the particle theory were further encouraged by an accidental discovery made by Étienne Malus, a French engineer, in 1809. Malus discovered that light is polarized when it is reflected from a surface at a certain angle. He found that for a specific angle of incidence (53° for water, 56° for glass) the reflected beam would *not* be transmitted through a tourmaline crystal if the axis of the crystal was perpendicular to the glass surface (Figure 8–32b).

These findings posed grave difficulties for the wave theorists. Finally, in 1817

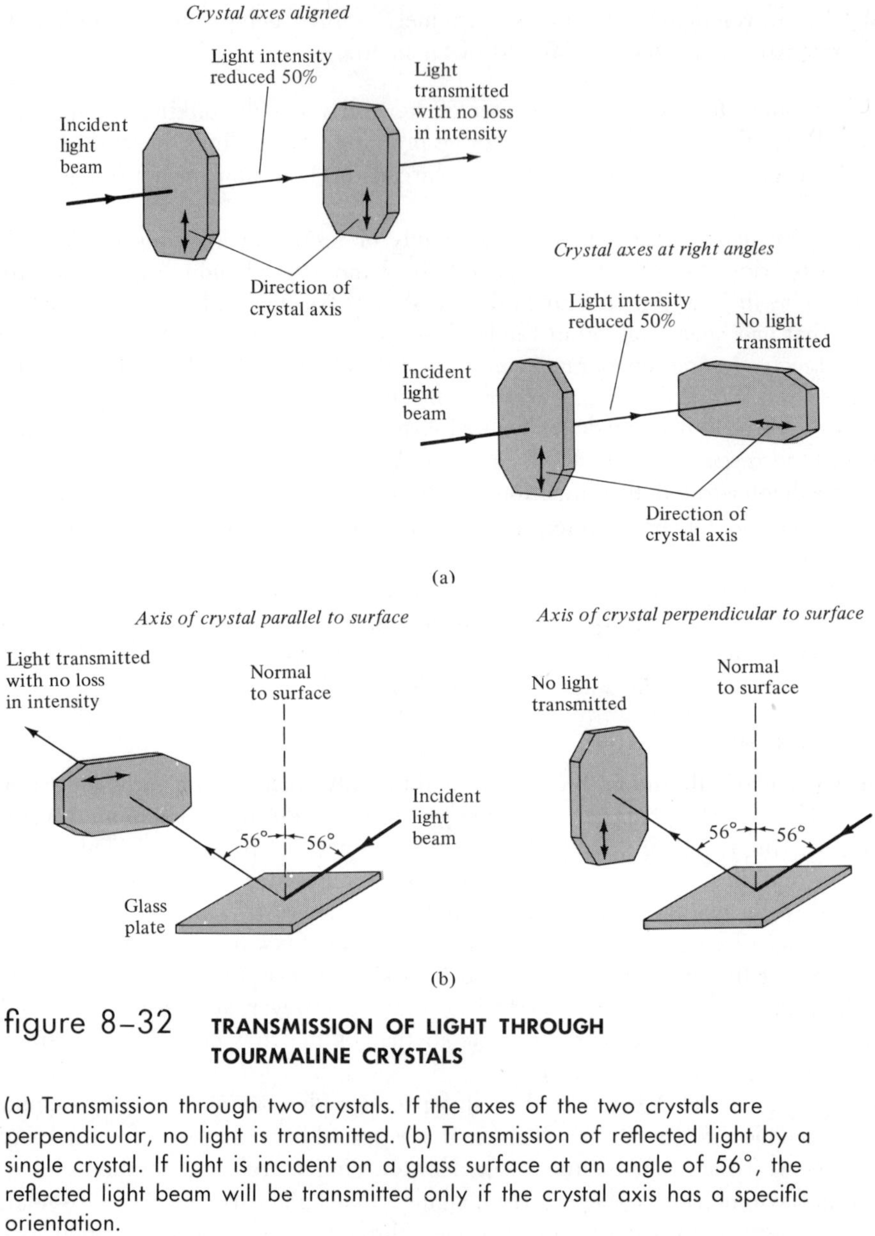

figure 8–32 **TRANSMISSION OF LIGHT THROUGH TOURMALINE CRYSTALS**

(a) Transmission through two crystals. If the axes of the two crystals are perpendicular, no light is transmitted. (b) Transmission of reflected light by a single crystal. If light is incident on a glass surface at an angle of 56°, the reflected light beam will be transmitted only if the crystal axis has a specific orientation.

an explanation was arrived at. From the time of Huygens it had been tacitly assumed that light is a *longitudinal* wave. Young, Fresnel, and Arago realized that if light was a *transverse* wave, all of these strange phenomena could be readily accounted for.

In a transverse wave the direction of vibration of the medium is perpendicular to the path of the wave. In three dimensions there is an infinite number of directions that are perpendicular to a given path; it is assumed that in an ordinary light beam all directions of vibration are equally probable. In other words, each particular wave front has a specific direction of vibration, which remains fixed as the wave front travels. However, successive wave fronts do not necessarily vibrate in the same direction. A light beam from an ordinary source consists of wave fronts emitted at different times from different parts of the source. Thus on a statistical basis all directions of vibration should be equally represented in the light beam. A light beam in which there is no preferred direction of vibration is said to be ***unpolarized light.***

A substance such as tourmaline is internally constructed so that its atoms vibrate in one particular direction with greater amplitude than in any other. When light enters, only those waves that are vibrating in this direction will be transmitted with unchanged intensity. If the direction of vibration of the wave is inclined at an angle to the preferred direction of vibration of the atoms of the crystal, then only the component of the wave vibration along the preferred direction will be transmitted. (See Figure 8–33a.) A wave vibrating perpendicular to the preferred direction of the crystal will accordingly not be transmitted at all. Thus each wave will have its parallel component transmitted and its perpendicular component absorbed. On the average, then, the intensity of an unpolarized beam will be reduced by about 50 percent. Moreover, the light waves that are transmitted all emerge vibrating in the same direction. Such a light beam, in which there is a preferred direction of vibration, is said to be ***linearly polarized light.***

The effects illustrated in Figure 8–32a can now be easily understood. The light emerging from the first tourmaline crystal is linearly polarized; it will be completely transmitted by the second crystal if the direction of polarization is aligned with the preferred direction of the crystal. As the second crystal is rotated through 90°, the transmission will decrease from 100 to 0 percent. The effect shown in Figure 8–32b can be explained if it is assumed that the light reflected at the particular angle (called the ***polarizing angle***) is completely linearly polarized in a direction parallel to the reflecting surface. (See Figure 8–33c.)

Substances such as tourmaline are known as ***polarizers,*** nearly all of which are naturally occurring crystals. One material of particular interest is ***herapathite*** (iodoquinine sulfate), which forms into tiny needle-shaped crystals. In 1935 the American inventor Edwin H. Land perfected a technique for orienting a large number of these crystals in the same direction and embedding them in plastic. The result is a polarizer that can be made into sheets of any size. This material is more familiarly known by its brand name, Polaroid. Polaroid sunglasses will transmit only 50 percent of unpolarized light. Moreover, since much reflected light—including the light of the sky—is polarized to some degree, it will be absorbed even more strongly by Polaroid glass. Light reflected off horizontal surfaces will be horizontally polarized; therefore by orienting the preferred direction of Polaroid in a vertical direction, glare is considerably reduced.

THE ACCEPTANCE OF THE WAVE THEORY OF LIGHT

By assuming that light is a transverse wave, Young, Fresnel, and Arago were able to account for all such polarization phenomena in terms of the wave theory of light. Fresnel and Arago provided further evidence for the theory when they demonstrated that two beams of light polarized at right angles to each other could not be made to produce an interference pattern (for if the amplitudes are perpendicular to one another, cancellation cannot be attained, even if the waves are 180° out of phase). Thus another barrier to the acceptance of the wave theory was removed.

In 1850 the *coup de grace* was delivered to Newton's particle theory by Léon

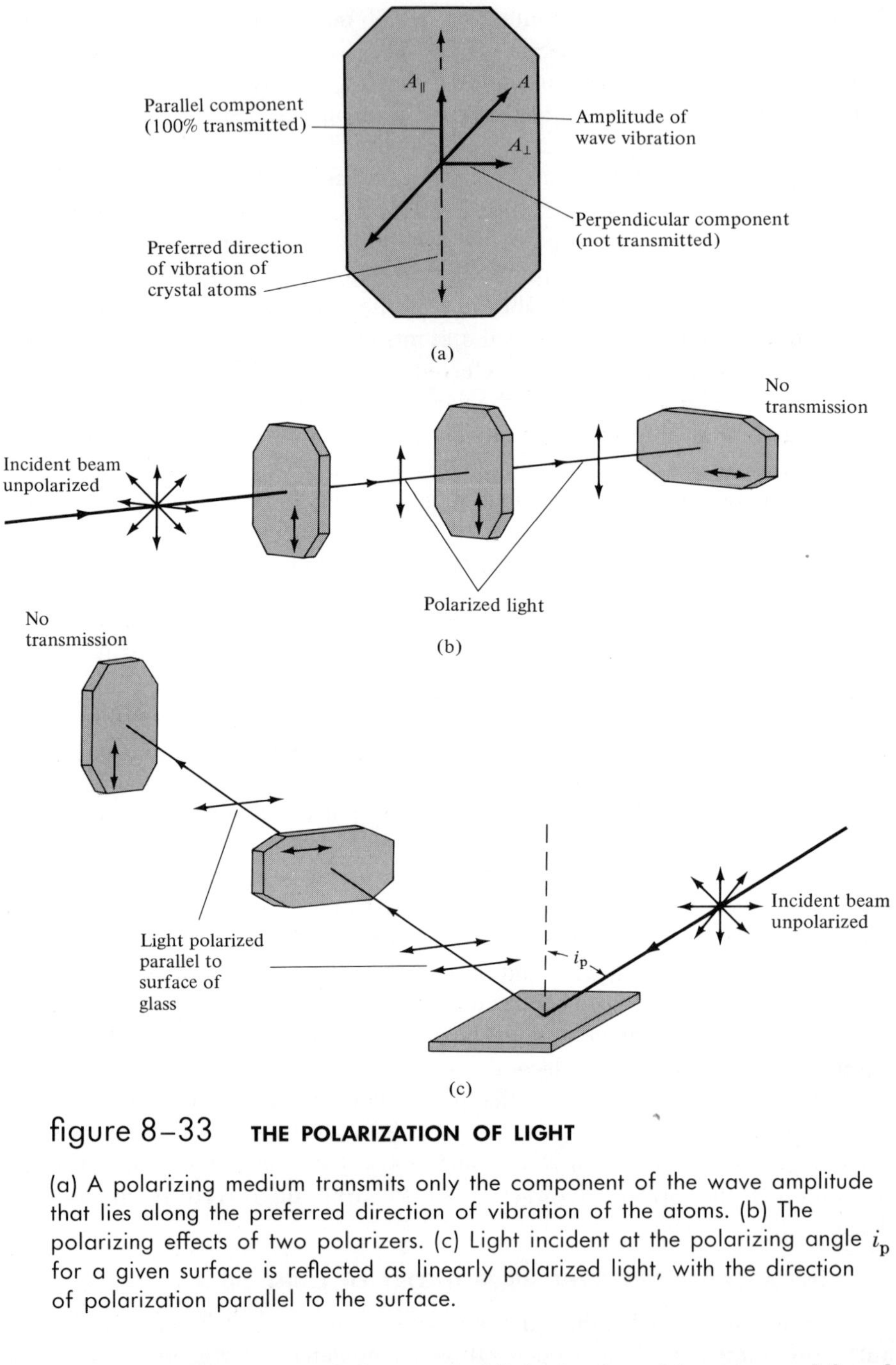

figure 8–33 **THE POLARIZATION OF LIGHT**

(a) A polarizing medium transmits only the component of the wave amplitude that lies along the preferred direction of vibration of the atoms. (b) The polarizing effects of two polarizers. (c) Light incident at the polarizing angle i_p for a given surface is reflected as linearly polarized light, with the direction of polarization parallel to the surface.

Foucault, who successfully measured the speed of light in air and in water and found that the speed was 1.3 times greater in air than in water. We previously noted that Snell's law could be derived from both the particle and the wave theories, but the particle theory predicted that light would *speed up* upon moving from air to water, and the wave theory predicted the opposite. Thus Foucault's measurement clearly supported the wave theory and showed the fundamental assumptions of the particle theory to be in error. Upon the announcement of Foucault's results, support of the particle theory crumbled rapidly.

Huygens' wave theory derivation of Snell's law can be used to obtain a relationship between the speed of light in a particular medium and the index of refraction of the medium. The index of refraction n can be determined experimentally from Snell's law (Eq. 8–1a):

$$\frac{\sin i}{\sin r} = \frac{n_r}{n_i}$$

Let us compare this equation with the equation for Snell's law obtained theoretically by Huygens (Eq. 8–7):

$$\frac{\sin i}{\sin r} = \frac{v_i}{v_r}$$

We find that the index of refraction is inversely proportional to the speed:

$$\frac{n_r}{n_i} = \frac{v_i}{v_r}$$

We therefore define the index of refraction n for any medium by the equation

$$n = \frac{c}{v} \tag{8–9}$$

where c is the speed of light in free space and v the speed of light in the medium. Since it is found experimentally that $n > 1$ for every known medium, it follows that light travels with maximum speed in free space and slows down upon entering any given material medium.

Let us now turn to the phenomenon of the dispersion of light by a prism into component colors. Newton showed that substances have different indexes of refraction for different colors, with n being smallest for red and greatest for violet. It follows from Eq. 8–9 that different colors must travel at different speeds in a medium; red has the greatest speed and violet the slowest. In free space, however, all colors travel with the same speed.

Newton proposed that each color was represented by a different type of particle, but examination of a spectrum reveals that it is not made up of six or seven distinct colors. Rather, the colors change gradually in hue and blend into one another without any clear line of demarcation between them. It would take an infinitude of different particles to represent all the different colors seen in the spectrum! According to the wave theory, on the other hand, differences in color correspond to differences in wavelength. Thus the index of refraction of a medium is dependent on the wavelength of the incident light. Each of the six basic colors is represented not by a single wavelength, but by a *range* of wavelengths (the boundaries of which are approximate). These ranges are listed in Table 8–2.

At this point we will introduce a unit called the ***angstrom,**** which is commonly

*In honor of the Swedish physicist Anders Ångström (1814–1874), who helped to standardize the measurement of the wavelengths of light.

table 8–2 **APPROXIMATE WAVELENGTH RANGES OF MAJOR COLORS**

(1 angstrom = 10^{-8} cm)

Color	Wavelength (angstroms)
Red	6100–7000
Orange	5900–6100
Yellow	5700–5900
Green	5000–5700
Blue	4500–5000
Violet	4000–4500

used to express the wavelength of visible light (except in calculations, when it must be converted back into cm). The angstrom (Å) is defined as follows:

$$1\ \text{Å} = 10^{-8}\ \text{cm}$$

Thus a wavelength of 5×10^{-5} cm is expressed in angstroms as follows:

$$5 \times 10^{-5}\ \text{cm} = 5000 \times 10^{-8}\ \text{cm} = 5000\ \text{Å}$$

The wavelengths of visible light range from 4000 Å to 7000 Å, approximately.

If light is a wave, then it obeys the equation $v = f\lambda$. In free space $v = c = 3 \times 10^{10}$ cm/sec. It is now possible to calculate the frequency of a typical light wave. If we assume $\lambda = 5000\ \text{Å} = 5 \times 10^{-5}$ cm,

$$f = \frac{v}{\lambda} = \frac{3 \times 10^{10}\ \text{cm/sec}}{5 \times 10^{-5}\ \text{cm}} = 6 \times 10^{14}\ \text{Hz}$$

This represents a vibration rate of 600 trillion waves per second! Hence the ultimate source of visible light must be vibrating so rapidly as to generate several hundred trillion waves per second, and the receptors in the human eye can not only respond to this vibration rate, but can also distinguish between relatively small differences in the vibration rate so as to make us consciously aware of slight differences in color.

In 1848 it was proposed that if light is a wave, it should exhibit the Doppler effect. That is, there should be an apparent change in the frequency of the light emitted by a light source when the source is in motion with respect to the observer. If the source is moving *toward* the observer, the frequency should appear to increase; hence the color of the emitted light should shift to the *violet* (high-frequency) end of the spectrum. If the source is *receding*, the observer should detect a lower frequency than is being emitted; that is, the color should shift toward the *red* (low-frequency) end of the spectrum. Thus if the frequency of the light emitted by a source is compared with the frequency seen by the observer, a *violet shift* indicates that the source is approaching, and a *red shift* indicates that the source is receding from the observer. The size of the shift is proportional to the speed of approach or recession.

Because of the high frequency and speed of light waves, the Doppler shift of light is extremely difficult to detect in terrestrial experiments. On an astronomical scale, however, the shift is more noticeable. In the 1870s and 1880s astronomers began to detect the Doppler effect in starlight, providing further evidence for the wave nature of light. To this day the Doppler effect is one of the astronomer's most important tools. It is used to measure the speeds of motion and rotation of planets, stars, and galaxies.

THE PROBLEM OF THE ETHER

By the 1870s the wave theory of light had been solidly established. The demonstrations of interference and diffraction showed that light has wave properties. Color phenomena could be simply interpreted in terms of differences in wavelength or frequency. The Doppler effect for light had been observed. In fact, all phenomena relating to the transmission of light could be readily accounted for with the wave theory.

However, one major difficulty remained. To account for polarization effects, it was necessary to postulate that light is a transverse wave. Yet transverse waves will travel only through solids; therefore the ether—which is the medium through which light waves travel when in free space—must have the properties of a solid. Furthermore, the speed of propagation of a transverse wave depends on the rigidity of the medium, and considering the extremely high speed of light, the ether must have a rigidity many times greater than steel or diamond. Thus if light is to propagate as a transverse wave, the ether must have the properties of an extremely rigid, extremely elastic solid. Yet the planets can move through the same ether without experiencing the slightest resistance or retardation!

This puzzle completely baffled a whole generation of scientists. If the ether does not exist, light cannot propagate through free space, and if the ether does exist, it must have properties that are quite unbelievable. Until a resolution of the dilemma could be found, scientists could not claim to have a full understanding of the nature of light.

To see how this problem was solved, we now turn to an apparently unrelated field of study, which came into its own in the early nineteenth century—electricity and magnetism.

QUESTIONS FOR REVIEW

1 Describe Michelson's method for measuring the speed of light.

2 State and explain (a) the law of reflection and (b) the law of refraction.

3 What kind of lens is found in the human eye? How is the focal length of this lens varied? Discuss the defects known as nearsightedness and farsightedness, and explain how they can be corrected for.

4 Describe the experiment devised by Newton to show (a) that white light is divisible into component colors, and (b) that each of the components cannot be further subdivided.

5 What is the phenomenon known as dispersion?

6 Outline the basic assumptions of Newton's particle theory of light. How did Newton explain (a) reflection, (b) refraction, (c) color, and (d) dispersion?

7 Discuss in general terms what is meant by a wave. List all the different kinds of wave motion you can think of.

8 What is the difference between a transverse wave and a longitudinal wave? How does each kind of wave propagate through a medium?

9 What factors influence the speed of a wave as it travels through a medium? Are these factors the same for both transverse and longitudinal waves?

10 What is a sound wave? Explain why not all compressional waves are sound waves.

11 Define each of the following properties of a wave or wave train: (a) period, (b) frequency, (c) wavelength, and (d) amplitude.

12 Describe and explain the Doppler effect. Under what conditions will the frequency measured by the observer be greater than the source frequency? Less than the source frequency?

13 Explain what is meant by the wave function of a wave. Can any wave form be represented by a wave function?

14 Distinguish between a finite and an infinite wave train. Explain why a finite wave train cannot be represented by a single sine or cosine wave function. How is it represented?

15 Discuss the phenomenon of wave interference. Under what conditions do reinforcement and cancellation take place?

16 What is a standing wave? How is it produced? What are the nodes and antinodes of a standing wave pattern?

17 What is Huygens' principle? Why is it a useful physical principle?

18 Describe the phenomenon of diffraction. How is diffraction accounted for using (a) Newton's particle theory and (b) the wave theory of light?

19 Outline the basic assumptions of the wave theory derivation of Snell's law of refraction.

20 Discuss Young's experiment. Describe the experimental setup, explaining the purpose of each of the three slits. Summarize the results of the experiment.

21 What is the difference between polarized and unpolarized light? What wave property of light can be deduced from the observation that light can be polarized?

22 Explain (a) why a beam of light will not be transmitted through two tourmaline crystals oriented with their axes at right angles and (b) why light reflected from glass at a certain angle will not be transmitted through a tourmaline crystal with its axis perpendicular to the glass surface.

23 Discuss the significance of Foucault's discovery that the speed of light is less in glass or water than in air.

24 Describe how the Doppler effect applies to light. List the conditions that lead to (a) a red shift of the emitted light and (b) a violet shift of the emitted light.

25 Discuss the theoretical difficulties that arise when one assumes that light is a transverse wave that propagates through the ether. What must be the properties of the ether if it is the medium for light waves? How do these proposed properties conflict with observations?

EXERCISES

1 Explain how the speed of light can be calculated from observations of stellar aberration. How does the length of the astronomical unit enter into this calculation?

2 A ray of light strikes a plane piece of glass ($n = 1.50$) at an angle of 30° to the perpendicular. What is the angle of refraction?

3 A ray of light travels from glass ($n = 1.50$) to air. For what angle of incidence will the refracted ray emerge tangent to the surface of the glass (i.e., the angle of refraction = 90°)? What happens if the angle of incidence is *greater* than this value?

4 Explain why a convex lens acts as a diverging lens (i.e., tends to spread the light rather than bring it to a focus) if the lens is immersed in a material with a greater index of refraction.

5 Explain why a transverse wave cannot be transmitted through a liquid or gas and a compressional wave can be transmitted through these media.

6 A string 40 cm long, with a mass of 20 grams, is stretched by a 5000-dyne force. What is the speed of a wave traveling along the string?

7 If a tree falls in a forest and no one is around to hear the crash, is there a sound?

8 Calculate the wavelength of (a) a sound wave in air of frequency 5000 Hz and (b) a wave train of 8 waves/sec traveling on the string described in question 6.

9 Discuss whether or not it is possible for wave interference to take place between two waves of different frequencies. Is complete reinforcement or cancellation possible in this case?

10 A string 60 cm long is fixed at both ends. Calculate the wavelengths of the standing wave patterns produced when there are (a) two nodes and (b) five nodes on the string, not including the ends.

11 A string 20 cm long is fixed at both ends and tightened until the wave speed is 4000 cm/sec.
(a) Calculate the fundamental frequency.
(b) Calculate the frequencies of the first three overtones.
(c) Explain what happens if a wave having a frequency not equal to the fundamental frequency or any of the overtones is produced on the string.

12 In terms of the basic principles of wave generation, explain why the pitch of the sound produced by a guitar string is changed by (a) tightening the string and (b) pressing down on the string at some point with one's finger.

13 Explain why two identical strings with the same applied tension produce different sounds when one is attached to a piano and the other to a bass fiddle.

14 In a setup of Young's experiment, the slits are 0.02 cm apart, and the screen is 100 cm from the slits.
(a) If the third bright line is found to be 0.6 cm from the center line, what is the wavelength of the light being used?
(b) If light of wavelength 6×10^{-5} cm is used, how far from the central line will the second bright line be located?

15 Discuss whether or not it is possible to produce polarized sound waves.

16 Young's experiment is modified by covering each of the two slits with a tourmaline crystal such that the two crystals are oriented at right angles to one another. How will this affect the outcome of the experiment? What will be the effect of placing a single crystal in front of the first slit only?

17 A ray of light incident on a piece of glass ($n = 1.50$) is partially reflected and partially refracted. It has been found that the reflected ray will be polarized if the reflected ray and the refracted ray are at right angles to each other. Calculate the angle of incidence required for this to happen. Compare your answer with the discovery made by Malus.

18 Calculate the frequency of light having a wavelength of (a) 6000 Å and (b) 3000 Å.

19 Explain how the Doppler effect for the light from a distant star can be used to demonstrate the annual orbital motion of the earth.

ELECTRICITY AND MAGNETISM

9

Nor is it of much importance to us to know the manner in which nature executes her laws; it is enough if we know the laws themselves. It is of real use to know that china left in the air unsupported with fall and break; but how it comes to fall, and why it breaks, are matters of speculation. It is a pleasure indeed to know them, but we can preserve our china without it.

Benjamin Franklin

THE PACE OF LIFE in a highly technological modern-day society can be overbearing at times, so it is not surprising to find that a growing number of people, both young and old, are calling for a return to the simple life, free of all technological influences, as a means of achieving personal freedom. Although this view is admirable in certain respects, it bespeaks a certain blindness to the idea that scientific and technological advancements *can* be beneficial to society.

In this respect let us consider the consequences of the discoveries of the fundamental laws of electricity and magnetism and the technological advances that followed. It is not an exaggeration to say that every facet of modern living—whether it be transportation, communication, health services, the manufacture and distribution of consumer goods—is at least partly dependent on electromagnetic devices. Our most popular forms of entertainment—radio and television, phonograph records, movies—fall into this category. One is therefore led to wonder whether life without electricity—without electric lighting, telephones, automobiles, airplanes, refrigerators, washing machines, air conditioners, and so on ad infinitum—is in the long run a life of freedom or a life of hardship and drudgery.

Dependent as we are on these devices, it is difficult to realize that they have all been developed within the past hundred years. Only the past two or three generations of mankind have been able to enjoy their benefits. In this chapter we shall consider the historical development of the theories and applications of electricity and magnetism—discoveries that in a relatively short time have resulted in truly profound changes in the scientist's understanding of nature and in the very evolution of human society.

9–1

ELECTRICITY AND MAGNETISM BEFORE 1800

Not surprisingly, the recorded history of electricity and magnetism begins with the ancient Greeks, who were aware of two phenomena that we would classify today as electromagnetic. The first is the power of the mineral amber, when rubbed, to attract small, light objects. The second is the power of the ***lodestone*** to attract iron. A lodestone is actually a piece of the mineral magnetite,* so named because it was commonly found in Magnesia, a region of Asia Minor. Thus in ancient times the word "magnet" was used synonymously with "lodestone." It was also known that certain other materials, such as iron, could be magnetized by bringing them into contact with lodestones.

The properties of magnets were studied during the Middle Ages. One important development of this period was the invention of the magnetic compass—a needle-shaped body made from a lodestone or some magnetizable material that always points toward the geographic north when allowed to pivot freely about its center. It is not known who was responsible for this discovery; it appears, however, that the Chinese and the Arabs began to use the magnetic compass for navigation during the eleventh and twelfth centuries.

The first systematic studies of these phenomena were made by William Gilbert, court physician to Queen Elizabeth I. Gilbert, more or less a contemporary of Galileo, was a strong defender of the scientific method and was also one of the first supporters of the Copernican theory in England. His great work *De Magnete* (*On the Magnet*), published in 1600, added considerably to the accumulated knowledge of electric and magnetic effects and served to clear up many misconceptions about these phenomena.

Gilbert showed that many other materials have the power of attraction previously attributed only to amber, and he was the first to use the word "electric" (from *elektron,* the Greek word for amber) to describe such effects. He experimented extensively with magnetism, making many important discoveries in this area. Gilbert proposed that the earth itself is a giant magnet and used this concept to account for the action of a magnetic compass. According to Gilbert, a magnetic compass—in fact, *every* magnet—has two opposite poles, which can be labeled "north" and "south." The north pole of a magnet is actually a north-*seeking* pole; the earth's magnetism causes that end of the magnet to point toward the geographic north. Similarly, the south pole of a magnet is, in fact, a south-seeking pole. Gilbert demonstrated that similar poles of different magnets will repel one another (i.e., north poles repel all other north poles, and south poles repel all other south poles), and opposite poles will always attract each other. Thus the tendency of a compass to line up in a north–south direction can be readily explained in terms of the attraction of the magnetic poles of the earth for their opposite poles on the compass. In *De Magnete* Gilbert could not resist a jab at those whose explanations of the action of the compass had been less successful:

> The crowd of philosophers, in order to discover the reason of the magnetical motions, called up causes lying remote and far away. And one man seems to me beyond all others worthy of censure, Martin Cortes, who, since there was no cause which could satisfy him in the whole of nature, dreamed that there was a point of magnetical attraction beyond the heavens, which attracted iron. Peter Peregrinus thinks that the direction arises from the poles of the sky. Cardan thought that the turning of iron was caused by a star in the tail of the Great Bear; Bessard, the Frenchman, opines that a magnet turns toward the pole of the zodiac. . . . Others have recourse to I know not what magnetic rocks and mountains. Thus it is always customary with mortals, that they despise things near home whilst foreign and distant things are dear and prized.[1]

* Magnetite is now known to be a compound of iron and oxygen having the chemical formula Fe_3O_4.

FLUID THEORIES OF ELECTRICITY

Gilbert's discovery that many substances could be electrified by friction (i.e., by rubbing them with some other material such as fur or cloth) led him to propose a theory of electrical attraction. This theory held that friction causes the substances to release a subtle fluid or "effluvium," which, upon returning to the original substance, will draw with it any light objects in the surrounding area. Gilbert's ideas on electricity were widely accepted, and they stimulated much experimentation in the 1600s and 1700s. (Strangely, neither Galileo nor Newton expressed more than passing interest in electric or magnetic phenomena.)

By the middle of the eighteenth century, the study of electric forces had become a rather popular pastime. A great many new phenomena were discovered that could not be accounted for under Gilbert's theory. Of particular interest was the discovery that the power of electric attraction could be transferred to certain materials by bringing them into contact with previously electrified bodies. These materials, most of which were metals, were given the name of ***conductors.*** Thus it was shown for the first time that bodies could be electrified *without* friction. As a result, it was more reasonable to think in terms of an ***electric fluid,*** which could be transferred from one body to another; an electrified body was thus considered as being "charged" with electric fluid. Furthermore, a clear distinction could now be made between conductors and ***insulators.*** Conductors readily accept electric fluid and allow the fluid to move freely through them, whereas insulators do not permit the free motion of electric fluid. In general, insulators are nonmetallic substances such as glass, rubber, plastic, and air. When an insulator is electrically charged, the fluid tends to remain in place, whereas when a conductor is charged, the electric fluid moves and is distributed over the body of the conductor. When a conductor is brought into contact with a charged insulator, the fluid on the surface of the insulator moves freely onto the conductor, and the conductor becomes charged as a result. A rather substantial charge can be built up on a conductor, provided the conductor is insulated to prevent the charge from leaking off. Experimenters were able to construct numerous devices that stored a charge in this manner.* Frequent public exhibitions were given in which scientists took great delight in drawing sparks from these devices or in delivering electric shocks to members of the audience. In one famous demonstration, which took place around 1750,

> The Carthusian monks at the convent in Paris were formed into a line 900 feet long, by means of iron wires between every two persons, and the whole company upon the discharge . . . gave a sudden spring at the same instant.[2]

The sight must have been quite impressive.

In the 1730s Charles Du Fay, a former soldier attached to the court of Louis XV of France, proposed, as a consequence of his experiments, that there are *two* kinds of electric fluid. One type of fluid, which Du Fay called ***vitreous electricity,*** appears on certain materials, such as glass, that have been electrified; the other type, known as ***resinous electricity,*** appears on other substances of which electrified amber is the most notable example. Du Fay discovered that two bodies possessing the same kind of electric fluid, whether vitreous or resinous, repel one another, whereas two

* A device used to store an electric charge is known as an ***electric condenser*** or ***capacitor.*** Condensers are widely used in modern electronic equipment.

bodies possessing different kinds of fluid attract each other. In addition, a body containing vitreous electricity becomes electrically neutral upon absorbing an equal quantity of resinous electricity and vice versa.

An alternative explanation of these phenomena was offered by Benjamin Franklin (1706–1790). Franklin is, of course, best known for his role as one of the founding fathers of the American republic, overshadowing the fact that he was also one of the most well-known and highly respected scientists of his times. Franklin proposed that there is only *one* electric fluid, which is present in all bodies, electrified or not. A body becomes electrified when it attains through some process an excess or deficiency of electric fluid. Franklin assumed on rather arbitrary grounds that an electrified glass rod has an *excess* of electric fluid. The glass is then said to be ***positively charged.*** Thus Franklin's single electric fluid is identical with Du Fay's "vitreous electricity." On the other hand, electrified amber has a *deficiency* of electric fluid and is considered to be ***negatively charged.*** Thus Du Fay's "resinous electricity" is not a separate substance, but rather a deficiency of vitreous electricity.

According to Franklin's theory, when a glass rod is rubbed with a piece of cloth, electric fluid is transferred from the cloth to the glass. Thus while the glass becomes positively charged, the piece of cloth loses fluid and therefore becomes negatively charged. On the other hand, when a piece of amber is rubbed with fur, electric fluid is transferred from the amber to the fur. Hence the amber is negatively charged (i.e., is deficient in electric fluid), while the fur has become positively charged. Thus Franklin was proposing that electric fluid is never created or destroyed; it is merely *transferred* from one body to another. This was in contrast to Du Fay's assumption that vitreous and resinous electricity were *created* by friction between the appropriate bodies. In modern terminology we recognize Franklin's concept as a conservation law; namely, the total amount of electric charge remains constant. This is known as the ***principle of the conservation of charge;*** to the present day no violations of this principle have ever been found.

In accordance with observations, Franklin assumed that similarly charged bodies repel one another, and oppositely charged bodies attract one another. (See Figure 9–1.) He even attempted to give an explanation of why this was so. He proposed that particles of the electric fluid repel one another, but are attracted to ordinary matter. Thus two positively charged bodies (each containing an excess of electric fluid) ought to repel, and oppositely charged bodies should attract (because of the attraction of the matter in the negatively charged, fluid-deficient body for the excess fluid in the positively charged body). However, it was difficult to understand why two negatively charged bodies should repel (since both have a deficiency of electric fluid), and Franklin and others put forth several rather complicated explanations to account for this particular effect.

The repulsion of like charges can be put to advantage in the ***electroscope,*** which is a device used to detect an electric charge. The construction of one common type of electroscope is shown in Figure 9–2. A conducting metal rod with a knob at the top extends into an enclosed, insulated chamber. At the end of the rod, two lightweight metal leaves are attached. Typically, the leaves are made of gold, which is so malleable that it can be flattened to a thickness of a few thousandths of an inch. When a charge of either sign is transferred to the metal knob, the charge will be distributed over the entire metal rod as well as the gold leaves (since they are conductors). Thus both leaves will acquire the same charge and will repel each other. The leaves are light in weight, so they will spread apart as a result of the electric repulsion (Figure 9–2b).

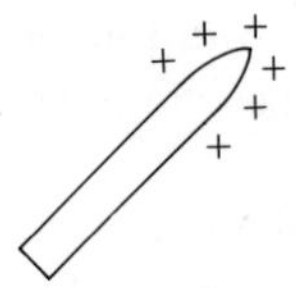

(a) A glass rod rubbed with cloth attains a positive charge.

(b) An amber rod rubbed with fur attains a negative charge.

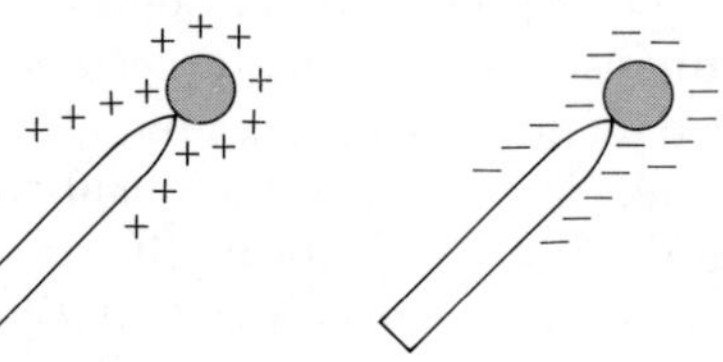

(c) Charge may be transferred to a conductor by bringing the rod in contact with the conducting body.

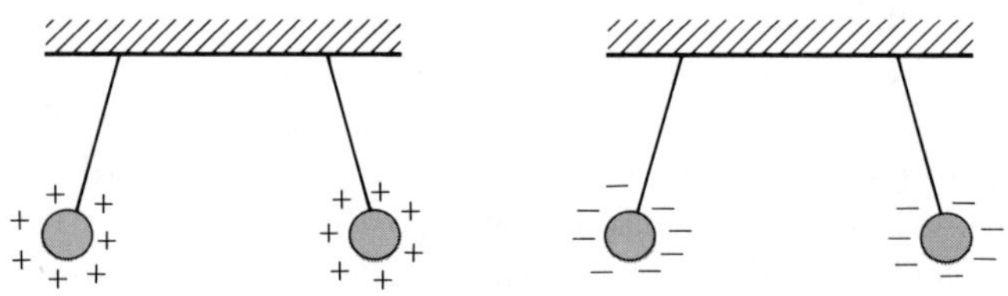

(d) Two lightweight conductors (small balls of tinfoil suspended from strings are suitable) will repel when given the same charge.

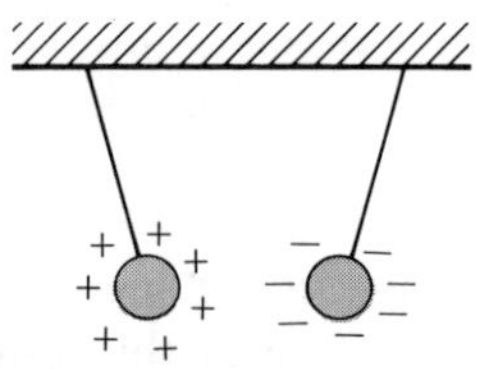

(e) When given opposite charges, the conductors will attract one another.

figure 9–1 **FORCES BETWEEN CHARGED BODIES**

Experiments show that there are two kinds of charge, called positive and negative. Bodies with the same charge repel one another, whereas bodies with opposite charges attract.

Because of the peculiar properties of conductors, an uncharged electroscope will react to a charged body that is near to, but not in contact with, the knob. For example, the presence of a positively charged body will repel the positive charges in the electroscope. In the conducting parts of the electroscope, where charges are free to move, a separation of charges will result, with the knob having a preponderantly negative charge and the leaves having an excess positive charge (Figure 9–2c). Consequently, even though the electroscope remains electrically neutral as a whole, the leaves will diverge. When the positively charged body is removed, the leaves will then return to their normal position. Thus an electroscope can be used to detect the presence of both positive and negative charges. It can even be calibrated to measure the *amount* of charge: the greater the charge, the farther apart the leaves will be.

The action of the electroscope can be readily understood in terms of the one-fluid theory, but this is not a decisive argument in favor of the theory. Actually, the electroscope and all the other electric phenomena known at that time could have been successfully accounted for by either the one-fluid or the two-fluid theory. The one-fluid theory certainly did not disprove the two-fluid theory by any means. However, the one-fluid theory was much simpler, and it gained wide acceptance. Nevertheless, the two-fluid theory retained many adherents.

It is worthwhile at this point to stop and consider the terminology that developed, particularly in the light of modern usage. We commonly speak of a body as having "a positive charge" or "a negative charge," but as we have seen, this was originally a shorthand way of saying that the body is either charged (in the sense of *loaded* or *imbued*) with excess electric fluid or else has a deficiency of fluid. The terms "positive charge" and "negative charge" are still used today, even though the concept of an electric fluid distinct from matter has long been discarded. It is presently believed that electricity is a basic property of the fundamental particles of matter itself. In

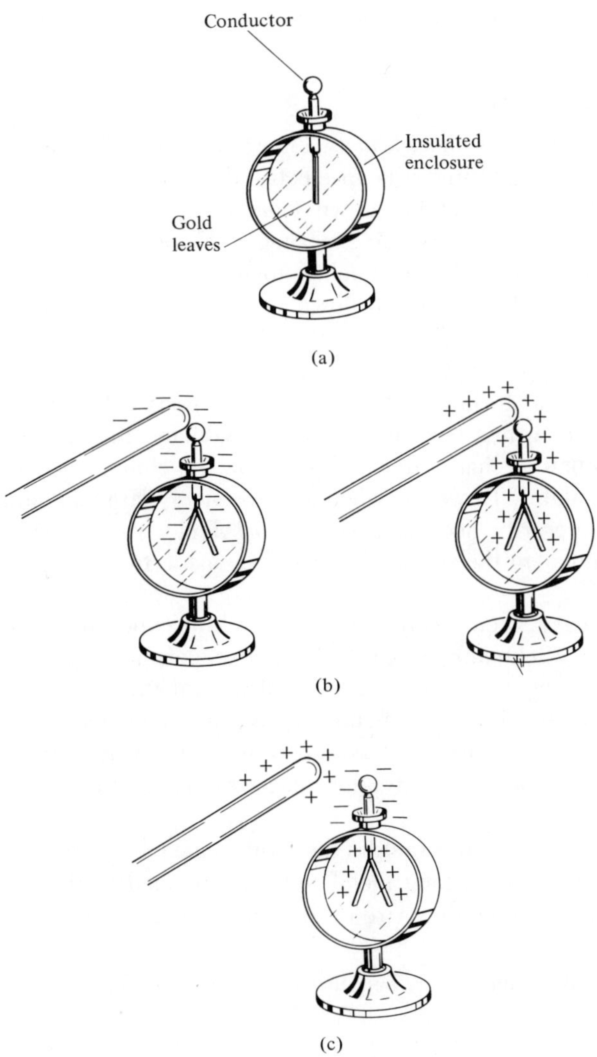

figure 9–2

THE GOLD–LEAF ELECTROSCOPE

(a) In an uncharged electroscope the gold leaves hang vertically. (b) When either a positive or a negative charge is touched to the conducting knob, the leaves acquire the same charge and repel one another. (c) The leaves will repel each other even when a charged body is brought near to, but not in contact with, the conducting knob. The nearby positive charge will attract the negative charges and repel the positive charges to the leaves, causing them to separate.

that case one may justifiably ask: What is a positive (or negative) charge charged with? At present, there is no answer to this question; the nature of electric charge is every bit as mysterious as the nature of gravity. We must regard charge, like gravity, as a fundamental and unexplained theoretical concept that is nevertheless useful in describing certain natural phenomena.

Furthermore, we find that the terms "positive" and "negative," as applied to charge, convey less meaning than they would appear to. To Franklin, a positive charge meant an excess, and a negative charge a deficiency, of electric fluid; to modern-day physicists (who reject the idea of an electric fluid), the terms "positive" and "negative" are arbitrary terms, which signify nothing more than that there are two different kinds of charges.* The present-day theory offers no explanation as to *why* there are two kinds of charges, or why like charges repel and opposite charges attract.

Finally, mention should be made of Franklin's arbitrarily drawn assumption that glass (rather than amber) attains an excess of electric fluid by friction. We now know that his guess was wrong, for glass *loses* negatively charged particles when electrified. Thus glass has a positive charge because of a deficiency of negative charge. On the other hand, amber is negatively charged because it *gains* negatively charged particles through friction. Franklin's incorrect guess, as we shall see, is sometimes a source of some confusion in electric theory.

THE ELECTRIC NATURE OF LIGHTNING

As a scientist, Benjamin Franklin is most widely known for his experiments with lightning. Franklin believed that lightning was an electric discharge and sought to demonstrate that an electric charge could be drawn from the clouds during a thunderstorm. In 1752 he performed his famous kite experiment. Franklin's kite was made of silk cloth and had a pointed metal rod attached to it to draw the charge from the air. The charge acquired by the kite when flown in a thunderstorm would then be conducted along the kite string. Franklin demonstrated this by drawing sparks from a metal key attached to the string near the ground. In this way he showed that the clouds become electrified during a storm. Franklin therefore proposed that lightning was an electric phenomenon in which electric charge is conducted through the air to reduce the excess or deficiency of charge that builds up on the clouds. (Under normal circumstances air is an insulator. However, when an excessively large amount of charge builds up, air will temporarily become a conductor of electricity.) Thus lightning was considered to be similar to the sparks that could be made to jump from charged electric condensers.

Franklin was lucky not to have been electrocuted by his kite experiment. In 1753 a scientist by the name of Richmann was killed while experimenting with lightning. One of Richmann's contemporaries was moved to comment that "it is not given to every electrician to die in so glorious a manner as the justly envied Richmann." A posthumous contribution to the cause of science was made by Richmann when his body was thoroughly dissected to study the effects of his last and best experiment on his vital organs.

Following his successful kite experiment, Franklin hit upon an idea for protecting buildings from being struck by lightning. A long metal rod was attached to the top of a building and was connected by a conducting wire to the ground. Lightning would

*In the same way the terms "north" and "south" signify only that there are two distinct kinds of magnetic poles.

then be attracted to the rod, and the electric charge would be conducted directly to the ground without causing any damage to the building. Franklin's invention was quite successful, and from the 1750s onward these ***lightning rods*** were erected on most of the important structures of the world.

Remarkably, there was considerable opposition to the use of lightning rods. According to some theologians, lightning was an instrument of God's wrath, and it was presumptuous of man to attempt to circumvent it. However, others found it difficult to reconcile this view with the fact that church spires were the objects that were most likely to be struck by lightning. Eventually, opposition to the use of lightning rods dissipated when it was seen how effective they were. We may cite the following example:

> A typical case in Italy was that of the tower of St. Mark's at Venice. In spite of the angel at its summit and the bells consecrated to ward off the power of the air . . . the tower was frequently injured and even ruined by lightning. In 1388 it was badly shattered; in 1417 and again in 1489, the wooden spire surmounting it was utterly consumed; it was again greatly injured in 1548, 1565, 1653, and in 1745. . . . The tower was again badly struck in 1761 and 1762; and not until 1766—fourteen years after Franklin's discovery—was a lightning-rod placed upon it; and it has never been struck since.[3]

THE INVERSE-SQUARE FORCE LAW

By 1760 the study of electric phenomena had resulted in some rather notable attainments. Yet the theories that had been put forth were entirely descriptive, and no serious attempts had been made to put them into mathematical form. Between 1760 and 1790, however, a number of scientists succeeded in establishing the law that describes the dependence of the force between two charges on the distance between them. It was found that the electric force, like the gravitational force, is inversely proportional to the square of the distance.

The inverse-square force law was first proposed by Joseph Priestley in 1767. Priestley was an English minister and amateur scientist, who is best known for his contributions to chemistry. (He is credited with the discovery of oxygen.) Because of his outspoken support of both the American and French revolutions, Priestley was forced to leave England; in 1794 he emigrated to the United States—the first of a long line of scientists who came to America to seek political freedom. Priestley deduced the inverse-square force law in a most ingenious way. He had learned from his close friend Franklin that when a small charged body is placed anywhere inside a hollow charged conducting sphere, no electric force is exerted on that body. Priestley recalled that in the *Principia* Newton had shown mathematically that the gravitational force is zero everywhere inside a hollow spherical shell because of cancellation of opposing forces, and that this is true only if the gravitational force is inversely proportional to the square of the distance. Priestley therefore reasoned that since the same effect was observed with charges, the force between charged bodies must also follow an inverse-square law.

Priestley's proposal was given a firm experimental basis through the careful experiments of the French scientist Charles Coulomb* (1736–1806). Coulomb made

*Coulomb's attitude toward political upheavals was in sharp contrast to that of Priestley. With the storming of the Bastille in 1789, Coulomb resigned all his positions and retired to his country estate to continue his experiments. He returned to Paris in 1799 when Napoleon came to power.

direct measurements of the forces between two charges by means of a ***torsion balance,*** a device similar to that used by Cavendish to measure the gravitational force. (See Figure 5–2.) As shown in Figure 9–3, a test charge is located at one end of a carefully balanced and insulated horizontal rod, which is suspended from a long steel wire. When a second charge of the same sign is brought near the first, the repulsive force causes the wire to twist away to a new equilibrium position in which the torsion in the wire balances the electric repulsion. By applying a counter-twist to the wire, Coulomb was able to vary the distance between the two charges. Using the fact that the angle of twist of a wire is proportional to the applied torque, he was able to calculate the magnitude of the force in each case. In this way he managed to show with some precision that the force diminishes according to the square of the distance between the charges.

One difficulty that had limited previous experimenters was the inability to measure precisely the amount of charge on a body. The only device available at the time was the electroscope, which was much too crude for Coulomb's purposes. He overcame this problem by using two identical bodies as his test charges. Bodies A and B are first brought into contact, then touched by some charged object. Since they are identical, A and B will share the charge equally, and thus they will have like charges. When B is touched to a *third* identical body, B will share its charge with the body, and the charge on B will be reduced by half. In this manner Coulomb

figure 9–3 **COULOMB'S TORSION–BALANCE MEASUREMENT OF ELECTROSTATIC FORCE**

(a) Bodies A and B are uncharged; the wire is at equilibrium. (b) A and B are given the same charge; a new equilibrium is established when the torsion force in the wire exactly balances the force between the charges. (c) By applying measurable torsion forces to the wire, the distance between the two charges can be varied. Coulomb discovered that the electrostatic force is inversely proportional to the square of the distance between the charges.

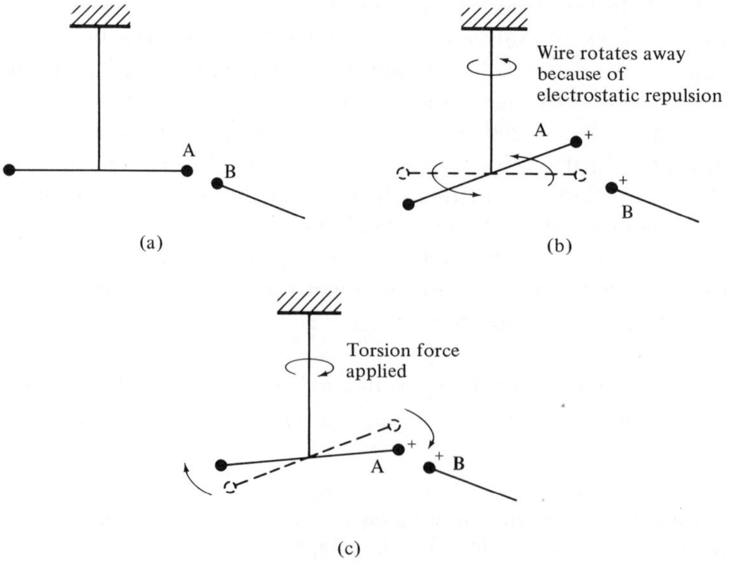

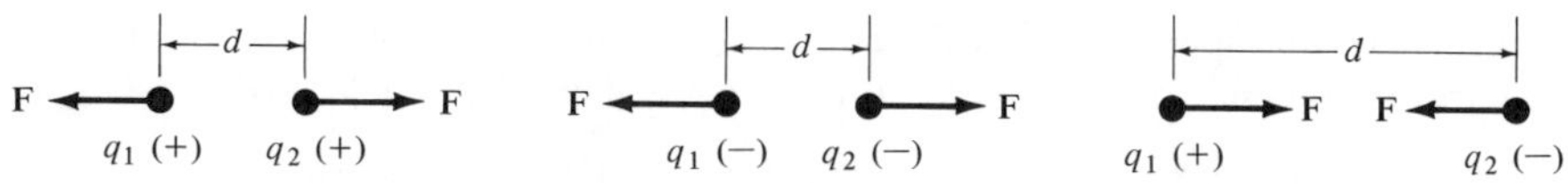

figure 9–4 **ELECTROSTATIC FORCES ON LIKE AND UNLIKE CHARGES**

The mutual force is repulsive if q_1 and q_2 have the same sign (F is positive); the force is attractive if they have opposite signs (F is negative).

was able to control the amount of charge on his test bodies, without having to measure them directly. He could then demonstrate that *the force between two stationary charges is directly proportional to the product of their charges.* Because the charges are stationary, this force is referred to as the ***electrostatic*** force. We will discover later that additional effects may appear when the charges are in motion. In mathematical form the electrostatic force law becomes

$$F = K_e \frac{q_1 q_2}{d^2} \tag{9–1}$$

where d is the distance between the charges, q_1 and q_2 are the magnitudes of the charges, and K_e is a constant of proportionality to be determined. In this equation F represents the magnitude of the mutual, equal, and opposite electrostatic forces exerted by the charges on one another (in accordance with the action–reaction principle). The forces will be repulsive or attractive according to whether the charges are like or unlike. If the charges have opposite signs, F will have a negative sign according to Eq. 9–1. If the charges have like signs, F will have a positive sign. Therefore we can establish the rule that a negative F represents an *attractive* force, and a positive F indicates a *repulsive* force. (See Figure 9–4.)

Equation 9–1 is known as ***Coulomb's law of electrostatic force*** and is one of the fundamental laws of electromagnetism. Coulomb presented the results of his experiments in a series of papers published between 1785 and 1789. Apparently, Coulomb's research was not at all hindered by the fact that he was a staunch proponent of the two-fluid theory.

Let us now consider the units with which a quantity of charge is measured. In the CGS system the unit of charge is the ***esu*** (short for electrostatic unit). It is defined as an amount of charge such that if two like charges of 1 esu each are placed 1 cm apart in free space, they will experience a mutual repulsive force of 1 dyne. Note that if we substitute $q_1 = q_2 = 1$ esu, $d = 1$ cm, and $F = 1$ dyne into Eq. 9–1, we obtain $K_e = 1$ dyne-cm^2/esu^2. Thus the esu is defined in such a way as to make $K_e = 1$ in the CGS system of units. If a medium is placed between the two charges, the mutual electrostatic force is found to diminish in strength. Consequently, the effect of an intervening material medium is to make the proportionality constant less than the free-space value.

In the MKS system the unit of charge is the ***coulomb.*** It is defined as follows:

$$1 \text{ coulomb} = \frac{c}{10} \text{ esu} = 2.9979 \times 10^9 \text{ esu}$$

where c is the speed of light.* The proportionality constant K_e in Coulomb's law has the following value in MKS units (using $c = 2.9979 \times 10^8$ m/sec):

$$K_e = 10^{-7}c^2 = 8.9874 \times 10^9 \text{ N-m}^2/\text{coulomb}^2$$

For simple calculations we can take

$$1 \text{ coulomb} = 3 \times 10^9 \text{ esu} \qquad \text{and} \qquad K_e = 9 \times 10^9 \text{ N-m}^2/\text{coulomb}^2$$

To summarize, Coulomb's law in metric units is

$$CGS: \quad F = \frac{q_1 q_2}{d^2} \qquad (q \text{ in esu, } d \text{ in cm, } F \text{ in dynes}) \qquad (9\text{–}2a)$$

$$MKS: \quad F = (9 \times 10^9)\frac{q_1 q_2}{d^2} \qquad (q \text{ in coulombs, } d \text{ in meters, } F \text{ in newtons}) \qquad (9\text{–}2b)$$

Fortunately, there are no special British units for electromagnetic quantities; MKS units are generally used with the British system.

As it turns out, 1 coulomb is a rather immense quantity of charge. Suppose that two 1-coulomb charges are placed 1 meter apart. The repulsive force, according to Eq. 9–2b, will be equal to 9×10^9 newtons—approximately a million tons! For most practical applications, the ***microcoulomb*** (1 microcoulomb = 10^{-6} coulomb) is commonly used.

EXAMPLE 9–1: Two charges of +5 microcoulombs and −4 microcoulombs are 50 cm apart. Calculate the electrostatic force between them.

SOLUTION: It is more convenient to solve this problem in MKS units. We first convert to appropriate units:

$$q_1 = 5 \text{ microcoulombs} = 5 \times 10^{-6} \text{ coulomb}$$
$$q_2 = -4 \text{ microcoulombs} = -4 \times 10^{-6} \text{ coulomb}$$
$$d = 50 \text{ cm} = 0.5 \text{ meter}$$

Substituting into Eq. 9–2b gives

$$F = \left(9 \times 10^9 \frac{\text{N-m}^2}{\text{coulomb}^2}\right) \frac{(5 \times 10^{-6} \text{ coulomb})(-4 \times 10^{-6} \text{ coulomb})}{(0.5 \text{ meter})^2}$$
$$= \underline{-0.72 \text{ newton}}$$

The minus sign, according to the rule established previously, indicates that the force is attractive. Thus each charge attracts the other with a force of 0.72 newton.

Although history credits Coulomb with the experimental verification of the inverse-square law, he was not the first to verify the law, for Henry Cavendish had performed the same experiments some ten years before Coulomb. Cavendish, however, did not see fit to publish his findings. Nearly a hundred years after his death, Cavendish's papers were discovered by Lord Kelvin, and they were eventually edited and

*The student may wonder why c, the speed of light, should appear in an equation pertaining to electric force. The answer to this question will be given in the next chapter.

published by Maxwell in 1879. At that time it became known that Cavendish had anticipated many of the important principles of electricity and magnetism.*

GRAVITATION, ELECTRICITY, AND MAGNETISM

Coulomb also used his torsion balance to measure magnetic forces and was able to confirm the observation that the force between two stationary magnetic poles also obeys an inverse-square law. That is,

$$F = k\frac{p_1 p_2}{d^2} \tag{9–3}$$

where p_1 and p_2 represent the strengths of the magnetic poles; d is the distance between them; and k is a suitable proportionality constant.

The remarkable similarities among the laws describing the gravitational force between two masses, the electrostatic force between two charges, and the magnetostatic force between two poles should not go unnoticed. The fact that all three of these fundamental natural forces obey an inverse-square law is indicative of an underlying relationship. There are, of course, significant differences; electric and magnetic forces can be attractive or repulsive, whereas gravitation is always attractive. (More precisely, no incidences of gravitational repulsion have ever been observed.) We do not find gravitational "conductors" or "insulators" in nature. Whereas all bodies in the universe are affected by gravitation, many material substances exhibit no apparent electric or magnetic properties. Finally, gravitation is immensely weaker than either electric or magnetic forces. (For example, the electrostatic repulsion between two electrons is approximately 10^{43} times larger than their mutual gravitational attraction.)

In spite of these differences, the fact that all three forces obey inverse-square laws remains highly significant and invites an explanation. Unfortunately, the attempts of physicists to discover a deeper unity that would connect gravitation, electricity, and magnetism in a single theoretical framework have been largely unsuccessful. This mysterious similarity will undoubtedly remain one of the great unsolved problems of physics for many years to come.

9–2

CURRENT ELECTRICITY

Before 1800 the study of electricity was the study of ***static*** (i.e., stationary) electricity. The separation of electric charges could be brought about through friction, but once separated the charges tended to remain in place. Charges could be made to move by transferring them to conductors, and electric sparks and lightning discharges represented situations in which charges were in motion. However, in these cases motion takes place very rapidly, and it could not be maintained for any length of time.

* Cavendish's desire to avoid his fellow man was legendary. A man of immense wealth, Cavendish lived in total seclusion; he ordered his meals by leaving a note on his hall table and required his female servants to keep out of his sight. He nevertheless attended regularly the weekly meetings of the Royal Society, at which he rarely spoke.

THE VOLTAIC PILE

The chain of events leading to the discovery of a way to produce a steady flow of charge began with a most unusual accident. In 1780 a dissected frog was being studied in the laboratory of Luigi Galvani, professor of anatomy at the University of Bologna. The frog lay on a table not far from an "electric machine" (a device used to generate static electricity by friction). As Galvani described it:

> It happened by chance that one of my assistants touched the inner crural nerve of the frog with the point of a scalpel; whereupon at once the muscles of the limbs were violently convulsed. Another of those who used to help me in electrical experiments thought he had noticed that at this instant a spark was drawn from the conductor of the machine. I myself was at the time occupied with a totally different matter; but when he drew my attention to this, I greatly desired to try it for myself and discover its hidden principle. So I, too, touched one or other of the crural nerves with the point of the scalpel, at the same time that one of those present drew a spark; and the same phenomenon was repeated exactly as before.[4]

Galvani began to experiment in order to find other conditions under which the frog's muscles could be made to contract. Eventually, he discovered that the presence of a spark or other electric discharge was not necessary. The same effect could be produced by using two dissimilar metals, one in contact with the nerve, the other touching the muscle. When the two metals were brought in contact with one another (thus completing the circuit), prolonged muscular contractions resulted.

Galvani theorized that his observations gave evidence for what he called "animal electricity." He proposed that the nerves contained electric fluid, which was transmitted to the muscles along the path provided by the two metallic conductors. Thus the source of the electric impulse was thought to be in the animal tissue itself.

Galvani's observations, published in 1791, stimulated enormous interest and controversy. Considerable opposition arose to Galvani's theory of animal electricity, under the leadership of Alessandro Volta, a professor of physics at the University of Pavia in Italy. It was Volta's contention that the electric impulse arose from the contact of the two dissimilar metals, and that the frog's leg served merely to conduct the electricity. Volta completely rejected the idea of any sort of animal electricity (or "galvanism," as it came to be known) and proposed instead the concept of "metallic electricity."

A spirited dispute between Galvani and Volta went on for several years afterward. But the political turmoil that reigned in Europe during the latter part of the 1700s and early part of the 1800s raised serious difficulties for scientists. Not all suffered the fate of Lavoisier, who lost his head to the French Revolution, but many renowned scientists were obliged to swear allegiance first to one conqueror, then to the next. In 1798 Galvani refused to take the oath of allegiance to Napoleon's Cisalpine Republic. He was forced to give up his professorship; disgraced and in poverty, Galvani died that same year. Volta, an Italian by birth, lived under Austrian, then French, then Austrian rule again and managed to flourish; he received honors from Napoleon and from the Austrian emperor as well.

After the death of Galvani, Volta began to accumulate more and more evidence for his theory that metallic electricity arose from the contact of two dissimilar metals. He showed that when two insulated pieces of zinc and copper are placed in contact and then separated, the zinc acquires a positive charge and the copper a negative charge.

Volta's major problem was that the electric discharge produced by a pair of dissimilar metals was extremely feeble and difficult to demonstrate. He began to experiment with ways to intensify the effect. Finally, Volta decided to try stacking a series of copper-zinc pairs, each one separated from the next by a piece of wet cardboard, as shown in Figure 9–5. Volta discovered that in this way a much stronger electric discharge could be produced, and the more pairs he added to the "pile," the stronger the effect. The most significant result was that *the pile was able to produce a steady flow of electricity,* which did not dissipate until after many, many hours of operation.

Thus the ***voltaic pile*** (as it was called) became the first source of a steady ***electric current.*** (In modern terminology the voltaic pile is called a ***battery.*** Each copper-zinc pair represents one ***cell*** of the battery.) Volta himself did not entirely understand how his pile worked. In one of his papers he wrote, "This endless circulation or perpetual motion of the electric fluid may seem paradoxical and may prove inexplicable, but is nevertheless real."[5] To Volta, the piece of wet cardboard served merely as a conductor; that is, it took the place of the frog's leg in completing the circuit. Volta considered this to be a disproof of the idea of animal electricity, since the animal was no longer needed. Later, scientists showed that the wet cardboard permits the chemical reactions that serve to maintain a continual electric current. We shall later consider more fully how a pile operates.

ELECTROCHEMISTRY

Many significant discoveries followed rapidly upon Volta's announcement (in March 1800) of his invention. Within six weeks of the announcement, the first dividends were paid; in England William Nicholson and Anthony Carlisle constructed a voltaic pile, and by putting the terminals of the pile into water, they were able to decompose water into hydrogen and oxygen gas. This method of chemical decomposition by the action of an electric current, known as ***electrolysis,*** was soon extended to other chemical compounds. For many years the substances potash and soda (now known as potassium hydroxide and sodium hydroxide, respectively) were thought to be pure chemical elements. In 1807 the English chemist Humphry Davy succeeded in decomposing both of them by electrolysis, thereby producing potassium and sodium as free elements. Similarly, common salt (NaCl) could be decomposed into pure sodium and chlorine gas. Thus the composition of many chemical compounds could now be ascertained.

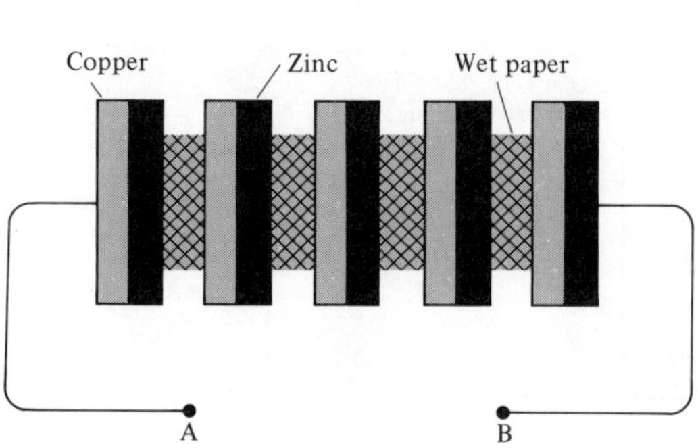

figure 9–5 **THE VOLTAIC PILE**

The first voltaic pile consisted of pairs of zinc and copper disks in contact, each pair separated from the next by a wet piece of paper. When the zinc disk at one end of the pile is connected to the copper disk at the other end (by connecting terminals A and B), a steady electric current is developed.

Greatly stimulated by the invention of the voltaic pile and the availability of current electricity, chemistry advanced rapidly in the nineteenth century. The fact that chemical compounds could be decomposed by passing an electric current through them gave chemists the first idea that chemical forces are electric in nature. In 1812 the Swedish chemist J. J. Berzelius made the proposal that all chemical compounds are made up of oppositely charged parts, which are held together by the electrical attraction between them. This picture of chemical structure—although now considered much too simplified and inadequate—provided a means of understanding many of the phenomena of chemical combination.

The electrolysis of chemical compounds can be easily understood in terms of this electrochemical picture of matter. The procedure in electrolysis is to connect the opposite terminals of a battery to conducting strips or rods known as ***electrodes,*** which are inserted into a liquid solution containing the compound dissolved in water (Figure 9–6). If the compound conducts electricity in water solution (not all compounds do), it is known as the ***electrolyte.*** Because of the chemical action inside the battery, one of the electrodes will be positively charged, the other negatively charged. In solution the molecules of the electrolyte are presumed to break up into positively and negatively charged particles called ***ions.*** The electrical attraction and repulsion of the oppositely charged electrodes will cause the positive ions to be attracted to the negatively charged electrode, called the ***cathode,*** while the negative ions will move to the positive electrode, or the ***anode.*** This flow of ions through the solution constitutes an electric current. Once the ions arrive at their respective electrodes, they become electrically neutralized and are deposited on the electrode itself. (Any gases that are produced will bubble out of the solution.)

The electrode at which the element is deposited identifies the ionic state of that element when in solution. For example, in the electrolysis of NaCl, sodium collects

figure 9–6 **ELECTROLYSIS OF A CHEMICAL COMPOUND**

In solution the molecules of the compound dissociate into positive and negative ions. The ions are attracted to the oppositely charged electrodes, where they are neutralized. The battery supplies a continual flow of charge to the electrodes.

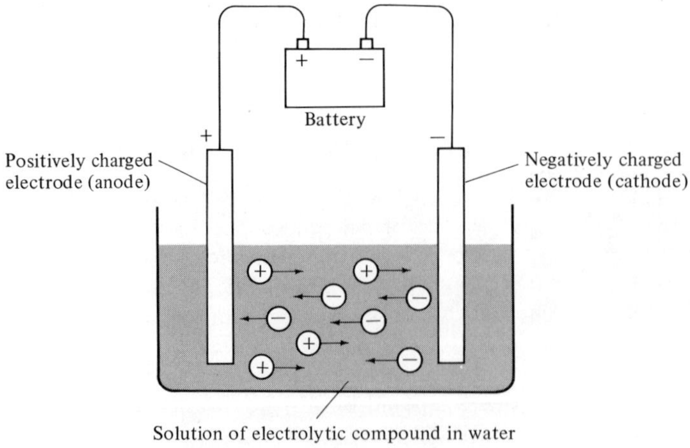

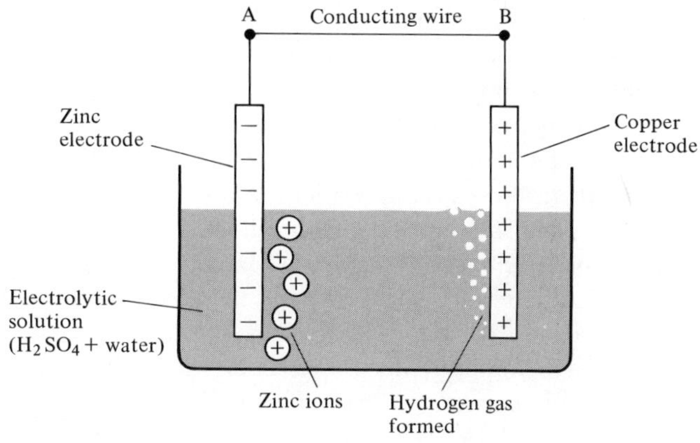

figure 9–7 **THE PRIMARY VOLTAIC CELL**

At the zinc electrode, positive zinc ions go into solution and the electrode acquires a negative charge. At the copper electrode, hydrogen ions in the solution remove negative charges from the copper atoms. The copper electrode becomes positively charged, and hydrogen gas is formed. When terminals A and B are connected together, a flow of electrons from A to B is established.

at the negatively charged cathode, so the sodium ions must have been positively charged. Similarly, chlorine collects at the positively charged anode, so chlorine ions are thereby identified as negative ions. Such information was of considerable use to nineteenth-century chemists in understanding the nature of chemical combination.

The electrical picture of chemical combination helps us to understand the operation of a voltaic pile. Volta had incorrectly believed that the current is generated at the zinc–copper interface, where the two dissimilar metals come into contact. It is now understood that the basic interactions occur where the two metals come into contact with the wet cardboard. Thus the basic unit, or primary cell, of the voltaic pile consists of two dissimilar metals separated by an electrolytic (conducting) solution (Figure 9–7). Since pure water is by itself a poor conductor, a substance such as sulfuric acid (H_2SO_4), which ionizes easily, is usually added. (Ordinary tap water is a good conductor because of the many impurities in it.)

In the process of electrolysis, the electrolytic solution is chemically decomposed. The electrodes serve to conduct current from the battery, and they do not react chemically. But in the voltaic cell the zinc and copper electrodes undergo chemical reactions with the electrolyte. The reaction between the zinc and the electrolyte causes zinc atoms to enter the solution as positively charged ions. The zinc electrode itself becomes negatively charged from the loss of positive ions. A different reaction occurs at the copper electrode; here positively charged hydrogen ions from the solution react with the copper. The hydrogen ions remove negative charges from the copper atoms and then combine to produce hydrogen gas. Because of the loss of negative charge to the hydrogen, the copper electrode attains a positive charge.

As a result of the separate chemical reactions occurring at the zinc and the copper electrodes, a situation is established in which the zinc electrode has a net negative charge and the copper electrode has a net positive charge. Now suppose an external path is provided between the two electrodes by connecting them with some conductor (such as a metal wire). A flow of charge will be established between the two electrodes along this path. According to modern-day theory, negatively charged particles (electrons) will move from the zinc to the copper through the external conducting path (from A to B in Figure 9–7). The negative charges arriving at the copper electrode serve to replenish the supply needed for the formation of more hydrogen gas. The resultant removal of hydrogen ions from the solution makes the solution less positively charged, permitting more zinc ions to enter the solution. Thus the chemical reactions continue, and the charges on the two electrodes are maintained. The chemical reactions (and thereby the action of the cell) will stop when the zinc electrode is completely decomposed or when all the hydrogen in the solution is used up. Until this occurs, a continual flow of charge can be provided by the cell.

Because of Benjamin Franklin's wrong guess described previously, it was assumed for many years that an electric current consists of a flow of positive charge. For a variety of reasons, modern-day physicists still define current as the flow of positive charge, even though it is now known that the negative charges (and not the positives) are in motion in a conducting wire. Thus by convention the current in the external wire is assumed to flow *from* the positive (copper) electrode *to* the negative (zinc) electrode—even though the actual motion consists of negative charges moving from the negative to the positive electrode.

Electric cells and batteries are extremely useful as portable sources of electricity, and they can be found in a wide variety of modern devices, including automobiles, portable televisions and radios, flashlights, and so on. The widely used "dry" cell (Figure 9–8) operates on much the same principle as the original voltaic cell, except that in the most common type the positive electrode is made of carbon instead of copper. The conducting liquid is ammonium chloride (NH_4Cl) in solution. This liquid is not in the cell as a free liquid, but is soaked up in some absorbent substance such as sawdust, blotting paper, or plaster to make a sort of paste and is carefully sealed

figure 9–8 **THE COMMON ZINC–CARBON DRY CELL**

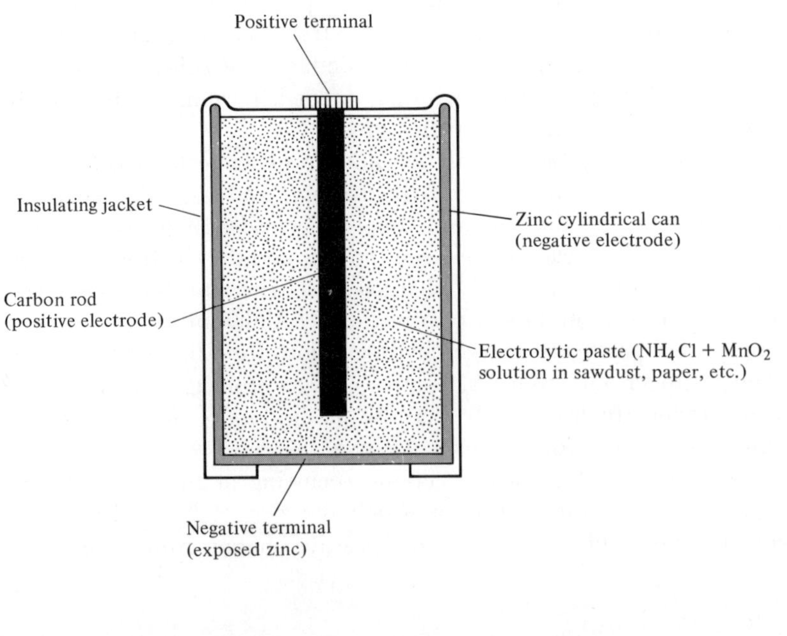

inside the cell. Thus a dry cell is not truly dry, only relatively so. Manganese dioxide (MnO_2) is also added to absorb the hydrogen gas that is produced at the positive electrode; this serves to prolong the life of the cell. Other kinds of dry cells commercially available (such as the mercury or alkaline cells) use other combinations of electrodes and electrolytes.

A ***storage battery*** is one in which the chemical reactions that tend to break down the electrodes are reversible; that is, when a current is driven through the battery in the opposite direction (from negative to positive), the reverse chemical reactions occur, restoring the electrodes to their original state. This is the process commonly known as charging a battery. It can be done by connecting the battery to a more powerful battery, positive terminal to positive terminal and negative to negative.

The lead–acid storage battery used in most automobiles consists of several cells having negative electrodes of lead (Pb) and positive electrodes of lead dioxide (PbO_2). The electrolyte is sulfuric acid. The chemical reactions tend to change both electrodes to lead sulfate ($PbSO_4$); however, when the battery is being charged, the reverse chemical reactions will occur, and the $PbSO_4$ will be changed back into Pb and PbO_2 at the proper electrodes. In this way the life of a storage battery can be prolonged considerably. In an automobile the battery is being charged (by a current from the generator or alternator) whenever the engine is running.

ELECTRIC ENERGY

We have seen that a voltaic cell or battery can generate a steady current of electric charge in a circuit as a consequence of the chemical reactions that occur internally. During the first half of the nineteenth century, there was some confusion as to how to describe the impetus provided to electric charge by this action; scientists generally used such terms as "electric tension" or "electromotive force" to identify the output of a battery. Fortunately, the situation was greatly improved by the development of the principle of conservation of energy in the 1850s. Although this principle was originally applied to mechanical energy only, it was quickly expanded to include electric and magnetic effects as well.

In order to bring the electrostatic force within the framework of the conservation of energy principle, ***electric potential energy*** must be defined. This can be done rather easily, in view of the strong similarity between Newton's law of gravitation and Coulomb's law. In Chapter 6 we learned that the gravitational potential energy of two attracting masses is represented by the equation (Eq. 6–19):

$$PE_G = -\frac{GMm}{d}$$

where d is the distance between the two masses M and m. If we use the same mathematical techniques, it becomes a relatively simple matter to show that the electrostatic force is a conservative force and that the following equation represents the electric potential energy of a system of two charges:*

$$PE_e = \frac{K_e q_1 q_2}{d} \tag{9–4}$$

*In comparing Eqs. 6–19 and 9–4, we notice that the minus sign does not appear in the latter. This deletion takes into account the minus sign that appears when q_1 and q_2 have opposite signs and the electrostatic force is attractive.

where d is the distance between the two charges q_1 and q_2. The units of electrostatic potential energy are the same as for any other kind of potential energy, namely ergs (CGS) and joules (MKS).

The following will illustrate how the concept of electric potential energy is related to the general principle of conservation of energy. Suppose that two oppositely charged bodies, A and B, with charges of $+8$ esu and -6 esu, respectively, are 4 cm apart, as shown in Figure 9–9a. Suppose further that body A is fixed in place and cannot move. There is an attractive electrostatic force acting on each body, so that if B is allowed to move freely, it will "fall" in the direction of A; that is, B will accelerate along the straight line joining the bodies. As B moves, its velocity and kinetic energy will steadily increase.

According to the principle of conservation of energy, this increase in kinetic

figure 9–9 **ELECTRIC POTENTIAL ENERGY OF TWO CHARGED BODIES**

Body A is fixed, while body B moves under the influence of the mutual electrostatic force **F.** In (a) the bodies are oppositely charged; the electric potential energy decreases as B moves toward A. In (b) the bodies have like charges; the electric potential energy decreases as B moves away from A.

$$PE_e = \frac{K_e q_A q_B}{d}$$

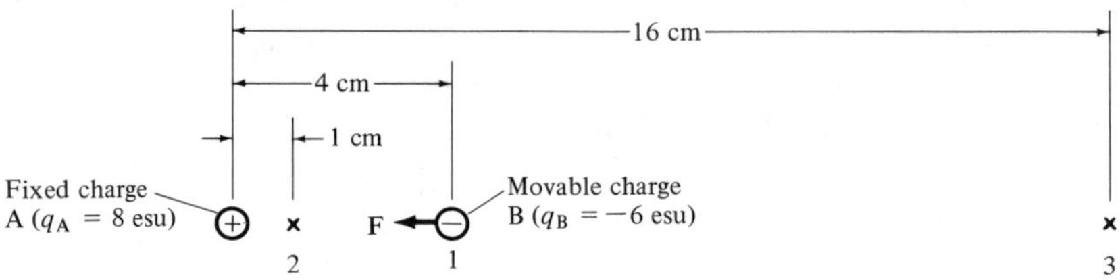

At point 2:

$$PE_{e_2} = \frac{(8 \text{ esu})(-6 \text{ esu})}{(1 \text{ cm})} = -48 \text{ ergs}$$

At point 1:

$$PE_{e_1} = \frac{(8 \text{ esu})(-6 \text{ esu})}{(4 \text{ cm})} = -12 \text{ ergs}$$

At point 3:

$$PE_{e_3} = \frac{(8 \text{ esu})(-6 \text{ esu})}{(16 \text{ cm})} = -3 \text{ ergs}$$

(a) Charges Have Opposite Signs.

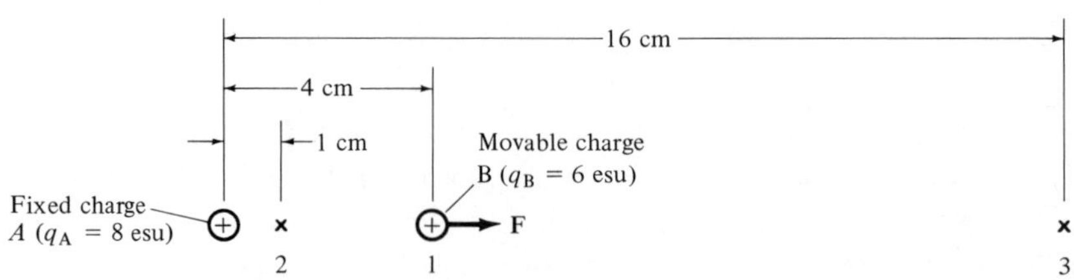

At point 2:

$$PE_{e_2} = \frac{(8 \text{ esu})(6 \text{ esu})}{(1 \text{ cm})} = 48 \text{ ergs}$$

At point 1:

$$PE_{e_1} = \frac{(8 \text{ esu})(6 \text{ esu})}{(4 \text{ cm})} = 12 \text{ ergs}$$

At point 3:

$$PE_{e_3} = \frac{(8 \text{ esu})(6 \text{ esu})}{(16 \text{ cm})} = 3 \text{ ergs}$$

(b) Charges Have Like Signs.

energy must be compensated by an equivalent decrease in potential energy. Using Eq. 9–4, we calculate that the electric potential energy is initially -12 ergs. However, as B moves toward A, B's electric potential energy decreases (becomes more negative). When the bodies are 1 cm apart, the electric potential energy will be -48 ergs. Thus when B moves from point 1 to point 2, its electric potential energy *decreases* by 36 ergs; consequently, its kinetic energy will have *increased* by the same amount.

Now let us suppose that body B has a *positive* charge of 6 esu (Figure 9–9b). In this case B is acted on by a *repulsive* force and will accelerate *away* from A if allowed to move freely. At point 1 body B has a *positive* electric potential energy of 12 ergs; as B moves to point 3, its electric potential energy decreases to 3 ergs. Hence there is a *decrease* of 9 ergs in the electric potential energy of B, which will be matched by an *increase* of 9 ergs in the kinetic energy of B as it accelerates away from A.

In Chapter 6 gravitational potential energy was defined such that the work done by a gravitational force would be equal to the decrease in the gravitational potential energy. (See Eq. 6–14.) The work done by an electrostatic force is similarly related to the electric potential energy:

$$W = PE_{e_i} - PE_{e_f} \tag{9–5}$$

Let us apply this equation to the examples shown in Figure 9–9. In Figure 9–9a the work done by the electrostatic attractive force in moving B from 1 to 2 is

$$W = PE_{e_1} - PE_{e_2} = -12 - (-48) = 36 \text{ ergs}$$

In Figure 9–9b the work done by the electrostatic repulsive force in moving B from 1 to 3 is

$$W = PE_{e_1} - PE_{e_3} = 12 - 3 = 9 \text{ ergs}$$

Note that the electric potential energy has been defined so that it is always negative for a pair of opposite charges and always positive for a pair of like charges. Consequently, the work done by an electrostatic force—whether attractive or repulsive—will result in a decrease in potential energy. In this way both the attractive and repulsive aspects of the electrostatic force can be encompassed in the conservation of energy principle.

Suppose we wanted to move B *against* the electrostatic force—that is, to point 3 in Figure 9–9a or to point 2 in Figure 9–9b. It would be necessary to apply some outside force that would increase the potential energy of B by doing work on the system. The action of this outside force is quite analogous to that of the force required to lift a book from a table, for in this case the lifting force does work against the gravitational pull of the earth on the book, and the potential energy of the book is increased accordingly.

We will now attempt to gain a clearer understanding of what a battery does, in terms of the energy concept. The net effect of the chemical reactions that occur in a battery is to separate electric charges, so that one electrode has an excess of positive charges and the other has an excess of negative charges. In order to separate oppositely charged bodies, outside forces must be exerted to act against the electrostatic attraction. In a battery these forces are generally exerted by the ions of the electrolyte on the atoms of the electrodes, resulting in the ionization of the latter. The work done by these forces tends to raise the electric potential energy of the system of charges. Thus the effect of the battery action is to separate positive and negative charges and, in so doing, to provide these charges with electric potential energy. If the charges are now provided with a path along which they can move freely (i.e., by connecting the electrodes with some kind of conductor), the charges will convert this potential energy into kinetic energy—that is, the charges will be set in motion, resulting in an electric current. (Specifically, the negative charges will be set in motion by the attraction of the fixed positive charges at the opposite electrode.) Electric devices such as light bulbs, toasters, and radios can be included in the conducting path, so that the kinetic energy of the moving charges can be further converted into light, heat, sound waves, or other energy forms.

One should not be misled into believing that a battery is a device that creates electric energy. More precisely, a battery is an *energy converter.* It converts the electrochemical energy stored in the atoms and molecules of the electrolyte and electrodes into the electric potential energy of separated charges. Furthermore, a battery cannot continue to produce electric potential energy indefinitely, for once the active chemical substances have been used up, there is no more available electrochemical energy to convert, and the battery will cease to operate. Thus a given battery can supply only a certain fixed amount of electric energy over its entire lifetime, unless the battery is resupplied with energy through a charging process.

It might seem that the best basis for comparing different types of batteries would be the total electric energy that the battery supplies over its lifetime. However, this is not the most suitable criterion, for we must also take into account the total amount of charge that has been set in motion by the battery. The total electric potential energy generated by a battery can be distributed either to a large number of charges (in which case each charge gets a small amount of energy) or to a smaller number of charges, so that each charge has a relatively large amount of energy. What we are interested in is the amount of energy supplied to each charge, or more specifically, the *electric potential energy per unit charge* provided by the battery. This quantity is known as the ***electric potential*** U:

$$U = \frac{PE_e}{q} \qquad (9\text{–}6)$$

In MKS units energy is measured in joules and charge in coulombs. Therefore the unit of electric potential is the joule/coulomb. This unit is called the ***volt*** (in honor of Volta). Since the MKS electric units are used universally for most practical applications, the CGS electric units will not be identified from here on.

A relationship between work and electric potential can be obtained by combining Eqs. 9–5 and 9–6:

$$W = PE_{e_i} - PE_{e_f} = qU_i - qU_f = q(U_i - U_f) \qquad (9\text{–}7)$$

This equation can be interpreted in the following way: when a quantity of charge q moves from one point to another, the work done on that charge is equal to the quantity of charge moved times the difference in potential between the two points. Note that the amount of work done depends on the *difference* in electric potential, and not the actual electric potentials themselves. The ***potential difference*** $(U_i - U_f)$ is also commonly referred to as the ***voltage*** or ***voltage difference.*** For simplicity, we will use the symbol V to represent potential difference; in other words, if two points have electric potentials U_1 and U_2, then the potential difference (or voltage) between the two points is $V = U_1 - U_2$.

It is standard practice to compare batteries (and other sources of electric energy) in terms of their voltage. For example, a 1.5-volt dry cell has a potential difference of 1.5 volts between its positive and negative terminals, and therefore supplies 1.5 joules of electric energy to each coulomb of charge that flows from one terminal to the other when the terminals are connected by a conducting path. Similarly, a common 120-volt wall outlet is connected to an energy source that provides a potential difference of 120 volts between the two terminals of the outlet; thus 120 joules of energy are supplied to each coulomb of charge.

The potential difference provided by a battery depends entirely on the specific chemical reactions that take place and not on the amount of material in the battery. That is, the standard zinc–carbon dry cell provides a potential difference of 1.5 volts between its terminals, *regardless of its size*. Similarly, a single cell of a lead–acid storage battery has a potential difference of about 2 volts under normal conditions. Since most automobiles are designed with a 12-volt electrical system, a standard automobile battery contains six lead–acid cells connected together in series.

ELECTRIC CURRENT AND RESISTANCE

When the positive and negative terminals of a battery are connected by a conducting path (making a complete electric circuit), electric charges are set in motion. The electric current in the circuit is defined as the *rate of flow of electric charge*. Specifically, if q is the net amount of charge that flows past a given point in the circuit in a time interval t, then the electric current I is

$$I = \frac{q}{t} \tag{9–8}$$

As defined, the current I represents the *net* flow of *positive* charge across any cross section of a conducting wire or electric device. Thus current is defined as the flow of *positive* charge, even though we now know that the positive charges remain fixed while the negative charges move. However, from a mathematical point of view, a flow of negative charge in one direction is identical to a flow of the same quantity of positive charge in the opposite direction. Thus the fiction of a current of positive charge can be maintained without deleterious effect (except to one's aesthetic sensibilities, perhaps).

The MKS unit of current is the coulomb/sec, which has been named the ***ampere*** (commonly abbreviated to amp). A current of 1 amp in a given direction through a conductor means that there is a net flow of 1 coulomb of charge crossing any section of the conductor in 1 second (or a flow of 1 coulomb/sec of negative charge in the opposite direction).

In 1827 a German high-school teacher named Georg Ohm discovered a simple and extremely useful relationship between the voltage and current in a metallic conductor. Ohm found through a combination of experiment and theoretical analysis that the current through a given conducting body is directly proportional to the voltage difference across the conductor. In other words, an increase or decrease in the voltage produces a proportional increase or decrease in the current. In mathematical terms the relationship is written

$$\frac{V}{I} = \text{constant}$$

Ohm called this constant the ***resistance*** (denoted by the symbol R) of the body. The above equation then becomes

$$V = IR \tag{9–9}$$

which is known as ***Ohm's law.*** Not surprisingly, the MKS unit of electric resistance is called the ***ohm****[*] (1 ohm = 1 volt/amp). The resistance of a given body depends on the body's size and shape, the type of material, and the temperature. Good conductors, such as copper, silver, and aluminum, tend to have low resistances, and substances usually classified as insulators have extremely high resistances. Thus the distinction between conductors and insulators can be understood in terms of Ohm's law. The resistance of an insulator is considerably higher than that of a conductor of the same size and shape. For the same potential difference, the current is much lower through an insulator than through a conductor. A perfect insulator would therefore be a body having an infinite resistance.

For most metallic conductors, the resistance tends to increase with temperature. (There are a number of substances, such as the elements germanium, silicon, and selenium, for which the resistance *decreases* with temperature. Such substances are called ***semiconductors*** and are widely used in modern electronic technology.) A simple explanation of this increase with temperature can be constructed from the kinetic theory of matter. In a solid conductor the atoms are arranged in a reasonably orderly array or lattice. When an electric potential difference is applied, some of these atoms will yield their negatively charged electrons readily, and a current of moving electrons will be established. However, the atoms are in vibration, and electrons will collide with the atoms and lose some of their energy. In this sense electric resistance is viewed as a kind of "electric friction," which arises as a consequence of the collisions between the moving electrons and the atoms of the conductor itself. As the temperature of the conductor increases, the conductor's atoms will vibrate more violently, and the number of collisions with electrons should increase. Thus the resistance would be expected to increase with temperature.

Conversely, as the temperature of a conductor is lowered, its resistance will decrease. We would expect some residual resistance to be present even at extremely low temperatures. In 1911 the Dutch physicist Heike Kamerlingh-Onnes was conducting experiments on the resistance of bodies at temperatures near absolute zero. He made the surprising discovery that the resistances of certain materials will disappear completely at temperatures a few degrees above absolute zero. In this state a substance is said to be a ***superconductor.*** Several hundred substances (metallic elements or compounds) are now known that will become superconductors at temperatures of 20°K or less.

This puzzling phenomenon was not satisfactorily explained until a theory was presented in 1957 by three American physicists, John Bardeen, Leon Cooper, and J. R. Schreiffer. In greatly simplified terms, their theory shows that at low temperatures the electrons interact in pairs with the atomic lattice in such a way that the energy

[*] The ohm is very often represented by the symbol Ω (the Greek capital letter omega); that is, 5 Ω = 5 ohms.

lost by one electron will be transferred by the lattice to the second electron. Hence there is no *net* loss of energy by the electrons to the atoms. Thus in essence the resistance is zero.

Once a current is established in a completely superconducting circuit, it can continue to flow indefinitely—even after the voltage source has been removed. Superconductivity and its associated effects have many potential applications in modern technology, the major drawback being the very low temperatures that must be produced in order for superconductivity to be maintained.

ELECTRIC POWER

An important consideration in the design and construction of mechanical or electric devices is the *rate* at which they generate or use up energy. To measure this, we define a quantity called ***power*** as *the amount of energy generated or dissipated per unit time.* In an electric device the amount of electric energy generated or dissipated is equal to the work done by the electric forces. Let us suppose that a potential difference, $V = U_1 - U_2$, exists across the terminals of some electric device. (See Figure 9–10.) According to Eq. 9–7, the work done (and hence the electric energy generated or dissipated) when a charge q moves through the device is

$$W = q(U_1 - U_2) = qV$$

To obtain an equation for the power (the *rate* of energy generation or dissipation), we divide by the time t required for the charge to move from one terminal to the other. We obtain

$$P = \frac{W}{t} = \frac{qV}{t} = V\left(\frac{q}{t}\right)$$

But according to Eq. 9–8, q/t is the current I through the device. Therefore the equation becomes

$$P = VI \tag{9–10}$$

This equation tells us that the power generated or dissipated by an electric device is equal to the potential difference across the terminals of the device multiplied by the current through the device. Power has the units of energy/time. In MKS units power is measured in joules/sec. The joule/sec is more familiarly known as the ***watt.*** (The unit of power in the British system is also familiar. It is the ***horsepower,***

figure 9–10 **VOLTAGE–CURRENT RELATIONSHIPS**

V represents the potential difference (voltage) between the two terminals of an electric device. The current I is the net flow of positive charge through the device.

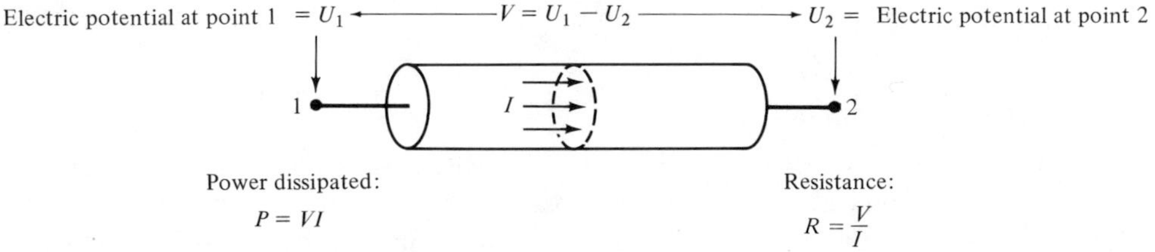

originally defined as the rate of work that can be done by a horse; 1 hp = 550 ft-lb/sec = 746 watts.)

Most common electric devices are rated in terms of their "wattage," which we now recognize as the rate at which they dissipate energy. For example, a 60-watt bulb converts electric energy into light and heat at the rate of 60 joules/sec. Other devices designed to generate heat (such as toasters, irons, and broilers) usually give off energy at a rate of about 1000 watts or more.

EXAMPLE 9–2: A lamp containing a 60-watt bulb is plugged into a 120-volt outlet. Calculate the current in the lamp and the resistance of the lamp.

SOLUTION: The current can be found directly by using Eq. 9–10:

$$P = VI$$

$$60 \text{ watts} = (120 \text{ volts})I$$

$$I = \frac{60}{120} = \underline{0.5 \text{ amp}}$$

The resistance can now be calculated from Eq. 9–9:

$$V = IR$$

$$120 \text{ volts} = (0.5 \text{ amp})R$$

$$R = \frac{120}{0.5} = \underline{240 \text{ ohms}}$$

9–3

MAGNETISM FROM ELECTRICITY

When viewed in historical perspective, the invention of the battery turns out to be of pivotal importance in the development of physics. An easily constructed source of a strong and steady electric current enabled the scientists of the nineteenth century to discover in rapid succession a number of interrelated electric and magnetic phenomena. The practical applications of these discoveries were quickly recognized, and a technological revolution ensued in which the electric devices now so essential to modern living were invented one after another. We can only speculate as to whether current electricity would have been discovered by some other means, but history records the fact that the entire course of human civilization was dramatically altered as the result of the accidental discovery in an anatomy laboratory of the twitching of a frog's leg.

At the beginning of the nineteenth century, electricity and magnetism were viewed as two separate and independent aspects of nature. Yet the idea that electricity and magnetism were related in some way was an appealing one to the scientists of that era, and many experiments were devised in an effort to find some connection between electric and magnetic phenomena.

OERSTED'S EXPERIMENT

The first success was obtained by the Danish physicist Hans Christian Oersted in 1820. Oersted placed a magnetic compass needle next to a current-carrying wire and discovered that the needle was deflected from its normal north–south orientation when the current was turned on (Figure 9–11). Oersted showed further that the magnitude and direction of this deflection depended on the strength and direction of the current. These results clearly demonstrated that a magnetic force could be generated by an electric current.

One aspect of this newly discovered force was particularly unique. The direction of the force was found to be *perpendicular* to the line joining the wire and the magnetic pole. This was unlike all the other forces previously studied, which were known to act *along* the line joining the interacting bodies. The pattern of the magnetic force

figure 9–11 **OERSTED'S EXPERIMENT**

(a) In the experiment a current is generated in a conducting wire by a voltaic cell. The wire is oriented in a north–south direction so that it is parallel to and directly above a magnetic compass. When the switch is closed, the compass is deflected from its normal north–south orientation. (b and c) Top views, showing the wire over the compass. When the current is from south the north, the north pole is deflected westward (b); when the current is reversed, the deflection is also reversed in direction (c). (d) When the compass is placed above the wire, the direction of deflection is again reversed.

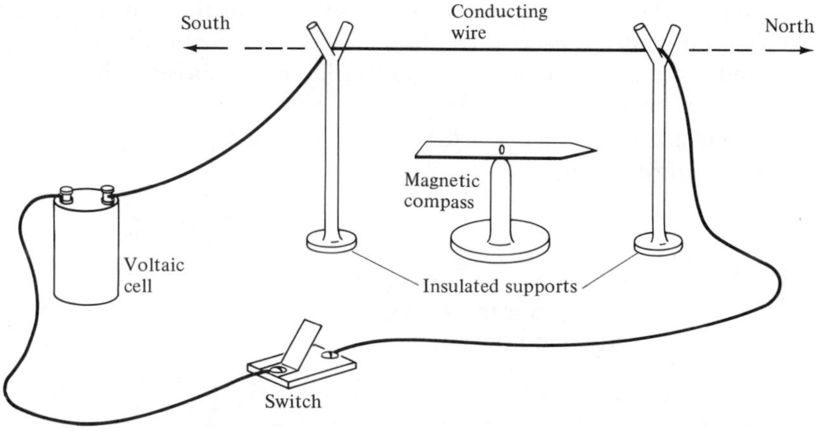

(a)

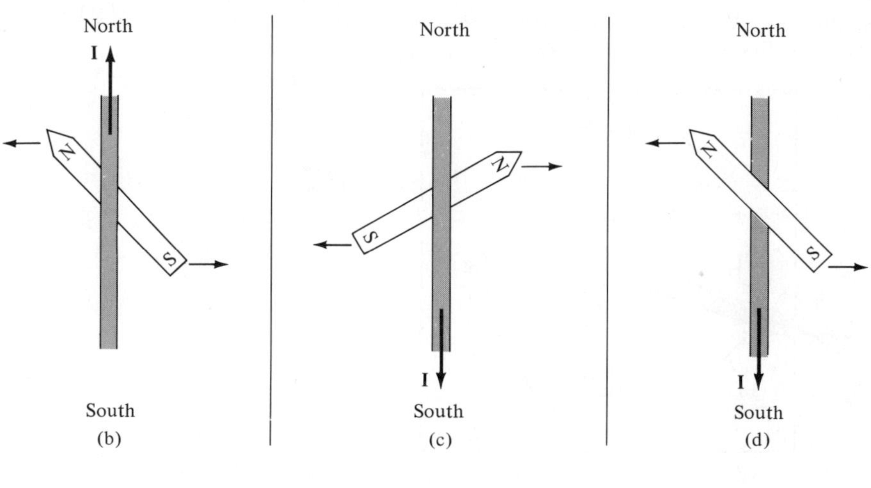

(b) (c) (d)

of an electric current is illustrated in Figure 9–12. Here the force exerted on an imaginary north magnetic pole is shown at various points in the plane perpendicular to the wire. (The force on a south magnetic pole would point in the opposite direction.) Note that the force direction is always perpendicular to the line joining the wire and the magnetic pole. In addition, a more precise measurement of the magnetic force shows that the force is directly proportional to the strength of the current and inversely proportional to the distance from the wire; that is, $F \propto I/d$.

An important consequence of Oersted's discovery derives from the application of the action–reaction principle: *if a current exerts a force on a magnet, a magnet should exert an equal and opposite force on a current.* Thus by placing a strong magnet near a current-carrying wire, one can establish a force that will set the wire in motion. This effect, as will be shown later, is the basic principle of operation of the electric motor. A further extension of this principle was reasoned out by the French scientist André-Marie Ampère. If a current generates a force similar to that produced by a magnet, and a magnet exerts a force on a current, then, Ampère proposed, two current-carrying wires should exert forces on one another. In a series of carefully designed experiments, Ampère succeeded in showing that two parallel wires will attract one another if the currents are in the same direction, and will repel one another if the currents are in opposite directions (Figure 9–13a).

Ampère chose to describe this effect in terms of the mutual forces between the individual elements of each wire. That is, he mathematically subdivided each wire into infinitesimal pieces or elements and considered the forces acting on these elements. As a result of his experiments, Ampère was able to show that the force between any pair of elements can be described by the following equation, which we will refer to here as ***Ampère's law:***

figure 9–12 **DIRECTION OF MAGNETIC FORCE EXERTED BY AN ELECTRIC CURRENT**

The diagram shows the direction of the force in a plane perpendicular to the current-carrying wire. The arrows show the direction of the force exerted on a north magnetic pole (the direction is opposite for a south pole). When the current direction is reversed, the direction of the magnetic force is also reversed.

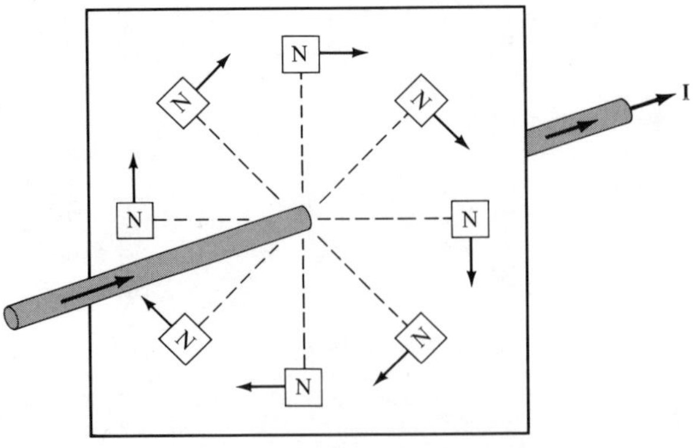

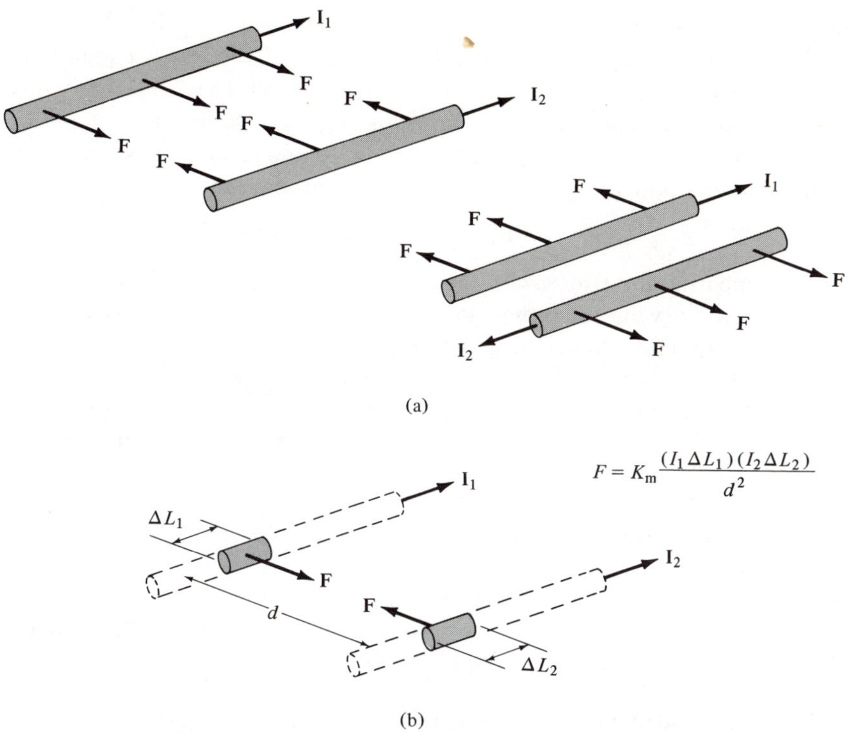

figure 9–13 **FORCES BETWEEN CURRENT–CARRYING WIRES**

(a) Two parallel current-carrying wires attract each other if the currents are in the same direction and repel if the currents are in opposite directions. (b) The force between two small elements of the two wires is inversely proportional to the square of the distance between them.

$$F = K_m \frac{(I_1\, \Delta L_1)(I_2\, \Delta L_2)}{d^2} \tag{9–11}$$

In the equation I_1 and I_2 represent the currents in the two wires; ΔL_1 and ΔL_2 represent length elements of the wires; d is the distance between the two elements (see Figure 9–13b); and K_m is a constant whose value depends on the units used.* From this equation the force between two wires of any given length can be found by mathematically summing all the forces between the elements of one wire and the elements of the second wire. For example, given two long, parallel wires of finite length (by "long" we mean that the length of each wire is large compared to the distance between them), it can be determined from Eq. 9–11 that the force on a length L of either wire is

$$F = K_m \frac{2I_1 I_2 L}{d} \tag{9–12}$$

Equation 9–12 is used to define the units of electric current. In CGS units L

*It is assumed that the two length elements are parallel to one another. The equation given here actually is a simplified form of a more general equation given by Ampère.

and d are in cm; F is in dynes; and the constant K_m is assigned a numerical value of 1. The CGS unit of current (defined in terms of the current required to produce a mutual force of 2 dynes/cm on two parallel wires 1 cm apart) is called the ***emu*** (electromagnetic unit). Thus in CGS units $K_m = 1$ dyne/emu^2.

In MKS units (with F in newtons and L and d in meters) the unit of current is the ampere, and the constant K_m is assigned a value

$$K_m = 1 \times 10^{-7}\ \text{N/amp}^2$$

Comparing the two systems of units, one can show that 1 emu = 10 amp.* The CGS and MKS values assigned to K_m are the *free-space* values; in material media K_m will have different values depending on the medium.

In Ampère's formulation the force described by Eq. 9–11 acts along the line joining the current elements. Note also that this force depends on the inverse square of the distance, and hence has the same mathematical form as the law of gravitation and Coulomb's laws of electrostatic and magnetostatic force. Ampère was convinced that he had identified the basic electromagnetic interaction. He was one of a number of nineteenth-century physicists who were as devoted to the Newtonian principle that all fundamental forces follow an inverse-square law as the ancient Greeks were to the Platonic concept that all celestial motions are circular. Hence Ampère and his colleagues were especially pleased that the harmony of the Newtonian system of mechanics had been preserved through Ampère's law.

ELECTRIC AND MAGNETIC FIELDS

Several years after Oersted's and Ampère's discoveries, an important conceptual advance was made through the introduction of the field representation of electric and magnetic forces. This development was a consequence of the work of one of the most remarkable men of science, Michael Faraday (1791–1867). Born under conditions of extreme poverty, Faraday received no more than a grade school education. At the age of fourteen he went to work as a bookbinder's apprentice in London and educated himself by reading some of the books that were brought in to be bound. Stimulated by an article on electricity in the *Encyclopaedia Britannica*, Faraday's interest in science was awakened, and he continued his self-education in this direction. Finally, at the age of twenty-two he managed to obtain a position as an assistant to Humphry Davy. At that time Davy was the head of the Royal Institution, a center for science education and research that had been founded by the ubiquitous Count Rumford in 1799. Here Faraday remained for the rest of his life.

Faraday was an extremely gifted experimenter, and he made many important discoveries in chemistry as well as in electromagnetism. Because of his complete lack of mathematical training, Faraday tended to think in terms of simple pictorial models. He was not impressed by Ampère's description of the magnetic force in terms of an inverse-square law, for Oersted's experiments had shown that the force exerted by a current on a magnetic pole acts at right angles to the line joining them. In Faraday's view this was the basic electromagnetic interaction, and no amount of mathematical manipulation could obscure the fact that the force clearly appeared to act at right angles.

*The student should not allow himself to be unduly confused by the two sets of units described here. The MKS units are the ones most commonly used; the CGS units have been presented here for historical reasons only. The relationship between the two systems will be explained more fully in the next chapter.

Faraday chose to represent the magnetic action by drawing *lines of force*. The line of force passing through a given point depicts the direction of the force that would be exerted on a particle at that point. More precisely, a line of force is drawn such that at every point on the line the force vector is tangent to the line. In this manner the influence of a given force can be represented pictorially. The lines of force thus drawn are said to represent the ***field*** of the given force.

To illustrate Faraday's concept, we will first construct the lines of force representing the electric field of a single negatively charged body. Let us imagine a small, positive "test charge," which we can place at various points in the vicinity. At every point the force of electrostatic attraction is directed along the straight line joining the charges. If the test charge were released from rest and allowed to move freely, it would always move directly in a straight line toward the negative charge. Thus the lines of force are straight lines radiating toward the negatively charged body (Figure 9–14a). By convention, the arrows on each line are drawn to point in the direction in which a positive charge will move. Hence the field of a positively charged body is the same as that of a negatively charged body, except that the arrows point away from the source of the field (Figure 9–14b). Because of the similarities between the electric and gravitational forces, the same pattern of lines of force will also represent the gravitational field of a single mass (Figure 9–14c).

Strictly speaking, there should be a line of force passing through every point of the surrounding region. Hence there should be an infinite number of lines emanating from the source of force. To make the diagram intelligible, however, only a finite number of equally spaced lines are drawn. One commonly used procedure is to make the number of lines emanating from a given charge or mass proportional to the magnitude of the charge or mass. For example, twice as many lines would be drawn emanating from a 10-esu charge as from a 5-esu charge. When this procedure is used, the density of the lines (i.e., the number of lines passing through some given region) is a measure of the intensity of the force in that region. The force is most intense where the lines are most closely spaced.

When the force is exerted by a single charge or mass, little insight or understanding is gained by constructing the lines of force. The advantage of this procedure is more apparent with more complicated arrangements. For example, consider two

figure 9–14 **LINES OF FORCE OF A FIELD**

Shown below are the lines of force of (a) the electric field of a negative charge, (b) the electric field of a positive charge, (c) the gravitational field of a body of mass M. The arrows indicate the direction of the force on a positive charge or a mass at any point on the line of force.

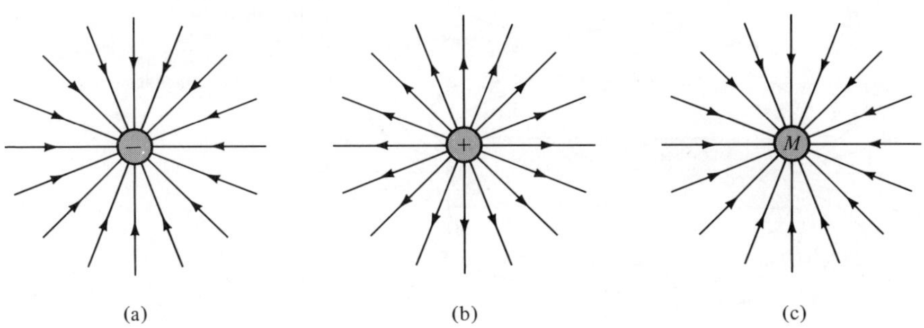

parallel conducting plates that have been given equal and opposite charges (Figure 9–15a). Using calculus, one can show that the net electric force due to all the charges is zero above and below the plates, and in the region between the plates the force is directed from the positive to the negative plate. This result could be stated in terms of a set of mathematical formulas, but it is much more clearly represented by the field diagram showing the lines of force.

figure 9–15 **ELECTRIC AND MAGNETIC FIELDS**

(a) Two conducting parallel plates have equal and opposite charges uniformly distributed over them. An electric field exists between the plates, but not outside of them. (b) In the electric field of two equal positive charges, a small positive test charge would be repelled along the lines of force under the influence of the net repulsive force of the two charges. (c) In the magnetic field of a bar magnet, an imaginary "test north pole" would be repelled by the north pole and attracted to the south pole of the magnet. The direction of the net force is along the line of force as shown.

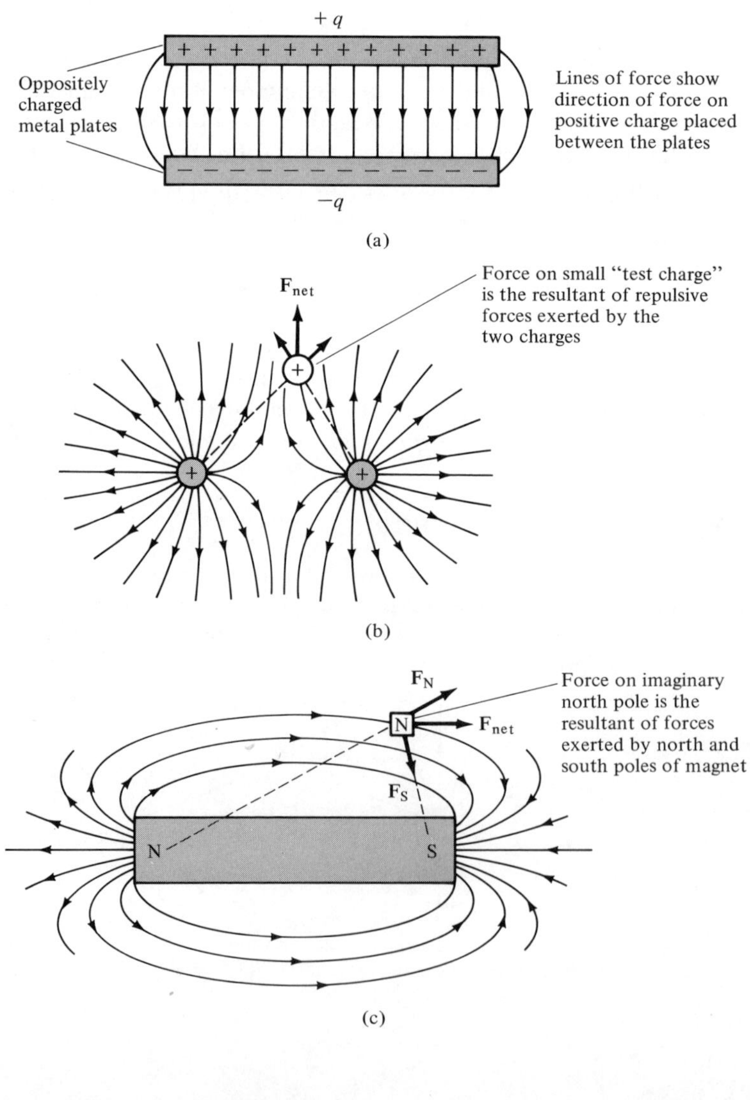

When one constructs the lines of force of the field of some complicated system of charges, magnetic sources, or masses, the guiding principle is that the force at any point is equal to the vector sum of the individual forces acting at that point. The vector representing the net force will be tangent to the line of force passing through the point in question. The diagrams showing the electric field of two equal positive charges (Figure 9–15b) and the magnetic field of a bar magnet (Figure 9–15c) illustrate this principle.

It is useful to have a numerical measure of the intensity of an electric or magnetic field. The ***electric field intensity*** **E** at a given point is defined as the electric force exerted on a unit charge located at that point. That is,

$$\mathbf{E} = \frac{\mathbf{F}_{\text{elec}}}{q} \qquad (9\text{–}13a)$$

In the MKS system the electric field intensity is measured in units of newtons/coulomb. Similarly, the ***magnetic field strength*** **B** (more formally called the ***magnetic induction***) measures the magnetic force acting on a unit pole located at the point in question:

$$\mathbf{B} = \frac{\mathbf{F}_{\text{mag}}}{p} \qquad (9\text{–}13b)$$

The MKS unit of magnetic field intensity is the ***tesla.*** In practice, the CGS unit is more commonly used; this unit is called the ***gauss*** (1 tesla = 10^4 gauss).* The magnetic field strength of an ordinary bar magnet has a typical value of a few hundred gauss near its poles; the magnetic field of the earth (responsible for the deflection of compass needles) has an average strength of about 0.5 gauss.

THE MAGNETIC FIELD OF A CURRENT-CARRYING WIRE

Now let us use the field representation to describe Oersted's discovery. As shown in Figure 9–12, the force exerted by a current-carrying wire on a magnetic pole is always at right angles to the line drawn from the pole to the wire. Thus the lines of force of the magnetic field are drawn as concentric circles lying in a plane perpendicular to the wire (Figure 9–16). The arrows indicating the direction of the force are drawn to show that the direction is clockwise when the observer is looking along the direction of current flow (i.e., when the current direction is into the plane) and counterclockwise when the current direction is pointing toward the observer (out of the plane).

The same rule can be applied when the current is moving in a circular loop rather than along a straight wire. At any point on the loop the lines of force are concentric circles lying in a plane perpendicular to the wire. However, because of the curvature of the wire, the magnetic fields caused by each small length element of the wire will overlap. Thus the net magnetic field at any point will be the vector sum of the fields produced by each portion of the wire. In particular, at the center of the loop the individual forces will all point in the same direction, so that the net magnetic field will be intensified at the center (Figure 9–17a). To further intensify the field at the center, one can place two or more loops side by side by winding

*These units are named after Nikola Tesla (1856–1943), who made many important contributions to the development of the modern-day system of electric power generation and distribution, and Karl Friedrich Gauss (1777–1855), one of the greatest mathematicians in history.

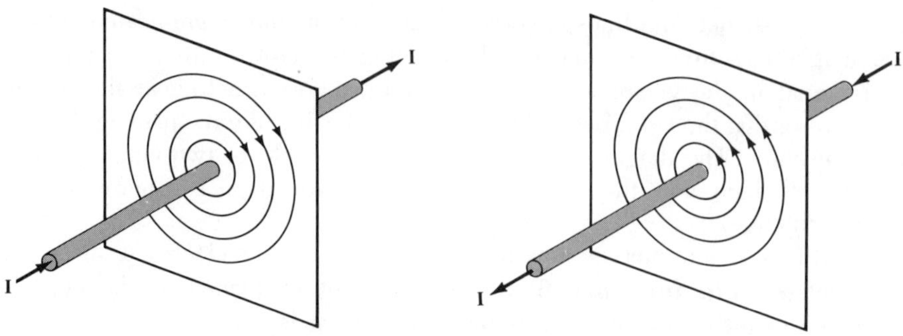

figure 9–16 **THE MAGNETIC FIELD OF A STRAIGHT CURRENT–CARRYING WIRE**

The lines of force are concentric circles in a plane perpendicular to the wire. When one sights along the wire in the direction of current flow, the direction of the force is clockwise.

the wire into a coil. The pattern of the lines of force for a coiled, current-carrying wire (sometimes called a ***solenoid***) is shown in Figure 9–17b.

If we examine Figure 9–17b more closely, we notice that the magnetic field of a solenoid (as represented by the solenoid's lines of force) bears a remarkable similarity to that of an ordinary bar magnet (Figure 9–15c). Since their fields are similar, it is not unreasonable to expect that a solenoid and a bar magnet will have other properties in common. In fact, experiments show that one end of a solenoid will behave like a north magnetic pole—that is, it will be repelled by the north pole of a bar magnet or another solenoid—and the other end will act like a south magnetic pole.* When allowed to pivot freely, a solenoid will align itself with the earth's magnetic field just as a magnetic compass does. The remarkable similarity of the properties of two seemingly unrelated objects—a bar of magnetized metal and a coil of wire through which a current is passing—is easily visualized through the technique of drawing lines of force. This is but one example of the value of the field concept.

The discovery of the magnetic properties of a solenoid is of great importance from a practical point of view, for it enables scientists and engineers to produce magnetic forces without recourse to natural magnets. One major advantage of a solenoid is that its magnetic field can be turned on or off simply by connecting or disconnecting the source of current. When the current direction in a solenoid is reversed, the polarity of the solenoid changes—that is, its north pole changes to a south pole and vice versa. Furthermore, whereas the magnetic field strength of a natural magnet is fixed, the field strength of a solenoid can be varied in the following ways:

(1) By increasing or decreasing the current.
(2) By increasing or decreasing the number of turns of wire in the coil.
(3) By coiling the wire around a core of iron or some other magnetizable metal. The magnetic field of the wire will magnetize the metal core, and the magnetic field of the core will be added to the field of the wire, thereby producing a more intensified overall field. A device of this type is commonly known as an ***electromagnet.***

*To determine which end of a solenoid is the north pole, the following rule can be used. Curl the fingers of the *right* hand around the coil of wire in the direction of the current. The thumb then points to the north pole.

With these methods magnetic fields of up to several hundred thousand gauss can be produced. One difficulty is that the extremely high currents needed to generate such fields produce so much heat that the conducting wire begins to melt. By using superconductors, one can negate the heating effects. It is believed that steady magnetic fields of as much as 500,000 gauss may be possible with superconducting solenoids.

THE MAGNETOMOTIVE FORCE

Let us now return to Ampère's discovery that current-carrying wires exert forces on one another. The distinction between Ampère's and Faraday's conceptions of this effect can now be made. In Ampère's view there is a direct and fundamental mutual force between the respective current elements of the two wires. In Faraday's picture the basic interaction is between a current-carrying wire and a magnetic field. Thus according to Faraday's point of view, two current-carrying wires exert forces on one another because *each wire lies in the magnetic field of the other.*

figure 9–17 **MAGNETIC FIELD OF A SOLENOID**

(a) The magnetic field of a single current loop. (b) The magnetic field obtained when a series of loops are coiled side by side to form a solenoid. Note the similarity to the field of a bar magnet shown in Figure 9–15c.

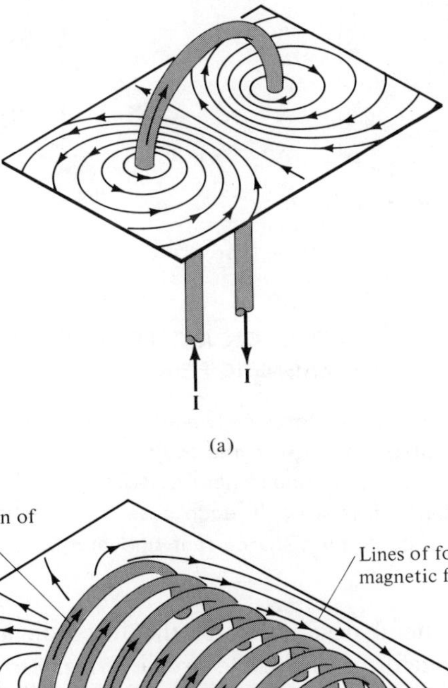

(a)

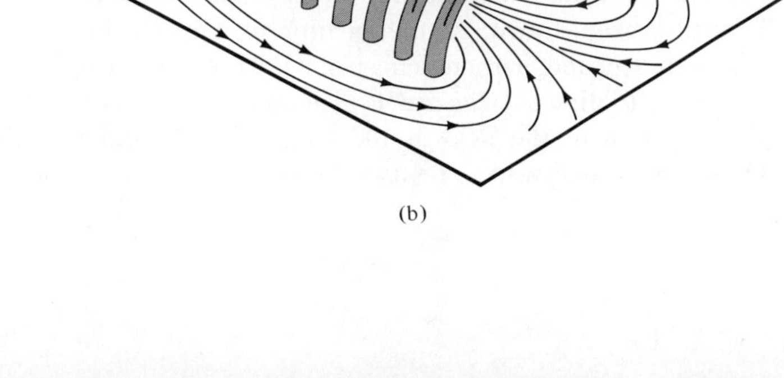

(b)

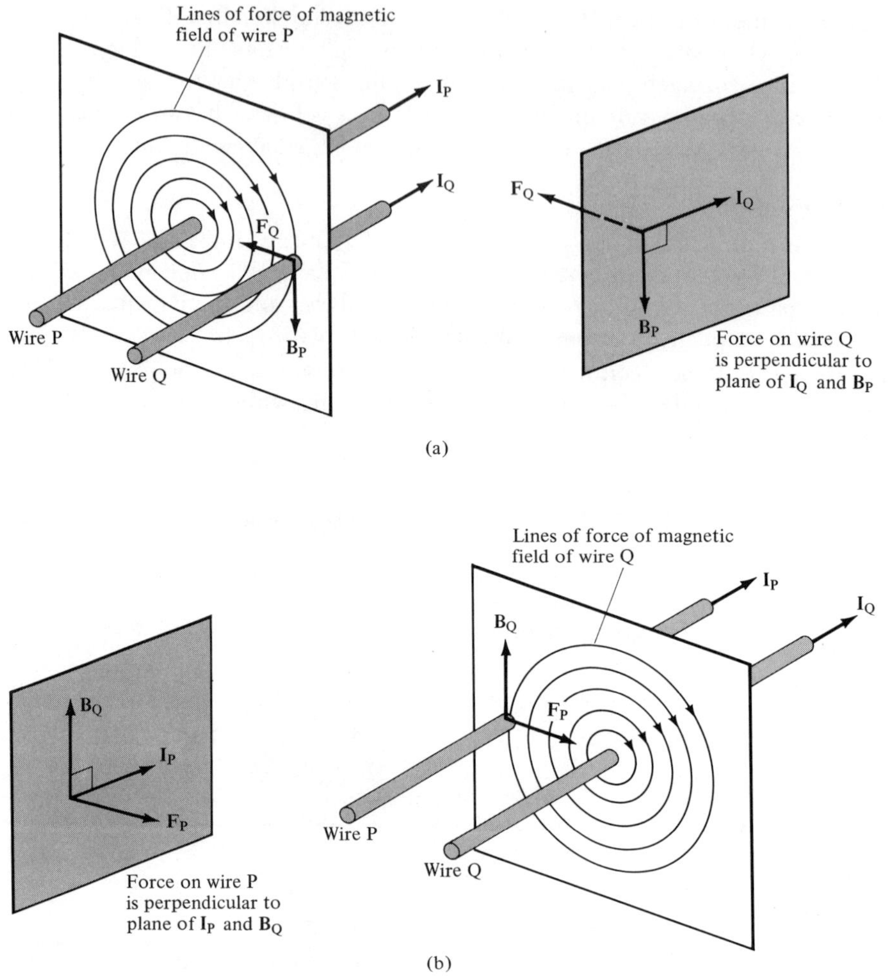

figure 9–18 **FORCE ON A CURRENT-CARRYING WIRE IN A MAGNETIC FIELD**

The mutual attractive force between two parallel wires is depicted in terms of the interaction of one wire with the magnetic field of the other. (a) The force on wire Q due to the magnetic field of wire P. (b) The force on wire P due to the magnetic field of wire Q. In each case the direction of the force is perpendicular to the plane containing the current and magnetic field directions.

A general description of the force on a current-carrying wire in a magnetic field—henceforth referred to as the ***magnetomotive force***—can be obtained from a consideration of the results of Ampère's experiments. Figure 9–18a shows two parallel wires carrying currents in the same direction. The lines of force of wire P have also been drawn. At the location of wire Q, the direction of P's magnetic field is downward and perpendicular to Q. However, Ampère's experiment showed that the force is directed *toward* P. Therefore the force on Q is perpendicular to both the direction of the field at the location of Q and the direction of the current in Q. A similar analysis can be used to show that the force on wire P due to the magnetic field of Q is perpendicular to both the direction of the magnetic field and the direction of the current in P. (See Figure 9–18b.)

The magnetomotive force exerted on a given length of wire is found to be proportional to the current in the wire and to the magnetic field strength. The magnitude of the force is also found to depend on the orientation of the wire with respect to the lines of force of the magnetic field. The equation for the magnitude of the magnetomotive force in MKS units* is

$$F = ILB \sin \theta \qquad (9\text{–}14)$$

where I is the current in amperes; L is the length of the wire in meters; B is the magnetic field strength in tesla; and θ is the angle between the wire and the magnetic field vector. (See Figure 9–19a.) The dependence of the force on $\sin \theta$ indicates that no force is exerted when the wire is parallel to the lines of force ($\theta = 0°$). Thus in order for this force to exist, the wire must cut across the lines of force of the magnetic field.

Experiments show that the direction of the force is perpendicular to the plane containing both the wire and the magnetic field strength vector **B.** There is a strong analogy between the description of the magnitude and direction of the magnetomotive force and the definition of the angular momentum vector **L** in Chapter 6. Just as **L** is perpendicular to the plane containing **v** and **r,** the direction of the magnetomotive force vector **F** is perpendicular to the plane containing **B** and the current-carrying wire. The direction of **F** is such that if the current direction is rotated toward **B** in a clockwise direction, then **F** points away from the viewer (Figure 9–19a).

A simpler way to identify the direction of **F** is through the "right-hand rule." The right hand is held with the thumb pointing straight up, the index finger pointing outward, and the middle finger pointing sideways, as shown in Figure 9–19b. If the *M*iddle finger is aligned with the direction of **B** (the *M*agnetic field), and the *I*ndex finger is aligned with the direction of the current (**I**), then the *T*humb will automatically point in the direction of the force (or *T*hrust).

Note that so far no explanation has been given as to *why* an electric current produces a magnetic force. To this day, the effects described by Oersted, Ampère, and Faraday remain as fundamental and unexplained principles of nature. Yet significant technological advances have been made, despite our lack of a real understanding of the relationship between electricity and magnetism. For example, the operation of the electric motor is based on the fundamental principle that a force is exerted on an electric current in a magnetic field. In the electric motor the force exerted on a coil of wire in a magnetic field is used to set the coil into continual rotation, thereby converting electric energy into the mechanical energy of motion.

The essentials of this action are depicted in Figure 9–20. A coil of wire, connected to a source of electric current, is placed in a magnetic field between opposite poles of a natural magnet or electromagnet. The coil is constrained to pivot about its axis. (In the diagram only a single loop of the coil is shown; however, the greater the number of loops, the greater the force on the coil as a whole.) The forces exerted by the magnetic field on the coil will be perpendicular to the wire, as shown in Figure 9–20a; these forces will produce a torque on the coil, causing it to rotate. When

*More precisely, the equation is $F = kILB \sin \theta$, where k is a constant. The MKS unit of magnetic field strength (the tesla) is defined so as to make $k = 1$ in this system of units. In CGS units (F in dynes, I in emu, L in cm, and B in gauss) k is also 1.

the plane of the coil is perpendicular to the field (Figure 9–20b), the forces will lie in the plane. Consequently, they will act as if to stretch the coil instead of to rotate it—that is, there will be no torque on the coil in this position. However, because of its inertia, the moving coil will rotate past the no-torque position. Once the coil is beyond this point, the forces will act to reverse the motion (Figure 9–20c) and return the coil to the no-torque position. But if the current direction is changed before the coil reverses its motion, the direction of the forces will reverse, and the coil will continue to rotate in the same direction. Thus in order to keep the coil in continual rotation, the current direction must be reversed at precisely the instant when the coil overshoots the no-torque position. This reversal can be performed automatically by a ***commutator*** (shown in Figure 9–20d).

The electric motor is perhaps the most widely used technological device in the modern world. Motors are essential to the operation of electric fans, air conditioners, power tools, washing machines, electric toothbrushes—in short, any device in which motion is produced from electricity.

figure 9–19 **THE MAGNETOMOTIVE FORCE**

Shown below are two methods for finding the direction of the magnetomotive force **F** on a wire of length L carrying a current **I** in a magnetic field **B.**

$$F = ILB \sin \theta$$

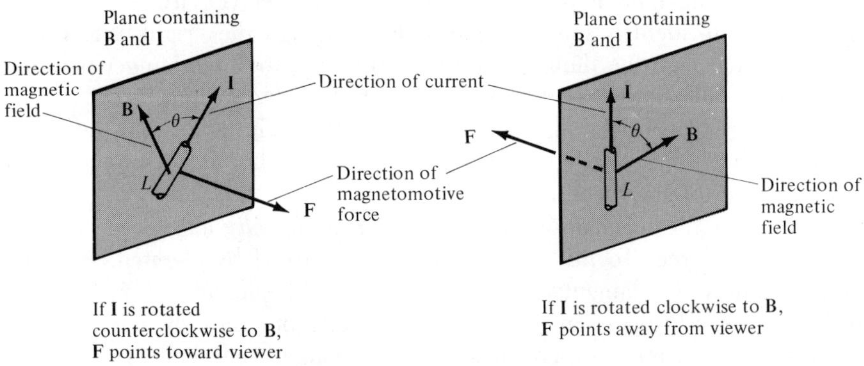

(a) The direction of **F** is always perpendicular to the plane containing **B** and **I**.

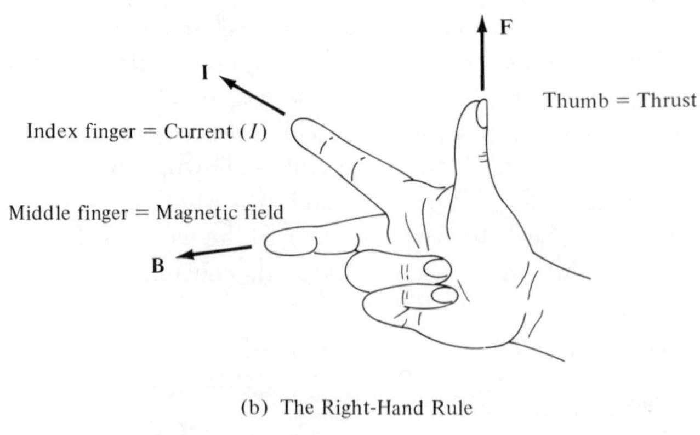

(b) The Right-Hand Rule

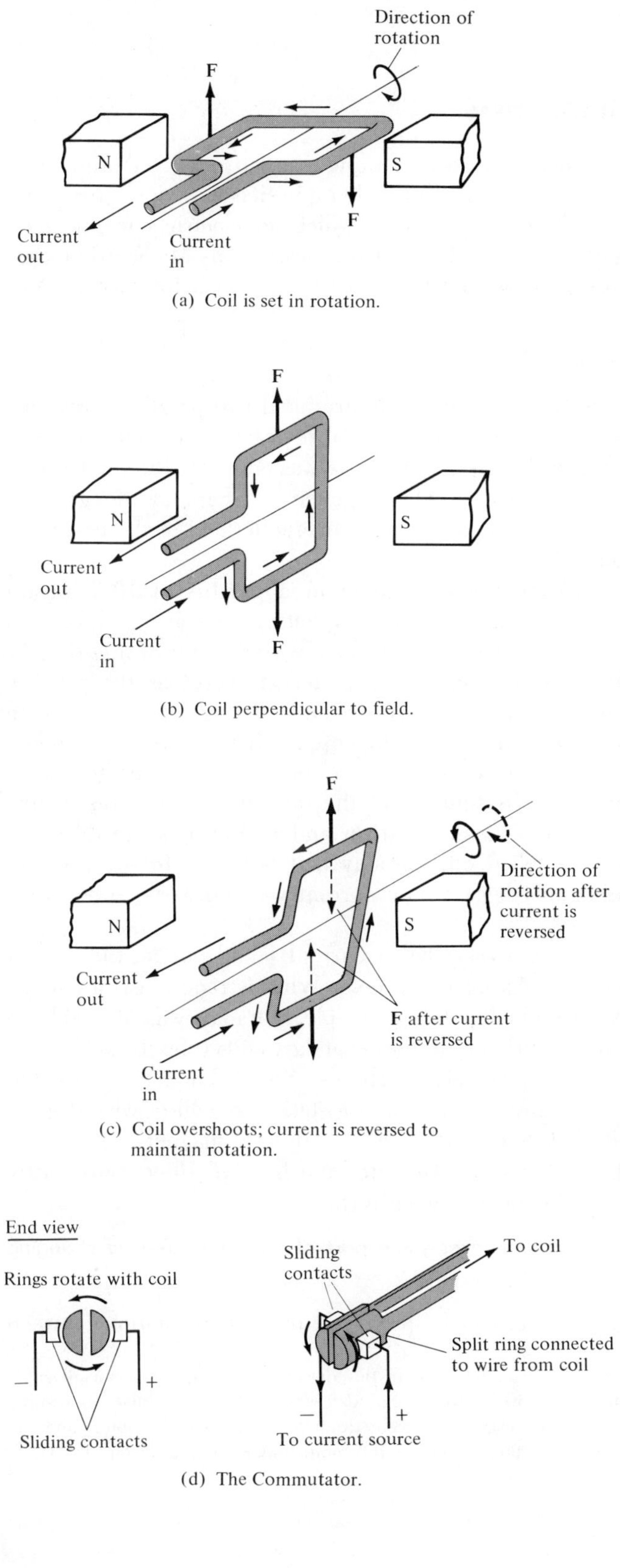

figure 9–20

PRINCIPLES OF THE ELECTRIC MOTOR

A current-carrying coil of wire placed in a magnetic field is set into continual rotation by the torques produced by the magnetomotive forces on the current.

9–4

ELECTRICITY FROM MAGNETISM

The discovery of a coupling between electricity and magnetism via the production of magnetic forces by electric currents naturally led physicists to look for other connections. In particular, was there some way in which an electric current could be induced through magnetism? At the forefront of this search was Michael Faraday, whose researches into this problem were to span a six-year period before success was attained.

ELECTROMAGNETIC INDUCTION

Faraday's first attempt was made in 1825. He arranged two parallel wires, one connected to a battery, the other to a galvanometer (a device that measures electric current), as shown in Figure 9–21a. When the first wire was connected to the battery, a current was established, and a magnetic field was generated. Faraday thought that a current would be induced in the second wire by the magnetic field of the first wire. However, he observed no such effect.

Faraday repeated this experiment with a different apparatus in 1828, again without detecting an induced current. Finally, in 1831 Faraday was successful. Instead of using two parallel wires, he wound two separate coils of wire on an iron ring (Figure 9–21b). Once again, one wire was connected to a battery to serve as the primary source of the magnetic field (which was intensified by the iron ring); the second wire was connected to a galvanometer. This time, Faraday noticed that a current did appear in the second wire, but only *momentarily*—whenever the current in the first wire was switched on or off. In addition, he found that the induced current was in one direction when the primary current was switched on and in the opposite direction when the primary current was switched off. Faraday had not seen these effects in his previous experiments, presumably because the current was too weak to be measured.*

Faraday surmised from this discovery that a current could not be induced in a circuit just by the mere presence of a magnetic field. What is required is a *change* in the magnetic field strength. Employing this principle, Faraday was also able to induce currents by moving either the primary or the secondary coils (thereby subjecting the secondary coil to a changing magnetic field). He also found that a current could be induced by moving a natural magnet in and out of a stationary coil of wire (Figure 9–21c) or by moving the coil back and forth over a stationary magnet.

Summarizing these findings, Faraday formulated the ***law of electromagnetic induction,*** which can be stated in the following form:

A current will be induced in a circuit whenever some part of the circuit is in a changing magnetic field.

The change in the magnetic field can be produced by relative motion between

*It is now generally accepted that Faraday was preceded in this discovery by the American physicist Joseph Henry, who observed induction in 1830. However, Faraday was the first to publish his results. Henry made important contributions to the design of the electromagnet, the electric motor, and the telegraph. He was also instrumental in the establishment of the Smithsonian Institution and the U. S. Weather Bureau.

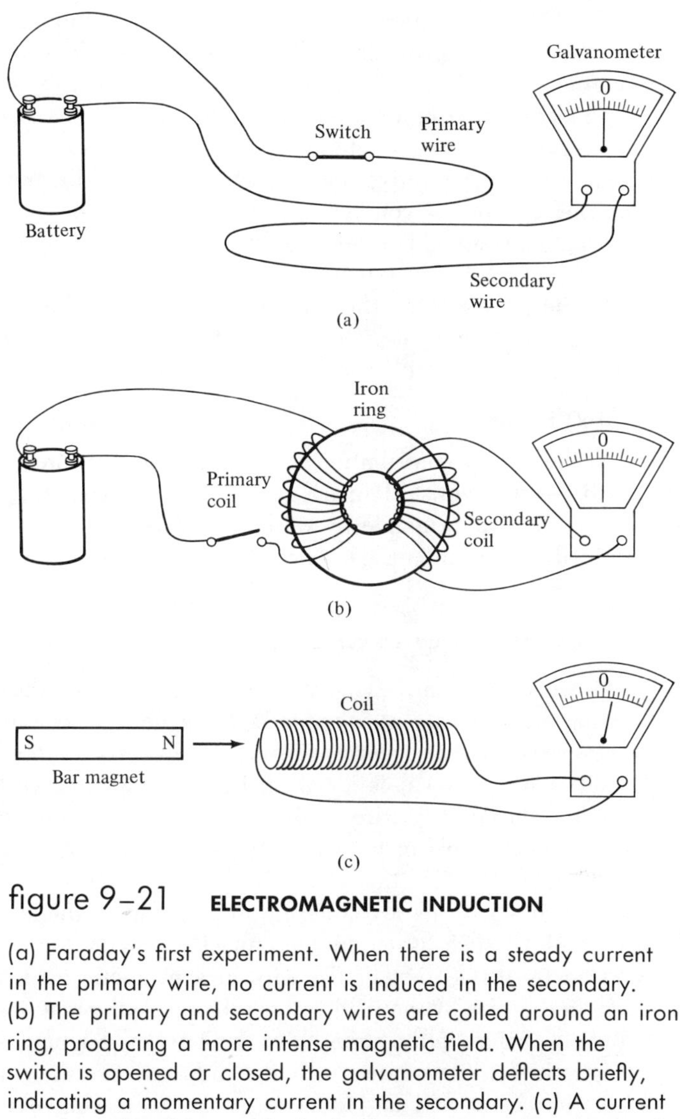

figure 9–21 **ELECTROMAGNETIC INDUCTION**

(a) Faraday's first experiment. When there is a steady current in the primary wire, no current is induced in the secondary. (b) The primary and secondary wires are coiled around an iron ring, producing a more intense magnetic field. When the switch is opened or closed, the galvanometer deflects briefly, indicating a momentary current in the secondary. (c) A current may be induced in a coil of wire by thrusting a bar magnet in and out of the coil. A current will appear whenever the magnet is in motion.

the circuit and the source of the magnetic field (either by moving the circuit through the field or by moving the field past a stationary circuit). The change can also be effected *without any motion at all* simply by increasing or decreasing the magnetic field strength—for example, by changing the current in a solenoid or electromagnet.

In any case, the induced current will appear *only* while the field is changing; when the change stops, the current stops also. In general, the magnitude of the induced current depends on the rate of change of the magnetic field with respect to the circuit.*

*The law of electromagnetic induction can be stated in precise mathematical terms; however, for the purposes of this text there is nothing to be gained by writing down the law in equation form.

The basic interaction in electromagnetic induction is between electric charges and a *changing* magnetic field. When the charges in a conductor are subjected to a changing magnetic field, a force is exerted on them that sets them in motion. If the conductor is part of a closed circuit, the moving charges will constitute a current in the circuit. If the conductor is not part of a closed circuit, the charges will still experience a force; positive and negative charges will move in opposite directions, so that one end of the conductor will have a net positive charge and the other end a net negative charge. This separation of charges will produce a potential difference across the terminals of the conductor.

At the present time there is no real understanding of why charges are set in motion by a changing magnetic field. Thus electromagnetic induction must also be regarded as an unexplained law of nature.

LENZ'S LAW

An important contribution to the study of electromagnetic induction was made by the Russian scientist Heinrich Lenz in 1834. Expanding on Faraday's work, Lenz established a clear principle by which the *direction* of the induced current could be determined in any circuit. This principle, commonly known as ***Lenz's law,*** can be stated thus:

The direction of the induced current is such as to oppose the action that induces it.

As we know from Oersted's discovery, an electric current produces a magnetic field. Thus when a current is induced, a magnetic field is induced along with it. We have also learned that a current-carrying wire in an external magnetic field (i.e., a field that is not its own) is acted on by a magnetomotive force. Therefore, when a current is induced in a wire, a force on the wire is created. The directions of the induced magnetic field of the wire and the induced magnetomotive force on the wire both depend on the direction of the current. According to Lenz's law, when a current is induced in a wire by a change in an external magnetic field, the direction of the current will be such as to produce an effect to oppose the change. If the current is induced by the motion of the wire in a magnetic field, the magnetomotive force exerted on the wire by the field will act to oppose the motion. If the current is induced simply by a change in the strength of the external magnetic field without any motion, the magnetic field associated with the induced current will tend to counteract that change; that is, it will act to decrease an increasing external field or increase a decreasing field.

Some simple illustrations of this principle can be given. Consider two parallel wires—the primary connected to a battery, the secondary connected to a galvanometer, as shown in Figure 9–22a. Initially, both wires are stationary; there will be a steady current in the primary wire but no current in the secondary, since the secondary is not in a changing magnetic field. Now suppose that the secondary wire is pushed toward the primary. We expect that a current will be induced, since the wire is now moving through a magnetic field (the field of the primary wire). Lenz's law predicts that the direction of the induced current will be *opposite* to that in the primary, so that the two wires will repel each other. (Remember from Ampère's experiments that parallel wires will repel each other if they are carrying currents in opposite directions.) The repulsive force on the secondary wire thus acts in opposition to the force that is trying to push the secondary closer to the primary. Similarly, if we tried to pull the secondary away from the primary, the induced current would be in the *same* figure 9–22 **ILLUSTRATIONS OF LENZ'S LAW**

(a) When the secondary wire is moved toward the primary, the induced current will be opposite to the direction of the primary current. The repulsive force between the wires resists the motion of the secondary toward the primary.
(b) When a bar magnet is moved toward a coil, the induced current moves in a direction so as to create an opposing magnetic field. When the magnet is moved away, the current in the coil reverses so as to generate an attractive force between the coil and the magnet.

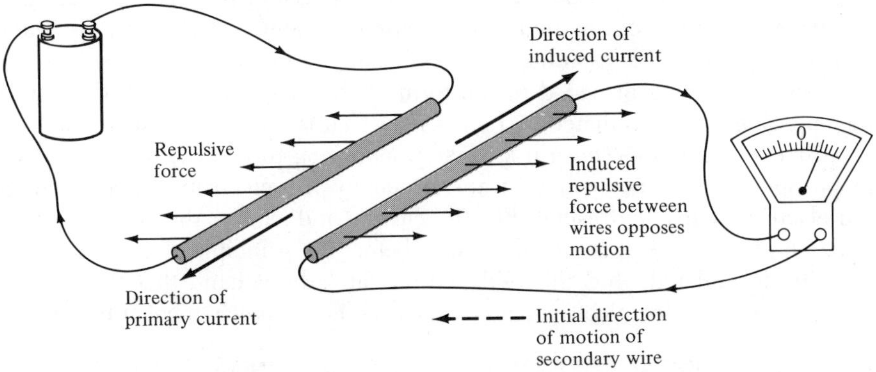

(a)

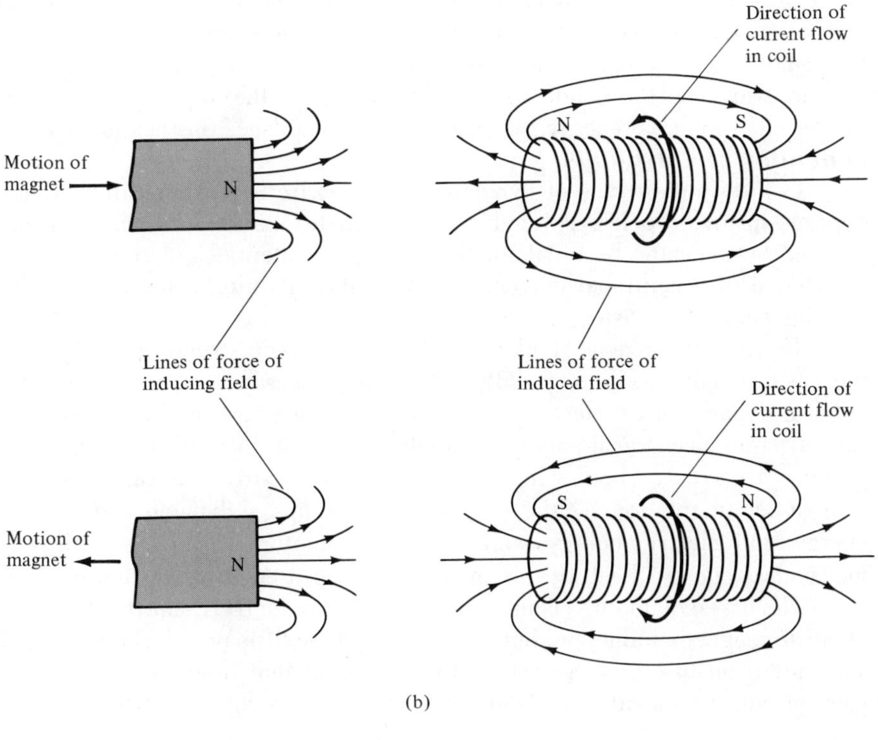

(b)

direction as in the primary, producing an *attractive* force between the wires that would then oppose the attempt to separate them.

As a second example, consider the current induced by the motion of a natural magnet toward a coil of wire. A current-carrying coil or solenoid will have a magnetic field similar to that of a bar magnet and will have its own north and south poles. Suppose the north pole of a natural magnet is pushed toward a coil of wire, inducing a current in the coil (Figure 9–22b). The direction of the induced current will be such as to make the end of the coil nearest the magnet into a north pole. The like poles of the coil and magnet will repel each other, thereby opposing the approaching motion of the magnet. When the magnet is withdrawn, the induced current reverses direction; the near end of the coil becomes a south pole. Now there is an attractive force between coil and magnet that again opposes the motion of the magnet.

The significance of Lenz's law is that it is really the principle of conservation of energy in disguise. An induced current represents energy; that is, energy must be supplied in order to establish an electric current in a circuit. We can now recognize that energy is being provided in the form of the mechanical work done to overcome the opposing forces predicted by Lenz's law. That is, one cannot simply *move* a magnet toward a coil; one must *push* it against an opposing force. The mechanical work done in pushing the magnet will be transformed by the electromagnetic interaction into the electric energy that establishes the current induced in the coil. It is notable that Lenz's law was stated some thirteen years before the principle of conservation of energy was first formulated; thus the evidence was already at hand that the conservation of energy was valid for electromagnetic as well as for mechanical interactions.

TECHNOLOGICAL APPLICATIONS OF ELECTROMAGNETIC INDUCTION

One consequence of the law of induction is that if a conductor is kept continually in motion in a magnetic field, a potential difference will be maintained across the conductor, and a current will be continually induced in a closed circuit of which the conductor is a part. The effect will be the conversion of mechanical energy (required to keep the conductor in motion against the opposing Lenz's law forces) into electric potential energy. A device that accomplishes this is known as an ***electric generator*** or a ***dynamo.***

In the simplest type of dynamo, a coil of wire is mechanically rotated between opposite magnetic poles. Through electromagnetic induction a potential difference is induced between the terminals of the coil. This potential difference can then act as the source of a continual current in a circuit containing some electric device such as a light bulb or toaster.

There is an obvious similarity between the basic structures of the dynamo and the electric motor (Figure 9–23). Essentially, the same machine could be used as either a dynamo or an electric motor. In effect, they are the inverse of one another. In a dynamo mechanical energy is supplied to rotate the coil in a magnetic field and produce a current, whereas in an electric motor a current is supplied to a coil that is then set in rotation by the magnetic field. Thus a dynamo converts mechanical energy into electric energy, and an electric motor converts electric energy into mechanical energy. Each device, in fact, can be used to drive the other.

The first dynamo was constructed by Faraday in 1831 shortly after his discovery of electromagnetic induction, but Faraday could see little practical use for his invention and moved on to other researches. To be sure, at that time there was little need for electric current outside the laboratory, and the voltages produced by the earliest dynamos were rather feeble compared to those obtainable from voltaic batteries. However, many scientists recognized the intrinsic advantage of dynamos over batteries—namely, that batteries have a limited lifetime because of the chemical breakdown that takes place, whereas dynamos can supply current indefinitely so long as mechanical energy is being supplied. In the 1850s and 1860s, as more and more uses for electric current were found, improvements were made in the design of dynamos. By this time larger dynamos were being constructed in which a steam-driven turbine was used to keep the coil in continual rotation (thereby effecting the conversion of heat energy into electric energy). By the 1870s dynamos were being used to supply current for the first electric railways and electric lighting systems.

The first commercial lighting systems employed the carbon arc lamp, in which a large electric current passing through a block of carbon caused the carbon to burn with a brilliant white light. This was suitable for the illumination of streets or public buildings, but was much too bright for use in the home. In 1878 the American inventor Thomas A. Edison set out to devise a system of electric lighting in which current could be delivered from a central power station to private homes to illuminate lamps of much less brilliance than those of the carbon arc. This system had to be designed to compete economically with the gas lighting system then in common use in homes.

If electric power is to be delivered from a central power station to private homes, a certain amount of power will be dissipated as heat in the transmission lines because of their electric resistance. Combining Eqs. 9–9 and 9–10, we find that the power dissipated in a conductor having resistance R is

$$P = VI = (IR)I = I^2R$$

That is, the power loss in a transmission line is proportional to the square of

figure 9–23 **COMPARISON OF THE ELECTRIC MOTOR AND THE DYNAMO**

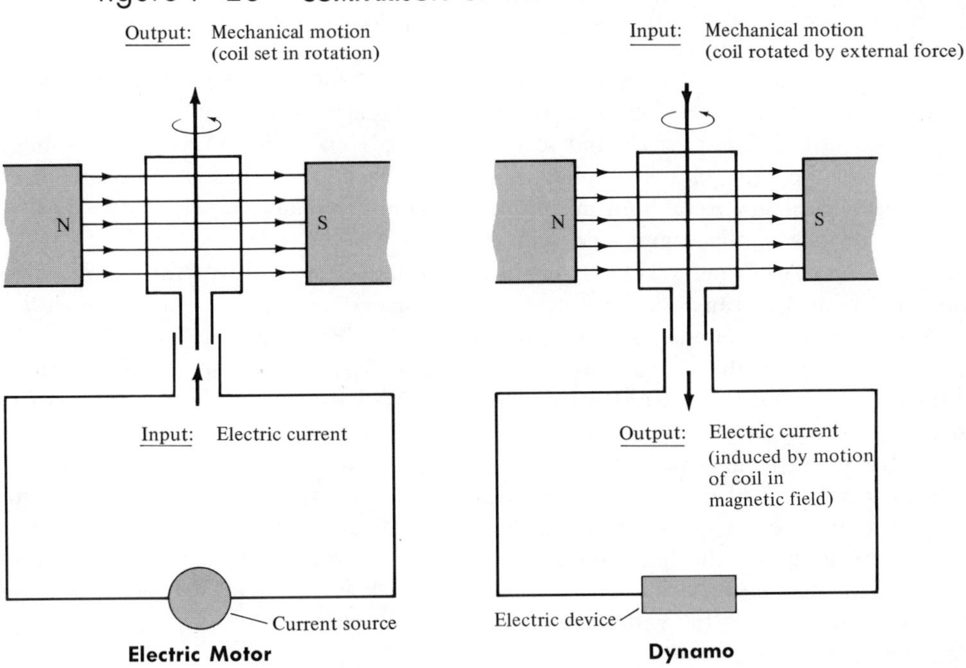

the current in the line. Edison recognized that in order for his system to be economically competitive, it would have to operate at a low current. He therefore planned a system in which a high-voltage dynamo would supply a small current to high-resistance incandescent lamps. Edison and his associates then designed a highly efficient dynamo that provided a potential difference of 110 volts, and after considerable difficulty, they also succeeded in constructing a long-life incandescent bulb suitable for home use. By 1882 the first of Edison's proposed central power stations was in operation in New York City.

A dynamo does not produce a steady current. As the coil rotates, the induced current reverses direction with each half revolution. Hence an ***alternating current*** (AC) results. However, the output of a dynamo can be converted to a ***direct current*** (DC) by means of a commutator similar to that used in the electric motor. Although the first electric power systems were designed to supply DC rather than AC, it was soon recognized that AC could be supplied more efficiently than DC, particularly over large distances.

The key to the many advantages of AC is the ***transformer.*** In its essentials the transformer is the same as the device Faraday used to discover electromagnetic induction. (See Figure 9–21b.) Two coils of wire are kept in close proximity to one another (usually by winding them on the same iron core); the primary coil is connected to some source of current, and the secondary may be connected to some electric device such as a light bulb or electric motor.

Faraday had observed that a current will be induced in the secondary only when the primary current is switched on or off—that is, when the magnetic field of the primary is changing in magnitude. Thus if a direct current is established in the primary, no current will be induced in the secondary because the secondary is in a *constant* magnetic field. On the other hand, if an alternating current is supplied to the primary coil, the coil's magnetic field will be continually changing in intensity. This continually changing magnetic field will in turn induce a continually changing current in the secondary. Hence, an alternating current in the primary will result in an alternating current in the secondary.

A transformer is useful because the voltage developed in the secondary can be made either larger or smaller than the primary voltage, depending on the number of turns of wire in the primary and secondary coils. Generally, the secondary voltage depends on the ratio of the number of turns in each coil. Thus if there are more turns in the secondary coil than in the primary, the secondary voltage will be greater than the primary voltage and vice versa.

With a transformer a wide variety of different electric devices, requiring different operating voltages, can be driven by the same primary voltage source. Each device contains a transformer that is specifically designed to "step up" or "step down" the primary voltage to the required operating voltage. Hence a TV set (some of whose components require over 1000 volts) and a toy train (requiring about 10 volts) can be plugged into the same wall outlet.

Transformers also permit the efficient transmission of electric power from a distant generating plant to homes and offices. The voltage output of the generator is first stepped up by means of a transformer to about 10,000 volts. Electric power is then transmitted at this high voltage and with a very small current. Since the current is small, the heating losses in the wires will be very low. At a local substation another transformer decreases the voltage (and increases the current) to values more suitable for common electric devices.

Electric power producers in North America conventionally supply 60-cycle alternating current* of 110–120 volts potential difference to their consumers, and manufacturers of electric devices design their products on the basis of these figures. In most other countries, however, 220–240 volt, 50-cycle AC is supplied. Hence most American-made electric appliances cannot be used in foreign countries without a specially designed "converter."

Throughout the world, electric energy is created from mechanical energy by the process of electromagnetic induction. To produce sufficient electric energy for a city, large-scale sources of mechanical energy must be available for conversion. One source that is quite suitable for this purpose is the kinetic energy contained in waterfalls or in large, rapidly moving rivers. When such sources are available near populated areas, hydroelectric power plants can be constructed in which the kinetic energy of moving water is used to keep the dynamo coil in continual rotation. However, many nations do not have adequate water resources to meet their electric power demands. They must rely on *thermal* power generation, in which the dynamo coil is turned by the force of steam. The required heat energy is obtained from the burning of coal or oil, creating two serious problems: (1) Thermal power plants must be supplied with a continual flow of fuel. However, there is a finite amount of fuel resources available in the earth, and these resources may be completely used up in a few hundred years. (2) When fuel is burned, the waste products are given off into the atmosphere, creating a serious health hazard.

As a consequence, the producers of thermal electric power are faced with the dilemma of trying to meet a mushrooming demand for more electricity with a dwindling supply of high-grade fuel, while keeping atmospheric pollution down to acceptable levels. Hydroelectric power production is not affected by these problems, but there are few places in the world where adequate water resources are located near regions of high population density. Hydroelectrically produced power must therefore be transmitted over long distances, and the power losses in transmission become prohibitive. There is little optimism at the present time with regard to alternative ways of generating electricity. Nuclear power plants (in which the heat used to produce steam is obtained from the energy released in nuclear reactions) pose radiation hazards that have not as yet been satisfactorily overcome. In any case, the cost of producing electricity seems destined to continue to spiral upward for years to come.

QUESTIONS FOR REVIEW

1 Discuss the two-fluid theory of electricity. What evidence indicated the existence of two kinds of electricity? What was the distinction between vitreous and resinous electricity?

2 How can the evidence for two kinds of electricity be explained in terms of a one-fluid theory? How did Franklin distinguish between vitreous and resinous electricity? Do we agree with his explanation today?

3 Describe how an electroscope works. Explain why the leaves of an electroscope diverge (a) if a charged body is brought in contact with the electroscope and (b) if a charged body is brought near to, but not in contact with, the electroscope.

4 What does the modern-day physicist mean by the term "electric charge"? Can we explain why there are two kinds of charge (positive and negative)?

* A "cycle" represents *two* changes in the current direction, say, from forward to backward to forward again. A 60-cycle current has 60 such cycles per second.

5 Explain how the inverse-square law of electrostatic force was discovered (a) by Priestley and (b) by Coulomb.

6 State Coulomb's law of electrostatic force. Under what conditions will the force between two charged bodies be attractive or repulsive?

7 Identify the CGS and MKS units of electric charge, and explain how they are defined.

8 Describe Galvani's "frog-leg" discovery. Discuss the conflicting theories that Galvani and Volta proposed to explain this effect. What was the historical importance of the discovery?

9 Explain how chemical compounds are decomposed by the process of electrolysis. How did the use of electrolysis help scientists to understand the nature of chemical forces?

10 Explain how a battery produces electric current. What effects eventually cause a battery to weaken and fail altogether?

11 Give the equation for the electric potential energy of two charges. What are the units of electric potential energy? Is it a vector or a scalar quantity? Under what conditions will it be positive or negative?

12 Define and explain (a) electric potential and (b) voltage. In what units are these quantities measured? How are they related to electric energy?

13 State and explain Ohm's law.

14 Give a physical explanation of electric resistance. Explain why the resistance of a metallic conductor increases with temperature.

15 Give the equation for electric power. What are the units of this quantity?

16 Describe Oersted's experiment. Summarize his results and discuss their significance.

17 Describe the force between two parallel current-carrying wires. Discuss the law formulated by Ampère to represent this force. Why did Ampère choose to describe the force in this particular way? How did Faraday describe the same phenomenon?

18 What is a line of force? What is the purpose of drawing the lines of force for a particular arrangement of charges, currents, and so on?

19 What is meant by a field? What distinguishes an electric field from a magnetic or gravitational field? Give definitions for the electric field intensity **E** and the magnetic field strength **B.** In what units are they measured?

20 Describe the magnetic field of (a) a long, straight, current-carrying wire and (b) a circular loop of current.

21 What is a solenoid? What are the advantages of a solenoid over a natural magnet as a source of magnetic force?

22 Describe the magnetomotive force acting on a current-carrying wire in a magnetic field. Explain how the magnitude and direction of this force is determined.

23 State and explain the law of electromagnetic induction. Discuss the various conditions under which an electric current can be induced in a circuit.

24 State and explain Lenz's law, giving examples of its application. How is it related to the principle of conservation of energy?

25 Explain in terms of fundamental principles how each of the following devices works: (a) an electric motor, (b) a dynamo (electric generator), and (c) a transformer.

EXERCISES

1 Explain why each of the following occurs.
(a) A person gets an electric shock when he touches a doorknob after walking on a rug.
(b) A comb sometimes "crackles" when it is being used.

(c) A person's hair is sometimes "fuzzy" after it has been dried with a towel.

2 Explain how an electroscope could be used to determine (a) the amount of charge on a body and (b) whether the charge on a body is positive or negative.

3 Determine the magnitude and direction of the electrostatic force (a) between a positive 5-esu charge and a negative 8-esu charge placed 4 cm apart and (b) between two positive charges of 10 microcoulombs and 20 microcoulombs placed 3 meters apart.

4 An electron has a negative charge of 4.8×10^{-10} esu and a mass of 9.11×10^{-28} gram. Compare the electrostatic repulsion with the gravitational force of attraction for two electrons 10^{-8} cm apart.

5 Explain what happens when water undergoes electrolysis. What substances will be formed at each electrode?

6 Explain whether the electric potential energy of a system of two charges increases or decreases when (a) the distance between two positive charges is increased, (b) the distance between two negative charges is decreased, and (c) the distance between two opposite charges is increased.

7 Let A be a fixed, positive 8-esu charge and B a movable, negative 5-esu charge.
(a) Calculate the electric potential energy of the system when B is 5 cm, 10 cm, and 20 cm away from A.
(b) Calculate the change in electric potential energy when B is moved from 10 cm to 5 cm from A and when B is moved from 10 cm to 20 cm from A.
(c) B is initially at rest at a distance of 20 cm from A. B is released and allowed to move freely. What will its speed be when it is 5 cm from A? The mass of B is 3 grams.

8 Two positive charges of 10 microcoulombs each are 4 meters apart. How much work must be done in order to decrease this distance to 2 meters?

9 Show that the electric potential at a point a distance d away from a fixed charge Q is given by the equation $U = K_e Q/d$. (Hint: Combine Eqs. 9–4 and 9–6.)

10 Using the equation derived in Exercise 9,
(a) Calculate U at distances of 5 m and 20 m from a positive 0.1-microcoulomb charge.
(b) What is the voltage difference between these points?
(c) How much work is done on a negative 10-microcoulomb charge in moving it from 5 m to 20 m from the 0.1-microcoulomb charge?

11 Explain how the output of a 6-volt dry cell differs from that of a 12-volt automobile battery.

12 A broiler has a power output of 1200 watts. What is the current in the broiler coil if it is connected to a 120-volt source? What is the resistance of the broiler coil?

13 A wire with a resistance of 30 ohms is connected to the terminals of a 120-volt battery. How many watts of power will be delivered to the wire?

14 An electric kiln draws a current of 30 amp when connected to a 220-volt source.
(a) What is the power output of the kiln in watts?
(b) If this energy is completely converted to heat, how many calories/sec will be generated?
(c) If electricity costs 5¢/kwhr (5 cents per kilowatt per hour), how much will it cost to run the kiln for 4 hours?

15 How does an electric fuse work, and what is its purpose? What happens to a 10-amp fuse when a current of more than 10 amp passes through it? How do you explain this effect?

16 If a magnet is oriented at right angles to a straight current-carrying wire, the magnet will not rotate. Explain why this is so.

17 In the text electric current is defined in terms of a flow of charge (Eq. 9–8) and also in terms of the force between parallel wires (Eq. 9–12). Give two definitions of the ampere based on these equations. Which of these two is the better definition? (Consider the practical aspects of measurement as well as the theoretical.)

18 Two long parallel wires, 10 cm apart, carry currents of 5 amp and 10 amp in opposite directions. Calculate the magnitude and direction of the force exerted on 1 meter of each wire.

19 A current of 2 amp flows in a rectangular loop that lies in the plane of a constant magnetic field of 10 tesla (Figure 9–24). Calculate the magnitude and direction of the force on each of the four sides of the rectangle.

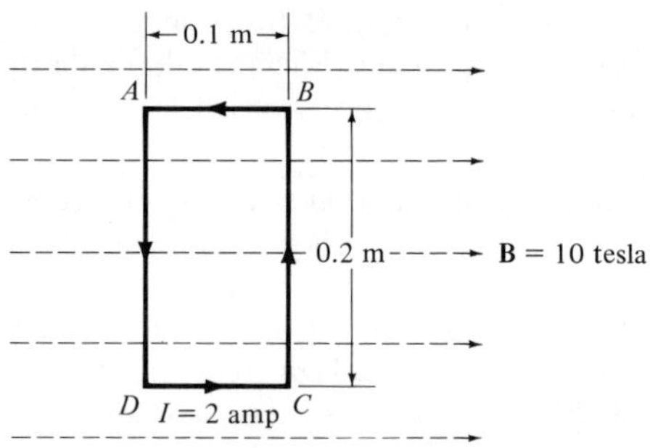

figure 9–24

20 Two coils of wire are placed in close proximity to one another. The primary coil has been connected to a battery that provides a constant current. The secondary coil is connected to a galvanometer, which measures the induced current in the coil.
(a) Explain why there is no current in the secondary.
(b) Explain why a current will be induced if (i) the secondary coil is moved away, (ii) an iron bar is thrust into the primary coil, and (iii) the connection to the battery is broken.

21 Two parallel wires carry equal currents in the same direction. An attempt is made to pull the two wires away from each other, increasing the distance between them.
(a) Using the law of electromagnetic induction, explain why this should cause both currents to change.
(b) Using Lenz's law, explain why this should cause both currents to increase rather than decrease.

THE ELECTROMAGNETIC FIELD THEORY

10

One cannot escape the feeling that these mathematical formulae have an independent existence and an intelligence of their own, that they are wiser than we are, wiser even than their discoverers, that we get more out of them than was originally put into them.

Heinrich Hertz

IN THE PREVIOUS CHAPTER we concentrated our attention on the fundamental electromagnetic phenomena. The aim was to identify the basic laws of electricity and magnetism and to describe their technological applications. Now we shall proceed to the final step of the scientific method: namely, the construction of a theoretical model encompassing these basic laws. The value of such a theory, as has been emphasized, is in the deeper understanding it gives us of the basic phenomena and its ability to predict the existence of undiscovered phenomena.

In this chapter we shall learn how the discoveries of Oersted, Ampère, Faraday, and others were brought together by James Clerk Maxwell's electromagnetic field theory. In terms of its comprehensiveness and predictive power, Maxwell's theory is on a par with Newtonian mechanics and Einstein's theory of relativity, and is justly considered one of the great masterpieces of physics.

10–1

EARLY ELECTROMAGNETIC THEORIES

Before we examine Maxwell's theory, let us first consider some of its important antecedents. One of the first attempts to provide a theoretical understanding of the basic electromagnetic phenomena was made by Ampère in the 1820s. Working from the results of his own and Oersted's experiments, Ampère surmised that if a magnetic force could be produced by a current-carrying wire, perhaps *all* magnetic forces could be attributed to the effects of electric currents.

Ampère proposed that the fundamental elements of magnetism are microscopic current loops. He noted that a single current loop—as the basic unit of a solenoid—has properties similar to those of a bar magnet. For example, a current loop orients itself in a magnetic field in the same way that a bar magnet does and in general acts as if it had its own north and south poles. (In terms of Faraday's representation, a current loop resembles a bar magnet in that both have approximately the same pattern of force lines—see Figure 9–17a.) From these observations Ampère constructed a theory of magnetism in which the magnetic properties of matter are attributed to microscopic current loops—presumably existing as circular currents within the atom.

It is assumed that in normal nonmagnetic matter the microscopic current loops are randomly oriented, so that such matter exhibits no *net* magnetic properties. In natural magnets the current loops tend to be lined up in one particular direction, so that the magnetic fields of the individual current loops add together. Certain substances are not naturally magnetic but are easily magnetized—a phenomenon that can be explained in the following way: normally the current loops in such substances are randomly oriented; but under the influence of an external magnetic field, the current loops become aligned, and the substance is found to exhibit a net magnetism. Experiments have also shown that the heating of a magnet causes it to lose its magnetism; according to the kinetic theory heating increases the random motion of the particles of the magnet, and sufficient heating can cause the current loops to become so randomly oriented that no net magnetism can remain.

Ampère's theory proposed that the ultimate source of magnetism resides in current loops rather than in individual north and south magnetic poles. This accounts for one of the fundamental differences between electricity and magnetism. Whereas is is possible to isolate single positive or negative charges, separated north or south magnetic poles (known as ***magnetic monopoles***) have never been observed in nature. North magnetic poles are always found attached to south poles. If a bar magnet is cut in half, each half becomes a new bar magnet with its own north and south poles. If the current loop is the basic unit of magnetism, there should be no way to produce an isolated north or south pole. However, if the existence of magnetic monopoles could be demonstrated experimentally, it would represent strong evidence against Ampère's theory. From time to time experiments designed to observe magnetic monopoles are performed, but to date none have been successful.

In general, we can state that there are no magnetic phenomena that cannot be accounted for in terms of electric currents. Thus Ampère's theory provided an important link in connecting electricity and magnetism, for it presents magnetism as a special class of phenomena related to the effects of electric currents.

A theory of electromagnetism that was given considerable importance in the middle of the nineteenth century was proposed by the German physicist Wilhelm Weber (1804–1890). Weber's theory was based on the assumption that electric currents consist of moving, discrete charges—in contrast to the more prevalent belief held at that time that electric currents represent a continuous flow of electric fluid. Weber derived an equation to describe the force between two moving charges. From this equation he was able to show that the force between two current-carrying wires and the law of electromagnetic induction could both be explained in terms of an interaction between moving charges. As an adherent of Ampère's theory of magnetism, Weber believed that all magnetic effects were due to current loops, which in turn consisted of charges moving in microscopic circular paths. Hence all magnetic effects could be attributed in principle to moving charges; for example, the force exerted by a current-carrying wire on a magnet could be explained as an interaction between the moving charges in the wire and the moving charges in the magnet.

According to Weber's theory, Coulomb's law of electrostatic force (Eq. 9–1) is valid only for *stationary* charges. When charges are in motion, there are additional forces that depend on the velocities and accelerations of the charges. In this manner Weber attempted to account for all electromagnetic interactions in terms of direct, action-at-a-distance forces between charges. This represented an entirely different philosophy from that of Faraday and Maxwell, who adopted the view that particles interact directly with electric and magnetic fields, and not with each other. In the end the field approach proved to be much more fruitful, and Weber's theory was discarded in favor of Maxwell's. Nevertheless, a few of Weber's ideas remain—most notably, the concept of an electric current as a stream of moving charges. We shall return to this matter later in the chapter.

10–2

MAXWELL'S FIELD THEORY

We come now to the electromagnetic field theory of James Clerk Maxwell. This monumental work is significant not only for the advances it stimulated in our understanding of electricity, magnetism, and light, but also for its philosophical importance. The theory led to the first significant divergence from Newtonian mechanics and provided an entirely new way to view the physical world.

JAMES CLERK MAXWELL

The author of the electromagnetic field theory, James Clerk Maxwell, is generally considered one of the greatest physicists of history—in a class with Galileo, Newton, and Einstein. His work on the kinetic theory of gases was discussed in Chapter 7; this contribution alone would have been sufficient to ensure his greatness. His researches into the theory of color vision are also of considerable importance.

Born in 1831 into a well-to-do Scottish family, Maxwell was educated at the University of Edinburgh and then at Cambridge. As a youngster Maxwell was given to solitude and was rather shy and uncomfortable with people. But the company of his fellow students at Cambridge brought Maxwell out of his shell; here he formed numerous friendships, which were to last through his lifetime. One of these was with William Thomson (later Lord Kelvin), with whom Maxwell carried on an extensive correspondence.

After receiving his degree in 1854, Maxwell became a professor of physics, first at the University of Aberdeen and then at King's College in London. In 1865 he retired from teaching to devote more time to his scientific research and to his favorite pastimes—the study of English literature and the writing of poetry. In 1879, while Maxwell was still actively engaged in his research, his health began to fail, and he died that year at the age of forty-eight.

Maxwell's first paper on electromagnetism, *On Faraday's Lines of Force,* was published in 1855. His electromagnetic theory gradually developed over a period of many years, culminating in the publication in 1873 of the *Treatise on Electricity and Magnetism.* This great and difficult book—beyond the understanding of many of Maxwell's contemporaries—has come to be regarded as one of the classics of physics, worthy of comparison with the *Principia.*

THE NATURE OF ELECTROMAGNETIC FIELDS

The starting point of Maxwell's theory was the problem of action at a distance. Along with many scientists since the time of Newton, Maxwell could not accept the idea of a force being transmitted through space without any medium. Hence Maxwell proposed his field theory as an alternative to Weber's theory of electromagnetic forces. A basic assumption of Maxwell's theory was that all electromagnetic interactions between bodies are transmitted by actions in the intervening medium. Maxwell described this principle in the introduction to one of his earlier papers:

> The mechanical difficulties . . . involved in the assumption of particles acting at a distance with forces which depend on their velocities are such as to prevent me from considering this theory as an ultimate one. . . . I have therefore preferred to seek an explanation of the fact in another direction, by supposing them to be produced by actions which go on in the surrounding medium as well as in the excited bodies, and endeavoring to explain the action between distant bodies without assuming the existence of forces capable of acting directly at sensible distances.
>
> The theory I propose may therefore be called a theory of the *electromagnetic field,* because it has to do with the space in the neighborhood of the electric or magnetic bodies, and it may be called a *dynamical* theory, because it assumes that in space there is matter in motion, by which the observed electromagnetic phenomena are produced. The electromagnetic field is that part of space which contains and surrounds bodies in electric or magnetic conditions.[1]

"It assumes that in space there is matter in motion"—thus a central feature of Maxwell's theory is *the existence of the ether as the medium through which electromagnetic forces are transmitted.* In Maxwell's time the existence of the ether was generally accepted, and it was thought to be the medium through which light waves propagated. Maxwell surmised that the same medium might also transmit electromagnetic interactions.

This idea was suggested to Maxwell by Faraday's work, which Maxwell had studied thoroughly. Faraday had proposed that the presence of electric charges or currents caused the molecules of the surrounding medium to be displaced from their equilibrium positions, thereby setting up "strains" in the medium. These strains, according to Faraday, coincided with the lines of force. Using Faraday's theory as a starting point, Maxwell first found a suitable mathematical representation for the lines of force—something Faraday had never been able to do. Maxwell then proceeded to derive equations by which the electric and magnetic field strengths (the quantities **E** and **B** described in Section 9–3) could be calculated at any point in the medium for a given arrangement of electric charges or currents. In Maxwell's theory the quantities **E** and **B** were supposed to represent the state of electric or magnetic strain of the medium. Thus these quantities were thought to describe specific properties of the ether.

In its original form Maxwell's theory was a completely mechanical theory; that is, Maxwell conceived of the electromagnetic force as being transmitted through the medium by the action of one particle on the next. He worked out an elaborate theory in which he proposed that small rotary motions or ***vortices*** were established in the ether by the presence of electric or magnetic sources, and that the action of one vortex on the next effected the transfer of electromagnetic energy from one place to another. Maxwell continued to speculate on the nature of the ether throughout his life; the ether remained, in his mind, an integral part of his theory.

MAXWELL'S EQUATIONS

The essence of Maxwell's theory is Maxwell's equations. These highly complex equations—which can be written down in a deceptively simple form—describe the variation of the electric and magnetic field strengths through space and time for a given arrangement of media, charges, currents, and so on. Maxwell's equations do not, in the main, represent Maxwell's own discoveries; rather they are the basic laws of electromagnetism (the discoveries of Coulomb, Oersted, Ampère, Faraday, and Lenz) transformed into a suitable mathematical form. For free space (using MKS units), Maxwell's equations can be written as follows:

$$\nabla \cdot \mathbf{E} = \rho/\epsilon_0 \tag{10–1}$$

$$\nabla \cdot \mathbf{B} = 0 \tag{10–2}$$

$$\nabla \times \mathbf{E} = -\frac{\partial \mathbf{B}}{\partial t} \tag{10–3}$$

$$\nabla \times \mathbf{B} = \mu_0 \mathbf{j} + \mu_0 \epsilon_0 \frac{\partial \mathbf{E}}{\partial t} \tag{10–4}$$

These equations have been written here not because the student is expected to understand or be able to use them; they are presented merely for reference purposes and also to show how concisely the basic equations of electromagnetism can be expressed. In the equations the symbols $\nabla \cdot$ and $\nabla \times$ are shorthand for rather complicated vector operations involving first and second derivatives of the vector components of **E** and **B.** The quantities ϵ_0 (called the ***dielectric constant***) and μ_0 (the ***magnetic permeability***) are related to the electric and magnetic proportionality constants K_e and K_m by the equations

$$\epsilon_0 = \frac{1}{4\pi K_e} \qquad \text{and} \qquad \mu_0 = 4\pi K_m \tag{10–5}$$

The quantity **j** in Eq. 10–4 is the ***current density,*** defined as the current flow per unit area (I/A) measured in amp/m^2.

Equations 10–1 and 10–2 are equivalent to Coulomb's laws of electrostatic force (Eq. 9–1) and magnetostatic force (Eq. 9–3). In Eq. 10–1 ρ stands for the density of stationary electric charge. The zero on the right-hand side of Eq. 10–2 expresses the fact that there are no isolated magnetic poles. (Should magnetic monopoles ever be discovered, this equation could be modified rather easily.)

Equation 10–3 is Faraday's law of electromagnetic induction. The term on the right-hand side is the mathematical symbol for the rate of change with time of the magnetic field. The left-hand side gives the mathematical representation for the change in the electric field, which can be observed in the form of an induced current or potential difference. The minus sign in the equation is a consequence of Lenz's law.

Equation 10–4 is derived from Oersted's and Ampère's discoveries and includes a significant contribution made by Maxwell himself. When Ampère's law (Eq. 9–11), as originally formulated, was placed in the appropriate mathematical form, only the first term ($\mu_0 \mathbf{j}$) appeared on the right-hand side of Eq. 10–4. In this form the equation defines the connection between electric currents and their associated magnetic fields. Maxwell discovered, however, that Ampère's law is not valid when the electric field is not constant. He found through mathematical analysis that a contradiction arises between Ampère's law and the law of conservation of charge when changing electric fields are taken into consideration. To remove this contradiction, Maxwell proposed that when an electric field is changing, the molecules of the medium readjust their positions in response to this change, and this motion of the molecules (actually, of the charges within the molecules) constitutes an electric current. Thus by Maxwell's principle a changing electric field causes an electric current, which in turn generates a magnetic field. Maxwell modified Eq. 10–4 by adding the term $\mu_0 \epsilon_0 (\partial \mathbf{E}/\partial t)$ to take this effect into account.

The effect proposed by Maxwell is quite difficult to demonstrate experimentally;

nevertheless its inclusion is clearly justified by the mathematical requirements of the electromagnetic theory. Most notably, it establishes a not unwelcome symmetry between electric and magnetic fields. According to Faraday's law, a changing magnetic field induces an electric field; according to Maxwell's principle, a changing electric field induces a magnetic field.

ELECTROMAGNETIC WAVES

The last two of Maxwell's equations are concerned with the effects of changes in electric and magnetic fields. The question that now presents itself is: How rapidly do such changes make themselves felt? The problem can be illustrated in the following way.

Two charged bodies, A and B, are located some distance apart, as shown in Figure 10–1. The two bodies are stationary, so that the mutual force is strictly an electrostatic one. Each body lies in the constant electric field of the other; there are no magnetic fields. When body A is suddenly moved to a new position, the direction of the electrostatic force on B must change. Moreover, the motion of body A constitutes a current, which produces a magnetic field. Because the current is not steady, the magnetic field will not be constant. According to Faraday's law, a changing magnetic field generates an electric field. Hence the motion of body A should cause body B to experience both a changing magnetic field and a changing electric field. Does body B experience these effects instantaneously, or is there some time delay before the effects of the motion of A are felt by B?

On this question the action-at-a-distance theory and the field theory part company. According to Weber's theory, body B should experience the above-mentioned effects immediately, for there is nothing in the theory that leads us to expect a time delay. On the other hand, the field theory gives quite a different answer. According to Maxwell's model, electromagnetic effects are transmitted through the ether by mechanical action. Therefore the effects of A's motion *cannot* be felt instantaneously by B because they must propagate through the intervening medium. Thus until the disturbance caused by the motion of A arrives at B, the force on B acts *just as if A were still at its initial position.* During this time, then, the electrostatic forces on A and B *do not* act along the line joining them.

Maxwell was able to demonstrate his assertion with mathematical rigor. Starting

figure 10–1 **ACTION–AT–A–DISTANCE VS. THE FIELD THEORY**

Two bodies, A and B, have equal positive charges q. As A is moved from 1 to 2, the direction of the electrostatic force on B changes from $\mathbf{F}_1$ to $\mathbf{F}_2$. Changing magnetic and electric fields are also generated by the motion of A. Are these changes felt instantaneously by B?

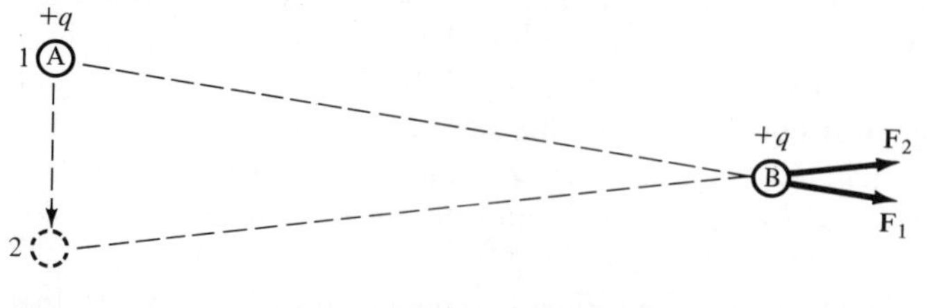

from the basic equations, he was able to prove that a disturbance in an electric or magnetic field will travel through a medium with exactly the properties of a wave. Maxwell believed that the propagation took place in an entirely mechanical way; that is, electric or magnetic disturbances pass through the medium from particle to particle in the same way a wave pulse travels along a rope or a sound wave travels through the air. Since an ordinary mechanical wave involves an interchange between kinetic and potential energy in the medium, the ether must therefore be capable of storing and transmitting electromagnetic energy.

Maxwell's equations specify exactly the nature of an electric or magnetic wave disturbance. Maxwell's equations predict that a changing magnetic field will induce an electric field, and a changing electric field will induce a magnetic field. Thus a changing magnetic field creates a changing electric field, which creates a changing magnetic field, which creates a changing electric field, and so on. It follows that the magnetic and electric disturbances propagate together; in other words, *a magnetic wave must always be accompanied by an electric wave and vice versa.* Hence the phenomenon is more properly described as an ***electromagnetic wave.***

If the charged body depicted in Figure 10–1 is moved back and forth, a train of electromagnetic waves will be generated with a frequency equal to the frequency of oscillation of the charged body. It can be shown that the electric and magnetic fields change in a direction that is perpendicular to the direction of propagation of the wave; therefore *an electromagnetic wave is a transverse wave.* Moreover, the electric and magnetic field vectors are perpendicular to, and in phase with, one another. (See Figure 10–2.)

In general, any accelerating charge will generate electromagnetic waves. A charge moving with constant velocity is equivalent to a steady current and creates a constant magnetic field. Hence a steady current does not generate electromagnetic waves. However, any *changes* in the current—for example, when the current is turned on or off—will result in the generation of electromagnetic waves.

Thus Maxwell's equations predict that electric and magnetic disturbances will propagate as electromagnetic waves with a finite speed. The equations can also be

figure 10–2 **AN ELECTROMAGNETIC WAVE**

An oscillating charge will generate a train of electromagnetic waves. The magnetic and electric field vectors are always perpendicular to one another and to the direction of propagation of the wave. Maxwell regarded an electromagnetic wave as a disturbance propagating through the ether.

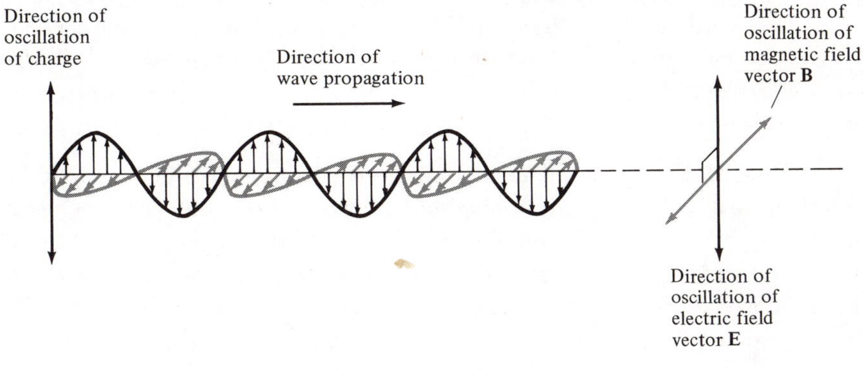

used to determine the numerical value of the speed of an electromagnetic wave. Maxwell showed that the speed in the ether can be calculated from the equation

$$v = \sqrt{\frac{K_e}{K_m}} \tag{10–6}$$

where K_e and K_m are the constants found in Coulomb's electrostatic force law (Eq. 9–1) and Ampère's law (Eq. 9–11), respectively. It will be recalled that in the CGS system of units (in use in Maxwell's time), $K_e = 1$ dyne-cm^2/esu^2 and $K_m = 1$ dyne/emu^2, where the esu and the emu are the CGS units of charge and current, respectively. Substituting these values into Eq. 10–6 gives

$$v = \sqrt{\frac{K_e}{K_m}} = \sqrt{\frac{1 \text{ dyne-cm}^2/\text{esu}^2}{1 \text{ dyne/emu}^2}} = \frac{1 \text{ cm-emu}}{\text{esu}}$$

In order to convert this value of the speed of electromagnetic waves into meaningful units (such as cm/sec), the relationship between the emu and esu must be determined.

Since current measures the amount of charge moving per unit time ($I = q/t$), the problem becomes one of determining how many esu of charge per second constitute 1 emu of current. That is, 1 emu = x esu/sec, where x is the quantity to be determined. Fortunately for Maxwell, the value of x had been determined by Wilhelm Weber and Rudolph Kohlrausch in 1856. Their method was to charge a condenser with a known quantity of charge, then measure the current that appears when the condenser is discharged. They reported a value of

$$1 \text{ emu} = 3.1 \times 10^{10} \text{ esu/sec}$$

The speed of propagation of electromagnetic waves through the ether could then be calculated by Maxwell:

$$v = \frac{1 \text{ cm-emu}}{\text{esu}} = \frac{1 \text{ cm}(3.1 \times 10^{10} \text{ esu/sec})}{\text{esu}}$$
$$= 3.1 \times 10^{10} \text{ cm/sec}$$

Maxwell noted the similarity between this speed and the values of the speed of light reported by Fizeau and Foucault a few years before. With brilliant intuition, Maxwell concluded:

> The velocity is so nearly that of light that it seems we have strong reason to conclude that light itself (including radiant heat, and other radiations, if any) is an electromagnetic disturbance in the form of waves propagated through the electromagnetic field.[2]

Thus Maxwell was led to the most remarkable consequence of this theory: the assertion that *light is an electromagnetic phenomenon.* Another of the great syntheses of physics was thereby accomplished: light, electricity, and magnetism were found to be connected phenomena, and a unified theory had been formulated to describe their interrelations. The seemingly unrelated and independent researches of Young and Fresnel, on one hand, and of Oersted, Ampère, and Faraday, on the other, had all been brought to a focus by the mathematical genius of Maxwell.

A BRIEF DIGRESSION ON UNITS

At this point let us stop for a moment to consider the relationships between the CGS and MKS electrical units given in Chapter 9. In free space the speed of an electromagnetic wave is equal to c, the speed of light. Thus we can write Eq. 10–6 as

$$c = \sqrt{\frac{K_e}{K_m}}$$

In the MKS system K_m *has been arbitrarily assigned a numerical value of* 10^{-7}. Solving the above equation for K_e gives*

$$K_e = K_m c^2 = 10^{-7} c^2 \approx 9 \times 10^9 \text{ N-m}^2/\text{coulomb}^2$$

Setting $K_m = 10^{-7}$ in Ampère's law (Eq. 9–11) establishes the MKS current unit, the ampere, as equal to $\frac{1}{10}$ emu. Since 1 emu of current $= 3 \times 10^{10}$ esu/sec $= c$ esu/sec,

$$1 \text{ amp} = \frac{1}{10} \text{ emu} = \frac{c}{10} \text{ esu/sec}$$

But

$$1 \text{ amp} = 1 \text{ coulomb/sec} = \frac{c}{10} \text{ esu/sec}$$

Therefore

$$1 \text{ coulomb} = \frac{c}{10} \text{ esu} \approx 3 \times 10^9 \text{ esu}$$

Thus the relationship between the electric and magnetic constants and the speed of light is seen to be the reason for the mysterious appearance of c in the MKS form of Coulomb's law. (See page 298.)

10–3

EXPERIMENTAL EVIDENCE FOR MAXWELL'S THEORY

The test of any theory, as we know, is whether or not its predictions can be verified. In this respect, acceptance of Maxwell's theory was slow in coming because of a lack of hard experimental evidence in favor of the theory. Maxwell's most important predictions dealt with the electromagnetic nature of light and the existence of electromagnetic waves. Clearly, the validity of Maxwell's theory would stand or fall on the verification of these predictions.

Experimental evidence for the electromagnetic nature of light was not especially abundant. An important discovery in this area had been made by Faraday in 1845. He found that when a beam of polarized light passes through a piece of ordinary glass, the direction of polarization changes when the glass is placed in a magnetic field. Maxwell was able to give a full theoretical explanation of this effect. However, this phenomenon showed only that the passage of light through a medium could be affected by a magnetic field; it did not clearly demonstrate the electromagnetic nature of light itself.

Many of the effects that Maxwell predicted were extremely subtle and difficult

*To calculate K_e in MKS units, we substitute $c = 3 \times 10^8$ m/sec.

to verify. One such prediction had to do with the energy and momentum that an electromagnetic wave transfers to any object with which it interacts. Like any other wave disturbance, an electromagnetic wave carries energy and momentum. Maxwell was able to show that the ***electromagnetic momentum*** p of a wave is given by the equation

$$p = \frac{E}{c} \tag{10–7}$$

where E is the energy carried by the wave, and c is the speed of light.

When an electromagnetic wave strikes the surface of a conducting body, it should transfer energy and momentum to the body. Because of the transfer of momentum, the body should experience a force that can set it in motion. In other words, a body should "recoil" when struck by light. This effect is commonly known as ***light pressure*** or ***radiation pressure.*** However, the radiation pressure of an electromagnetic wave is extremely tiny. For example, the pressure exerted by the sunlight received at the earth's surface amounts to only about 7×10^{-10} lb/in.^{2}. Because of the experimental difficulties involved in unequivocally detecting such a small effect, confirmation of the existence of radiation pressure did not come until 1901, more than thirty years after its existence had been proposed.

HERTZ'S EXPERIMENTS

Clearly, the evidence was not sufficient to substantiate Maxwell's theory fully. There was some experimental support for the concept of light as an electromagnetic phenomenon, but in general the existence of electromagnetic waves had not been demonstrated. In theoretical terms light was thought of as an electromagnetic wave having a particular range of frequencies to which the human eye responds. Young's experiments had shown that this frequency range is roughly from 4×10^{14} to 7×10^{14} waves/sec. Thus electromagnetic waves of higher or lower frequencies could theoretically exist, but they would not stimulate vision. Was it experimentally possible to generate and detect such waves?

The researches of the German physicist Heinrich Hertz provided an affirmative answer to this question. In 1886 Hertz devised a means of generating and detecting electromagnetic waves of known frequency. Hertz's transmitter was an electric circuit with a small gap in it. When a large voltage was built up, sparks jumped back and forth across the gap. These sparks constituted an alternating electric current, which—according to Maxwell's theory—should radiate electromagnetic waves. From the properties of the circuit, Hertz was able to estimate the frequency of the waves being emitted.

To detect the electromagnetic waves, Hertz used a metal wire bent into a circle with a small gap between the ends (Figure 10–3a). Hertz found that whenever sparks were produced in the transmitter circuit, sparks also appeared in the gap in the circular wire. Sparks could be induced in the detecting wire when the wire was as far as 6 feet from the primary circuit. To explain this phenomenon, Hertz asserted that the electromagnetic waves radiated by the sparks in the transmitter circuit induced a current in the detecting wire, causing a spark to jump across the gap in that wire.*

*The first transmission and detection of electromagnetic waves in a laboratory was performed, quite unwittingly, by Galvani. In Galvani's original experiment the spark drawn at the "electric machine" generated electromagnetic waves. The frog's leg acted as the detector.

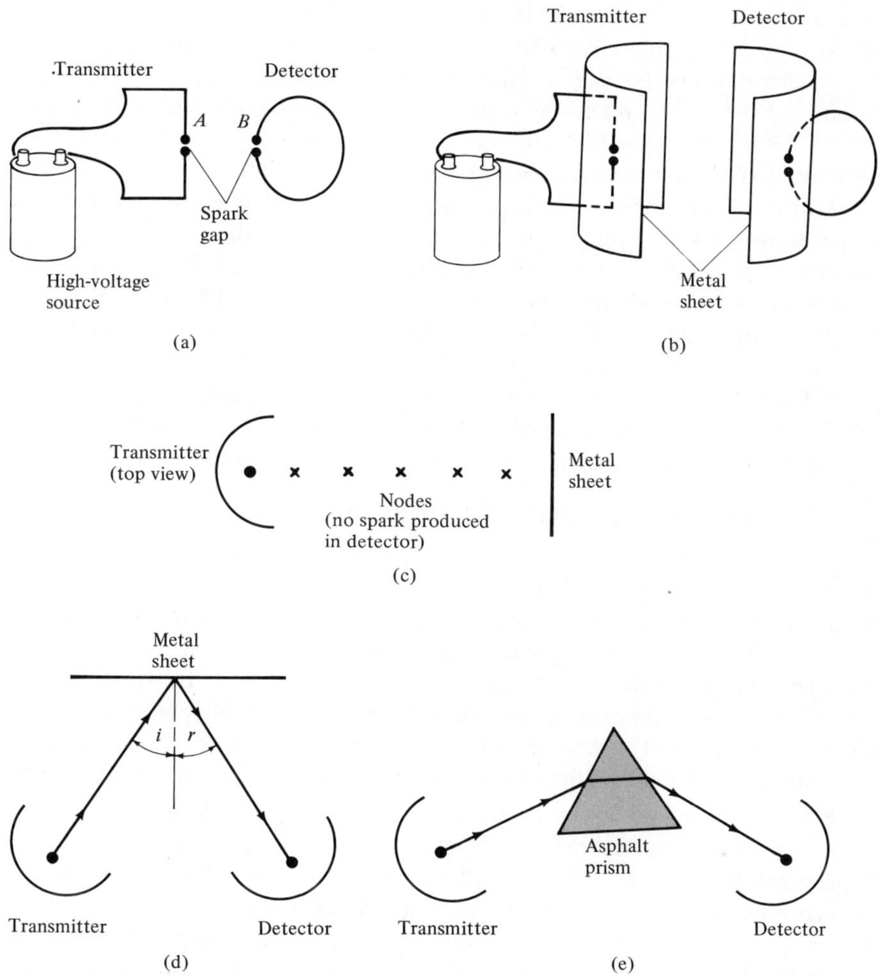

figure 10–3 **HERTZ'S EXPERIMENTS**

(a) A high voltage across the gap at A causes sparks to jump across the gap. The appearance of sparks at A induces sparks to jump the gap at B in the detector. (b) The effect can be enhanced by putting the transmitter and detector at the focus of a concave metal "mirror." (c) A standing wave pattern can be produced by using a metal reflecting sheet. (d) Reflection of emitted waves. When the detector is moved a few degrees to either side of the position shown, the sparking stops. (e) Refraction by a "prism" may be demonstrated in the same manner.

Hertz found that he could intensify the effect by bending a metal sheet into a concave shape and placing either the transmitting spark gap or the detecting wire at the focus (Figure 10–3b). The metal sheet thus acted as a reflecting "mirror" that tended to concentrate the emitted waves, just as a concave optical mirror concentrates light into a beam if the source is at the focus. Hertz then employed his apparatus to study the properties of the waves being emitted. He found that a flat metal sheet, placed some distance from the source, reflected the advancing waves, and a pattern of standing waves was set up (Figure 10–3c). Hertz used his detector to locate the nodes of the standing wave pattern. These were the points between the source and the reflector where no sparks could be induced in the detector. By measuring the distances between the nodes, he could determine the wavelength of the emitted waves. By combining this number with his calculated emission frequency, Hertz was able to show that the emitted waves had a speed equal to the speed of light.

In the course of further experimentation, Hertz demonstrated that the emitted waves were transverse and could be polarized. He also showed that the waves obeyed the laws of reflection and refraction (Figures 10–3d and e). To demonstrate the latter, he constructed a large triangular prism 3 feet on a side and 4 feet tall, made of asphalt (an insulator), and weighing nearly 1200 lb. Electromagnetic waves were refracted by the prism, just as light is refracted by a glass prism.

Thus Hertz successfully demonstrated that electromagnetic waves exist and have all the properties usually attributed to light. The results of Hertz's experiments, published in 1889, provided a most convincing verification of Maxwell's theory.

FUNDAMENTALS OF RADIO AND TV BROADCASTING

No less impressive than Hertz's demonstration of the existence of electromagnetic waves were the technological developments that rapidly followed this discovery. Soon after Hertz announced the results of his experiments, it was suggested that electric signals could be transmitted from place to place by electromagnetic waves traveling through free space rather than by electric currents in wires. By 1894 the first wireless telegraphy system was in operation, and in 1901 transatlantic wireless communications were established.* The first decades of the twentieth century saw the development of more powerful transmitters and more sensitive receivers; by the 1920s radio programs were being broadcast on a regularly scheduled basis.

For most people a radio is a rather commonplace object, so they do not find it particularly mysterious or wonderful to hear the voice of another person, who may be many hundreds of miles away, coming out of a small box that can be carried in a shirt pocket. Nevertheless, the entire system of radio and TV broadcasting is an extremely complex one, based on the contributions of hundreds of scientists and engineers made over a span of several decades. Let us consider briefly the basics of a radio communication system.

We cannot hear a radio wave; our auditory sense responds only to compressional (sound) waves of relatively low frequency. Sound waves are mechanical disturbances in a medium (air) and are not electromagnetic in nature. Hence the purpose of a radio receiver is to intercept the radio waves given off by the transmitter and to convert them into audible sound waves. These radio waves must be transmitted in a manner that allows the radio receiver to generate sound waves that faithfully reproduce the original sounds made at the source. The radio waves carry the information that describes the sounds to be reproduced through the process of ***modulation.*** That is, the transmitted radio waves are *modulated,* or varied in a way that is related to the variations in loudness and pitch of the original sound waves.

There are two basic methods for modulating radio waves. In the ***amplitude modulation (AM)*** system the transmitter broadcasts a radio wave of a fixed frequency (the carrier frequency), but the amplitude of the wave is varied according to the

*The transmission of radio waves around the curvature of the earth is made possible by a region of charged particles in the atmosphere, called the ionosphere, that reflects radio waves.

intensity (loudness) and frequency (pitch) of the source sound waves. Thus the variation in the amplitude of the carrier wave is itself a wave form that duplicates the wave pattern of the source sound (Figure 10–4a). Since the frequency range of audible sound is from 20 to 20,000 Hz and the frequency of the carrier signal is typically around 10^6 Hz, the amplitude variations have a much longer wavelength than the carrier waves.

Frequency modulation (FM) achieves the same effect by transmitting a signal of constant amplitude in which the frequency is varied with respect to the carrier frequency according to the intensities and frequencies of the sounds to be reproduced (Figure 10–4b). Thus in radio communication it is not the sounds themselves that are transmitted; instead, radio waves that are encoded with information describing the sounds to be reproduced are transmitted.

A simplified block diagram of a transmitting–receiving system is shown in Figure 10–5. The conversion of sound energy into electric energy is effected by a microphone, which is a device specifically designed so that one of its electric properties (such as its resistance) will vary according to the pressure of the surrounding air. Thus the electric output of a microphone varies in strength according to the intensities and frequencies of the sounds being picked up.

figure 10–4 **MODULATION OF RADIO WAVES**

The sound wave form may be superimposed on the carrier-frequency radio wave in two ways. (a) Amplitude modulation. The amplitude of the carrier wave is varied according to the intensity and frequency of the sound wave. (b) Frequency modulation. The frequency of the carrier wave is varied from the unmodulated carrier frequency according to the intensity and frequency of the sound wave.

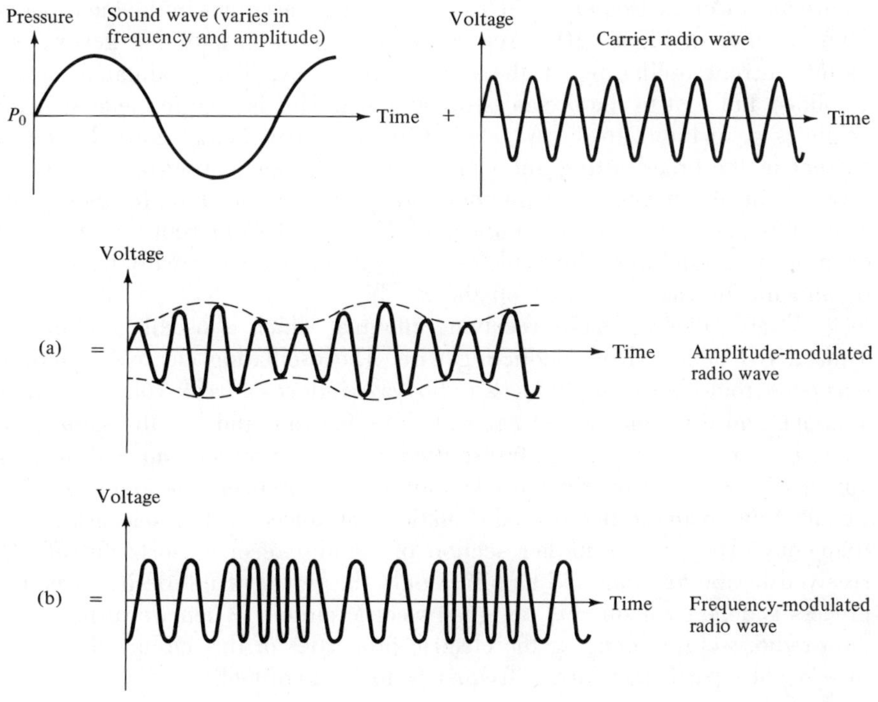

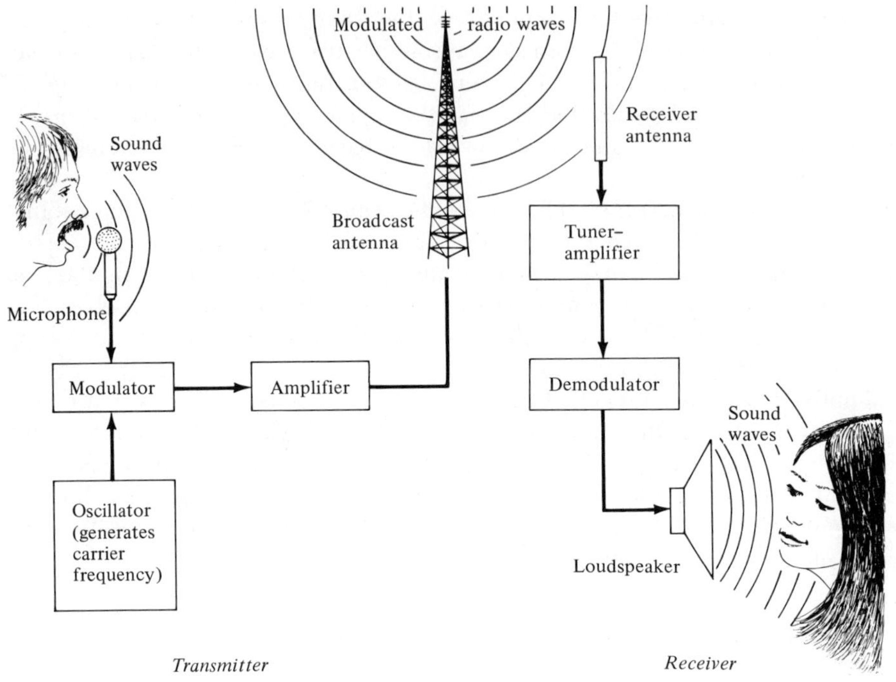

figure 10–5 **BASIC ELEMENTS OF A RADIO COMMUNICATION SYSTEM**

The signal from the microphone is fed into the modulator section of the transmitter. Here the carrier wave is modulated (by varying either the amplitude or the frequency of the carrier wave) according to the variations in the microphone output. The oscillator in Figure 10–5 represents an electric circuit that generates an alternating current oscillating at the carrier frequency. The modulated signal is then amplified and sent to the broadcasting antenna. This is a large metal structure whose height is comparable to the wavelength of the waves being emitted. The oscillating current in the broadcasting antenna results in the emanation of radio waves. These waves, being electromagnetic, do not require a material medium for their transmission. Thus it is possible to receive radio and TV signals from transmitters on the moon or in orbiting satellites. (Nevertheless, most radio stations usually begin broadcasting by announcing that they are "on the air"!)

A radio receiver has a receiving antenna, which is usually nothing more than a metal rod or a coil of conducting wire. The oscillating electromagnetic field that emanates from the transmitter as a radio wave induces a weak voltage in the receiving antenna, and this voltage alternates at the same rate and has the same variations in amplitude or frequency as the transmitted wave. The voltage induced in the receiving antenna is rather weak and must be amplified. Moreover, the antenna is picking up signals from many different radio stations at once, each broadcast at a different frequency. The tuner-amplifier section of a radio receiver sorts out all the signals received by the antenna and amplifies only the signal of the desired frequency. This process is done by a suitably designed electric circuit. When we turn the tuning dial on a radio, we are changing the electric properties of this circuit. Thus we are able to select the particular carrier frequency to be amplified.

The amplified signal is now "demodulated"; that is, the variations in amplitude or frequency that are related to the source sounds are in effect separated from the carrier frequency. The demodulator produces an "audio-frequency" electric signal—that is, a signal that oscillates with the same amplitudes and frequencies as the original sound waves. This signal directly drives the loudspeaker, which converts the electric signals into sound waves of the same amplitudes and frequencies. A standard loudspeaker consists of a cone-shaped paper or cloth diaphragm connected to a small motor. The motor is driven by the audio-frequency signal, so that the diaphragm vibrates according to the variations in this signal. The diaphragm's vibrations cause the air in contact with the diaphragm to vibrate, generating sound waves duplicating those originally produced.

Television broadcasting operates in much the same way as radio broadcasting. In TV broadcasting the essential problem is to modulate the transmitted radio waves in such a way as to provide information about the picture being televised. The modulating signal is provided by the television camera, which scans each small region of the picture and generates an electric current proportional to the light intensity in that region. A set of signals describing each region of the picture in sequence (i.e., representing one complete scan of the picture) can be generated in 1/30 sec. The same sequence of signals is demodulated from the carrier wave by the television receiver and is used to reproduce the picture on the television screen.

In the early days of radio broadcasting, one of the major problems was interference between two or more stations operating at approximately the same carrier frequency. Before long government regulation of radio broadcasting was necessary. At the present time all commercial radio and TV stations in the United States must be licensed by the government, and carrier frequencies are assigned in accordance with international agreements. All AM radio stations are assigned frequencies between 550 and 1600 kHz* (stations are spaced 10 kHz apart), and FM radio stations operate in the frequency range from 88 to 108 MHz (spaced 200 kHz apart). Television stations are assigned various "channels" ranging from 44 to 216 MHz. (In the 1960s a new band of frequencies—the UHF or ultrahigh-frequency region from 470 to 890 MHz—was opened up to commercial television.) Interspersed between the AM, FM, and TV bands are specific frequencies (the "shortwave" frequencies) set aside for government, scientific, industrial, and amateur use.

A rather brief outline of the basic principles of radio and TV transmission and reception has been given here. A description of the electric circuitry and devices required to perform such functions as modulation or amplification would be much too involved for the purposes of this text. Historically, the most impressive thing is how rapidly these technological developments followed upon the work of Maxwell and Hertz, and the enormous effect they have had on civilization. One of the most characteristic features of modern-day life is the speed with which information is transmitted from place to place on the earth. This "communication explosion" has profoundly affected economic, social, and political developments in the twentieth century. These changes are all the more remarkable when we trace their origin back to Maxwell's equations. Few people in Maxwell's time could have been visionary enough to predict that this highly abstract theory would so thoroughly revolutionize nearly all aspects of human life.

* 1 kHz = 1 kilohertz = 10^3 cycles/sec; 1 MHz = 1 megahertz = 10^6 cycles/sec; 1 GHz = 1 gigahertz = 10^9 cycles/sec.

10–4

THE ELECTROMAGNETIC SPECTRUM

The electromagnetic theory places no restrictions on the frequency of an electromagnetic wave. Theoretically, an electromagnetic wave can be generated at any frequency between zero and infinity (although the production of waves of certain frequencies—particularly the higher ones—may be a *practical* impossibility). Regardless of frequency, all electromagnetic waves have certain properties in common: they all travel at the same speed in free space, exhibit interference and diffraction phenomena, and obey the laws of reflection and refraction.

Frequency differences, however, have certain effects on the properties of electromagnetic waves. For example, the spacing between the regions of reinforcement and cancellation in an interference or diffraction pattern depends on the frequency. The index of refraction for a given medium varies according to the frequency of the wave. As a general rule, the rates at which matter emits, transmits, and absorbs electromagnetic waves are all frequency-dependent. Hence the processes used to generate and detect electromagnetic waves differ according to the frequencies of the waves. In this sense visible light is distinguished from other electromagnetic waves in that the frequencies in the visible light range trigger the nerve responses involved in the process of vision, whereas waves with frequencies outside this range do not trigger these responses. Even within the visible range, waves of different frequencies trigger different responses, and the result is that we experience the phenomenon of color vision. Hence the colors of the visible spectrum represent differences in the response of the eye–brain system to electromagnetic waves of different frequencies.

It is useful to extend the spectrum concept to cover the entire range of electromagnetic wave frequencies. Just as the visible spectrum is divided into several main colors with no distinct boundaries between them, the ***electromagnetic spectrum*** is divided into several regions, each representing a different manifestation of electromagnetic radiation (Figure 10–6). Electromagnetic waves are primarily classified into different types according to the ways in which they are generated and/or detected. The boundaries between the different regions of the electromagnetic spectrum are somewhat arbitrary, and there is considerable overlap between regions, since waves of certain frequencies can be generated or detected by more than one process. The basic data on the different regions of the electromagnetic spectrum are summarized in Table 10–1. We shall now consider briefly some of the more interesting properties of the different types of electromagnetic waves.

Radio waves can be generated and detected by electric circuits. Frequencies from 20 kHz to 1 GHz are used for broadcasting AM, FM, shortwave, and television signals. Radio waves in the 1 to 30 GHz region (corresponding to wavelengths of 1 to 30 cm) are known as ***microwaves*** and are used for radar and satellite communications. The upper limit of frequency that can be generated electrically is about 3000 GHz.

Radio waves are also emitted by natural objects. The sun, several planets, the interstellar gases, and galaxies give off radio waves. Huge radio telescopes that detect and focus these radiations have been constructed. Many significant discoveries pertaining to the nature of the universe have been made on the basis of the measurements obtained with these telescopes.

figure 10–6 **THE ELECTROMAGNETIC SPECTRUM**

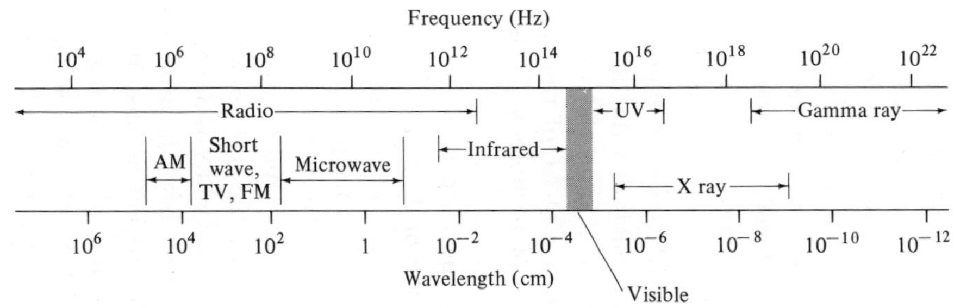

table 10–1 PROPERTIES OF THE ELECTROMAGNETIC SPECTRUM

Type of wave	Frequency (Hz)	Wavelength (cm)	How usually described	How usually generated
Radio	0 to 3×10^{12}	$\infty \longrightarrow 0.01$	By frequency, measured in Hz, kHz, MHz, or GHz (1 Hz = 1 cycle/sec)	Oscillations in electric circuits
Infrared	6×10^{11} to 4×10^{14}	0.05 to 7.5×10^{-5} ($500\ \mu$ to $0.75\ \mu$)	By wavelength, measured in microns ($1\ \mu = 10^{-4}$ cm)	Vibrations of atoms within molecules, molecular vibrations in solids
Visible	4×10^{14} to 7.5×10^{14}	7.5×10^{-5} to 4×10^{-5} (7500 Å to 4000 Å)	By wavelength, measured in angstroms (1 Å $= 10^{-8}$ cm)	Motions of electrons in atoms
Ultraviolet	7.5×10^{14} to 3×10^{16}	4×10^{-5} to 1×10^{-6} (4000 Å to 100 Å)	By wavelength, measured in angstroms	Motions of electrons in atoms
X ray	2×10^{15} to 3×10^{19}	1.5×10^{-5} to 1×10^{-9} (1500 Å to 0.1 Å)	By wavelength, measured in angstroms	Rapid acceleration or deceleration of charged particles
Gamma ray	$3 \times 10^{18} \longrightarrow \infty$	$1 \times 10^{-8} \longrightarrow 0$ (1 Å $\longrightarrow$ 0)	By wavelength, measured in angstroms	Nuclear reactions, radioactive disintegration

The gap between the radio and visible regions of the spectrum is encompassed by the ***infrared.*** Infrared radiation is emitted as a consequence of the vibrations of atoms within molecules and the vibrations of atoms and molecules about their equilibrium positions in solid matter. Since vibratory motion is associated with heat energy, infrared radiation is sometimes referred to as "heat radiation." In the interaction of infrared with matter this radiation is absorbed directly by the atoms and molecules in the matter, causing them to vibrate faster. The energy thus gained is converted to heat energy.

Infrared radiation was discovered by William Herschel in 1800. Herschel placed a thermometer in various parts of the spectrum of sunlight and found that the highest temperatures were recorded in the region just beyond the red end of the visible spectrum. In the 1840s Fizeau and Foucault were able to produce infrared interference patterns, using microscopic thermometers to detect the bands of reinforcement and cancellation.*

The range of the ***visible*** region of the electromagnetic spectrum is delineated by the sensitivity of its detector—the human eye. The narrowness of the visible region is notable, for it tells us that vision is an extremely specific sense, responding to a very small range of wavelengths out of the entire spectrum of electromagnetic waves. The gases in the earth's atmosphere absorb or reflect virtually the entire electromagnetic spectrum *except* for the visible wavelengths, which are transmitted with no appreciable attenuation by the atmosphere. (The sunlight reaching the surface of the earth has a wavelength range of 3000 to 11,000 Å, which is only slightly wider than the 4000 to 7000 Å range of human vision.) Thus it would seem that human vision evolved to be most sensitive to the wavelengths in the visible region because these are the only wavelengths available at the earth's surface. On the other hand, some biochemists have pointed out that the visible wavelengths are the only ones suitable for vision; radio waves and infrared do not carry enough energy to trigger the chemical reactions involved in vision, and ultraviolet and x rays carry too much energy. In this respect, "x-ray vision" and "radio-wave vision" are biological impossibilities. Thus the coincidence between the spectrum of available sunlight and the range of sensitivity of the human eye may be fortuitous. If the atmosphere had been so constituted as to not transmit the visible wavelengths, the sense of vision might not have evolved.

Ultraviolet radiation is both beneficial and harmful to life. Ultraviolet radiation stimulates the production of vitamin D in human skin and is also responsible for sunburn (which is caused specifically by ultraviolet in the 2900 to 3100 Å range). The shorter wavelength ultraviolet (2600 to 2900 Å) will be strongly absorbed by cell proteins and nucleic acids and is highly lethal. If these wavelengths were not absorbed by the ozone in the earth's atmosphere, large-scale life would probably not exist on the surface of the earth.

Ultraviolet and visible light are emitted and absorbed by the electrons in atoms. There are certain substances that absorb energy in the form of ultraviolet radiation and re-emit this energy as visible light. Such substances, known as ***fluorescent*** substances, are very useful as detectors of ultraviolet radiation.

X rays were accidentally discovered by the German-born scientist Wilhelm Roentgen in 1895. Roentgen was experimenting with a device known as a gas-

*Infrared radiation is normally referred to in terms of its wavelength instead of its frequency. The unit most commonly used is the ***micron,*** symbolized by the Greek letter μ; 1 micron (1 μ) = 10^{-6} meter = 10^{-4} cm.

discharge tube, which consists of a gas-filled glass tube in which two electrodes are implanted so that electric discharges can be passed through the gas (Figure 10–7). When a high voltage is established across the electrodes and the gas is evacuated from the tube, a stream of high-speed particles emanates from the cathode. These particles were known as ***cathode rays*** and were the subject of intense study in the late nineteenth century. (We will learn later in the chapter that these particles are actually high-speed electrons.)

While studying the properties of cathode rays, Roentgen noticed that a screen coated with a fluorescent material that happened to be located several feet from the tube began to glow whenever the cathode rays were generated. Roentgen found that the fluorescence was not caused by the cathode rays themselves, but rather by other radiations of an unknown nature that were emitted from the point where the cathode rays were stopped. Roentgen found that the mysterious "x rays," as he called them, could penetrate through most solid substances and would darken photographic film. The announcement of Roentgen's discovery stirred considerable public interest. Some people, not fully understanding the phenomenon, were afraid x rays would be a tremendous boon to voyeurs, who would be able to see through clothing! On a more serious level, the medical value of x rays was immediately recognized, and within a matter of months x rays were being commonly used as a diagnostic tool.

Roentgen was convinced that x rays are a form of electromagnetic radiation, since some of their properties are similar to those of ultraviolet light. However, he was unable to show that x rays could be refracted or made to exhibit interference effects. The difficulty lay in the extremely short wavelengths of x rays. In order to produce a measurable x ray interference pattern in a double-slit apparatus similar to that used in Young's experiment, the slits would have to be 10^{-6} to 10^{-7} cm apart. Eventually, experimental techniques were developed that made it possible to produce

figure 10–7 **THE PRODUCTION OF X RAYS**

When cathode rays strike the anode or any target placed in their path, x rays are given off. These x rays are high-frequency electromagnetic waves emitted as a consequence of the rapid deceleration of the cathode-ray electrons as they strike the target.

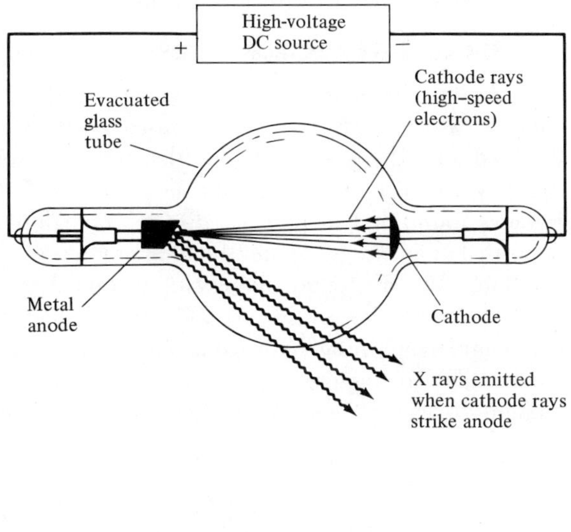

observable diffraction of x rays. By 1920 the electromagnetic nature of x rays had been clearly established.

According to Maxwell's theory, electromagnetic waves are radiated by accelerating (or decelerating) electric charges. When the high-speed electrons that constitute cathode rays strike a target, they are rapidly decelerated. The result is the emission of high-frequency electromagnetic waves that are in the x-ray region of the spectrum. The interaction of high-speed electrons with atoms can also result in the emission of x rays by the electrons within those atoms.

Electromagnetic radiations at the extreme high-frequency end of the spectrum are classified as ***gamma rays.*** They were first discovered in 1900 as one of the emanations given off by radioactive materials and were shown to be even more penetrating than x rays. In general, gamma rays are emitted and absorbed in interactions involving the nuclei of atoms. There is no theoretical upper limit to the frequency of gamma rays; frequencies as high as 10^{25} Hz have been produced by artificial means.

Between 1890 and 1920 all the existing gaps in the electromagnetic spectrum were filled in, and the existence of electromagnetic radiation across the entire range of frequencies was firmly established. Hence the most significant prediction of Maxwell's theory—that electromagnetic disturbances propagate through space at finite speed in the form of waves—had been thoroughly verified.

10–5

THE DISCOVERY OF THE ELECTRON

In spite of the brilliant success of Maxwell's electromagnetic theory with respect to the existence of electromagnetic waves, there were certain conceptual aspects of the theory that required modification. Foremost among these was the nature of electric charges and currents and their relationship to electric and magnetic fields. Maxwell's theory was based on the premise that all electromagnetic interactions occur in the ether rather than in matter itself. Maxwell speculated that charges and currents did not exist independently in matter, but were instead consequences of the interactions between matter and the ether. In short, charges and currents were assumed to be distortions or displacements of the ether produced by the presence of matter. Maxwell did not accept the existence of discrete, point charges; he preferred to relate his theory to the older concept of an electric fluid.

THE LORENTZ ELECTRON THEORY

In the 1880s an alternative viewpoint developed that held that discrete electric charges exist and are intimately connected with the atoms of matter rather than with the medium. The source of this idea was an experimental finding made by Michael Faraday in 1834. In his experiments on the electrolysis of chemical compounds, Faraday found that there was a direct proportion between the mass of a given element deposited at an electrode and the total amount of charge passed through the electrolytic solution. Faraday found, for example, that 1 gram of hydrogen would be deposited at the cathode for every 96,500 coulombs passed through the solution, for *any* compound containing hydrogen. This direct relationship between charge and mass led Faraday to suggest that if matter consists of atoms, each atom must have a fixed amount of charge associated with it. But in 1834 the atomic theory was not that widely accepted, and Faraday himself was not entirely convinced of the validity of his hypothesis.

Once the atomic theory was firmly established, the implications of Faraday's discovery became more obvious. In 1881 Helmholtz declared:

> If we accept the hypothesis that the elementary substances are composed of atoms, we cannot avoid concluding that electricity also, positive as well as negative, is divided into definite elementary portions which behave as atoms of electricity.[3]

Thus Helmholtz suggested that the "electric fluid" contained in matter is not completely continuous. Instead, there is a fundamental quantity or unit of charge (both positive and negative) that cannot be subdivided; all aggregations of charge consist of multiples of this fundamental unit. It was not clear whether the fundamental unit was concentrated in a specific particle of electricity or was a property of atoms themselves and hence not separable from atoms. The word ***electron*** (coined in 1891 by G. J. Stoney) was used at first to signify only the fundamental quantity of charge and did not presuppose the existence of an actual particle of electricity. Thus in the original sense of the word "electron," it was proper to speak of a body as containing "one or more electrons (i.e., units) of positive charge" or "one or more electrons of negative charge."

In 1892 the Dutch physicist Hendrik Lorentz proposed a significant modification of Maxwell's electromagnetic theory, which took into account the presumed existence of discrete units of charge. The ***Lorentz electron theory,*** as it came to be known, was in effect a compromise between Maxwell's theory and the old, discredited theory of Weber. According to the Lorentz theory, all electromagnetic effects are caused by the presence and motions of discrete electric charges. (In calling his theory the "electron theory," Lorentz was referring to the fundamental unit of charge and not to the particle itself.) Charges are assumed to have a real existence independent of the ether, in contrast to Maxwell's concept of charges and currents as disturbances produced in the ether by matter. In the Lorentz theory Maxwell's equations are retained intact, but are given a somewhat different physical interpretation. To Maxwell, the equations simply related mathematically the properties of electric and magnetic fields to the charges and currents that appear, without establishing any causal precedence. In Lorentz's view electric and magnetic fields are *caused* by stationary and moving charges; Maxwell's equations enable one to calculate the electric and magnetic fields that are produced by a given arrangement of charges.

From Weber's theory Lorentz adopted the idea that all magnetic effects are produced by electric currents, which are composed of discrete electric charges in motion. In the Lorentz theory, however, charges do not interact directly with one another; each charge interacts with the field produced by other charges. Thus the concept of action at a distance, which led to the downfall of Weber's theory, is absent from the Lorentz electron theory. Any effects produced by the motion of a charge will propagate through the intervening medium in the form of an electromagnetic wave, and other charges will be affected only when the wave reaches them.

Lorentz explained how to determine the effect of an electric or magnetic field on a charged body. In an electric field the force exerted on a charge q is

$$\mathbf{F} = \mathbf{E}q \tag{10–8}$$

where $\mathbf{E}$ is the electric field intensity. This equation follows directly from the definition of electric field intensity given previously (Eq. 9–13a). It describes the purely electrostatic force acting on a stationary charge in an electric field. Note that the direction of the force on a positive charge is the same as the direction of the electric field, whereas the force on a negative charge is in the opposite direction.

When a stationary electric charge is placed in a *magnetic* field, it experiences no force. However, if the electric charge is in motion, it will be affected by a magnetic field. A moving electric charge constitutes an electric current, and (as Ampère and Faraday discovered) an electric current in a magnetic field experiences a magnetomotive force. Thus a force should act on a *moving* charge in a magnetic field. According to Lorentz, therefore, the force on a moving charge in a magnetic field is the source of the magnetomotive force on a current-carrying wire.

We learned previously that the magnetomotive force on a current-carrying wire is given by the equation (Eq. 9–14):

$$F = ILB \sin \theta$$

where I is the current through a wire of length L, and θ is the angle between the current direction and the direction of the magnetic field **B.** The direction of the magnetomotive force, it will be recalled, is perpendicular to both the current and magnetic field directions. Lorentz modified this equation to represent the force on a freely moving charged particle. Suppose that a single charge q passes through a wire of length L in a time t. The current is $I = q/t$. Also, the speed v of the particle is $v = L/t$. Therefore

$$IL = \left(\frac{q}{t}\right)L = q\left(\frac{L}{t}\right) = qv$$

so that

$$F = ILB \sin \theta = qvB \sin \theta$$

Thus the force acting on a charge q moving with a velocity **v** in a magnetic field **B** is described by the equation*

$$F = qvB \sin \theta \tag{10–9}$$

where θ is the angle between **v** and **B.** This force is sometimes referred to as the ***Lorentz force.*** The direction of the Lorentz force is perpendicular to the plane containing **v** and **B;** that is, the force is perpendicular to both the direction of motion and the direction of the magnetic field. The direction of **F** can be obtained by using the right-hand rule for the magnetomotive force. (See Figure 9–19.) It should be noted, however, that the direction of the force on a negatively charged particle will be the opposite of that on a positively charged particle, since the q in Eq. 10–9 will have a negative sign.

MOTION OF A CHARGED PARTICLE IN A MAGNETIC FIELD

Suppose that a freely moving, positively charged particle of charge q and mass m enters a uniform magnetic field at right angles to the lines of force, as shown in Figure 10–8. (The crosses in the diagram represent the lines of force of the magnetic

*It is assumed here that all quantities are measured in MKS units. In CGS units I is measured in emu and q in esu, and the conversion factor of 1 emu $= c$ esu/sec must be taken into account. Thus in CGS units $IL = qv/c$, so that Eq. 10–9 becomes

$$F = \frac{1}{c} qvB \sin \theta$$

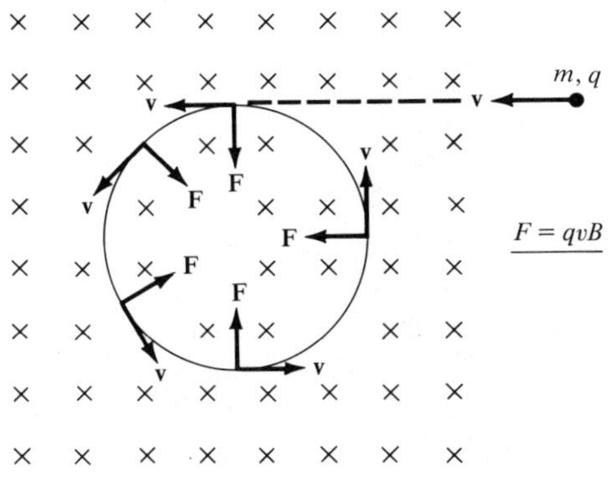

Magnetic field **B** is directed into paper; angle between **v** and **B** = 90°.

figure 10–8

MOTION OF A CHARGED PARTICLE IN A UNIFORM MAGNETIC FIELD

A particle entering a uniform magnetic field at right angles to the direction of the field lines will move in a circle at constant speed under the influence of the Lorentz force. The crosses represent the lines of force of the magnetic field, which are directed away from the reader perpendicular to the plane of the paper.

field, which point into the plane of the paper.) The force exerted on the particle will be perpendicular to both the direction of motion and the direction of the magnetic field, and the particle will be deflected away from its straight-line motion, as shown. As the particle changes direction, the direction (but not the magnitude) of the force also changes. Since the force is always perpendicular to the direction of motion, the particle will move in a circular path. Furthermore, no work is done on the particle, so that its speed and energy remain constant. Thus in this situation the Lorentz force acts as a centripetal force that keeps the particle moving in a circle at constant speed.

The magnitude of the Lorentz force on the particle is

$$F = qvB \sin \theta = qvB$$

because **v** and **B** are always perpendicular to one another. Since the Lorentz force provides the centripetal force that keeps the particle in circular motion, we have

$$F = \frac{mv^2}{r} = qvB$$

where m is the mass of the particle, and r is the radius of the particle's circular path. Solving for r, we obtain

$$r = \frac{mv}{qB} \tag{10–10}$$

Hence the radius of the circle is directly proportional to the speed of the particle.

An important practical application of this effect is made in the ***cyclotron,*** a device used to accelerate charged particles to very high speeds.* As illustrated in Figure 10–9, the cyclotron consists of two semicylindrical chambers separated by a small gap. A uniform magnetic field is impressed perpendicular to the faces of the two chambers, and an electric field (in the form of a large voltage difference) acts horizontally across the gap. A charged particle is injected at the center and is accelerated across the gap by the electric field. The magnetic field causes the particle to

*The person most responsible for the development of the cyclotron was Ernest O. Lawrence, who supervised the construction of the first cyclotron at the University of California in 1931. One important use of the cyclotron is in the production of artificial elements (to be discussed in Chapter 19) and for this reason element 103 was named lawrencium.

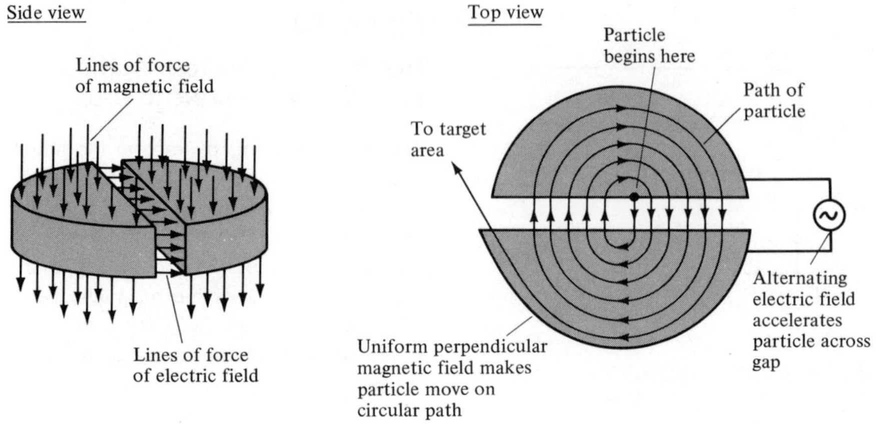

figure 10–9 **THE CYCLOTRON**

A charged particle is injected at the center. It is accelerated by an electric field across the gap and deflected onto a circular path by a uniform magnetic field. With each crossing of the gap, the particle gains energy and the circles grow wider.

move in a circular path without increasing the particle's speed. After a half-turn the particle has returned to the gap, where it is again accelerated. The electric field is made to reverse its direction (by means of an alternating voltage) every half-turn, so that the particle is always *accelerated* (and never decelerated) by the electric field.

When the particle enters the other chamber, it again moves in a circular path. However, the speed of the particle has been increased by the electric field, so that the radius of the circle is greater than before. (See Eq. 10–10.) With each crossing of the gap the particle gains speed, so it spirals around in ever-widening circles. But the increase in radius is exactly compensated by the increase in speed, so that the particle always takes the same time to complete a half-turn. Thus the electric field can be alternated at a steady rate. When the particle reaches the outer edge of the cyclotron, it is directed toward an area where it can be used to bombard a target or to perform other experiments.

Note that in a cyclotron a particle gains energy from the electric field only; the magnetic field serves merely to move the particle in a circular path without any increase in speed. To accelerate particles in an electric field alone would require an extremely long, straight electric field (since circular electric fields are extremely difficult to construct). Thus the main virtue of a cyclotron is its compactness, for the magnetic bending allows the particle to be accelerated by the same electric field over and over again and confines the path of the particle to a relatively small area. It is possible to accelerate particles up to speeds close to the speed of light in a cyclotron 10 to 20 feet in diameter.

The cyclotron and similar devices that use electric and magnetic fields to accelerate particles have played an important role in the development of nuclear physics. The impact of a high-speed charged particle on a target atom can result in the disintegration of the atomic nucleus. Much has been learned about nuclear forces, the nature of the nucleus, and the fundamental particles of matter by studying the results of these collisions. These topics will be discussed more fully in Chapter 20.

THOMSON'S CATHODE-RAY EXPERIMENT

The Lorentz electron theory represented little more than a reinterpretation of electromagnetic phenomena based on the assumption that discrete charges exist. In 1892, when the theory was proposed, there was ample evidence to indicate that discrete charges did indeed exist, but their relationship to matter had not been ascertained. Is charge an innate property of matter or is it something distinct and separable from matter?

Scientists generally believed that research on the conduction of electricity by matter would be most likely to provide the answer to this question. As we have learned, Faraday's study of the conduction of electricity in liquids provided the first evidence of a direct relationship between mass and charge. Attention was turned to the conduction of electricity in gases, with the recently developed kinetic theory serving as a useful analytic tool.

The device used to study conduction in gases was the gas-discharge tube, one

figure 10–10 **CONDUCTION OF ELECTRICITY IN GASES**

(a) At a sufficiently high voltage and low pressure, a gas will ionize and conduct electric current. (b) When the gas is evacuated, cathode rays are emitted. The cathode rays themselves are not visible. However, if they are allowed to pass through a hole in the anode and strike the wall of the glass tube, the glass will fluoresce and x rays will be emitted.

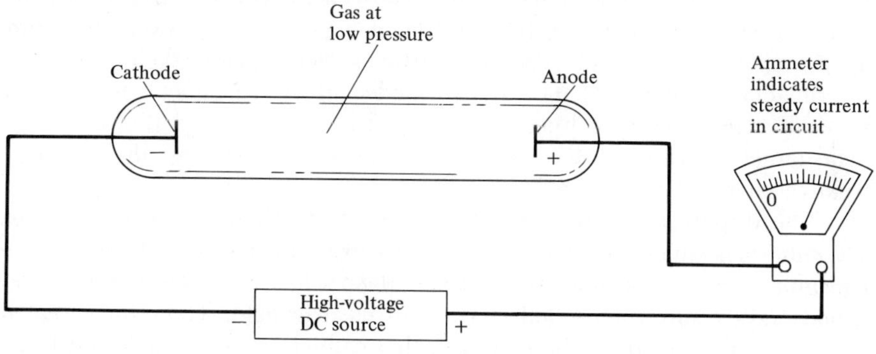

(a)

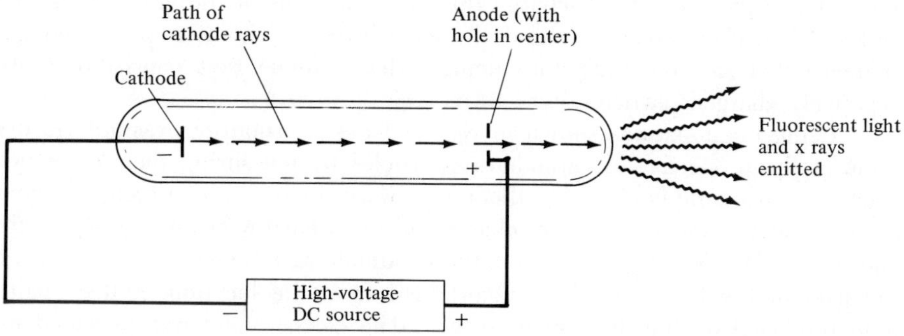

(b)

form of which is shown in Figure 10–10a. Under normal conditions a gas is a poor conductor of electricity. However, a large voltage difference between the cathode and the anode causes the gas atoms in the tube to ionize. A current through the gas is established as positive ions move to the cathode and negative ions to the anode. Scientists who were studying this phenomenon were interested to find out what happens to the current as the amount of gas in the tube decreases. It was found that as gas is removed, the current at first increases. Even though there are fewer atoms to be ionized, there is less likelihood that an ion will encounter an oppositely charged ion and be neutralized, so there is a net increase in the ion population. However, with continued evacuation of gas there are fewer and fewer atoms to be ionized, and the current decreases.

When the gas is almost completely evacuated, the current surprisingly begins to increase again. If a hole is drilled in the anode, the glass directly behind the anode begins to fluoresce with a characteristic green color, and x rays are given off (Figure 10–10b). The first studies of this phenomenon, made by the German scientists W. Hittorf and E. Goldstein in the 1870s, indicated that an invisible stream of radiation was emanating from the cathode itself. It was shown that the "cathode rays" travel in straight lines, since an object placed in their path casts a distinct shadow on a fluorescent screen.

Two schools of thought developed concerning the nature of cathode rays. The German scientists believed that cathode rays are a form of electromagnetic radiation. The most notable proponent of this view was Heinrich Hertz, who announced that he had found that cathode rays could penetrate through a thin sheet of gold leaf and that they could not be deflected in an electric field. On the other hand, the English scientists who were studying the phenomenon generally believed that cathode rays are a stream of negatively charged particles. They based their argument on two experimental findings: (1) the rays are deflected by a magnetic field in a direction perpendicular to the field lines, as predicted by Lorentz, and (2) when an insulated metal target is placed in the path of a beam of cathode rays, the target acquires a negative charge.

The dispute was settled in 1897 when J. J. Thomson, a professor of physics at Cambridge, announced the results of his experiments on cathode rays. The most damaging argument against the particle theory had been Hertz's observation that cathode rays could not be deflected in an electric field. Thomson repeated Hertz's experiment and found, to the contrary, that cathode rays *are* deflected in an electric field, in the direction that a negatively charged particle would be expected to move. Thomson attributed Hertz's result to faulty experimental technique; Hertz had not sufficiently evacuated the discharge tube, and the ions of the gas that had remained in the tube effectively neutralized the electric field. Thomson presented additional evidence that showed quite convincingly that cathode rays consist of a stream of negatively charged particles.

Using the apparatus shown in Figure 10–11a, Thomson was able to determine some of the properties of cathode-ray particles by measuring their deflection in an electric or magnetic field. The cathode rays were generated by a high voltage between the cathode and anode, and were channeled into a narrow beam by holes in the anode and in a second metal plug. Although the cathode rays themselves could not be seen, the path of the beam could be ascertained from the location of the small spot of light produced on the fluorescent screen. The electric field was provided by a pair

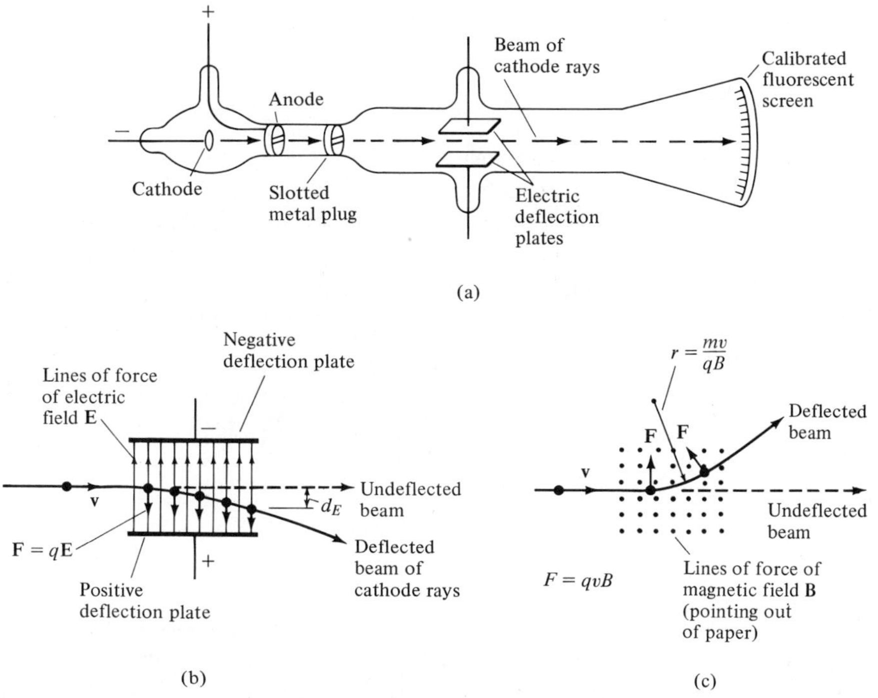

figure 10–11 **THOMSON'S EXPERIMENT**

(a) In the experiment a narrow beam of cathode rays is formed by the slots in the anode and metal plug. A spot of light appears at the point where the beam strikes the fluorescent screen. Deflection of the beam by an electric or magnetic field can be measured from the change in position of the spot on the screen. An electric field may be produced between the deflection plates; a magnetic field may be produced by an electromagnet (not shown). (b) Deflection of cathode rays in an electric field. (c) Deflection of cathode rays in a magnetic field. The direction of deflection indicates that the particles are negatively charged.

of oppositely charged metal deflection plates, and an electromagnet was used to produce the magnetic field.*

As shown in Figure 10–11b, the cathode-ray particles passing between the deflection plates are acted on by an electric force directed along the lines of force toward the positively charged plate. The particles experience an acceleration perpendicular to their initial velocity, so that their motion is similar to that of a projectile in a gravitational field. The magnitude of the acceleration can be found by applying Newton's second principle to the electric force:

$$F = Eq = ma$$

*Thomson's apparatus is the forerunner of the modern-day ***cathode-ray tube*** used in television sets. A rapidly alternating voltage on the deflection plates makes the cathode-ray beam sweep back and forth across the fluorescent screen, and a picture is produced by varying the intensity of the beam to give light and dark spots of different intensities.

so that

$$a = \frac{Eq}{m}$$

where E is the electric field intensity, m is the mass, and q is the magnitude of the charge on the cathode-ray particle.

Since the particle has a constant acceleration perpendicular to its initial motion, its vertical deflection will be equal to $d_E = \frac{1}{2}at^2$, where t is the time required for the particle to pass through the electric field. If l is the length of the field and v the initial speed of the particle, then $t = l/v$. Substituting for the acceleration and time, we obtain

$$d_E = \frac{1}{2}at^2 = \frac{1}{2}\left(\frac{Eq}{m}\right)\left(\frac{l}{v}\right)^2$$

Rearranging terms gives us

$$\frac{q}{m} = \frac{2d_E v^2}{El^2} \tag{10–11}$$

E and l can be readily measured, and d_E can be determined from the change in position of the spot of light on the fluorescent screen. Thus the q/m (charge-to-mass) ratio for a cathode-ray particle can be calculated if its speed v is known. Thomson was able to determine the speed by simultaneously applying an electric and a magnetic field and adjusting their strengths so that the electric and magnetic forces exactly cancel each other. (When the forces balance, there will be no net deflection of the beam.) Then

$$F_{\text{elec}} = F_{\text{mag}}$$
$$qE = qvB$$

and

$$v = E/B$$

Thus Thomson was able to calculate the particle speed from the magnitudes of the field strengths required to balance the electric and magnetic forces exactly. He found that the cathode rays in his experiment were traveling at speeds of approximately one-tenth the speed of light. (The speed of a cathode-ray particle depends on the voltage between cathode and anode.) With v known, Thomson could calculate the q/m ratio from Eq. 10–11. He obtained a value of

$$\frac{q}{m} = 1.76 \times 10^{11} \text{ coulombs/kg}$$

Thomson devised several alternative methods for determining the q/m ratio of the cathode-ray particles and obtained the same value in each case. He demonstrated that the value of q/m is independent of the gas, the type of metal used for the cathode, and the voltage between the cathode and anode.

Thomson's experimental methods permitted him to determine q/m for the cathode-ray particles, but not q or m separately. However, Thomson could compare his result with the q/m ratios of various ions obtained from electrolysis experiments. Faraday had found that 96,500 coulombs of charge were required to liberate 1 gram of hydrogen. Assuming that this charge is equally distributed over all the hydrogen ions in 1 gram (= 0.001 kg) of hydrogen gives a q/m value for the hydrogen ion of

$$\frac{q}{m} = \frac{96{,}500 \text{ coulombs}}{0.001 \text{ kg}} = 9.65 \times 10^7 \text{ coulombs/kg}$$

The charge on a hydrogen ion could also be estimated. Using the kinetic theory, we can determine that the mass of a hydrogen atom is 1.67×10^{-24} gram = 1.67×10^{-27} kg. (See page 230.) Multiplying the q/m ratio for hydrogen times the mass of the hydrogen atom gives

$$\left(\frac{q}{m}\right)(m) = q = (9.65 \times 10^7 \text{ coulombs/kg})(1.67 \times 10^{-27} \text{ kg})$$
$$= 1.6 \times 10^{-19} \text{ coulomb}$$

Hence the average positive charge on a hydrogen ion is 1.6×10^{-19} coulomb.

Thomson did not know the charge of a cathode-ray particle, but he did know that its q/m ratio (1.76×10^{11} coulombs/kg) is more than 1800 times larger than that of a hydrogen ion (9.65×10^7 coulombs/kg). He could at least conclude that a cathode-ray particle must have either a much larger charge or a much smaller mass than a hydrogen ion.

Additional experiments performed by Thomson in 1898 and 1899 settled the question. It had been known for many years that ultraviolet light striking a metal plate causes the emission of negatively charged particles from the metal. (This phenomenon, known as the ***photoelectric effect,*** will be discussed in detail in Chapter 14.) Thomson was able to measure the average charge carried by these particles as well as their q/m ratio. He discovered that the average charge was -1.6×10^{-19} coulomb, equal in magnitude to that of the hydrogen ion, and that the q/m ratio was the same as that of a cathode-ray particle. Thomson therefore concluded that the emitted particles were identical to cathode-ray particles, so that he now knew the charge of a cathode-ray particle. Since the charge is equal in magnitude to the charge on a hydrogen ion, it followed from a comparison of the q/m ratios that the mass of a cathode-ray particle is 1/1800 the mass of a hydrogen ion. The following values could now be established for the cathode-ray particle:

$$q = -1.6 \times 10^{-19} \text{ coulomb}$$
$$m = 9.1 \times 10^{-31} \text{ kg}$$

Thomson was now convinced that the cathode-ray particle was a fundamental particle of matter that could be found in the atoms of all elements. These particles, or "corpuscles," as Thomson called them, were assumed to carry one fundamental unit or "electron" of negative charge. Before long, the corpuscles themselves became known as electrons. (Thus in the modern usage the word "electron" is used to denote a specific particle of matter, although originally the word was used to define only the charge on that particle.)

Thomson then went on to propose a new theory of the atom. He regarded the atom as a sphere of positively charged matter in which enough electrons are embedded to make the atom electrically neutral. Ionization consists of the separation of one or more electrons from the atom, so that the remaining atom becomes a positively charged ion. Because the negatively charged electrons are lighter and more mobile than the positively charged ions, an electric current in a conductor consists of electrons in motion. (Thus we see that Benjamin Franklin made the wrong guess when he assumed that electric current is the flow of positive charge.)

THE NATURE OF THE ELECTRON

Further developments during the more than seventy-five years that have elapsed since the discovery of the electron have served to confirm the status of the electron as one of the fundamental constituents of matter. The charge and mass of the electron have been measured with great precision; the most recently announced values (1971) are as follows:*

$$e = 1.6021917 \pm 0.0000070 \times 10^{-19} \text{ coulomb}$$
$$= 4.803250 \pm 0.000021 \times 10^{-10} \text{ esu}$$

and

$$m = 9.109558 \pm 0.000054 \times 10^{-31} \text{ kg}$$

There is no evidence to indicate that the electron can be subdivided or that it has an internal structure. Yet the nature of the electron is only vaguely comprehended. The size of an electron remains unknown; its radius is certainly less than 10^{-13} cm, and it gives the appearance of being a point particle. The gravitational mass of an electron is unknown; the deflection of an electron in a gravitational field is much too small to be measured experimentally. The measured mass of an electron is its *inertial* mass, or more specifically, its resistance to acceleration in an electric or magnetic field. This resistance can, in principle, be attributed to the interaction between the electron and its own magnetic field, which is generated when the electron is in motion. Hence it may be somewhat naïve even to consider the electron as a material particle. In spite of these uncertainties, it is possible to predict the interactions between electrons and other matter with great accuracy. At the present time, however, the exact nature of the electron remains one of the unsolved problems of physics.

THE NATURE OF CHARGE

Thomson's discovery of the electron also served to provide the Lorentz electron theory with clear evidence for the existence of discrete charges that act as the sources of electric and magnetic fields. It remained to be shown, however, that the charge on the electron was truly the fundamental unit of charge, and that smaller quantities of charge do not appear in nature.

Thomson's methods for determining the electronic charge were entirely statistical. The total charge on a large aggregate of ions was measured, and an estimate was made of the total number of ions present. Thus only the *average* charge per ion could be calculated—and not to any high degree of accuracy. This situation was remedied by the American physicist Robert A. Millikan, who announced in 1911 that he had devised a method for measuring the charges on *individual* ions.

The experimental apparatus that Millikan used is illustrated in Figure 10–12. Tiny oil drops sprayed from an atomizer were allowed to enter a small cylindrical chamber, where their motions could be studied individually. The oil drops usually acquired a charge, either by friction as they were sprayed out of the atomizer nozzle or by capturing ionized oxygen or nitrogen molecules from the surrounding air. Inside

*The symbol e is commonly used to symbolize the magnitude of the electronic charge; thus $q_{\text{electron}} = -e$.

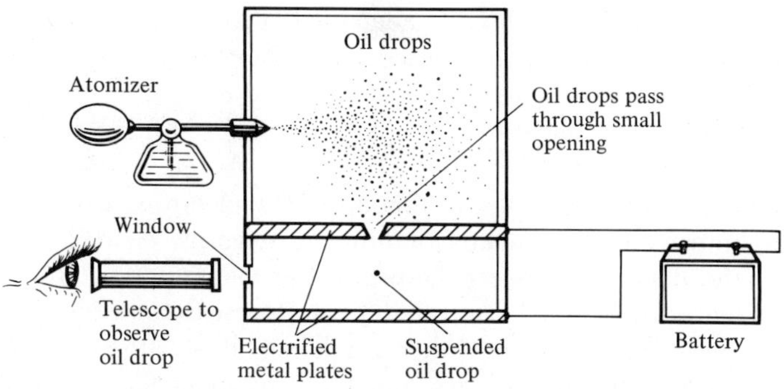

figure 10–12 **THE MILLIKAN OIL DROP EXPERIMENT**

Individual oil drops can be suspended between the two electrified plates by counteracting gravity with an upward electric force. The charge on the oil drop can then be calculated.

the chamber the oil drops were subjected to an electric field that could be made to counteract the force of gravity. Thus when the upward electric force ($F = Eq$) on the oil drop is exactly equal to the oil drop's weight ($w = mg$), the drop will be suspended in midair, or will move up or down with a constant speed. When the forces are balanced,

$$Eq = mg$$

Thus if the mass m of an oil drop can be determined, its acquired charge q can be calculated, since E and g are known quantities.

There is an important distinction between Millikan's method and Thomson's experiment with cathode rays. Thomson was working with *free electrons,* which are too light to be measurably deflected by gravity; in Eq. 10–11 q and m represent the charge and mass of the electron itself. In Millikan's experiment the oil drop acts as a *carrier* of electric charge; one or more electrons or positive ions have been attached to (or removed from) the oil drop, which would otherwise be electrically neutral. Hence q represents the net charge of all the attached ions, and m represents the mass of the entire oil drop, which is large enough for its mass to be measured.

Millikan was able to calculate the mass of a single oil drop by observing its motion when the electric field was turned off. The drops do not fall freely in a gravitational field; because of air resistance and their very small size, the drops fall with a constant speed which depends on their mass and the resistance properties of air. Since the drops fall rather slowly, Millikan was able to measure the speed of an individual oil drop as it fell between two reference cross hairs. The data thus obtained enabled Millikan to calculate the mass of the drop. Hence by comparing the motion of an oil drop under gravity alone to its motion when the electric field is turned on, Millikan was able to determine the charge on an individual oil drop.

Millikan discovered, in the course of observing the motions of hundreds of oil drops, that the charges carried by the oil drops were always exact multiples of 1.6×10^{-19} coulomb. He observed oil drops with both positive and negative charges of anywhere from 1 to 150 times this number, without finding any exception to the rule. On several occasions an oil drop under observation was seen to change speed suddenly, indicating that it had gained or lost charge. In all of these instances Millikan found that the amount of charge gained or lost was an exact multiple of 1.6×10^{-19} coulomb.

With this beautifully simple experiment Millikan convincingly demonstrated that there is an elementary unit of electric charge, and that the charges on individual ions are exact multiples of this unit. The electron, a fundamental subatomic particle, is known to carry exactly one unit of negative charge. Thus we say that the charge of the electron is $-e$ (where e is the magnitude of the fundamental charge unit), or more simply, -1. Twentieth-century research on the nature of matter has revealed the existence of many other fundamental subatomic particles—protons, neutrons, pions, mesons, and so on—and these particles have all been found to have charges of either -1, 0, or $+1$. The fundamental units of positive and negative electricity (the charges on the proton and electron, respectively) have been compared and have been found to be identical to an accuracy of 1 part in 10^{20}.

The data amassed since the time of Faraday clearly indicate that the concept of an electric fluid is untenable. Electric charge exists in discrete, indivisible units or packets; it is never found in a disembodied or free state, but is always attached to some material particle such as an electron. Yet in spite of these discoveries, we cannot profess to know what electric charge is. The existence of *two* kinds of charge simply compounds the mystery. It will be recalled that "electric" originally meant "having the properties of amber," and the word "charge" is a relic from the electric fluid theory. We know at least that all matter possesses electric properties; electric neutrality results from equalizing effects of positive and negative charges. Hence we must think of electricity, like gravity, as a fundamental, unexplained property of matter.

10–6

THE ETHER AND THE ELECTROMAGNETIC FIELD

As we have seen, the existence of the ether was a central assumption in the development of the theories of light and electromagnetism. In the wave theory of light, the ether acts as the medium through which light propagates as a transverse mechanical wave. In Maxwell's electromagnetic theory, it was assumed that all electromagnetic forces are transmitted by mechanical action through the particles of the ether. An electromagnetic field was thought to represent a state of strain or tension (i.e., a disturbance from equilibrium) in the ether. Thus an electromagnetic wave was depicted as a kind of mechanical wave that propagates through the "strained" medium in much the same way as a transverse wave propagates along a taut string. With the identification of light as an electromagnetic wave, the theories of light, electromagnetism, and the ether were conveniently merged.

The major unsolved problem in this system had to do with the properties of the ether itself. During the last half of the nineteenth century numerous theories were put forth in an attempt to describe the internal structure of the ether and to explain, in terms of this structure, how forces could be transmitted through the ether. Foremost among these was the vortex theory proposed by Maxwell as part of his electromagnetic theory. The vortex theory was quite popular for a time; in the 1860s and 1870s the properties of vortices were analyzed in considerable mathematical detail by such luminaries as Helmholtz and Lord Kelvin as a possible way toward understanding the nature of the ether. Eventually, the vortex theory lost favor, however. In later editions of his *Treatise on Electricity and Magnetism,* Maxwell abandoned the vortex theory, although he persisted in his belief that electromagnetic forces are transmitted by mechanical action from particle to particle through the ether. In Maxwell's thought the electromagnetic theory remained as a purely mechanical theory in the Newtonian tradition.

Thus the ether provided the backbone of the most important theoretical developments of nineteenth-century physics. But at the end of the century, a series of experiments were performed that led to the inescapable conclusion that *the ether did not exist.* In the following chapter we will consider these experiments more fully and study the wider implications of this discovery. At this point, however, it will be appropriate to consider just the consequences for electromagnetic theory.

Early in the twentieth century it became quite evident that a model of the ether as a medium composed of particles or possessing an internal structure or any properties describable in traditional mechanical terms could no longer be held. Obviously, an etherless theory of electromagnetism would have to be formulated. In doing so, physicists had to keep in mind that even though the mechanical electromagnetic theory was formulated on an incorrect premise, *its conclusions turned out to be correct.* That is, the predictions made on the basis of Maxwell's and Lorentz's equations—such as the existence of electromagnetic waves—had been accurately verified. Moreover, the concept of an electromagnetic field seemed to be a generally useful and convenient way of depicting electromagnetic interactions. Hence physicists were faced with the task of tearing out the foundation of the electromagnetic theory without disturbing the superstructure.

Thus the etherless electromagnetic theory retains Maxwell's and Lorentz's equations without alteration, but subjects them to an entirely different interpretation. In the mechanical form of the electromagnetic theory, Maxwell attempted to show that his equations followed from the general mechanical properties of the ether. In the etherless theory no such interpretation can be made. Maxwell's and Lorentz's equations become the *fundamental postulates* of the new electromagnetic theory; that is, they are not explainable in terms of other concepts. In the mechanical theory the electric and magnetic field intensities **E** and **B** were presumed to measure the amount of strain or tension in the medium. In the etherless theory **E** and **B** have no such physical significance. They are considered to be abstract mathematical quantities defined entirely by Maxwell's equations.

In the face of this drastic restructuring of the postulates of electromagnetic field theory, how are we now to understand the meaning of the term "field"? To Faraday and Maxwell, the field represented the state of the ether, but in the absence of an ether such an interpretation must be rejected. Yet it would be foolish to discard the field concept entirely in view of its considerable value to physics as a practical tool.

One way to understand the field concept is to think of it as a mathematical abstraction. For a given arrangement of stationary and moving charges, Maxwell's four equations enable us to calculate the magnitudes and directions of two vector quantities, **E** and **B,** at every point in space. These "field intensities" are to be considered as nothing more than mathematical functions, and they are not directly measurable through experiment. Maxwell's equations attain physical significance only when related to a charged particle located at some point in space. The calculated values of **E** and **B** at that particular point can now be substituted into Lorentz's equations, and the *force* on the charged particle can be calculated. The force on a particle can be measured directly; hence an electromagnetic field can be detected and measured only indirectly in terms of the force exerted.

Consequently, in this interpretation the "field" is merely a mathematical scaffold that enables us to calculate the magnitudes and directions of electromagnetic forces. The fundamental equations describe the basic physical laws of electromagnetism; they are derived from experiments (by Oersted, Ampère, Faraday, and so on). No theoretical attempts are made to explain these laws. Nor is any attempt made to explain how forces are transmitted between bodies. Thus the term "field" acts as a verbal shorthand signifying the mathematical system that enables us to describe and predict various electromagnetic interactions, without understanding the underlying causes. As such, "field" is somewhat akin to "gravity" in Newtonian physics, wherein the motions of planets and satellites can be accurately predicted from a mathematical formula (the law of gravitation), even though the nature of gravity is not understood.

This mathematical interpretation of field represents a step backward in many respects. It becomes necessary to accept as postulates certain aspects of electromagnetism that could formerly be explained in terms of an ether. We must now assume that electromagnetic forces can be transmitted between particles that are separated by empty space. We can no longer visualize an electromagnetic wave as a form of mechanical wave proceeding through the ether; we must now postulate that an electromagnetic wave exists as a unique phenomenon, able to carry energy and momentum while proceeding at finite speed from place to place without the need for an intervening medium.

Hence we are forced to reject the fundamental Newtonian principle that all physical phenomena are ultimately explainable in terms of interactions between particles. The transfer of energy by electromagnetic waves in the absence of any medium or particles to carry this energy stands in direct conflict to Newtonian mechanics. Considering the enormous advances in physics that had been wrought from the mechanical view (not to mention its influence on thought in areas outside physics), its abandonment in this case is perhaps the most disturbing aspect of the new interpretation of the electromagnetic field.

Yet this new picture is sufficient, consistent, and above all, practical. It enables physicists to describe and predict electromagnetic phenomena just as Kepler's laws enabled seventeenth-century astronomers to describe planetary motions. But it is intellectually unsatisfying and aesthetically unpleasant to many physicists, so it is not surprising to find that several new theories have been formulated to fill the gap created by the demise of the mechanical electromagnetic theory. At the present time none of these theories has proved to be singularly successful, and they remain as mostly speculative ventures. In general, two main lines of thought have developed.

One approach involves a rethinking of the concept of a vacuum, by which we mean completely empty space devoid of matter and ether. In the older Newtonian view, it was presumed that free space is filled with the ether, which is composed of particles having properties quite different from those of ordinary matter. Although the exact nature of the ether was difficult to ascertain, it was nevertheless clear that the properties of free space—that is, space devoid of ordinary matter—are the properties of the ether. Where there is neither matter nor ether, a vacuum exists, and the space is completely featureless and without properties.

If the ether does not exist, free space is actually a vacuum. Since it is difficult to divorce ourselves completely from Newtonian concepts, it seems most reasonable to assume that empty space, being entirely devoid of substance, cannot possess any properties of its own and ought to be incapable of transmitting forces or supporting electromagnetic fields or waves.

The modern concept of a vacuum departs from the Newtonian picture by assuming that a vacuum, although featureless, can nevertheless possess properties. In particular, a field represents a *state of the vacuum.* That is, the electric or magnetic field intensity at some point in space *is considered to be a property of the space itself.* Since electromagnetic fields originate in charged particles, it follows that the properties of the vacuum are determined by the presence of matter. As before, Maxwell's and Lorentz's equations provide the connecting link between matter (stationary or moving charged particles) and field (the electric and magnetic intensities at points in space).

In this picture, the vacuum—empty space—becomes a medium, taking on the role of the ether in the Newtonian system. Electromagnetic fields can exist in this medium, and electromagnetic waves can travel through it. But the vacuum is a different kind of medium, for it is featureless and without structure. Unlike the ether, the vacuum is nonsubstance, and its nature is unknowable—or at least it is beyond the realm of ordinary matter. Because we are so engrained in the Newtonian tradition, it is difficult to resist the temptation to describe the vacuum in terms of some mechanical picture. Thus the vacuum is sometimes pictured as a kind of all-pervading fluid in which matter is embedded, with the field representing the distortions produced in the fluid by the embedded matter.

Some modern-day theoretical physicists have attempted, with some success, to set up a theory in which particles of matter are seen as condensations or concentrations in the field. Thus their aim is to do away with the concept of matter entirely! Hence only fields exist, with the vacuum as the medium. Nothing is substance; all is field.

An entirely opposite approach has been taken by other theorists in answer to the problem of an etherless electromagnetic theory. Their idea is to do away with the field concept entirely and to deal only with particles. In this scheme of things, electromagnetic and all other forces arise from an extremely rapid and steady interchange of particles between the interacting bodies. These interchanges take place so rapidly that they cannot be observed by any measuring instruments. Hence the interchanges are referred to as "virtual" processes.

The conflict between the matter and field concepts of nature has pervaded twentieth-century physics and is as yet unresolved. It is just one aspect of the revolution, fomented by the death of the ether hypothesis, that shattered classical physics. In the next chapter we will begin our study of the new physics that arose from the ashes.

QUESTIONS FOR REVIEW

1 Discuss Ampère's theory of magnetism. How is the theory used to explain the following?
(a) How naturally magnetic and nonmagnetic substances differ.
(b) Why no isolated north or south poles are found in nature.
(c) Why some nonmagnetic substances can be magnetized.
(d) Why heating destroys magnetism.

2 Contrast Weber's theory of electromagnetic interactions with the field concept of Faraday and Maxwell.

3 Outline the basic principles of Maxwell's electromagnetic field theory. According to Maxwell, what does an electric or magnetic field represent? How are electromagnetic forces transmitted?

4 List the basic laws of electromagnetism that are encompassed in Maxwell's equations. What was Maxwell's contribution to the laws of electromagnetism?

5 What is an electromagnetic wave? What are its properties? How is it generated?

6 According to Maxwell, how does an electromagnetic wave propagate? Why do an electric wave and a magnetic wave always accompany one another?

7 Explain how Maxwell calculated the speed of an electromagnetic wave. What value did he arrive at? What did Maxwell conclude from this result?

8 What is electromagnetic momentum? Give the equation for this quantity. What is the evidence that light carries momentum?

9 Describe Hertz's experiment. How did Hertz generate and detect electromagnetic waves? What properties of electromagnetic waves did Hertz demonstrate?

10 Explain what is meant by the modulation of a radio wave. What is the difference between an AM and an FM radio signal?

11 Explain how the speaker on a radio converts an electric signal into a sound wave.

12 List in order of increasing frequency the major regions of the electromagnetic spectrum. What distinguishes one region from another? How is each type of wave produced?

13 Why is ultraviolet radiation harmful? How is life on earth protected against solar ultraviolet radiation?

14 Explain how x rays were discovered.

15 Summarize the basic principles of the Lorentz electron theory. Contrast Lorentz's theory with the theories proposed by Weber and by Maxwell.

16 Give the equation for the force on a charged particle in an electric field. How does the direction of the force depend on the sign of the charge?

17 Under what conditions will an electric charge experience a force in a magnetic field? What is the equation that describes the magnitude of this force? How is the direction of this force determined?

18 Explain how a cyclotron works. What purpose is served by (a) the electric field; (b) the magnetic field?

19 How are cathode rays produced experimentally? Before Thomson's experiment what were the two main theories as to their nature?

20 Describe Thomson's cathode-ray experiment. Why are an electric field and a magnetic field both needed? What properties of cathode-ray particles did Thomson measure with his apparatus?

21 How did Thomson conclude that cathode-ray particles have the same charge as a hydrogen ion, but a smaller mass?

22 What was the original meaning of the word "electron"? What has it come to mean today? What are the known properties of the electron?

23 Describe Millikan's oil-drop experiment. What was the purpose of the experiment? What were Millikan's conclusions?

24 Why was the ether a necessary component of the electromagnetic field theory? Why wasn't the field theory rejected entirely when the concept of the ether was discarded?

25 What role do Maxwell's and Lorentz's equations play in the etherless electromagnetic field theory? How are the existence of electric and magnetic fields accounted for in this theory?

EXERCISES

1 Explain how the concept of action at a distance is dealt with in Maxwell's theory.

2 In Maxwell's theory what is the distinction between an electromagnetic field and an electromagnetic wave?

3 The dielectric constant ϵ_0 and the magnetic permeability μ_0 are defined by Eqs. 10–5.
(a) Give the numerical values of these constants in CGS and MKS units.
(b) Show that they are related to the speed of light by the equation $c = \sqrt{1/\mu_0\epsilon_0}$.

4 Explain how one can demonstrate, using Hertz's experimental setup, that radio waves are transverse and can be polarized.

5 In his experiment Hertz produced a standing-wave pattern in which the distance between adjacent nodes was 30 cm. Show that the radio wave was being generated with a frequency of 500 megacycles/sec.

6 Explain how an AM radio signal is modulated to carry both the pitch and loudness of a given sound. Do the same for an FM signal.

7 AM radio stations are spaced 10 kHz apart, and FM stations are spaced 200 kHz apart. Why must FM stations be kept farther apart to avoid interference?

8 What is the purpose of an antenna on a radio or TV set? What is the basic law of electromagnetism that explains how an antenna picks up a radio signal?

9 Review the operation of a transformer (described in Chapter 9). Show that the primary circuit acts as a transmitter of radio waves, and the secondary acts as a receiver.

10 Calculate the frequency of (a) 20-cm radio waves, (b) 20-μ infrared radiation, and (c) 100-Å x rays.

11 Discuss whether or not Superman's "x-ray vision" is in conflict with the presently accepted laws of physics.

12 A hydrogen ion ($q = 1.6 \times 10^{-19}$ coulomb, $m = 1.67 \times 10^{-27}$ kg) is traveling at a speed of 3×10^7 m/sec in a magnetic field of 1000 tesla. The ion is moving perpendicular to the magnetic field.
(a) Calculate the magnitude and direction of the force acting on the ion.
(b) Calculate the radius of the circular orbit of the ion in the magnetic field.
(c) How would your answers to (a) and (b) be changed if the particle were an electron, traveling at the same speed?

13 Explain why no work is done by the magnetic force acting on a moving charged particle.

14 In Thomson's experiment to measure the q/m of a cathode-ray particle, the electric and magnetic field directions had to be oriented at right angles to one another. Explain why this is the case.

15 How does the modern-day theory of electric charge differ from the theories advanced by Du Fay and Franklin in the 1700s? Does the evidence tend to support the two-fluid theory or the one-fluid theory?

16 In Millikan's experiment how was it possible to tell whether the charge on a given oil drop was positive or negative?

17 In the modern-day version of the electromagnetic field theory (in which the concept of the ether no longer appears) is it still possible to think of an electromagnetic wave as a mechanical wave? If not, how is the propagation of electromagnetic waves explained?

Four

THE EINSTEINIAN REVOLUTION

Albert Einstein

With the theory of relativity, Einstein completely revolutionized physics. By successfully challenging the most fundamental concepts of the nature of space and time, he brought about the downfall of the Newtonian system. A philosopher and humanitarian as well as a scientist, and perhaps the most influential figure of the twentieth century, Einstein has come to be regarded as the symbol of creative genius.

THE ORIGINS OF RELATIVITY

11

The evidence for the existence of the ether has accumulated as additional phenomena of light and other radiations have been discovered. . . . Whatever difficulties we may have in forming a consistent idea of the constitution of the ether, there can be no doubt that the interplanetary and interstellar spaces are not empty, but are occupied by a material substance.

James Clerk Maxwell

WE NOW BEGIN the study of one of the greatest intellectual achievements in the history of mankind—Einstein's theory of relativity. Historians of science generally discuss the evolution of physics in terms of two eras—the "classical" and the "modern"—with the first decade of the twentieth century as the dividing line between the two. Newton, as the progenitor of the classical, and Einstein, as the symbol of the modern, represent the two dominant figures in the history of post-Renaissance physics. Just as Newtonian thought pervaded the eighteenth and nineteenth centuries, the spirit of Einstein has come to dominate twentieth-century physics and has influenced developments in many other fields of intellectual, social, and political endeavor.

One of the major reasons for the demise of classical physics was the breakdown of the ether hypothesis. For Newton, Maxwell, and the other supporters of the classical view, the ether was a necessary element of the physical universe. It served as the universal medium through which gravitational and electromagnetic forces could be mechanically transmitted; in fact, nearly all of the theories of classical physics were connected in one way or another to the existence of the ether. Hence the rejection of the ether hypothesis meant that many seemingly well-established physical theories had to be drastically revised.

In this chapter we will consider the role of the ether in the context of the Newtonian system of the universe. We will learn about the discoveries that led to the failure of the ether hypothesis and see how this failure contained the seeds of Einstein's radically new view of the nature of the physical universe.

11–1

NEWTONIAN CONCEPTS OF SPACE AND TIME

The theory of relativity is concerned with the ***relativity of motion.*** Position and velocity are quantities that are measured relative to some frame of reference; hence the values of these quantities depend on the choice of reference frame. A student who is asked the question "What is your velocity?" may respond, "Zero—I'm sitting in a chair." But the chair is attached to the earth, which is rotating on its axis and revolving about the sun; the sun is revolving about the center of the galaxy; and the galaxy itself appears to be in motion with respect to other galaxies. Thus the student's velocity is zero relative to the chair; several hundred miles an hour relative to the center of the earth; 18 mi/sec relative to the sun, and so on. The question "What is your velocity?" is meaningless unless a frame of reference is specified. The student's response should have been "Relative to what?"

The student may believe that by adding together all the relative velocities—the velocity of the earth, the sun, the galaxy, and so on—he will arrive at his "true" velocity. But a "true" or absolute velocity is one that is measured relative to some fixed point in the universe. Thus the concept of an absolute velocity implies the existence of a reference frame that is truly at rest in the universe. Whether such a reference frame exists and can be located is a question of considerable importance in physics.

The concept of the relativity of motion did not originate with Einstein; it appears in the writings of Galileo and was given considerable attention in the *Principia.* Both Newton and Einstein used the principle of the relativity of motion as a point of departure to develop comprehensive theories in which the problems involved in the measurement of motion are related to fundamental concepts of space and time.

To appreciate the significance of Einstein's theory of relativity, we must first examine Newton's theory of the nature of space and time. The concepts proposed by Newton were the cornerstones of his entire mechanical theory and (in view of the success of the Newtonian system) came to be considered by some scientists as almost self-evident. It was these concepts that Einstein challenged.

We will begin by examining the mathematical methods used in the Newtonian system to measure the positions and motions of bodies in space.

COORDINATE SYSTEMS AND REFERENCE FRAMES

In order to study the motion of a body, we must have a means of clearly identifying the position of a body in space at a given instant of time. Coordinate systems, therefore, play an important role in physics, even at the most elementary level. Essentially, a coordinate system assigns to each point in space a unique set of numbers or ***coordinates*** which locate that point with respect to a chosen reference point, known as the ***origin.*** The coordinates represent distances measured from a set of ***coordinate axes*** that are fixed to the origin. The student should already be familiar with the ***Cartesian*** or ***rectangular coordinate system.*** In this system a pair of numbers (x_0, y_0) locates the position of any point in a plane with respect to the origin in terms of two coordinate axes set at right angles to one another. The ***polar coordinate system*** locates points in a plane by using two coordinates (r, θ) such that r_0 is the distance measured along the line joining the origin and the point in question, and θ_0 is the angle between this line and a given reference line (Figure 11–1a). These two systems represent the most convenient methods for locating points in a plane, but they are by no means unique. An infinite number of two-coordinate systems could be devised.

To locate points in physical space, three coordinates are needed, corresponding to the three dimensions of our visual sense perceptions. A three-dimensional rectangular coordinate system can be constructed by defining a third axis (the z axis) perpendicular to the x and y axes. Similarly, the polar coordinate system can be

figure 11–1 **COORDINATE SYSTEMS**

Shown are the most common two-dimensional and three-dimensional coordinate systems used to locate a point P with respect to the origin O.

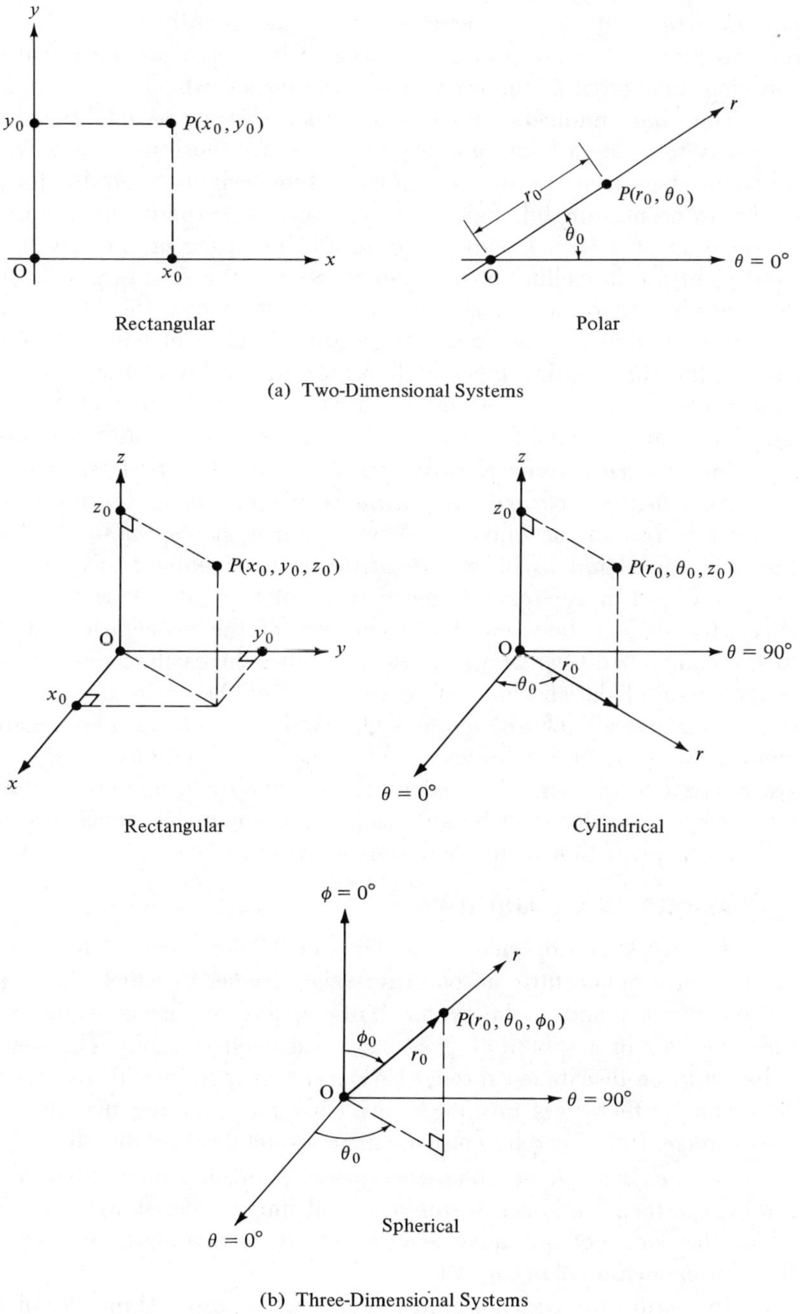

modified by the addition of either a z axis (the cylindrical coordinate system) or a second angle, ϕ (the spherical coordinate system). These systems are illustrated in Figure 11–1b.

Three coordinates are sufficient to locate the *position* of a body in space, but they are not enough to describe the *motion* of a body. For this purpose, a ***time coordinate*** must also be specified. The time coordinate must be measured with reference to some "time origin." This is an arbitrarily chosen event at which we begin counting time—that is, the event marks the instant when we set our clocks at $t = 0$.

Thus four numbers—three space coordinates and one time coordinate—are necessary to locate a body in space and time. We must also specify *two* origins—the origin of the space coordinates and the time origin—in order for the coordinate system to be meaningful. For a simple laboratory experiment, say one in which the acceleration of a falling body is measured, the space origin may be taken as some fixed point on the ceiling of the laboratory, and the time origin as the instant when the timer is turned on. For astronomical measurements, the space origin may be the center of the earth or the sun, and the time origin can be some astronomical event such as the sun crossing the meridian on a given day of the year. In any case, the choice of space and time origins is based mainly on convenience. Once the origins are chosen and a suitable set of space and time coordinates is defined, the system represents the ***reference frame*** against which all motions are measured.

An ***absolute reference frame*** is one in which the origin is considered to be truly at rest in the universe. Any motion with respect to an absolute reference frame is considered to be an ***absolute*** or true motion; any motion with respect to some object or reference frame that is not truly at rest is a ***relative*** motion. In effect, the dispute between the proponents of the geocentric and the heliocentric theories amounted to an argument as to whether the earth or the sun should be chosen as the origin of the absolute reference frame for the entire universe. In a geocentric reference frame all celestial motions observed from the earth are regarded as absolute motions, whereas in a heliocentric reference frame motions are absolute only if they are referred to the sun. (For example, the retrograde motion of Mars as seen from the earth is considered to be an absolute motion in the geocentric reference frame and a relative motion in the heliocentric system.)

ABSOLUTE SPACE AND TIME

The work of Copernicus, Kepler, and Galileo seemed to settle the argument in favor of a heliocentric absolute reference frame, in which the sun was presumed to be stationary and at the center of the sphere of "fixed" stars. But by Newton's time the idea of a sphere of fixed stars had been rejected. The sun and stars were believed to be distributed through space and in motion with respect to one another. Accordingly, there was no longer any basis for assuming that the sun is at rest in the universe. In his search to provide a more suitable absolute reference frame, Newton proposed the concept of ***absolute space.*** Newton conceived of absolute space as infinite, perfectly uniform, featureless, and immovable. It was, in effect, the void in which the bodies of the universe move about. Hence absolute space itself could not be seen or measured in any way.

In conjunction with absolute space, Newton proposed the idea of ***absolute time.*** In the *Principia* Newton described this concept:

> Absolute, true, and mathematical time, of itself, and from its own nature, flows equably without regard to anything external. . . . relative, apparent, and common

> time is some sensible and external measure of duration by the means of motion, which is commonly used instead of true time.[1]

Thus for Newton absolute time was something that flowed uniformly onward, independent of any clock; "common" time, as measured by the most accurate time pieces, could only approximate the true flow of time.

It is surprising, in view of his repeated injunctions against the framing of hypotheses not directly supported by empirical evidence, that Newton should propose such concepts. But absolute space and time were elements of Newton's philosophy more than of his physical theory. To Newton, absolute space and time were metaphysical entities that served to relate the universe to its Creator. A description of this relationship appears in the *Principia:**

> [God] is eternal and infinite, omnipotent and omniscient; that is, his duration reaches from infinity to infinity. . . . He endures forever, and is everywhere present; and by existing always and everywhere, he constitutes duration and space. . . . In him all things are contained and moved; yet neither affects the other: God suffers nothing from the motion of bodies; bodies find no resistance from the omnipresence of God.[2]

Hence in Newton's view absolute space and time were manifestations of the oneness of God and the universe. An omnipresent God must exist forever throughout the infinite reaches of absolute space, and an omniscient God must know what exists at each point in absolute space. Man, on the other hand, is unable to determine the position of any object with respect to absolute space—for absolute space is perfectly uniform and infinite; every portion of it is identical to every other, and there is no way to distinguish physically one location from any other location. Thus there is no way to select or recognize any point of absolute space as the origin of coordinates of an absolute reference frame. There is no point that can be certified as "the center of the universe." Hence because of its uniformity, absolute space has certain disadvantages as an absolute reference frame. We cannot specify the absolute position of a body in absolute space; we can determine only the *relative* positions of bodies with respect to other bodies. Neither is it possible to determine the *absolute velocity* of a body (i.e., its velocity with respect to absolute space); only the *relative velocity* of one body with respect to another body can be ascertained.

For example, let us imagine two spaceships, A and B, located in space far away from any other material bodies. For our purposes we can consider them to be the only objects in an otherwise empty universe. Given the uniformity of absolute space, the absolute positions of these bodies cannot be known. It is possible, however, to determine the distance between A and B with normal measuring instruments. That is, the spaceships' *positions relative to one another* can be measured. Moreover, it is possible to determine whether the distance between them is increasing, decreasing, or constant. Using a clock, observers on A and B can determine the *rate* at which the distance between the two spaceships is changing. In other words, one can measure the *relative velocity* and *relative acceleration* of one body with respect to another.

Let us suppose that the distance between the two spaceships is observed to be

*This is an excerpt from a section entitled "General Scholium," which concludes the second edition of the *Principia.* Newton added this section in response to criticism that his system of the universe was atheistic. Newton carefully detailed his belief in an omniscient, omnipotent God who controls the universe and occasionally intervenes in human affairs.

decreasing at a uniform rate of 10 ft/sec. This could be accounted for in any of the following ways:

(1) A is stationary in absolute space, and B is approaching at an absolute speed of 10 ft/sec.
(2) B is stationary, and A is approaching at an absolute speed of 10 ft/sec.
(3) A and B are moving toward one another, such that the sum of their absolute speeds is 10 ft/sec.
(4) A and B are moving in the same direction, but the trailing spaceship is moving 10 ft/sec faster than the other and is catching up.

Given the properties of absolute space, *there is no way to distinguish between any of these possibilities.* Without a visible origin of coordinates, there is no way to determine one's absolute velocity. Thus we can conclude only that the two spaceships have a relative velocity of 10 ft/sec with respect to each other; we cannot determine the absolute velocity of either one.

We cannot perform any experiments that might enable us to calculate the absolute velocity of a body. This restriction is a direct consequence of the fundamental principles of mechanics—namely, Newton's principles of motion. According to the principle of inertia, no force is required to maintain a body in constant-velocity motion. This is the case whether the body is moving slowly or rapidly or is at rest. Hence inertial motion is independent of velocity; the observation that $F = 0$ on a given body tells us nothing about the absolute velocity of that body.* Similarly, accelerated motion is also independent of velocity. A given force acting on a body always produces the same acceleration ($\mathbf{F} = m\mathbf{a}$) whether the body is at rest or in motion. That is, a given force always produces the same *change* in the velocity of a body, regardless of the initial velocity of the body. Also, neither the principle of action and reaction nor the principle of gravitation depends in any way on the velocity of the body being acted on. Hence the Newtonian system of mechanics offers no method for determining experimentally the absolute velocity of a body, for none of Newton's principles depends in any way on the absolute velocity.

Remarkably, the Newtonian system does provide a means for determining the *absolute acceleration* of a body. In any circumstance in which the force acting on a body can be measured or calculated, the equation $\mathbf{F} = m\mathbf{a}$ can be used to determine the acceleration. Thus it is possible to know whether or not a body is accelerating in absolute space. Consider the following examples:

(1) Two spaceships are *accelerating* relative to one another. (That is, their relative velocity is changing.) It is possible that only one of the spaceships is accelerating while the other is moving at constant velocity, or that both are accelerating but at different rates. With the aid of the second principle we can distinguish between the two possibilities. In each spaceship the passengers can determine, through observation and experiment, whether or not they are accelerating. In a spaceship moving at constant velocity through empty space, an astronaut will float freely. If the spaceship accelerates forward, however, the freely floating astronaut (who continues to move forward at constant velocity while the ship accelerates) will collide with the rear of the ship, as if he were being thrown backward. If a spaceship is accelerating, seated astronauts will feel themselves being pressed against their seats, whereas they will

* We could imagine a hypothetical universe in which a force *is* required to maintain constant-velocity motion, such that $F \propto v$. Then by measuring the force on a body, one could calculate its absolute velocity.

feel no such force if the spaceship is moving at constant velocity (Figure 11–2a). The force can be measured, and the absolute acceleration of each spaceship can then be calculated.

(2) A large celestial body, such as a planet or a star, is located in space far from any other matter. As Newton demonstrated in the *Principia,* a large body held together by gravitation assumes a spherical shape, but if it is rotating it tends to flatten across the poles and bulge at the equator because of inertia. The amount of flattening (the disparity between the polar and equatorial diameters) is directly related to the rotation rate. (See page 156.) Rotational motion is accelerated motion; hence direct observation of the shape of a body can reveal whether or not the body is rotating (and thereby accelerating) in absolute space (Figure 11–2b).

Numerous similar examples show that in the Newtonian system it is possible to determine the acceleration of a body in absolute space. In fact, the detectability of absolute acceleration fortified Newton's belief that absolute space was more than

figure 11–2 **ABSOLUTE ACCELERATION**

(a) When the spaceship has no acceleration, all objects exhibit inertial motion and float freely. Astronauts in an accelerating spaceship will be able to detect the acceleration of the ship. (b) Rotation in absolute space may be detected by the extent of flattening of a spherical body.

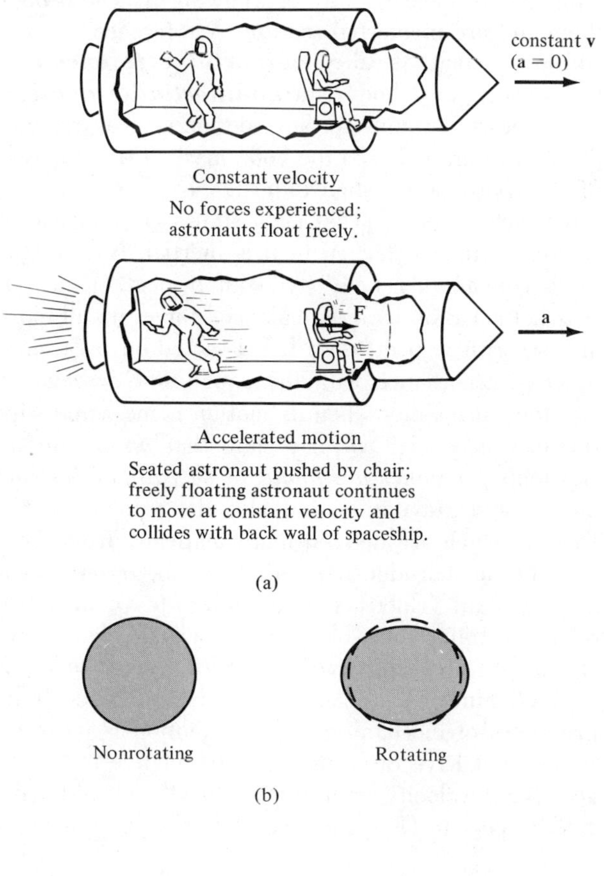

just a metaphysical principle, for he reasoned that if absolute acceleration can be measured, absolute space must exist. But acceleration is a measure of the *change* in the velocity or position of a body, so that even if we know the absolute acceleration of a body, *we still cannot determine its velocity or position in absolute space.*

INERTIAL REFERENCE FRAMES

Absolute space may have been a philosophically satisfying and even necessary element of Newton's cosmology, but as a scientific concept it left much to be desired. Given the perfect uniformity of absolute space, it follows that no object—not the sun, the earth, or the fixed stars, or the galaxy—can be considered to be absolutely at rest and suitable as the origin of coordinates for an absolute reference frame. Hence we must accept as a basic property of the Newtonian system that *it is impossible to define an absolute reference frame.* The Newtonian universe is neither heliocentric nor geocentric—in fact, it does not have a center at all.

Nevertheless, we must establish some frame of reference in order to study the motions of bodies, so we are forced to select an origin of coordinates from among the bodies that exist in absolute space. Once we have selected one of these bodies as an origin, we attach a set of coordinate axes to the body and consider this reference frame to be "stationary" for mathematical purposes. We then measure the positions and motions of other bodies with respect to this "stationary" reference frame.

Bodies in absolute space can be divided into two groups—those that exhibit absolute acceleration and those that exhibit no acceleration. The members of the latter group are moving with constant velocity in absolute space—and may even be at rest, although we have no way of knowing it. These bodies have no net force acting on them and are in inertial motion. A reference frame attached to a body moving with constant velocity is called an ***inertial reference frame.*** A reference frame attached to an accelerated body is a ***noninertial*** or ***accelerated reference frame.***

Let us now consider how motion is measured with respect to an inertial reference frame. In Figure 11–3a the body marked O is the reference body chosen as the origin of the coordinate system indicated by the x, y, z axes. Body O has no unbalanced forces acting on it. (That is, it shows no evidence of accelerated motion.) Thus the reference frame attached to it is inertial. Now suppose that we wish to describe the motion of another body (A) with respect to the reference frame attached to O. If A is not acted on by an unbalanced force, it will move at constant velocity (or remain at rest) with respect to O. If it is acted on by a force, A will accelerate with respect to O in accordance with the equation $\mathbf{F} = m\mathbf{a}$. In general, a body always obeys Newton's principles when its motion is measured with respect to an inertial reference frame. Conversely, we may state that *an inertial reference frame is one in which Newton's principles are found to be valid.* (This includes not only the principles of motion and gravitation, but also the principles of conservation of momentum and energy, which are derived mathematically from the principles of motion.)

Let us consider a second reference frame attached to body O′. If O′ is moving at a constant velocity $\mathbf{v}$ with respect to O, then O′ cannot be accelerating, and its reference frame must also be an inertial reference frame. Thus *any reference frame moving at a constant velocity with respect to an inertial reference frame is itself inertial.* Since O′ is also an inertial reference frame, body A will obey Newton's principles of mechanics when its motion is measured with respect to O′. Although A does not have the same velocity relative to O′ as to O, A nevertheless maintains a constant velocity with respect to O′ unless acted on by a force. If A accelerates with respect to O, it must accelerate with respect to O′ as well.

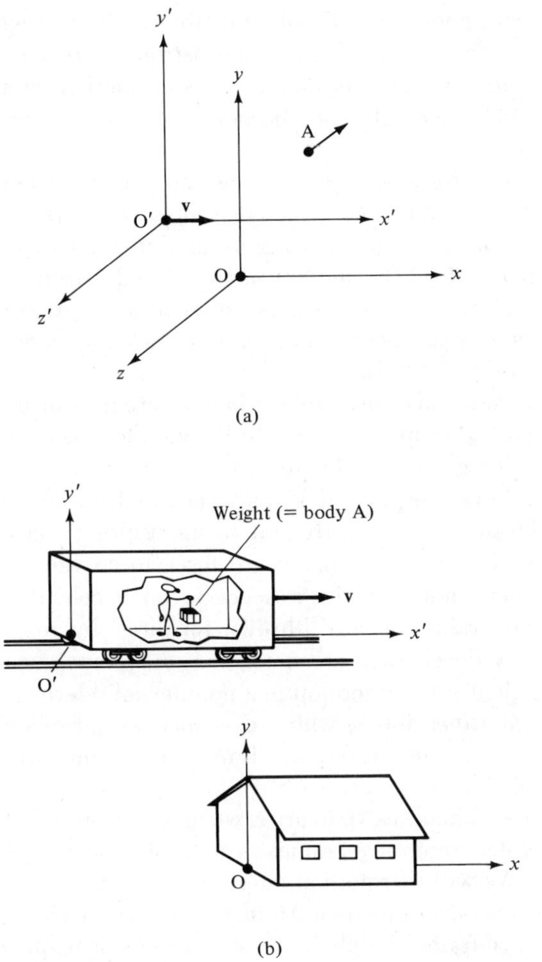

figure 11–3

INERTIAL REFERENCE FRAMES

(a) O and O′ represent the origins of two inertial reference frames. O′ has a constant velocity **v** with respect to O. The motion of body A will follow Newton's principles when measured relative to either coordinate system. (b) O and O′ are inertial reference frames attached to a station and a moving train, respectively. A weight carried in the train will obey Newton's principles in both reference frames.

As a more concrete example of these principles, let O′ be a train moving at constant velocity, O a train station, and A a weight carried by a scientist riding on the train (Figure 11–3b). The train and the station can both be considered inertial reference frames. If the weight remains in the scientist's lap, it is at rest with respect to the train and is moving at constant velocity with respect to the station. In other words, it is in inertial motion relative to both O and O′. If the weight is released and allowed to fall to the floor, it will be accelerated vertically by gravity. It will have the same acceleration (32 ft/sec^2 downward) relative to each reference frame. Thus the weight obeys the same principles of mechanics in both reference frames. The only difference is that the weight has a different velocity relative to each reference frame.

Newton's principles of motion are *not* obeyed, however, when the motion of a body is measured with respect to an *accelerated* reference frame. Suppose O is an inertial frame, and O′ is an accelerated frame. A is not acted on by any forces and is moving at constant velocity with respect to O. But with respect to O′, A *appears* to be accelerating. In other words, A has a measured acceleration relative to the O′ coordinate axes *even though A is not being acted on by a force.* Thus the motion of A seems to violate the principle of inertia.

As an illustration of such a situation, consider the motion of the freely floating astronaut in Figure 11–2a. When the spaceship accelerates, the astronaut *appears* to accelerate backward. In other words, the astronaut has an accelerated motion with respect to the spaceship (which is an accelerated reference frame) even though there are no identifiable forces acting on him.

Suppose that body A is acted on by a force that gives it the same acceleration as O′. Now A has a constant velocity *relative to O′*. That is, A *appears* to be in inertial motion, even though it is acted on by a force. Thus its motion relative to O′ seems to violate Newton's second principle. (The motion of the seated astronaut in Figure 11–2a when the spaceship accelerates is an example of such a situation.) Hence *when the motion of a body is viewed from the origin of an accelerated reference frame, Newton's principles of motion are not obeyed.*

In order to "explain" the failure of Newtonian principles when we are measuring motions with respect to accelerated reference frames, we tend to invoke the existence of unexplained forces. The "centrifugal force" that seems to throw us outward when the car we are in turns a corner appears because the car is an accelerated reference frame. A person in a freely falling elevator and an astronaut in an orbiting space capsule experience "weightlessness," as if there were an upward force to counteract gravity, because they are measuring their motions with respect to an accelerated reference frame. Passengers in a subway train who are "thrown forward" by some mysterious force when the train suddenly decelerates are not acted on by any such force at all; they are simply maintaining their inertial motion in a noninertial reference frame. In effect, such forces are really *fictitious* forces whose existence we postulate in order to make our motion in an accelerated reference frame consistent with Newtonian principles.

The simplest way to deal with the difficulties that arise when the motions of bodies are measured relative to accelerated reference frames is to avoid the use of accelerated reference frames altogether. We will therefore attempt to measure motions with respect to *inertial* reference frames whenever possible. An inertial reference frame is more useful than an accelerated reference frame, simply because Newton's principles of motion are valid in all inertial reference frames.

Since Newton's principles are equally valid in all inertial reference frames and cannot be used to distinguish absolute motion, there can be no "preferred" inertial reference frame. The choice of one inertial reference frame as "stationary" and another as "moving" is made solely for mathematical convenience. There is no physical justification for making such a distinction, for no mechanical experiment can be done that might distinguish an inertial reference frame as being at rest in absolute space. Thus from the point of view of Newtonian mechanics, *all inertial reference frames are equally valid.*

Of the bodies that populate the universe, which ones may be chosen as inertial reference frames? The earth cannot be an inertial reference frame, for it is rotating on its axis and revolving about the sun—both accelerated motions. The sun and other stars revolve about the center of the galaxy and are therefore not in inertial motion. Finally, we might try to establish an inertial reference frame in the form of a spaceship cruising through intergalactic space. But this spaceship will still be subject to the acceleration-producing gravitational forces (however small) exerted by even the most distant galaxies. As far as we know, there is no place in the universe that is completely free of external accelerating forces. Thus there are no truly inertial reference frames in the visible universe.

When the external forces on a body are small, however, it is possible to establish a reference frame that is *nearly* inertial. The forces exerted on the sun by the other stars are sufficiently small so that the sun can be considered a good inertial reference frame for the solar system. A laboratory attached to the surface of the earth serves as an adequate inertial reference frame for experiments of moderate accuracy in which the effects of the earth's rotation and revolution are considered negligible. Thus it is almost always possible to select a reference frame that is inertial to a desired degree of accuracy.

If the motion of a body is measured with respect to one inertial reference frame, it is a simple matter to describe the body's motion with respect to any other inertial reference frame. In Figure 11–4a we find two inertial reference frames, O and O′, in motion relative to one another. Let us assume that O′ has a velocity $\mathbf{v}$ relative to O. Let us further assume that the body A is in motion and has a velocity $\mathbf{u}'$ measured with respect to O′. To find the velocity $\mathbf{u}$ of A relative to O, we simply add together the two velocity vectors:

$$\mathbf{u} = \mathbf{u}' + \mathbf{v} \tag{11–1}$$

Or, in other words,

$$\begin{bmatrix}\text{velocity of A} \\ \text{relative to O}\end{bmatrix} = \begin{bmatrix}\text{velocity of A} \\ \text{relative to O}'\end{bmatrix} + \begin{bmatrix}\text{velocity of O}' \\ \text{relative to O}\end{bmatrix}$$

Equation 11–1 expresses in mathematical form the familiar rule of *addition of velocities*. For example, let O be a train station and O′ a train moving at $v = 60$ mi/hr (Figure 11–4b). A bee (body A) flies through the train from back to front at a speed of 5 mi/hr relative to the train ($u' = 5$ mi/hr). Since both $\mathbf{u}'$ and $\mathbf{v}$ point in the same direction, the bee's velocity relative to the train station is

$$\mathbf{u} = \mathbf{u}' + \mathbf{v} = 5 \longrightarrow + 60 \longrightarrow = 65 \text{ mi/hr} \longrightarrow$$

If, however, the bee flies in the opposite direction through the train, his velocity relative to the station is

$$\mathbf{u} = \mathbf{u}' + \mathbf{v} = 5 \longleftarrow + 60 \longrightarrow = 55 \text{ mi/hr} \longrightarrow$$

One consequence of Equation 11–1 is that a *change in the velocity of a body* will have the same value measured in two different inertial reference frames. If the bee increases his speed relative to the train from 5 mi/hr to 8 mi/hr, he also increases his speed relative to the station from 65 mi/hr to 68 mi/hr. But in either reference frame the observed *change* in speed is 3 mi/hr.

The acceleration of a body is, as we learned in Chapter 3, the rate of change of velocity with respect to time:

$$\mathbf{a} = \frac{\Delta \mathbf{v}}{\Delta t}$$

The measurement of time in different reference frames must now be considered. Given the nature of absolute time, which—according to Newton—flows on without relation to anything external, it seems natural to assume that time will elapse at the same rate in all reference frames. That is, the time interval Δt between two events will be of the same duration in all reference frames. It follows mathematically that if the change of velocity and the time interval over which the change occurs have

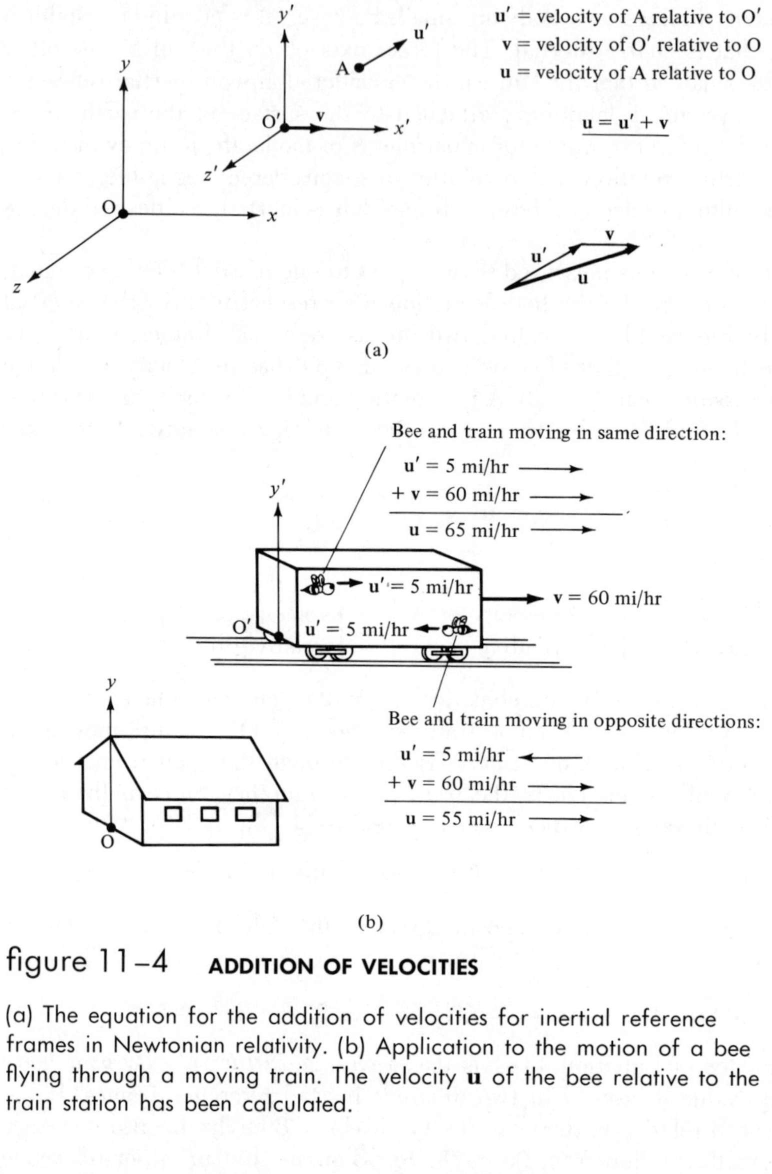

figure 11–4 **ADDITION OF VELOCITIES**

(a) The equation for the addition of velocities for inertial reference frames in Newtonian relativity. (b) Application to the motion of a bee flying through a moving train. The velocity **u** of the bee relative to the train station has been calculated.

the same value in all inertial reference frames, *the acceleration of a body has the same value in all inertial reference frames.*

If we assume further that the mass of a body has the same value in all reference frames, the accelerating force ($\mathbf{F} = m\mathbf{a}$) must also have the same magnitude in all inertial reference frames.

NEWTONIAN RELATIVITY

This concludes our summary of what may justifiably be called "Newton's theory of relativity." The basis of Newtonian relativity is the concept of absolute space. The properties of absolute space are such that we are unable to detect absolute motion. Thus we cannot make measurements with respect to absolute space; we can make measurements only with respect to other bodies. To make such measurements, we define an inertial reference frame, which is a reference frame attached to a body moving at a constant velocity. Measurements can be made with respect to any inertial reference frame, since Newton's principles of motion are equally valid in all inertial reference frames. We find that the magnitudes of certain physical quantities depend on the choice of inertial reference frame; specifically, position and velocity are examples of such *relative* quantities. On the other hand, certain physical quantities, such as time, mass, acceleration, and force, have the same value in all inertial reference frames, so they are *absolute* quantities.

Newtonian relativity has few surprises. Most of its assumptions and conclusions seem eminently reasonable and rather obvious. Unfortunately, in physics what seems obvious is not always what is true, as the geocentrists discovered during the Renaissance. Eventually, certain observations were made that seemed to contradict the Newtonian picture. These contradictions could not be resolved within the Newtonian framework, and Newtonian relativity had to be rejected. As we shall learn in the next section, the rock on which the Newtonian system foundered was the ether hypothesis.

11–2

THE FAILURE OF THE ETHER HYPOTHESIS

In previous chapters we learned about the role played by the ether in the development of the theories of physics. Newton believed the ether to be a medium composed of tiny particles, similar to air but considerably more rarefied, and offering no measurable resistance to the motion of bodies. Therefore it would be extremely difficult, if not impossible, to determine whether a given body is at rest or in motion with respect to the ether. Moreover, the ether could have no special relationship with respect to absolute space—for the ether is considered to be a *material* substance; like other matter, it is something that *exists in absolute space*. Consequently, there could be no way to ascertain whether the ether is at rest or in motion relative to absolute space. Although it is tempting to assume that the ether is stationary in absolute space, we realize that there is no logical justification for doing so. Hence from a *mechanical* point of view, the ether cannot be certified as a special reference frame in the universe.

RELATIVITY AND ELECTROMAGNETIC THEORY

We have learned that there is no *mechanical* way to detect absolute motion because Newton's principles are independent of the velocity of a body. But with Maxwell's successful development of the electromagnetic field theory in the 1860s, the possibility of detecting absolute motion had to be re-examined, because some of the laws of electromagnetism are velocity-dependent. We have learned, for example, that the Lorentz force (Equation 10–9) acting on a moving charged particle in a magnetic field depends on the velocity of the particle. Of even greater significance is the fact that Maxwell's equations predict a free-space speed of $c = 3 \times 10^{10}$ cm/sec for electromagnetic waves. Newtonian relativity shows that velocities must be measured with respect to some particular reference frame, and indeed Maxwell's theory states that c is the speed of electromagnetic waves *relative to the ether*.

Electromagnetic theory thus establishes the ether as a "special" reference frame:

The ether is the reference frame in which electromagnetic waves have a free-space speed equal to c. Given the rule for addition of velocities, it follows that the speed of light relative to a reference frame moving with respect to the ether must be *different* from c. That is, if we let

$$\begin{aligned} O &= \text{"stationary" frame attached to the ether} \\ O' &= \text{"moving" frame that has velocity } \mathbf{v} \text{ relative to ether} \\ \mathbf{c} &= \text{velocity of light relative to O (ether)} \\ \mathbf{c}' &= \text{velocity of light relative to O}' \text{ (moving observer)} \end{aligned}$$

then, according to Equation 11–1, $\mathbf{c} = \mathbf{c}' + \mathbf{v}$, or if we solve for $\mathbf{c}'$,

$$\mathbf{c}' = \mathbf{c} - \mathbf{v} \tag{11–2}$$

In other words, the *observed* velocity of light is equal to the *vector difference* of $\mathbf{c}$ (the velocity of light in the ether) and $\mathbf{v}$ (the velocity of the observer in the ether).

Referring to Figure 11–5, we can consider two simple examples of this principle. Let O′ be a spaceship that is moving through the ether with a velocity $\mathbf{v}$. In Figure 11–5a, a light ray is shown moving toward the spaceship, which is moving away from the light ray in the same direction. Since $\mathbf{c}$ and $\mathbf{v}$ point in the same direction, then $\mathbf{c}'$, the velocity of light relative to the spaceship, will have a magnitude of

$$c' = c - v \tag{11–3a}$$

figure 11–5 **THE SPEED OF LIGHT AS MEASURED BY A MOVING OBSERVER: PARALLEL MOTION**

According to the ether theory, an observer moving relative to the ether will measure a speed different from c depending on his direction of motion. (a) Light ray and observer moving in the same direction. (b) Light ray and observer moving in opposite directions.

O = stationary reference frame fixed to the ether
O' = reference frame moving through the ether with velocity $\mathbf{v}$ relative to the ether
$\mathbf{c}$ = velocity of light in the ether
$\mathbf{c}'$ = velocity of light measured by moving observer at O'

$$\mathbf{c}' = \mathbf{c} - \mathbf{v}$$

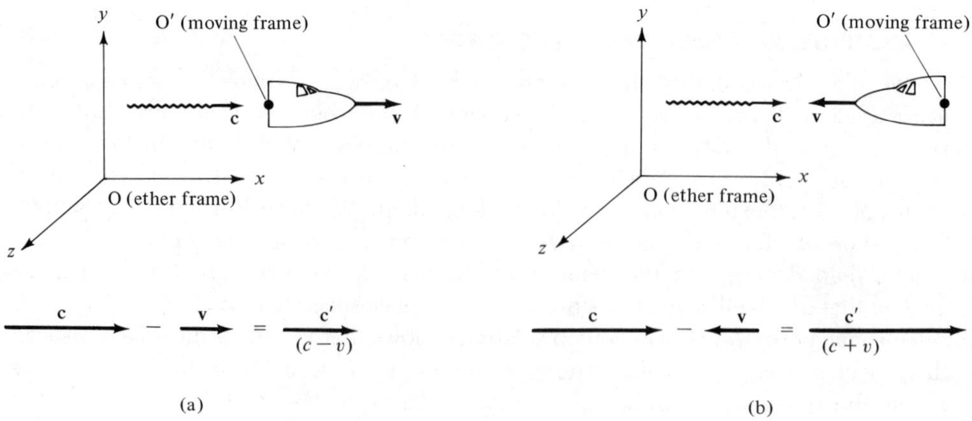

In Figure 11–5b the spaceship is moving *toward* the oncoming light ray; in this case the observed speed of light is

$$c' = c + v \tag{11–3b}$$

Thus for motion in parallel directions, the *observed* velocity of light $\mathbf{c}'$ in a moving reference frame is expected to have a magnitude of $c + v$ or $c - v$, depending on whether the observer is moving toward or away from the light ray.

Figure 11–6 illustrates a more complicated situation. In the moving reference frame the light ray appears to be moving at right angles to the motion of the reference frame itself (i.e., parallel to the y' axis). But to an observer at O fixed in the ether,

figure 11–6 **THE SPEED OF LIGHT AS MEASURED BY A MOVING OBSERVER: PERPENDICULAR MOTION**

(a) A light ray is seen to move parallel to the y' axis by a moving observer. In the ether frame the same light ray appears to move at an angle to the y axis. (b) The relation between the observed velocity $\mathbf{c}'$ and the ether velocity $\mathbf{c}$ is obtained by vector subtraction.

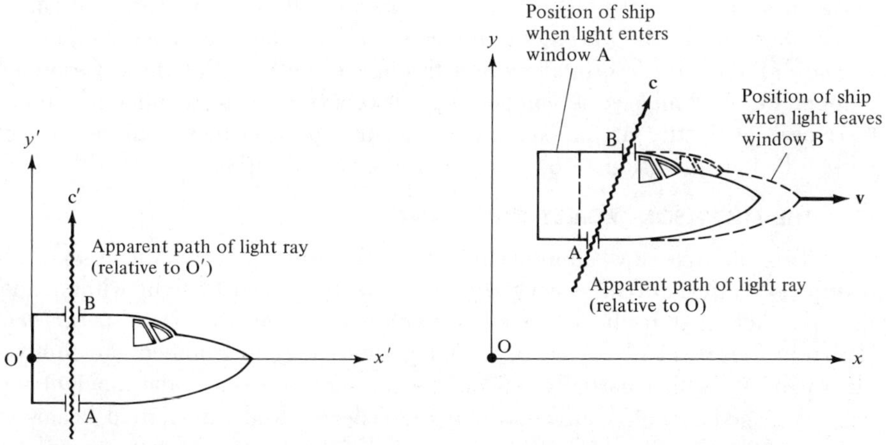

Moving frame. Observer inside ship sees light ray enter window A and leave window B. Its path is parallel to the y' axis (which is attached to the spaceship and is moving relative to the ether).

Ether frame. Observer stationary in the ether sees light ray moving at an angle to compensate for motion of spaceship as light passes from A to B. The path of the light ray is not parallel to the y axis.

(a)

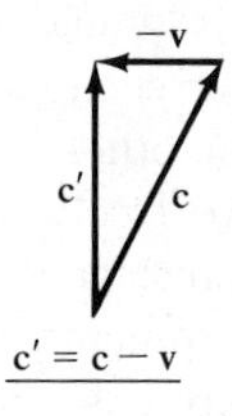

$$c^2 = c'^2 + v^2$$
$$c'^2 = c^2 - v^2$$
$$\therefore\ c' = \sqrt{c^2 - v^2}$$

(b)

the same light ray appears to be moving at an angle with respect to the y axis. For example, let O′ be a spaceship with two symmetrically placed windows. An observer inside the spaceship sees a light ray enter by one window and leave by the other; hence the light ray appears to be traveling in a path perpendicular to the motion of the spaceship. But an observer who is stationary in the ether has a different point of view. If the light ray is to pass through the two windows, it must be moving at an angle to compensate for the motion of the spaceship while the light is passing from one window to the next (Figure 11–6a). The magnitude of the observed velocity $\mathbf{c}'$ measured in the spaceship can be calculated by taking into account the fact that $\mathbf{v}$ and $\mathbf{c}'$ are now perpendicular to one another. As shown in Figure 11–6b, the velocity $\mathbf{c}'$ measured in the moving reference frame has a magnitude

$$c' = \sqrt{c^2 - v^2} \qquad (11\text{–}4)$$

In general, the velocity of light in a given reference frame should depend on the motion of the reference frame relative to the ether. It should then be possible to detect the magnitude and direction of the velocity of a body relative to the ether by comparing the observed velocity of light $\mathbf{c}'$ to the "ether" velocity $\mathbf{c}$. In practice, however, such a comparison is extremely difficult to make. The speeds of large material bodies are extremely small compared to the speed of light, so that the differences between c' and c are rather small. For example, if the sun were stationary in the ether, there would be variations in the velocity of light *as measured on the earth* because of the earth's orbital motion through the ether. But these variations would amount to ± 18 mi/sec as compared to 186,300 mi/sec—a difference of less than 0.01 percent. In the late nineteenth century the speed of light could not be measured with sufficient precision to observe such a small deviation.

THE MICHELSON–MORLEY EXPERIMENT

This obstacle was circumvented in 1879 when Albert Michelson devised an ingenious method for measuring *differences* in the speed of light without measuring the speed of light itself. The method centered on the use of the ***interferometer,*** shown in Figure 11–7a. A beam of monochromatic light is aimed at a thinly silvered glass plate G, which partially reflects and partially transmits the incident light. The glass is placed at a 45° angle so that the reflected and transmitted beams go off at right angles to one another. Each beam is then reflected back to the plate by the mirrors at M_1 and M_2. Through further reflection and transmission, portions of each beam reach the telescope at T. Since the two beams are originally from the same wave front, they start out in phase; but any difference in path length or velocity will cause the beams to arrive at the telescope out of phase. In that case an interference pattern of alternating bright and dark lines should be seen in the telescope.

Michelson planned to use his interferometer to detect the motion of the earth through the ether. Suppose that the interferometer (which is attached to the earth) is moving with a speed v through the ether such that one arm of the interferometer is parallel to the direction of motion and the other is perpendicular to it. It is assumed that the earth and the interferometer move freely in the ether. The light beams are not "dragged along" with the interferometer; once emitted, they move in the ether with a speed c, independently of the interferometer. However, when measured *relative to the interferometer* (which can be considered as a moving reference frame), the light beams will have speeds different from c. According to Equation 11–3, a light beam has a speed of $c - v$ relative to the interferometer when traveling from G to M_1

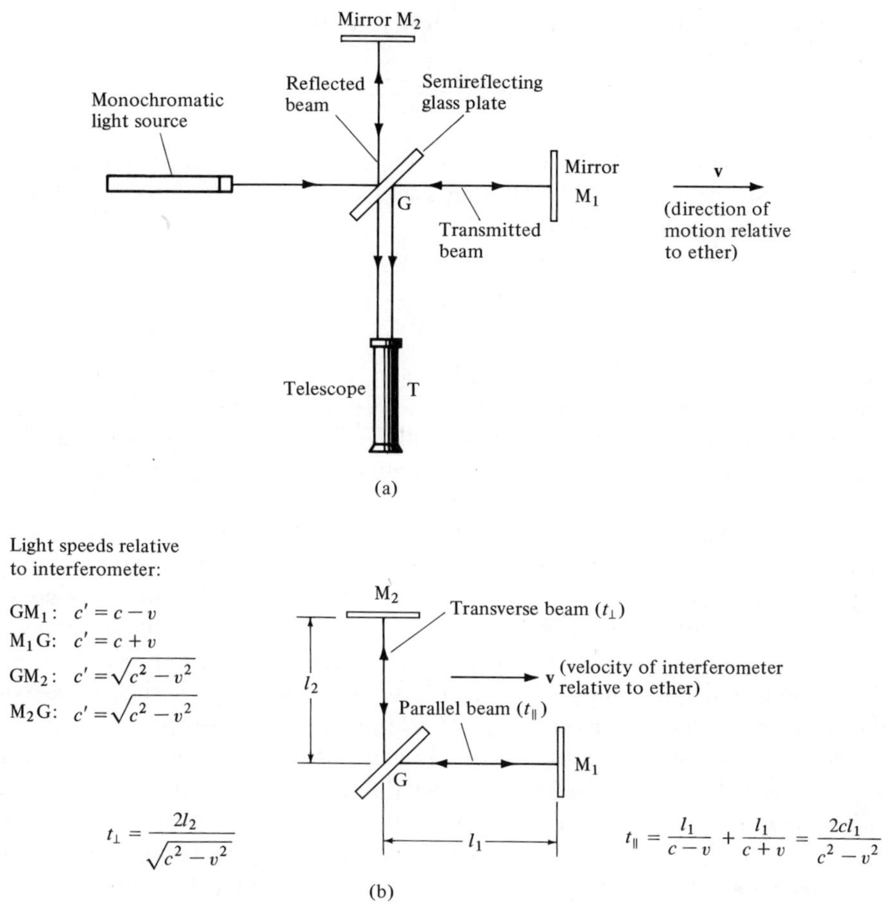

figure 11–7 **THE MICHELSON–MORLEY EXPERIMENT**

(a) In the interferometer, the partially reflecting glass plate G splits the incident light beam into a reflected beam and a transmitted beam. These beams are reflected back into a telescope. If the two beams travel different path lengths or have different velocities, an interference pattern will be seen in the telescope. (b) Comparison of travel times of perpendicular and parallel beams for interferometer moving through the ether with a velocity $\mathbf{v}$.

(see Figure 11–7b) and a speed of $c + v$ when traveling from M_1 back to G. A light beam traveling from G to M_2 and back is moving perpendicular to the motion of the interferometer through the ether; therefore the light beam has a measured speed of $\sqrt{c^2 - v^2}$ relative to the interferometer (Equation 11–4).

Let us now compute the time $t_\parallel$ required for the light beam moving parallel to the motion of the interferometer to travel from G to M_1 and back. Let l_1 be the distance between G and M_1. The light beam travels from G to M_1 with a speed $c - v$, and from M_1 back to G with a speed $c + v$. Using time = distance/speed, we have

$$t_\parallel = t_{GM_1} + t_{M_1G} = \frac{l_1}{c - v} + \frac{l_1}{c + v}$$

Combining the two terms algebraically gives

$$t_{\parallel} = \frac{2cl_1}{c^2 - v^2} \tag{11–5}$$

On the other hand, the light beam that travels perpendicular to the motion of the interferometer (from G to M_2 and back) covers a distance of $2l_2$ at a speed of $\sqrt{c^2 - v^2}$. Hence its travel time is

$$t_{\perp} = \frac{2l_2}{\sqrt{c^2 - v^2}} \tag{11–6}$$

When we compare Equations 11–5 and 11–6, it is evident that the two travel times will not in general be equal. Because of the difference in travel times, the two beams that start out from G in phase will be out of phase with one another when they recombine at the telescope. Hence a pattern of bright and dark interference lines should be observed.

The interference pattern that appears can be traced to two factors: (1) the arms of the interferometer are not the same length ($l_1 \neq l_2$), and (2) the interferometer is moving through the ether with a speed v. When an interferometer is constructed, it is extremely difficult to make the two arm lengths equal, to the accuracy desired. Since both factors contribute to the interference pattern, it is impossible to determine the motion of the interferometer relative to the ether from a single observation of the interference pattern. Even if the interferometer were *stationary* in the ether, an interference pattern would still appear because of the difference in the arm lengths.

Nevertheless, it is possible to detect the motion of the interferometer by other means. It was assumed in deriving Equations 11–5 and 11–6 that the interferometer is aligned with l_1 parallel and l_2 perpendicular to the interferometer's motion through the ether. But if the positions are interchanged (by rotating the interferometer 90°), a slightly different interference pattern should result. In other words, the nature of the interference pattern depends on how the interferometer is aligned with respect to its direction of motion through the ether. If the interferometer is *not* moving through the ether, the interference pattern will remain static as the interferometer is rotated. But if the interferometer *is* moving through the ether, the interference pattern will change when the interferometer is rotated, as first one arm and then the other lines up with the direction of the interferometer's motion through the ether. These changes will appear as a shift in the position of the bright and dark lines as seen in the telescope (Figure 11–8).

Michelson began thinking about the ether-motion experiment shortly after he completed his first measurement of the speed of light. In 1880 he went to Germany. where the best optical equipment was available, and obtained a position in the laboratory of Hermann von Helmholtz. Here Michelson constructed his first interferometer, in which the optical parts were mounted on long metal arms that could be pivoted about the center. The arms provided a path length of 120 cm for the parallel and perpendicular light beams. Michelson calculated that if the earth's speed through the ether were equal to its orbital speed about the sun (18.6 mi/sec), he should see a shift of 0.04 line in the interference pattern as the interferometer was rotated. He considered his instrument accurate enough for him to observe this small displacement.

But Michelson ran into unexpected difficulties. Vibrations caused by traffic

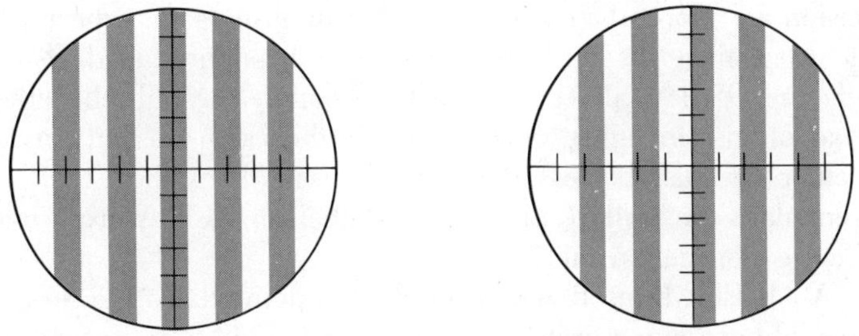

(a) Interference pattern observed in telescope. Note position of lines with respect to cross hairs.

(b) Expected pattern after interferometer is rotated. Pattern should shift by about a half a line.

figure 11–8 **INTERFERENCE PATTERN OBTAINED IN MICHELSON–MORLEY EXPERIMENT**

If the interferometer is moving through the ether, a rotation of the interferometer should cause the pattern to shift from (a) to (b).

outside the laboratory made it difficult for him to observe a steady interference pattern. In addition, the interferometer's arms were not rigid enough to maintain a steady interference pattern while the interferometer was being rotated. Consequently, Michelson could not attain the accuracy he desired. Nevertheless, he did not observe any large displacement of the interference pattern. If taken at face value, his results indicated no motion through the ether, but Michelson noted that any actual displacement could have easily been masked by experimental errors.

Considering his results to be inconclusive, Michelson abandoned his experiment, returned to the United States, and went back to work on his measurement of the speed of light. He obtained a professorship at Case Institute in Cleveland; here he became a close friend of Edward W. Morley, a chemistry professor at nearby Western Reserve University.

The importance of Michelson's ether-motion experiment was not lost on the physicists of the day, and he was urged to repeat the experiment. Finally, in 1884, with Morley as a collaborator, Michelson set out to construct a better interferometer. To do away with vibrations, Michelson and Morley mounted the light source, mirrors, and telescope on a stone slab 5 feet square and 14 inches thick. The slab was floated in a pool of mercury so that it could be rotated slowly and easily without vibration. By using multiple reflections of the light beams, they increased the path length to 11 meters. This time a displacement of 0.4 line was expected as the interferometer was rotated.

In 1887, after more than two years of preparation, Michelson and Morley made their measurement. They observed the interference pattern as the interferometer was set into rotation and found that the pattern did not change. They concluded that if there was any displacement of the lines, it had to be less than 0.01 line. This corresponded to a speed relative to the ether of less than one-fifth the earth's orbital speed. Michelson and Morley continued their observations for several months, and at no time did they ever observe a displacement of the lines larger than the experimental error. This "null result" indicated that *the earth has no measurable velocity with respect to the ether.*

Since 1887 the Michelson–Morley experiment has been repeated with increasing precision by more than a dozen different groups of experimenters. No convincing evidence against the null result has ever been produced. The most recent attempt (made in 1958) to detect the earth's motion through the ether (using microwaves instead of visible light) showed that if the earth has any motion at all relative to the ether, it must be less than one-thousandth of the earth's orbital speed. To the present day, the null result of the Michelson–Morley experiment stands as a well-verified experimental fact.

Michelson himself was keenly disappointed by the results of his experiment. He would have much preferred to have been able to report a positive result representing a direct measurement of the earth's velocity through the ether. He believed the null result indicated the failure of his experiment to detect the ether (in which he firmly believed). There is an interesting parallel between Michelson and Copernicus—both were dedicated to the classical tradition in physics, considered their work to be a failure, and were the unwitting founders of significant revolutions in physics.

INTERPRETATIONS OF THE NULL RESULT

A single observation of the null result would simply indicate that the earth had no velocity relative to the ether at the time when the observation was made. That is, when the observation was made, the earth and the ether could have happened to be moving in the same direction with the same speed. But the earth, because of its rotation and revolution, does not always move in the same direction; therefore, if the earth is moving with the ether at one time it must be moving *against* the ether at another time. Yet observations made over a period of months show no apparent effect resulting from the earth's change in direction. *We are forced to conclude on this basis that the earth is always stationary with respect to the ether.* Since the sun, planets, and stars move relative to the earth, they cannot also be stationary in the ether; thus it would seem that of all the bodies in the universe, only the earth is fixed to the ether. Were the geocentrists correct after all? If the Michelson–Morley experiment had been performed in 1600, it would have probably stifled all further attempts to advance the heliocentric theory; Copernicus and Galileo would have been regarded as prime examples of the aberrations of mind produced by the Renaissance, and Newton might well have stuck to alchemy. But in 1887 the heliocentric theory was too well entrenched. A return to geocentrism would have required a complete dismantling of physics. No one wished to entertain such thoughts.

An alternative explanation was offered. Suppose a material body does not move freely through the ether, but drags along a relatively thin layer of ether that adheres to the body's surface. If so, then the earth drags a portion of the ether around with it; hence the earth is not in motion with respect to the "dragged" ether. This hypothesis had in fact been proposed in the 1840s, but it did not gain much acceptance. The main argument against it was the observation of stellar aberration, which depends on the relative motion of the earth with respect to light received from the stars. If a layer of ether is dragged along by the earth, then the dragged ether is in motion with respect to the ether further away from the earth. Light reaching the earth from the stars would have to pass through layers of ether moving at different speeds. This would alter the path of the light rays with respect to the earth, producing noticeable distortions in the appearance of stellar aberration. But no such effects are observed. Other experiments were performed to test the idea that the ether close to the earth moves relative to the more distant ether. The Michelson–Morley experiment was done at various altitudes above the earth's surface, but a null result was always obtained.

Thus the weight of the observational evidence was strongly against the ether-drag hypothesis.

The null result of the Michelson–Morley experiment, coupled with the inadequacy of the ether-drag hypothesis, posed a formidable dilemma. If the earth is stationary in the ether, then the heliocentric theory, and with it all of Newtonian mechanics, is incorrect. A less shattering but still unpalatable alternative was that the ether did not exist at all. Even though there were some difficulties in accounting for its mechanical properties, the ether had also become a well-entrenched physical concept. Without the ether as a medium, how could electromagnetic waves travel from place to place? Maxwell's description of electromagnetic waves in terms of mechanical vibrations in the ether had been quite successful in all other respects. It appeared that one (or more!) of the well-established theories of physics had to be incorrect. Physics was facing a crisis of epic proportions.

ATTEMPTS TO SAVE THE ETHER

In order to preserve the ether concept, several attempts were made to modify Maxwell's electromagnetic theory. Two of them are worth mentioning: the Ritz emission theory and the Fitzgerald–Lorentz contraction hypothesis.

In 1908 the Swiss physicist Walther Ritz showed that the null result of the Michelson–Morley experiment could be accounted for by assuming that the velocity of light depends on the motion of the source. The velocity of light in the ether is presumed to be equal to the vector sum of the velocity of emission ($\mathbf{c}$) plus the velocity of the light source. Thus if the light source has a velocity $\mathbf{v}$ relative to the ether, a light beam emitted in the direction of motion of the source will have a speed of $c + v$, and a light beam emitted in the opposite direction will have a speed of $c - v$ relative to the ether. Hence according to this theory, light emission is analogous to a bullet fired from a gun; if the gun is moving, the velocity of the gun will be added to the velocity of the bullet itself. The emission theory thus assumes that light travels with a speed of c relative to the *source,* not relative to the ether as proposed by Maxwell. In the Michelson–Morley experiment the light source is part of the interferometer; hence light beams move through the interferometer with a speed of c relative to the interferometer, regardless of the motion of the interferometer. As a consequence, no interference pattern shift is to be expected.

However, strong evidence against the emission theory was provided in 1913 by Willem de Sitter's observations of double stars. In a double star system the two stars are revolving about their center of mass, so that when one star is approaching the earth, the other is moving away. If v is the orbital speed of the star, the light leaving the star when it is approaching the earth should have a speed of $c + v$ relative to the earth, and the light leaving the star when it is receding should have a speed $c - v$ relative to the earth (Figure 11–9). If the emission theory is correct, the observed light velocities would change continually as the stars revolve. De Sitter noted that such changes should produce certain irregularities in the motions of the stars as they are seen from the earth. But no such irregularities are ever observed. All the observations indicate that the light rays from both stars reach the earth with the same speed, regardless of the motions of the stars.

Further evidence contradicting the emission theory appeared when the Michelson–Morley experiment was performed using starlight (by R. Tomaschek in 1924) and sunlight (by D. C. Miller in 1925). In both cases a displacement of the interference pattern is expected because the interferometer is moving with respect to the light

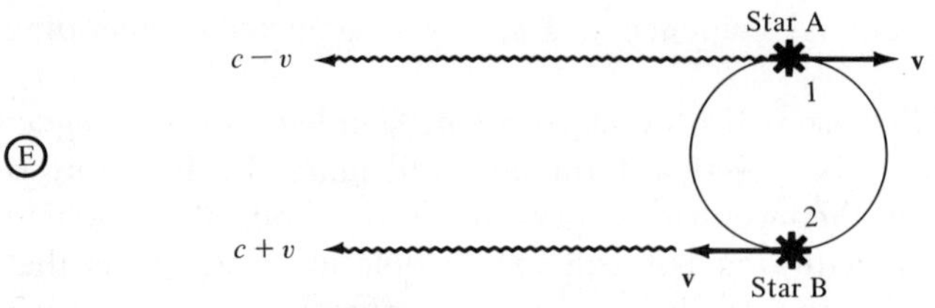

figure 11–9 **THE SPEED OF LIGHT FROM A DOUBLE STAR SYSTEM**

According to the emission theory, the light from star A should approach the earth with a speed of $c - v$, while the light from star B should approach with a speed of $c + v$. Because of the difference in light speeds, B should appear to reach position 2 before A arrives at position 1, even though they actually arrive simultaneously. De Sitter showed that irregularities of this sort do not appear in the observed motions of double stars.

source. However, a null result was found in both experiments. These observations clearly showed that the emission theory was untenable.

A completely different explanation of the null result was suggested by G. F. Fitzgerald in 1892. Examining the equations used in the Michelson–Morley experiment, Fitzgerald noted that the null result could be accounted for by assuming that the arm of the interferometer parallel to the motion *physically contracts* by a factor of $\sqrt{1 - v^2/c^2}$ (i.e., l_1 is replaced by $l_1 \sqrt{1 - v^2/c^2}$ in Equation 11–5). Fitzgerald proposed the hypothesis that a moving body contracts by a factor of $\sqrt{1 - v^2/c^2}$ along the direction of motion through the ether, where v is the speed of the body relative to the ether. He pointed out that a contraction of only 1 part in 200 million would be sufficient to account for the null result. But Fitzgerald's proposal was purely an ad hoc hypothesis that served only to explain the results of the Michelson–Morley experiment, and Fitzgerald did not attempt to give any physical reason why a body should contract in the direction of its motion.

Nevertheless, Fitzgerald's hypothesis was adopted by Lorentz, who attempted to give it a physical basis in terms of his electron theory. Lorentz noted that bodies are held together by the electric forces between atoms, and since the ether is the medium for the transmission of electric and magnetic interactions, it might not be unreasonable to assume that electric forces increase in strength along the direction of motion of a body moving through the ether. This increase in force would cause the body to contract in that direction. But Lorentz did not explain *why* electric forces should increase with motion, and to support his argument Lorentz had to assume that *all* forces, including gravitation, increase in strength along the direction of motion. Lorentz also found it necessary to assume that the electron itself flattens into an ellipsoidal shape when in motion through the ether. Many physicists found these assumptions difficult to accept. Moreover, other effects of the proposed contraction (such as a change in the index of refraction for light traveling through a moving medium) were not observed. The Fitzgerald–Lorentz contraction hypothesis did not seem to hold the answer to the puzzle of the null result.

THE ETHER REJECTED

Frustrated by their inability to find a rational explanation for the null result, many physicists came to believe that it was intrinsically impossible to detect motion through the ether. In 1899 the French mathematician-philosopher Henri Poincaré proposed the principle that *absolute motion with respect to the ether is in principle not detectable by any experimental means.* It will be recalled that according to the principles of Newtonian relativity, absolute motion cannot be detected by any *mechanical* means because the laws of mechanics are the same in all inertial reference frames. That is, Newton's laws do not distinguish between different inertial reference frames. But Poincaré's principle asserts that absolute motion cannot be detected by electromagnetic or optical experiments either. It follows that the laws of electromagnetism should not distinguish between different inertial reference frames. On this basis Poincaré proposed a more general ***principle of relativity,*** which can be stated as follows: *all laws of physics are the same in all inertial reference frames.*

But a rather glaring contradiction arises. Maxwell's equations do distinguish between inertial reference frames; they select the ether as the special reference frame in which electromagnetic waves travel with a speed of c. In an inertial reference frame that is moving through the ether, electromagnetic waves should appear to have a speed different from c. But the null result of the Michelson–Morley experiment and the failure of the emission theory clearly indicate that *electromagnetic waves appear to travel with a speed c with respect to all inertial reference frames, regardless of the motion of the source or the observer.*

From the midst of the chaos created by the null result of the Michelson–Morley experiment, a few clear ideas were beginning to emerge. It was becoming more and more evident that the concept of the ether would have to be modified radically or even rejected altogether. If it cannot be shown by experiment that an electromagnetic wave propagates as a mechanical disturbance with a unique velocity relative to the ether, the idea of the ether as a mechanical medium becomes untenable. But in that case electromagnetic waves could no longer be considered as mechanical waves. This alternative seemed so unacceptable that a number of physicists (especially Lorentz) preferred to hang grimly onto the ether hypothesis, in spite of its obvious contradictions.

Such was the state of affairs when Albert Einstein burst upon the scene in 1905.

11–3

ALBERT EINSTEIN

The man who was to revolutionize physics gave no outward indication of any special brilliance until he was twenty-six years old. Born in Ulm, Germany, in 1879, Albert Einstein was such a slow learner as a child that his parents were afraid he might be retarded. Einstein's aversion to discipline and his distaste for the rigid, authoritarian teaching methods used in the German schools made him less than a model student. He was not without his intellectual interests, however, for he took a liking to geometry at the age of twelve and pursued the study of mathematics entirely on his own. Nevertheless, in high school Einstein's teachers considered him a disruptive influence, and he was asked to leave. At the age of fifteen Einstein became a high-school dropout.

At this point Einstein moved to Milan with his family and spent a year loafing and roaming about northern Italy. He realized, however, that he would have to become self-supporting, and he applied for admission to the Institute of Technology in Zurich. Not having a high-school diploma, Einstein had to take an entrance examination, which he failed. He returned to high school, obtained his diploma, and was then admitted to the university.

Even as a college student Einstein refused to follow any externally imposed discipline. He rarely attended lectures, preferring to learn physics entirely on his own. When examination time came around, he borrowed the notes of a fellow student and crammed from them. In this manner Einstein managed to pass his examinations, but he later reported, "It had such a deterring effect on me that I found the consideration of any scientific problems distasteful to me for an entire year."[3]

Einstein graduated in 1900, but he had made such a slight impression on his professors that he was unable to obtain an academic post. Finally, he was hired as an examiner in the Swiss Patent Office in Bern. Einstein was particularly happy with this position, for he found the work interesting, and it gave him plenty of time to ponder the problems in physics that caught his interest. He remained in Bern for seven years, examining patent applications by day and the foundations of physics by night. During this time Einstein discussed his new ideas with only his wife and a few friends, and he had practically no contact with any professional physicists. He remained, by choice, outside the scientific and academic establishment.

Einstein's first investigations were in the field of thermodynamics. Between 1902 and 1904 he published a series of papers in which he independently worked out the basic principles of statistical mechanics—not realizing that many of his results had been obtained previously by other physicists. Einstein's interest in statistical mechanics led him to consider the interaction between light and matter, and in 1905 he proposed a revolutionary theory in which he showed that under certain conditions light appears to behave like a stream of particles. This photon theory of light, as it is now called, was to have a profound effect on the development of physics. (We will consider it more fully in a later chapter.) A few months later Einstein's first paper on the special theory of relativity appeared in print.

The year 1905 was truly a *wunderjahr* for Einstein. Besides publishing the epochal papers on special relativity and the photon theory, he also completed his doctoral thesis (on the relationship between molecular sizes and statistical fluctuations in the properties of gases), gave a theoretical analysis of molecular motion, and proposed the relationship between mass and energy. Like Newton 240 years before him, Einstein was definitely "in the prime of his age for invention." The significance of Einstein's achievements was soon recognized, and before long he was being offered academic positions at a number of leading universities. In 1908 he left the Patent Office with some reluctance to accept a professorship at the University of Zurich. Five years later he was appointed to a highly prestigious position at the University of Berlin. By this time Einstein was considered one of the outstanding physicists of his day.

In 1916 Einstein published the general theory of relativity, which represented a complete reformulation of the theory of gravitation and an entirely new concept of the nature of space and time. This work was highly controversial, for it represented a complete overthrow of Newtonian mechanics, and its mathematical abstractness made it quite difficult to comprehend. But as part of the theory, Einstein proposed that light should bend in a gravitational field, and in 1919 this effect was strikingly confirmed experimentally by a team of English scientists.

The fact that the theory of a German scientist was confirmed in England shortly after the end of World War I attests to the international spirit of science. But not all scientists can transcend the bounds of nationalism. When Einstein was invited to speak in England and France, there was considerable opposition, and many professors chose to boycott his talks. Einstein's own position in Germany was somewhat tenuous, for he was an avowed pacifist and steadfastly refused to support the German war effort. But his scientific reputation and his acquired Swiss citizenship afforded him protection in wartime Germany.*

With the experimental verification of general relativity, Einstein's fame rose to unparalleled heights. His name became a household word as a synonym for genius. He was showered with honors wherever he went. In spite of this adulation Einstein remained an unpretentious, good-natured individual, who was equally at home with royalty and tradesmen. He was amused by the fact that there could be so much public enthusiasm for a mathematical physicist. Yet the theory of general relativity, with its strange new concepts of space and time so contrary to common sense (i.e., firmly imbedded Newtonian concepts), was discussed and debated endlessly by physicists, philosophers, and theologians.

Einstein did take advantage of his preeminence to speak out in favor of political causes—notably world government and disarmament. For this he was strongly attacked by the more extreme German nationalists. These attacks came in the guise of scientifically and philosophically muddled arguments against the theory of relativity. It was quite apparent, however, that Einstein was being criticized for his political rather than his scientific beliefs, and his opponents made little headway. Most Germans regarded Einstein as a national treasure and a symbol of German culture.

But even Einstein's prestige could not withstand the rising tide of anti-Semitism. The attacks against him grew more and more vitriolic, and his supporters were effectively silenced. To the Nazis, Einstein became a national enemy, and his writings on relativity were burned publicly. He was reviled as the leading proponent of "Jewish physics," which was viewed as a conscious attempt to obscure the foundations of physics and to confuse Aryan scientists. Foremost among Einstein's attackers was Philipp Lenard, a world-renowned physicist and a winner of the Nobel Prize, who wrote the following words:

> Jewish physics can best and most justly be characterized by recalling the activity of one who is probably its most prominent representative, the pureblooded Jew Albert Einstein. His relativity theory was to transform and dominate all physics; but when faced with reality, it no longer had a leg to stand on. Nor was it intended to be true. In contrast to the equally intractable and solicitous desire for truth of the Aryan scientist, the Jew lacks to a striking degree any comprehension of truth.[5]

Seeing his homeland plunged into such irrationality, Einstein left Germany, never to return. In 1933 he emigrated to the United States. He continued his research on the general theory of relativity and also devoted much energy to the formulation of a unified field theory in which the theories of gravitation and electromagnetism were to be combined. (Unfortunately, his efforts in this area were largely unsuccessful.) Einstein continued to speak forcefully on the major political and social questions of

*In 1919 Einstein offered the following description of his situation: "Today in Germany I am called a German man of science and in England I am represented as a Swiss Jew. If [my theory is overthrown] the description will be reversed, and I shall become a Swiss Jew for the Germans, and a German for the English."[4]

the times. At his death in 1955 Einstein was revered as an outstanding humanitarian as well as one of the greatest thinkers of all recorded history.

QUESTIONS FOR REVIEW

1 Explain the distinction between absolute motion and relative motion.

2 What is a coordinate system? What is meant by the origin of a coordinate system?

3 Discuss the Newtonian concept of absolute time. What distinction did Newton make between absolute time and common time?

4 Discuss the Newtonian concept of absolute space. What properties did Newton attribute to absolute space? How do these properties prevent us from establishing an absolute reference frame?

5 Why is it impossible to measure the absolute velocity of a body in space? In particular, explain why Newton's principles of motion cannot be used to calculate the absolute velocity of a body.

6 Explain in terms of the basic Newtonian principles why it is possible to determine the absolute acceleration of a body.

7 What is an inertial reference frame? What properties must a body have to be suitable as the origin of an inertial reference frame?

8 How is it possible to distinguish between an inertial and a noninertial reference frame?

9 Give examples to show how each of the following is violated in an accelerated reference frame:
(a) The principle of inertia.
(b) $\mathbf{F} = m\mathbf{a}$

10 State and explain the equation for the addition of velocities in the Newtonian system.

11 Why must all inertial reference frames be considered equally valid in Newtonian mechanics?

12 Which physical properties of a moving body have the same values in all inertial reference frames in the Newtonian system? Which properties have different values in different inertial reference frames?

13 Explain why Maxwell's theory of electromagnetism establishes the ether as a special inertial reference frame. Does the theory make the ether an *absolute* reference frame?

14 According to Maxwell's theory, how does the observed speed of light depend on the motion of the observer? Under what conditions will an observer measure a light speed exactly equal to c?

15 Describe the construction and operation of Michelson's interferometer. Under what conditions will the two beams interfere with each other? Why is it necessary to rotate the interferometer in order to detect the motion of the earth through the ether?

16 What was the purpose of the Michelson–Morley experiment? What was the result of the experiment? What problems were created by this result?

17 What was the ether-drag hypothesis? How does it account for the null result? Why was it rejected?

18 Outline Ritz's emission theory. According to this theory, how is the speed of light determined? How does this differ from Maxwell's theory? Outline the evidence that led to the rejection of the emission theory.

19 Discuss the Lorentz–Fitzgerald contraction hypothesis. How did Lorentz justify this hypothesis theoretically?

20 What principle did Poincaré propose with respect to the ether? How did he justify this principle?

EXERCISES

1 List the astronomical motions that are considered to be *absolute* motions in (a) the geocentric system of the universe and (b) the heliocentric system.

2 Two spaceships, A and B, are approaching one another such that the distance between them is decreasing at a constant rate of 5 mi/sec.
(a) What is the relative velocity of B with respect to A?
(b) What is the relative acceleration of B with respect to A?
(c) Is it possible to determine the absolute acceleration of either spaceship?
(d) An astronaut on a third spaceship observes that A is stationary. How does this information change the answers to (a) and (b)?

3 Explain how each of the following experiments can be used to test whether a given reference frame is accelerated or inertial:
(a) A rock is thrown vertically upward.
(b) A ball is placed on a smooth, level table.
(c) A weight is hung from the ceiling on a long string, and the angle with the vertical is measured.

4 Explain why each of the following statements is correct:
(a) A body moving with a constant velocity relative to an inertial reference frame can be used as the origin of another inertial reference frame.
(b) A body moving with a constant velocity relative to one inertial reference frame is moving with a constant velocity relative to *all* inertial reference frames.
(c) A body accelerating in an inertial reference frame has the same acceleration relative to all inertial reference frames.
(d) A body moving with constant velocity relative to an accelerated reference frame cannot be chosen as the origin of an inertial reference frame.
(e) It is possible for a body to have the same velocity relative to two or more inertial reference frames.

5 Is it reasonable to consider a laboratory on the earth as an inertial reference frame? Can you think of any effects of the earth's motion that might appear in a simple laboratory experiment?

6 Three spaceships, A, B, and C, are all traveling in the same straight line, with B in the middle. Spaceship A is moving eastward toward B with a relative speed of 3 mi/sec. Spaceship C is moving westward toward B (from the opposite direction) with a relative speed of 4 mi/sec.
(a) Using Equation 11–1, calculate the velocity of C relative to A.
(b) Suppose that spaceship C is moving at 4 mi/sec northward along a straight line *perpendicular* to the line joining A, B, and C. What is the velocity of C relative to A?

7 The Lorentz force on a charged particle ($F = qvB \sin \theta$) contains a term v that represents the speed of the charged particle. Can v be considered an absolute speed? If not, relative to what reference frame is it measured?

8 According to the ether theory, light travels with a speed c relative to the ether. Suppose the light *source* is moving through the ether with a speed v. How will the speed of light be affected? What effect will the motion of the source have on the emitted light?

9 A light source is stationary in the ether. Spaceship A is moving directly toward the light source with a speed $c/2$. Spaceship B is moving directly away from the light source with a speed $c/2$. What is the speed of light as measured on each spaceship, according to the ether theory?

10 A spaceship is moving northward through the ether at a speed of $c/2$. Its path is perpendicular to that of a light ray emitted in an eastward direction by a stationary source. Specify the magnitude and direction of the velocity of the light ray relative to the spaceship.

11 A spaceship, moving through the ether at a speed c, is traveling parallel to a light ray going in the same direction. According to the ether theory, what is the speed of light relative to the spaceship? Does a stationary light wave seem physically possible?

12 Beginning with Equations 11–5 and 11–6, show that

$$t_{\|} = \frac{2l_1/c}{(1 - v^2/c^2)} \qquad \text{and} \qquad t_{\perp} = \frac{2l_2/c}{\sqrt{1 - v^2/c^2}}$$

Write the equation for the time delay $\Delta t = t_{\|} - t_{\perp}$. Using this equation, show that

(a) If $l_1 \neq l_2$ (the arms are unequal in length) and $v = 0$ (the interferometer is stationary in the ether), an interference pattern will result.

(b) If $l_1 = l_2$ (the arms are exactly equal) and $v \neq 0$ (the interferometer is moving through the ether), an interference pattern will result.

(c) If $l_1 = l_2$ and $v = 0$, there will be no interference pattern.

13 Using the result of exercise 12(b), show that if l_1 is replaced by $l_1\sqrt{1 - v^2/c^2}$ in the direction parallel to the motion, there will be no interference. (Note: This is the Fitzgerald–Lorentz hypothesis.)

THE SPECIAL THEORY OF RELATIVITY

12

Oh, that Einstein, always skipping lectures. . . .
I certainly never would have thought he could do it.

Hermann Minkowski

AS A STUDENT, Einstein was not especially interested in learning the mathematical techniques and procedures used in solving specific problems in physics. Instead, his major concern was to probe more deeply in an effort to ferret out and examine the fundamental principles underlying the major physical theories. His goal was nothing less than to find the simplest set of assumptions from which all physical theories could be derived. Buttressed by an unbounded faith in the essential unity and simplicity of physics, Einstein was not afraid to challenge the most fundamental aspects of the classical Newtonian system. Radically new concepts of inertia, gravitation, energy, light, and the very nature of space and time emerged from his vision, and Einstein rose to a position of preeminence in physics attained previously only by Newton.

The special theory of relativity, published in 1905, represents a thorough reformulation of space and time as physical concepts. This theory marked the first major break with the Newtonian tradition and served as a cornerstone of the transition from classical to modern physics.

12–1

BASIC PRINCIPLES OF RELATIVITY

Upon examining the fundamental principles of mechanics and electromagnetism, Einstein became particularly disturbed by the conflict between Newtonian mechanics and Maxwell's electromagnetic theory as they relate to the ether hypothesis. In mechanics there is no way to distinguish between inertial reference frames; they are all equivalent. But in Maxwell's theory the ether is a special reference frame. Einstein's faith in the unity of physics led him to believe that if all reference frames are equivalent with respect to mechanical experiments, they should be equivalent with respect to electromagnetic experiments as well. The ether was, after all, supposed to be a *mechanical* medium. Why should it have a special status in electromagnetic theory and not in mechanics? This situation signified to Einstein a fundamental incompatibility between the then-existing theories of mechanics and electromagnetism. One, or possibly both, had to be incorrectly constructed. The fact that physicists were experimentally unable to detect motion through the ether served to fortify Einstein's belief.*

Einstein therefore proposed that the theories of mechanics and electromagnetism should be reformulated so that they are compatible with the principle that *all laws of physics have the same form in all reference frames.* This principle is quite similar to Poincaré's principle of relativity, which was described previously. (Einstein was apparently unaware of Poincaré's work and formulated the principle independently.) However, Poincaré proposed the principle of relativity in response to the apparent inability to detect the ether experimentally, whereas Einstein proposed this principle because he believed that the laws of physics should not depend on a choice of reference frame.

A strong influence on Einstein's thought was exerted by the scientific philosophy of Ernst Mach (1838–1916). It was Mach's goal to rid physics of its metaphysical foundations. The early development of scientific thought had been founded on the belief that scientific principles could be based on philosophical arguments. The Platonic principle that all celestial motions are composed of combinations of circular motions and Aristotle's arguments on the motions of falling bodies are prime examples of this approach. Mach argued that philosophical explanations of scientific facts have no place in science. He believed that one should deal only in terms of quantities or physical entities that can be observed and measured.

Mach's approach represented the application to science of the philosophical doctrine known as ***positivism.*** In simple terms the positivist believes that questions that are inherently unanswerable through our senses are "meaningless" questions. (A good example of such a question is the famous "How many angels can fit on the head of a pin?"—a question that was endlessly debated by medieval philosophers.) As applied to scientific thought, the positivist doctrine asserts that entities that are *inherently unmeasurable*—that is, that cannot be defined in terms of physically observable quantities or events—should not be employed in physical theories. The Newtonian concept of absolute space represents such an entity, for it is inherently undetectable; one cannot determine the position or velocity of a body in absolute space. Thus to Mach, absolute space is a meaningless physical concept, and it should play no role in physics.

Einstein wholeheartedly adopted the positivist doctrine and made it an integral part of his own scientific philosophy. Within this framework the inability to detect the existence of the ether by experimental means takes on a special significance—for if the ether is *experimentally undetectable in principle,* it is a meaningless and superfluous concept that has no scientific value. If there is no means of verification, the question of whether or not the ether exists becomes a purely academic one.

THE AXIOMS OF SPECIAL RELATIVITY

Two fundamental principles—the equality of the laws of physics in all reference frames (the principle of relativity) and the impossibility of detecting the existence of the ether—form the foundations of Einstein's theory of relativity. It was Einstein's intention to reformulate the fundamental equations of mechanics and electromagnetism so that they would be consistent with these principles. This task proved at first to

* Paradoxically, Einstein was not directly aware of the Michelson–Morley experiment in 1905. Instead, he had learned (by reading the works of Lorentz) of the failure to detect the ether, without knowing of the details of specific experiments.

be beyond his ability. Poincaré's form of the principle of relativity applied to inertial reference frames only; accelerated reference frames were excluded. But Einstein believed that a truly comprehensive theory should obey the principle that all laws of physics are identical in *all* reference frames, inertial or accelerated. The construction of such a theory proved to be an extremely difficult problem—one that Einstein could not solve at first.

Einstein therefore had to restrict his theory to include *inertial* reference frames only. This proved to be a much simpler problem. In 1905 he published a paper entitled "On the Electrodynamics of Moving Bodies" presenting the ***special theory of relativity,*** which is applicable to inertial reference frames only.* Finally, in 1916 Einstein succeeded in his original intention and proposed the ***general theory of relativity,*** which extends the principle of relativity to accelerated reference frames. In this chapter we shall confine our attention to the special theory.

Einstein stated the basic principles of special relativity in the form of two axioms:

(1) All laws of physics have the same form in all inertial reference frames.

(2) Light propagates in free space with a speed of c relative to all observers.

The first axiom is, of course, the restricted form of the principle of relativity. The second axiom is an expression of the inability of physicists to detect motion with respect to the ether experimentally. In effect, it raises the null result of the Michelson–Morley experiment to the status of a law of nature. Applying positivist doctrine, Einstein declared that the ether is a meaningless physical concept. Hence the idea that light travels with a speed c relative to the ether must be rejected. In Newtonian theory the velocity of a mechanical wave is measured *with respect to the medium,* and the magnitude of the wave velocity depends on the rigidity of the medium. According to Einstein, however, light (or any other type of electromagnetic wave) can no longer be regarded as a mechanical disturbance in the ether.

The second axiom also rejects categorically any *emission* theory of light; the speed of light is declared to be independent of the motion of its *source.* Neither does the motion of the observer affect the measured speed of light. An observer moving toward an approaching light beam will measure the same speed (c) as a stationary observer or an observer moving away from the light beam. The assertion that *the observed speed of light does not depend on the motion of the source or the observer* is a denial of the validity of Newtonian mechanics as it applies to the addition of velocities. Einstein did not give any explanation as to why light has this peculiar property; he merely stated that this is the way light behaves in nature.

We have found that the axioms of relativity—particularly the second—violate "commonsense" notions with respect to the nature and motion of light. Why, then, should we be inclined to accept these axioms? We must remember that a physical theory is judged not on the reasonableness of its axioms, but on the correctness of the conclusions that follow logically from the axioms. Our attitude must be to accept temporarily the validity of the axioms and to see what conclusions can be drawn from them. If these conclusions disagree with the results of observations and experiments, then we can legitimately reject the axioms. If the predictions based on the axioms turn out to be true, then our "commonsense" notions must be rejected.

*This does not mean that accelerated motions cannot be analyzed within the framework of special relativity. Accelerated motions measured with respect to *inertial* reference frames can be discussed. *Any* motions with respect to *accelerated* reference frames are excluded.

THE MEASUREMENT OF TIME

The concept of absolute time is one of the most fundamental ideas of Newtonian physics. It is assumed that time flows at the same uniform rate for all observers, independent of the choice of coordinate system. It follows that if two events occur at the same instant for one observer, they should be simultaneous for all observers, regardless of the observers' motion. But Einstein discovered that this seemingly obvious assertion is in conflict with the axioms of special relativity.

To understand why there is a conflict, let us carefully examine how one determines the time of an event and whether two events have occurred simultaneously. To measure time, scientists have constructed clocks that advance at a uniform rate. The units of time measurement can be defined in terms of certain readily observable astronomical motions, and any clock can be compared against these motions to ensure that it is properly synchronized and advancing at the proper rate.

It would appear, then, that determining the time of an event should be a simple matter. One merely records the reading of the clock when the event occurs. But in fact the observer is recording the time when he *sees* the event occur and not the actual time of the event itself. If an event occurs at some point in space at a great distance from an observer, the observer will not know about the event until he receives some *signal* (a transfer of energy) from the event. The "observer" does not have to be a living person; it can be any mechanical or electric device that is triggered by the receipt of a signal. In any case the observer measures the time when a signal from the event is received (the ***time of observation***) rather than the time when the signal was emitted (the ***time of occurrence***). The difference between the two times is simply the time required for the signal to travel to the observer from the point in space where the event took place.

The signal being transmitted may be a sound wave, a particle—anything that transfers energy from the source to the observer. The most effective means of transmission, however, is by electromagnetic waves, which travel through free space with a constant speed of c. Unlike all other forms of energy transmission, electromagnetic waves require no material medium and travel with a speed that has the same value for all observers (according to the second axiom). Moreover, at the present time no signal that travels with a speed greater than that of an electromagnetic wave is known to exist. For these reasons we will assume henceforth that the signals transmitted from source to observer are *light waves* (by "light," we mean any electromagnetic wave).

The difference between the time of occurrence of an event and the time of observation of the event is the ***light travel time*** from source to observer. If the source and observer are a fixed distance d apart, the light travel time is equal to d/c (the distance traveled divided by the speed). Obviously, the closer the observer is to the point of the event, the shorter the light travel time will be. If the observer is located directly at (or extremely close to) the point of the event, the light travel time becomes arbitrarily small. Hence the actual time of occurrence of an event can be measured directly by a clock *only* if the clock is located exactly at the point where the event takes place. Otherwise, a correction must be made for the light travel time. (See Figure 12–1.)

Given two events occurring at two different points in space, the order in which they are *seen* to occur depends on the location of the observer. As shown in Figure 12–2, if two events occur simultaneously at points P and Q, an observer who is closer

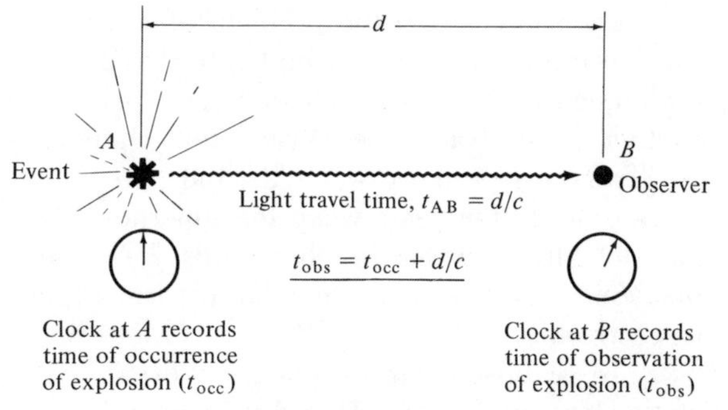

figure 12–1 **TIME OF OBSERVATION AND TIME OF OCCURRENCE**

An event (say, an explosion) occurs at point A. A clock located at A records the time of occurrence of the explosion. An observer at B records the time at which he receives a light signal from the explosion (the time of observation). The difference is the time required for light to travel from A to B (the light travel time).

to P will see the event at P *before* he sees the event at Q (because of the different light travel times). The sequence of events will appear to be reversed for an observer located closer to Q, and an observer located precisely midway between P and Q will see the two events simultaneously. Hence whether or not two events *appear* to be simultaneous depends on the location of the observer. In order to determine whether the events are truly simultaneous—that is, whether they have the same *time of occurrence*—an observer must correct for the light travel times to the points where the events took place.

figure 12–2 **THE SEQUENCE OF TWO EVENTS**

Two explosions occur simultaneously at P and Q. Whether or not they *appear* to be simultaneous events depends upon the location of the observer.

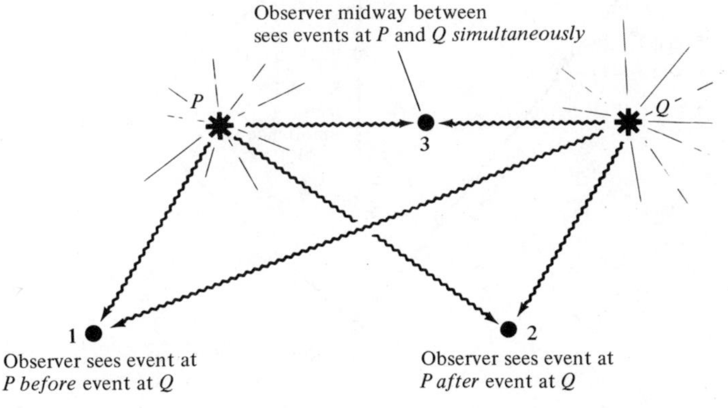

Let us illustrate the problem of establishing the true sequence of events by considering a historical example. In 1572 Tycho Brahe observed a supernova (an exploding star) in the constellation Cassiopeia. Then in 1604 Johannes Kepler observed another supernova in the constellation Ophiuchus. These two events are referred to as "Tycho's supernova of 1572" and "Kepler's supernova of 1604." However, the dates indicate the years when the supernovae were *seen* on the earth, and not when they actually occurred. In other words, 1572 and 1604 represent the *times of observation* (measured by observers on the earth) of Tycho's and Kepler's supernovae, respectively.

To establish the true sequence of events, we must calculate the *times of occurrence* of the two supernovae. In order to do this, we have to know their distances from the earth when they exploded. Astronomers are able to estimate (through methods too complex to go into here) that Tycho's supernova was 10,000 light years from the earth, and Kepler's supernova was 15,000 light years away.

A light year (LY), which is a unit of *distance*, is a most convenient unit for our purposes. It is the distance that a light ray, proceeding with a speed c, travels in one year (1 LY $= 5.88 \times 10^{12}$ miles $= 9.46 \times 10^{17}$ cm). Hence a light ray travels at a rate of 1 LY per year, or in other words, *the light travel time in years is equal to the distance in light years.*

The light travel times to the earth for Tycho's and Kepler's supernovae are, respectively, 10,000 and 15,000 years. Since the supernovae were *seen* on the earth in 1572 and 1604, it follows that Tycho's supernova occurred in approximately 8400 B.C., and Kepler's supernova occurred in about 13,400 B.C. Thus Kepler's supernova actually occurred some 5000 years *before* Tycho's supernova, but observers on the earth saw the events in *reverse* order, only 32 years apart!

Hence the *observed* time interval between two events that occur at different points in space depends on the location of the observer. As shown in Figure 12–3,

figure 12–3 **THE TIME BETWEEN EVENTS**

The apparent time interval between Tycho's and Kepler's supernovae will be different for different observers, depending on the observer's distance from the two supernovae.

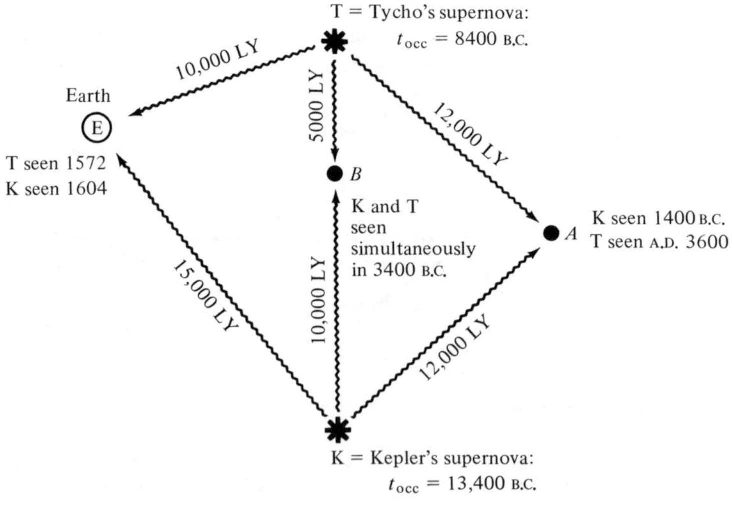

an observer located at point A, equidistant from both supernovae, would see the two explosions in the proper sequence, 5000 years apart in time. An observer located at B (5000 LY from Kepler's and 10,000 LY from Tycho's) would see both objects explode simultaneously.

Obviously, these observed time intervals are *apparent* time intervals. To find the "true" time interval, an observer determines the distances to the two events, corrects for the different light travel times, and finds the time of occurrence for each event. (In this case the time of occurrence is 13,400 B.C. for Kepler and 8400 B.C. for Tycho.) Hence the "true" time interval between the two explosions is *calculated* (but not directly observed) to be 5000 years. It seems reasonable to assume that *all* observers, after they have properly corrected for the light travel time, would calculate the same times of occurrence. These times of occurrence should all agree with the times recorded by observers located at the point of each event (the only observers who can *directly measure* the time of occurrence). Thus if their clocks and calendars are synchronized, all observers ought to agree that the events *occurred* in 13,400 B.C. and 8400 B.C. In effect, we are asserting that each event occurs at a precise instant on an *absolute time scale* to which all observers agree.

THE RELATIVITY OF TIME

Now let us consider the problems that arise when a *moving* observer attempts to determine the time of occurrence of an event. Suppose an observer M receives a signal from some distant event—say, the light from an exploding star. The observer is moving toward the star with a velocity **v.** He is able to calculate that the star is a distance d away from him at the instant when he receives the signal.*

To determine the time of occurrence of the explosion, the observer must calculate the light travel time. *But there is more than one way to make this calculation!* Let us first assume that the star is stationary in space at point P, and the observer is moving toward it, as shown in Figure 12–4a. When the light from the explosion was emitted, the observer was at M′. But while the light is traveling, the observer moves toward it, and he meets it at point M. At this point the observer is a distance d from the star. Hence the light travel time is simply d/c.

Since all motion is relative, however, we cannot rightfully assume that the star is stationary. Let us assume that the *observer* is stationary and the star is moving toward the observer with a velocity **v.** (See Figure 12–4b.) It follows that the star has been moving since the explosion, so that the explosion took place *not* at point P (where the star is now located) but at a more distant point, P'. Hence the light from the explosion has had to travel a distance d' (which is greater than d), and the light travel time is d'/c.

The dilemma can be summarized as follows. If the star is stationary and the observer is moving, the light travel time is d/c. If the observer is stationary and the star is moving, the light travel time is d'/c (which is greater than d/c). Since the light travel times differ, in each case *a different time of occurrence will be calculated.* Thus the answer to the question "When did the explosion occur?" depends on whether the star or the observer is moving. Since all motion is relative, there is no way to tell whether the observer is moving or not.

To clarify the problem, let us introduce another observer, S, who is stationary

*The speed can be determined from the Doppler shift of the light from the star. From the known speed and distance at the time of the explosion, the present position of the star can be calculated.

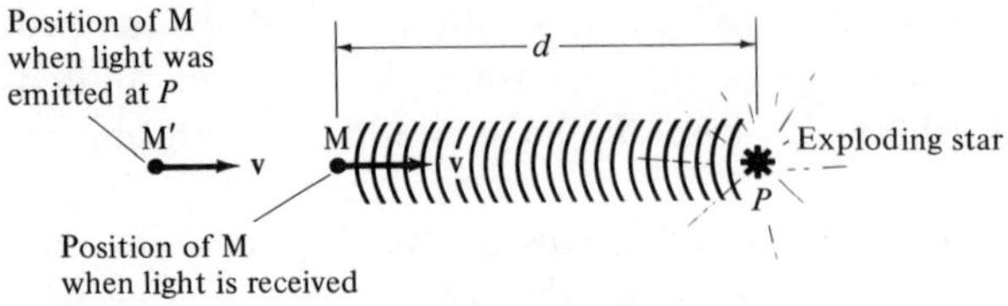

Distance traveled by light = d
Light travel time, $t = d/c$

$t_{\text{occc}} = t_{\text{obs}} - d/c$

(a)

Observer M moving. Star is stationary at point P; M moves toward P with velocity **v**.

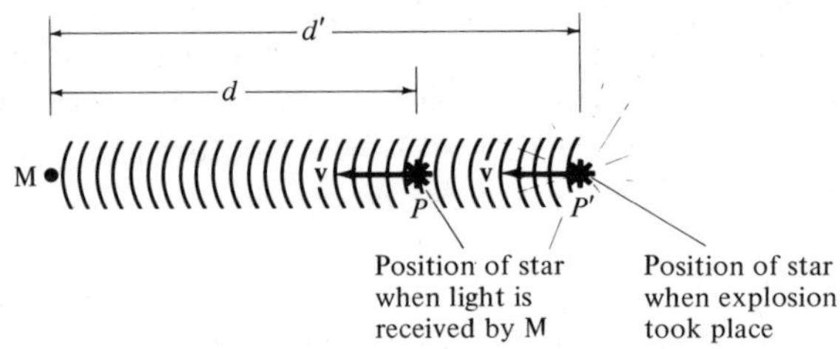

Distance traveled by light = d'
Light travel time, $t = d'/c$

$t_{\text{occ}} = t_{\text{obs}} - d'/c$

(b)

Observer M stationary. Star is moving to left with velocity **v**; explosion took place at P'.

figure 12–4 **RELATIVE MOTION AND THE MEASUREMENT OF TIME**

An observer M is moving relative to a star that exploded in the past. When the observer receives the light from the explosion, he is a distance d from the star. The light travel time depends on whether the observer or the star is moving.

with respect to the star. Both observers (S and M) see the explosion simultaneously (their times of observation are the same), and S is also a distance d from the star. In the reference frame of observer S (Figure 12–5a), the star is fixed at a point P, and the observer M is moving toward P at a velocity **v**. The light travel time in this reference frame is d/c. If the explosion was observed in the year 1970 and the star is known to be 100 LY away, the explosion must have occurred in 1870, according to observer S.

Now let us consider the reference frame of observer M (Figure 12–5b). In this reference frame M is stationary, and observer S and the star are both moving with a velocity **v** in the same direction. The star exploded when it was at point P'; hence the light travel time in this reference frame is d'/c. Suppose that $d' = 110$ LY. (That is, the star has moved 10 LY closer while the light from the explosion was traveling to the two observers.) Observer M also sees the explosion in 1970, and he calculates that the light travel time is 110 years; so according to M, the explosion must have occurred in 1860.

Thus the two observers do not agree as to the time of occurrence of the explosion. Did the explosion take place in 1870 (as S contends) or in 1860 (as M calculates)? Common sense tells us that one of the observers is right and the other is wrong. How do we settle the disagreement? We might think that since S is the stationary observer, his time of occurrence is the correct one. But we have stipulated only that S *is stationary with respect to the star;* we cannot be certain that S's reference frame is stationary in absolute space. Observer M has an equally valid claim to being the "stationary" observer.

Clearly, the answer hinges on which of the two observers is stationary. If S is stationary and M is in motion, the light travel time from the explosion is d/c, and the explosion occurred in 1870. If M is stationary and S (and the star) are moving, the light travel time is d'/c, and the explosion occurred in 1860. Thus in order to settle the disagreement, we must determine which of the two observers is truly stationary. If there were an observable ether, we could determine whether M or S is at rest with respect to the ether, and the argument could then be settled. If S were stationary in the ether, his time of occurrence (1870) would be the true (absolute) time, and if M were stationary in the ether and S were moving, M's time (1860) would be the true time of occurrence. However, we have learned that motion relative to the ether is undetectable, so the argument cannot be settled in this manner.

The disagreement could also be settled if the emission theory of light were true. If the star were stationary and M were moving, light would proceed from the star with a speed c, and if the star were moving, the light waves emitted by the star would travel with a speed $c + v$, according to the emission theory. The greater speed would

figure 12–5 **LIGHT TRAVEL TIME AS CALCULATED IN DIFFERENT REFERENCE FRAMES**

Observer S is stationary with respect to the star; observer M is moving relative to the star.

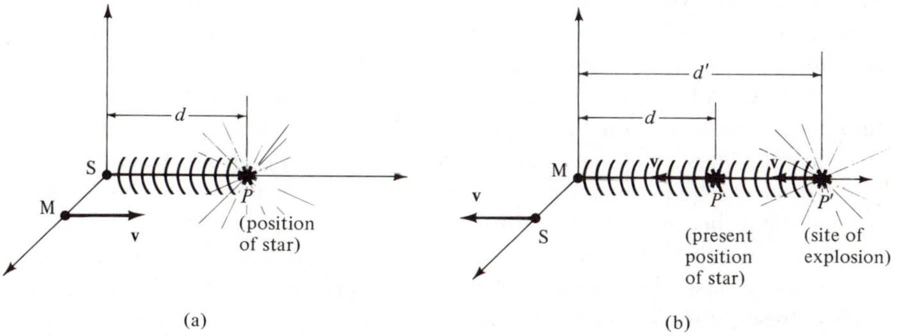

Reference frame of observer S. The star is located at P, a fixed distance d from observer S. M is moving toward P with a velocity **v**. The light travel time in this reference frame is d/c.

Reference frame of observer M. The star is moving toward M with a velocity **v**. The explosion took place when the star was at P'; the star has since moved to its present position at P. The light from the explosion was emitted at P'; hence the light travel time is d'/c.

compensate exactly for the greater distance, and the discrepancy would disappear. However, the emission theory is not supported by experimental evidence and is rejected by the axioms of special relativity.

We noted previously that the time of occurrence of an event could be measured *directly* by a clock placed at the point of the event. Thus it would seem that the disagreement could be settled by referring to the time recorded by a clock located at the star at the time of the explosion. (We are assuming, of course, that the hypothetical clock is not destroyed by the explosion.) But if the time recorded by the star clock is to have any meaning when compared to the times recorded by observers S and M, the clock must be synchronized to the same calendar and time system that the two observers use.

The synchronization of the star clock could be achieved by taking three clocks, setting them to the same time, and then locating them at S, M, and the star, respectively. We cannot be certain, however, that the clocks will remain synchronized while they are being moved. When a clock is moved from one place to another, it must be accelerated and then decelerated, and the forces required could easily speed up or slow down the clock's mechanism. Therefore we must check that the clocks are still synchronized after they are in place.

The synchronization of the clocks located at S and the star could be checked by the procedure illustrated in Figure 12–6. A light pulse is sent from S to the star clock. If the pulse leaves S when the clock at S reads a time t_S, the star clock should read $t_S + d/c$ when the pulse arrives. Similarly, a pulse leaving the star at time t_P should arrive at S when S's clock reads $t_P + d/c$. If this is so, S would consider his clock and the star clock to be synchronized. However, *the observer at M would not agree that the clocks are properly synchronized.* In his reference frame S and the star are both in motion with a velocity $\mathbf{v}$. A pulse traveling from S to the star travels a distance shorter than d, because the star is moving toward the oncoming light pulse. According to observer M, the light pulse leaving S at time t_S should reach the approaching star at a time *sooner* than $t_S + d/c$. Similarly, a pulse leaving the star at time t_P should be received by S at a time *later* than $t_P + d/c$, since S is moving *away* from the oncoming light pulse. The synchronization check proposed by observer M, which is based on the assumption that S and the star are moving, would, of course, be unacceptable to observer S. Hence the two observers are not able to agree on a procedure for synchronizing the star clock.

Thus two observers moving relative to one another and measuring the same light signal from some distant event cannot agree on the light travel time of the signal, so each calculates a different time of occurrence for the event. The system of synchronized clocks, which presumably measure occurrence times directly, does not solve the problem because light pulses must be used to check the clocks' synchronization—and since the two observers cannot agree on light travel times, each considers the other's clock to have been improperly synchronized. There seems to be no way to settle the disagreement unless we can determine which of the two observers is truly stationary. According to the principles of relativity, however, absolute motion cannot be detected by any means whatsoever. Consequently, the true or absolute time of occurrence of the explosion cannot be determined.

On the basis of such considerations, Einstein questioned the validity of the fundamental Newtonian concept of absolute time. The measurement of absolute time requires knowledge of absolute motion; hence absolute time and absolute space are fundamentally linked. Einstein argued that if absolute space (or absolute motion) is

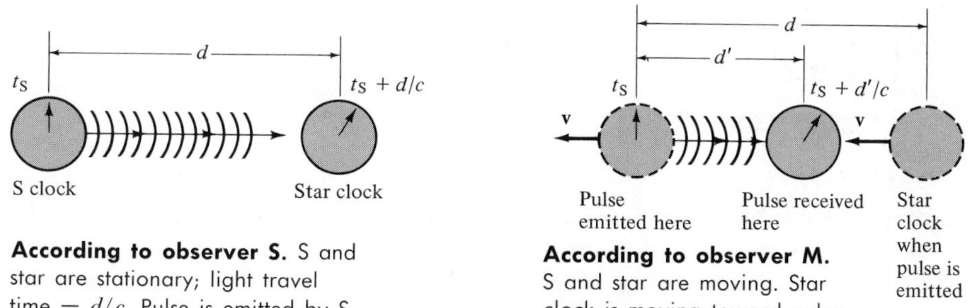

According to observer S. S and star are stationary; light travel time $= d/c$. Pulse is emitted by S clock at time t_s; star clock should read $t_s + d/c$ when pulse arrives.

According to observer M. S and star are moving. Star clock is moving toward pulse; therefore light travel time is d'/c. Star clock should read $t_s + d'/c$ when pulse arrives.

(a) Light Pulse Sent from S Clock to Star Clock

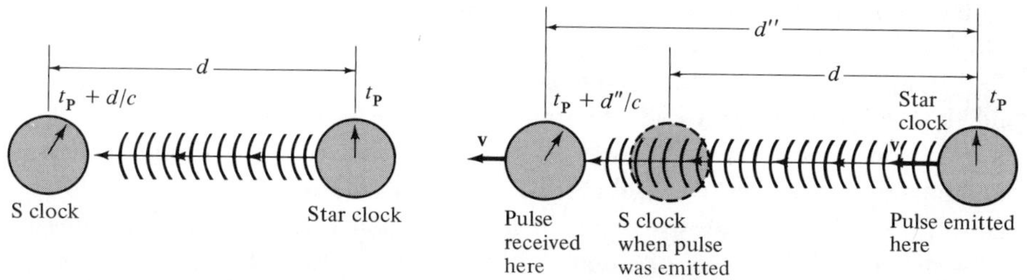

According to observer S. S and star are stationary; light travel time $= d/c$. Pulse is emitted by star clock at time t_P; S clock should read $t_P + d/c$ when pulse arrives.

According to observer M. S and star are moving. S clock is moving away from pulse; therefore light travel time is d''/c. S clock should read $t_P + d''/c$ when pulse arrives.

(b) Light Pulse Sent from Star Clock to S Clock

figure 12–6 **THE SYNCHRONIZATION OF A CLOCK IN DIFFERENT REFERENCE FRAMES**

Two observers moving relative to each other are unable to agree on a procedure for checking the synchronization of a distant clock.

inherently unmeasurable, absolute time must also be unmeasurable. *In the absence of any absolute reference frame, there can be no absolute time.* Time measurements are relative and differ from one reference frame to the next; each observer measures the *time with respect to his own reference frame.* Thus observers S and M are *both* right; the explosion occurred in 1870 relative to S's reference frame, and it occurred in 1860 relative to M's reference frame.

It is a tribute to Einstein's genius (and a characteristic of his anti-authoritarian nature) that he alone of the many brilliant thinkers who were struggling with the same problems was bold enough to challenge the concept of absolute time. We should recognize, though, that at this point Einstein had not actually proved the concept to be false; he had shown only that the idea of absolute time is inconsistent with the axioms of special relativity. Rather than reject the axioms because of this conflict, Einstein chose to exclude absolute time from physical theory.

12–2

COORDINATE TRANSFORMATIONS

We have learned that it is impossible to determine one's absolute position or absolute velocity in space. One can detect only *relative* motion or position; therefore it becomes necessary to measure one's position and velocity with respect to some arbitrarily chosen reference frame. Inertial reference frames are specifically used for this purpose because the principles of Newtonian mechanics are the same in all inertial reference frames. In this respect all inertial reference frames moving relative to one another with constant velocity are equally valid.

To these principles Einstein added the following: (1) *all* laws of physics (not just the laws of mechanics) should be the same in all inertial reference frames, and (2) the absolute time of an event is not measurable. Thus time, like position and velocity, must be measured relative to a given reference frame.

It is useful to develop a set of equations that enable one to relate the space and time coordinates of a body relative to one inertial reference frame to the coordinates of the same body relative to any other inertial reference frame. Such a set of equations is known as a ***coordinate transformation,*** because it allows one to transfer from one set of space and time coordinates to another. The measurement of the position and velocity of a body with respect to different inertial reference frames can then be described in terms of the transformation equations.

To illustrate the idea of a coordinate transformation, let us take the simple case of two reference frames that are stationary with respect to one another, but whose origins do not coincide, as shown in Figure 12–7. Point A has coordinates $(x = 3, y = 4)$ with respect to the x–y axes attached to O and coordinates $(x' = 5, y' = 7)$ relative to the x'–y' axes attached to O′. The difference in coordinates stems from the fact that the origins of the two coordinate systems differ by two units in the x direction and three units in the y direction. Thus the x' coordinate of any point will be greater than the x coordinate by two units, and the y' coordinate will be three units greater than the y coordinate. In other words, if a point P has coordinates (x, y) relative to O, its coordinates (x', y') relative to O′ can be obtained by using the equations

$$x' = x + 2$$
$$y' = y + 3$$

These two equations comprise the coordinate transformation that connects the two reference frames O and O′. If the coordinates of any point in one reference frame are known, the coordinates of that point in the other reference frame can be calculated with these equations.

THE GALILEAN TRANSFORMATION

Now let us derive a coordinate transformation for two inertial reference frames that are moving relative to one another. Two inertial reference frames attached to bodies O and O′ are shown in Figure 12–8. We will consider O to be the "stationary" reference frame and O′ to be the "moving" reference frame; it is assumed that O′ is moving in the x direction with a speed v relative to O. Point P is a fixed point in reference frame O. (That is, a body located at point P would be stationary relative

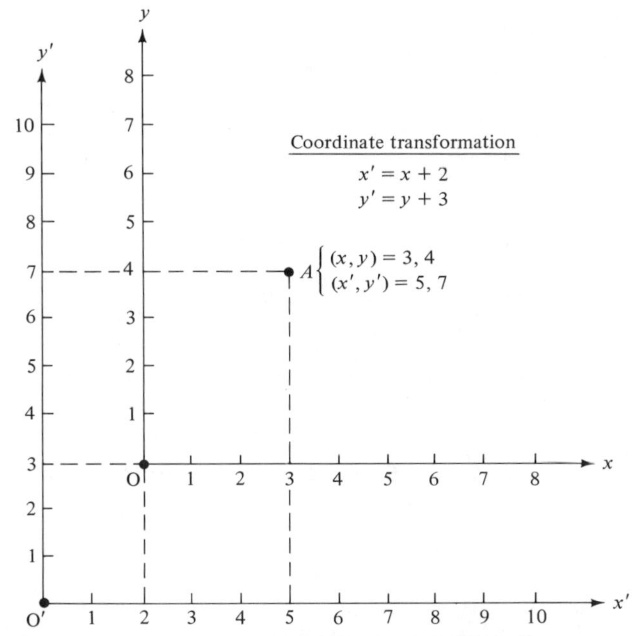

figure 12–7 **COORDINATE TRANSFORMATION FOR STATIONARY REFERENCE FRAMES O AND O′**

The coordinates (x, y) of any point in reference frame O are related to the coordinates (x', y') of the same point in reference frame O′ by the coordinate transformation equations.

to O.) Point P has space coordinates x_1, y_1, z_1 with respect to the x–y–z axes attached to O and space coordinates x_1', y_1', z_1' relative to the x'–y'–z' axes attached to O′. For simplicity, we have assumed that the x and x' axes are collinear and that the y and z axes are parallel to the y' and z' axes, respectively.

Since O′ is in motion relative to O, the x'–y'–z' coordinates will change with time. The time origin ($t = 0$) has been chosen as the instant when O and O′ coincide. Figure 12–8 shows the positions of O and O′ after a time t_1 has elapsed. By this time O′ has moved along the x axis a distance vt_1 relative to its initial position at $t = 0$. Now we will determine the relationships between the coordinates x_1, y_1, z_1 and x_1', y_1', z_1' of point P. At $t = 0$, when O and O′ (and their associated axes) coincide, the coordinates will be the same ($x_1 = x_1'$, $y_1 = y_1'$, $z_1 = z_1'$). But as O′ moves along the x axis, the x' coordinate of the point P will change with time. (Remember that point P is stationary with respect to O.) It is easy to see from the diagram that after a time t_1 has elapsed, $x_1' = x_1 - vt_1$. Furthermore, if the motion of O′ is entirely in the x direction, the y and z coordinates will not change; thus at all times $y_1' = y_1$ and $z_1' = z_1$.

The measurement of time in the two reference frames must also be considered. According to Einstein, time is measured differently relative to different reference frames, but here we will adopt the Newtonian premise that there is an absolute time of occurrence for any given event and that this time is the *same* for all observers.

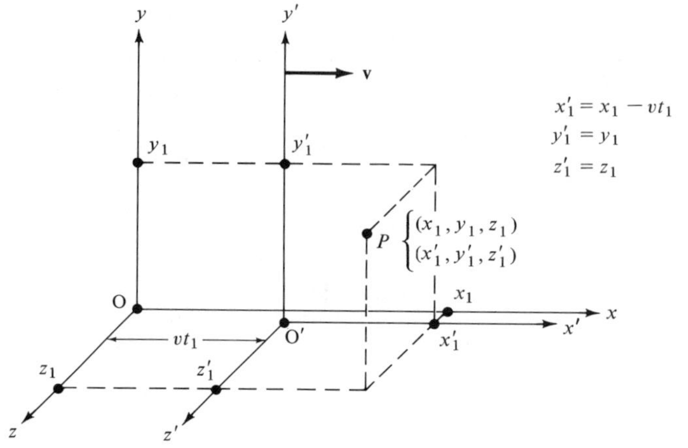

figure 12–8 **THE GALILEAN COORDINATE TRANSFORMATION**

O and O′ are both inertial reference frames. O′ is moving in the x direction with a speed v relative to O. At $t = 0$, the origins coincide. The diagram shows the two reference frames after a time t_1 has elapsed. The equations relate the coordinates of the point P in the two reference frames.

Hence time will elapse at the same rate in both reference frames. If we establish the same event as the time origin for both reference frames (i.e., if we synchronize the clocks by setting $t = t' = 0$ at the same instant), the time coordinate will be the same in both reference frames. Thus at any instant, $t = t'$. Hence t and t' represent the *times of occurrence* of a given event as calculated by the two observers. Although the observers' times of observation of a distant event may differ, they should calculate the same time of occurrence, according to the Newtonian system.

Thus the three equations $x'_1 = x_1 - vt_1$, $y'_1 = y_1$, and $z'_1 = z_1$ relate the three space coordinates of the point P with respect to the two reference frames at a given instant of time t_1 $(= t'_1)$. In general, if x, y, z, t are the space and time coordinates of any point relative to reference frame O and x', y', z', t' are the space and time coordinates of the same point relative to O′, the coordinate transformation equations relating the two sets of coordinates can be written as follows:

$$x' = x - vt \qquad (12\text{–}1a)$$
$$y' = y \qquad (12\text{–}1b)$$
$$z' = z \qquad (12\text{–}1c)$$
$$t' = t \qquad (12\text{–}1d)$$

where v is the speed of O′ relative to O. This set of equations is commonly known as the ***Galilean transformation*** in honor of Galileo, who was one of the first scientists to analyze the nature of relative motion. These equations permit us to determine the coordinates of a point in one inertial reference frame in terms of the coordinates of the same point in another inertial frame, when the inertial frames are moving relative to one another and the same time origin is chosen. Equations 12–1 apply specifically to situations in which the motion is in the x direction only, but they can easily be modified to include motion in the y and z directions as well. Note that the Galilean transformation assumes the existence of *absolute time* ($t' = t$); hence *the Galilean transformation is the coordinate transformation of Newtonian relativity.*

Now we shall consider some of the mathematical consequences of the Galilean transformation. To ascertain the length of some object in a given reference frame, we must know the coordinates of the end points of the object. But the two end points must be examined *simultaneously*, for the object may be in motion, and if the coordinates of the end points are measured at two different times, the length will be incorrect.

Suppose an object lies on the x axis, and at a given instant of time t_0 one end is at point x_A and the other end is at x_B (Figure 12–9). The length of the object in the x–y–z reference frame is the difference between the coordinates of the object's end points; that is,

$$l = x_B - x_A$$

In the x'–y'–z' reference frame, one end is at x'_A and the other end is at x'_B at the time $t'_0 = t_0$. Thus in this reference frame

$$l' = x'_B - x'_A$$

Using Equation 12–1a of the Galilean transformation, we substitute $x'_B = x_B - vt_0$ and $x'_A = x_A - vt_0$ to obtain the following:

$$l' = x'_B - x'_A = (x_B - vt_0) - (x_A - vt_0) = x_B - x_A = l$$

That is, the object has the same length ($l = l'$) in both reference frames. This is an eminently reasonable result, since we would certainly expect the length of an object to be independent of the choice of coordinate system.

The transformation equations can also be used to determine the velocity of a body measured with respect to different inertial reference frames. Let us assume that a body is in motion in the x direction. At time t_1 it is at point 1 (x_1, y_1, z_1); at time t_2 it has moved to point 2 (x_2, y_1, z_1). (Note that since the body is moving in the x direction, only the x coordinate changes). Thus in the time interval $\Delta t = t_2 - t_1$,

figure 12–9 **MEASUREMENT OF THE LENGTH OF A BODY**

In a given reference frame, the length of a body is determined from the coordinates of its end points. In a moving reference frame, the coordinates must be measured simultaneously.

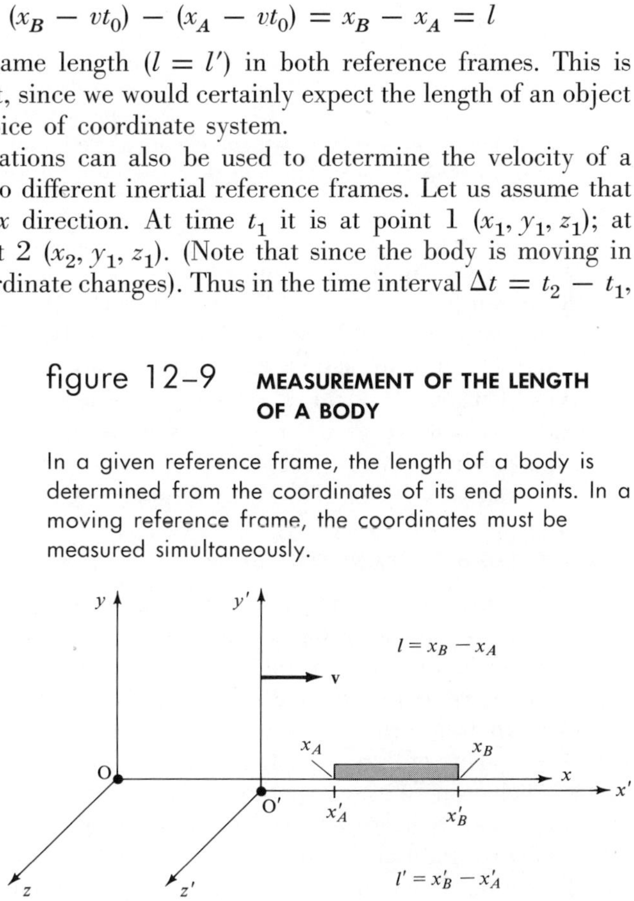

the body travels a distance $x_2 - x_1$. Hence the speed u of the body with respect to O is:

$$u = \frac{\text{distance}}{\text{time}} = \frac{x_2 - x_1}{t_2 - t_1}$$

In the same manner we can determine the speed u' of the body as measured in reference frame O′. Relative to O′, the body has moved from the point x_1', y_1', z_1' at time t_1' to the point x_2', y_1', z_1' at time t_2'. Therefore its speed relative to O′ is

$$u' = \frac{\text{distance}}{\text{time}} = \frac{x_2' - x_1'}{t_2' - t_1'}$$

An equation relating u and u' can be obtained by using the equations of the Galilean transformation for x_1', x_2', t_1' and t_2' in terms of x_1, x_2, t_1, and t_2. Substituting and rearranging terms yields

$$u' = \frac{(x_2 - vt_2) - (x_1 - vt_1)}{t_2 - t_1} = \frac{(x_2 - x_1) - v(t_2 - t_1)}{t_2 - t_1}$$

$$= \frac{x_2 - x_1}{t_2 - t_1} - v\frac{t_2 - t_1}{t_2 - t_1} = u - v$$

Therefore

$$u = u' + v$$

In deriving this result, we have considered motion in the x direction only. If motion in all directions is allowed, a more general vector equation results:

$$\mathbf{u} = \mathbf{u}' + \mathbf{v}$$

This is the equation for the addition of velocities (Equation 11–1), which was stated without proof—because it seemed almost self-evident—in the previous chapter. Now we see that the rule of addition of velocities is a mathematical consequence of the Galilean transformation, which in turn is based on the principles of Newtonian relativity. Thus Equation 11–1 is the equation for the ***velocity transformation*** that relates the velocity $\mathbf{u}'$ of a moving body measured in the "moving" reference frame O′ to the velocity $\mathbf{u}$ of that body relative to the "stationary" reference frame O, where $\mathbf{v}$ is the relative velocity of O′ with respect to O.

THE LORENTZ TRANSFORMATION

In the previous chapter we learned that the equation for the addition of velocities ($\mathbf{u} = \mathbf{u}' + \mathbf{v}$) leads to a contradiction when it is applied to light waves. Specifically, the equation predicts that an observer moving through the ether or moving relative to a light source will measure a velocity $\mathbf{c}' = \mathbf{c} - \mathbf{v}$ for a light wave. Consequently, the Galilean transformation (from which the equation is derived) predicts that the observed velocity of light should depend on the motion of the observer, but the failures of the Michelson–Morley experiment, the emission theory, and so on indicate that this prediction is disproved by the experimental evidence.

Poincaré and Lorentz were among the first to recognize the failure of the Galilean transformation. Poincaré proposed that it was necessary to extend the principle that absolute motion is undetectable by mechanical experiments to include electromagnetic experiments. But Maxwell's equations select the ether as a special reference frame in which electromagnetic waves travel with a velocity $\mathbf{c}$. When the velocity $\mathbf{c}$ is transformed to a different reference frame, it is changed to $\mathbf{c} - \mathbf{v}$ by the Galilean transformation. According to Poincaré's principle, the theory must be modified so that electromagnetic waves have the same velocity ($\mathbf{c}$) relative to *all* inertial frames.

Lorentz posed the mathematical problem of finding a coordinate transformation that would make Maxwell's equations independent of the choice of reference frame, so that the same velocity would be predicted for electromagnetic waves in all inertial reference frames. In 1904 Lorentz derived the following set of equations, now known as the ***Lorentz transformation:***

$$x' = \frac{1}{\sqrt{1 - v^2/c^2}}(x - vt) \tag{12–2a}$$

$$y' = y \tag{12–2b}$$

$$z' = z \tag{12–2c}$$

$$t' = \frac{1}{\sqrt{1 - v^2/c^2}}\left(t - \frac{v}{c^2}x\right) \tag{12–2d}$$

Like the Galilean transformation, these equations relate the coordinates (x, y, z, t) of a fixed point in one inertial reference frame to the coordinates (x', y', z', t') of the same point relative to a second inertial reference frame, which is moving relative to the first with a speed v in the x direction. However, these equations were derived specifically to make Maxwell's equations retain the same mathematical form in both coordinate systems. Hence the Lorentz transformation replaces the Galilean transformation in electromagnetic theory.

Of particular interest is the time equation (Equation 12–2d), which predicts that t' and t are different—that is, that time proceeds at different rates in the two coordinate systems. Lorentz could not bring himself to accept this idea as a physical fact. He continued to believe that there was a "true" time (similar to Newton's absolute time) irrespective of the choice of coordinate system, and he viewed the time equation as a purely mathematical device with no physical meaning. Lorentz refused to abandon his belief in the existence of the ether (even though it seemed to be undetectable). He believed that an observer who is stationary in the ether measures the true or absolute time (t), and an observer who is moving relative to the ether measures an "apparent time" (t').

Einstein did not accept Lorentz's interpretation of the new transformation equations. Like Lorentz, Einstein believed that the Galilean transformation was invalid, but Einstein based his rejection on the grounds that the Galilean transformation was in conflict with the axioms of special relativity. One conflict is, of course, that the Galilean transformation equations require the speed of light to vary from one reference frame to the next, whereas the axioms of relativity require the speed of light to be the same for all observers. A second source of conflict is that the Galilean transformation assumes that time is the same for all observers, whereas Einstein had shown that the time of occurrence of an event depends on the observer's motion.

Einstein set out to derive a new set of transformation equations for space and time coordinates that would be completely consistent with the axioms of special relativity. These equations were to maintain the constancy of the speed of light for all observers and to make the laws of physics the same in all inertial reference frames—in agreement with the axioms of special relativity. The transformation equations were to relate coordinates in inertial reference frames moving relative to one another without any reference to the ether or any other special reference frame.

The derivation of these transformation equations is not a difficult one, but it is much too long to include at this point in the text. The derivation can be found in Section C–2 of the Appendix. The mathematically inclined student will find it rewarding to read and study Einstein's method for obtaining the relativistic transformation equations.

The result of Einstein's derivation turned out to be the following set of equations:

$$x' = \frac{1}{\sqrt{1 - v^2/c^2}}(x - vt) \tag{12–2a}$$

$$y' = y \tag{12–2b}$$

$$z' = z \tag{12–2c}$$

$$t' = \frac{1}{\sqrt{1 - v^2/c^2}}\left(t - \frac{v}{c^2}x\right) \tag{12–2d}$$

These equations should be recognized as the equations of the Lorentz transformation! Einstein was able to demonstrate that the Lorentz transformation is consistent with the axioms of special relativity and is, in fact, the correct relativistic coordinate transformation.

Although Lorentz and Einstein arrived at the same set of equations, their respective interpretations of the significance and applicability of these equations were quite different. Lorentz derived the equations by attempting to find a transformation specifically for Maxwell's equations that would maintain c as the speed of electromagnetic waves in all reference frames. In Lorentz's interpretation *O is a reference frame fixed to the ether, and O' is a reference frame moving with velocity* $\mathbf{v}$ *relative to the ether.* Lorentz also considered t to be the true or absolute time and t' to be an "apparent" time that had no physical significance.

Einstein, on the other hand, derived the transformation equations directly from the axioms of special relativity without reference to any specific laws of physics. Thus the equations are sufficiently general to be applied to *all* laws of physics. That is, they are valid for mechanics as well as for electromagnetic theory. In Einstein's interpretation $\mathbf{v}$ is the relative velocity between any two inertial reference frames. In the absence of a detectable ether, there is no justification for considering any reference frame to be at rest. Reference frame O is referred to as "stationary" purely as a matter of convenience; it is equally valid to call O′ the stationary frame and O the moving frame. Furthermore, one should not assume that either t ot t' measures the absolute time. Each represents the time coordinate for some particular event in space as measured in different reference frames; hence each is a physically meaningful measure of the time *relative to a given inertial reference frame.*

The fact that Einstein was able to derive the Lorentz transformation from the axioms of special relativity strengthened his faith in the validity of these axioms and in the concept of the relativity of time. Lorentz, on the other hand, was too tied to the classical tradition to accept Einstein's interpretation. Ironically, through the Lorentz transformation the name of Lorentz has become inextricably tied to the theory of relativity—a theory that he could never bring himself to believe.

In examining the Lorentz transformation, we note that if the relative speed v is very small compared to the speed of light ($v \ll c$), then $\sqrt{1 - v^2/c^2} \approx 1$ and $v/c^2 \approx 0$. When we make these approximations, we find that the Lorentz transformation equations reduce to the equations of the Galilean transformation. Thus the Galilean transformation, which follows from Newtonian mechanics, represents an *approximation* to the Lorentz transformation when the speed of the "moving" frame is much smaller than the speed of light. What constitutes "much smaller than"? At what speed will the Galilean transformation begin to deviate significantly from the Lorentz transformation, so that the former can no longer be considered a "good" approximation? These questions can be answered by examining the function $\sqrt{1 - v^2/c^2}$, which appears in nearly all relativistic formulas. A graph of this quantity is shown in Figure 12–10. Here we see that the value of $\sqrt{1 - v^2/c^2}$ begins to deviate appreciably from 1 when v/c is about 0.2. Numerically, $v/c = 0.2$ corresponds to a speed of 37,200 mi/sec! An orbiting satellite travels at a speed of about 7 mi/sec ($v/c = 0.000037$); a commercial jet cruises at 600 mi/hr ($v/c = 0.0000008$). Hence within the range of speeds available to human experience, the value of $\sqrt{1 - v^2/c^2}$ does not differ from 1 by more than 1 part in 10^{10}. Thus when our moving reference frames are common objects, such as cars, trains, or airplanes, the Galilean transformation is quite suitable and is virtually indistinguishable from the Lorentz transformation. We would expect relativistic effects to be noticeable only for objects moving at speeds comparable to the speed of light. Note also that $\sqrt{1 - v^2/c^2}$ equals zero at $v = c$ and is an imaginary number (a negative square root) for $v > c$. Hence the Lorentz transformation is inapplicable for speeds greater than the speed of light.

We have learned that the Lorentz transformation relates the coordinates (x, y, z, t) of a fixed point in a given inertial reference frame O to the coordinates (x', y', z', t') of the same point measured in another inertial reference frame O′ that is moving

figure 12–10 **GRAPH OF $\sqrt{1 - v^2/c^2}$ VS. v/c**

Note that the value of the function approaches zero as v approaches c.

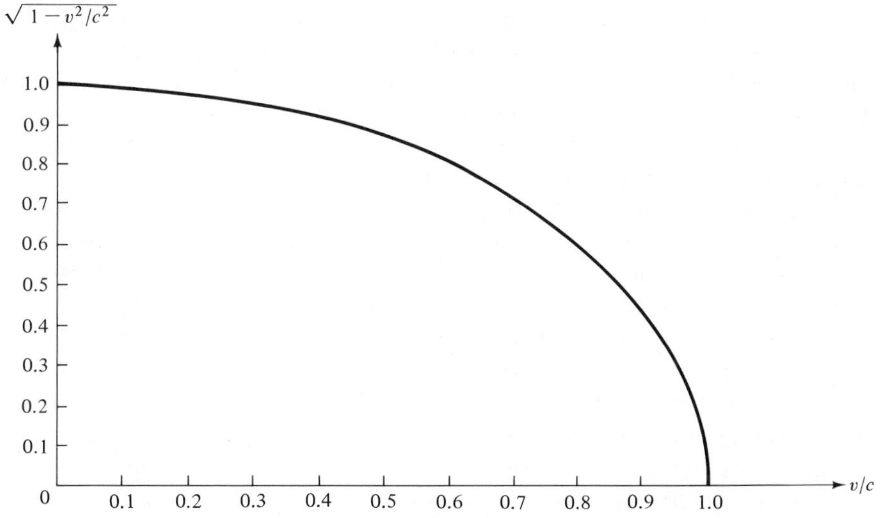

relative to the first. It is customary to refer to the first reference frame as the "stationary" reference frame and to the second as the "moving" reference frame, in the sense that O *is stationary with respect to the point in question,* and O′ is moving relative to the same point. It is useful to write the Lorentz transformation in a form that relates the coordinates of a point that is fixed in the O′ reference frame. If we solve the equations of the Lorentz transformation (Equations 12–2) for x, y, z, and t in terms of x', y', z', and t', we obtain the ***inverse Lorentz transformation:***

$$x = \frac{1}{\sqrt{1 - v^2/c^2}}(x' + v't) \tag{12–3a}$$

$$y = y' \tag{12–3b}$$

$$z = z' \tag{12–3c}$$

$$t = \frac{1}{\sqrt{1 - v^2/c^2}}\left(t' + \frac{v}{c^2}x'\right) \tag{12–3d}$$

These equations are used when the point (x', y', z', t') is considered to be fixed in the O′ frame. The equations give the coordinates of the same point (x, y, z, t) relative to the O frame, which is moving relative to this point with a speed v.

With these considerations in mind, let us now examine some of the consequences of the use of the Lorentz transformation.

12–3

TIME AND SPACE IN SPECIAL RELATIVITY

One of the most striking features of the Lorentz transformation is the time coordinate transformation (Equations 12–2d and 12–3d), which gives the relationship between the times determined by two observers moving relative to one another. These equations indicate that the passage of time, as measured by the time interval between specific events, will be different for the two observers. Does this mean that time physically slows down or speeds up in one observer's reference frame as compared to the other observer? Or is the difference in time only an apparent effect? It is evident that the interpretation of the Lorentz time transformation will significantly affect our physical and philosophical concepts of the nature of time. It is therefore important that the meaning of the time transformation be clearly understood.

Let us compare the findings of two observers, one in O and one in O′, moving relative to one another. The observer in O knows the coordinates (x, y, z) of all points in space with respect to the origin of his coordinate system. He therefore knows the distances and light travel times between any two points in his reference frame, and can thus calculate the time of occurrence of any distant event by subtracting the light travel time from the time of observation. He may also have placed any number of clocks at various points in his reference frame, all synchronized to read the same time, t. Each of these clocks will directly read the time of occurrence of any event at its location. The observer in O′ can do the same in his own reference frame, by calculating light travel times based on the x', y', z' coordinates or by synchronizing his clocks to read the same time, t'. Thus each observer is capable of determining the time of occurrence of any event with respect to his reference frame. However, the two observers will calculate different occurrence times for the same event. Each observer will consider the other to have improperly synchronized clocks and incorrectly calculated light travel times. Given the axioms of special relativity, there is no way to resolve the disagreement. Because absolute motion is undetectable and the speed of light is the same for all observers (i.e., all inertial reference frames are equally valid), there is no way to decide which observer has measured a "correct" travel time or has properly synchronized his clocks.

The conclusion that two observers moving relative to one another measure different times is therefore seen to be a consequence of *a fundamental and unresolvable disagreement as to how times of occurrence should be determined.* Thus the discrepancy is an *apparent* one. In special relativity time does not physically slow down or speed up; the mechanism of a clock is not affected in some mysterious way that causes it to turn at a different rate when it is moving relative to an observer. One can still accept as a philosophical concept the idea of an absolute time that flows on without reference to anything external, but in relativity theory the absolute time is *intrinsically unmeasurable* and therefore serves no purpose as a physical concept. Therefore all times measured in different inertial reference frames must be considered equally valid. The Lorentz transformation provides the mathematical link between the different times of occurrence measured in different reference frames that are moving relative to one another.

THE RELATIVISTIC TIME DILATION

Suppose two observers in different inertial reference frames calculate the time interval between two given events. The Lorentz transformation can be used to determine the relationship between the time intervals calculated by each observer. To simplify matters, let us assume that these events occur at a fixed point in one reference frame. For illustrative purposes, let us assume that we are determining the time interval between two successive flashes of a blinking light located at the point P on the x axis a distance d from the origin of reference frame O, as shown in Figure 12–11a. A stationary observer located anywhere in O records the time at which he observes the first flash, the time at which he observes the second flash, and then subtracts the light travel times from each to obtain the times of occurrence, t_1 and t_2. (Since the observer is stationary with respect to the light source, the light travel times will be the same for both flashes.) He then calculates $\Delta t = t_2 - t_1$, which is the calculated time interval between the occurrence of the first and second flashes for any observer who is stationary relative to the light source at P.

Now consider an observer in O′, the "moving" reference frame (Figure 12–11b). Relative to reference frame O′ the light source appears to be in motion between successive flashes. The observer in O′ records the times when he observes the successive flashes. He then corrects for the light travel times, which are different for each flash because the light source is moving relative to the observer. He then subtracts the calculated light travel times from the respective observation times to obtain the successive times of occurrence, t_1' and t_2', and the time interval, $\Delta t' = t_2' - t_1'$. However, the "stationary" observer in O will disagree with the moving observer's light travel times. (From the stationary observer's point of view the moving observer's light travel times are too large.) Hence the two observers will calculate different time intervals.

The Lorentz transformation mathematically takes into account the disagreement between the two observers as to light travel times, so that it becomes possible to

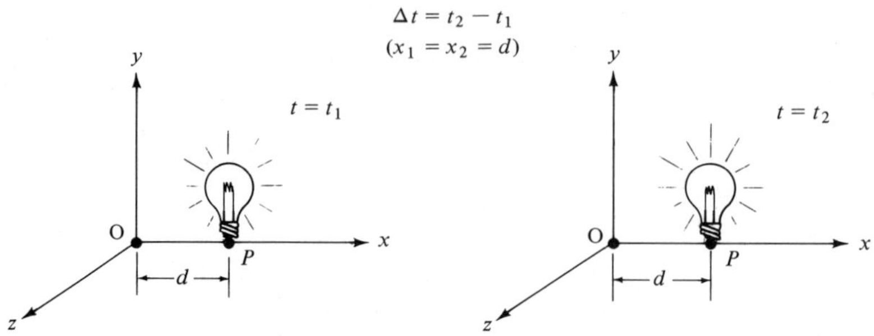

(a)

Stationary observer. Two events (two successive flashes of light) take place at a fixed point in the observer's reference frame.

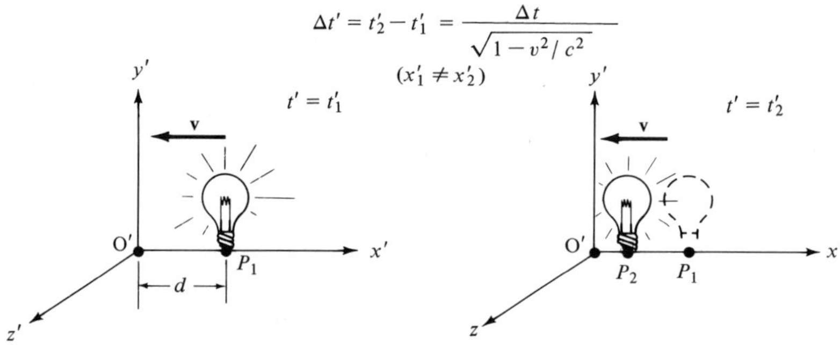

(b)

Moving observer. Two events occur at different points with respect to the observer's reference frame.

figure 12–11 RELATIVISTIC TIME DILATION

Stationary and moving observers will measure different time intervals between successive flashes of light. The time interval measured by the moving observer will be longer than that of the stationary observer.

obtain an equation directly relating the time intervals calculated by the two observers in this situation. Specifically, the times of occurrence, t_1' and t_2', of the two light flashes relative to the observer in O′ can be related to t_1 and t_2 (the times relative to O) by Equation 12–2d. We obtain

$$t_2' - t_1' = \left[\frac{1}{\sqrt{1 - v^2/c^2}}\left(t_2 - \frac{v}{c^2}x_2\right)\right] - \left[\frac{1}{\sqrt{1 - v^2/c^2}}\left(t_1 - \frac{v}{c^2}x_1\right)\right]$$

$$= \frac{1}{\sqrt{1 - v^2/c^2}}\left[(t_2 - t_1) - \frac{v}{c^2}(x_2 - x_1)\right]$$

Here x_1 and x_2 represent the coordinates of the point P at t_1 and t_2, respectively, relative to O. But P is stationary in O; that is, $x_1 = x_2 = d$, so that $x_2 - x_1 = 0$. Substituting $(x_2 - x_1) = 0$, $\Delta t' = t_2' - t_1'$, and $\Delta t = t_2 - t_1$ into the above equation, we arrive at the expression

$$\Delta t' = \frac{\Delta t}{\sqrt{1 - v^2/c^2}} \qquad (12\text{–}4a)$$

Since $\sqrt{1 - v^2/c^2}$ is always less than 1, it follows that $\Delta t'$ will always be greater than Δt. In other words, the moving observer calculates a *longer* time interval than does the stationary observer. The greater the relative speed v is, the greater the disparity between $\Delta t'$ and Δt.

Now let us suppose that the light source is fixed at a point that is stationary in the O′ system. Then the observer in O becomes the "moving" observer; the two flashes occur at two different points with respect to his origin. However, the position of the light source in O′ is fixed; thus $x_1' = x_2'$. We now use Equation 12–3d to relate the calculated time intervals Δt and $\Delta t'$:

$$t_2 - t_1 = \frac{1}{\sqrt{1 - v^2/c^2}}\left[(t_2' - t_1') + \frac{v}{c^2}(x_2' - x_1')\right] = \frac{t_2' - t_1'}{\sqrt{1 - v^2/c^2}}$$

or

$$\Delta t = \frac{\Delta t'}{\sqrt{1 - v^2/c^2}} \qquad (12\text{–}4b)$$

Now the observer in O calculates a longer time interval! These two results appear to be contradictory, but they are not because they refer to two different situations. On closer inspection we find that Equations 12–4a and 12–4b actually express the same result. In each case *the observer who is moving with respect to the point at which the events take place calculates a longer time interval than the observer who is stationary with respect to this point.* We can combine Equations 12–4a and 12–4b into a single ***relativistic time dilation equation:***

$$\Delta t = \frac{\Delta t_0}{\sqrt{1 - v^2/c^2}} \qquad (12\text{–}5)$$

In this equation Δt_0, the *proper time interval,* is the time interval between two events measured by an observer who is stationary with respect to the place at which the events occur, and Δt is the time interval between the same two events as determined by a moving observer. (Here "moving" and "stationary" mean with respect to the events taking place.) Since $\Delta t > \Delta t_0$, it follows that a moving observer always calculates a longer time interval than a stationary observer. Thus time seems drawn out or dilated to a moving observer. This should not be regarded, however, as a true slowing down of time; it is a consequence of the disagreement between the two observers as to how time should be measured. The following example illustrates how Equation 12–5 is used.

EXAMPLE 12–1: Spaceship X is traveling with a speed of 0.6 of the speed of light with respect to spaceship Y. Light pulses are sent from X to Y at 10-sec intervals, as timed by a clock on board spaceship X. What is the rate of emission of these pulses as calculated by an observer on Y?

SOLUTION: In this situation X is the stationary observer, since the clock used to time the pulses is stationary with respect to the light source. Thus the proper time interval $\Delta t_0 = 10$ sec. The observer on Y is moving at $v = 0.6c$ $(v/c = 0.6)$ relative to the light source. According to Equation 12–5, the time interval between light pulses relative to Y is

$$\Delta t = \frac{\Delta t_0}{\sqrt{1 - v^2/c^2}} = \frac{10 \text{ sec}}{\sqrt{1 - (0.6)^2}} = \frac{10}{\sqrt{0.64}} = \frac{10}{0.8} = \underline{12.5 \text{ sec}}$$

In this example Δt (= 12.5 sec) is *not* the time between Y's *observation* of successive pulses. Rather it is the time between the occurrence of successive pulses as *calculated* by Y after subtracting the light travel time for each pulse. X and Y measure different pulse rates because of their disagreement over the values of the light travel time from X to Y. If the pulses were being emitted by Y at 10-sec intervals, X would be the "moving" observer and would calculate a time interval of 12.5 sec between pulses.

THE RELATIVISTIC LENGTH CONTRACTION

In Section 12–2 we used the Galilean transformation to show that the length of an object or the distance between two points is independent of the motion of the reference frame. Let us now repeat the calculation using the Lorentz transformation.

An object lies along the x axis of reference frame O and is at rest with respect to O (Figure 12–12a). If one end lies at point x_A and the other end at x_B, the object has a length $l = x_B - x_A$. Suppose an observer in O′ wishes to measure the length of the object with respect to his own reference frame. Since the object is in motion relative to his reference frame, the observer must be certain to measure *simultaneously* the coordinates on the x' axis of the end points of the object. (If the end points are not measured simultaneously, the object will move during the measurement process.) Thus the observer in O′ must measure x'_A and x'_B at the same instant of time, $t'_A = t'_B = t'_0$. The length l' measured in O′ will be $l' = x'_B - x'_A$.

The relationship between l' and l can be obtained from the inverse Lorentz transformation. Substituting for x_A and x_B from Equation 12–3a and subtracting gives

$$x_B - x_A = \frac{1}{\sqrt{1 - v^2/c^2}}[(x'_B - x'_A) + v(t'_B - t'_A)]$$

But $t'_A = t'_B$; therefore $t'_B - t'_A = 0$. Substituting $x_B - x_A = l$ and $x'_B - x'_A = l'$, we obtain

$$l' = l\sqrt{1 - v^2/c^2} \qquad \text{(12–6a)}$$

Therefore the "moving" observer in O′ measures a *shorter* length than the "stationary" observer in O. However, if the object is at rest in O′ (Figure 12–12b), then O is the "moving" observer and must simultaneously ($t_A = t_B = t_0$) measure the end-point coordinates. In this case it can be shown, using the Lorentz transformation, that

$$l = l'\sqrt{1 - v^2/c^2} \qquad \text{(12–6b)}$$

Once again the moving observer measures the shorter length. We can combine these results into the ***relativistic length contraction equation:***

$$l = l_0\sqrt{1 - v^2/c^2} \qquad \text{(12–7)}$$

In this equation l_0, the ***rest length,*** is the length of an object (or distance between two points) as measured by an observer at rest with respect to the object or points in question; l, the ***moving length,*** is the length of the object measured by an observer in whose reference frame the object is moving with a speed v.

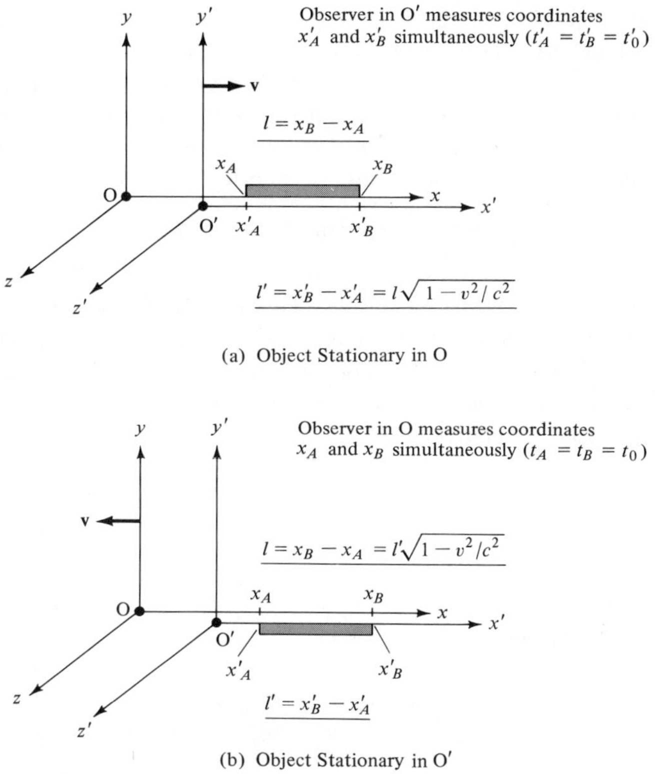

figure 12–12 **RELATIVISTIC LENGTH CONTRACTION**

(a) An object lying on the x axis is at rest in O, and its length is measured by the moving observer in O′. (b) The object is at rest in O′, and its length is measured by the moving observer in O. In each case the moving observer measures a shorter length than the stationary observer does.

The source of the disagreement between stationary and moving observers as to the length of an object can be traced to the problem of the relativity of time. The moving observer must measure the end points simultaneously, but he and the stationary observer cannot agree as to whether or not this has been done properly. The moving observer believes that he has made the measurements simultaneously ($t'_A = t'_B$), but the stationary observer believes that the moving observer has made the measurements at two different times. From the stationary observer's point of view, the moving observer has measured the length improperly. However, the first axiom of special relativity tells us that both measurements are equally valid. Hence length, like time, becomes a *relative* quantity that has different values when measured in different reference frames.

In the previous chapter we discussed the Fitzgerald–Lorentz contraction hypothesis, which was offered as an explanation for the null result of the Michelson–Morley experiment. The hypothesis assumed that a body contracts by a factor of $\sqrt{1 - v^2/c^2}$ along its direction of motion. It would seem that our findings have confirmed this hypothesis, but this is not the case. The Fitzgerald–Lorentz hypothesis proposed that a body *physically* contracts (because of changes in the intermolecular forces) as a result of its motion with respect to the ether. But in special relativity motion relative to the ether is undetectable. In Einstein's interpretation the relativistic length contraction is an *apparent* contraction that arises from a fundamental disagreement between observers as to how a moving length is measured.

THE RELATIVISTIC ADDITION OF VELOCITIES

We will now derive an equation to relate the velocity of a body with respect to different reference frames. This equation will be the relativistic analogue of the velocity addition equation $\mathbf{u} = \mathbf{u}' + \mathbf{v}$ (Equation 11–1).

We assume that a body is moving along the x axis of reference frame O with a constant speed. Suppose that the body moves from x_1 at time t_1 to x_2 at time t_2. Its speed u relative to O is:

$$u = \frac{\text{distance}}{\text{time}} = \frac{x_2 - x_1}{t_2 - t_1} \tag{12–8a}$$

We wish to determine the speed u' of the same body measured relative to a second reference frame O′, which is moving with a speed v relative to O. Relative to O′, the body has a speed

$$u' = \frac{x_2' - x_1'}{t_2' - t_1'} \tag{12–8b}$$

Substituting the Lorentz transformation equations for x_1, x_2, t_1, and t_2 in Equation 12–8a and canceling out the common factor $\sqrt{1 - v^2/c^2}$, we obtain

$$u = \frac{(x_2' + vt_2') - (x_1' + vt_1')}{\left(t_2' + \frac{v}{c^2}x_2'\right) - \left(t_1' + \frac{v}{c^2}x_1'\right)} = \frac{(x_2' - x_1') + v(t_2' - t_1')}{(t_2' - t_1') + \frac{v}{c^2}(x_2' - x_1')}$$

Dividing through by the factor $t_2' - t_1'$ yields

$$u = \frac{\left(\frac{x_2' - x_1'}{t_2' - t_1'}\right) + v}{1 + \frac{v}{c^2}\left(\frac{x_2' - x_1'}{t_2' - t_1'}\right)}$$

Substituting for u' from Equation 12–8b, we obtain

$$u = \frac{u' + v}{1 + \frac{u'v}{c^2}} \tag{12–9}$$

This equation is the relativistic formula for the addition of velocities.* Note that for $v \ll c$, $u \approx u' + v$; thus Equation 11–1 (derived from the Galilean transformation) is an *approximation* to the relativistic equation at small velocities.

* In this form Equation 12–9 applies only when $\mathbf{u}'$ and $\mathbf{v}$ point in the same direction. When $\mathbf{u}'$ and $\mathbf{v}$ point in opposite directions, the plus signs are changed to minuses. There is no simple relativistic vector equation (equivalent to $\mathbf{u} = \mathbf{u}' + \mathbf{v}$) when $\mathbf{u}'$ and $\mathbf{v}$ are not parallel.

EXAMPLE 12–2: Three spaceships, X, Y, and Z, are all moving in the same direction. Z is traveling with a speed of $0.8c$ relative to Y, and Y is traveling with a speed of $0.6c$ relative to X. What is the speed of Z relative to X?

SOLUTION: Let Y be reference frame O′ and X be reference frame O. Then u' is the speed of Z relative to Y ($0.8c$); v is the speed of Y relative to X ($0.6c$); and u is the speed of Z relative to X. Using the nonrelativistic formula, we would obtain $u = 1.4c$ by simple addition. Using Equation 12–9, however, we obtain

$$u = \frac{u' + v}{1 + \dfrac{u'v}{c^2}} = \frac{0.8c + 0.6c}{1 + \dfrac{(0.8c)(0.6c)}{c^2}} = \frac{1.40}{1.48}c$$

or

$$\underline{u = 0.946c}$$

The following application of Equation 12–9 is particularly interesting. Suppose a light pulse is emitted from the origin of the O′ system along the x' axis. The light pulse will have a speed $u' = c$ relative to O′. What is its speed u with respect to O? (We assume that O′ has a speed v relative to O):

$$u = \frac{c + v}{1 + \dfrac{cv}{c^2}} = \frac{c + v}{1 + \dfrac{v}{c}} = \frac{c(c + v)}{c + v} = c$$

Thus *any speed added to the speed of light gives the speed of light.* This means that all observers measure a speed of c for any light pulse regardless of their motion. Equation 12–9 is therefore consistent with the axioms of special relativity—as it should be, since the formula was derived from these axioms.

Let us summarize what has been accomplished thus far. At the start Einstein proposed for the sake of argument the acceptance of the two axioms of special relativity. Proceeding logically from these axioms, he was led to the idea that time and length are *relative* quantities that are different when measured by observers in different inertial reference frames. Using the Lorentz transformation equations (which follow from the axioms), he developed formulas to relate the different values of length, time, and velocity measured by two observers moving uniformly relative to one another.

These formulas (Equations 12–5, 12–7, and 12–9) are different from those obtained from Newtonian concepts, according to which length and time are absolute quantities. We therefore have two rival theories—the Einsteinian (relativistic) and the Newtonian (nonrelativistic)—which make conflicting predictions with respect to the measurement of length, time, and velocity. In principle we should be able to decide between these theories on the basis of an experimental test. However, the predictions of the two theories differ significantly only at speeds comparable to the speed of light. At the speeds common to everyday human experience, the discrepancies between the Einsteinian and Newtonian results are so minute as to be virtually unmeasurable. To its credit, the theory of relativity fully accounted for the failure of all attempts to detect the ether. Many scientists were sure, however, that the null result of the Michelson–Morley experiment would eventually be explained in a more conventional manner within the confines of Newtonian mechanics.

In 1905 no definite experimental evidence was available that could be used to decide between the relativistic and nonrelativistic theories. In the absence of any such evidence, special relativity could not be considered as either proved or disproved;

it remained a purely hypothetical enterprise. Einstein, however, was convinced that he was on the right track. He plunged further into the theory in the hope of finding some result that could be put to a clear-cut experimental test.

12–4

MASS AND ENERGY

A few months after the appearance of his first paper on special relativity, Einstein published a second article entitled "Does the Inertia of a Body Depend upon Its Energy Content?" In this brief paper Einstein demonstrated the possibility of a direct relationship between mass and energy, which represented a radical departure from the ideas of Newtonian physics.

Einstein derived this relationship by devising a "thought experiment." He considered a body, at rest in an inertial reference frame, that emits two pulses of light (or any other kind of electromagnetic radiation) of equal energy E in opposite directions (Figure 12–13). Since energy is conserved, the total energy of the body should decrease by $2E$, equal to the amount of light energy radiated. After the emission, the body should remain at rest, for if the pulses are emitted with equal intensity in opposite directions, the body should not recoil in one direction or the other. Hence a stationary observer does not see any change in the velocity of the body.

Now we consider the same situation as seen by an observer moving relative to the body. With respect to this observer's reference frame, the body appears to be moving with a velocity $\mathbf{v}$. Since the *stationary* observer sees no change in velocity ($\mathbf{v} = 0$ before and after emission of the pulses), the *moving* observer must also observe no change in velocity. Hence in the moving observer's reference frame the body continues to move with the same velocity $\mathbf{v}$ after emitting the light pulses.

The moving observer sees the pulses being emitted by a moving body. According to the principles of wave motion (and in particular because of the Doppler effect), he should observe that the forward-emitted light pulse has an energy E_f that is greater than E, and the backward-emitted light pulse has an energy E_b less than E. Einstein used the equations of special relativity to show that the sum of $E_f + E_b$ is greater than $2E$.* Thus *the amount of light energy emitted by the body appears greater to the moving observer than to the stationary observer.* It follows that the body appears to lose a greater amount of energy in the moving reference frame than it does in the stationary reference frame. Unless this extra energy can be accounted for, the moving observer will have seen a violation of the principle of conservation of energy.

Einstein showed that energy conservation could be preserved in the moving observer's reference frame *only if the body loses kinetic energy in the process.* (In other words, the extra energy lost by the body in the moving reference frame comes from its kinetic energy.) But the velocity of the body did not change! How could the kinetic energy decrease if the velocity remained constant? Since $KE = \frac{1}{2}mv^2$, it follows that if v is constant, the kinetic energy can decrease only if the *mass* of the body decreases. Thus Einstein came to the conclusion that *when a body emits electromagnetic radiation, it loses mass in the process.* In the stationary reference

* Specifically, Einstein found that $E_f + E_b = \dfrac{2E}{\sqrt{1 - v^2/c^2}}$.

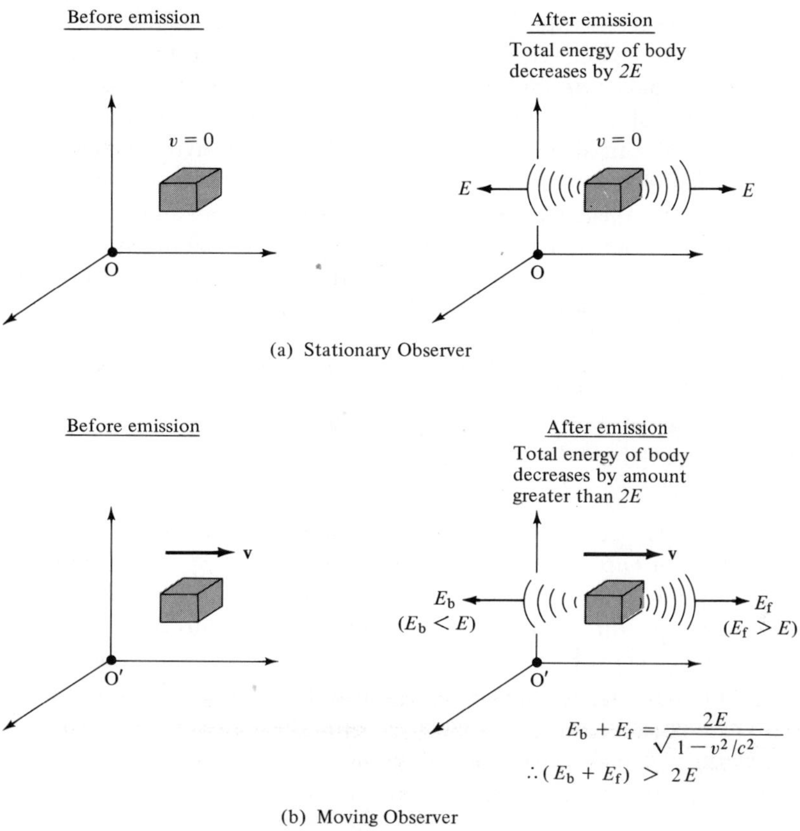

figure 12–13 **EINSTEIN'S MASS-ENERGY "THOUGHT EXPERIMENT"**

(a) A body emits two light pulses of energy E in opposite directions; its speed remains constant. (b) The same event is observed by a moving observer. In the moving reference frame the forward light pulse appears to have an energy E_f, which is greater than E, while the backward light pulse appears to have an energy E_b, which is less than E. The body appears to lose more energy in the moving reference frame than in the stationary reference frame. In order to maintain conservation of energy, the mass of the body must decrease.

frame, where $v = 0$, the body can lose mass without losing kinetic energy, so Einstein's conclusion applies whether the body is moving or at rest in the observer's reference frame.

In Newtonian physics the mass of a body is a constant quantity. Hence Einstein's thought experiment leads to a serious dilemma. If we assume that the mass decreases, we are rejecting the principle of conservation of mass. If we assert that the mass remains constant, the moving observer has observed a violation of conservation of energy. If we continue to accept both conservation of energy *and* conservation of mass, we have a conflict with the axioms of special relativity.

Since no violations of either conservation of energy or conservation of mass had been observed experimentally in 1905, Einstein might well have reasoned at this point that the axioms of special relativity would have to be abandoned, but his faith in these axioms was much too strong. Instead, he chose to reject the principle of conservation of mass, while retaining conservation of energy as a fundamental physical principle valid in all reference frames. Thus in relativity theory it is taken as a basic *assumption* that energy is conserved in all reference frames, and the equations of relativity are constructed so as to agree with this assumption.

In his original paper Einstein arrived at a mathematical relationship between the energy emitted or absorbed by a body and the mass change required to maintain conservation of energy. He stated his conclusions as follows:

> If a body gives off the energy E in the form of radiation, its mass diminishes by E/c^2. . . . The mass of a body is a measure of its energy-content; if the energy changes by E, the mass changes in the same sense by E/c^2. . . . If the theory corresponds to the facts, radiation conveys inertia between the emitting and absorbing bodies.[1]

Here Einstein suggested for the first time that mass and energy are directly related in an entirely non-Newtonian way. According to Einstein, a body's emission or absorption of energy in the form of electromagnetic radiation must be accompanied by a corresponding change in the body's mass. Thus if body A emits light that is absorbed by body B, the mass of A decreases, and the mass of B increases by the same amount. In effect, mass is being transferred from the emitter to the absorber. The change in mass Δm associated with the emission or absorption of a quantity of energy ΔE is given by the equation

$$\Delta m = \frac{\Delta E}{c^2} \tag{12–10}$$

This is the relationship between mass and energy that is required for the principle of energy conservation to be compatible with the axioms of special relativity.

A fundamental assumption of Newtonian physics is that the mass of a body is a constant quantity. Newton himself believed that mass is a measure of the amount of matter in a body. It therefore becomes difficult to understand how the mass of a body can change with the emission or absorption of electromagnetic waves—which are nonsubstance—or how light can transfer mass from one body to another. But note Einstein's statement that radiation conveys *inertia* between bodies. We recall that inertia denotes the tendency of a body to retain its state of motion (to resist acceleration), and that mass is specifically a measure of the inertia of a body and not of the amount of matter in that body. Hence it is the inertia, and not the substance, of a body that changes when light is emitted or absorbed; a body that has absorbed light becomes more resistant to acceleration.

The weight of a body is, as we know, directly proportional to its mass ($w = mg$). It follows that if Einstein's prediction about mass and energy is true, light bulbs should weigh less the longer they shine, and a devoted sun bather should gain weight along with a suntan. However, because of the c^2 factor in Equation 12–10, the mass of a body would be expected to change by only 1 gram for every 9×10^{20} ergs (= 25 million kilowatt-hr!) of energy absorbed or emitted. Thus the nature of the mass–energy relationship is such that the mass (and weight) changes associated with light emission or absorption should be immeasurably small.

The equation $\Delta m = \Delta E/c^2$ is obviously related to the more famous $E = mc^2$, which will be discussed more fully in the next section. The mass–energy relationship was, in fact, first derived by Poincaré in 1900. We learned in Chapter 10 that electromagnetic waves carry momentum ($p = E/c$). Poincaré calculated that the "mass" associated with this momentum is equal to E/c^2. However, Poincaré considered this to be a fictitious mass and went no further. Einstein, on the other hand, believed that the mass associated with electromagnetic radiation is real and can be converted to or from the inertial mass of a material body.

Einstein's first derivation of the mass–energy relationship contained a serious mathematical error, which was not noticed until many years later. Nevertheless, his conclusion was still correct. He was able to derive the same result by several other "thought experiments," which he presented in 1906 and 1907.

RELATIVISTIC MECHANICS

In the first published paper dealing with the mass–energy relationship, Einstein was concerned only with the mass changes that should accompany the absorption or emission of electromagnetic radiation. Next he sought to find a more general and mathematically rigorous formulation of the mass–energy relationship. Many other physicists also became interested in the problems raised by Einstein's first papers. The German physicists Max Planck and Hermann Minkowski* and the American Gilbert N. Lewis, as well as Einstein, made many important contributions to the theory of special relativity between 1906 and 1910.

The main purpose of the research done by these investigators was to construct a ***relativistic mechanics*** to replace the Newtonian theory of mechanics. The rejection of the Newtonian system was a matter to be considered very carefully. On one hand, it had been shown that Newtonian mechanics was inconsistent with the axioms of special relativity. On the other hand, the Newtonian system had accounted for the motions of the earth, moon, and planets with considerable accuracy. These bodies, however, move with speeds that are small compared to the speed of light. (For example, $v/c = 0.0001$ for the earth's orbital motion.) Newtonian mechanics had, in fact, never been tested experimentally at speeds comparable to c.

Hence the new relativistic mechanics would constitute a description of force and motion logically consistent with the axioms of special relativity. But the relativistic equations would have to resemble closely the Newtonian equations in the low-velocity realm ($v \ll c$), where Newtonian mechanics was known to be accurate. Thus Newtonian mechanics should appear as a low-velocity approximation to relativistic mechanics.

The formulation of the new mechanics pivoted on the conservation principles of energy and momentum. Newtonian mechanics began with Newton's three principles of motion as the fundamental axioms, and many years later the conservation principles were derived from them. By 1900 most physicists had come around to the point of view that the conservation principles were more fundamental. Accordingly, in the formulation of relativistic mechanics, the conservation principles regarding energy and momentum were adopted as *fundamental assumptions* of the theory, along with the axioms of special relativity.

In the previous section we saw that Einstein arrived at the mass–energy relationship by assuming that conservation of energy must be observed in all inertial

*Minkowski was Einstein's mathematics professor when Einstein was an indifferent undergraduate at the University of Zurich. Within a decade teacher and pupil exchanged roles rather dramatically.

reference frames. Now a similar stipulation was made with respect to the conservation of *linear momentum.* Various thought experiments were devised in which a collision between bodies or the recoil of one body upon emission of another body was examined from the points of view of different observers moving relative to one another. It was found that if the Newtonian definition of linear momentum, $\mathbf{p} = m\mathbf{v}$, was used, linear momentum would not appear to be conserved in all inertial reference frames. It was shown mathematically that in order for conservation of linear momentum to be compatible with the axioms of special relativity, linear momentum would have to be redefined in the following way:

$$\mathbf{p} = \frac{m_0 \mathbf{v}}{\sqrt{1 - v^2/c^2}} \tag{12–11}$$

This equation defines the ***relativistic linear momentum,*** a quantity that is conserved in all interactions. From this equation it can be deduced that the ***relativistic mass*** m of a body varies with the body's speed, according to the equation

$$m = \frac{m_0}{\sqrt{1 - v^2/c^2}} \tag{12–12}$$

The term m_0 appearing in Equations 12–11 and 12–12 is the ***rest mass***—the mass of a body measured by an observer when the body is stationary in his reference frame. When the body is in motion with respect to the observer, the measured mass m will be greater than the rest mass m_0, and will increase as the speed v of the body relative to the observer increases. Thus in relativity theory mass becomes a *relative* quantity, like length and time, and is different for different observers moving relative to one another. By contrast, in Newtonian mechanics the mass of a body is always equal to the rest mass m_0, regardless of the body's speed. At low speeds, however, Equations 12–11 and 12–12 give the approximate results $\mathbf{p} \approx m_0\mathbf{v}$ and $m \approx m_0$. Thus the relativistic formulas reduce to the Newtonian results for $v \ll c$, as required.

In relativity theory Newton's principle of inertia remains essentially intact. All bodies possess inertia; that is, they tend to move with constant velocity unless acted on by a force. In relativity theory, however, a body does not have a constant inertia as in Newtonian mechanics. The inertia of a body is measured by the body's mass, and in relativistic mechanics the mass increases with speed. Hence the tendency of a body to maintain constant velocity becomes greater as the body moves faster; all things being equal, it is more difficult to accelerate a fast-moving body than a slow-moving one. For instance, it takes a greater force to accelerate a given body from 100,000,010 mi/hr to 100,000,020 mi/hr than from 10 mi/hr to 20 mi/hr in a given time interval. Even though the acceleration is the same in both cases, a greater force is required at the greater speed because the body has more inertia at 100,000,010 mi/hr than at 10 mi/hr.

An even more remarkable conclusion can be drawn from the specific way in which mass increases with speed. According to Equation 12–12, the mass of a body becomes infinite as the speed of the body approaches c. Hence it is impossible to accelerate a body up to the speed of light, for an infinitely strong force would be required to accelerate an infinite mass. Therefore relativity theory predicts that no material object can have a speed equal to or greater than c; the speed of light in a vacuum* becomes the limiting speed for motion in the universe. This is in stark contrast to Newtonian mechanics, in which it is presumed that by continued application of a constant force a body can be accelerated up to any speed without limit.

In 1906 Planck showed how to formulate a relativistic analogue for Newton's second principle that would yield the familiar $\mathbf{F} = m\mathbf{a}$ as a low-velocity approximation. In Chapter 6 we demonstrated that Newton's second principle could be written in the form (Equation 6–2)

$$\mathbf{F} = \frac{\Delta \mathbf{p}}{\Delta t} = \frac{\Delta(m\mathbf{v})}{\Delta t}$$

That is, *force equals the rate of change of linear momentum.* In Newtonian mechanics the mass m is constant ($m = m_0$), and the equation $\mathbf{F} = m\mathbf{a}$ follows directly from Equation 6–2. This statement of the second principle can also be used in relativistic mechanics. But now the mass is no longer a constant; the mass increases with speed as described by Equation 12–12. Thus it is still possible to state that the *force equals the rate of change of linear momentum,* provided we use the *relativistic* linear momentum as defined in Equation 12–11. Thus in relativistic mechanics Newton's second principle becomes

$$\mathbf{F} = \frac{\Delta \mathbf{p}}{\Delta t} = \frac{\Delta(m_0\mathbf{v}/\sqrt{1 - v^2/c^2})}{\Delta t} \tag{12–13}$$

and the Newtonian formula $\mathbf{F} = m\mathbf{a}$ appears only as a low-velocity approximation to this equation.

CONSERVATION OF MASS–ENERGY

In relativistic mechanics, as in Newtonian mechanics, a quantity called work is defined as the product of force and displacement ($W = Fd\cos\theta$). In Chapter 6 it was shown that the work done on a body by any force is equal to the change in the kinetic energy of the body, provided kinetic energy is defined by Equation 6–10:

$$KE_{\text{Newtonian}} = \tfrac{1}{2}m_0v^2$$

A similar derivation relating kinetic energy to work can be done in relativity theory. However, the relativistic force equation (Equation 12–13) must now be used. A rather complicated calculation leads to the following:

$$KE_{\text{relativistic}} = m_0c^2\left(\frac{1}{\sqrt{1 - v^2/c^2}} - 1\right) \tag{12–14}$$

This is the equation for the ***relativistic kinetic energy*** of a body of rest mass m_0. Although it is not at all obvious from the equation's appearance, Equation 12–14 reduces to $KE = \frac{1}{2}m_0v^2$ as a low-velocity approximation. The mathematical proof of this relationship appears in Appendix C–3.

*When moving through a material medium, light travels with a speed less than c, so it is possible for a particle to travel faster than light in a medium. The theory stipulates that no particle can have a speed equal to or greater than c ($= 3 \times 10^{10}$ cm/sec), which is the speed of light in a vacuum.

Now let us rewrite Equation 12–14 and substitute for the relativistic mass (Equation 12–12) as follows:

$$KE = \frac{m_0c^2}{\sqrt{1 - v^2/c^2}} - m_0c^2 = mc^2 - m_0c^2$$

which can be rearranged as follows:

$$mc^2 = m_0c^2 + KE \tag{12–15}$$

This equation was particularly interesting to Einstein. In his previous work Einstein had shown that the mass of a body changes when electromagnetic energy is emitted or absorbed, and the equation $\Delta E = c^2 \, \Delta m$ correlates the change in mass with the amount of energy emitted or absorbed. This interchange between mass and energy suggested to Einstein that *mass is a form of energy*, and the appearance of the m_0c^2 term in conjunction with kinetic energy in Equation 12–15 strengthened this belief. He therefore proposed that a body possesses an intrinsic energy proportional to its mass, over and above all other forms of energy. Einstein defined the ***rest energy*** E_0 of a stationary body by the equation

$$E_0 = m_0c^2 \tag{12–16}$$

The rest energy is the intrinsic energy contained in a body by virtue of its mass alone. When the body is in motion, it has kinetic energy in addition to its rest energy. The total energy E is then

$$E = mc^2 = m_0c^2 + KE = E_0 + KE$$

By comparing this equation with Equations 12–12 and 12–15, we see that the total energy of a moving body (rest energy plus kinetic energy) is

$$E = mc^2 = \frac{m_0c^2}{\sqrt{1 - v^2/c^2}} = \frac{E_0}{\sqrt{1 - v^2/c^2}} \tag{12–17}$$

In addition, the body can possess potential energy (gravitational, electromagnetic, etc.) and heat energy.

With the identification of mass as a form of energy, the separate principles of conservation of mass and conservation of energy become merged. Mass conservation by itself can no longer be maintained. Moreover, the principle of conservation of energy must again be expanded to include the rest energy (m_0c^2) of a body. Hence mass and energy can be viewed as two forms of the same thing, which can be called ***mass–energy.*** Mass and energy are encompassed under a single conservation principle, which we shall call the ***principle of conservation of mass–energy.*** The theoretical possibility exists for the conversion of mass into various other forms of energy and vice versa,* just as kinetic and potential energy can be converted into one another. However, the theory gave no indication as to *how* mass–energy conversion might be effected. Hence Einstein and his co-workers did not speculate on the possibility or practicality of producing huge quantities of energy from the conversion of small amounts of mass. The technological value of $E = mc^2$ could at best be only

*In this respect the change in mass of a body upon emission or absorption of radiation represents a conversion between mass and electromagnetic energy.

dimly foreseen in 1910. (It has become customary to attribute the development of nuclear weapons and nuclear reactors to the theory of relativity. Actually, the discovery that nuclear energy could be released in great quantities was made quite incidentally to the theory and should more properly be seen as the outgrowth of the nuclear physics research done in the 1930s. Although the existence of nuclear energy can be accounted for in terms of mass–energy conversion, it is not directly predictable from relativity theory.)

The equation $E = mc^2$ is undoubtedly the most familiar equation in all of science. The beginning student may be puzzled as to why c, the speed of light, appears in this equation. There is no simple explanation of this fact, except to say that it is required by the mathematics of the theory. One should recognize, however, that in this equation c does *not* represent the speed of the body in question. Rather, c (or c^2) should be thought of as a conversion factor, relating energy to mass. Since mass and energy are two forms of the same thing (mass–energy), they must have the same units. The c^2 term is the proper factor to convert mass units (grams, kilograms) to energy units (ergs, joules). In this sense c^2 is akin to the "mechanical equivalent of heat" (4.18 joules = 1 calorie), which converts heat units to mechanical energy units.

By combining Equations 12–11 and 12–17, we obtain the following equation relating energy and momentum:

$$E^2 = m_0{}^2c^4 + p^2c^2 = (m_0c^2)^2 + (pc)^2 \qquad (12\text{–}18)$$

Note that for a body at rest ($p = 0$) Equation 12–18 reduces to Equation 12–16. Of more interest is the result obtained by setting $m_0 = 0$—namely, that $E = pc$. The equation therefore predicts that if a massless body exists, it will possess momentum $p = E/c$, even though it has no mass. (Previously, we learned that an electromagnetic wave possesses a momentum $p = E/c$.) It can also be shown (by solving Equation 12–17 for v, then setting $m_0 = 0$) that a massless particle must necessarily travel at the speed of light. This result will be of some interest to us in later chapters.

A FOUR-DIMENSIONAL UNIVERSE?

The equations of special relativity clearly indicate that the space and time coordinates of any event are *not* independent of one another as they are in Newtonian mechanics. The concept of the relativity of time, for example, is based on the premise that the time coordinate of any event depends on the observer's position and motion with respect to that event. In the same manner the measured distances between points in space are found to depend on the times when the measurements are made. This interrelationship is made evident by the equations of the Lorentz transformation, in which the space coordinates (x', x) and time coordinates (t', t) are seen to be mathematically interrelated.

This interdependence of space and time in relativity theory led Hermann Minkowski to construct a new mathematical formulation of the equations of physics. In their new form these equations are written in terms of *four* coordinates; the time coordinate is placed on an essentially equal footing with the three space coordinates x, y, and z to emphasize the relationship between space and time. Since most of the quantities that appear in the equations of physics are vector quantities (force, acceleration, momentum, etc.), it becomes necessary to have ***four-component vectors*** (usually called ***four-vectors***) in the Minkowskian formulation. In addition to the normal three components (corresponding to the three space directions) of pre-relativistic physics, each vector quantity must have a fourth component to be consistent with a four-coordinate system. For example, Minkowski found it useful to choose the time coordinate as the fourth component of the displacement vector and energy as the fourth component of the momentum vector.*

Minkowski discovered that the relativistic equations of mechanics and electromagnetism could be written with considerable mathematical simplicity in terms of a four-coordinate system. The rules of vector arithmetic, extended to four components instead of three, could be directly applied to four-vectors, revealing numerous mathematical and geometric properties of the equations of relativity that had hitherto gone unnoticed. Most notably, it was shown that the magnitude of the four-component displacement vector, which represents the "distance" between two points in space–time, has the same value in all inertial reference frames. Thus in special relativity the space interval between two points and the time interval between two events are both relative—that is, they have different magnitudes for different observers—but the combined space–time interval is an *absolute* or *invariant* quantity for all observers. Contrary to what special relativity seemed to indicate—that all physical quantities are relative—the four-coordinate formulation demonstrated the existence of numerous absolute quantities such as the space–time interval.

Yet the importance of Minkowski's achievement goes beyond its mathematical effectiveness, for it raises some rather intriguing questions about the nature of the physical universe. The world appears to our senses to be three dimensional. Objects appear to have length, width, and depth. Three coordinates are needed to locate the position of an object in space, and vectors have three components to correspond to the three directions of space. Consequently, the equations of pre-relativistic physics were written in terms of three coordinates so as to describe the three-dimensional world of our senses. But Minkowski showed that the equations of physics can be written more conveniently in a four-coordinate form. Do we interpret this to mean that we exist in a *four-dimensional* universe? By way of analogy, let us imagine a civilization of flat, two-dimensional creatures constrained to move on a single plane. They would not be able to perceive a third dimension and would probably be unaware of its existence. Are we, by the same token, three-dimensional creatures who are prevented by our imperfect senses from perceiving the fourth dimension?

In the absence of any hard experimental evidence, we can only speculate as to the answers to such questions. The success of the four-coordinate mathematics does not logically require a belief in a four-dimensional universe; mathematical simplicity need not be equated with physical reality. Thus the concept of a four-dimensional universe does not necessarily follow from the Minkowskian formulation, but neither is there any reason to rule out such an interpretation.

On the other hand, the choice of time as the fourth coordinate has led to much unwarranted speculation about time as the "fourth dimension." Modern science-fiction writing abounds with references to time travel, time warps, and the like. There is nothing in the theory of relativity, however, to indicate that one can travel through

*Actually, the fourth component of displacement is defined as ct, so that it will have the dimensions of length and be dimensionally equivalent to the other three components, x, y, and z.

time the way one travels through space. Such ideas lie outside the bounds of our present-day scientific knowledge.

12–5

EXPERIMENTAL EVIDENCE

It is the essence of the scientific method that a scientific theory stands or falls on whether or not its predictions correspond with the facts. As noted previously, the major difficulty in testing the predictions of the theory of relativity was that the relativistic and nonrelativistic equations are significantly different from one another only for bodies moving at speeds comparable to c. At first such speeds seemed to be experimentally unobtainable. This barrier was eventually overcome, however, by the development of techniques for accelerating charged subatomic particles up to relativistic speeds. It became possible to measure the masses and kinetic energies of these particles, so that the mass and energy relationships predicted by relativity theory could be put to experimental test.

THE MASSES OF HIGH-SPEED PARTICLES

In 1900 it was found that certain radioactive substances spontaneously emit high-speed electrons. Using Thomson's method of deflecting these particles in electric and magnetic fields, physicists could determine the speeds and charge-to-mass ratios of these electrons. In 1901 Walter Kaufmann discovered that the masses of the electrons tended to increase with velocity. Attempts were made to explain this effect in terms of the electromagnetic theory of electrons. In 1908 Alfred H. Bucherer repeated Kaufmann's measurements with greater precision. Using electrons with speeds ranging from $0.32c$ to $0.69c$, Bucherer found that the electron mass increased exactly as predicted by the relativistic mass formula (Equation 12–12).

Since the 1930s a variety of particle accelerators have been constructed in which electrons and other charged particles can be accelerated up to very high speeds by electric and magnetic fields. (The most familiar of these devices is the cyclotron, described in Chapter 10.) The results obtained over the years from particle accelerator experiments tended to confirm fully the relativistic mass formula. The experience with the cyclotron was typical. As higher and higher energies were obtained, it was found that the accelerated particles began to deviate from Newtonian behavior. According to Equation 10–10, the ratio of the path radius to particle speed ($r/v = m/qB$) should be a constant, but in fact it was found to increase at higher and higher speeds. This effect can be completely accounted for in terms of the mass increase predicted by the relativistic mass formula. To the present day the use of relativity theory is essential in the design of all high-energy particle accelerators.

Experiments using particle accelerators have also confirmed Einstein's prediction that bodies cannot attain speeds equal to or greater than the speed of light. All attempts to accelerate particles up to the speed of light have failed. Particles have been brought very close to the speed of light, but the addition of more kinetic energy results in an increase in the *mass* rather than the speed of the particle—as predicted by the relativistic mass–energy relationships. The most powerful accelerators available today can accelerate electrons to speeds of $0.9999999997c$. These electrons have gained so much kinetic energy that they should be traveling at several hundred times the speed of light if Newtonian mechanics were correct. Instead, observations agree closely with the relativistic kinetic energy formula (Equation 12–14). All available evidence to date agrees with the assertion that c is the ultimate speed for material bodies.

MASS–ENERGY CONVERSION

Einstein's mass–energy relationship $E = mc^2$, which represents the most radical departure from Newtonian principles in relativity theory, received its first experimental verification in 1932. In that year John Cockcroft and Ernest Walton of England used a particle accelerator to produce a beam of high-speed hydrogen ions. These ions were aimed at a target made of lithium. The collision of a hydrogen ion with a lithium atom resulted in a nuclear reaction in which two helium ions were produced. Cockcroft and Walton found that mass was not conserved in this reaction; the total mass of the product helium ions turned out to be slightly less than the original total mass of the hydrogen ion and lithium atom. The energy equivalent of the missing mass (calculated from $E = mc^2$) was found to be equal to the kinetic energy of the product helium ions.

Thus Cockcroft and Walton produced the first direct evidence for the conversion of mass to an equivalent amount of kinetic energy. Since 1932 many thousands of different nuclear reactions have been studied. In all cases a conversion of either mass to energy or energy to mass in accordance with $E = mc^2$ was observed. Einstein's mass–energy relationship stands as the most thoroughly verified prediction of relativity theory.

TIME DILATION

Many unstable subatomic particles are generated in high-energy nuclear reactions. One such particle is the ***muon,*** which is a negatively charged particle with a mass about 207 times that of an electron. A muon at rest or traveling at low speed will have, on the average, a lifetime of 2×10^{-6} sec before disintegrating into other particles. In 1941 Bruno Rossi and D. B. Hall discovered that the lifetime of a muon depends on its speed; namely, that slow muons decay more rapidly than fast ones. This observation can be explained in terms of the relativistic time dilation. The time interval between the birth and death of a muon has a different value relative to different reference frames. For a stationary observer (at rest with respect to the muon), the lifetime is 2×10^{-6} sec. For an observer moving relative to the muon (i.e., an experimenter measuring the properties of a moving muon), the calculated lifetime T will be

$$T = \frac{2 \times 10^{-6}\ \text{sec}}{\sqrt{1 - v^2/c^2}}$$

where v is the speed of the muon relative to the observer. Hence the greater the speed of the muon, the longer its observed lifetime. The time dilation equation thus predicts the results observed by Rossi and Hall.

Just a few examples of the experimental evidence that supports the theory of special relativity have been cited here. In fact, virtually all the predictions of the theory have been verified to considerable accuracy, and there is an overwhelming amount of evidence to support the theory. At the present time there is no serious doubt among physicists as to the validity of special relativity.

12–6

THE SIGNIFICANCE OF RELATIVITY

The wealth of experimental evidence supporting the theory of relativity also serves to document the failure of Newtonian mechanics. The Newtonian system can be regarded as a good approximation to relativity at low velocities, but it breaks down completely at velocities near the speed of light. As a practical tool, however, Newtonian mechanics still remains as an integral part of modern-day physics. The high-velocity realm in which relativity theory reigns is one that is seldom encountered in the natural world. There has been virtually no need to employ relativity theory in the analysis of planetary or stellar motions;* the discrepancies between Newtonian theory and observations of celestial motions are, at best, barely measurable. The classical (Newtonian-based) theories in other branches of physics such as thermodynamics, optics, and electromagnetism have proved to be quite adequate for most practical purposes, and reformulation of these theories along relativistic lines would be of mainly academic interest. The practical uses of relativity are nonexistent at the present time. There is certainly no need to take into account the change in mass with velocity in the design of an automobile, jet plane, or even a space vehicle. It is only in the fields of atomic and nuclear physics that relativity theory has been of value. Here the introduction of the relativistic equations of motion and energy has served to improve our understanding of the details of atomic and nuclear structure.

Hence, at least from a practical standpoint, the impact of the theory of special relativity on physics has not been that substantial. It is the philosophical implications of relativity that have made such a deep impression on the course of physical theory. Einstein was responsible for introducing several important new concepts into the mainstream of physical thought. One of these was the idea (expressed in the first axiom) that the laws of physics are the same in all reference frames. This principle has proved to be of considerable importance in many aspects of modern physics. Of even greater significance is the concept that physics should deal with quantities in terms of the way they are measured. Thus the Newtonian concepts of absolute space and absolute time, so central to physical theory for many years, were expunged from physics because Einstein showed that they were inherently unmeasurable. The measurement of time involves space, and the measurement of space involves time; hence it follows that space and time can no longer be dealt with as separate concepts. In the words of Hermann Minkowski,

> Henceforth space by itself, and time by itself, are doomed to fade away into mere shadows, and only a kind of union of the two will preserve an independent reality.[2]

Thus new concepts of space and time—of which the "four-dimensional universe" is but one example—evolved as a consequence of the theory of relativity.

Certainly, one of the most profound effects of the theory was the overthrow of the ether hypothesis. It had been shown that the ether hypothesis led to unresolvable

*The major exception is the motion of the planet Mercury, in which there is a small but clearly measurable deviation from the Newtonian theory. This effect was explained by Einstein using the general theory of relativity and will be discussed in the next chapter.

contradictions with respect to the constancy of the speed of light and that the ether itself was inherently undetectable; therefore it could have no place in physical theory. Yet the ether had served an extremely important role in physics as the medium for the transmission of electromagnetic interactions, including light. With the rejection of the ether hypothesis, the question of how light is transmitted once again became a mystery. The solution lay in the reformulation of the concept of a vacuum as a featureless medium capable of supporting electromagnetic fields. Hence the Newtonian picture of "empty" space, devoid of any properties, could no longer be an acceptable part of the physicist's picture of the universe. Another puzzle raised by Einstein's successes is the special role played by c in the equations of relativity. Why should there be a limiting speed to material motion, and why should it be the speed of electromagnetic waves? This result possibly reflects a deeper connection between inertia and electromagnetism than is found in the present-day equations of physics.

Thus the theory of special relativity has been responsible for many subtle yet profound changes in man's concepts of his own universe. Even more significant changes in these concepts have arisen from the general theory of relativity, as we shall discover in the following chapter.

QUESTIONS FOR REVIEW

1 State and explain the two axioms of special relativity. Discuss the reasons why Einstein adopted each of them.

2 Explain the distinction between the time of observation and the time of occurrence of an event. How are these two quantities related?

3 In relativity theory what distinction is made between a stationary observer and a moving observer?

4 Explain why the light travel time from a distant event depends on whether we consider the light source or the observer to be moving. Why can't the "correct" light travel time be determined?

5 Describe the procedure used to synchronize an observer's clock with a distant clock (a) in a stationary reference frame and (b) in a moving reference frame. Explain why it is impossible to decide which synchronization procedure is the proper one.

6 Discuss the reasons why Einstein rejected the Newtonian concept of absolute time. According to Einstein, how is time to be measured? Can we say definitely that absolute time does not exist?

7 Explain what is meant by a coordinate transformation.

8 Explain what the Galilean transformation is and what it is used for. What assumptions are made in the derivation of the transformation equations? Why is the Galilean transformation associated with Newtonian mechanics?

9 Explain why the Galilean transformation conflicts with (a) electromagnetic theory and (b) the axioms of special relativity.

10 State and explain the equations of the Lorentz transformation. Compare the assumptions made by Lorentz and by Einstein in deriving these equations.

11 Compare Lorentz's and Einstein's interpretations of the following aspects of the Lorentz transformation:

(a) The difference between the stationary observer at O and the moving observer at O′.

(b) The meaning of the speed v.

(c) The length contraction equation.
(d) The time dilation equation.

12 What is the inverse Lorentz transformation? In what situation is it applicable?

13 State and explain the time dilation equation. For which observer does time seem to run more slowly? Discuss whether this effect is a real or an apparent one.

14 State and explain the length contraction equation. According to Einstein, why does a moving observer measure a shorter length? Is this effect real or apparent?

15 State the relativistic equation for the addition of velocities. Identify all terms in the equation, and explain the conditions under which the equation is valid. How does the equation compare with the Newtonian form?

16 Describe Einstein's thought experiment on the relationship between inertia and energy. What conclusion did he come to? What conservation principle was rejected, and why did Einstein find it necessary to reject it?

17 Outline the fundamental assumptions that were made in the formulation of relativistic mechanics.

18 State and explain the equation for the relativistic mass of a body. What does the speed v in this equation refer to? What does it mean when we say that the mass of a body increases with speed?

19 What modifications of Newton's three principles of motion are made in relativistic mechanics?

20 What is meant by the rest energy of a body? How is the concept of rest energy incorporated into the principle of conservation of energy?

21 Discuss the main advantages of Minkowski's four-coordinate representation of the equations of physics.

EXERCISES

1 Outline the basic principles of the philosophy of positivism, as it applies to scientific thought. List as many Newtonian or other physical concepts as you can think of that would be rejected by a positivist.

2 An observer on the earth simultaneously sees two supernova explosions. He determines that one of them (A) is 500 LY away and the other (B) is 800 LY away.
(a) What is the true time sequence of the two supernovae for the observer on the earth?
(b) Where would an observer have to be located with respect to the two supernovae in order to see A occur 100 years before B? In order to see B occur 200 years before A?

3 Show that the *calculated* time interval between two distant events is equal to the *observed* time interval for a stationary observer if (a) the events occur at the same point in space; (b) the observer is equidistant from the locations of the two events.

4 A light source and an observer are a distance d apart when a light pulse from the source is received by the observer. Show that the light travel time is (a) d/c if the observer is stationary relative to the light source, (b) d/c if the observer is assumed to be moving toward the source with a speed v, (c) $d/(c - v)$ if the source is assumed to be moving toward the observer with a speed v, and (d) $d/(c + v)$ if the source is assumed to be moving away from the observer with a speed v.

5 The following table gives the coordinates of three points with respect to two stationary reference frames, O and O′. Determine the equation of the coordinate transformation connecting the two reference frames.

Point	(x, y)	(x', y')
1	(5, 6)	(2, 10)
2	(3, 4)	(0, 8)
3	(−2, −4)	(−5, 0)

6 Starting with the Lorentz transformation equations, derive the equations of the inverse Lorentz transformation.

7 A beacon attached to the nose of a spaceship gives off light flashes at a rate of one flash per minute. A second spaceship approaches the first at a speed of $0.8c$.
(a) What is the rate of the emission of the light flashes according to an observer on the second spaceship?
(b) If an identical beacon is attached to the nose of the second spaceship, what will the rate of the flashes be as measured by an observer on the first spaceship?

8 A scientist measures the period of a pendulum and finds it to be 1 sec. A second scientist, who is moving relative to the first, calculates the period of the same pendulum to be 2 sec. What is the speed of the second scientist relative to the first?

9 Two observers, moving relative to one another, each possess a stick that is 1 meter long. When each observer looks at the other observer's meter stick, he calculates it to be 80 cm long.
(a) What is their relative speed?
(b) Why does each observer think that the other's meter stick is shorter? Is this a logical contradiction? Explain your answer.

10 Derive Equation 12–6b from the Lorentz transformation. Specify the assumptions made in your derivation.

11 A spaceship moves away from the earth at a constant speed of $0.9c$. A rocket is launched from the spaceship in the same direction at a speed of $0.8c$ relative to the spaceship. What is the speed of the rocket relative to the earth? Compare your answer with the nonrelativistic answer.

12 Equation 12–9 for the relativistic addition of velocities is written for the case when $\mathbf{u}'$ and $\mathbf{v}$ point in the *same* direction. What form will the equation take if $\mathbf{u}'$ and $\mathbf{v}$ point in opposite directions?

13 Using the equation obtained for exercise 12, calculate the speed of the rocket of exercise 11 relative to the earth when the rocket is launched *toward* the earth instead of away from it. Compare your answer with the nonrelativistic answer.

14 An observer on a space station sees two spaceships moving toward each other. Each has a speed of $0.9c$ relative to the observer. What is the speed of the two spaceships relative to each other?

15 Throughout his work Einstein made considerable use of the idea of a "thought experiment".
(a) What is the purpose of a thought experiment, and what are its advantages and disadvantages?
(b) Outline a thought experiment that could be used to show that conservation of linear momentum is incompatible with special relativity if the Newtonian definition of linear momentum ($\mathbf{p} = m_0\mathbf{v}$) is used.

16 The sun radiates energy at a rate of 3.9×10^{33} ergs/sec.
(a) Assuming that this energy comes from the conversion of mass, how many grams per second of material are converted to energy in the sun?
(b) If the mass of the sun is 2×10^{33} grams, for how many years could it continue to convert mass to energy at this rate? (1 year $= 3.15 \times 10^7$ sec)

17 A body has a rest mass of 10 grams. It is traveling at a speed of $0.8c$ relative to an observer.
(a) What is the mass of the body relative to the observer?
(b) What is the mass of the body relative to a second observer who is moving parallel to the body at the same speed?

18 A proton has a rest mass of 1.67×10^{-24} gram. It is traveling at six-tenths the speed of light. Calculate its kinetic energy according to both the relativistic and the nonrelativistic formulas.

19 How fast must a body move in order for its relativistic kinetic energy to be equal to its rest mass?

20 Suppose that 1 gram of spinach is completely converted to energy. What would be its caloric value? (Note: 1 Calorie = 1 kilocalorie = 4185 joules).

21 A pion is an unstable subnuclear particle with a rest lifetime of 2.5×10^{-8} sec. An experimenter produces a beam of pions having a speed of $0.6c$.
(a) What lifetime will the experimenter measure for the moving pions?
(b) What distance will the pions travel during their lifetime?

22 Using Equations 12–11 and 12–17, derive the equation $E^2 = m_0{}^2c^4 + p^2c^2$.

23 Solve Equation 12–17 for v in terms of E, m_0, and c. Then show that if a particle has energy but no mass, it must travel at the speed of light.

24 Use the equations of the Lorentz transformation to show that $x'^2 + y'^2 + z'^2 - (ct')^2 = x^2 + y^2 + z^2 - (ct)^2$. This proves that the space–time interval (the four-dimensional length) is an invariant quantity for all observers.

THE GENERAL THEORY OF RELATIVITY

13

Of the general theory of relativity you will be convinced once you have studied it. Therefore I am not going to defend it with a single word.

Albert Einstein

WITHIN A FEW YEARS after the publication of the special theory of relativity in 1905, many of Einstein's contemporaries came to regard the theory as a contribution of major significance. But Einstein himself looked on it as an incomplete theory, because he had found it necessary to restrict the validity of the special theory to inertial reference frames. By 1908 Einstein was actively seeking a way in which the theory of relativity could be generalized to encompass accelerated reference frames as well.

The difficulties encountered in extending the principle of relativity to accelerated reference frames were, in effect, first recognized by Newton. As we learned in Chapter 11, it is not possible in the Newtonian system to determine the position or velocity of a body with respect to absolute space. Newton's principles of motion cannot be used to distinguish between inertial frames because the principles have the same mathematical form and are equally valid in all inertial reference frames moving relative to one another. However, it is possible in the Newtonian system to determine the *acceleration* of a body in absolute space, since Newton's principles appear to be violated for motions measured relative to accelerated reference frames. Any deviation from Newtonian behavior can in principle be used to measure the absolute acceleration of the reference frame. For example, the amount of flattening of a rotating sphere is directly related to the rotation rate (i.e., the acceleration) of the sphere in absolute space.

Accelerated reference frames therefore take on a special status. Because of the apparently absolute nature of accelerated motion, Einstein was faced with a formidable barrier to his attempts to make *all* motion relative and *all* reference frames equal. He had been able to reject immediately the concepts of absolute velocity and absolute time because they were inherently unmeasurable, but he could not dismiss the notion of absolute acceleration on the same basis. For this reason Einstein had to exclude accelerated reference frames from consideration when he developed the special theory of relativity. In order to construct a truly general theory of relativity, Einstein clearly had to demonstrate in some manner that all accelerated motions are *relative* rather than absolute. Only in this way could the concepts of an absolute space and an absolute reference frame be completely discarded once and for all.

13–1

THE RELATIVITY OF ACCELERATED MOTION

In order to appreciate fully the difficulties involved in "relativizing" accelerated motions, let us remind ourselves of some of the effects of motion in accelerated reference frames. To begin with, we shall consider the conditions that arise in a rotating reference frame.

Suppose we are in a circular, windowless room and observe that the wall and floor are rotating with respect to one another. We will assume that one or the other is stationary; that is, either the floor is stationary and the wall is turning, or the wall is stationary and the floor is turning. Can we do an experiment that will enable us to determine which of these possibilities is correct?

There are a number of ways to solve this problem. An observer can often *sense* that he is rotating by the feeling of "dizziness" produced by rapid rotation. (This is the physiological effect caused by the shifting of the fluid in the ear canals when one's head is rotated.) However, if the rotation is slow, the effect may not be noticeable. We are not, for instance, made dizzy by the rotation of the earth!

More objective experiments can be performed to determine whether or not the floor is rotating. A ball thrown outward from the center of the room will travel in a straight line if the floor is stationary, but will appear to curve away from the observer if the floor is rotating (Figure 13–1a). Another simple test involves placing an object on the floor (Figure 13–1b). If the floor is stationary, the object will remain motionless, but if the floor is rotating (and there is negligible friction), the object should slide out toward the wall—just as passengers in a subway train are "thrown outward" when the train makes a sharp turn. Any such effect that gives the appearance of the action of a fictitious "centrifugal force" is an indication of rotational motion.

Clearly, it is a simple matter to determine whether the wall or the floor is rotating, so that it is possible to account for the relative motion between the floor and the wall in terms of the *absolute* rotation of one or the other. Hence relative rotational motion is of quite a different nature from the relative motion of inertial reference frames (say two spaceships moving at constant velocity with respect to one another), for no experiment can be performed that will determine the absolute motion of an inertial reference frame.

These considerations seem to demonstrate clearly the impossibility of encompassing accelerated motion (particularly rotation) under the principle of relativity. It is quite difficult to comprehend how rotational motion can be viewed as a relative motion between equivalent reference frames if it is possible to do experiments that tell whether the observer is or is not rotating. As we shall now discover, the key to the solution of this problem becomes more apparent when we turn our attention to the effects of *straight-line* accelerated motion.

Figure 13–2 depicts a spaceship that is accelerating through empty, gravity-free space. (The acceleration is provided by the continuous firing of the rockets at the back of the spaceship.) Inside the ship are two freely floating objects, A and B, of different sizes and masses. If these objects are not acted on by any forces, they will continue to move with constant velocity. However, the spaceship is accelerating forward, so that these bodies must eventually collide with the rear wall of the spaceship. If the motion of the two bodies is observed by someone who is fixed to the spaceship

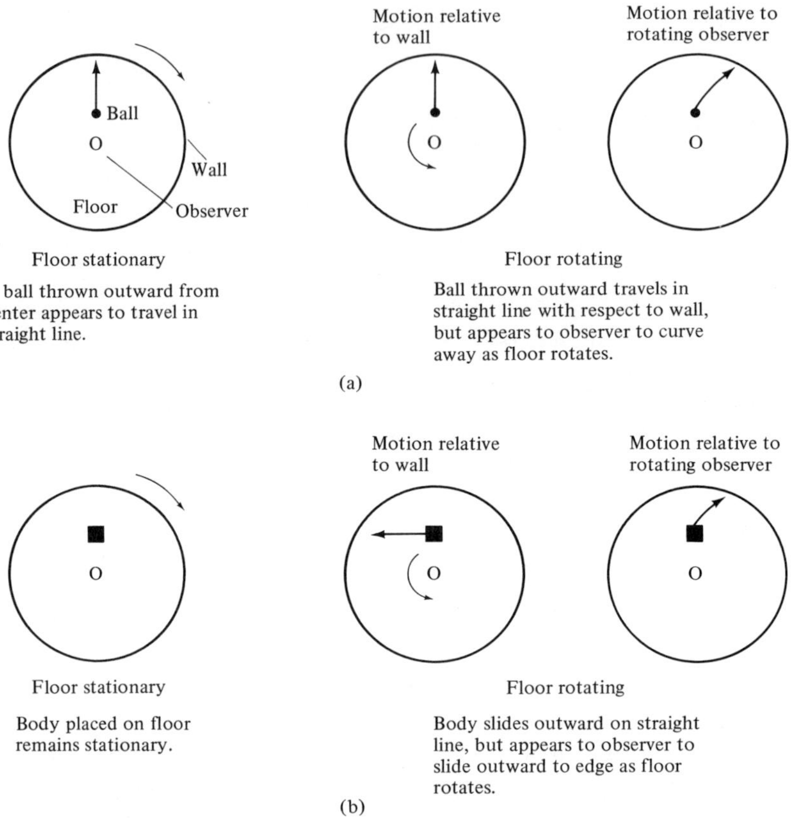

figure 13–1 **THE DETECTION OF ABSOLUTE ROTATION**

An observer is located at the center of a room in which either the floor or the wall is rotating. The observer may distinguish whether or not he is rotating by (a) throwing a ball outward or (b) placing an object on the floor (assuming negligible friction). In each case the motion of the body will depend upon whether the floor is or is not rotating.

(i.e., an observer who is stationary with respect to the accelerated reference frame), the two bodies appear to be accelerating backward—that is, they appear to be drawn by some unknown force toward the rear of the ship. However, an outside (nonaccelerating) observer would identify the unknown force as a "fictitious" force that results not from any interaction between bodies but from the accelerated motion of the reference frame.

It is particularly interesting that bodies A and B appear to have the same acceleration with respect to the spaceship, despite their different masses. This is, of course, a consequence of the fact that it is the *reference frame,* and not the bodies themselves, that is accelerating in the opposite direction. In general, we can assert that *to an observer who is fixed in an accelerated reference frame, all freely moving bodies will appear to exhibit the same acceleration, regardless of their masses.* Evidently, an observer ought to be able to use this phenomenon to test whether or not he is in an accelerated reference frame and then to determine his absolute acceleration. For instance, an astronaut in a spaceship can decide whether or not his ship is accelerating by observing the motions of freely floating bodies inside the ship. If they all appear to accelerate at the same rate and in the same direction, the astronaut can conclude that his spaceship is accelerating in the opposite direction. Once again we find a situation in which relative motion (between the freely floating bodies and the spaceship) can seemingly be accounted for in terms of an absolute acceleration.

THE PRINCIPLE OF EQUIVALENCE

Remarkably, this phenomenon provided Einstein with the key to the solution of the problem of relativizing accelerated motion. The most significant clue is the fact that all freely moving objects have the same apparent acceleration, *regardless of their masses,* with respect to an accelerated reference frame. This phenomenon also occurs under entirely different conditions—namely, in a gravitational field! As Galileo had demonstrated three centuries before, all objects fall with the *same* acceleration, *regardless of their masses,* under the influence of a gravitational force. It follows that the motion of a freely moving body is the same in an accelerated reference frame

figure 13–2 **MOTION OF BODIES IN AN ACCELERATED REFERENCE FRAME**

(a) Bodies A and B are freely moving objects of different sizes and masses floating inside a spaceship. (b) As the spaceship accelerates forward, the two bodies (which are moving with constant velocity) will be struck from behind by the rear of the spaceship. (c) To an observer fixed to the spaceship, it appears as if both bodies are accelerating toward the rear of the ship. Both A and B will have the same apparent acceleration, despite their different masses.

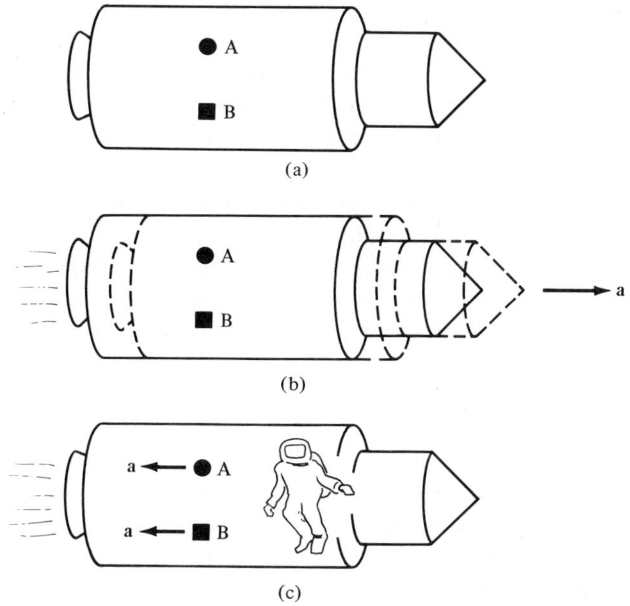

as in a gravitational field. On this basis Einstein proposed in 1911 the ***principle of equivalence,*** which can be stated as follows:

It is impossible to distinguish by experiment between the effects of accelerated motion and the effects of a gravitational field.

According to the principle of equivalence, one cannot distinguish between gravitational forces and the "fictitious" forces produced by accelerated motion. Since the effects of accelerated motion are identical to the effects of gravitation, it is not possible to determine whether a body has an absolute acceleration or is simply being acted on by gravity. Hence the concept of absolute acceleration must be rejected. In Einstein's words,

> This assumption of equivalence makes it impossible for us to speak of the absolute acceleration of the system of reference, just as the usual theory of relativity forbids us to talk of the absolute velocity of a system.[1]

To illustrate how the concept of absolute acceleration is done away with through the principle of equivalence, let us consider the experience of a physicist who finds himself in a sealed, windowless elevator. He has no contact with the outside world and does not know where the elevator is located or what its motion is. The physicist decides that he might get some idea as to his location by measuring the acceleration of gravity, so he takes out a coin and a stopwatch. He releases the coin and finds that it falls to the floor of the elevator with a constant acceleration of 32 ft/sec^2. Can the physicist thereby conclude that he is located at the surface of the earth? He *cannot,* because there are *two* ways to explain this result (see Figure 13–3a):

(1) The elevator is stationary in a gravitational field where $g = 32$ ft/sec^2.
(2) The elevator is in a gravity-free region and is being accelerated upward at a uniform rate of 32 ft/sec^2.

In the latter case the coin would remain suspended in space when released, but the floor of the elevator would accelerate up to meet it, giving the appearance that the coin is falling. The two explanations are entirely equivalent; the physicist is unable to determine whether he is in an accelerated reference frame or in a gravitational field. According to the principle of equivalence, there is no experiment that will allow the physicist to decide between the two possibilities. He is able to measure an acceleration, but cannot decide whether this acceleration is caused by gravitation or by the absolute motion of his reference frame. Thus he is unable to determine whether or not he has an acceleration with respect to absolute space.

Now let us consider a different situation. Suppose that when the physicist releases the coin, it does not appear to fall, but remains suspended in midair. The physicist could explain this phenomenon in two ways (Figure 13–3b):

(1) The elevator is in free fall, so that the physicist, the coin, and the elevator are all falling with the same acceleration.
(2) The elevator is located at a place where there are no net gravitational forces.

Once again, the equivalence of acceleration and gravitation prevents the physicist from distinguishing between the two explanations. In this case a reference frame falling with constant acceleration is entirely equivalent to a gravity-free inertial reference frame. This situation is quite similar to the phenomenon of weightlessness experienced

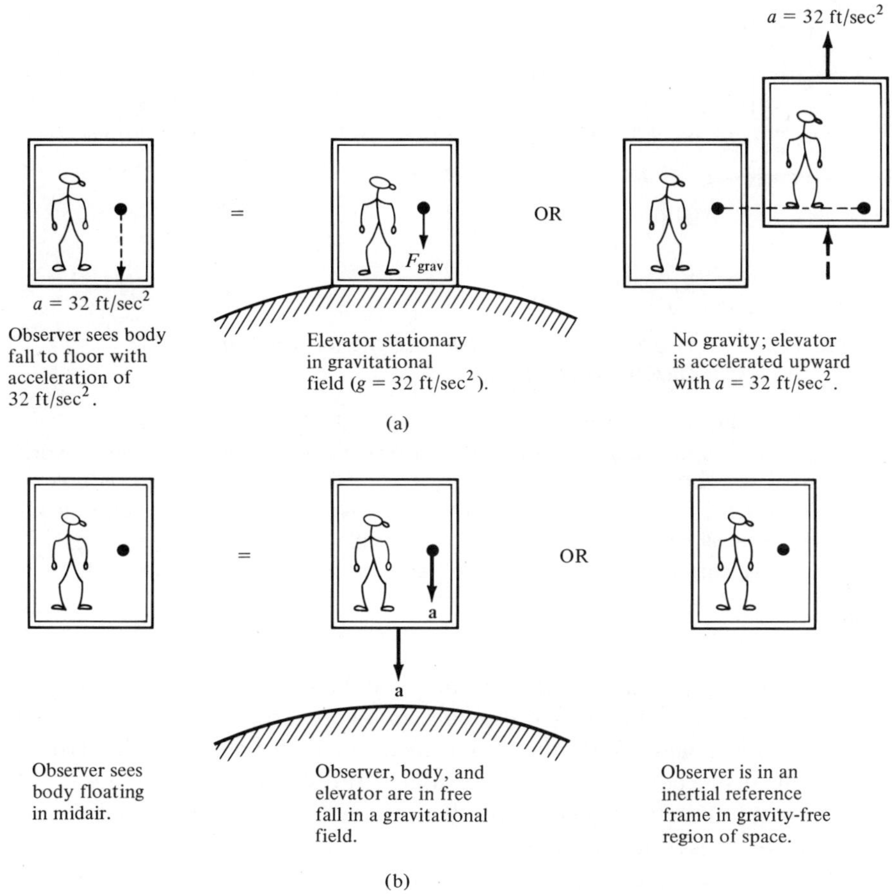

figure 13–3 **THE PRINCIPLE OF EQUIVALENCE**

In each situation the observer is unable to distinguish between the effects of accelerated motion and the effects of a gravitational field.

by astronauts in an orbiting satellite (discussed in Chapter 5). Because the satellite and its contents are in free fall toward the earth, conditions inside the satellite are identical to those of a gravity-free environment. There are no experiments that will allow an astronaut in a satellite (or the physicist in the elevator) to distinguish between free fall and the absence of gravity. Hence the absolute acceleration of the reference frame cannot be known.

An important restriction with regard to the principle of equivalence is that the principle is strictly true only with respect to a *uniform* gravitational field—that is, a field in which the gravitational force has the same magnitude and direction at all points. The earth's gravitational field is one example of a nonuniform field. The gravitational force decreases with increasing distance from the earth and does not have a constant direction. (The direction at any point is toward the center of the earth.) The difficulty raised by nonuniform gravitational fields is that it *is* possible in principle to distinguish between a nonuniform gravitational field and an accelerated reference frame. In a nonuniform gravitational field objects have different accelerations

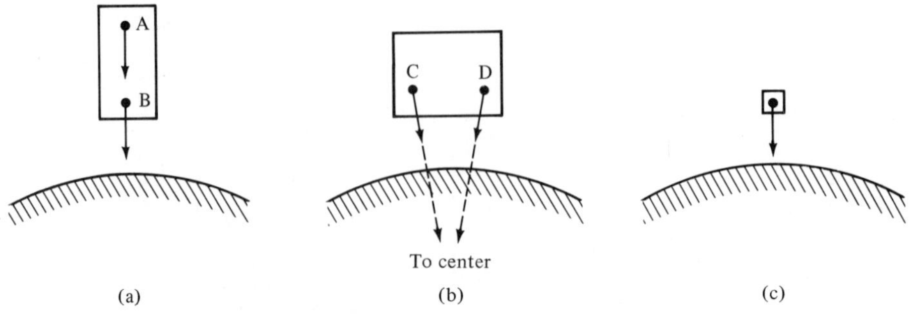

figure 13–4 **THE PRINCIPLE OF EQUIVALENCE IN A NONUNIFORM GRAVITATIONAL FIELD**

(a) Body B will have a greater acceleration than body A toward the earth. (b) Bodies C and D will move toward one another, since they are attracted toward the center of the earth. An observer inside either reference frame would be able to detect that he is in a nonuniform gravitational field because in an accelerated reference frame all bodies have the same acceleration. (c) The principle of equivalence remains valid in a region small enough so that the nonuniformity of the field cannot be detected.

depending on their location, whereas in an accelerated reference frame all objects exhibit the same acceleration. For example, in a very large satellite such as that shown in Figure 13–4a, body A (farther from the earth than body B) will have a *smaller* acceleration than body B. This effect enables an observer to determine that he is in a nonuniform gravitational field rather than in an accelerated reference frame (where A and B would have the same acceleration).

A similar example is shown in Figure 13–4b. Since all objects fall toward the center of the earth, bodies C and D will move toward one another as they fall. If C and D are sufficiently far apart for this effect to be measurable, an observer will be able to detect the nonuniformity of the gravitational field.

Thus the principle of equivalence is not valid throughout a nonuniform gravitational field. The principle remains valid, however, *if the observer is confined to a single point in space.* Under this restriction the observer is not able to discern the nonuniformity of the field. In practical terms a "point in space" corresponds to a region so small that the nonuniformity of the field cannot be detected experimentally. Thus within a small satellite orbiting the earth or in an elevator or a laboratory at the earth's surface, the gravitational field is effectively uniform, and the principle of equivalence remains valid.

To summarize the principle of equivalence, *an observer located at a given point in space is unable to distinguish by experiment between the effects of a gravitational field and the effects of motion relative to an accelerated reference frame.* In both situations all objects will appear to move with the same acceleration regardless of their sizes, compositions, masses, and so on. Hence it is impossible to certify an absolute acceleration, because every instance of accelerated motion can also be attributed to the action of a gravitational field.

Thus by recognizing the equivalence of acceleration and gravitation, Einstein was able to reject the notion that one could unequivocally detect absolute acceleration. It is interesting to speculate as to why Newton—who was, after all, well aware of the nature of gravitation—did not recognize the principle of equivalence before Einstein did. We must remember, however, that it was Newton's aim to establish the existence of *absolute* acceleration as a means of justifying his concept of absolute space. The experiments performed in rotating systems by which an observer is apparently able to determine whether or not he is rotating seemed to Newton to demonstrate clearly the existence of absolute acceleration. But Einstein pointed out that the equivalence principle holds in a rotating system also. That is, the experiments used to detect rotation can also be explained in terms of gravitational forces. For example, the curved motion of the ball thrown outward in Figure 13–1a could conceivably be attributed to the presence of a suitably large distribution of mass beyond the wall that gravitationally attracts the ball. (See Figure 13–5.) Given the possibility that gravitational forces are being exerted by objects beyond the wall of the room, an observer located at a point inside *cannot* truly determine whether or not he has an absolute acceleration.

Before going on to consider some of the consequences of the principle of equivalence, we should take note of an aspect of the principle that is of considerable interest to physicists. *The principle of equivalence depends for its validity on an unexplained fact of nature*—namely, that all objects fall with the same acceleration in a gravitational field. In Chapter 5 we saw that this phenomenon requires the inertial mass of a body (the m in $F = ma$) to be identical to the gravitational mass (the m in $F = GmM/d^2$). Conversely, if the inertial and gravitational masses were *not* exactly equal to one another, objects of different mass or composition could conceivably fall at different rates in a uniform gravitational field. In that case an observer would be able to determine whether or not he was in a gravitational field or in an accelerated reference frame (where all objects always have the same acceleration). The principle of equivalence would therefore be invalid.

Hence the principle of equivalence, which is one of the fundamental postulates of general relativity, requires the equality of inertial and gravitational mass. This equality is unaccounted for theoretically and is based entirely on experimental observations. Should someone ever perform an experiment that clearly demonstrates that the inertial and gravitational masses of some substance or particle are *not* equal to one another, the entire theory of general relativity would have to be radically revised

figure 13–5 **THE PRINCIPLE OF EQUIVALENCE IN A ROTATING ROOM**

The curved path of a ball thrown outward may be attributed to either (a) the rotation of the floor or (b) the gravitational attraction of a body outside the room.

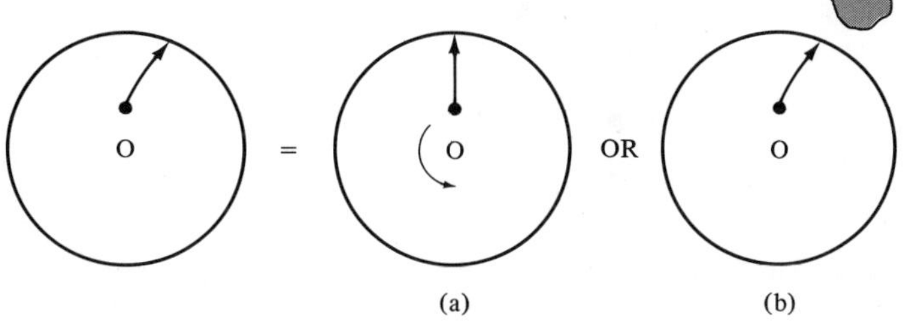

or even rejected outright. At the present time, however, there is no reason to doubt the validity of the principle of equivalence.*

13–2

LIGHT AND GRAVITATION

In view of the principle of equivalence, Einstein asserted that the special status previously accorded to accelerated reference frames could now be rejected. Consequently, the principle of relativity could be expressed in its most general form, as follows:

The laws of physics are the same in all reference frames.

Stated in this manner the principle of relativity has considerable philosophical and aesthetic merit, for it proposes that the principles that govern the order of the physical universe are independent of such an arbitrary factor as the reference frame or coordinate system chosen by the observer. But physical theories require more than aesthetic appeal for their justification; they must, of course, be firmly grounded in observation and experiment. Since the adoption of the principle of relativity is made possible by the principle of equivalence, the first step is to give the principle of equivalence some experimental validity. Accordingly, in his first major publication on general relativity (presented in 1911), Einstein took note of some consequences of the principle of equivalence that could be put to experimental test.

The experiments that Einstein suggested dealt specifically with the behavior of light in a gravitational field. If the principle of equivalence holds true, the behavior of light should be the same in an accelerated reference frame as in a gravitational field—for according to the principle of equivalence, an observer cannot distinguish between the two by experiment, and this should include experiments involving the properties of light.

THE BENDING OF LIGHT IN A GRAVITATIONAL FIELD

What happens to a beam of light when it passes through a gravitational field? According to classical theory, light is an electromagnetic wave disturbance and hence should not be affected by gravitational forces. Consequently, light would be expected to follow its usual straight-line path even in a gravitational field. Einstein demonstrated, however, that the principle of equivalence would be violated if this were the case. That is, *if the principle of equivalence is valid, light must bend from a straight-line path when passing through a gravitational field.*

To see how the bending of light is required by the principle of equivalence, let us return to our scientist in the elevator. Consider a hypothetical experiment in which a narrow beam of light enters parallel to the floor of the elevator through a small window on one wall and strikes a detector located on the opposite wall, as shown in Figure 13–6. Now suppose that the elevator is accelerating upward. During the time the light beam is crossing the elevator, the elevator is accelerating upward, so that the beam should strike a point *below* the level of the window (Figure 13–6b).

*The most recent attempt to find a difference between the inertial and gravitational mass was made in 1964. The equality was verified to a precision of 1 part in 100 billion.

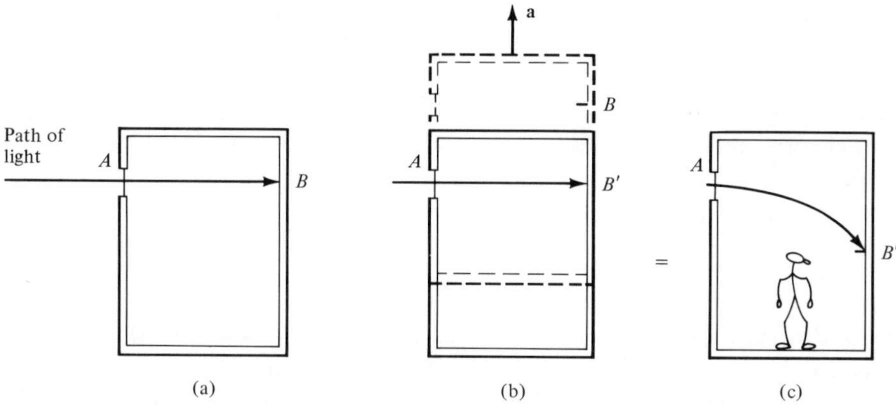

figure 13–6 **THE DEFLECTION OF LIGHT IN AN ACCELERATED REFERENCE FRAME**

(a) A light beam enters the elevator through a window at A. If the elevator is stationary, the beam will strike a detector on the opposite wall at point B. (b) If the elevator has an upward acceleration, the beam will strike the wall at point B', located below B. (c) To an observer inside the elevator, the light beam appears to have been deflected downward.

Thus as a consequence of the upward acceleration of the elevator, the light should appear to curve downward as seen by an observer inside the elevator (Figure 13–6c). Of course, the amount of curvature would be so minute as to be unobservable to the naked eye, but in principle the curvature ought to exist and could conceivably be detected if sufficiently precise measuring instruments were available.

Thus a light beam should appear to have a curved path in an accelerated reference frame. According to the principle of equivalence, however, no experiment can distinguish between the effects of an accelerated reference frame and the effects of a gravitational field. Therefore if the principle of equivalence is valid, *a beam of light must bend in a gravitational field.* If light did *not* travel in a curved path in a gravitational field, it would be possible to distinguish between an accelerated reference frame and a gravitational field simply by observing the path of a light beam.

Einstein noted particularly that this proposed influence of gravitation on light was consistent with his theories on the relationship between mass and energy. Einstein had previously demonstrated that the inertia of a body depends on its energy content, and that light conveys inertia from one body to another. Furthermore, all the available experimental evidence indicates that the inertial mass and gravitational mass of a body are identical. Hence if light conveys inertial mass, it must also convey gravitational mass. It therefore seems reasonable to assume that light must be affected by gravitation in some way. In short, given the relationships connecting (1) mass and energy and (2) inertia and gravitation, it is possible to conclude that energy (in the form of electromagnetic radiation) will be influenced by gravitational fields.

Einstein therefore proposed that the path of a beam of light is altered by gravity, so that a light beam passing near a massive object ought to be deviated from a straight-line path just as if it were "falling" toward the object. (See Figure 13–7.) It had been suggested as far back as 1801—when the corpuscular theory of light was still the accepted theory—that light corpuscles might be subject to gravitational forces. Hence a light beam passing near the sun should travel in a hyperbolic orbit about the sun, as predicted by the Newtonian theory of gravitation. However, Einstein's detailed calculations of the path of the light beam showed that the deflection of the beam should be twice as large as that predicted by the Newtonian theory. In particular, Einstein was able to calculate that a light beam passing near a massive object will be deflected through an angle θ such that

$$\tan \theta = \frac{4GM}{rc^2} \tag{13–1}$$

where θ is the angle of deflection, G the gravitational constant, M the mass of the object, c the speed of light, and r the distance from the center of the object to the light beam, as shown in Figure 13–7. Note that the angle of deflection is *inversely* proportional to the distance r, so that the amount of deflection is greater for light beams passing closer to the center of mass.

Calculations made with Equation 13–1 reveal that the amount of deflection will be exceedingly small, even for light beams passing close to an object as massive as the sun. The equation predicts, for example, that a beam of light that just grazes the edge of the sun (r = 430,000 miles) will be deflected through an angle of 1.75″—about 1/2000 of a degree! A beam of light passing near the earth would be deflected through a much smaller angle of no more than 0.0006″. Nevertheless, Einstein believed that the deflection of light by the sun might be just large enough to be observable, so he proposed to astronomers that they make an experimental test of his theory. If the light from a distant star is deflected when passing near the sun, an observer on the earth ought to notice a shift in the apparent position of the star in the sky during that time of year when the light from the star must pass near the sun before reaching the earth (Figure 13–8). Specifically, Einstein proposed that two sets of observations of the positions of stars be made several months apart, as shown in Figure 13–8. When the earth is at position A, the light from the star should travel in a straight line, thus indicating its true position. When the earth is at position B in its orbit, however, the starlight should bend slightly and appear to come from a different position in the sky. The shift in the apparent position of the star should be less than or equal to 1.75″, depending on how close the path of the starlight comes to the sun.

However, when the earth is at position B (and the starlight is presumably

figure 13–7 **THE DEFLECTION OF A LIGHT BEAM IN A GRAVITATIONAL FIELD**

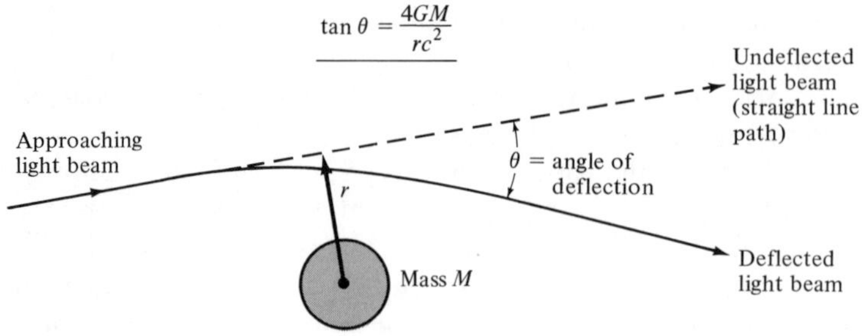

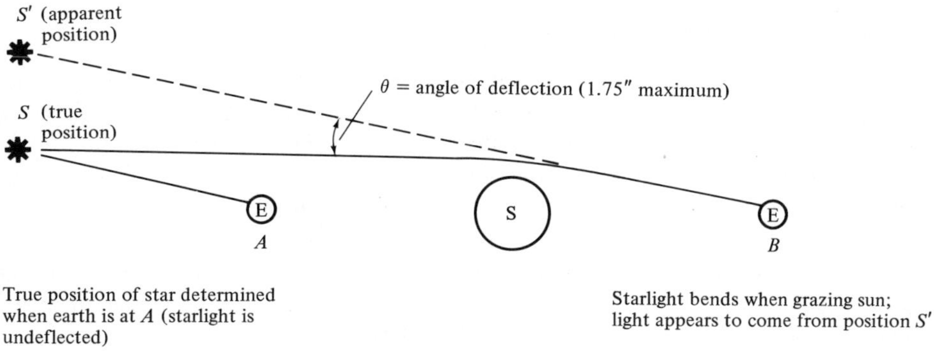

figure 13–8 **EXPERIMENTAL TEST OF THE DEFLECTION OF STARLIGHT BY THE SUN**

Two observations are made from the earth of the position of a given star. When the star is observed from point A in the earth's orbit, the starlight travels in a straight line from its true position S. When the star is observed from point B in the earth's orbit, the starlight is deflected by the sun and appears to come from S'.

deflected), the star and the sun are almost directly in line with each other. Thus the star will not be visible because it will be daytime, and its light will be lost in the sunlight! So Einstein suggested that the measurement could be made during a total eclipse of the sun, when the sky is momentarily dark and the stars are visible.

The first crucial test of Einstein's theory was made during the total solar eclipse of May 1919. A group of British astronomers organized expeditions to Brazil and West Africa, where the eclipse would be visible, and took photographs of the darkened sky at the height of the eclipse. The stars that could be seen near the eclipsed sun on the photograph were found to be shifted from their true positions by the amounts predicted by Einstein, within the limits of experimental error.

The existence of the deflection of starlight has been confirmed by approximately ten separate observations made since 1919 during favorable solar eclipses. In every instance a positive deflection was observed. The measured maximum deflections have ranged from 1.33″ to 2.73″, as compared to Einstein's predicted theoretical value of 1.75″. The wide range of experimental values reflects the difficulties involved in making precise measurements of the deflection of starlight. At best, the experimental uncertainty at the present time is considered to be no better than 10 percent.* Thus the evidence clearly indicates that light is deflected by the sun as predicted by Einstein, although it is not entirely certain that the observed amount of deflection is in precise agreement with the predicted numerical value. (As noted previously, if light beams were deflected according to the Newtonian theory of gravitation, they would bend through angles half the size of those predicted by Einstein. Hence the maximum deflection for light passing the sun would be 0.87″. The observational results thus clearly support the theory of relativity over Newtonian gravitation.)

*The most recent (1970) measurement was based on the deflection by the sun of radio waves emitted by distant galaxies, rather than of visible starlight. The maximum deflection observed was $1.70 \pm 0.16''$ for radio waves of about 10 cm wavelength.

THE GRAVITATIONAL FREQUENCY SHIFT

We have seen how the equivalence principle requires the alteration of the direction of a light beam when the light beam passes through a gravitational field. We shall now examine a second application of the equivalence principle. Through this application Einstein showed that the *frequency* of light should also be changed under the influence of gravitation. As before, we will perform a "thought experiment" in an accelerating elevator to demonstrate this effect.

Let us imagine a light source, mounted in the ceiling of an elevator, that emits light of a specific frequency, f_0. In the floor of the elevator is a detector that measures the frequency of the light it receives (Figure 13–9a). Presumably, if there is no relative motion between the source and the detector, the light received by the detector should have the same frequency as the light emitted by the source. But let us suppose that the elevator is accelerating upward (Figure 13–9b). The detector is moving toward the oncoming light beam with a steadily increasing velocity. Because of this accelerated motion, the velocity of the detector at the instant it receives the light is greater than the velocity of the source at the time the light was emitted. It will be as if the source were stationary and the detector were moving toward the oncoming light waves. As we learned in Chapter 8, relative motion between a wave source and an observer creates the Doppler effect. Therefore the frequency of the light measured by the moving detector will be different from the frequency of the light emitted by the source. If the detector is moving *toward* the source, the observed frequency will be *greater* than the emitted frequency. That is, there will be a *violet shift* of the emitted light toward the higher frequency (lower wavelength) end of the spectrum. Similarly, if the source and detector are interchanged, so that the detector is accelerating *away* from the light (Figure 13–9c), there will be an apparent *decrease* in the observed frequency (corresponding to a *red shift* of the emitted light). The greater the acceleration of the elevator, the greater will be the velocity of the detector toward or away from the light. Therefore the size of the frequency shift will be directly proportional to the acceleration.

Now let us consider the consequences of the equivalence principle as it applies to this situation. If an accelerated reference frame is indistinguishable from a gravitational field, the frequency shift of light observed in an accelerated reference frame must also be observed in a gravitational field. Thus the principle of equivalence predicts that *the frequency of light changes in a gravitational field.* As shown in Figure 13–9d, a beam of light moving downward in an upward-accelerating elevator is equivalent to a light beam that is "falling" in a gravitational field (i.e., is moving toward the center of mass). According to the principle of equivalence, an *increase* in frequency must be observed in both circumstances. Similarly, a light beam moving upward in an upward-accelerating elevator is equivalent to a light beam "rising" in a gravitational field (Figure 13–9e), and a *decrease* in frequency should be observed in both situations. Thus it follows that if the principle of equivalence holds, light falling in a gravitational field must exhibit a frequency increase (violet shift) and light rising in a gravitational field must exhibit a frequency decrease (red shift).

Einstein's detailed mathematical formulation of the principle of equivalence leads to the following equations for the magnitude of the frequency shift. Consider a light beam that travels from point 1 to point 2 in a gravitational field of a point mass (or uniform sphere of mass), as shown in Figure 13–10. Let f_0 be the frequency figure 13–9 **THE GRAVITATIONAL FREQUENCY SHIFT OF LIGHT**

Elevator experiment. The light source emits light of frequency f_0; the detector measures the frequency f of the light received.

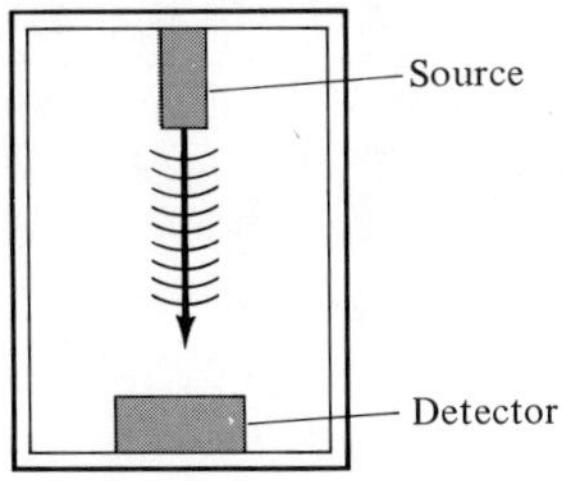

(a)

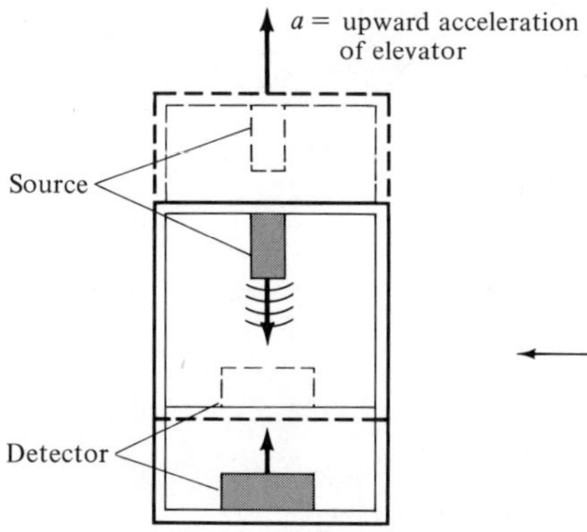

(b)

Detector accelerates toward light; increase in frequency observed due to Doppler effect ($f > f_0 \rightarrow$ violet shift).

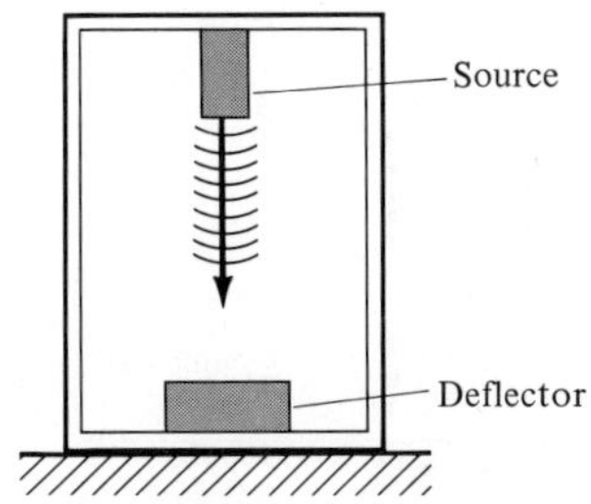

(d)

Equivalent to light "falling" in gravitational field: by equivalence principle, violet shift must be observed.

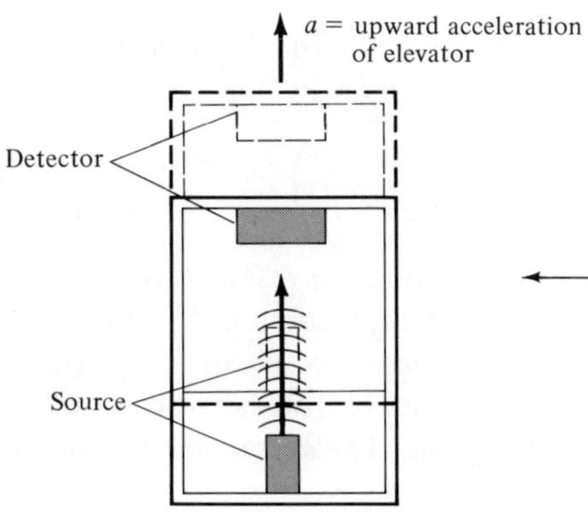

(c)

Detector accelerates away from light; decrease in frequency observed due to Doppler effect ($f < f_0 \rightarrow$ red shift).

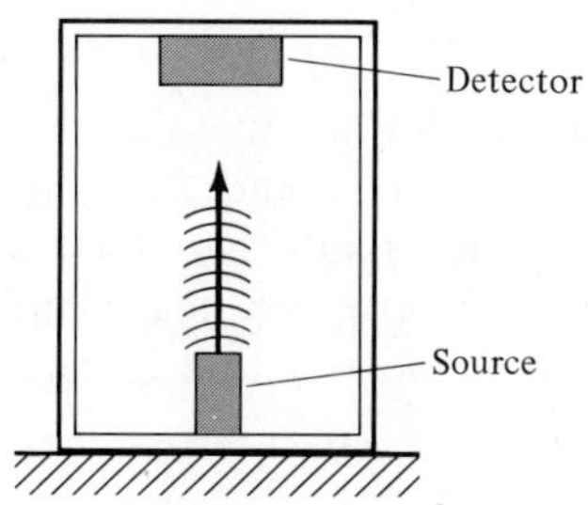

(e)

Equivalent to light "rising" in gravitational field; by equivalence principle, red shift must be observed.

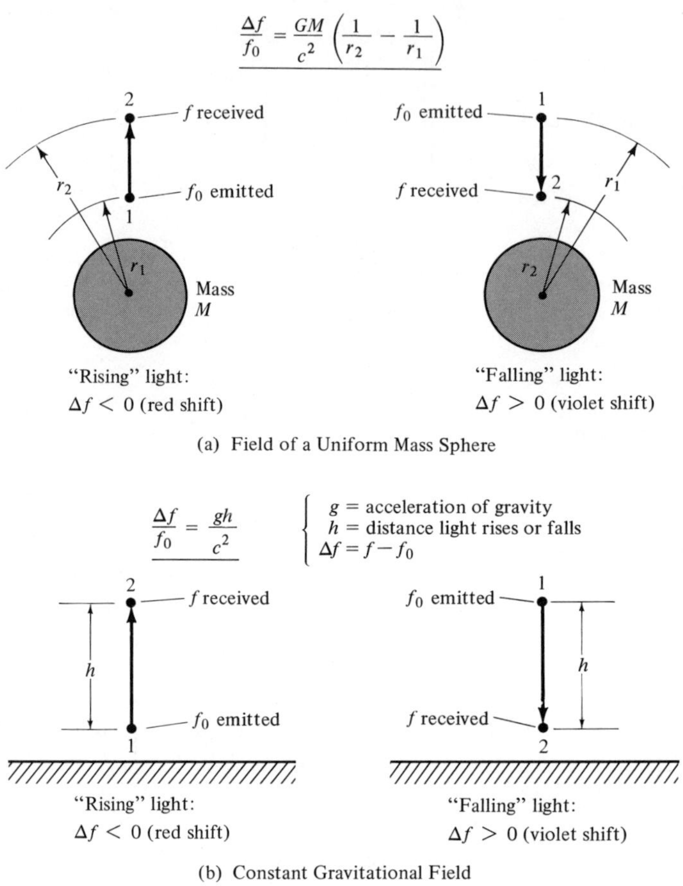

figure 13–10 **EINSTEIN'S EQUATIONS FOR THE GRAVITATIONAL FREQUENCY SHIFT OF LIGHT**

of the light emitted at point 1 and f the frequency of the light received at point 2. The *change* in the frequency Δf $(= f - f_0)$ is given by the equation

$$\frac{f - f_0}{f_0} = \frac{\Delta f}{f_0} = \frac{GM}{c^2}\left(\frac{1}{r_2} - \frac{1}{r_1}\right) \tag{13–2a}$$

where M is the mass of the central body, and r_1 and r_2 are the respective distances from points 1 and 2 to the mass center. Here Δf represents either the *increase* in frequency that occurs for "falling" light or the *decrease* that occurs for "rising" light. In a uniform (or nearly uniform) gravitational field of constant acceleration (such as is found near the surface of the earth), Equation 13–2a can be approximated as follows:

$$\frac{\Delta f}{f_0} = \frac{gh}{c^2} \tag{13–2b}$$

where g is the local acceleration of gravity, and h is the distance that the light rises or falls.

Einstein's prediction of a frequency shift of light in a gravitational field provides what appears to be an opportunity for a clear-cut experimental test of the theory of general relativity and of the validity of the principle of equivalence. That is, one ought to be able to send a beam of light up or down in a gravitational field to determine whether or not the frequency of the light beam changes in accordance with Einstein's prediction. As a practical matter, however, an experiment of this type is extremely difficult to perform with sufficient accuracy, primarily because the magnitude of the gravitational frequency shift is extremely small. For example, a light beam rising or falling a distance of 1000 feet (the approximate height of the Empire State Building) near the surface of the earth should exhibit a frequency shift of only 1 part in 30 trillion!

A larger shift is to be expected for a light beam traveling from the sun to earth. In this instance the light can be considered to be rising in the sun's gravitational field. According to Equation 13–2a, light that leaves the surface of the sun (r_1 = radius of sun = 430,000 miles) and rises in the sun's gravitational field to the vicinity of the earth (r_2 = distance to earth = 93,000,000 miles) should experience a *decrease* in frequency of about 2 parts in a million (i.e., $\Delta f/f_0 = 2 \times 10^{-6}$). Hence light leaving the sun and arriving at the earth should exhibit a small red shift. One would expect to find similar red shifts in the light reaching the earth from any star.

Einstein had hoped that astronomers would be able to detect the proposed gravitational red shift of sunlight or starlight, thereby providing experimental verification of his theory. Indeed, a number of observers have found gravitational red shifts in the light from the sun and several stars. Unfortunately, the experimental uncertainties have thus far proved to be too large to allow for a clear-cut *numerical* verification of Einstein's prediction. Thus at the present time the available evidence clearly indicates that a gravitational red shift of sunlight and starlight does exist, but the observations are not accurate enough to show that the magnitude of the observed red shifts are in agreement with the values predicted by Einstein.

Remarkably, the most convincing evidence for the gravitational frequency shift comes from a measurement of the change in frequency of light falling through a distance of less than 100 feet at the surface of the earth. The measurement of this extremely small frequency shift was made possible as a consequence of an important scientific breakthrough achieved in 1957 by Rudolf Mossbauer of Germany.

Mossbauer's work dealt with the emission and absorption of gamma rays by radioactive substances. It was previously known that a substance that *emits* gamma rays of a particular frequency will also *absorb* gamma rays of the *same* frequency. Suppose, for example, that a certain radioactive substance emits gamma rays of a given frequency, f_0. When the gamma rays from another source are aimed at this substance, it will absorb the gamma rays having this same frequency. The absorber is said to "resonate" to gamma rays of the frequency f_0, and f_0 is referred to as the ***resonant frequency*** of the absorbing substance. Under ideal theoretical conditions the absorber should resonate only to those gamma rays having *precisely* the resonant frequency; there should be very little absorption of gamma rays having slightly higher or lower frequencies.

Since a radioactive substance generally absorbs gamma rays of the same frequency that it emits, the substance should resonate to its own gamma radiation. Thus precise resonance could in principle be attained with two samples of the same radioactive substance. One sample would serve as the emitter, giving off gamma rays of the resonant frequency f_0; the other sample would act as the absorber that resonates to these gamma rays. Under normal laboratory conditions, however, such precise resonance cannot be achieved. The natural motions of the absorbing and emitting atoms tend to produce a Doppler shift, which changes the frequency of the gamma rays with respect to the absorber. As a consequence, the resonance effect is destroyed.

Mossbauer's accomplishment was to develop techniques for attaining precise resonance. By selecting the appropriate radioactive substances and operating at very low temperatures (to reduce the thermal motions of the atoms), Mossbauer was able to achieve such precise resonance that a deviation from the resonant frequency of more than 1 part in 10^{14} would result in the destruction of resonance. Consequently, when Mossbauer's techniques are used, the resonance absorption of gamma rays virtually disappears as a result of any effect that makes the emitted gamma-ray frequency differ from the resonant frequency by more than 1 part in 100 trillion! Thus the ***Mossbauer effect*** can be used to detect extremely small changes in frequency. For example, the motion of either the source or absorber at a rate of *less than an inch per minute* will create a Doppler shift sufficient to destroy resonance.

Soon after Mossbauer published his findings, a number of physicists recognized that the Mossbauer effect was sensitive enough to permit the detection of the gravitational frequency shift. According to Einstein's prediction, if the gamma rays from the source are allowed to fall (or rise) in the earth's gravitational field, they should experience a change in frequency that will destroy or at least weaken the resonance between the source and absorber. In 1960 Robert Pound and Glen Rebka of Harvard University made the first successful measurement of the gravitational frequency shift by this method. They placed the gamma-ray source at the top of a 74-foot tower and the absorber at the bottom. They observed a decrease in gamma-ray absorption that was equivalent to a gravitational frequency shift of $\Delta f/f_0 = 2.5 \times 10^{-15}$. This experimental value is within 5 percent of the value that Einstein predicted theoretically. Subsequent refinements of this experiment confirmed Einstein's prediction to an accuracy of better than 1 percent. Thus the existence of the gravitational frequency shift and the validity of Equation 13–2 are now quite solidly confirmed.

13–3

TIME AND GRAVITATION

In the special theory of relativity Einstein demonstrated that time is measured differently in different reference frames. He noted that since it is impossible to establish an absolute reference frame, an absolute time cannot be established; hence all time measurements are equally valid. In the special theory, however, Einstein restricted his considerations to inertial reference frames. In the general theory accelerated reference frames are "relativized" through the equivalence principle. It should come as no surprise that in the general theory Einstein has extended the principle of the relativity of time to apply to accelerated reference frames and gravitational fields.

In special relativity the measurement of time depends on the relative motion of the observer. If measurements are being made of the time interval between two events, an observer who is moving with respect to these events will measure a longer time interval than an observer who is stationary with respect to the events. As was shown in Chapter 12, the time interval Δt_0 measured by the "stationary" observer will be related to the time interval Δt measured by the "moving" observer through Equation 12–5:

$$\Delta t = \frac{\Delta t_0}{\sqrt{1 - v^2/c^2}}$$

This equation shows that Δt is always greater than Δt_0; that is, the moving observer always measures a longer time interval than the stationary observer.

We expect this same principle to hold with respect to accelerated reference frames. That is, if the "moving" observer is accelerating with respect to the events he is measuring, he should also calculate a longer time interval between events as compared to an observer who is stationary with respect to the events. Thus time dilation should occur in an accelerated reference frame as well as in an inertial reference frame.

According to the principle of equivalence, an accelerated reference frame is indistinguishable from a gravitational field. Therefore we expect to find a time dilation effect in a gravitational field. The "moving" observer is equated with an observer in a gravitational field, and the "stationary" observer becomes the observer who is in a field-free region. Thus *the time interval between two events should appear to be longer when the events occur in a gravitational field.* Suppose, for example, that there are identical flashing beacons, one located in a field-free region and one in a gravitational field. An observer in the field-free region and an observer in the gravitational field compare the time between successive flashes of the two beacons. Both observers will agree that the time between flashes is *longer* for the beacon in the gravitational field than for the beacon in the field-free region.

The size of the time dilation in a gravitational field depends, of course, on the strength of the field. The equation describing the time dilation is as follows. Let Δt_0 be the time interval measured by an observer in a field-free region, and Δt_G the time interval measured by an observer in a gravitational field. Assume that the field is produced by a body of mass M. Then

$$\Delta t_G = \frac{\Delta t_0}{\sqrt{1 - 2GM/rc^2}} \tag{13-3}$$

where r is the distance from the observer to the center of mass of the body.

The gravitational frequency shift, described in the previous section, can be considered as a manifestation of the gravitational time dilation. To an observer in a gravitational field, the period of any vibratory motion should be longer because of time dilation; hence the *frequency* of vibration (the number of vibrations per second) should be less. Thus light emitted from a source in a strong gravitational field should exhibit a *lower* frequency (i.e., a red shift) as compared to light emitted by the same source located in a weaker gravitational field. As we have learned, this prediction has been confirmed experimentally through the use of the Mossbauer effect.*

Unlike the time dilation of special relativity, the gravitational time dilation is considered to represent a real dilation of time. A clock or any other vibrating system will continue to vibrate with a smaller frequency for as long as it remains in a gravitational field. In particular, atoms vibrate at a slower rate in a gravitational field, and many physicists speculate that as a consequence biological processes should tend to proceed at a slower rate. That is, the rate of aging of a biological system should

*The Mossbauer effect has also been used to confirm the time dilation in an accelerated reference frame. An experiment was performed in which the gamma-ray source was placed at the center of a turntable, with the absorber at the edge. When the turntable was set in rotation (so that the edge was accelerated with respect to the center), a decrease in resonance was observed.

be slower in a gravitational field. This leads us to a most intriguing speculation. Imagine twin brothers who both have jobs in the Empire State Building. One twin works on the ground floor, the other on the 102nd floor. The twin who works on the ground floor is closer to the center of the earth and is therefore in a stronger gravitational field. He should age at a slower rate, thereby becoming younger than his brother! (The difference in their ages will, of course, be extremely small, amounting to about 1 second in 100 million years.)

13–4

GEOMETRY AND GRAVITATION

We have seen that there is rather substantial experimental evidence attesting to the validity of the principle of equivalence. Once again we stand in awe of Einstein's uncanny ability to identify the fundamental principles of physics and to deduce from them the existence of such extremely subtle phenomena as the deflection of light and its frequency shift in a gravitational field. Yet the principle of equivalence represents merely the starting point of the general theory of relativity, for Einstein's main goal was to establish the principle of relativity—that is, to show that the laws of physics must take the same form in all reference frames. At first, however, he found that the principle of equivalence and the principle of relativity seemed to be incompatible. His efforts to resolve this contradiction led to what is perhaps his most spectacular achievement—a complete reformulation of the theory of gravitation.

To begin, let us see how the principle of equivalence and the principle of relativity appear to conflict with each other. The principle of relativity, in its most general form, states that the laws of physics have the same form in *all* reference frames, accelerated or not. Let us consider as an example the law of physics that describes the path of light through space. It is a fundamental principle of classical physics that a light beam travels from one point in space to another along the shortest possible path joining the points, or in other words, light travels through space in a straight line.* This principle has been found to be valid in any *inertial* reference frame. However, we know (as a consequence of the principle of equivalence) that a light beam *does not* appear to travel in a straight line when seen in an accelerated reference frame or in a gravitational field. Thus it appears that the principle describing the path of a light beam cannot have the same form in inertial and accelerated reference frames.

A similar difficulty is provided by Newton's principle of inertia. This principle states that the natural motion of a body is in a straight line at a constant speed. Yet, as we learned in Chapter 11, this principle appears to be violated in an accelerated reference frame. That is, to an observer attached to an accelerated reference frame, an object may appear to have a nonuniform motion, even though it is not acted on by any forces. Similarly, in a gravitational field the "natural" motion of a body is acceleration (free fall), rather than constant-velocity motion. Once again we find that it is apparently impossible to state a physical principle—in this case the principle of inertia—so that it is uniformly valid in all reference frames.

*Actually, the most exact form of this principle (first proposed by Pierre de Fermat around 1640) is that light travels along the path of minimum *time*. (This takes into account the motion of light through a medium in which it changes velocity.) The principle as stated above is a special case that applies to the motion of light through empty space.

Einstein seemed to be forced to choose between the principle of equivalence and the principle of relativity, since both apparently could not be satisfied simultaneously. Yet his faith in the validity of both of these principles led him to seek a resolution of the dilemma that would preserve both of them. Eventually, he was able to find a solution to the problem. Characteristically, he did so by challenging a concept that was essential to the very foundations of physics, *for Einstein called into question the validity of the system of geometry used to describe objects and events in the universe.*

THE ROLE OF GEOMETRY IN PHYSICS

The branch of mathematics known as geometry (literally translated "the measurement of the earth") had its earliest beginnings in ancient Babylon and Egypt. The Egyptians were the first to devise methods for measuring angles and for calculating the areas of various geometric figures. Their interest in geometry, however, was limited to its practical applications, particularly its use in land surveying. The ancient Greeks, on the other hand, studied geometry as an abstract form of mathematics. In the schools led by Thales, Pythagoras, Plato, and Aristotle, considerable effort was devoted to the systematic organization and development of geometry as a logical system. Most of the familiar theorems having to do with the properties of triangles, circles, and other geometric figures were proved by the Greeks. This effort reached its culmination in the work of Euclid (about 350–280 B.C.), an Athenian mathematician of the Platonic school. Euclid compiled and reorganized all of the previously available material, which he combined with his own contributions into a comprehensive and logically rigorous formulation of geometry. His great treatise *The Elements* was recognized as the fundamental text on geometry until the end of the nineteenth century.* The system of geometry presented in *The Elements* is now commonly referred to as ***Euclidean geometry.***

Although geometry was developed and formalized as a form of pure mathematics, it has always been firmly rooted in the observable world. Such abstractions as the circle and the straight line represent idealizations of the shapes of various objects found in nature; hence the properties of these natural objects are found to obey geometric theorems. Thus, for example, a cylindrical soda can will turn out to have a volume of $\pi r^2 h$ (where r is the radius and h is the height of the cylinder), and a rectangle marked off by four tightly stretched strings will be found by direct measurement to enclose an area equal to the length times the width, in agreement with the theorems of geometry. Consequently, from the earliest times to the present day, scientists have found it quite fruitful to express physical principles in terms of geometric concepts. The ancient Greeks themselves came to regard Euclidean geometry as more than just an abstract mathematical system; it was *the* mathematical system on which the organization of the universe was based. We learned in Chapter 2 that the geocentric model of the universe developed by Pythagoras and Plato was constructed of spheres and circles, and the Platonic doctrine of "circular motion at constant speed" dominated astronomical thought for 2000 years, until it was overthrown by Kepler in favor of a different geometric figure (the ellipse). The Euclidean straight line plays an important role in Newtonian physics and optics as the "natural path" of moving bodies and light beams.

* Euclid's *Elements* was used to teach beginning geometry at the high-school level until the twentieth century, when it fell out of favor with educators. Thus the *Elements* was in almost continuous use as a textbook for more than 2200 years.

Thus it appears that the shapes and motions of natural objects can be described successfully and usefully in terms of the concepts and theorems of Euclidean geometry. Physicists have come to speak of the ***geometry of space,*** meaning the geometry that represents the properties of natural bodies as they exist and move in space. Inasmuch as these properties seem to be represented properly in terms of the concepts of Euclidean geometry, one can say that *the geometry of space is Euclidean.*

NON-EUCLIDEAN GEOMETRY

Indeed, the statement that the geometry of space is Euclidean would have been regarded as redundant by all scientists and mathematicians living before the nineteenth century, for it seemed self-evident to them that there was no other kind of geometry but Euclidean geometry. The notion that a consistent, logical system of geometry might exist in which, let us say, the sum of the angles of a triangle could be more than 180° was regarded as absurd. Yet nineteenth-century mathematicians successfully demonstrated that Euclidean geometry was *not* unique, and that a number of logically consistent "non-Euclidean" geometries could be formulated. We will now consider how these non-Euclidean geometries came to be discovered, and we will examine briefly some of the important properties of a non-Euclidean geometry.

Geometry, as a mathematical system, is based on certain defined entities (the straight line, the circle, etc.) and certain unproved propositions dealing with these entities. These propositions are known as ***axioms*** or ***postulates.*** Their truth cannot be verified in terms of any other axioms or postulates; they are simply *accepted* as true and are considered to be self-evident.* There are in geometry additional propositions known as ***theorems,*** which can be proved to be true by demonstrating that they follow logically from the axioms and postulates. Thus a geometric theorem such as "The sum of the angles of a triangle is equal to 180°" is true in the sense that it is consistent with the axioms and postulates of Euclidean geometry.

A careful analysis of Euclidean geometry has shown that the axioms and postulates chosen by Euclid do not lead to any logical inconsistencies, and there are no redundancies among them. (That is, no axioms or postulates can be proved in terms of other axioms or postulates.) There is one commonly recognized weakness, however, in that part of Euclidean geometry that deals with the properties of parallel lines. Euclid defined parallel lines as straight lines lying in the same plane that do not meet if extended indefinitely in any direction. In order to prove certain theorems involving other properties of parallel lines, Euclid found it necessary to propose an additional postulate. This postulate, commonly known as the ***parallel postulate,*** can be stated in a number of different ways. The most familiar form is: *Through a given point, only one straight line can be drawn parallel to a given straight line.* (Euclid's original statement of the parallel postulate, which is entirely equivalent to the one stated here, is depicted in Figure 13–11.)

Many geometers (including Euclid himself) were not satisfied with the parallel postulate because they felt that it was not as self-evident as the other postulates of geometry. They attempted to show that the parallel postulate was really a theorem—that is, that it could be shown to follow as a logical consequence of the other postulates of geometry. However, none of these attempts was ever successful. By the early

* Some examples include "When equal things are added to equals, the results are equal," and "Things equal to the same thing are equal to each other." The difference between an axiom and a postulate in geometry is too complicated to go into here.

1. Through a given point, only one straight line may be drawn parallel to a given straight line.

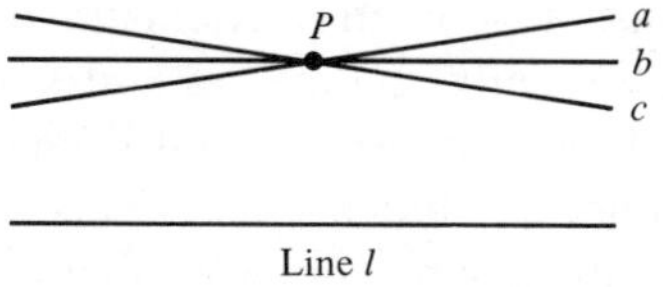

Lines a and c, if extended, will intersect line l. Line b is the *only* line through P that does not intersect l.

2. If a straight line falling on two straight lines makes the interior angles on the same side less than two right angles, the two straight lines, if extended indefinitely, meet on that side on which the angles are less than two right angles.

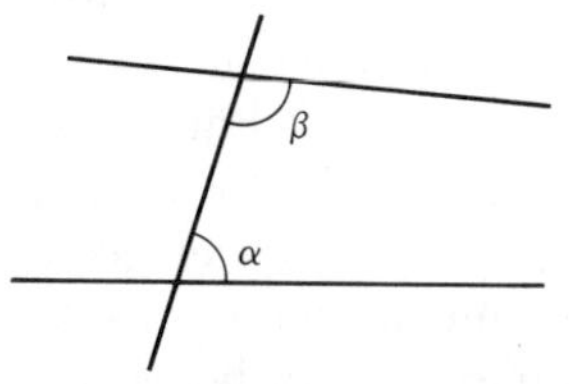

If $\alpha + \beta < 180°$, lines will intersect.

figure 13–11 **EQUIVALENT STATEMENTS OF THE PARALLEL POSTULATE**

Statement 2 is the one originally presented by Euclid. The parallel postulate was considered by geometers not to be self-evident; yet it could not be proved as a theorem.

1800s—after more than 2000 years of effort—geometers were forced to admit that the parallel postulate was completely independent of the other postulates of geometry and was truly a postulate in its own right.

A number of geometers now posed the following question: If one denies the truth of the parallel postulate and replaces it with a different postulate, is it possible to construct another system of geometry, different from Euclidean geometry? In other words, could it be shown that there are other kinds of geometry in addition to Euclidean geometry that are also logically consistent and completely free from contradictions?

The search for a ***non-Euclidean geometry*** (a geometry in which Euclid's parallel postulate is rejected) was begun by Karl Friedrich Gauss in the early 1800s. Gauss, who is considered by many to be the greatest mathematician in history, developed some general ideas about non-Euclidean geometry. Most of his work was unpublished, but in 1830 Gauss did report that it was possible to construct a non-Euclidean geometry free from contradictions. In the same year the first complete non-Euclidean geometry was published by Nikolai Lobachevsky of Russia. In 1832 János Bolyai, a Hungarian, independently completed a system of non-Euclidean geometry similar to that developed by Lobachevsky. (Although Lobachevsky and Bolyai were unaware of each other's work, both were influenced indirectly by Gauss. Bolyai's father and Lobachevsky's mathematics teacher both studied with Gauss and were aware of his interest in non-Euclidean geometry.)

Little attention was paid to the work of Lobachevsky and Bolyai until the 1850s, when the next significant advance was made. In 1851 the German mathematician Bernhard Riemann completed his doctoral thesis (under the supervision of Gauss) on the mathematical foundations of geometry. Riemann described geometry in the most general way possible. Traditionally, geometers had worked in three dimensions (since we perceive the world in three dimensions), but Riemann showed how geometry could be extended to any number of dimensions. Even though it is impossible to visualize or draw four- or five-dimensional objects, they can still be considered as abstract entities; through Riemann's work it became possible to describe the mathematical and geometric properties of such objects. Furthermore, he did not restrict his attention to Euclidean geometry alone. He demonstrated that there were at least two types of logically consistent non-Euclidean geometries that—along with Euclidean geometry itself—could be encompassed within the general system of geometry that he had developed.

Riemann showed that different geometries arise when one deals with the geometric properties of the various two-dimensional surfaces—planes, spheres, cylinders, and so on.* Euclidean geometry is commonly known as ***plane geometry,*** which signifies that it is the geometry dealing specifically with entities that exist in a plane (flat) surface. In Riemann's terminology, Euclidean geometry is the geometry for a surface of ***zero curvature.*** The familiar definitions and theorems of Euclidean geometry—the straight line is the shortest path between two points; the sum of the angles of a triangle equals 180°; the circumference of a circle equals $2\pi R$; and so on—are valid on such a surface.

Now suppose we consider the geometry that applies to a curved surface, such as the surface of a sphere. Since all geometric figures must lie on the surface, the familiar Euclidean straight line cannot exist in this geometry. It can be shown, however, that the most direct path between two points on the surface of a sphere is on the ***great circle*** connecting the points. A great circle is one whose plane passes through the center of the sphere (see Figure 13–12a) and whose radius is equal to the radius of the sphere. Thus all circles of longitude on the earth are great circles, but circles of latitude (except for the equator) are not great circles (Figure 13–12b).

Hence in the geometry of the spherical surface, the shortest distance between any two points is always measured along the great circle joining the points. On the earth, for example, the most direct path between New York and Paris cannot be obtained by connecting them with a Euclidean straight line, because this path would have to be a tunnel through the earth. Since we are constrained to move on the surface of the earth, the shortest path between the two cities is along the great circle connecting them. Thus on an airplane flight between two cities that lie on the same line of latitude, the pilot does not fly directly along the line of latitude. Instead, he takes the "great circle route" between the two cities. Although it is difficult to visualize, it is nevertheless true that the distance measured on a great circle path is always *shorter* than the distance measured on a line of latitude, as shown in Figure 13–12c.

Consequently, we shall consider great circles to be the "straight lines" of spherical geometry. (Lines of latitude, since they are not great circles, are equivalent to the "curved lines" of plane geometry.) On the surface of a sphere, all great circles

*The student should note that although a sphere or a cylinder is a three-dimensional object, its *surface* has two dimensions. For example, the surface of the earth is two-dimensional; it takes only *two* coordinates—latitude and longitude—to locate exactly a point on the earth.

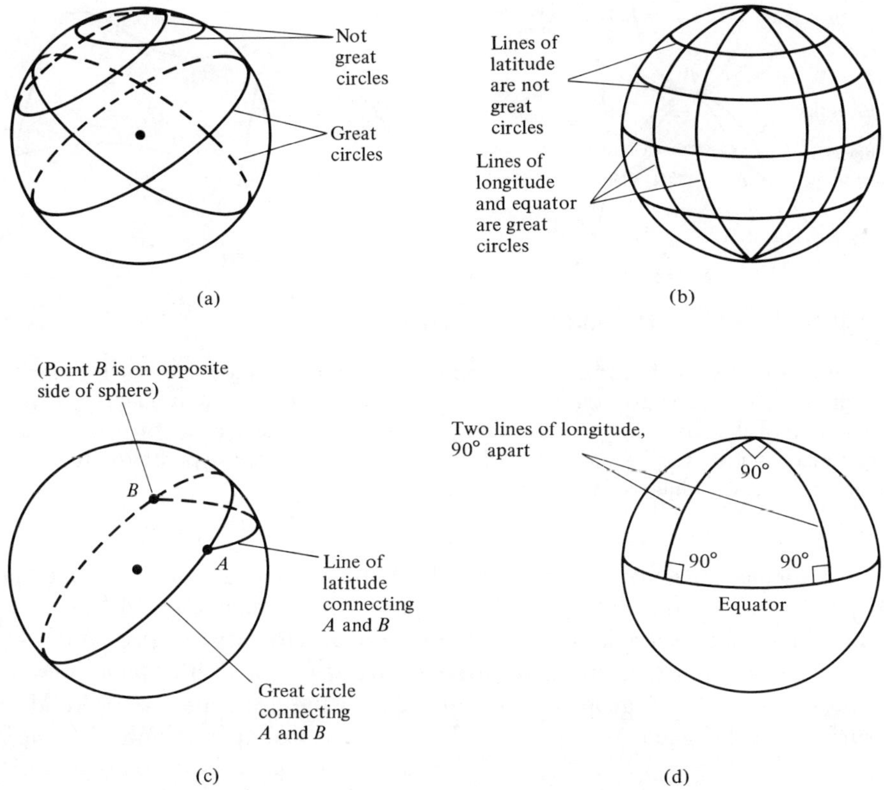

figure 13–12 **GEOMETRY ON THE SURFACE OF A SPHERE**

(a) Great circles are circles whose centers coincide with the center of the sphere. (b) On the earth, all lines of longitude are great circles, but lines of latitude are not great circles (with the exception of the equator). (c) The great circle connecting points A and B is the shortest path between the two points. (d) A triangle constructed from three intersecting great circles (in this case two lines of longitude 90° apart and the equator) will have the sum of its angles greater than 180°.

must intersect; conversely, it is impossible to draw two nonintersecting (parallel) great circles. Since the great circle is the "straight line" of spherical geometry, it follows that *there can be no parallel straight lines in this geometry*. Thus the geometry of a spherical surface represents a non-Euclidean geometry in which the parallel postulate is rejected. It is the geometry that results when the Euclidean parallel postulate is replaced by the postulate that *through a given point, no straight lines can be drawn parallel to a given straight line*. The geometry of a spherical surface (sometimes called ***Riemannian geometry***) can be shown to be a complete and logically consistent non-Euclidean geometry. The standard geometric figures can be defined in this geometry, but they have different properties as compared to their Euclidean counterparts. For example, a triangle can be constructed out of three intersecting great circles (see Figure 13–12d), but the sum of the angles of the triangle is always *greater than* 180°.

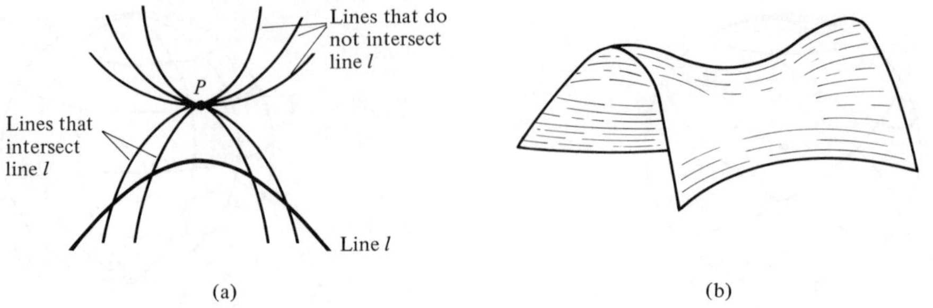

figure 13–13 **HYPERBOLIC GEOMETRY**

In this geometry the hyperbola is the equivalent of the straight line. (a) Through a given point P it is possible to construct more than one "straight line" (hyperbola) that does not intersect the given line l. (b) The surface of hyperbolic geometry is the "saddle surface." All cross sections of it are hyperbolas. The surface extends infinitely in all directions.

Riemann also demonstrated that the non-Euclidean geometry obtained previously by Lobachevsky and Bolyai could be depicted in a similar way. In Riemann's system the surface of a sphere is considered to have a ***positive curvature.*** It is also possible to define a surface having a ***negative curvature,*** and Riemann showed that the Lobachevsky–Bolyai geometry is equivalent to the geometry that is valid on such a surface. In this geometry the parallel postulate is stated as follows: *Through a given point, more than one line can be drawn parallel to a given line.* In this case the equivalent of the Euclidean straight line becomes the hyperbola, and the surface on which the geometry is valid is the "saddle surface," which is curved in such a way that every cross section of it is a hyperbola (Figure 13–13). In this ***hyperbolic geometry*** the sum of the angles of a triangle (constructed out of three intersecting hyperbolas) is always *less than* 180°.*

Thus Riemann showed that there are at least two non-Euclidean geometries, corresponding to the geometries of positively or negatively curved surfaces. In this respect Euclidean geometry represents the special case of a geometry on a flat surface. In each of these geometries the shortest path connecting any two points on a surface—the "straight line" of the geometry—will be a different kind of geometric line. In general, this line is known as the ***geodesic*** of the given geometry. Thus in Euclidean (plane) geometry the geodesic is the Euclidean straight line; in Riemannian (spherical) geometry the geodesic is the great circle; and in Lobachevsky–Bolyai (hyperbolic) geometry the geodesic is the hyperbola. The major properties of the three kinds of geometry are summarized in Table 13–1.

WHAT IS THE GEOMETRY OF SPACE?

The discovery of the existence of non-Euclidean geometry was naturally of great interest to mathematicians. Moreover, this discovery turned out to be of revolutionary significance for physics. At the beginning of this section the notion of the "geometry

*In 1733, nearly a hundred years before Lobachevsky and Bolyai, Geralamo Saccheri arrived at this conclusion in an attempt to validate Euclid's parallel postulate. Saccheri considered it so absurd for the angles of a triangle to add up to less than 180°, that he considered it to be a proof that non-Euclidean geometries could not exist!

of space"—the geometry that describes the properties of bodies in the physical world—was mentioned. We noted that it had long been assumed that the geometry of space is Euclidean geometry, and that this assumption had been made in part because Euclidean geometry was thought to be the only possible kind of geometry. In the light of the discovery of other systems of geometry, however, the question of the geometry of space must be raised anew, for as Riemann himself pointed out, there was no longer any reason to assume that space *must* be Euclidean. The question could be answered only by performing experiments. These experiments would involve the measurement of the geometric properties of natural objects in order to determine which of the various geometries best describes them.

As has already been noted, natural objects do appear to follow the theorems of Euclidean geometry. However, it is possible that more precise experiments involving very large distances (perhaps on an astronomical scale) might well reveal the non-Euclidean nature of space. For example, one could construct a large triangle out of three long pieces of stretched string, and then by careful surveying determine whether the sum of the angles of the triangle is less than, greater than, or equal to 180°. Yet there was no experimental evidence of any sort to indicate that space was anything but Euclidean. It was not until Einstein turned his attention to the problem that the non-Euclidean nature of space was finally revealed.

EINSTEIN'S THEORY OF GRAVITATION

At this point let us recall Einstein's dilemma in reconciling the principles of equivalence and relativity. We saw that the principle of equivalence requires light

table 13–1 **PROPERTIES OF THE THREE DIFFERENT TYPES OF GEOMETRY**

	Euclidean	Non-Euclidean	
		Spherical (Riemann)	Hyperbolic (Lobachevsky–Bolyai)
Type of surface	Plane	Sphere	Saddle
Curvature of surface	Zero	Positive	Negative
Geodesic	Straight line	Great circle	Hyperbola
Parallel postulate	Only one line parallel to a given line	No lines parallel to a given line	More than one line parallel to a given line
Sum of angles of a triangle	180°	Greater than 180°	Less than 180°
Circumference of circle of radius R	$2\pi R$	Less than $2\pi R$	Greater than $2\pi R$
Area of circle of radius R	πR^2	Less than πR^2	Greater than πR^2

beams and freely moving bodies to travel on curved paths in an accelerated reference frame or gravitational field. In an inertial reference frame, however, their paths are straight lines. Hence the principle of relativity, which proposes that the laws of physics take the same form in all reference frames, seemingly cannot be satisfied.

Einstein realized that the two principles could be made compatible through the use of non-Euclidean geometry. He proposed that in an accelerated reference frame—or in its equivalent, a gravitational field—*the path of a light beam is curved because the geometry of space in the region of the gravitational field is non-Euclidean.* The light beam travels on the most direct path between any two points (the geodesic of the particular geometry), but if the geometry is non-Euclidean, this path will be curved. In an inertial reference frame, where there are no gravitational fields, the geometry will be Euclidean and light will travel in a straight line. In any case *a light beam always travels on a geodesic.* In this manner the law of physics describing the path of light has been stated in a form that is the same for all reference frames. The principle of relativity is thereby satisfied, since the law describing the path of light is the same for all reference frames. It is the *geometry* that differs from one reference frame to the next.

Einstein could then account for the bending of a light beam in the sun's gravitational field in the following way. Far from the sun the geometry of space is Euclidean, and light travels in a straight line (the geodesic of Euclidean geometry). As the light beam approaches the sun, it moves through regions where the geometry gradually becomes more non-Euclidean, and the geodesic becomes more of a curved path. The bending of the light beam will be greatest near the sun, where the geometry is most non-Euclidean. As the light beam recedes from the sun, its path gradually returns to a straight line. (The geometry becomes more and more Euclidean.)

To put his ideas in more precise mathematical form, Einstein turned to the generalized geometry developed by Riemann. From 1911 to 1915 Einstein devoted himself to mastering the intricacies of Riemann's mathematics and then constructed his theory of gravitation in full mathematical detail. Einstein adopted the "space–time" description of relativity (first proposed by Minkowski) in which time is placed on an equal mathematical footing with the three space dimensions to provide a four-coordinate system. He then matched this four-coordinate representation of space–time to Riemann's generalized geometry. The result was that in the general theory of relativity the motions of bodies in space and time are described in terms of the mathematics of a four-dimensional non-Euclidean geometry.

In its full detail the general theory of relativity is mathematically quite complicated and extremely abstract, particularly because of the four-dimensional geometry that is used. In general, a four-dimensional object has a three-dimensional surface, and in a four-dimensional non-Euclidean geometry these three-dimensional surfaces are curved. In this context it is common usage to speak of the ***curvature of space–time.*** This curvature exists in an abstract sense and is impossible to visualize. What is meant by this terminology is that the properties and motions of bodies in the physical world correspond to the theorems of a four-dimensional non-Euclidean geometry, and as a consequence the natural paths of bodies moving through space and time are curved paths. In general relativity the principle of inertia is stated as follows:

The natural motion of a freely moving body is along a geodesic in space–time.

Stated in this form, the principle of inertia satisfies the principle of relativity. In the absence of a gravitational field, space–time is flat, and the natural paths of bodies are Euclidean straight lines (the geodesics of a Euclidean space–time geometry).

The presence of a gravitational field causes the geometry of space–time to be non-Euclidean; the stronger the gravitational field, the more non-Euclidean or "curved" the space–time geometry becomes. The natural paths of bodies will be the geodesics of the geometry. Einstein derived an equation of motion describing the space–time motion of a particle that possesses inertia and is under the influence of a gravitational field. If there is no gravitational field, the equation of motion reduces to the equation of a straight line, in accordance with the Newtonian principle of inertia.

The theory of special relativity is included as a special case of general relativity. Special relativity, it will be recalled, is valid only in inertial reference frames, where there are no gravitational fields. Hence the geometry of an inertial reference frame is Euclidean, and special relativity corresponds to general relativity as it applies to inertial reference frames.

To go one step further, we note that gravitational fields are always associated with masses; that is, any body that possesses mass will have a gravitational influence on other bodies. Ultimately, therefore, *the geometry of space–time is determined by the presence of matter.* In the absence of matter (no gravitational fields) space–time is "flat"; that is, the geometry describing the motions of bodies is Euclidean. The presence of matter in a region causes the geometry of that region to become non-Euclidean; the closer one is to a concentration of matter, the more non-Euclidean the geometry becomes, or in other words, the greater the curvature of space–time. The keystone of general relativity is Einstein's ***field equation*** (see Figure 13–14), which mathematically links the presence of matter in a region of space to the properties of the space–time geometry of the region.

Hence in Einstein's view *matter determines the properties of space.* This is in keeping with Einstein's positivistic philosophy, for without matter the properties of space and the passage of time cannot be measured. Thus Einstein once again rejected as meaningless the Newtonian concepts of absolute space and absolute time. Even

figure 13–14 **EINSTEIN'S FIELD EQUATION**

The equation (actually a related family of ten equations) relates the distribution of matter in a region of space to the curvature of the four-dimensional geometry of the region.

$$R_{\mu\nu} - \tfrac{1}{2}g_{\mu\nu}R = -\frac{8\pi G}{c^4}T_{\mu\nu}$$

(Geometry of space-time) $\longleftrightarrow$ (Distribution of matter)

$R_{\mu\nu}$, $g_{\mu\nu}$, and $T_{\mu\nu}$ are mathematical quantities called *tensors;* each represents an array of ten independent quantities.

$R_{\mu\nu}$, $g_{\mu\nu}$, R: quantities that define the properties (particularly the curvature) of the space-time geometry. For a flat space-time, $R_{\mu\nu} = 0$ and $R = 0$.

$R_{\mu\nu}$ = Ricci tensor

$g_{\mu\nu}$ = metric tensor

R = curvature scalar

$T_{\mu\nu}$: the "stress-energy tensor": a quantity that describes the distribution of matter and energy in space. In empty space, $T_{\mu\nu} = 0$.

G = universal constant of gravitation

c = speed of light

more significantly, Einstein provided a new concept of gravity. The Newtonian view of gravitation as a mechanical force exerted by one body on another was cast aside. The Einsteinian view is that *gravity is not a force, but a manifestation of the curvature of space–time*. The Newtonian and Einsteinian theories of gravitation can be compared as follows. We observe that a small body moving in the direction of a much larger body will be deflected from a straight-line path. According to Newton, this deflection is the result of the gravitational force exerted by the larger body, which causes the small body to deviate from its natural straight-line (inertial) motion. According to Einstein, the deflection is merely the natural motion of the small body as it travels along the curved geodesics of the non-Euclidean geometry of the space–time in the vicinity of the larger body.

Thus gravitational forces in the Euclidean space of Newtonian mechanics are replaced by the "curved space–time" of general relativity. With this step Einstein completed his dismantling of the seemingly indestructible edifice of Newtonian physics. Through the special and general theories of relativity, Einstein replaced Newton's principles of motion and gravitation with entirely new concepts.

As a test of his theory, Einstein considered the problem of a planet moving in the gravitational field of the sun. Einstein was aware that Newton's theory of gravitation accounted for the motions of the planets rather accurately, so that the equations of planetary motion derived from general relativity should not differ widely from Newton's results. In other words, the space–time geometry in the vicinity of the sun should be at most only slightly non-Euclidean.

The problem of planetary motion, as it turned out, is one of the two or three problems for which an exact solution of the equations of general relativity can be obtained. Einstein's solution, as expected, is quite similar to Newton's; that is, it predicts as a first approximation that a planet in the gravitational field of the sun will travel in an elliptical orbit about the sun at one focus.* Einstein's solution differs slightly from Newton's, however, in that it predicts that the orbit of the planet does not close on itself. Instead, the planet moves such that the axis of the orbit (the line from aphelion to perihelion) rotates or ***precesses*** slowly around the sun. (See Figure 13–15.) Thus at each successive perihelion the planet is in a slightly different position in space than at the previous one. This effect is known in astronomy as the ***precession of the perihelion*** of a planetary orbit.

Astronomers had been aware for many years of the precession of planetary orbits; in fact, it had been observed that all the planets exhibit this effect to some degree. However, astronomers had been able to account for precession using Newtonian theory by showing that it was a consequence of the much smaller gravitational forces exerted on a planet by the other planets of the solar system. Thus, for example, the perihelion of the orbit of the earth was known to precess by an angle of about 11″ per year, an amount that could be accounted for (to the limits of observational error) in terms of the effects of the gravitational forces exerted on the earth by Jupiter, Saturn, and Venus.

Among the planets only Mercury posed a problem. The perihelion of the orbit of Mercury precesses 574″ in a century. In the 1840s Urbain Leverrier (the co-

*The student should be careful not to be led to believe that the ellipse represents the geodesic in the gravitational field of the sun. According to general relativity, a planet travels on a geodesic in *four-dimensional space–time*. An elliptical orbit represents the three-dimensional space portion of the four-dimensional geodesic.

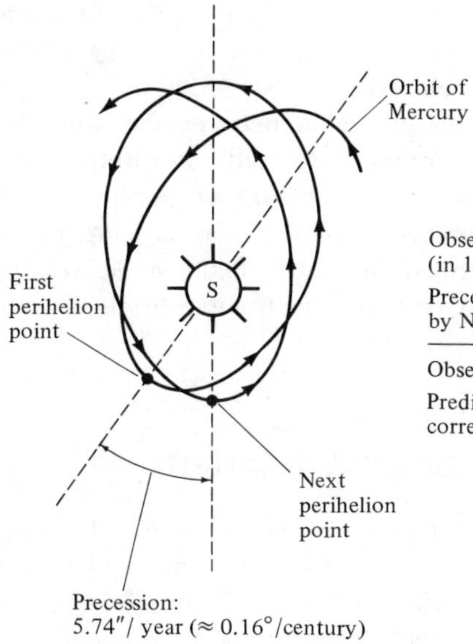

Observed precession (in 100 yr)	= 574.2 ± 0.4″
Precession predicted by Newtonian theory	= 531.1 ± 0.2″
Observed discrepancy	= 43.1 ± 0.5″
Predicted relativistic correction	= 43.03″

figure 13–15 **THE PRECESSION OF THE PERIHELION OF MERCURY**

As Mercury orbits the sun, its perihelion point moves slowly through space. This effect could not be completely explained by Newton's theory of gravitation. Using general relativity, Einstein was able to account for the discrepancy.

discoverer of Neptune) demonstrated that the precession of the perihelion of Mercury could *not* be entirely accounted for by Newtonian theory. Through careful astronomical observations, the precession was found to be $43.1 \pm 0.5''$ per century *larger* than was predicted. This deviation, although extremely small, represented a noticeable failure of Newton's theory of gravitation. Some astronomers attributed the unexplained precession to the existence of another planet located between Mercury and the sun. This hypothetical planet was given the name Vulcan in anticipation of its discovery. Astronomers looked for Vulcan for many years, but it was never seen.

Einstein's solution in 1915 of the general relativistic equations of planetary motion demonstrated that there is actually a *second* cause of the precession of the perihelion of a planetary orbit. In addition to the "Newtonian" precession produced by the gravitational effects of the other planets, there is also a "relativistic" precession generated as a consequence of the non-Euclidean nature of space–time in the sun's gravitational field. Einstein predicted that the relativistic precession is most noticeable closest to the sun, where the space–time geometry is most non-Euclidean. The relativistic precession should also depend on the eccentricity of the orbit, for a planet in an eccentric orbit will move from regions of greater curvature to lesser curvature and vice versa as it orbits the sun. Thus for most of the planets the relativistic part of the precession should be virtually negligible in comparison to the Newtonian part. The relativistic precession would be greatest for the planet Mercury, which is closest to the sun and has a relatively eccentric orbit. Einstein's calculations predicted a relativistic precession of $43.03''$ per century for Mercury's orbit—in remarkably good agreement with the observed discrepancy of $43.1 \pm 0.5''$ found by astronomers. Upon achieving this success after years of drudgery, Einstein wrote to a friend: "Imagine my joy . . . at the result that the equations yield the correct perihelion motion of Mercury. I was beside myself with ecstasy for days."[2]

Thus the disturbing disagreement between Newton's theory of gravitation and the observed motions of Mercury was finally accounted for. Newton's theory could not account completely for the precession of the perihelion of Mercury, whereas Einstein's could account precisely for the observed precession. Moreover, Einstein's theory of gravitation contained Newton's theory as a first approximation. Therefore, Einstein rightfully claimed that his was a superior theory on all counts.

13–5

THE VALIDITY OF GENERAL RELATIVITY

The most severe test of a theory, as we have noted many times, is whether its predictions agree with the results of experiments and observations. In this respect the general theory of relativity rests on a much less secure foundation than many other theories, for there is a rather limited body of experimental evidence in support of general relativity.

EXPERIMENTAL TESTS OF GENERAL RELATIVITY

Three of Einstein's predictions that have been verified by observation and experiment have already been discussed. The first is the precession of the perihelion of Mercury. The second and third are the bending of light and the frequency shift of light in a gravitational field. These latter two effects were described in general terms in the context of the principle of equivalence. However, Einstein required the full detail of the theory of general relativity to make precise mathematical predictions of the magnitudes of these effects. We noted previously that the observed bending of starlight has a maximum predicted value of 1.75″, which represents the angle by which a ray of light that just grazes the edge of the sun is deflected from a straight line. This extremely small deflection indicates that the space–time geometry even very close to the sun is indeed very slightly non-Euclidean.

In recent years two new confirmations of Einstein's predictions have been achieved. The first of these had to do with a measurement of time dilation in a gravitational field. According to the theory, a light beam passing through a gravitational field will appear to slow down to a distant observer; hence there will be a time delay in the passage of the light through the field. This apparent slowing down of a light ray occurs in addition to the deflection of the ray's path in a gravitational field, and the two effects are considered to be different phenomena.

In order to observe this effect, one must be able to time the passage of a known signal through a gravitational field. (The time delay cannot be observed for starlight passing the sun, for example, since the time of emission of the starlight is not known.) In 1970 a group of American scientists announced that they had detected the time-delay effect by bouncing radio waves off Venus and Mercury as they passed behind the sun (Figure 13–16). In the experiment radio pulses were emitted at a precisely known time from the earth. The pulses passed near the sun, reflected off Venus or Mercury on the other side of their orbits, and returned to the earth (passing near the sun a second time). Since the distance from the earth to the planet is precisely known, the round-trip travel time of the signal from emission to receipt of the echo

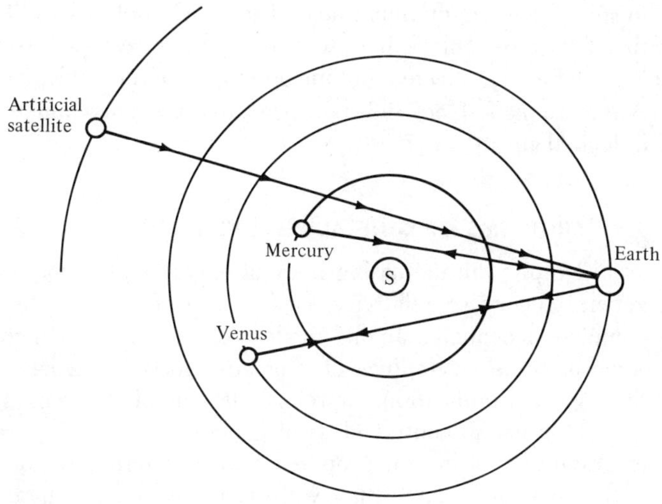

figure 13-16 **THE RELATIVISTIC TIME DELAY EFFECT**

When Venus or Mercury is on the opposite side of the sun from the earth, a radio signal from earth echoing off the planet will exhibit a time delay because of the time dilation effect in the gravitational field of the sun. The signals emitted by an artificial satellite in a solar orbit will be similarly affected.

can be calculated. The predicted time delay in the receipt of the signal was observed and was found to be within 6 percent of the theoretical value.

A confirmation of the time-delay effect was provided in 1970 by the signals from Mariner 6 and Mariner 7. These two artificial satellites were originally launched several years before for the purpose of exploring the planet Mars, but both had since gone into orbits around the sun. When the satellites were on the opposite side of the sun from the earth, their radio signals were found to be time-delayed, as predicted. The observed time delay was within about 5 percent of the predicted value.

The testing of the general theory of relativity is one of the most active areas of modern-day physics. Several experiments are now being planned or performed to test other predictions of the theory or to improve the accuracy of the experimental tests that have already been made. The main source of difficulty in obtaining experimental verification of general relativity is that the differences between the Newtonian and Einsteinian predictions are extremely minute and at the limits of detectability of modern experimental equipment. Extremely intense gravitational fields—many times stronger than that of the sun—are needed to produce easily observable general relativistic effects. Where such conditions prevail, extremely strange phenomena would be expected. For example, astrophysicists have shown that a star can collapse rapidly under its own gravitation until the entire mass of the star is contained within a sphere no more than a few miles in diameter. With such a great concentration of mass, the space–time geometry should become so curved that the geodesic of a light beam will be a circle or an ellipse. Thus any light emitted or reflected by the collapsed star cannot escape. The star is therefore completely invisible; it becomes a "black hole" in space. The gravitational field of the black hole will still be effective, and any matter that falls into a black hole will be unable to escape. As of 1974 no black holes had been definitely detected, but the possibility of their existence has created great interest.* Are there many black holes in space, and if so, how much of the matter of the universe is locked up in them?

CRITICISMS OF GENERAL RELATIVITY

Despite the increasing body of experimental evidence seemingly supporting the general theory of relativity, certain aspects of the theory have not as yet gained complete acceptance among modern-day physicists. In this section we shall consider some of the shortcomings of Einstein's theory that have led a number of physicists to propose modifications or reformulations of the general theory.

One of the central ideas of general relativity is a reformulation of the concept of gravitation. Einstein proposed that gravitation is not a force, but a consequence of the non-Euclidean geometry of space–time created by the presence of matter. Hence the curved paths of bodies moving through gravitational fields represent the natural motions of these bodies in a curved space–time geometry. After some reflection, one realizes that Einstein has not *explained* gravitation any more than Newton did; he has simply reinterpreted it in terms of different unexplained concepts. In the words of physicist G. C. McVittie, "To say instead that gravitation is a manifestation of the curvature of four-dimensional geometrical manifolds is to account for a mystery by means of an enigma."[3]

It is especially notable that Einstein was not able to extend his geometric concept to explain the existence of other forces—particularly, the electromagnetic forces. In general relativity the idea of gravitation as a mechanical force between bodies is abolished, but the idea of electromagnetic forces between bodies remains. Einstein and numerous other physicists tried for many years to construct a ***unified field theory*** in which *all* forces could be explained in geometric terms, but a satisfactory theory of this nature has not been produced to date.

The general theory of relativity has also been criticized for its failure to explain the concept of inertia. In Newtonian mechanics it is stated as a fundamental principle that bodies possess inertia—the tendency to maintain constant-velocity motion—but it is not explained *why* bodies possess this property. In the 1870s Ernst Mach (who had a significant philosophical influence on Einstein) theorized that inertia may be the result of an *interaction* between bodies—that is, the inertia of a body is somehow due to the presence of other bodies in the universe. This proposal is commonly known as ***Mach's principle.***

It has seemed to many physicists that Mach's principle should be an integral part of the general theory of relativity. The significance of Mach's principle becomes more evident when one considers the application of general relativity to rotating reference frames. The earth itself can be taken as a case in point. The earth is presumed to rotate on its axis with respect to the fixed stars, and the flattening of the earth

*Astronomers have recently located a star whose motion indicates that a massive, unseen object is in orbit around it. Spectral analysis indicates that gas is flowing out of the star. The unseen object is known to emit x rays. Some astronomers believe that the object is a black hole, and that x rays are being given off by the outflowing gas as it falls into the black hole.

is attributed to this rotation. More specifically, the flattening is explained in terms of the inertial tendency of the material at the equator to fly off in a straight line. That is, if we consider the earth to be rotating, its flattened shape is a consequence of inertia.

Now let us consider a reference frame in which the earth is stationary, and the "fixed stars" (i.e., the entire universe) rotate around the earth. This reference frame may seem to be an absurd one, but the principle of relativity tells us that it is just as valid as any other reference frame. Yet changing the reference frame does not change the shape of the earth; it is still observed to be flattened. Since the earth is not rotating in this reference frame, the flattening of the earth has to be attributed to a net gravitational interaction between the rotating universe and the earth. That is, in a reference frame in which the earth is stationary, the flattening of the poles must be a consequence of the rotation of the universe about the earth.

It would seem to follow from this line of reasoning that if there were fewer bodies in the universe, the interaction between the earth and the rotating universe would be weaker, and the earth would not be as flattened; in an otherwise empty universe the earth would not be flattened at all. Thus in an empty universe (if we return to the reference frame in which the earth is rotating), we find a rotating earth that is not flattened. Therefore, inertia (which causes a rotating earth to flatten) must depend on the presence of other bodies in the universe.

The point of this example is that if one carefully applies the principles of relativity to a rotating system, one arrives at Mach's principle. Yet Mach's principle does not appear in the general theory of relativity. For a variety of reasons, Einstein decided not to include it in the theory, and so inertia remains as an unexplained property of matter.

In 1961 Carl Brans and Robert H. Dicke of Princeton University proposed a modified form of general relativity that takes Mach's principle into account. In the Brans–Dicke theory additional terms have been added to Einstein's field equation to account for the dependence of inertia on the presence of other matter. Because of these additional terms, the predictions of the Brans–Dicke relativity theory differ slightly from those of Einstein's theory; for example, the Brans–Dicke theory predicts a deflection of light in the sun's gravitational field of 1.6″ (as compared to Einstein's 1.75″), and at the present time the experimental evidence is not good enough to decide between the two. It appears likely, however, that within the next few years experiments of sufficient accuracy for us to decide between the two theories will be performed.

Whatever the outcome of these experiments, the greatness of Einstein's achievement cannot be denied. The fundamental concepts and basic framework of general relativity are no longer doubted; it is the details of the theory that are still open to question. (The Brans–Dicke theory, for example, represents a modification rather than a complete overthrow of Einstein's theory.) What is perhaps most remarkable about the theory of relativity is that it grew out of Einstein's faith in a few fundamental principles. Unlike other great theories in physics, which were developed to account for observed phenomena (e.g., Newton's theory of gravitation was formulated with Kepler's laws as a guide), the general theory of relativity was created from basic principles, and experimental confirmation came afterward. It is a unique achievement in the annals of physics, and one of the greatest intellectual accomplishments of human history.

QUESTIONS FOR REVIEW

1 State and explain the principle of equivalence. Why did Einstein consider this principle to be the key to the development of general relativity?

2 Illustrate the principle of equivalence by describing measurements and observations made (a) in a upwardly accelerating laboratory and (b) in a freely falling laboratory.

3 Explain why the principle of equivalence is restricted to measurements made at a single point (or in a small region) of space. In particular, explain how a nonuniform gravitational field can be distinguished from an accelerated reference frame.

4 What is the unexplained law of nature that makes the principle of equivalence valid? Explain your answer.

5 Explain why the bending of light in a gravitational field is a necessary consequence of the principle of equivalence.

6 Describe the astronomical test of Einstein's prediction of the gravitational deflection of light. Why do the observations have to be made during a total solar eclipse? How well does Einstein's prediction conform to the results of these observations?

7 Describe the gravitational frequency shift of light. Under what conditions will the frequency of light increase or decrease in a gravitational field?

8 Explain why the gravitational frequency shift is required by the principle of equivalence.

9 Describe the Mossbauer effect. Explain how it was used to test for the gravitational frequency shift.

10 Discuss time dilation in a gravitational field. Is this effect apparent or real? Why is it a consequence of the principle of equivalence?

11 State the parallel postulate of Euclidean geometry. Why was it of special interest to geometers?

12 What is the difference between Euclidean and non-Euclidean geometry? Why is the existence of more than one kind of geometry particularly interesting to physicists?

13 Describe the basic properties of Riemannian (spherical) geometry and hyperbolic geometry. Explain why each geometry is non-Euclidean.

14 What is a geodesic?

15 Discuss the apparent conflict between the principle of relativity and the principle of equivalence. How did the existence of non-Euclidean geometries enable Einstein to resolve this dilemma?

16 State in a form that is valid in all reference frames (a) the principle governing the motion of light and (b) the principle of inertia.

17 Explain what is meant by the geometry of space. What does it mean to say that the geometry of space is Euclidean or non-Euclidean? According to Einstein, what determines the geometry of space?

18 Compare Einstein's theory of gravitation with Newton's theory. In what sense does gravity become a "fictitious force" in Einstein's theory?

19 Discuss the precession of the perihelion of Mercury. What are the causes of this effect according to Newton and to Einstein? Why was this effect considered to be an important test of general relativity?

20 Summarize the main experimental tests of the general theory of relativity.

21 State and explain Mach's principle. Discuss its connection to the theory of relativity.

CA$H For BOOKS

buy back dates

Dec.	19	9-3	
	20	9-3	6-8
	21	9-3	

EXERCISES

1 How does the principle of equivalence apply to experiments performed (a) in a laboratory on the earth and (b) in an orbiting satellite?

2 Suppose evidence is found that the inertial mass of a certain body is *not* equal to its gravitational mass. Explain how this discovery would invalidate the principle of equivalence.

3 Show that the existence of antigravity (a body "falling up" in a gravitational field or being repelled by gravity) would invalidate the principle of equivalence.

4 Suppose that the elevator of Figure 13–6 moves upward with a *constant velocity,* rather than a constant acceleration.
(a) How would the apparent path of the light beam be different in this case?
(b) Is a gravitational field required to explain the motion of the light beam?
(c) Is it possible for a light beam to travel in a straight line in a gravitational field? Explain your answer.

5 Using Equation 13–2a, show that the light reaching the earth from the surface of a distant star of mass M and radius R exhibits a gravitational frequency shift $\Delta f/f_0 = GM/Rc^2$. Explain why this frequency shift always appears as a red shift.

6 A white dwarf star is a star that has condensed under its own gravitation. A typical white dwarf star may have a mass equal to that of the sun, but a radius of 5000 to 10,000 miles. Explain why the gravitational frequency shift of the light from a white dwarf should be noticeably larger than that of the sun.

7 Using Equation 13–2b, calculate the frequency shift ($\Delta f/f_0$) expected for a gamma ray falling a distance of 100 feet at the surface of the earth. (Note: Make sure your units are consistent.)

8 In the Pound–Rebka experiment to detect the gravitational frequency shift, the resonance absorption of the Mossbauer effect is decreased when the gamma rays are falling or rising in a gravitational field. To restore resonance (and to measure the magnitude of the frequency shift), Pound and Rebka set the source in motion.
(a) Explain why the motion of the source restores the resonance effect.
(b) In what direction was the source moved to restore resonance when the source was at the top of the tower and the detector at the bottom? Explain your answer.
(c) What happens when the positions of the source and detector are interchanged?

9 Two identical clocks are synchronized; then one is placed in an orbiting satellite, while the other remains on the ground. After a sufficiently long time, the orbiting clock is brought down and compared with the ground clock. Will the two clocks read the same time? If not, which clock will have measured a longer elapsed time? Explain your answer.

10 One member of a pair of identical twins goes off on a long journey through gravity-free space, while his brother remains on the earth.
(a) Explain why we expect the astronaut twin upon his return to have aged more than his earth-bound twin.
(b) A "twin paradox" can be formulated by noting that in the reference frame of the astronaut twin, the earth-bound twin has gone off on a long journey. Hence each twin thinks the other is older! Show that the two reference frames are *not* equivalent, and thereby resolve the paradox.

11 Prove the theorem of plane (Euclidean) geometry that the sum of the angles of a triangle is equal to 180°. Show that this proof depends on the Euclidean parallel postulate.

12 Explain why the circumference of a circle drawn on the surface of a sphere is *less* than $2\pi R$. (Note: R is *not* the radius of the sphere, but the radius of the *circle;* the radius is a curved line that lies *on* the surface of the sphere.)

13 Explain what is meant by the statement "Space is curved."

14 Why is the space–time of special relativity "flat"?

15 Explain why the problem of action at a distance, which was so difficult for Newton to accept, is dissolved by Einstein's theory of gravitation.

16 Is Einstein's explanation of gravitation better than Newton's? Why do modern-day physicists consider Einstein's theory to be the better theory?

17 Suppose that a *single* planet were in an elliptical orbit around a star. Would this planet exhibit a precession of its perihelion (a) according to Newton's theory and (b) according to Einstein's theory?

18 Einstein gave the following equation for the relativistic precession of the perihelion of a planetary orbit:

$$P = \frac{3.84''}{a^2\sqrt{a}(1 - e^2)}$$

where P is the precession in seconds of arc per century; a is the orbital radius in AU; and e is the orbital eccentricity. Calculate the relativistic precession of the orbits of Venus and the earth, and compare your results with the observed values obtained by astronomical measurement:

(a) Earth: $a = 1$ AU; $e = 0.017$; observed $P = 5.0 \pm 1.2''$/century.
(b) Venus: $a = 0.72$ AU; $e = 0.007$; observed $P = 8.4 \pm 4.8''$/century.

19 Discuss whether it is possible to determine if an isolated body in an otherwise empty universe does or does not possess inertia. How does this question relate to Mach's principle?

20 The Brans–Dicke modification of the general theory of relativity predicts that the universal constant of gravitation G is not truly constant, but is decreasing with time.

(a) What effect would a decreasing G have on the various bodies in the universe?
(b) What experimental tests could be made to determine whether G is decreasing or not?
(c) Why does the presently available evidence indicate that if G is decreasing at all, it must be decreasing at an extremely slow rate?

Five

LIGHT AND THE NATURE OF THE ATOM

COURTESY OF THE AMERICAN INSTITUTE OF PHYSICS

Niels Bohr and Max Planck

The physicists of the twentieth century have had remarkable success in their attempts to understand the nature of the atom. Two of the earliest important breakthroughs were made by Max Planck (shown on the right), who proposed the quantum theory of radiation, and Niels Bohr, who provided the first successful theory of atomic structure. The contributions made by these two men have greatly influenced the course of modern physics.

THE ORIGINS OF THE QUANTUM THEORY

14

It is undeniable that there is an extensive group of data concerning radiation which shows that light has certain fundamental properties that can be understood much more readily from the standpoint of the Newton particle theory than from the standpoint of the wave theory.

Albert Einstein

WE WILL NOW turn our attention back to a problem that remained in 1900, as it was in 1700, a central concern of many physicists: the nature of light. We recall that the scientists of the nineteenth century, beginning with Thomas Young and ending with Heinrich Hertz, succeeded in demonstrating that light and other forms of electromagnetic radiation travel from place to place with all the properties normally associated with the motion of a transverse wave. By 1900 considerable evidence had been amassed to show that all the various types of electromagnetic radiation exhibit the phenomena of interference, diffraction, and polarization—which are characteristic properties of waves. Although some serious unanswered questions as to the nature of the medium of propagation remained, the wave nature of electromagnetic radiation had been made quite evident.

On the other hand, when attention turned from the transmission of light to the processes by which light is emitted and absorbed, the wave theory no longer provided the proper description. The result was the construction by Planck and Einstein of a radically new theory concerning the nature of light and its interactions with matter. In this chapter we shall examine the discoveries that led to the formulation of the quantum theory of light.

14–1

SPECTRA

To begin to understand how light is emitted by matter, let us consider some of the common natural sources of light—the sun and stars; burning gases; and hot, glowing solids (such as the filament in an electric light bulb). Most natural light sources have a characteristic color; the sun appears yellowish white, candles burn with a yellow flame, natural gas gives a blue-colored flame, and so on. Often the color of an incandescent body will change with the temperature; for example, an iron bar will glow red at first, then become successively orange, yellow, and white as its temperature is raised higher and higher.

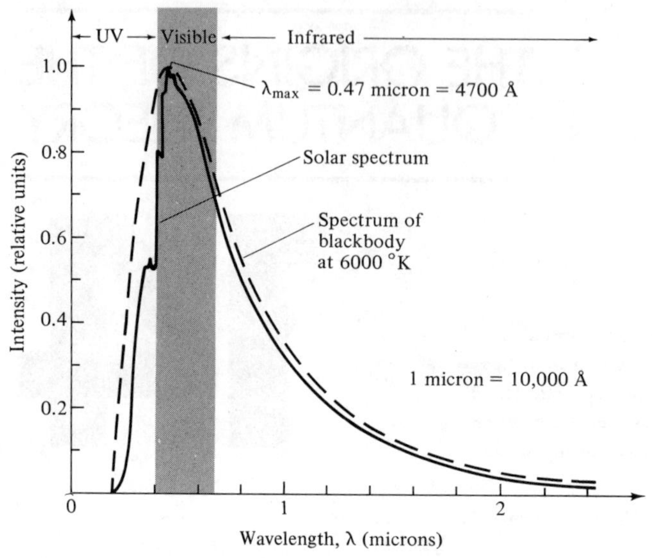

figure 14–1 **THE SOLAR SPECTRUM**

The graph shows the variation of intensity with wavelength of the light given off at the surface of the sun. The spectrum of a blackbody (a perfect radiator) at a temperature of 6000°K is shown for comparison.

The most effective way to analyze the light given off by these bodies is to obtain a ***spectrum.*** This can be done by passing the light through a prism so that it will be separated into its component colors by the process of dispersion. With suitable measuring instruments, the spectrum of any natural light source can be analyzed to determine what wavelengths of light are emitted and with what intensities.

The first spectral analysis of a natural light source was performed by Isaac Newton in 1666. Newton produced a spectrum of sunlight and showed that the white light of the sun is actually a combination of many different colors. In modern terminology the spectrum of sunlight is described as ***continuous;*** that is, electromagnetic radiation is being emitted at all wavelengths. Furthermore, it is found that the sun does not shine with equal intensity at all wavelengths. Actually, sunlight is most intense at a wavelength of about 4700 Å, which is in the blue-green part of the spectrum. (See Figure 14–1.) Although the peak of the solar spectrum is in the blue-green part of the spectrum, the earth's atmosphere selectively removes (by absorption and scattering) much of the blue light from the incoming radiation. Thus at the earth's surface, sunlight appears yellowish white.

In 1815 Newton's experiment with sunlight was repeated by a Bavarian optician, Joseph von Fraunhofer. To perform this analysis, Fraunhofer used a ***spectroscope*** (Figure 14–2). In the simplest form of this device, light from the source enters a narrow slit, is refracted by a prism, and is then brought to a focus by means of a converging lens. The dispersion of the different wavelengths by the prism causes each color to come to a focus at a slightly different position, and a spectrum is produced.

When Fraunhofer examined the spectrum of sunlight, he was surprised to find that the spectrum was crossed with thousands of narrow dark lines (Figure 14–3a).

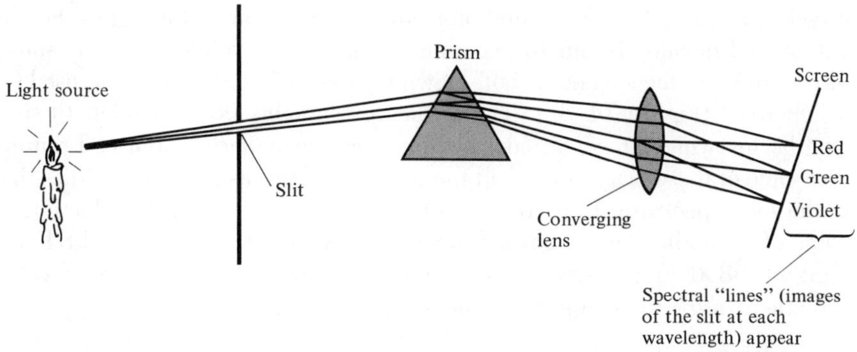

figure 14–2 **THE SIMPLE SPECTROSCOPE**

A narrow beam of light passing through the slit is dispersed by the prism into component colors according to wavelength. The converging lens brings each color to a focus at a different point on the screen. Each color will come to a focus as a narrow band or "line," which is an image of the slit.

Fraunhofer reasoned that each of these dark lines represents a wavelength of light that is absent from the solar spectrum (or at least is present at considerably reduced intensity). Thus it was discovered that the solar spectrum was not as smoothly continuous as had been believed.*

Fraunhofer could not offer any explanation for the appearance of these dark lines in the solar spectrum, but he continued to catalog them and measure their

figure 14–3 **TYPES OF LINE SPECTRA**

(a) In a dark-line spectrum certain wavelengths are missing from an otherwise continuous spectrum or are emitted with much lower intensity than neighboring wavelengths. Thus dark lines appear in the spectrum corresponding to these wavelengths. (b) In a bright-line spectrum only certain specific wavelengths are seen. The rest of the spectrum between these wavelengths is absent.

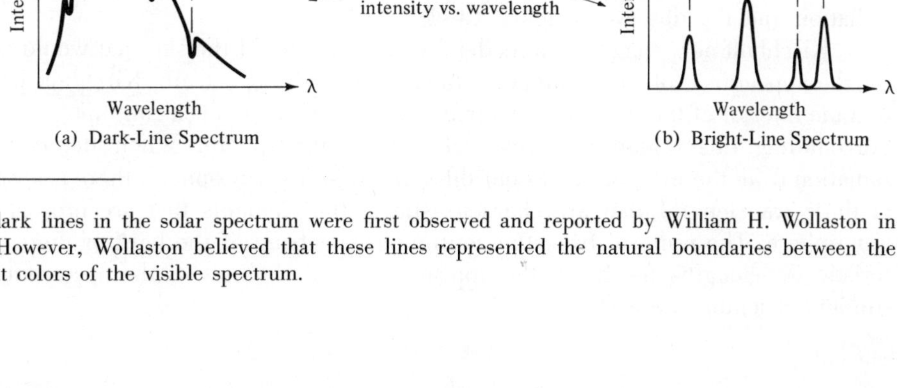

*The dark lines in the solar spectrum were first observed and reported by William H. Wollaston in 1802. However, Wollaston believed that these lines represented the natural boundaries between the different colors of the visible spectrum.

wavelengths. In addition, Fraunhofer (and independently, John Herschel and William Talbot in England) began to examine the spectra of other luminous sources. It was discovered that the spectra of hot, glowing gases or flames were composed of numerous bright lines (Figure 14–3b). These lines presumably corresponded to specific wavelengths for which the emitted radiation was particularly intense. Fraunhofer found that some of these bright lines had the same wavelengths as the dark lines that appeared in the solar spectrum. In particular, Fraunhofer observed that the characteristic yellow color of a candle flame comes from a pair of closely spaced bright lines of yellow light of 5890 and 5896 Å wavelength. At precisely the same wavelengths, two prominent *dark* lines appear in the solar spectrum.

SPECTRAL ANALYSIS

A comprehensive explanation of these effects was provided in 1859 by two German scientists, Gustav Kirchhoff and Robert Bunsen, who worked together at the University of Heidelberg. Bunsen is best known for his invention of the Bunsen burner, which played an important role in the work done by Kirchhoff and Bunsen. The hot, colorless flame was capable of vaporizing and igniting many common substances. In examining the spectra of various vaporized substances, Kirchhoff and Bunsen discovered that each element emits a unique bright-line spectrum, so that no two elements have the same set of spectral lines. This discovery opened up an entirely new method of chemical analysis, for the chemical composition of any unknown substance could be identified by vaporizing the unknown and comparing the resulting spectrum with the known bright-line spectra of the various elements. This method of ***spectral analysis*** proved to be successful in identifying the chemical composition of even very minute amounts of matter. Using this technique, Kirchhoff and Bunsen were in fact able to discover in certain minerals two new chemical elements, cesium and rubidium, because the spectra of these elements were different from those of any previously known elements.

Many of the dark "Fraunhofer" lines seen in the solar spectrum were found to have exactly the same wavelength as certain bright lines in the spectra of various elements. Kirchhoff was led to hypothesize that the dark lines are produced by substances that are absorbing the same wavelengths that they emit when hot. A hot, luminous vapor will normally *emit* a bright-line spectrum, but if light from a hotter, more intense source is passed through the same vapor, the vapor will *absorb* light of the same wavelengths. Kirchhoff was able to verify this hypothesis experimentally. He successfully produced dark-line spectra by passing the light from an intense source of continuous radiation through the vapors of various elements. In each case the dark lines that appeared matched the bright lines in the emission spectrum of the particular element. Thus Kirchhoff showed that a dark-line spectrum is actually an ***absorption*** spectrum that results from the presence of a vapor between a source of continuous radiation and the observer (Figure 14–4).

Kirchhoff next suggested a model for the structure of the sun that would account for the appearance of the Fraunhofer lines in the solar spectrum. Kirchhoff proposed that the interior of the sun is extremely dense and hot and emits a continuous spectrum. Surrounding this region is a relatively cooler atmosphere. When the continuous radiation from the interior passes out through the sun's atmosphere, there is absorption at those wavelengths that are characteristic of the elements that are present in the atmosphere. The removal from the continuous spectrum of the light of these characteristic wavelengths results in the appearance of dark lines in the spectrum of the sunlight reaching the earth.

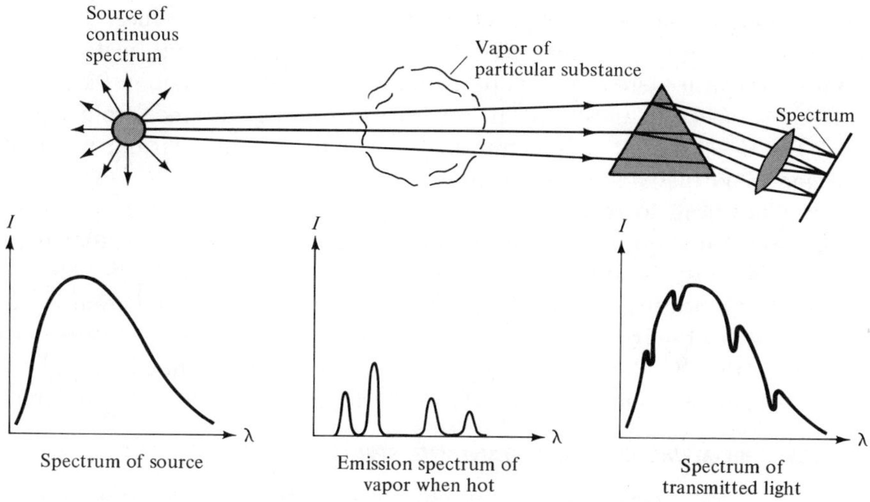

figure 14–4 **PRODUCTION OF A DARK–LINE (ABSORPTION) SPECTRUM**

Light from a source of continuous radiation passes through a vapor. The spectrum of the transmitted light will exhibit dark lines that correspond to the bright lines seen in the emission spectrum of the vapor.

By comparing the Fraunhofer lines with the bright-line spectra of known elements obtained in their laboratory, Kirchhoff and Bunsen were able to determine which elements are present in the atmosphere of the sun. For example, the two yellow lines discovered by Fraunhofer in candle flames turn out to be characteristic of sodium, and since these lines appear as dark lines in the solar spectrum, sodium must be present in the sun. Other prominent dark lines discovered by Fraunhofer could be definitely attributed to hydrogen, iron, calcium, magnesium, and silicon.

As better spectroscopes were constructed, more dark lines were discovered in the solar spectrum, and more elements were identified. At present, there are more than 25,000 known dark lines in the solar spectrum (many of which have yet to be identified). To date, more than sixty elements and several dozen different kinds of molecules have been found in the sun through spectral analysis. In fact, one element was discovered in the sun *before* it was found on the earth. In 1868 Norman Lockyer discovered in the solar spectrum a spectral line that could not be attributed to any known element. Lockyer proposed the name *helium* for this element (after *helios*, the Greek word for the sun). Finally, in 1895 William Ramsay positively identified helium as an inert gas present in minute quantities in terrestrial rocks and in atmospheric air.

In retrospect, the discoveries made by Kirchhoff and Bunsen have proved to be among the most fruitful in the history of science. In the next chapter we will learn how the modern theory of the atom evolved out of attempts to understand bright-line spectra. In chemistry the techniques of spectral analysis provided a powerful new analytic tool. In the field of astronomy, spectral analysis turned out to be the most significant advance since the discovery of the telescope, for the methods used to determine the elements present in the sun could also be used to ascertain the chemical composition of planetary atmospheres, stars, interstellar gases, and even the most distant galaxies. The spectra of the light from these objects exhibit absorption and/or emission lines. From a complete analysis of these lines, it is possible to determine not only the chemical composition of the absorbing material but also its physical properties (size, density, pressure, temperature), its motion (using the Doppler shift of the spectral lines), and in some cases even its distance from the earth.

The development of spectral analysis represents the starting point of modern astronomy. Virtually all that is known about the far-flung bodies of the celestial universe has been learned from their spectra. Through spectral analysis it has been established that the same chemical elements can be found throughout the universe, even in the most distant known objects several billion light years away. Moreover, all the discoveries made through spectral analysis have been consistent with the idea that the laws of physics are the same throughout the known universe. Hence the knowledge that has been gained in this manner has had significant philosophical as well as scientific value, for it attests to the essential physical unity of the universe.

THE PROBLEMS OF UNDERSTANDING SPECTRA

The use of spectra to identify chemical elements was a remarkably successful scientific advance, but it did not really come to grips with the main problem. It had been shown that each element emits a unique set of spectral lines, but it had not been explained *why* this is so. That is, Kirchhoff and Bunsen discovered several particularly useful *laws* pertaining to spectra, but they did not propose any *theories* to explain them.

Let us summarize the main phenomena that had to be accounted for theoretically. The spectra of luminous bodies can be classified into three main types:

1. Continuous Spectra. Electromagnetic radiation is emitted at all wavelengths. In general, hot solids, liquids, and dense gases emit continuous spectra. The spectral distribution of intensity versus wavelength will depend mainly on the temperature of the emitting body. Thus two different substances heated to the same temperature will have nearly identical spectra. As the temperature of a given body is increased, its spectral distribution changes; the peak of the distribution (the wavelength of maximum intensity) tends to shift to shorter and shorter wavelengths.

2. Bright-Line Spectra. Radiation is emitted only at specific wavelengths. (See Figure 14–3b.) Bright-line spectra are emitted only by gases at low density and pressure. Each element or compound emits a characteristic and unique set of wavelengths. As the temperature of a gas changes, its spectral lines change in intensity; some lines fade out completely and others appear. Closer inspection of an individual bright line shows that the emission is not truly at a single wavelength but is spread over a narrow band of wavelengths (Figure 14–5). The "width" of a spectral line is a measure of this spreading. It is found that an increase in pressure or temperature results in an increase in the line width.

3. Absorption (Dark-Line) Spectra. An absorption spectrum can be produced by passing a continuous spectrum through a low-density gas (Figure 14–4). The absorption lines that appear are characteristic of the particular gas and correspond to the lines seen in the bright-line spectrum of the gas. The properties of an absorption (dark) line are similar to those of an emission (bright) line; namely, dark lines vary in depth (intensity) and width according to the pressure and temperature. The depth of an absorption line also depends on the amount of gas between the source and the observer.

These are the major features of the three types of spectra. To explain the different kinds of spectra, we must identify and understand the processes by which bodies emit

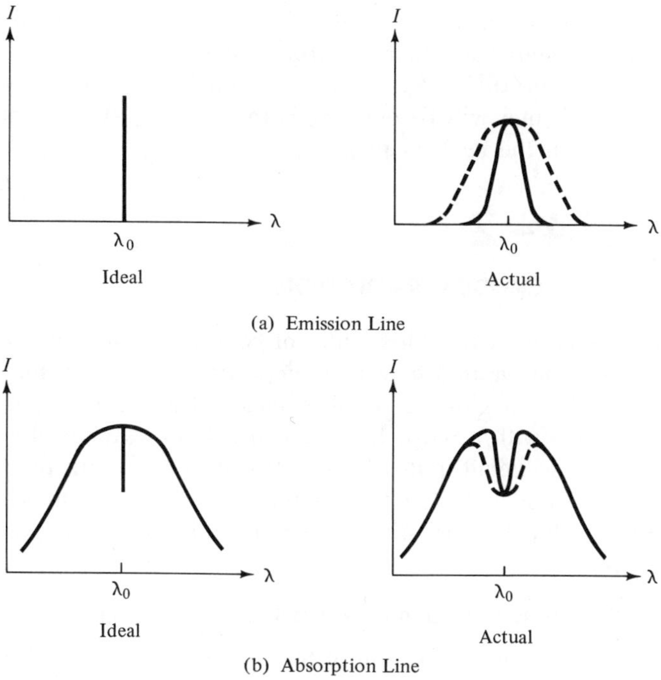

figure 14–5 THE WIDTH OF A SPECTRAL LINE

An ideal spectral line represents emission or absorption of a single wavelength, λ. Actual spectral lines involve the emission or absorption of light over a narrow range of wavelengths centered around λ. As the pressure or temperature increases, the width of a spectral line will broaden, as shown by the dashed lines.

and absorb light. The first question that arises is: Why is it that only low-density gases produce line spectra, whereas other states of matter emit continuous spectra? In attempting to answer this question, we note that in a low-density gas the atoms are widely separated and exert little influence on one another. It follows that in a gas light is being emitted and absorbed by *individual* atoms acting independently of each other. Therefore to understand line spectra, we should examine the characteristics of individual atoms. On the other hand, in solids, liquids, and dense gases the atoms are relatively close to one another, so that interatomic forces come into play. Under such conditions atoms do not act independently of one another; their behavior is strongly affected by the presence of neighboring atoms. Therefore the processes governing the emission and absorption of light that lead to the production of continuous spectra must depend more on the nature of the interatomic forces than on the individual atoms themselves.

In view of these considerations, we shall divide the study of spectra into two parts. To account for line spectra, we must understand the nature of the atom. We must explain why the atom emits or absorbs certain specific wavelengths and no others, and why these wavelengths differ from atom to atom. To account for continuous spectra, we must understand how light is emitted and absorbed by large numbers of atoms that are bound together by strong interatomic forces.

Historically, the problems of continuous spectra were solved first, so the rest of the chapter will be devoted to this subject. We will return to the question of line spectra in the next chapter.

14–2

BLACKBODY RADIATION

The most notable feature of continuous spectra is that the spectral distribution of the light emitted by a body depends primarily on the temperature of the emitting body. Thus a glowing piece of charcoal and a hot iron bar both at about 1000°F will glow with a red color and will exhibit virtually the same spectra, but the same iron bar will emit quite a different spectrum when its temperature is raised to 2000°F. Clearly, the first step toward understanding continuous spectra is to discover the relationship between the temperature of a body and the spectrum of radiation the body emits.

EMISSION AND ABSORPTION OF RADIATION

A simple application of the principle of conservation of energy tells us that the *emission* of radiation is a *cooling* process, and the *absorption* of radiation is a *heating* process. That is, in the absence of outside influences the temperature of a body will decrease as the body emits radiation and increase when the body absorbs radiation. It follows that for a body to maintain a constant temperature, it must absorb as much energy from its surroundings as it emits. Hence it is assumed that when a body attains thermal equilibrium with its surroundings, it does not cease to emit or absorb radiation; rather it has attained a state in which its rate of emission is exactly balanced by its rate of absorption. This principle is known as the ***law of exchanges*** and was first proposed by Pierre Prévost in 1809. The law of exchanges is assumed to hold for every wavelength; that is, when a body is at equilibrium, the rate of emission of radiation of any given wavelength is equal to the rate of absorption at the same wavelength. (It is now known that there are some exceptions to the law of exchanges, most notably the process of fluorescence, which is to be discussed in the next chapter. However, for most substances the law is essentially correct.)

Natural bodies can absorb, reflect, or transmit light. From the law of exchanges it follows that if a body at a given temperature reflects or transmits but does not absorb light of a particular wavelength, it will not emit that wavelength. Hence a perfect reflector or a perfectly transparent material (which reflects or transmits all wavelengths) will not emit *any* radiation. In nature, however, such materials do not exist. For example, glass, although transparent to visible light, is a good *absorber* of infrared radiation; therefore it must also be an efficient *emitter* of infrared.

Let us consider the perfect absorber. In order to be a perfect absorber, a material must absorb 100 percent of the radiation of all wavelengths incident upon it and must not reflect or transmit any radiation whatsoever. A body that does not reflect any visible light appears black to the eye; hence a perfect absorber is known as a ***blackbody.***

The temperature of a body is a function of its surroundings. If a cool body is exposed to radiation, it will absorb and heat up. As the body gets hotter, it will emit more and more radiation until an equilibrium is reached at some temperature at which the rates of emission and absorption are equal. A blackbody must necessarily absorb more radiation than any other body at the same temperature; therefore by the law of exchanges it must *emit* more radiation than any other body. Consequently, a blackbody—which is a perfect absorber—must also be a perfect emitter. *At a given temperature no body can emit more radiation at any wavelength than a blackbody.* In this respect the name blackbody is perhaps a bit confusing. Actually, a blackbody appears black only when it is cool; when it is hot enough to emit visible light, a blackbody will appear *brighter* than any other body at the same temperature!

Since a blackbody is the most efficient emitter at every wavelength, the spectrum of blackbody radiation gives at each wavelength the upper limit on the amount of radiation that could be emitted by any body at the same temperature. A blackbody absorbs and emits all wavelengths; therefore it must exhibit a continuous spectrum. Hence the blackbody radiation spectrum represents the ideal continuous spectrum for a body at a given temperature.

It would, of course, be extremely useful if the spectrum emitted by a blackbody at a given temperature could be measured experimentally. However, there are no truly black substances in nature; even black paint or black ink will reflect light to some extent. The best absorber known is lampblack (a fine soot produced by burning hydrocarbon compounds), which absorbs about 97 percent of any incident light. Thus the perfect blackbody is an idealized object whose properties can only be approximated by natural substances.

CAVITY RADIATION

Since there are no real blackbodies in nature, it would seem that an ideal blackbody spectrum could not be obtained. In 1859, however, Kirchhoff explained how blackbody radiation could be produced experimentally. He considered a completely enclosed and insulated cavity whose walls are maintained at a uniform temperature—in effect, an oven (Figure 14–6). The radiation emitted by the walls remains

figure 14–6 **CAVITY RADIATION**

The walls of the enclosed cavity are maintained at a constant temperature and emit radiation into the cavity. This radiation is reflected or absorbed and re-emitted by the walls. At equilibrium, the cavity is filled uniformly with radiation having the same spectrum as the radiation from a blackbody at the same temperature. A hole in one wall allows a minute amount of radiation to escape and be measured.

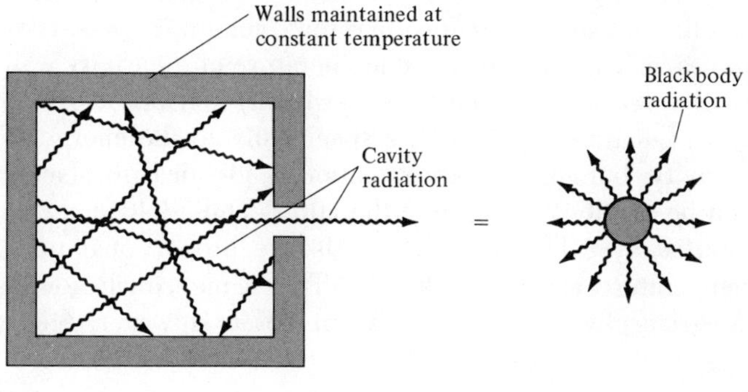

trapped in the cavity. Any radiation striking the walls will either be reflected back into the cavity or be absorbed and re-emitted by the walls.

Eventually, the cavity will become uniformly filled with electromagnetic radiation of all wavelengths. Through a series of theoretical arguments based on the laws of thermodynamics, Kirchhoff proved that when equilibrium is reached, *the radiation in the cavity will have the same spectral distribution as blackbody radiation.* That is, the cavity radiation for a given wall temperature has the same spectrum as the radiation from a blackbody at the same temperature. Moreover, Kirchhoff showed that the cavity would contain blackbody radiation *regardless of the size or shape of the cavity or of the chemical composition of the walls.* Even if the walls are made of different materials with different absorption and emission properties, the multiple reflection, absorption, and emission of radiation by the walls will produce the same net effect that would be produced if the walls themselves were blackbodies—hence at equilibrium, blackbody radiation will fill the cavity.

Thus through theoretical analysis Kirchhoff found a way to measure the spectrum of blackbody radiation. One has to build a well-insulated oven or kiln that can be brought to high temperatures. (Electric heating coils in the walls are one way to accomplish this.) When equilibrium is reached, the oven will be filled with blackbody radiation having a spectrum characteristic of the wall temperature. A small hole could be cut in the wall to allow a tiny portion of the radiation to escape. (A person who looks into a high-temperature oven or electric kiln can actually see the blackbody radiation in the form of a "glow" that fills the enclosure.) The spectral distribution of this radiation could then be measured. By repeating this process for different wall temperatures, one could determine the dependence of the blackbody radiation spectrum on temperature.

THE STEFAN–BOLTZMANN LAW

In Kirchhoff's time experimental techniques were not sufficiently advanced for this experiment to be performed. Nevertheless, scientists maintained a strong interest in the problem and continued to analyze blackbody radiation in theoretical terms. One especially interesting result was obtained by Ludwig Boltzmann in 1884. By combining the laws of electromagnetism with the principles of thermodynamics, Boltzmann was able to derive the following equation for the total energy of the electromagnetic radiation contained in a cavity:

$$u = \frac{E}{V} = aT^4 \tag{14–1}$$

In this equation E is the total electromagnetic energy of the radiation within the cavity, and V is the volume of the cavity. Thus E/V is the ***energy density*** u—the amount of energy per unit volume in a cavity, measured in ergs/cm^3 or joules/m^3. T is the absolute temperature of the cavity walls, and a is a constant whose CGS value (determined by experiment) is 7.56×10^{-15} erg/cm^3-°K^4.

Equation 14–1 applies specifically to the energy of the radiation contained in a cavity, but since cavity radiation is identical to blackbody radiation, this equation can be modified to apply to the latter. Thus Boltzmann proposed that the total energy emitted by a blackbody must also be proportional to T^4, where T is the absolute temperature of the blackbody. This same conclusion had been arrived at by the Austrian physicist Josef Stefan in 1879. However, Stefan's discovery seems to have been one of those "discoveries that should not have been made" that occasionally appear in the annals of physics. According to science historian Max Jammer,

> Having read in a textbook on heat about John Tyndall's experiments according to which the total emission of a heated platinum wire at 1200°C (1473°K) was 11.7 times that at 525°C (798°K) and realizing that $(1473/798)^4$ is approximately 11.7, Stefan concluded that the total radiation E is proportional to T^4. . . . Today we know that Stefan's inference, based after all on relatively meager experimental evidence, was to a high degree fortuitous. A modern repetition of Tyndall's experiment—Tyndall measured a radiation which was far from that of a blackbody—would yield a ratio of 18.6 rather than 11.7.[1]

Nevertheless, Stefan has been credited as a co-discoverer of what is now known as the ***Stefan–Boltzmann law*** for the rate of energy emission of a blackbody:

$$P = \frac{E}{A} = \sigma T^4 \tag{14–2}$$

In this equation E is the rate of energy emission (in ergs/sec in CGS units), and A is the surface area of the body. P is the ***emissive power,*** defined as the rate of energy emission per unit surface area. (In CGS units P is measured in ergs/sec-cm^2.) The constant σ (Greek sigma) is called the ***Stefan–Boltzmann constant*** and has a numerical value in CGS units of 5.67×10^{-5} ergs/cm^2-sec-°K^4.

Equation 14–2 is exactly correct only for ideal blackbodies and is used as an approximation for bodies that approach blackbody behavior. Note that this equation predicts a rather sharp rise in emissive power with increasing temperature. For instance, if the absolute temperature of a blackbody is doubled, the emissive power increases by $(2)^4 = 16$ times. Notice that the Stefan–Boltzmann law predicts that *all* bodies radiate some energy as long as they are not at a temperature of absolute zero.

WIEN'S RADIATION FORMULA

The Stefan–Boltzmann law enables us to calculate the *total* energy of the radiation contained in a cavity or emitted by a blackbody at a given temperature, but it gives no information about the *spectrum* of this radiation. It is considerably more useful to know the energy density at a single wavelength. This quantity is known as the ***monochromatic energy density*** u_λ (or monochromatic emissive power P_λ) for any given wavelength λ.* Taken together, the different values of u_λ for different wavelengths give the spectrum of the cavity radiation at a given temperature.

Many scientists recognized the value of having an equation for u_λ as a function of the wavelength and the temperature, for such an equation would represent a mathematical formula for the spectrum of blackbody radiation. As a first step, the German physicist Wilhelm Wien made an educated guess at a suitable formula. In 1896 Wien proposed the following equation for u_λ:

$$u_\lambda = \frac{b_1}{\lambda^5 e^{b_2/\lambda T}} \tag{14–3}$$

*Strictly speaking, one cannot measure the radiation being emitted at a single wavelength. Thus u_λ is more precisely defined as the energy density per unit wavelength interval—that is, the energy density of the radiation emitted in the wavelength range from λ to $\lambda + 1$.

where T is the absolute temperature of the cavity walls; $e = 2.718$ (the base of natural logarithms); and b_1 and b_2 are constants of undetermined value. Wien did not derive this formula rigorously, but rather pieced it together from several theoretical arguments. He assumed that the atoms in the walls of the cavity that are responsible for the absorption and emission of the radiation are in random motion, and that their speeds are distributed according to the Maxwell–Boltzmann distribution. (The $e^{b_2/\lambda T}$ term appearing in Wien's formula is a modification of the $e^{-mv^2/2kT}$ term found in the Maxwell–Boltzmann distribution formula—see Figure 7–14). Another consideration that Wien used to determine the mathematical form of the equation was that if the *total* energy density is calculated (by adding up the u_λs for every emitted wavelength), the result should be in agreement with the Stefan–Boltzmann law. Wien hoped to obtain numerical values of b_1 and b_2 by matching his formula against experimental results.

Wien's formula had little value in the absence of any experimental data to compare it with, so at Wien's suggestion a research group was organized in Berlin with the aim of making careful experimental measurements of cavity radiation. Techniques were developed for the construction of an oven whose walls could be maintained at constant temperatures, and instruments were devised to measure the intensity of the contained radiation at different wavelengths. The first results, announced in 1897 by Otto Lummer and Ernst Pringsheim, seemed to confirm Wien's formula rather nicely. When appropriate values were selected for the constants b_1 and b_2, Wien's formula could be made to fit closely to the Lummer–Pringsheim data, as shown in Figure 14–7.

When we examine the graphs shown in Figure 14–7 of the cavity radiation spectrum at different temperatures, we observe that there is a peak in each spectrum.

figure 14–7 **WIEN'S FORMULA FOR THE SPECTRUM OF BLACKBODY RADIATION**

The solid lines represent Wien's formula as fitted to the early data (dots) of Lummer and Pringsheim. The wavelength of maximum energy emission λ_{max} for each temperature obeys the Wien displacement law.

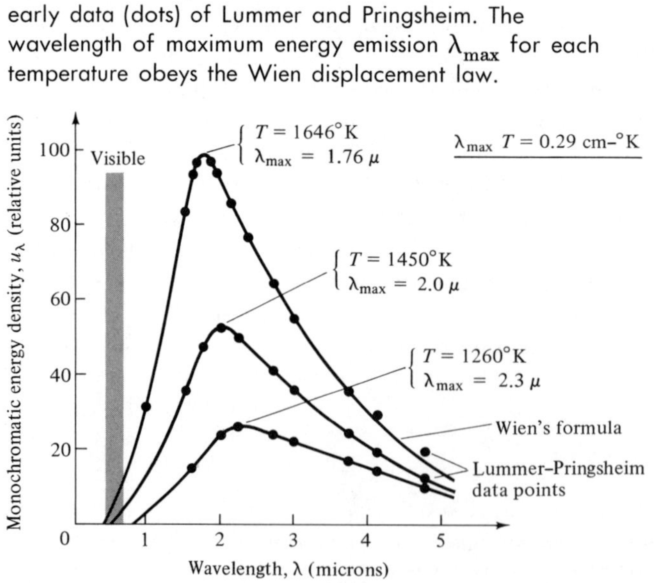

This peak represents the wavelength at which the energy density is most intense. If we compare the radiation curves for different temperatures, we find that the wavelength of maximum intensity λ_{max} tends to decrease as the temperature increases. This is in accordance with our basic observations. When a body is heated to incandescence, it first glows red (the color of the longest wavelength in the visible spectrum). As the temperature of the body is raised, its color changes to orange, yellow, and so on, corresponding to a shift in peak intensity to light of shorter wavelengths.

As a corollary to Wien's radiation formula, it can be shown that λ_{max} and the absolute temperature are inversely proportional; that is, $\lambda_{max}T$ = constant. The data of Lummer and Pringsheim verified this relationship and gave a numerical value for the constant. Thus

$$\lambda_{max}T = 0.29 \text{ cm-°K} \tag{14–4}$$

This equation is known as the ***Wien displacement law.*** It states that as the absolute temperature of a blackbody increases, the wavelength of peak energy emission is displaced toward shorter wavelengths. (λ_{max} decreases.)

Equation 14–4 is extremely useful for determining the temperature of a radiating body from its spectrum. If the spectrum is approximately that of a blackbody, the wavelength of maximum intensity can be determined and the temperature calculated directly. The Wien displacement law is particularly valuable in astronomy, where it can be used to estimate the surface temperatures of stars.

EXAMPLE 14–1: The solar spectrum has the approximate shape of a blackbody spectrum. The peak intensity (see Figure 14–1) is found at a wavelength of 4700 Å. Calculate the surface temperature of the sun.

SOLUTION: The Wien displacement law can be used. Solved for T, Equation 14–4 is

$$T = \frac{0.29}{\lambda_{max}}$$

with λ_{max} in cm and T in °K. Thus

$$\lambda_{max} = 4700 \text{ Å} = 4700 \times 10^{-8} \text{ cm} = 4.7 \times 10^{-5} \text{ cm}$$

and

$$T = \frac{0.29}{4.7 \times 10^{-5}} = \underline{6200\text{°K}}$$

THE FAILURE OF CLASSICAL THEORY

Wien's success was short-lived. Lummer and Pringsheim's first measurements were made over a fairly narrow range of relatively short infrared wavelengths (from 1 to 5 microns). As Lummer and Pringsheim extended their measurements to longer wavelengths, they found systematic discrepancies between their measured values of energy emission and Wien's radiation formula. By 1900 they had become quite convinced that the observed spectrum of blackbody radiation could not be represented by Wien's formula. The formula could be made to fit the data for shorter wavelengths up to and beyond the peak wavelength (so that the Wien displacement law remained

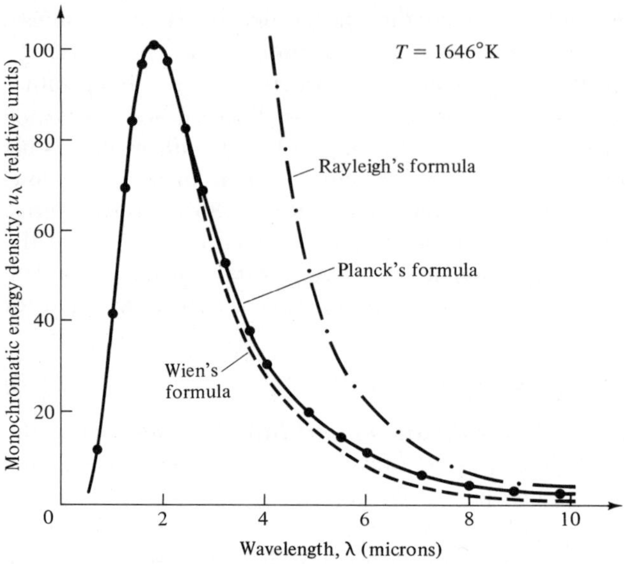

figure 14–8 **COMPARISON OF RADIATION FORMULAS**

The formulas proposed by Wien, Rayleigh, and Planck for the energy density of cavity radiation are compared with the experimental data. The dots represent the data obtained by Lummer, Pringsheim, Rubens, and Kurlbaum.

valid), but at longer wavelengths the values predicted by Wien's formula were consistently lower than the experimental values, as shown in Figure 14–8.

At the same time a new formula for the spectrum of blackbody radiation was advanced by John William Strutt, an English physicist and nobleman more familiarly known as Lord Rayleigh. While recognizing the partial success of Wien's formula, Rayleigh pointed out that it was not much more than an empirical formula with virtually no theoretical foundation. He therefore decided to ignore the available data and derive a formula for blackbody radiation that followed rigorously from the laws of physics.

Starting with Maxwell's equations, Rayleigh showed that at equilibrium any electromagnetic radiation in a cavity would exist as standing waves. The situation is somewhat analogous to the problem of standing mechanical waves on a string (discussed in Chapter 8), except that the wavelengths are determined by the dimensions of the cavity instead of the length of the string. For a string of length L, the allowed wavelengths for standing waves are given by the equation (8–6a)

$$\lambda_n = \frac{2L}{n} \qquad n = 1, 2, 3, 4, 5, \ldots$$

A similar (though somewhat more complicated) equation can be obtained for the allowed wavelengths of the standing electromagnetic waves in a three-dimensional cavity at equilibrium. We see in Equation 8–6a that as the integer n takes on larger values, the wavelength gets shorter and shorter; hence there will be a limited number of waves of long wavelength and a very large number of waves of short wavelength in the cavity.

Rayleigh next applied the principle of ***equipartition of energy,*** which states (in simplified form) that at equilibrium the available energy of a system will be equally distributed among all parts of the system. The principle of equipartition of energy can be derived as a direct consequence of Newtonian mechanics and is a fundamental concept of the kinetic theory. Rayleigh assumed that the equipartition principle applies to electromagnetic radiation as well; therefore all standing waves must have the same average energy, regardless of wavelength. Taking this assumption into account, he arrived at the following formula for the monochromatic energy density of cavity radiation:

$$u_\lambda = \frac{8\pi kT}{\lambda^4} \tag{14–5}$$

where k is the Boltzmann constant from the kinetic theory ($k = 1.38 \times 10^{-16}$ erg/°C). Rayleigh's formula* is shown graphically in Figure 14–8. Obviously, the formula does not agree with the data except at very long wavelengths. It predicts that the greatest amount of energy is emitted at the shortest wavelengths. Therefore any emitting body ought to give off intense amounts of x rays and ultraviolet, regardless of temperature. This is a consequence of the assumption that all waves have the same average energy; if there is a large number of short wavelengths (as predicted by Maxwell's equations), the short wavelength end of the spectrum must contain most of the energy! Furthermore, Rayleigh's formula contradicts both the Wien displacement law (since there is no maximum in the curve) and the Stefan–Boltzmann law. (The total energy density, according to Rayleigh's formula, would be infinite!) Yet both of these laws had been experimentally verified.

Rayleigh's point in deriving an obviously incorrect formula was to show that attempts to explain blackbody radiation represented a *failure of classical physics,* since Rayleigh's formula was derived rigorously from the classical electromagnetic and kinetic theories. This meant that either one (or both) of these well-established theories was partially incorrect—or at least that the two theories were incompatible with one another.

Thus at the beginning of the twentieth century, physicists were faced with two serious difficulties: the failure to account for the null result of the Michelson–Morley experiment and the failure to explain the spectrum of blackbody radiation. As we shall see, equally radical solutions were required to resolve both of these problems.

14–3

THE QUANTUM HYPOTHESIS

The key to the puzzle of blackbody radiation was provided by a scholarly physicist who was so thoroughly grounded in the classical tradition that he could not bring himself to accept his own conclusions. Max Planck, a member of a well-to-do Prussian family, studied physics under Kirchhoff, Helmholtz, and Clausius, and was an expert

*In his original work Rayleigh made a slight mathematical error, which was corrected by James Jeans in 1905. The equation given here is the correct one and is sometimes known as the Rayleigh–Jeans formula.

on thermodynamics. After several years as a professor at the University of Kiel, he was invited in 1889 to the University of Berlin to succeed to the professorship vacated by the death of Kirchhoff.

Planck's first success with the problem of blackbody radiation came when he managed to derive Wien's formula from thermodynamic principles. Naturally, Planck was disturbed when Wien's formula later failed to agree with the Lummer–Pringsheim data. Then in October 1900 Planck was informed by his two colleagues, Heinrich Rubens and Ferdinand Kurlbaum, that their experimental findings had confirmed the failure of Wien's formula and seemed to agree more closely with Rayleigh's law at long wavelengths. Planck set out to find a formula that would fit the new data. As the result of what he admitted was a "lucky guess," Planck made a simple modification of Wien's formula and obtained an equation that fit the Lummer–Pringsheim and Rubens–Kurlbaum data with considerable precision. (See Figure 14–8.) Planck's formula was

$$u_\lambda = \frac{b_1}{\lambda^5(e^{b_2/\lambda T} - 1)} \tag{14–6}$$

where b_1 and b_2 are the unknown constants whose values are chosen to fit the data. Despite his success, Planck was not satisfied at having merely guessed the correct formula, so he began immediately to seek a physical explanation for it. After less than two months of intense work, Planck arrived at a theoretical justification for his formula. He presented his findings at a meeting of the German Physical Society in December 1900.

Planck began with the assumption that the atoms in the cavity walls are fixed in place by the interatomic forces that exist in solid matter, and the atoms vibrate about their equilibrium positions. The vibration frequency for an atom depends on its mass, the interatomic forces, and so on, and is a fixed quantity. The addition of energy to an atom will cause the atom to vibrate with a greater amplitude, but at the same frequency (just as the frequency of oscillation of a pendulum is independent of the amplitude). It is expected that there will be a statistical distribution of vibration frequencies among all the atoms in the walls, so that for any given frequency there will be a large number of atoms vibrating at that frequency. Planck further assumed that the vibrating atoms contain electric charges. Maxwell's electromagnetic theory predicts that vibrating charges should emit or absorb electromagnetic radiation of the same frequency as the vibration frequency. Thus at equilibrium there will be a steady interchange of energy between the wall atoms that are vibrating at a given frequency and the electromagnetic waves of the same frequency in the cavity.

Planck's aim was to show that his successful formula for the spectrum of cavity radiation (Equation 14–6) could be derived from this theoretical model. His task was to find a mathematical relationship that described the interchange of energy between the vibrating atoms and the electromagnetic radiation emitted and absorbed by these atoms. Planck believed that this problem could be solved by the established methods of statistical mechanics. Up to this point Planck's assumptions were quite consistent with the classical concepts of electromagnetic theory, thermodynamics, and the kinetic theory, but now Planck discovered that *he could not derive his formula without violating classical theory.*

Planck had to assume that vibrating atoms can change their energies only by "jumps"; that is, by emitting or absorbing discrete units or ***quanta*** of energy. Planck found it necessary to assert that *a vibrating atom of frequency f can change its energy only by emitting or absorbing one or more quanta of energy such that*

$$E = hf \tag{14-7}$$

where h is a fundamental physical constant—now known as ***Planck's constant***—whose value is determined by comparison with experimental data. The value is $h = 6.63 \times 10^{-27}$ erg-sec.* Hence according to Planck, a vibrating atom can change its energy by hf, $2hf$, $3hf$, . . . , but not by any values in between.

Planck's "quantum hypothesis" was in violation of several precepts of classical theory. According to Maxwell's electromagnetic theory, a vibrating electric charge should *continually* emit electromagnetic radiation and can emit or absorb any arbitrarily small amount of radiation. Planck's hypothesis requires, however, that a vibrator emit or absorb a *minimum* of one quantum of energy at a time. Thus a vibrating atom must emit or absorb radiation in spurts. The relationship $E = hf$ between the quantum of energy and the frequency of the vibrator is also a break with tradition. In classical physics the amplitude of a simple vibration is entirely unrelated to the frequency. According to Planck, the size of the quantum of energy absorbed or emitted depends on the vibrator frequency, so that high-frequency vibrators must emit or absorb larger quanta of energy than low-frequency vibrators. Because of this, the high-frequency vibrators will be less likely to absorb or emit radiation and will consequently have a smaller average energy than the low-frequency vibrators. Hence the principle of equipartition of energy is also contradicted.

Planck, being a generally conservative thinker, was rather reluctant to break with classical physics, but he found that *if he did not accept the quantum hypothesis, he would end up with Rayleigh's formula.* That is, Rayleigh's formula follows as a necessary consequence of the assumption that vibrators can absorb or emit unrestricted amounts of radiation, rather than quanta. On the other hand, Planck's formula—which alone agrees with the experimental data—can be obtained only from the quantum hypothesis.

THE PLANCK RADIATION EQUATION

By assuming the quantum hypothesis, Planck was able to derive the following equation for the monochromatic energy density of cavity radiation:

$$u_\lambda = \frac{8\pi ch}{\lambda^5(e^{ch/k\lambda T} - 1)} \tag{14-8}$$

which is identical in form to Equation 14–6 except that the arbitrary constants b_1 and b_2 are now related to more fundamental physical constants ($b_1 = 8\pi ch$ and $b_2 = ch/k$). Since the values of b_1 and b_2 can be determined by experiment, a numerical value of h can be readily obtained. (The speed of light c appears in the equation because of the relationship $c = f\lambda$ between the frequency and wavelength of electromagnetic waves.) Both the Stefan–Boltzmann law (Equation 14–2) and the Wien displacement law (Equation 14–4) can be derived mathematically from Planck's formula, and the constants that appear in these equations can also be related to fundamental physical constants. (For example, it can be shown that the Stefan–Boltzmann constant σ is related to more fundamental constants by the equation $\sigma = 2\pi^5k^4/15c^2h^3$.)

*The most precise value obtained as of 1970 is $h = 6.626196 \pm 0.000050 \times 10^{-27}$ erg-sec.

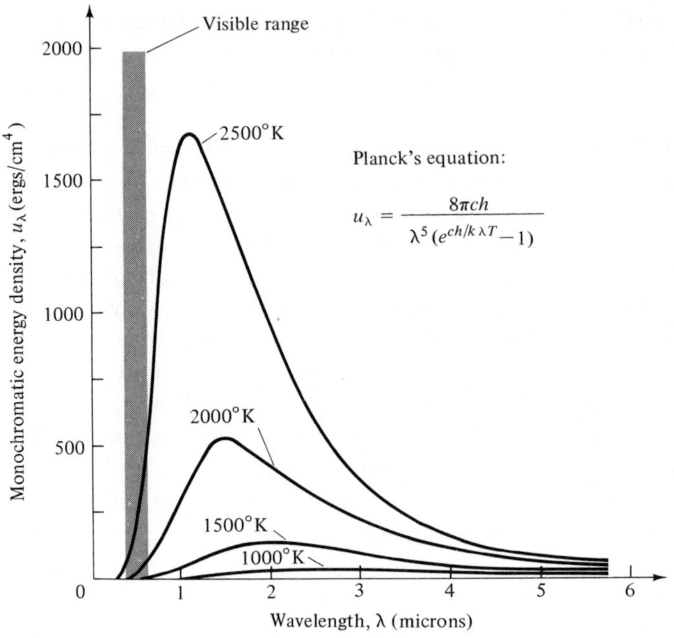

figure 14–9 **PLANCK'S RADIATION EQUATION**

Each curve shows the predicted spectrum of cavity radiation for the given temperature.

Further mathematical analysis shows that for very short wavelengths (high frequencies), Planck's equation reduces to Wien's formula. (See Figure 14–8.) Wien's formula fails at long wavelengths because it incorrectly implies that vibrators can absorb or emit any small amount of energy, and Rayleigh's formula fails at short wavelengths because it incorrectly assumes that high-frequency vibrations have the same capacity to absorb energy as low-frequency ones. Figure 14–9 shows the spectral distribution predicted by Planck's equation for a number of temperatures.

Thus the problem of blackbody radiation was apparently solved. Yet Planck's feelings about his success were rather ambivalent. He once confided to his son that he thought he had made "a discovery as important as the discoveries of Newton," but mainly Planck looked on the quantum hypothesis as "an act of desperation" that was required as a last resort to give physical meaning to his blackbody radiation formula. For many years afterward Planck labored to find a different explanation more consistent with classical physics, but he was not successful. Planck apparently suffered from the same "failure of nerve" that had afflicted Copernicus. Both were too rooted in the established physics of their times to be completely convinced of the truth of their discoveries, but just as Copernicus had his Galileo, Planck had his Einstein.*

*In fairness to Planck, we should note that he made several significant contributions to the theory of relativity and gave considerable encouragement to Einstein when the latter was still an unknown.

14–4

THE PHOTON THEORY OF LIGHT

Planck's quantum hypothesis received scant attention for several years after it was proposed. Physicists who were aware of Planck's proposal looked on it as merely a temporary hypothesis that served only to account for blackbody radiation and nothing else. It was a much too radical departure from classical physics to be taken seriously on the basis of its success with blackbody radiation alone. Planck himself was in almost complete agreement with this attitude.

Early in 1905, several months before the theory of relativity was proposed, Albert Einstein published a paper that not only supported Planck's hypothesis, but went far beyond it. According to Planck, electromagnetic radiation is continuous (it can consist of any arbitrarily small amount of energy), but it can be absorbed or emitted by atoms only in discrete amounts (quanta). Einstein considered this picture to be fundamentally inconsistent, so he adopted the point of view that if atoms can absorb or emit light only in the form of quanta of energy, *light itself must consist of quanta of energy.* To support this argument, Einstein showed that certain thermodynamic properties of cavity radiation are the same as those of an ideal gas made up of independent particles. He therefore proposed the hypothesis that *light consists of particles or light quanta,* each of which has a fixed amount of energy. The energy of a quantum of light is given by Planck's equation

$$E = hf$$

where f is the frequency of the light, and h is Planck's constant. Many years later the word "photon" was suggested as a name for Einstein's light quantum.* Hence the theory is often referred to as the ***photon theory of light.*** According to the photon theory, light energy is concentrated in individual photons, which travel through space with a speed c. Consequently, light propagates as a stream of particles rather than as an electromagnetic wave.

Einstein's proposal was, of course, extremely revolutionary, but then Einstein was certainly not one to be bound by the traditions of classical physics. In effect, the photon theory represented a throwback to Newton's corpuscular theory of light. It flew in the face of one hundred years of accumulated evidence that light behaves like a wave. How could Einstein ignore all the wave phenomena associated with light? Einstein replied that he was not denying the existence of these phenomena; he was simply stating that under certain conditions light exhibits the properties of a stream of particles, and under other circumstances light exhibits properties normally associated with the behavior of mechanical waves.

PROPERTIES OF PHOTONS

The equation $E = hf$ expresses the dual nature of light. This equation relates the energy of a photon to its frequency. But if a photon is an independent, single particle, how can it have a "frequency"? One should not have the mistaken impression

* This name was proposed in 1926 by G. N. Lewis on the ground that it would fit the pattern of "electron" and "proton"—at that time considered to be the fundamental constituents of all matter.

that a photon "rides along" on a wave, oscillating up and down as it advances through space; this is not the picture intended by the theory. The equation is best interpreted in the following manner. Light of a specific type (say blue light or infrared or x rays) exhibits both wave and particle characteristics. When the light behaves like a wave, it can be characterized by a particular frequency, f (which can be measured from its interference pattern). When the same type of light behaves like a particle, it appears to consist of discrete quanta of energy E. The equation $E = hf$ relates these two aspects of light; thus the energy of a photon of a particular kind of light is proportional to the frequency that characterizes the wave properties of the light.

According to the wave theory, different kinds of light are distinguished by having different frequencies. According to the photon theory, different kinds of light are distinguished by the fact that their quanta have different energies. Since blue light has a larger frequency than red light, a photon of blue light will have a correspondingly larger energy than a photon of red light. To illustrate this, let us compare blue light of wavelength 4800 Å with red light of wavelength 6500 Å. First we calculate the frequencies using the wave equation $c = f\lambda$:

Red: $\lambda = 6500\ \text{Å} = 6500 \times 10^{-8}\ \text{cm} = 6.5 \times 10^{-5}\ \text{cm}$

$$f = \frac{c}{\lambda} = \frac{3 \times 10^{10}\ \text{cm/sec}}{6.5 \times 10^{-5}\ \text{cm}} = \underline{4.62 \times 10^{14}\ \text{waves/sec}}$$

Blue: $\lambda = 4800\ \text{Å} = 4800 \times 10^{-8}\ \text{cm} = 4.8 \times 10^{-5}\ \text{cm}$

$$f = \frac{c}{\lambda} = \frac{3 \times 10^{10}\ \text{cm/sec}}{4.8 \times 10^{-5}\ \text{cm}} = \underline{6.25 \times 10^{14}\ \text{waves/sec}}$$

Now we calculate the energy of a photon of each:

Red: $E = hf = (6.63 \times 10^{-27}\ \text{erg-sec})(4.62 \times 10^{14}\ \text{waves/sec}) = \underline{3.06 \times 10^{-12}\ \text{erg}}$

Blue: $E = hf = (6.63 \times 10^{-27}\ \text{erg-sec})(6.25 \times 10^{14}\ \text{waves/sec}) = \underline{4.14 \times 10^{-12}\ \text{erg}}$

Hence red light consists of a stream of identical photons each having an energy of about 3×10^{-12} erg. Similarly, blue light consists of a stream of identical photons each having an energy of about 4×10^{-12} erg. The physiological response by which we sense color therefore depends on the size of the individual energy quanta that the eye receives. The *intensity* of a beam of light depends on the *number* of photons in the beam; the more photons of a particular kind that are received by the eye, the brighter the light will appear.

The modern-day theory pictures a photon as a particle that possesses energy and momentum,* but not mass. This concept of a massless particle is sometimes difficult for a beginning student to grasp, but the laws of physics do *not* require that a particle must have mass in order to exist. Remember that mass is a measure of the *inertia* of a body and not of the amount of substance. Hence a massless particle is simply one to which the principle of inertia does not apply. A photon certainly fits this description, since it cannot be accelerated or decelerated by normal forces.

It will be useful at this point to introduce two computational aids. In the realm

*According to Maxwell's theory, light possesses momentum $p = E/c$, where E is the energy of the electromagnetic wave. Therefore the momentum of a photon is given by the equation $p = E/c = hf/c$.

of atomic physics, the quantities of energy that are encountered are generally very small (a quantum of visible light, for example, has an energy of the order of 10^{-12} erg), and the standard units, the erg and the joule, are much too large and unwieldy. Therefore physicists have found it convenient to define a new unit of energy, the ***electron volt*** (abbreviated eV). The electron volt is defined as the amount of energy gained by an electron when it is accelerated by a potential difference of 1 volt. It can be shown that

$$1 \text{ eV} = 1.602 \times 10^{-19} \text{ joule} = 1.602 \times 10^{-12} \text{ erg}$$

Therefore to convert from ergs to eV, we divide as follows:

$$\text{energy (in eV)} = \frac{\text{energy (in ergs)}}{1.602 \times 10^{-12}}$$

The second computational aid is a convenient modification of the equation $E = hf$. It is customary to give a photon's wavelength in angstroms and its energy in electron volts. We will now change the equation into a form in which the wavelength in angstroms and the energy in eV can be directly related, taking into account the appropriate constants and conversion factors. We substitute for the frequency in terms of wavelength:

$$E = hf = \frac{hc}{\lambda}$$

If the wavelength is in angstroms, then

$$E = \frac{hc}{\lambda} = \frac{(6.626 \times 10^{-27} \text{ erg-sec})(2.998 \times 10^{10} \text{ cm/sec})}{\lambda(\text{in Å}) \times 10^{-8} \text{ cm}} = \frac{1.986 \times 10^{-8} \text{ erg}}{\lambda(\text{in Å})}$$

To obtain the energy in eV, we write

$$E(\text{in eV}) = \frac{1.986 \times 10^{-8} \text{ erg}}{\lambda(\text{in Å}) \times 1.602 \times 10^{-12} \text{ erg/eV}}$$

or

$$E(\text{in eV}) = \frac{12{,}400}{\lambda(\text{in Å})} \tag{14–9}$$

This equation allows us to calculate the energy in eV directly from the wavelength in angstroms or vice versa.

EXAMPLE 14–2: Calculate the energy in eV of a photon of (a) red light of 6500 Å wavelength and (b) blue light of 4800 Å wavelength.

SOLUTION: Substituting directly into Equation 14–9, we obtain

(a) $E = \dfrac{12{,}400}{6500} = \underline{1.91 \text{ eV}}$ (b) $E = \dfrac{12{,}400}{4800} = \underline{2.58 \text{ eV}}$

THE PHOTOELECTRIC EFFECT

To strengthen his case for the particle theory of light, Einstein used the theory to account for an extremely puzzling phenomenon known as the ***photoelectric effect,*** which could not be explained by the electromagnetic wave theory of light. The photoelectric effect was accidentally discovered in 1887 by Heinrich Hertz in the course of his research on the properties of electromagnetic waves. Hertz observed that ultraviolet light striking a metallic surface caused the surface to acquire a positive charge. Hertz did not pursue the matter further, but several other physicists studied the effect in more detail.

The photoelectric effect was investigated by J. J. Thomson in 1899. Having learned that ultraviolet light causes negatively charged particles to be ejected from certain metals, Thomson measured the charge-to-mass ratio of these particles and found that it was the same as that of the electron—which he had identified as a particle only two years before. However, the most substantial research on the photoelectric effect was done by Philipp Lenard.* In 1900 he reported that although ultraviolet light was necessary to eject electrons from most metals, visible light could also eject electrons from certain metals (among them sodium, potassium, cesium, and barium). Hence the photoelectric effect involves the ejection of electrons from metals by light.

Over the next few years Lenard and others made extensive studies of the photoelectric effect, relating the intensity and frequency of the incident light and the type of metal used to the energy of the ejected electrons. The major observations can be summarized as follows (Figure 14–10):

figure 14–10 THE PHOTOELECTRIC EFFECT

Light striking a metal plate causes electrons to be ejected from the plate. The graphs show (a) the relation between the frequency of the incident light and the maximum kinetic energy of the ejected electrons and (b) the relation between the intensity of the light and the number of electrons ejected.

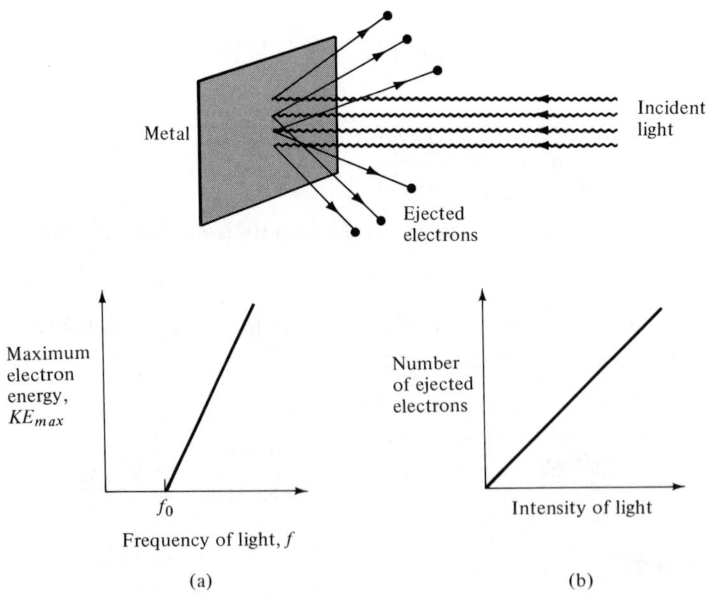

*This is the same Lenard who later espoused Hitler's racial policies and led the attack against Einstein and "Jewish physics." It is ironic that Lenard's research was so instrumental in Einstein's success.

(1) If the frequency of the incident light is less than a specific ***threshold frequency*** f_0, no electrons will be ejected; that is, the photoelectric effect will not occur. The threshold frequency is different for different metals.
(2) As the *frequency* of the incident light is increased, the *maximum kinetic energy* of the ejected electrons increases in proportion (Figure 14–10a). However, the number of ejected electrons remains constant.
(3) As the *intensity* of the incident light is increased, the *number* of ejected electrons increases proportionally (Figure 14–10b). However, the energies of the electrons remain constant.
(4) Electrons are ejected *instantaneously;* that is, there is no time delay between the light striking the metal and the appearance of electrons.

The photoelectric effect has many practical uses. It is the basis for the photoelectric photometer or "light meter," in which the current of ejected electrons is measured to determine the intensity of the incident light (Figure 14–11). The most important fact about the photoelectric effect, however, is that *it cannot be explained by the electromagnetic theory of light.*

The electrons ejected by the incident light are presumably close to the surface of the metal. These electrons are normally bound to the metal by the electric attraction of the positive charges in the metal. In order to escape, an electron must therefore absorb sufficient energy from the incident light to overcome the attractive force that binds it to the metal. According to the electromagnetic theory, the energies of the ejected electrons should depend on the *intensity* (related to the wave amplitude) of the light. But the electron energy is found to be independent of the intensity. The electromagnetic theory is unable to explain why the electron energy should be directly proportional to the *frequency*, rather than the intensity, of the light.

The theory cannot explain why there is a threshold frequency; light of any frequency ought to be able to eject electrons. Finally, the wave theory cannot explain why electrons are ejected instantaneously. According to the theory, the energy of the incident wave front should be distributed uniformly among all the electrons. For

figure 14–11 **THE PHOTOELECTRIC CELL**

Ejected electrons are attracted to the positively charged anode, completing the circuit. When the light is turned off, no current will flow in the circuit. The current is proportional to the intensity of the incident light.

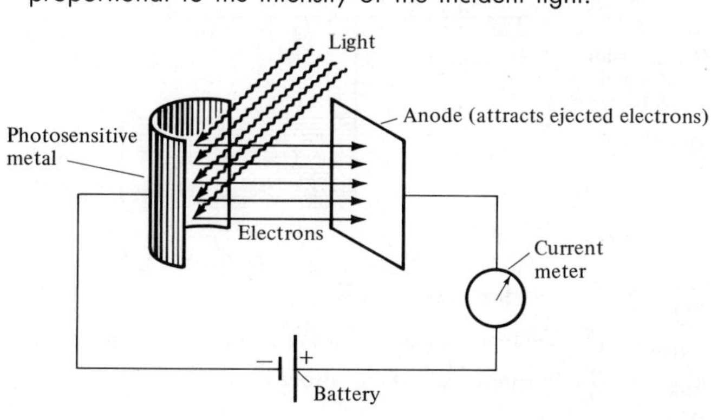

light of low intensity, there should be a considerable time lag before any single electron absorbs sufficient energy from the incoming waves to be able to escape. Yet observations show that ejection is instantaneous even for light of the lowest possible intensity (provided it is above the threshold frequency). Hence the predictions of the wave theory are all contradicted by the observations. Classical electromagnetic theory therefore fails completely with respect to the photoelectric effect.

Now we turn to Einstein's explanation. According to the particle theory, light consists of a stream of photons, each one possessing an energy proportional to its frequency ($E_{\text{phot}} = hf$). Einstein assumed that ejection results from a direct interaction between a single photon and a single electron, in which the photon is completely absorbed and its entire energy transferred to the electron. The electrons in the metal will have a range of different energies, so that each electron will need to absorb a different amount of energy in order to escape from the metal. Let W be the energy required to eject the most energetic electron. That is, W is the *minimum* amount of energy required to remove an electron. (See Figure 14–12.) If the energy of the incident photon is less than W ($E_{\text{phot}} < W$), no electrons will escape. If the photon energy is greater than W ($E_{\text{phot}} > W$), electrons will be ejected, and any excess energy they possess will appear as their kinetic energy of motion as free electrons. The electrons that require the minimum energy to escape (presumably those closest to the surface) will have the greatest kinetic energy. For these electrons

$$KE_{\text{max}} = E_{\text{phot}} - W \tag{14–10a}$$

In this equation KE_{max} is the maximum kinetic energy of an ejected electron and E_{phot} is the energy of the absorbed photon. W, which is commonly known as the ***work function,*** is a constant whose value depends on the type of metal used. (See Table 14–1.) All of the features of the photoelectric effect can now be readily explained.

figure 14–12 THE PHOTOELECTRIC EFFECT

Einstein's explanation in terms of the photon theory is shown in the form of an energy diagram.

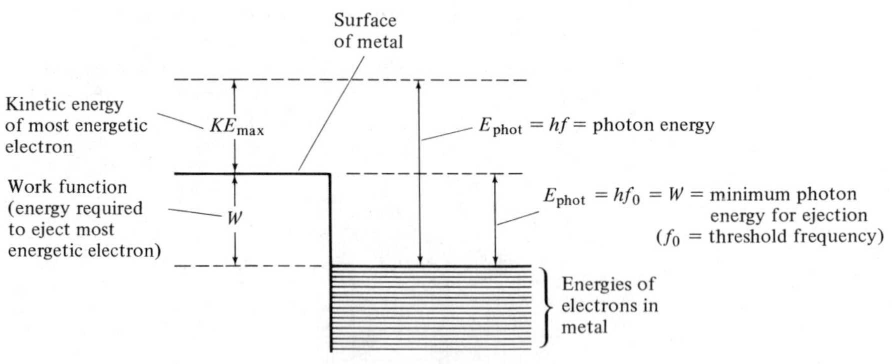

$$KE_{\text{max}} = E_{\text{phot}} - W$$

$E_{\text{phot}} < W$: no electrons are ejected.

$E_{\text{phot}} > W$: electrons ejected; excess energy becomes kinetic energy

$E_{\text{phot}} = W$: threshold at which effect begins

table 14–1 **WORK FUNCTIONS OF SOME COMMON METALS**

Metal	W (eV)	Metal	W (eV)
Aluminum	4.1	Platinum	5.4
Calcium	2.7	Potassium	2.3
Cesium	1.9	Rubidium	2.2
Copper	4.3	Silver	4.6
Gold	4.8	Sodium	2.3
Lithium	2.3	Titanium	4.1
Nickel	5.0	Tungsten	4.5

(1) A threshold frequency f_0 exists because the incident photon must supply at least enough energy to equal the work function of the metal (i.e., $E_{phot} \geq W$). The threshold for the effect is at $E_{phot} = hf_0 = W$. Therefore the threshold frequency is

$$f_0 = W/h$$

Hence if the frequency f of the incident light is less than f_0, the photoelectric effect cannot occur. The value of f_0 naturally depends on the type of metal used.

(2) As the frequency of the incident light is increased, the photon energy increases; therefore more energy is transferred to the electron. Since $E_{phot} = hf$, the photoelectric equation (Equation 14–10a) can be written as follows:

$$KE_{max} = hf - W \qquad (14\text{–}10b)$$

Or since $W = hf_0$ (where f_0 is the threshold frequency),

$$KE_{max} = h(f - f_0) \qquad (14\text{–}10c)$$

The photoelectric equation consequently predicts that the maximum kinetic energy of the ejected electrons should be proportional to the frequency of the incident light. Equation 14–10c is the equation of the experimentally observed straight line shown in Figure 14–10a.

(3) An increase in the light *intensity* means an increase in the *number* of incident photons. Since each photon interacts with a single electron, the number of ejected electrons will increase in direct proportion to the intensity, as shown in Figure 14–10b. However, there will be no increase in the electron energy so long as the light frequency remains the same.

(4) In the photon–electron interaction, all the photon energy is *instantaneously* transferred to the electron. Therefore the electron should be immediately ejected. No time delay would be expected.

Hence all of the observational features of the photoelectric effect can be correctly accounted for by the photon theory. The photoelectric effect therefore is a phenomenon that can be explained in terms of the photon theory of light but not in terms of the wave theory.

EXAMPLE 14–3: The work function for sodium is 2.30 eV. Calculate (a) the threshold frequency for the photoelectric effect and (b) the maximum kinetic energy of the electrons ejected by light of 4000 Å wavelength,

SOLUTION: (a) The threshold frequency for sodium is

$$f_0 = \frac{W}{h} = \frac{(2.30 \text{ eV})(1.6 \times 10^{-12} \text{ erg/eV})}{6.63 \times 10^{-27} \text{ erg-sec}} = \underline{5.55 \times 10^{14} \text{ waves/sec}}$$

Note that this corresponds to a threshold wavelength of

$$\lambda_0 = \frac{c}{f_0} = \frac{3 \times 10^{10} \text{ cm/sec}}{5.55 \times 10^{14} \text{ waves/sec}} = 5.41 \times 10^{-5} \text{ cm} = 5410 \text{ Å}$$

which is a wavelength of green light. Light with a wavelength *greater* than 5410 Å will have a frequency *less than* f_0 and will not trigger the photoelectric effect on sodium.

(b) The energy in eV of a photon of wavelength 4000 Å can be computed directly from Equation 14–9:

$$E = \frac{12{,}400}{\lambda} = \frac{12{,}400}{4000} = 3.10 \text{ eV}$$

Therefore (using Equation 14–10a) electrons that are ejected from sodium by light of this wavelength will have a maximum kinetic energy of

$$\begin{aligned} KE_{\text{max}} &= E_{\text{phot}} - W \\ &= (3.10 \text{ eV}) - (2.30 \text{ eV}) = \underline{0.80 \text{ eV}} \end{aligned}$$

THE ACCEPTANCE OF THE PHOTON THEORY

In spite of Einstein's successful explanation of the photoelectric effect, few people were willing to accept the photon theory of light. The evidence for the wave nature of light was much too comprehensive to be ignored. Led by Planck, physicists worked instead to find some way of explaining the photoelectric effect within the confines of classical electromagnetic theory. The general opinion held at the time of Einstein's theory can be illustrated by the following assessment, made in 1913, of Einstein's contribution to physics:*

> There is hardly one among the great problems, in which modern physics is so rich, to which Einstein has not made an important contribution. That he may sometimes have missed the target in his speculations, as, for example, in his hypothesis of light quanta, cannot really be held too much against him, for it is not possible to introduce fundamentally new ideas without occasionally taking a risk.[2]

The problem was compounded by the experimental verification of Einstein's photoelectric equation. The most precise measurements, made by Robert Millikan and published in 1916, completely confirmed the validity of Equation 14–10. Millikan found, as predicted, a straight-line relationship between the maximum electron energy and the light frequency, as shown in Figure 14–13. He was able to determine the value of Planck's constant h for several metals. (Mathematically, h is equal to the

* This statement was submitted as part of a recommendation made by Planck and others that Einstein be admitted to the Prussian Academy of Science.

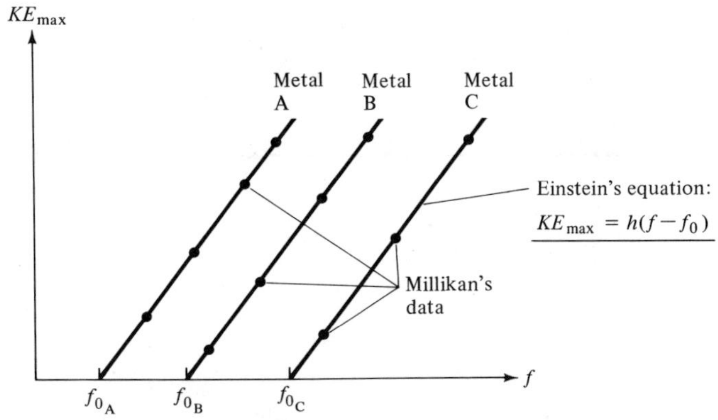

figure 14–13 **EXPERIMENTAL TEST OF THE PHOTOELECTRIC EQUATION**

The straight lines represent Einstein's equation relating the maximum electron kinetic energy KE_{max} to the frequency of the incident light f. Lines are shown for three metals having different threshold frequencies f_0. The slope of each line is h, Planck's constant. The equation was confirmed by the measurements of Millikan (shown by dots).

slope of the straight line.) Millikan found that the value of h was independent of the type of metal and was virtually the same as that obtained by Planck from blackbody radiation data. Describing his efforts, Millikan reported

> I spent ten years of my life testing that 1905 equation of Einstein's, and, contrary to all my expectations, I was compelled in 1915 to assert its unambiguous experimental verification in spite of its unreasonableness.[3]

Millikan's findings notwithstanding, Einstein's theory still did not win general acceptance. Most physicists reluctantly accepted Einstein's photoelectric equation as a valid formula, but they were not yet willing to believe in the reality of photons.

In 1923 more evidence in favor of the photon theory was provided by the experimental findings of the American physicist Arthur H. Compton. At the time there was much interest in the interaction between x rays and electrons. Classical electromagnetic theory predicts that when an x ray is absorbed by a free electron, it is immediately re-emitted with *the same wavelength* in a random direction. Hence a beam of x rays striking a target should be scattered in different directions by the electrons in the target without a change in wavelength. However, Compton discovered that an appreciable fraction of the scattered x rays have *longer wavelengths* than the incident ones (Figure 14–14a). This phenomenon became known as the ***Compton effect.***

Compton found that a complete description of the effect could be obtained by adopting the photon theory. He assumed that there is a direct "billiard-ball" collision between an x-ray photon and an electron (Figure 14–14b). The incident photon has a specific energy, $E = hf_0$, and momentum, $p = hf_0/c$, and in the collision some of this energy and momentum is transferred to the electron. Since the photon loses energy, its frequency must decrease, and therefore its wavelength increases. Compton

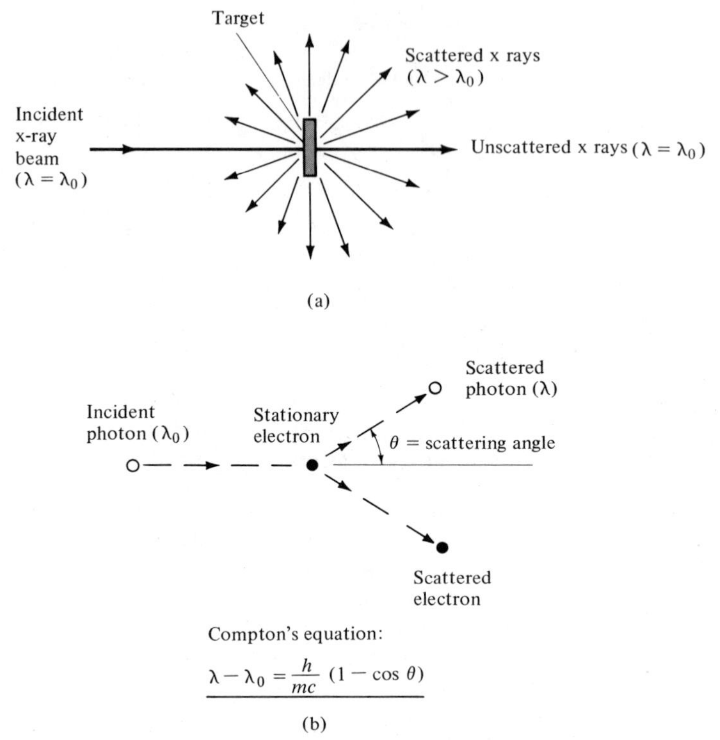

figure 14–14 **THE COMPTON EFFECT**

(a) A beam of x rays is scattered by a target. The scattered x rays have longer wavelengths than the x rays in the incident beam. (b) Compton explained the effect by assuming a billiard-ball collision between a photon and an electron.

applied the principles of conservation of energy and momentum to the collision and came up with an equation that matched rather accurately the experimental measurements of the wavelengths and scattering angles of the scattered x rays.

Hence the Compton effect, like the photoelectric effect, represents a phenomenon that could be accounted for by the photon theory, but not by the classical electromagnetic wave theory of light. With Compton's discovery the validity of the photon theory finally had to be taken seriously.

LIGHT: WAVE OR PARTICLE?

The success of the photon theory posed an extremely serious problem for physics, for it now became necessary to raise anew the question of the nature of light. There are certain phenomena (such as blackbody radiation and the photoelectric and Compton effects) that can be explained only by assuming that light is a particle; yet there is another class of phenomena (including interference, diffraction, polarization) that can be accounted for only in terms of wave concepts.* How can light have both wave

* An interesting historical manifestation of this dilemma is that Heinrich Hertz *in the same experiment* provided evidence for the wave nature of light (by producing radio waves) *and* its particle nature (by discovering the photoelectric effect).

and particle properties? Light seems almost capricious in expressing its duality; for example, in the Compton effect we find that x rays enter and leave as waves; yet they interact as particles! How can such behavior be accounted for theoretically?

Upon closer inspection, however, it is found that the behavior of light is not all that capricious. In general, light exhibits particle properties when it is interacting with matter and wave properties when it is traveling from place to place. This suggests the possibility that light does consist of particles, which influence one another such that their *group* behavior, averaged over space and time, is characteristic of wave motion. Einstein recognized the attractiveness of such a picture and set out to construct a theory along these lines from which one could deduce both the wave and particle characteristics of light. He worked for several years on the problem, but to no avail. In 1911 Einstein confessed his failure to a friend:

> I no longer ask whether these quanta really exist. Nor am I trying any longer to construct them, because I now know that my brain is incapable of accomplishing such a thing.[4]

The inability of Einstein and other physicists to find a suitable theoretical explanation for the dual nature of light has had an enormous effect on the development of twentieth-century physics. We will return to this problem in a later chapter.

QUESTIONS FOR REVIEW

1 What is a spectrum? How is one obtained?

2 Explain the difference between a continuous spectrum and a bright-line spectrum. Under what conditions is each produced?

3 What is an absorption spectrum? How is it produced? What is the relationship between the bright-line and absorption spectra of a given substance?

4 What are the Fraunhofer lines? What do they tell us about the structure and composition of the sun?

5 State and explain the law of exchanges. How does it relate to the attainment of thermal equilibrium by a radiating body?

6 Explain what is meant by a blackbody. Describe the properties of a blackbody. Explain why a blackbody is the most efficient emitter of radiation at a given temperature.

7 What is cavity radiation? How is it produced? Discuss the relationship between cavity radiation and blackbody radiation.

8 Define and explain the meaning of the following terms and give their CGS units:
(a) Energy density.
(b) Monochromatic energy density.
(c) Emissive power.

9 State and explain (a) Boltzmann's equation for the energy density of cavity radiation and (b) the Stefan–Boltzmann law for the emissive power of a blackbody. Discuss the conditions to which each equation applies. How are the two equations related?

10 What was the purpose of Wien's radiation formula? How was the formula derived? How did it compare with the experimental observations?

11 State and explain the Wien displacement law. Discuss its practical applications. How was the value of the constant (0 29 cm-°K) established?

12 Show that the Wien displacement law agrees with our observations of the relationship between the color of an incandescent body and its temperature.

13 Outline the assumptions that Rayleigh used to derive a formula for the spectrum of cavity radiation. How did Rayleigh's formula compare with Wien's formula? Discuss the implications of the failure of Rayleigh's formula to agree with observations.

14 Describe Planck's model for the production of cavity radiation. What is the relationship between the vibrating atoms in the oven walls and the electromagnetic waves in the cavity?

15 Explain Planck's quantum hypothesis. How did Planck justify this hypothesis? Discuss the ways in which it violates various aspects of classical physical theory.

16 What is Planck's constant? What role does it play in the quantum hypothesis? How can its numerical value be determined?

17 Outline the photon theory of light. What led Einstein to propose this theory? How did Einstein's interpretation of the quantum hypothesis differ from Planck's? In particular, compare their respective interpretations of the equation $E = hf$.

18 Describe the photoelectric effect, outlining the main experimental observations. In particular, describe how the properties of the ejected electrons depend on the intensity and frequency of the incident light.

19 Outline the predictions made by the electromagnetic wave theory with regard to the photoelectric effect. How do these predictions compare with observations?

20 Discuss Einstein's explanation of the photoelectric effect in terms of the photon theory. State and explain Einstein's photoelectric equation.

21 What does the work function represent? According to Einstein, how is it related to the threshold frequency?

22 Describe the Compton effect. Explain how it contradicts the predictions of the electromagnetic wave theory. What assumptions did Compton make in order to explain his observations?

23 Discuss the problems raised by the success of the photon theory in explaining the photoelectric and Compton effects.

24 Explain how the numerical value of Planck's constant can be obtained from experimental measurements of the photoelectric effect. What is the significance of the fact that h is found to have the same value as that obtained from measurements of blackbody radiation?

EXERCISES

1 List all the natural or artificial sources of light used in modern civilization that you can think of. In each case state whether the spectrum of the source is a continuous or a line spectrum.

2 Explain how you would determine the chemical composition of an unknown substance using the method of spectral analysis.

3 In general, stars exhibit absorption spectra.
(a) What information about the atmosphere of the star can be obtained from the star's spectrum?
(b) Explain why information about the chemical composition of the interior of a star cannot be obtained from the star's spectrum.
(c) What conclusion can be drawn if a star exhibits a purely continuous spectrum?

4 (a) What does the spectrum of the light from a planet look like? (Hint: Think about how planets shine.)
(b) If a planet has an atmosphere, how would it show up in the planet's spectrum?
(c) What kind of spectrum does moonlight have?

5 Suppose a glowing body is placed in completely empty space, far from any other sources of radiation. Will the body attain thermal equilibrium? Explain your answer.

6 The earth is continually receiving radiation from the sun. Explain why the temperature of the earth does not steadily rise as a consequence.

7 Two otherwise identical pieces of metal are painted with a light color and a dark color, respectively. They are left out in the sunlight for a long time. Explain why the dark-colored metal attains a noticeably higher temperature than the light-colored one, even though they both receive the same amount of sunlight.

8 Glass is considered to be a good absorber of infrared radiation. Use this fact to explain why the inside of a greenhouse (having glass walls and a roof) remains warm on a cold day.

9 By what factor will the energy density of an oven increase if the temperature changes (a) from 1000°K to 2000°K? (b) from 2000°K to 6000°K?

10 An oven is in the shape of a cube 1 meter on each side. Calculate the total electromagnetic energy contained in the oven at a temperature of 1000°K ($\approx$ 1340°F).

11 Refer to the solar spectrum shown in Figure 14–1. Is it a reasonable approximation to consider the sun as a blackbody? What errors would be made by applying (a) the Wien displacement law and (b) the Stefan–Boltzmann law to obtain information about the sun?

12 The surface temperature of the sun is approximately 6000°K. The surface temperature of Betelgeuse is only 3000°K. Both stars can be considered as blackbodies. Yet observations show that Betelgeuse is 10,000 times intrinsically brighter than the sun. (That is, Betelgeuse radiates 10^4 times more energy than the sun does.) How do you explain this? (Hint: What factors determine the *total* energy emitted by a blackbody?)

13 Calculate the temperature of a blackbody if its spectrum reaches its peak intensity at a wavelength of (a) 1450 Å in the ultraviolet, (b) 5.8 microns in the infrared, and (c) 29 cm in the radio range.

14 A blackbody is at room temperature (300°K).
(a) What is the wavelength of maximum intensity of its spectrum?
(b) In what part of the electromagnetic spectrum does this wavelength lie?
(c) If all bodies emit radiation at any temperature, why don't people or objects at room temperature glow in the dark?

15 An atom in a heated solid body is vibrating at a rate of 3×10^{14} vibrations/sec. According to Planck's hypothesis,
(a) what is the *minimum* amount of electromagnetic energy the atom can emit or absorb?
(b) what is the next largest amount?
(c) what is the wavelength of the light emitted or absorbed by this atom?

16 Calculate the energy in ergs and in eV of (a) an ultraviolet photon of $\lambda = 100$ Å, (b) an infrared photon of $\lambda = 10$ microns, and (c) a radio photon of $\lambda = 20$ cm.

17 Explain why ultraviolet and x rays tend to show their particle characteristics rather readily, whereas the particle properties of radio waves are much more difficult to observe. (That is, radio waves do not appear to be "quantized.")

18 Calculate the wavelength associated with each of the following photon energies, and state the region of the electromagnetic spectrum in which they lie (1 keV = 10^3 eV; 1 MeV = 10^6 eV).
(a) 13.6 eV
(b) 0.06 eV
(c) 30 keV
(d) 6 MeV

19 Discuss the principles of operation and applications of the device known as the "electric eye."

20 For a certain metal the threshold for the photoelectric effect occurs with photons of 3100 Å wavelength.

(a) What is the work function of the metal?

(b) Can visible light be used to eject electrons from this metal? Explain your answer.

(c) When light of a certain wavelength strikes the metal, electrons come off with a maximum kinetic energy of 4.8×10^{-12} erg. Calculate the wavelength of the light used.

(d) Calculate the maximum speed of the electrons ejected from the metal by ultraviolet light of 2000 Å wavelength. (The mass of an electron is 9.11×10^{-28} gram.)

21 Compton's equation for the Compton effect is given in Figure 14–14. Outline an experiment in which the Compton effect is used to determine the numerical value of Planck's constant.

THE STRUCTURE OF THE ATOM

15

Rutherford is a man you can rely on; he comes regularly and enquires how things are going and talks about the smallest details. . . . he is such an outstanding man and is really interested in the work of all the people around him.

Niels Bohr

AT THE BEGINNING of Chapter 14 spectra were classified into three basic types: continuous spectra, bright-line spectra, and absorption (dark-line) spectra. We learned that continuous spectra are emitted by hot solids, liquids, or dense gases, and that line spectra are produced by low-density gases. We noted that in low-density gases the atoms are far enough apart to be completely independent of one another; whereas in solids, liquids, and dense gases the motions of individual atoms are strongly influenced by the interatomic forces.

The problem of explaining the emission of continuous spectra was solved on the basis of this picture. It was presumed that electromagnetic radiation is emitted by vibrating atoms. Since there should be a complete range of vibration frequencies among the atoms in an emitting body, all frequencies should be represented in the spectrum of the emitted radiation. By adopting the quantum hypothesis—which states that vibrating atoms must absorb or emit discrete quanta of energy—Planck was able to account for the exact shape of the continuous spectrum of a blackbody.

With the problem of continuous spectra essentially solved, physicists next attempted to provide a theoretical explanation for line spectra. Since line spectra are produced by atoms acting independently of one another, it follows that spectral lines are characteristic of individual atoms. The problem was to identify the processes acting within the atoms that lead to the emission of specific frequencies of light and to explain why these frequencies differ from one element to the next.

In this chapter we will learn how the attempts to understand line spectra led to a completely new concept of the nature of the atom. We will find once again that the classical physical theories failed to provide a proper explanation of phenomena and had to be replaced by newer, more radical physical concepts.

15–1

THE PROBLEM OF LINE SPECTRA

The analysis of line spectra was begun in earnest by Kirchhoff and Bunsen in the 1860s. Over the next few decades the spectra of many substances were analyzed in considerable detail. Dozens of individual lines were found in the spectra of all the known elements, and the wavelengths of these lines were measured and cataloged. Scientists were now faced with the task of finding some mathematical formula that could predict the wavelengths of all the spectral lines of a given element.

The first successful formula of this type was provided by Johann Balmer (1825–1892). Balmer was a Swiss schoolteacher with an avid interest in numerology—but little knowledge of physics. One day a friend challenged him to find a mathematical formula relating the wavelengths of the four prominent spectral lines—at 6563, 4861, 4340, and 4101 Å—found in the visible emission spectrum of hydrogen. By trial and error Balmer found that these four numbers could be generated by multiplying 3645.6 by the ratios 9/5, 4/3, 25/21, and 9/8, respectively. This sequence can also be written as

$$\frac{9}{5} \qquad \frac{16}{12} \qquad \frac{25}{21} \qquad \frac{36}{32}$$

which is equivalent to

$$\frac{3^2}{3^2 - 4} \qquad \frac{4^2}{4^2 - 4} \qquad \frac{5^2}{5^2 - 4} \qquad \frac{6^2}{6^2 - 4}$$

Balmer therefore proposed the following formula:

$$\lambda = 3645.6\left(\frac{n^2}{n^2 - 4}\right) \qquad (\lambda \text{ in Å})$$

By letting $n = 3, 4, 5$, and 6 in turn, one could generate the numbers corresponding to the wavelengths of the four hydrogen lines.

Balmer announced his discovery in 1885. It was his first scientific contribution; he was sixty years old at the time. Balmer also suggested that additional lines might be observed at wavelengths calculated by letting $n = 7, 8, 9, \ldots,$ in his formula. Soon after, it came to his attention that five more hydrogen lines had been identified in the ultraviolet region of the spectrum. Balmer's formula predicted the right wavelengths of these lines by setting $n = 7, 8, 9, 10$, and 11 in sequence. The formula was then generalized as

$$\lambda = 3645.6\left(\frac{n^2}{n^2 - 4}\right) \qquad (n = 3, 4, 5, 6, \ldots, \infty) \tag{15–1}$$

This formula predicts that there is a spectral line corresponding to every integral value of n from 3 onward. Balmer noted that as n gets larger, the value of the fraction $n^2/(n^2 - 4)$ approaches 1; therefore the sequence of lines should converge to an upper wavelength limit of 3645.6 Å. (See Figure 15–1.) Subsequent measurements verified Balmer's predictions to an accuracy of better than 0.1 percent.

Balmer's discovery stimulated a number of attempts to find formulas for the spectral lines of other elements. In 1890 the Swedish physicist Johannes Rydberg found that a formula similar to Balmer's (but somewhat more complex) could be used to reproduce the spectral lines of the elements lithium, sodium, potassium, rubidium, and cesium. Rydberg found that the same numerical constant—now called the ***Rydberg constant***—appeared in all of these formulas, including Balmer's. The Rydberg constant appears when Balmer's formula for hydrogen (Equation 15–1) is written in the equivalent form

$$\frac{1}{\lambda} = R\left(\frac{1}{2^2} - \frac{1}{n^2}\right) \qquad (n = 3, 4, 5, \ldots, \infty) \tag{15–2}$$

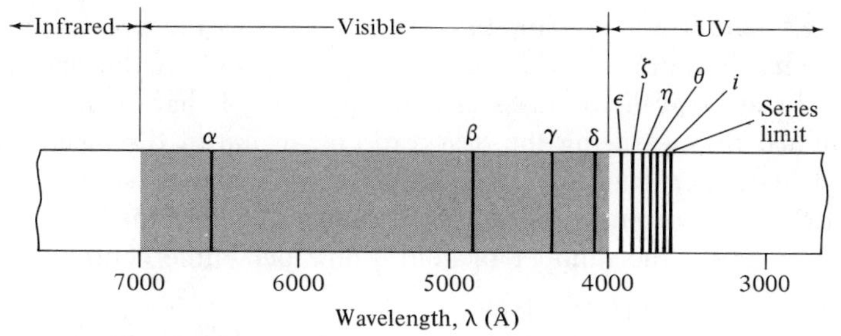

Spectral line	Measured wavelength (Å)	n	Spectral line	Measured wavelength (Å)	n
α	6563	3	ζ	3889	8
β	4861	4	η	3835	9
γ	4340	5	θ	3798	10
δ	4101	6	ι	3770	11
ϵ	3970	7	Limit	3646	∞

Balmer's formula:

$$\lambda = 3645.6\left(\frac{n^2}{n^2 - 4}\right) \qquad (n = 3, 4, 5, \ldots, \infty)$$

figure 15-1 **THE HYDROGEN SPECTRUM**

The lines are labeled in sequence by the letters of the Greek alphabet. Only the first four spectral lines are in the visible part of the spectrum.

The approximate numerical value of Rydberg's constant is found to be

$$R = 1.097 \times 10^5 \text{ cm}^{-1}$$

Although it was later found from precise spectroscopic measurements* that R shows slight variations (of no more than 0.06 percent) from element to element, it was considered to be a significant physical constant. Physicists were now aware that any theory of the atom purporting to account for the appearance of line spectra would have to explain the formulas obtained through trial and error by Balmer and Rydberg, and would have to account for the numerical value and universality of the Rydberg constant.

THE THOMSON MODEL OF THE ATOM

In 1910 the most widely accepted model of the structure of the atom was the one proposed by J. J. Thomson, the discoverer of the electron. Thomson pictured the neutral atom as containing equal quantities of positive and negative charge. The positive charge was evenly distributed throughout the atom, and negative charge was concentrated in electrons that were embedded in the positively charged matter.† Thomson considered the atom to be completely solid.

* As an example of the precision with which modern-day spectroscopic measurements can be made, the most recent determination of the Rydberg constant gave a value of $1.09737312 \pm 0.00000011 \times 10^5$ cm^{-1}—an accuracy of 1 part in 10^7.

† The Thomson model used to be called the "plum pudding" model, since the electrons were likened to plums stuck in a pudding (the pudding representing the positively charged matter). Since then, the popularity of plum pudding has declined considerably.

By this time it was possible to make estimates of the sizes of atoms. One method of determining atomic sizes follows from a knowledge of the density of an element in the solid state and of the mass of a single atom of that element. (The latter could be determined by measuring the q/m ratio of an ion of the element or from kinetic theory calculations.) Thus, for example, if the density of solid gold is found to be $19.3\ \text{g/cm}^3$ and the mass of a gold atom is determined to be approximately 3.3×10^{-22} gram, the number of gold atoms per cubic centimeter is

$$\frac{19.3\ \text{g/cm}^3}{3.3 \times 10^{-22}\ \text{g/atom}} = 5.8 \times 10^{22}\ \text{atoms/cm}^3$$

If 5.8×10^{22} atoms occupy $1\ \text{cm}^3$, the volume occupied by *one* atom is

$$\frac{1\ \text{cm}^3}{5.8 \times 10^{22}\ \text{atoms}} = 1.7 \times 10^{-23}\ \text{cm}^3/\text{atom}$$

That is, each atom is contained within a cube of $1.7 \times 10^{-23}\ \text{cm}^3$ volume. The volume of a cube is s^3, where s is the length of the side of the cube; therefore the cube in which the atom is contained has a side whose length equals $(1.7 \times 10^{-23})^{1/3} \approx 2.6 \times 10^{-8}$ cm. This number represents, in effect, the center-to-center distance between the atoms. It is reasonable to assume that in a dense solid such as gold the atoms are packed rather close together. Hence it follows that each gold atom has a radius of about half the center-to-center distance, or roughly 10^{-8} cm. Similar calculations for other elements established the fact that the radii of most atoms are no larger than about 10^{-8} cm.

Thomson and a group of his co-workers and students at Cambridge hoped to be able to explain line spectra on the basis of the Thomson atomic model. It was presumed that the embedded electrons maintain stationary positions within the atom, and calculations were made to determine how the electrons would distribute themselves within the atoms of different elements. It was assumed that when the atom was disturbed (by colliding with another atom, for example), the electrons would vibrate about their equilibrium positions. While vibrating they would emit electromagnetic radiation, in accordance with the principles of Maxwell's theory. Thomson's group tried to show that the frequencies of light appearing in line spectra could be related to the vibration frequencies of the embedded electrons, but they were totally unsuccessful. There seemed to be absolutely no way that the Thomson model could be made to yield Balmer's formula for hydrogen or to account for the patterns of spectral lines exhibited by the other elements.

15–2

THE DISCOVERY OF THE NUCLEUS

The failure of the Thomson model of the atom to explain line spectra raised serious doubts about its validity. The model was, after all, a rather speculative one, for it was not based on any *experimental* evidence as to the structure of the atom (other than the fact of the existence of electrons within the atom). But given the small size of the atom, how could it be possible to determine its internal structure experimentally?

The way out of this impasse was provided by Ernest Rutherford. Born and educated in New Zealand, Rutherford went to England in 1895 under a scholarship to study with Thomson. Here he took up an interest in two recently discovered phenomena: x rays and radioactivity. In 1898 he accepted a professorship at McGill University in Montreal, Canada, where he established a center for research on radioactivity. Here Rutherford and his colleagues discovered the laws governing radioactive decay and established the fact that radioactivity involves the natural transmutation of one element to another (to be discussed more fully in Chapter 18).

In 1899 Rutherford had discovered that radioactive substances emit two types of high-energy particles, to which he gave the names ***alpha*** (α) and ***beta*** (β). Rutherford studied the properties of these particles over a period of several years. By using Thomson's technique of electric and magnetic deflection, he was able to show that beta particles were actually high-speed electrons. The alpha particle proved to be more difficult to identify. Eventually, Rutherford showed that it had a mass four times greater than that of a hydrogen atom and a charge of two positive units of electricity (i.e., twice the magnitude of the charge of the electron). He was then able to establish that the alpha particle is actually a helium atom that has been doubly ionized. (That is, it has lost two electrons.)

THE ALPHA-SCATTERING EXPERIMENT

One of Rutherford's discoveries regarding alpha and beta particles was that these particles can penetrate through thin layers of solid matter. In doing so, the particles are deflected or scattered from a straight-line path, as a result of the forces exerted on them as they pass through the atoms. Rutherford hit on the idea that by studying the scattering of alpha and beta particles by atoms, it might be possible to learn something about the internal constitution of the atom.

In 1907 Rutherford moved to the University of Manchester in England. Here he devised an experiment to measure the scattering of alpha particles by matter. A narrow beam of alphas was produced by placing a radioactive alpha source inside a lead shield. (See Figure 15–2.) The beam was aimed at a thin sheet of gold that

figure 15–2 **RUTHERFORD'S ALPHA-SCATTERING EXPERIMENT**

Alpha particles from a radioactive source are scattered by the gold foil. The scattered alpha particles strike a zinc sulfide screen, causing light flashes that may be observed in the microscope.

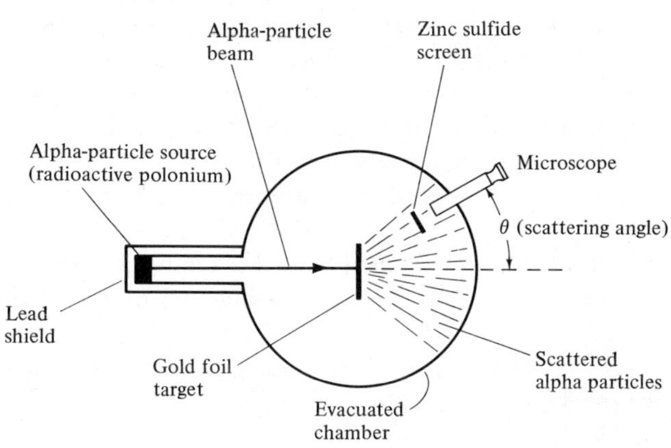

had been flattened to a thickness of about 10^{-5} cm. The alpha particles scattered by the gold atoms struck a screen coated with zinc sulfide, a substance that gives off a flash of light when struck. These flashes could be seen through a microscope, which could be moved to measure the number of particles scattered through different angles. The experimenter had to sit in total darkness so that he could see and keep an accurate count of the flashes produced by the scattered alphas. Rutherford was assisted in making his observations by Hans Geiger* and Ernest Marsden.

Rutherford's calculations, based on the Thomson model, predicted that the alphas would be scattered through angles of no more than 1 or 2°. Indeed, the first measurements seemed to agree with this prediction. Then Rutherford off-handedly suggested to Marsden (who was then an undergraduate) that he look for alphas being scattered backward (through angles greater than 90°). Rutherford did not expect any positive results; according to the Thomson model, the probability of an alpha being scattered at an angle greater than 90° was something like 1 in 10^{3500}! Rutherford was stunned by the result, as he recalled many years later:

> I remember two or three days later Geiger coming to me in great excitement and saying, "We have been able to get some of the alpha-particles coming backward." It was quite the most incredible event that has ever happened to me in my life. It was almost as incredible as if you fired a 15-inch shell at a piece of paper and it came back and hit you.[1]

Geiger and Marsden were finding that roughly one alpha in 10,000 was being back-scattered. Although this was an extremely small ratio, it was still much too large to be explained in terms of the Thomson model. The alphas were known to be traveling at very high speeds, so it would require a rather large force to turn them back. This force could certainly not be provided by the electrons in the atoms; the mass of an alpha particle is more than 7000 times the mass of the electron, so that in any collision between alphas and electrons the alpha would not be measurably deflected. Furthermore, since the positive charge of the atom was presumed to be distributed evenly over a sphere of approximately 10^{-8} cm radius, the repulsive force the atom exerts on the alpha (which is positively charged) could not be intense enough to cause an alpha to be back-scattered. In fact, it was calculated that if the Thomson model were correct, an alpha would have to pass through several thousand atoms even to be scattered by more than 1°! (See Figure 15–3a.)

In 1911, nearly two years after Geiger and Marsden reported their discovery, Rutherford proposed a new model of the atom, which could account for the alpha-scattering results. He reasoned that scattering through a large angle could come about only if *the entire positive charge and most of the mass of the atom is contained in a very small region,* which he called the ***nucleus.*** If the positive charge is concentrated in a small nucleus, the repulsive electric force between the alpha and the nucleus will be much greater than that expected for the Thomson model. And since the *mass* of the atom is also concentrated and is much larger than the mass of the alpha particle, the alpha particle, rather than the nucleus, will experience the greater deflection.

Rutherford estimated that the radius of the nucleus would have to be about 10^{-12} cm to account for the experimental findings. That is, an alpha particle would

*Geiger is best known for his invention of the Geiger counter, which electrically detects and counts high-speed charged particles.

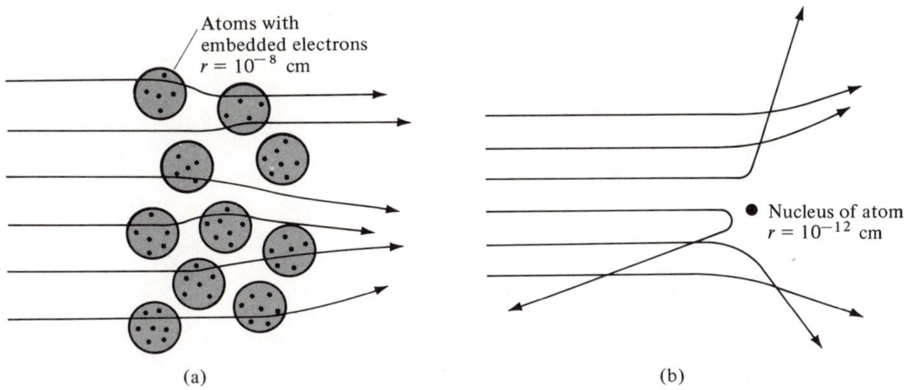

figure 15–3 **SCATTERING OF ALPHA PARTICLES ACCORDING TO THE THOMSON AND RUTHERFORD ATOMIC MODELS**

(a) Large deflections are highly unlikely with the Thomson model. (b) According to the Rutherford model, backward scattering occurs as a result of near collisions with the nucleus.

have to approach to within about 10^{-12} cm of a nucleus in order for the repulsive force to be strong enough to deflect the particle through a large angle. (Remember that the electrostatic force—described by Coulomb's law—decreases in proportion to the square of the distance.) But if the nucleus is so small, few alphas are likely to come this close to a nucleus; most will pass far away from a nucleus and will be deflected only slightly. It is only those few alphas that chance to have head-on collisions or near encounters with nuclei that exhibit large scattering angles (Figure 15–3b). Rutherford proceeded to derive from this model a formula to predict the relative number of alphas that will be scattered through a given angle as a function of the alpha particle energy and the thickness and composition of the target. Rutherford's formula was subsequently verified experimentally in full detail by Geiger and Marsden. A comparison of the experimental results with the predictions made from the Thomson and Rutherford models is shown in Figure 15–4. The superiority of Rutherford's model over Thomson's was clearly established.

THE NUCLEAR ATOM

Let us now examine some of the important features of the Rutherford model of the atom. The atom is assumed to consist of a tiny nucleus surrounded by electrons. The nucleus contains all the positive charge; an equal negative charge is distributed among the electrons to make the atom electrically neutral. Nearly all the mass of the atom must be contained in the nucleus, since the mass of an electron is known to be very small compared to the mass of an atom. Surely the most startling aspect of the model is the small size of the nucleus. The radius of the nucleus is on the order of 10^{-12} cm, and the radius of an atom is about 10^{-8} cm. Thus the nucleus is only *one ten-thousandth* (10^{-4}) the size of the atom. This is an extremely small ratio, equivalent to about 1 foot in 2 miles. On the scale of the solar system, if the nucleus were the size of the sun, the atom would be almost 50 AU in radius—well beyond the orbit of Pluto! Rutherford himself likened the nucleus to "a fly in a cathedral."

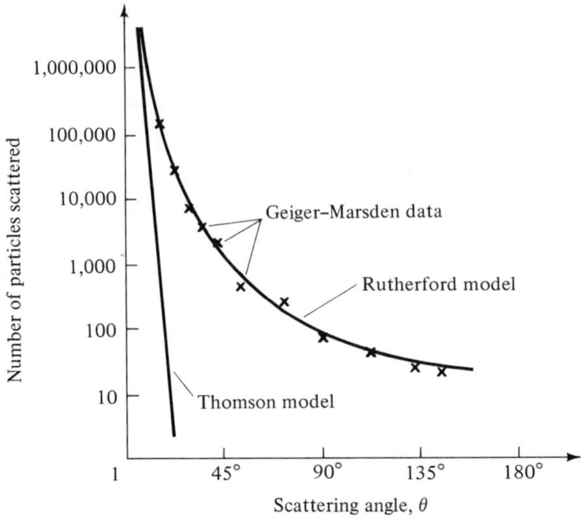

figure 15–4 **EXPERIMENTAL TEST OF THOMSON AND RUTHERFORD ATOMIC MODELS**

The crosses represent the data obtained by Geiger and Marsden. The solid lines show the theoretical predictions according to the Thomson and Rutherford models of the atom.

This picture of atomic structure seems quite contrary to our commonsense notions about the nature of the atom. Going back to the ancient Greeks, it was believed that atoms were solid, impenetrable bodies; now we find that the atom consists mainly of empty space! Thus we are being deceived by our senses when an object appears to be solid; actually, the only solid matter is contained in the nuclei, which all together occupy less than *one trillionth* of the total volume of a "solid" body.

When Rutherford proposed his atomic model, there was one serious difficulty that prevented its general acceptance. The problem had to do with the status of the electrons in the atom. The electrons could not be stationary, or else the electrostatic attraction would cause them to fall into the nucleus. Hence the electrons must be in orbit around the nucleus, with the electrical attraction providing the centripetal force. Since Coulomb's law and the law of gravitation have the same mathematical form, the electron orbits should be identical to planetary orbits. Thus we obtain a rather attractive picture of the atom as a miniature solar system, with the nucleus as the sun and the electrons as the planets.

Unfortunately, the "solar system" model is quite unstable according to classical physics. An orbiting electron is in accelerated motion, and Maxwell's theory requires that an accelerating charge must continually emit radiation. Hence an orbiting electron should steadily lose energy and spiral inward. In fact, the theory predicts that the electron should spiral into the nucleus within 10^{-8} sec! Furthermore, in doing so, the electron should emit a *continuous* spectrum of radiation, rather than the line spectrum that is actually observed. So although Rutherford's experiment points unequivocally to the existence of the nucleus, Maxwell's theory predicts that the nuclear atom is inherently unstable.

Characteristically, Rutherford made no attempts to resolve this dilemma. He considered himself to be purely an experimental physicist and often expressed his contempt for theoreticians who "play games with their symbols while we turn out the real solid facts of nature." Following the outbreak of World War I, Rutherford went into war research, working on the problem of detecting submarines. After the war he turned his attention to the study of collisions between alpha particles and lightweight atoms. In 1919 Rutherford reported that he had successfully disintegrated a nitrogen nucleus by alpha bombardment. This was the first artificial nuclear reaction resulting in the transmutation of one element to another—a discovery of major importance.

In view of Rutherford's seminal contributions to the study of radioactivity, atomic structure, and nuclear reactions, he may justifiably be considered the "father of nuclear physics." He was held in great esteem by his countrymen; in 1931 he was given the title of Lord Rutherford of Nelson (after his hometown in New Zealand), and on his death in 1937 he was buried in Westminster Abbey, not far from the grave of Newton.

15–3

THE BOHR THEORY OF THE ATOM

In 1912 Rutherford paid a visit to his former teacher Thomson at Cambridge and was introduced to a young Danish physicist named Niels Bohr. Bohr had obtained his Ph.D. at the University of Copenhagen and had come to Cambridge on a one-year fellowship to work with Thomson. Bohr was unhappy at Cambridge, partly because he was put off by the traditional academic formality and partly because Thomson had shown little interest in his work.

Bohr was a strong supporter of Planck's quantum theory—which at the time was considerably more popular in continental Europe than in England—and Bohr's attempts to employ quantum principles in his work met with resistance from Thomson. However, Rutherford took a liking to Bohr and invited him to visit Manchester, where Rutherford showed Bohr the results of the alpha-scattering experiments. Bohr returned to Cambridge quite convinced of the validity of the Rutherford atom. This did not go over well with Thomson, who finally suggested to Bohr that he complete the rest of his fellowship at Manchester.

POSTULATES OF THE BOHR THEORY

Once established at Manchester, Bohr—with Rutherford offering considerable encouragement—set out to resolve the problem of the instability of the nuclear atom. It was clear that Rutherford's atomic model and Maxwell's theory were incompatible; Bohr had no recourse but to postulate that Maxwell's theory does not apply on an atomic level. In defense of this hypothesis Bohr could point to Planck, who had to make the same assumption in order to account for the spectrum of blackbody radiation. Bohr therefore proposed that electrons in an atom exist in "stationary states"—now called ***energy levels***—of constant energy in which the electron moves in an orbit about the nucleus without radiating energy.

Bohr proposed further that an electron can change its energy only by jumping instantaneously from one energy level to another. In the process the electron gains or loses energy by absorbing or emitting a quantum of electromagnetic radiation. The frequency of this radiation is determined by the Planck–Einstein equation $E = hf$. Specifically, if ΔE is the energy gained or lost by the electron, the frequency of the absorbed or emitted radiation is

$$f = \frac{\Delta E}{h} \tag{15–3}$$

The use of this equation was in itself another break with Maxwell's theory, which requires that the frequency of the emitted radiation be equal to the frequency of revolution (the number of revolutions per second) of the orbiting electron. But according to Bohr, the frequency of the radiation depends on the *difference in energy between one energy level and the next.*

Although Bohr accepted Planck's quantum hypothesis, he did not accept Einstein's photon theory. (Very few people did in 1912.) A belief in photons is not necessary for the success of the Bohr theory; the "quantum of radiation" emitted or absorbed by an electron can be in the form of either a wave or a particle. Nevertheless, we will use photon terminology from here on, in keeping with the modern-day viewpoint.

Bohr turned to classical Newtonian mechanics to describe the orbit of a single electron. He assumed for simplicity that the electron travels in a circular orbit about the nucleus. Uniform circular motion requires a constant centripetal force, which in this case is provided by the mutual electric attraction between the positively charged nucleus and the negatively charged electron. The magnitude of this force is given by Coulomb's law, $F_{\text{elec}} = K_e q_1 q_2 / d^2$. The charge on the electron is numerically equal to the fundamental unit of charge, e $(= 4.8 \times 10^{-10}$ esu). The nucleus contains an integral number of positive charge units. Therefore we set $q_{\text{nucleus}} = Ze$, where Z is an integer representing the number of positive charge units in the nucleus. The distance d between the nucleus and the electron is r, the radius of the orbit. Hence the magnitude of the attractive force is (using CGS units with $K_e = 1$):

$$F_{\text{elec}} = \frac{(q_{\text{nucleus}})(q_{\text{electron}})}{d^2} = \frac{(Ze)(e)}{r^2} = \frac{Ze^2}{r^2}$$

Since F_{elec} provides the centripetal force, we have $F_{\text{elec}} = F_{\text{centripetal}}$, or

$$\frac{Ze^2}{r^2} = \frac{mv^2}{r}$$

where m and v are the mass and orbital speed, respectively, of the electron. Multiplying through by r gives

$$mv^2 = \frac{Ze^2}{r} \tag{15–4}$$

Bohr obtained a second equation from the principle of conservation of energy. (His assumption was that the energy of the electron remains constant while in a fixed orbit.) The total energy E of the electron is composed of its kinetic energy plus its electrostatic potential energy. That is,

$$E = KE + PE_e$$

The equation for electric potential energy (Equation 9–4) is $PE_e = K_e q_1 q_2 / d$; therefore for the orbiting electron (using CGS units),

$$PE_e = \frac{q_{\text{nuc}}q_{\text{elec}}}{r} = \frac{(Ze)(-e)}{r} = -\frac{Ze^2}{r}$$

The minus sign is necessary here because the charges are of opposite sign; hence the potential energy has a maximum value of zero when the charges are an infinite distance apart and *decreases* (becomes more and more negative) as the charges approach each other. The total energy of the electron is therefore

$$E = KE + PE_e = \frac{1}{2}mv^2 - \frac{Ze^2}{r}$$

Substituting for mv^2 from Equation 15–4 gives

$$E = \frac{1}{2}\left(\frac{Ze^2}{r}\right) - \frac{Ze^2}{r} = \frac{Ze^2}{r}\left(\frac{1}{2} - 1\right)$$

or

$$E = -\frac{Ze^2}{2r} \tag{15–5}$$

Note that the total energy is always negative, because the negative potential energy always outweighs the positive kinetic energy. A negative total energy is a necessary condition for a closed planetary orbit, as we learned in Chapter 6.

At this point Bohr was uncertain how to proceed. He had no idea how to modify Equation 15–5 so as to yield a distinct set of energy levels for the electron—and the existence of these energy levels was the cornerstone of his theory. In January 1913, after Bohr had returned to Copenhagen, this uncertainty was lifted. It was then that he learned for the first time of the existence of Balmer's formula. "As soon as I saw Balmer's formula," Bohr later recalled, "the whole thing was immediately clear to me."

Balmer's formula, as we know, gives the frequencies of the light emitted by glowing hydrogen gas. According to Bohr's theory, light is emitted by an atom when an electron in the atom changes from one energy level to another. The frequency of this light is related (through Equation 15–3) to the *differences* between the electron energy levels. Bohr realized that by working backward from Balmer's formula, he could determine what the energy levels were in the hydrogen atom. In doing so, he found that one more assumption had to be made to complete the theory. This assumption can be stated in the following form:

The angular momentum of the electron in its orbit must be an integral multiple of $h/2\pi$, where h is Planck's constant.

Let us express this condition mathematically. For a circular orbit the magnitude of the angular momentum $L = mvr$ (Equation 6–7). Therefore Bohr's assumption is that

$$L = mvr = \frac{nh}{2\pi} \qquad (n = 1, 2, 3, \ldots, \infty) \tag{15–6}$$

According to this principle, the orbital angular momentum of the electron is restricted to values of $h/2\pi$, $2h/2\pi$, $3h/2\pi$, and so on. Thus it is assumed that *there is a fundamental unit of angular momentum,* just as there is a fundamental unit of charge. The angular momentum is therefore said to be ***quantized.*** For the moment we shall not attempt to justify this rather strange assumption, other than to show that its adoption leads to Balmer's formula.

First we combine Equations 15–4 and 15–6 (the algebra will be left for the student to work out) to obtain

$$r_n = \frac{n^2h^2}{4\pi^2mZe^2} \qquad (n = 1, 2, 3, \ldots, \infty) \tag{15–7}$$

We will examine this equation later on. Now we substitute Equation 15–7 into Equation 15–5 to get

$$E_n = -\frac{2\pi^2mZ^2e^4}{n^2h^2} \qquad (n = 1, 2, 3, \ldots, \infty) \tag{15–8}$$

This equation gives, in mathematical form, the distinct set of energy levels required by Bohr. The ***quantum number*** n, which appears in the denominator, must be an integer; therefore the electron can have only those values of energy obtained by substituting $n = 1, 2, 3, \ldots$, in turn into Equation 15–8. Thus *to each value of n there corresponds a specific energy level E_n of the electron.* Consequently, the *energy* of the orbiting electron is also quantized.

Now we can proceed to derive Balmer's formula. Suppose an electron jumps from an energy level characterized by quantum number n_2 to a lower energy level with a quantum number n_1. The difference in energy between these two levels is:

$$\Delta E = E_{n_2} - E_{n_1} = \left(-\frac{2\pi^2mZ^2e^4}{{n_2}^2h^2}\right) - \left(-\frac{2\pi^2mZ^2e^4}{{n_1}^2h^2}\right)$$
$$= \frac{2\pi^2mZ^2e^4}{h^2}\left(\frac{1}{{n_1}^2} - \frac{1}{{n_2}^2}\right)$$

In jumping from a higher energy level to a lower one, the electron emits the energy difference ΔE in the form of radiation. The frequency of the radiation (according to Equation 15–3) is

$$f = \frac{\Delta E}{h} = \frac{2\pi^2mZ^2e^4}{h^3}\left(\frac{1}{{n_1}^2} - \frac{1}{{n_2}^2}\right)$$

Let us write this equation in terms of the wavelength rather than the frequency. Since $c = f\lambda$,

$$\frac{1}{\lambda} = \frac{f}{c}$$

Therefore

$$\frac{1}{\lambda} = \frac{2\pi^2mZ^2e^4}{ch^3}\left(\frac{1}{{n_1}^2} - \frac{1}{{n_2}^2}\right) \tag{15–9}$$

Let us compare this equation with Balmer's formula (Equation 15–2):

$$\frac{1}{\lambda} = R\left(\frac{1}{2^2} - \frac{1}{n^2}\right) \qquad (n = 3, 4, 5, \ldots, \infty)$$

These two equations are quite similar in form. They are identical when two conditions are met. First, n_1 must equal 2, which means that Balmer's formula presumably represents a special case in which electrons are jumping specifically to the $n = 2$ energy level. Second, we recall that the value of the Rydberg constant R, determined experimentally from the hydrogen spectrum, is 1.097×10^5 cm^{-1}. This number should be obtained from Equation 15–9. Since the hydrogen nucleus is known to contain a single unit of positive charge, we set $Z = 1$ in Equation 15–9. Comparing Equations 15–9 and 15–2, we see that the equation derived by Bohr agrees with Balmer's formula if

$$R = \frac{2\pi^2 me^4}{ch^3} \tag{15–10}$$

Let us now evaluate the quantity on the right side of this equation:

$$\frac{2\pi^2 me^4}{ch^3} = \frac{2\pi^2(9.108 \times 10^{-28}\ \text{gram})(4.803 \times 10^{-10}\ \text{esu})^4}{(2.998 \times 10^{10}\ \text{cm/sec})(6.626 \times 10^{-27}\ \text{erg-sec})^3}$$
$$= 1.097 \times 10^5\ \text{cm}^{-1}$$

which accurately matches the experimentally determined value of the Rydberg constant for hydrogen.

Bohr's achievement was most impressive. Balmer's formula had originally been derived to fit the hydrogen spectral lines by a trial-and-error procedure—in much the same way that Kepler had arrived at his laws of planetary motion. Just as Newton explained Kepler's laws through the theory of gravitation, Bohr provided a theory from which Balmer's formula could be obtained. Bohr's successful prediction of the value of the Rydberg constant in terms of a combination of fundamental physical constants was considered by many physicists to be as significant an achievement as Maxwell's calculation of the speed of light from the electric and magnetic constants.

Bohr's success in deriving Balmer's formula and explaining for the first time the peculiar properties of the hydrogen line spectrum naturally evoked considerable interest. But many physicists, including Bohr himself, were disturbed by the nature of the postulates required to attain this result. To see why, let us summarize Bohr's postulates:

(1) The Rutherford nuclear atomic model is assumed.
(2) Electrons traveling in stationary orbits about the nucleus *do not radiate electromagnetic energy*. Thus Maxwell's electromagnetic theory does not correctly predict the behavior of electrons on the atomic level.
(3) An electron can gain or lose energy only in a transition from one orbit (energy level) to another. When energy is gained or lost by absorption or emission of electromagnetic radiation, the frequency of that radiation is governed by the equation $f = \Delta E/h$.
(4) The orbital motion of the electron obeys the laws of Newtonian mechanics and Coulomb's law of electrostatic force. However, the laws of mechanics do not apply to transitions, in which the electron jumps *instantaneously* (in effect, with infinite velocity) from one orbit to another.
(5) The angular momentum of an orbiting electron must be an integral multiple of $h/2\pi$.

Bohr's theory, which began as an attempt to combine the Rutherford nuclear atom with the quantum hypothesis, thus involves a seemingly arbitrary and conflicting mixture of classical and quantum concepts. Bohr was especially disturbed about having to postulate that certain classical physical principles are obeyed within the atom and others are not. For many years afterward he devoted considerable effort to attempts to reconcile philosophically the validity of classical physics for large-scale objects with its failure at the atomic level.

15–4

THE SUCCESS OF THE BOHR THEORY

Yet whatever qualms physicists might have had about the peculiarities of the postulates, they could not argue with the tremendous power of the Bohr theory to account for a wide variety of atomic phenomena. Let us now examine some of the consequences of the theory.

THE PROPERTIES OF THE HYDROGEN ATOM

We begin with the formula for the radius r_n of the nth electron orbit (Equation 15–7), which we rewrite as follows:

$$r_n = \left(\frac{h^2}{4\pi^2 mZe^2}\right)n^2 \qquad (n = 1, 2, 3, \ldots, \infty)$$

The quantity in parentheses is composed of various constants. For hydrogen ($Z = 1$) the quantity has the value

$$r_0 = \frac{h^2}{4\pi^2 me^2} = 0.529 \times 10^{-8}\ \text{cm}$$

We can now write Equation 15–7 in a more concise form:

$$r_n = r_0 n^2 \qquad (n = 1, 2, 3, \ldots, \infty) \tag{15–11}$$

This equation states that in the hydrogen atom, the radius of an electron orbit must have one of the following values:

$$\begin{aligned} r_1 &= r_0 \\ r_2 &= 4r_0 \\ r_3 &= 9r_0 \\ &\vdots \end{aligned}$$

These represent the ***allowed orbits*** of the electron; no other radii are permitted. When an electron changes orbits, it goes instantaneously from one orbit to another—a process not describable by ordinary mechanical laws. The smallest orbit is the $n = 1$ orbit, which, as we shall see, is also the level of lowest energy. Hence the theory predicts that a hydrogen atom in its lowest energy state has a radius $r = r_0 = 0.529 \times 10^{-8}$ cm. This is exactly the right order of magnitude for the size of the atom, as compared with experimentally determined values. Thus *the Bohr theory correctly predicts the size of the hydrogen atom.* If r_0 had turned out to be, say 10^{-15} cm or 1 cm, the Bohr theory would have had to be rejected.

Now we turn to the electron energies. The energy equation (Equation 15–8) can be rewritten as follows:

$$E_n = -\frac{(2\pi^2 mZ^2e^4/h^2)}{n^2} \qquad (n = 1, 2, 3, \ldots, \infty)$$

For hydrogen ($Z = 1$) this equation simplifies to

$$E_n = -\frac{E_0}{n^2} \qquad (n = 1, 2, 3, \ldots, \infty) \tag{15-12}$$

where

$$E_0 = \frac{2\pi^2 me^4}{h^2} = 2.18 \times 10^{-11} \text{ erg} = 13.60 \text{ eV}$$

Hence the allowed electron energies for the hydrogen atom are:

$$\begin{aligned} E_1 &= -\frac{13.6}{(1)^2} = -13.60 \text{ eV} \\ E_2 &= -\frac{13.6}{(2)^2} = -3.40 \text{ eV} \\ E_3 &= -\frac{13.6}{(3)^2} = -1.51 \text{ eV} \\ E_4 &= -\frac{13.6}{(4)^2} = -0.85 \text{ eV} \\ &\vdots \\ E_\infty &= -\frac{13.6}{(\infty)^2} = 0 \end{aligned}$$

These represent the energy levels of the hydrogen atom. Each value corresponds to the total energy of an electron when it is in the orbit characterized by the given quantum number n. (See Figure 15-5.) Note that the $n = 1$ orbit (the smallest radius) has the *lowest* (i.e., most negative) energy. This orbit is referred to as the ***ground state*** of the hydrogen atom. Successive orbits, referred to as ***excited states,*** have higher and higher energies. The $n = \infty$ "orbit," which corresponds to zero energy,

figure 15-5 **ALLOWED ORBITS OF THE HYDROGEN ATOM**

The energies and radii of the first six allowed orbits are shown (the radii are drawn to scale; $r_0 = 0.529 \times 10^{-8}$ cm).

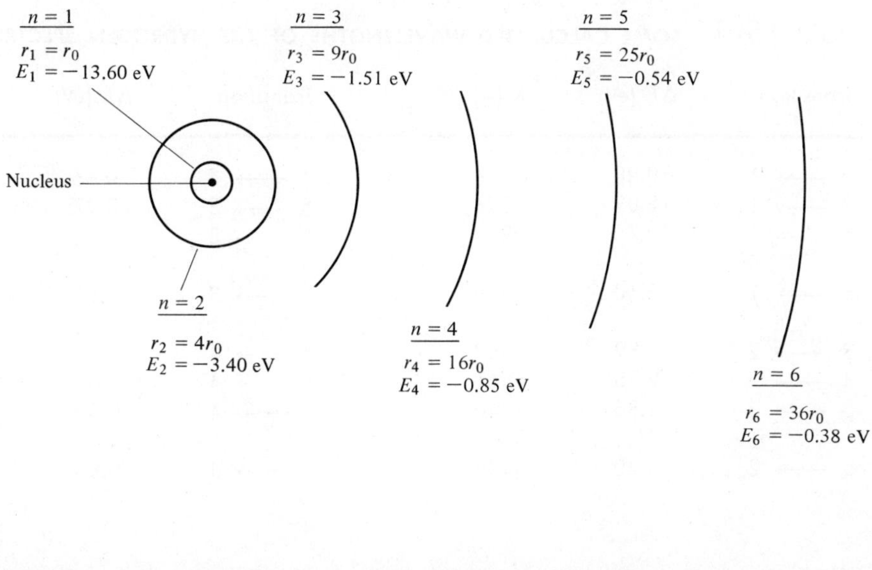

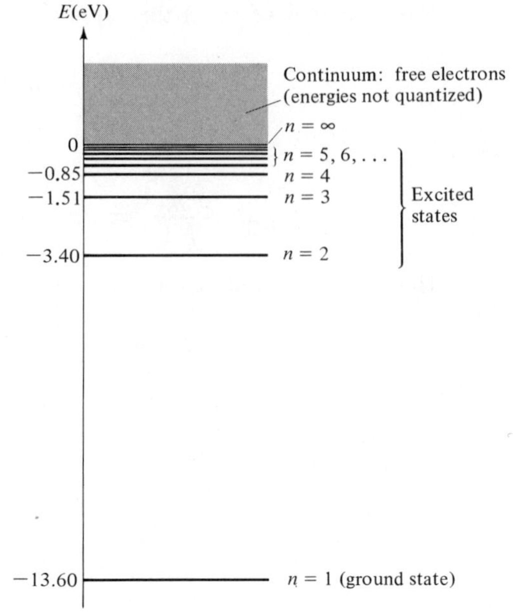

figure 15–6
ENERGY LEVEL DIAGRAM FOR HYDROGEN

The allowed energies for the electron in the hydrogen atom are plotted to scale on the vertical axis.

is actually a parabolic orbit and represents an electron no longer bound to the atom. Electrons having positive energies are "free" (unbound) electrons not bound to the atom (but which have been deflected into hyperbolic orbits by the electrical attraction of the nucleus). The energy of an unbound electron is not quantized; that is, it can change by any arbitrarily small amount, in agreement with Maxwell's theory.

It is customary to represent the allowed energy levels by an ***energy level diagram*** (Figure 15–6), which illustrates the energy differences between levels. Note that the levels get closer and closer together as n gets larger. The region above $E = 0$, which corresponds to the energy states of free electrons, is called the ***continuum.*** Since free electrons can have any positive energy and can change their energy by an arbitrarily small amount, there are no discrete energy levels in this region.

The energy level diagram provides a convenient scheme for determining the wavelengths of the lines in the emission spectrum of hydrogen. For a transition between any two given energy levels, we first find the energy difference between the two levels

table 15–1 SOME CALCULATED WAVELENGTHS OF THE HYDROGEN SPECTRUM

Transition	ΔE (eV)	λ (Å)	Transition	ΔE (eV)	λ (Å)
$2 \longrightarrow 1$	10.20	1216	$4 \longrightarrow 3$	0.66	18,750
$3 \longrightarrow 1$	12.09	1026	$5 \longrightarrow 3$	0.97	12,820
$4 \longrightarrow 1$	12.75	972	$6 \longrightarrow 3$	1.13	10,940
⋮	⋮	⋮	⋮	⋮	⋮
$\infty \longrightarrow 1$	13.60	912	$\infty \longrightarrow 3$	1.51	8,204
$3 \longrightarrow 2$	1.89	6563	$5 \longrightarrow 4$	0.31	40,510
$4 \longrightarrow 2$	2.55	4861	$6 \longrightarrow 4$	0.47	26,250
$5 \longrightarrow 2$	2.86	4340	$7 \longrightarrow 4$	0.57	21,660
⋮	⋮	⋮	⋮	⋮	⋮
$\infty \longrightarrow 2$	3.40	3646	$\infty \longrightarrow 4$	0.85	14,580

by subtracting the lower energy from the higher one. This energy difference will be equal to the energy of the emitted photon. To find the wavelength of the photon, we substitute the photon energy into the equation $\lambda = 12{,}400/E$ (with E in eV and λ in Å). For example, let us calculate the wavelength of the photon emitted when an electron falls from $n = 3$ to $n = 2$ in a hydrogen atom. For hydrogen the energies of the $n = 3$ and $n = 2$ levels are:

$$\underline{\text{n} = 3:} \quad E_3 = -\frac{13.6}{(3)^2} = -1.51 \text{ eV}$$

$$\underline{\text{n} = 2:} \quad E_2 = -\frac{13.6}{(2)^2} = -3.40 \text{ eV}$$

The energy difference is

$$E_3 - E_2 = (-1.51) - (-3.40) = 1.89 \text{ eV}$$

This is the energy of the emitted photon. The wavelength of the photon is

$$\lambda = \frac{12{,}400}{1.89} \approx 6560 \text{ Å}$$

By treating all the possible transitions in a similar manner, we can work out the wavelengths of all the spectral lines of hydrogen. A number of them are listed in Table 15–1. In examining the table, we notice that the spectral lines tend to fall into specific groups. The lines in the first set, having wavelengths between 912 and 1216 Å, all lie in the ultraviolet part of the spectrum and result from transitions from higher energy levels to the $n = 1$ level. The second group, between 3646 and 6563 Å, lies mainly in the visible spectrum; all these lines result from transitions to the $n = 2$ level. These are the spectral lines whose wavelengths are predicted by Balmer's formula. Other groups of lines, lying in the infrared, come about from transitions to the $n = 3$, $n = 4$, and higher energy levels. (See Figure 15–7.)

Thus the Bohr theory predicts the existence of more spectral lines than are accounted for by Balmer's original formula. The latter was derived at a time when the visible spectral lines (transitions to $n = 2$) were the only ones known. The $n = 1$ set of lines was discovered by Theodore Lyman in 1914—the year after the Bohr theory was proposed—thus providing strong evidence for the validity of the theory. This set of lines is now known as the ***Lyman series.*** The infrared lines of the $n = 3$ series were discovered by F. Paschen in 1908; additional series of infrared lines were identified by F. Brackett ($n = 4$) in 1922 and A. H. Pfund ($n = 5$) in 1924.

Hence from the Bohr theory we can derive a more general form of Balmer's formula, which includes all the possible hydrogen spectral lines. Combining Equations 15–9 and 15–10 (with $Z = 1$), we obtain

$$\frac{1}{\lambda} = R\left(\frac{1}{n_1^{\,2}} - \frac{1}{n_2^{\,2}}\right) \qquad (15\text{–}13)$$

The various lines are grouped as follows:

Lyman series:	$n_1 = 1$	$n_2 = 2, 3, 4, \ldots, \infty$	(ultraviolet)
Balmer series:	$n_1 = 2$	$n_2 = 3, 4, 5, \ldots, \infty$	(visible)
Paschen series:	$n_1 = 3$	$n_2 = 4, 5, 6, \ldots, \infty$	(near infrared)
Brackett series:	$n_1 = 4$	$n_2 = 5, 6, 7, \ldots, \infty$	(middle infrared)
Pfund series:	$n_1 = 5$	$n_2 = 6, 7, 8, \ldots, \infty$	(far infrared)

etc.

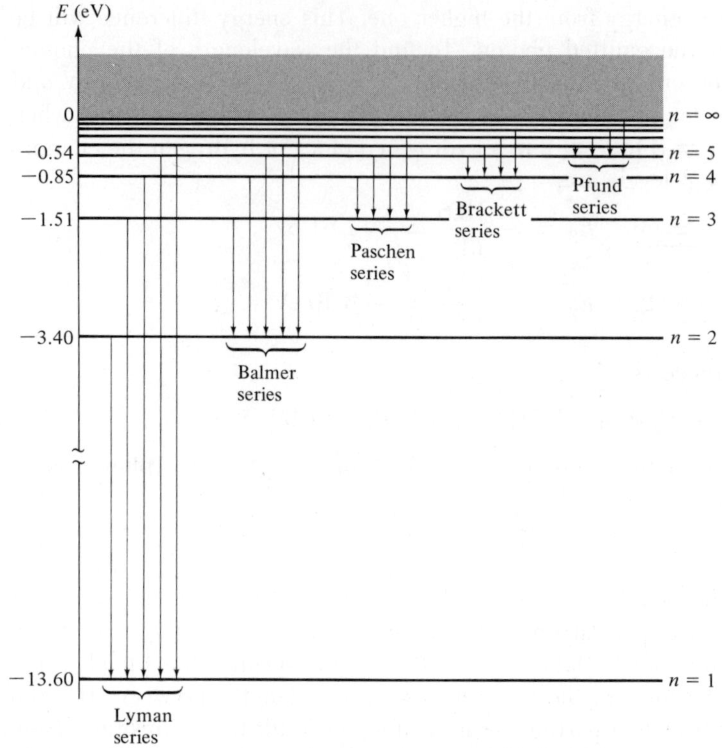

figure 15–7 **THE SPECTRAL SERIES OF HYDROGEN**

The energy level diagram for hydrogen shows the transitions that lead to the various spectral series of emission lines. The same lines appear as absorption lines when electrons are made to jump from lower levels to higher ones.

The wavelengths of any of these spectral lines can be calculated by direct substitution into Equation 15–13 of the appropriate values of n_1 and n_2.

ONE-ELECTRON ATOMS

One of the triumphs of the Bohr theory was the manner in which it accounted for the wavelengths of the spectral lines of hydrogen. Yet shortly after the theory was first published, Bohr was challenged on the ground that there were additional spectral lines in the hydrogen spectrum that could not be derived from the theory. Bohr successfully answered this challenge in a way that greatly strengthened the validity of his theory.

These extra lines were first observed as absorption lines in the spectrum of a star by the American astronomer E. C. Pickering in 1896. They were also seen in 1912 by the British physicist A. H. Fowler in the spectrum of the sun and in a heated mixture of hydrogen and helium gas. The wavelengths of these lines could be represented by a formula that bore a striking resemblance to the Balmer formula, and these lines also seemed to have the same wavelength limit—3645.6 Å—as the Balmer lines. For these reasons the lines were commonly thought to be a "second series" of hydrogen.

Bohr attributed these lines to the spectrum of singly ionized helium (He^+). The helium nucleus contains two positive charges; therefore there are two electrons orbiting the nucleus in a neutral helium atom. The Bohr energy level equation was derived on the assumption of a *single* electron orbiting the nucleus, so it is not applicable to neutral helium. If, however, the helium atom loses just one of its electrons, it becomes a ***one-electron atom*** and can be described by the Bohr theory. However, the double positive charge on the nucleus must be taken into account.

Bohr's energy formula (Equation 15–8)

$$E_n = -\frac{2\pi^2 m Z^2 e^4}{n^2 h^2} \qquad (n = 1, 2, 3, \ldots, \infty)$$

can be written as

$$E_n = -\frac{Z^2 E_0}{n^2} \qquad (n = 1, 2, 3, \ldots, \infty) \tag{15–14}$$

where $E_0 = 2\pi^2 me^4/h^2 = 13.60$ eV. Similarly, from Equations 15–9 and 15–10 it follows that the most general form of Balmer's formula becomes

$$\frac{1}{\lambda} = Z^2 R\left(\frac{1}{{n_1}^2} - \frac{1}{{n_2}^2}\right) \tag{15–15}$$

where R is the Rydberg constant. Equations 15–14 and 15–15 are valid for any one-electron atom, where Z represents the number of positive charges in the nucleus. For example, the energy levels of singly ionized helium (He^+), for which $Z = 2$, are given by the equation

$$E_n = \frac{-4E_0}{n^2}$$

Therefore the energy of each level in He^+ is four times greater (more negative) than the energy of the corresponding level in hydrogen. Thus $E_1 = -4E_0 = -54.4$ eV, $E_2 = -E_0 = -13.6$ eV, and so on. Consequently, there will be a certain similarity between the spectra of He^+ and hydrogen. Bohr was able to show that all the extra lines found by Pickering and Fowler were actually from the He^+ spectrum. In particular, the series resulting from transitions to $n = 4$ in He^+ has the same limit as the Balmer series ($n = 2$) in hydrogen. In fact, the lines resulting from the $6 \longrightarrow 4$, $8 \longrightarrow 4$, $10 \longrightarrow 4, \ldots$, transitions in He^+ closely match the lines of the hydrogen Balmer series ($3 \longrightarrow 2$, $4 \longrightarrow 2$, $5 \longrightarrow 2$, etc.) and the odd-numbered transitions in He^+ ($5 \longrightarrow 4$, $7 \longrightarrow 4$, $9 \longrightarrow 4$, etc.) represent the "extra" lines that had been previously attributed to hydrogen.

Fowler argued that Bohr's equation did not exactly predict the wavelengths of the extra lines, there being a discrepancy of nearly 3 Å. Bohr countered by showing that this error arose because he had assumed in the derivation of the equation that the electron travels in a circle about the nucleus at the center, when in fact both the electron and nucleus rotate about their common center of mass. Bohr found that center-of-mass motion could be taken into account simply by multiplying the Rydberg constant by the factor $M_{nuc}/(m + M_{nuc})$, where m is the electron mass, and M_{nuc} is the mass of the nucleus. With this correction—which is quite small, since $M_{nuc} \gg m$—Bohr was able to account for the errors between the calculated and measured wavelengths of the spectral lines of hydrogen and He^+. Bohr's center-of-mass correction also accounted for the small ($\approx$ 0.06 percent) variation in R from element to element that had been observed previously.

The Bohr theory was used to predict the spectra of other one-electron atoms such as doubly ionized lithium, Li^{2+} ($Z = 3$), triply ionized beryllium, Be^{3+} ($Z = 4$), and so on. The predicted lines have since been found in the spectra of the sun and other stars, where such species are expected to exist.

EXAMPLE 15–1: Calculate the wavelength of the photon emitted in the $4 \longrightarrow 3$ transition in Li^{2+} ($Z = 3$).

SOLUTION: Li^{2+} is a one-electron atom. Its energy levels are predicted by Equation 15–14:

$$E_n = \frac{-Z^2E_0}{n^2} = \frac{-(3)^2(13.60)}{n^2} = \frac{-122.4}{n^2}\text{ eV}$$

The electron energies of the $n = 4$ and $n = 3$ levels are therefore

$$E_4 = \frac{-122.4}{(4)^2} = -7.65\text{ eV} \qquad \text{and} \qquad E_3 = \frac{-122.4}{(3)^2} = -13.60\text{ eV}$$

Hence $\Delta E = E_4 - E_3 = (-7.65) - (-13.60) = 5.95$ eV. The wavelength of a photon of energy 5.95 eV is

$$\lambda = \frac{12{,}400}{5.95} = \underline{2084\text{ Å}}$$

which lies in the ultraviolet. The same result could also have been obtained by direct substitution into Equation 15–15:

$$\frac{1}{\lambda} = Z^2R\left(\frac{1}{{n_1}^2} - \frac{1}{{n_2}^2}\right) = (3)^2R\left(\frac{1}{(3)^2} - \frac{1}{(4)^2}\right) = 9R\frac{(16-9)}{(9)(16)} = \frac{7R}{16}$$

$$\lambda = \frac{16}{7R} = \frac{16}{(7)(1.097 \times 10^5)} = 2.084 \times 10^{-5}\text{ cm} = \underline{2084\text{ Å}}$$

ABSORPTION AND EMISSION PROCESSES

One of Bohr's postulates is that an electron can gain or lose energy only by a transition from one orbit to another. In order to jump from a lower energy level to a higher one, an electron must absorb from some outside source precisely the amount of energy needed for the transition. For example, a transition in the hydrogen atom from $n = 1$ ($E = -13.60$ eV) to $n = 2$ ($E = -3.40$ eV) requires that the electron absorb precisely 10.20 eV—no more, no less. Moreover, the electron cannot simply absorb small amounts of energy until it has accumulated the 10.20 eV needed for the transition. *It must absorb the energy all at once* and upon doing so instantaneously makes the transition.

An electron can absorb energy for a transition by a number of processes. One way is through a collision between the electron and another particle, in which some of the kinetic energy of the particle is transferred to the electron. Collisions between gas atoms in random motion can be sufficiently energetic to raise electrons to higher energy states, particularly if the gas is at a high temperature. Another way for the electron to gain enough energy for a transition is by absorbing a photon that possesses

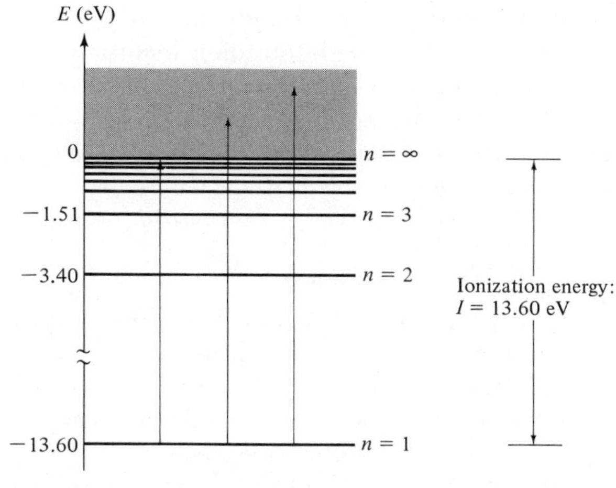

figure 15–8 **IONIZATION**

The hydrogen atom becomes ionized when an electron in the ground state absorbs sufficient energy to jump to an unbound state ($E \geq 0$).

exactly the energy needed. However, the electron must *completely* absorb the photon; a bound electron cannot absorb only a part of the photon energy.

Suppose that through collisions or photon absorption an electron gains sufficient energy to be raised to the $n = \infty$ level. Now the electron is no longer bound to the atom in a closed orbit, and it escapes. The atom has lost a negative charge and is no longer electrically neutral; *it is now a positively charged ion.* The energy level diagram (Figure 15–8) shows that an electron in the hydrogen ground state must absorb a minimum of 13.60 eV of energy for ionization to take place. This quantity is known as the ***ionization energy** I* of hydrogen. An electron can absorb more than 13.60 eV in the ionization process; any excess energy above 13.60 eV will appear as the kinetic energy of the removed electron.

When an electron jumps from a higher energy level to a lower one, it must emit a quantum of energy corresponding to the energy difference between the levels. Normally, this emission is in the form of a photon. (Under special circumstances the electron may lose the required amount of energy in a collision with another particle.) Thus the photons emitted by an atom must have energies that correspond to the energy differences between the electron energy levels. Furthermore, since the energy of the photon emitted or absorbed depends only on the energy difference between two energy levels, it follows that the energy of the photon *absorbed* by an electron jumping from a lower level A to a higher level B is the same as the energy of the photon *emitted* when the electron drops from B down to A.

An electron in the ground state cannot lose energy. Therefore before an atom can emit light, it must first absorb sufficient energy (through collisions or photon absorption) to raise an electron to an excited state. Once an electron is raised to an excited state, it will almost immediately return to the ground state. It may return *directly* to the ground state (emitting a single photon) or *indirectly* through intermediate states (emitting several photons). For example, an electron in the $n = 3$ state can return to the ground state either by a $3 \longrightarrow 1$ transition or by a $3 \longrightarrow 2$ followed by a $2 \longrightarrow 1$ transition (emitting two separate photons in the process).

In general, an electron remains in an excited state for an extremely brief period of time—typically, no more than 10^{-8} sec—before returning spontaneously to the ground state. In certain atoms, however, an electron may remain in an excited state for a rather long period of time before returning to the ground state. An excited state of this type is called a ***metastable state.***

LINE SPECTRA EXPLAINED

The properties of bright-line and absorption-line spectra, which were a mystery to physicists for many years, can be easily understood in terms of the absorption and emission processes described by Bohr.

A bright-line spectrum is produced when a low-density gas is heated to incandescence. Through collisions the electrons in the gas atoms are raised to excited states. In returning to the ground state, each electron emits a photon whose energy (and wavelength) corresponds to the energy difference between the levels. Since there is a limited number of electron levels in the atom, only certain wavelengths of light can be emitted. *This explains why bright-line spectra consist of relatively few wavelengths.* Each element has its own characteristic set of spectral lines, because the atoms of different elements have different arrangements of energy levels.

An absorption or dark-line spectrum is produced by passing a continuous spectrum through a gas. The atoms of the gas absorb only those photons having energies corresponding to the differences between electron energy levels. The light leaving the gas is deficient in those wavelengths of light that were absorbed. *This explains why dark lines appear in an absorption spectrum.* Moreover, since the photons absorbed by an atom in a given transition have the same energies as those emitted by the atom, it is easy to see why the wavelengths that are found in a bright-line spectrum are the same as the ones that are missing from the absorption spectrum of the same element.

To clear up one possible misconception, we should recognize that the lines seen in the spectrum of a gas do not all come from the same atom. A sample of gas contains an enormous number of atoms (roughly 10^{19} per cm^3 under normal conditions), each one of which is independently absorbing and/or emitting photons. At any given instant a certain fraction of these atoms may be emitting photons of one specific wavelength; another fraction may be emitting photons of another wavelength, and so on. What is seen in the spectroscope is the *total* light given off by all of these atoms at a given time.

All of the possible spectral lines of a given element do not necessarily appear with the same intensity, and some spectral lines may not be present at all under special circumstances. For example, suppose a sample of hydrogen gas is maintained at room temperature and a continuous spectrum is passed through it. The absorption spectrum that is produced will exhibit relatively few dark lines because at low temperature most of the hydrogen atoms have their electrons in the ground state ($n = 1$). If a hydrogen electron is in the $n = 1$ level, it can absorb only the wavelengths of the $1 \longrightarrow 2$, $1 \longrightarrow 3$, $1 \longrightarrow 4, \ldots,$ transitions. In order for any other lines to appear in an absorption spectrum, the electron must be in the $n = 2$ or higher levels *before* it absorbs a photon. If the temperature of the gas is raised to several thousand degrees, the random-motion collisions between atoms will be energetic enough to raise a significant number of electrons to the $n = 2$ level or higher. These electrons will, of course, immediately return to the ground state, but if the collision rate is high enough (depending on the density and temperature of the gas), there will always be at any instant a sizable fraction of atoms with their electrons in the $n = 2$ state or higher. These electrons are capable of absorbing photons corresponding to $2 \longrightarrow 3$, $2 \longrightarrow 4, \ldots$, transitions. Thus as the temperature of the gas changes, the different spectral lines change in intensity.

In Chapter 14 we saw that spectral lines have "width"; that is, they do not have a single wavelength but are spread out over a narrow range (Figure 14–5). This observation would seem to contradict the assertion that an electron must absorb or emit *exactly* the energy it needs to make a given transition. However, spectral line widening can be understood in terms of the Doppler effect. The gas atoms are in random motion as they absorb and emit light. Thus if an atom is moving toward the observer at the moment of emission, the emitted light will be slightly violet-shifted, and if the atom is moving away, the light will be red-shifted. Hence the total emitted light from a large group of atoms will exhibit Doppler shifts of various sizes in proportion to the speeds of the emitting atoms. The effect is that each spectral line appears to be spread over a range of wavelengths centered around the true wavelength emitted by a stationary atom.

Thus the basic features of line spectra can be readily explained by the Bohr theory. The theory can also be used to account for several other well-known phenomena.

FLUORESCENCE

Most objects, when illuminated by light of a particular wavelength, will either reflect the light or absorb it and re-emit the same wavelength. In either case the light emanating from the body has the same wavelength as the light that illuminates the body. Certain substances, however, will emit light of a particular wavelength *even though that wavelength is not present in the light incident on the body.* This phenomenon is known as ***fluorescence*** (named for the mineral fluor spar, which exhibits this property). Typically, *ultraviolet* light shining on a fluorescent material will result in the emission of *visible* light from the material.*

The process of fluorescence can be readily explained in terms of the Bohr theory. An electron in a lower energy state absorbs an ultraviolet photon and jumps to a high energy level. Instead of dropping directly back to the ground state by a single transition (emitting a photon of the same wavelength as the one absorbed), the electron returns to the ground state via two or more transitions involving intermediate states. With each transition, a photon is emitted. Hence the electron energy is released in the form of several lower-energy photons instead of a single high-energy one. In one of the intermediate steps, a photon of visible light may be given off. Thus the atom absorbs ultraviolet light and emits visible light in the process of returning to the ground state.

Let us use hydrogen to illustrate this process. (See Figure 15–9.) An electron in $n = 1$ absorbs a 12.09-eV ultraviolet photon and jumps to $n = 3$. The electron

*The so-called black-light lamps that cause objects to glow in the dark operate in this manner. A black-light lamp actually emits invisible ultraviolet light. An object made of or coated with fluorescent substances will absorb ultraviolet and emit visible light.

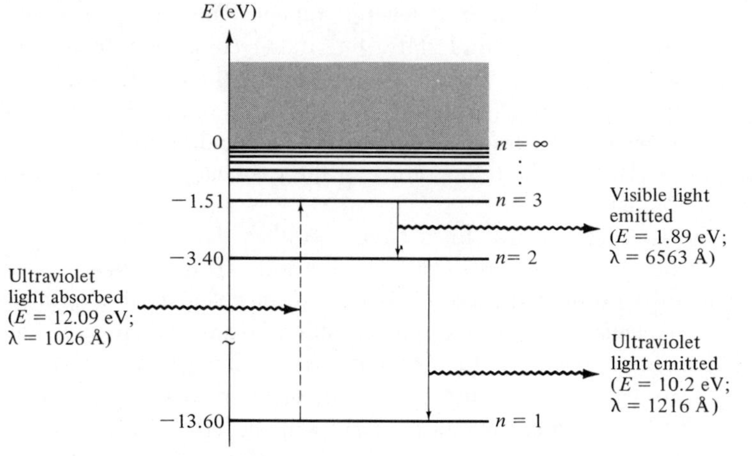

figure 15–9 **FLUORESCENCE**

An electron jumps to a high energy level in a single step by absorbing ultraviolet radiation. It returns to the ground state by successively dropping down through intermediate levels. Visible light is emitted in one of these transitions. Thus in the process of fluorescence, ultraviolet light is absorbed, and visible light is emitted.

returns to $n = 1$ by two transitions: $3 \longrightarrow 2$ (emitting a 1.89-eV photon), then $2 \longrightarrow 1$ (emitting a 10.20-eV photon). The 1.89-eV photon has a wavelength of 6563 Å, in the visible (red) part of the spectrum. Hence when ultraviolet light of sufficient energy is passed through hydrogen, the hydrogen will "fluoresce" by emitting red light.

A fluorescent light operates on this principle. The tube is filled with mercury vapor, and the glass wall is coated with a mixture of fluorescent materials. An electric discharge through the tube causes the mercury atoms to emit ultraviolet light. The ultraviolet light strikes the tube wall and causes visible light to be emitted by the fluorescent coating. An incandescent light bulb contains a high-temperature tungsten filament, which emits a continuous spectrum. Most of the energy emitted is in the infrared. A fluorescent tube, however, emits a line spectrum and little or no infrared. This is why a fluorescent tube is considerably cooler and a more efficient light source than an ordinary incandescent light bulb.

STIMULATED EMISSION: THE LASER PRINCIPLE

In 1916 Albert Einstein published a paper in which he derived Planck's radiation formula from the principles of the Bohr theory. The immediate significance of this work was to show that *continuous* spectra as well as line spectra could be accounted for within the framework of the Bohr theory. This could be achieved by assuming that the electrons in solid bodies also make quantum jumps between distinct energy levels—except that in solids the electron energy levels are determined by the forces exerted by a group of atoms, rather than by a single nucleus.

To arrive at Planck's formula, Einstein dealt with three processes by which an electron can emit or absorb radiation (Figure 15–10). The first is ***absorption,*** in which an electron in a lower energy state (E_1) jumps to a higher energy state (E_2) by absorbing a photon with a frequency $f_{12} = (E_2 - E_1)/h$. The second is ***spontaneous emission,*** in which an electron in a higher energy state (E_2) spontaneously and without outside influence drops to a lower energy state (E_1) and *emits* a photon of the same frequency, f_{12}. These two processes were described by Bohr. The third process, ***stimulated emission,*** was Einstein's contribution. Suppose an electron is in the higher energy state (E_2) when a photon of frequency f_{12} impinges on the atom. This photon will not be absorbed, since the electron is already in the higher energy state. However, the presence of this photon, Einstein proposed, can stimulate the electron to drop down to E_1 (*before* spontaneous emission occurs), *emitting in the process a second photon that is identical to the first.* The emitted photon will not only be identical in frequency, but will also be exactly in phase (producing reinforcement of the wave pattern) and will be traveling in the same direction as the incident photon that triggered its emission.

Einstein gave no theoretical justification for the existence of stimulated emission—other than to point out that if its existence is *not* assumed, the derivation leads to Wien's law, rather than Planck's. Characteristically, Einstein made no mention of any possible practical applications of the process. There the matter rested for nearly forty years.

Interest in stimulated emission was suddenly revived in the early 1950s when it was recognized that the process might be of value in communications. A given input signal (consisting of electromagnetic radiation of a given frequency) could trigger stimulated emission in an appropriate atom or molecule. Since the emission consists of radiation of the same frequency, phase, and direction as the input radiation, the effect is the amplification of the input signal to a much higher intensity. (One immedi-

figure 15–10 **STIMULATED EMISSION**

A photon of an energy exactly equal to the difference between the two energy levels E_1 and E_2 will stimulate an electron in the higher level to drop to the lower level and emit an identical photon. The incident and emitted photons will reinforce each other. The competing processes of absorption and spontaneous emission are shown for comparison.

$$f_{12} = \frac{E_2 - E_1}{h}$$

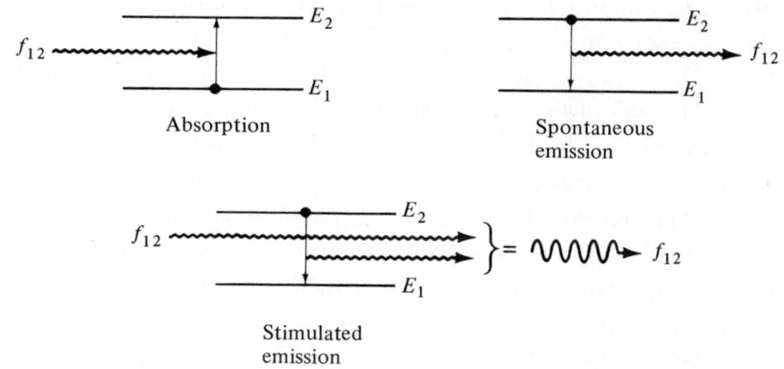

ate practical value is in the detection, through amplification, of extremely weak radio signals.) The theory of the amplification process was worked out independently by Charles Townes of the United States and N. G. Basov and A. M. Prokhorov of the U.S.S.R.

The first practical amplifier of this type was constructed by Townes and his co-workers at Columbia University in 1954. The device was named the ***maser*** (an acronym for Microwave Amplification by Stimulated Emission of Radiation). The first maser used ammonia (NH_3) molecules to amplify at the microwave frequency of 23,870 MHz ($\lambda = 1.25$ cm), which corresponds to the transition to the first excited state of NH_3. The amplification of visible light by the same principle was achieved in 1960, when the first ***laser*** (for Light Amplification by Stimulated Emission of Radiation) was constructed.

The fundamentals of the operation of a maser or laser can be easily understood in terms of the Bohr theory. In order for stimulated emission to occur, the electron must already be in an excited state when a photon of the proper frequency appears. At normal temperatures most electrons are in the ground state of the atom, where they will *absorb* the incident photon, rise to an excited state, and then re-emit another photon of the same frequency. In this case there will be no amplification. Therefore in order to minimize the competition between stimulated emission and absorption, there must be more atoms with their electrons in the excited state than in the ground state. (This is known as an ***inverted population.***) To obtain this condition, energy must be supplied to "pump" the electrons continually up to the excited state. This can be done by bombarding the laser material with an electric discharge or high-intensity light.

Once the electron is in an excited state, it should remain there until stimulated to emit a photon. If the electron returns to the ground state by spontaneous emission, the emitted photon will not be coupled to any incident photon. In other words, if there is any competition between *spontaneous* emission and *stimulated* emission, there will be little or no amplification. Such competition is avoided if the excited state is a metastable state. The substances chosen for use in masers or lasers are specifically selected on this basis. (For example, the ammonia molecule has a metastable excited state at 10^{-4} eV above the ground state.)

Radiation produced by stimulated emission is highly coherent (i.e., all waves are exactly in phase) and highly monochromatic. Under optimum conditions the frequency of the emitted radiation is constant to 1 part in 10^9. The emitted radiation can be concentrated into an extremely narrow beam, in which the energy concentration may be as high as 10^{12} watts/cm^2. A laser is not an especially efficient device, however, since a considerable amount of energy must be supplied initially to pump the electrons up to an excited state. The utility of the device lies in the fact that the output energy (which is only a few percent of the input) is highly concentrated.

The scientific and technical applications of masers and lasers are innumerable. Because of the high-frequency stability of the output signal, a maser can be used as an "atomic clock," with an error of no more than 1 sec per 1000 years. Because of the coherence of laser light, it is particularly suitable for high-precision interference experiments. A laser beam can be used for transmitting signals, measuring distances, precise surgical operations, cutting of metals or diamonds, ad infinitum. The invention of the laser has come to be regarded as one of the most significant technological advances of the 1960s.

15–5

LIMITATIONS OF THE BOHR THEORY

By 1914 the Bohr theory had reached its zenith. From the theory one could predict with remarkable accuracy the wavelengths of all the observed spectral lines of hydrogen and other one-electron atoms. The quantum nature of atomic energy levels had been clearly demonstrated. The century-old puzzle of the production of emission and absorption line spectra by gases had finally been explained through Bohr's radically new model of atomic structure. Yet it was not long before the Bohr theory began to display serious weaknesses. We will now briefly look at two of the most notable ones.

MANY–ELECTRON ATOMS

The first major shortcoming of the Bohr theory was that it could not be extended with any accuracy to any atom more complex than a one-electron atom. Bohr tried without success to derive a theoretical model for neutral helium, which has two electrons. The difficulty is that each electron experiences a repulsive force from the other electron in addition to the attractive force from the nucleus. The net force on each electron will be the vector sum of these two forces. Since both electrons are presumably in orbital motion, the distance between them will vary and so will the magnitude and direction of their mutual repulsive force. Because of these complications, it is impossible to get an exact solution for the radii and energy levels of the two electrons.

A similar problem had arisen once before in physics. In order to predict planetary orbits accurately, the gravitational forces exerted by the other planets had to be taken into account. However, the forces between planets are quite small compared to the sun–planet force, so the deviations from an elliptical orbit are quite minor and can be calculated by approximate methods. In the helium atom, however, the electron–electron force is nearly the same magnitude as the electron–nucleus force; hence the standard approximation techniques are not applicable. Bohr could not get his theory to give even a reasonable estimate of the energy levels or spectral lines of the helium atom.

FINE STRUCTURE

A second significant failure of the Bohr theory came about as the result of more precise measurements of spectral lines. As the resolution of the spectroscope was improved (i.e., spectra were spread out over a greater range), it was discovered that there were many more spectral lines in the hydrogen spectrum than just those predicted by the Bohr theory. What appeared to be a single line in a low resolution spectroscope turned out to be several closely spaced lines when seen in a spectroscope of higher resolution. These subdivisions of what were once thought to be single lines came to be known as the ***fine structure*** of atomic spectra. One notable example is the first Balmer line (the $3 \longrightarrow 2$ transition, $\lambda = 6563$ Å), which on closer inspection is found to be five separate lines (Figure 15–11). With the discovery that *all* hydrogen lines exhibit some fine structure, the Bohr theory no longer represented accurately the spectrum of hydrogen.

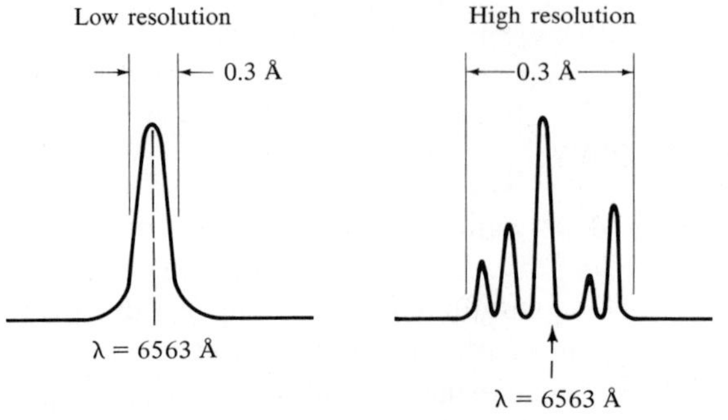

figure 15–11 **FINE STRUCTURE OF SPECTRAL LINES**

In a spectroscope of high resolution, the first line of the Balmer series of hydrogen is found to be composed of five separate, closely spaced lines.

In spite of these failures, the Bohr theory was still considered sufficiently successful so that few physicists believed it should be rejected completely. Instead, they preferred to make revisions of the theory, which would hopefully bring it into line with the more precise spectroscopic data. We shall assess the merits of these attempts to preserve the Bohr theory in the following chapter.

No one recognized the weaknesses of the Bohr theory more than Niels Bohr himself, and he remained at the focus of the attempts to devise a more comprehensive picture of atomic structure. Bohr was the inspirational force behind many of the significant developments of twentieth-century physics; his own contributions to the field span a period of nearly fifty years.

Not the least of Bohr's achievements was the establishment of the Institute for Theoretical Physics in Copenhagen. The Institute received most of its financial support from the estate of the founder of the Carlsberg brewery, and Bohr, as Denmark's leading scientist, was provided with his living quarters in the Carlsberg mansion. "Bohr's Institute" grew to be an international center for physics research, where young physicists could come to test out their ideas on one another. Bohr himself took a keen interest in the work of all of the scholars-in-residence and sought to maintain a convivial, informal atmosphere. One anecdote that perhaps typifies the spirit that Bohr fostered at the Institute was related by one of his former students, George Gamow:

> The evening work in the Institute's library was often interrupted by Bohr, who would say that he was very tired and would like to go to the movies. The only movies he liked were wild Westerns (Hollywood style), and he always needed a couple of students to go with him and explain the complicated plots. . . . But his theoretical mind showed even in these movie expeditions. He developed a theory to explain why although the villain always draws first, the hero is faster and manages to kill him. This Bohr theory was based on psychology. Since the hero never shoots first, the villain has to decide when to draw, which impedes his action. The hero, on the other hand, acts according to a conditioned reflex and grabs the gun automatically as soon as he sees the villain's hand move. We disagreed with this theory, and the next day I went to a toy store and bought two guns in Western holsters. We shot it out with Bohr, he playing the hero, and he "killed" all his students.[2]

Over the years the Institute came to acquire an almost legendary reputation. Among the well-known physicists who spent some time there were Wolfgang Pauli, Werner Heisenberg, Paul Dirac, Edward Teller, and J. Robert Oppenheimer. These were the men who shaped the history of twentieth-century physics. Their discoveries were to have a profound effect on the course of civilization. They all received some of their early encouragement from Niels Bohr.

QUESTIONS FOR REVIEW

1 What is Balmer's formula? What does it apply to specifically? How was it obtained?

2 Describe Thomson's model of the atom. In what two ways did it fail to agree with experimental findings about the properties of the atom?

3 Describe Rutherford's alpha-scattering experiment. What was the purpose of this experiment? What results were obtained?

4 Show how the results of Rutherford's experiment conflicted with the Thomson model of the atom. Explain how Rutherford used his experimental results to construct a different atomic model. How did Rutherford's model conflict with previously held views of the nature of atoms?

5 Why is Rutherford's atomic model unstable, according to the principles of classical physics? How did Bohr deal with this problem?

6 Explain what an energy level is. According to the Bohr theory, how does an electron in an atom change its energy?

7 State Bohr's assumption about the angular momentum of an orbiting electron. How did he justify this assumption? What is meant by the statement "The angular momentum of the electron is quantized."?

8 State and explain the equation governing the electron energy levels of the hydrogen atom. What is the significance of the quantum number n that appears in this equation?

9 Explain why E_n is a negative quantity. Under what conditions will E_n be positive? What is the lowest possible energy an electron can have in the hydrogen atom?

10 What is the significance of the fact that Balmer's formula can be derived from the Bohr theory? How did Bohr account for the numerical value of R (the Rydberg constant)?

11 Summarize the main postulates of the Bohr theory. Discuss whether each of these postulates does or does not agree with classical physics.

12 What prediction does the Bohr theory make about the radius of an electron orbit in the hydrogen atom?

13 Explain what is meant by each of the following:
(a) An energy level diagram.
(b) The ground state.
(c) An excited state.
(d) The continuum.

14 Identify each of the various series (Lyman, Balmer, Brackett, etc.) of spectral lines found in the hydrogen spectrum. In what parts of the spectrum do these lines appear? To what transitions in the hydrogen atom do they correspond?

15 Describe the process of ionization in terms of the Bohr theory. What is meant by the ionization energy? What is the ionization energy for hydrogen?

16 Explain how the main features of bright-line spectra can be accounted for in terms of the Bohr theory. Why must a gas be heated in order to give off an emission spectrum?

17 Use the Bohr theory to explain how an absorption spectrum is produced. Why do the dark lines in an absorption spectrum have the same wavelengths as the bright lines in the emission spectrum of the same gas?

18 Describe the phenomenon of fluorescence, and explain it in terms of the Bohr theory. How does a fluorescent light work, and what are its advantages?

19 Describe the process of stimulated emission. How does it differ from spontaneous emission?

20 Discuss the basic principles of the operation of a laser. Be sure to explain (a) what an inverted population is, and why one must be obtained, (b) why the substance used must have a metastable excited state, (c) how laser light differs from ordinary light, and (d) what the difference is between a laser and a maser.

21 Explain why the Bohr theory could not be used for anything more complicated than a one-electron atom.

22 What is fine structure? How was it discovered? How did the existence of fine structure affect the validity of the Bohr theory?

EXERCISES

1 What is the wavelength predicted by Balmer's formula for the spectral line corresponding to $n = 7$? How does this value compare with the observed wavelength of 3970 Å? Why wasn't the existence of this spectral line known before 1890?

2 Show that Equation 15–2 for Balmer's formula is identical to Equation 15–1.

3 What is an alpha particle? If $q/m = 9.65 \times 10^7$ coulomb/kg for a hydrogen ion, what is q/m for an alpha particle?

4 By what factor will the electrostatic repulsion between an alpha particle and an atomic nucleus increase or decrease if (a) the distance between them decreases from 10^{-8} to 10^{-12} cm? (b) silver ($Z = 47$) is used instead of gold ($Z = 79$)?

5 (a) Compare the electrical attraction between an electron and a hydrogen nucleus with their gravitational attraction when they are 10^{-8} cm apart. (The masses of the electron and the hydrogen nucleus are 9.11×10^{-28} gram and 1.67×10^{-24} gram, respectively; $e = 4.8 \times 10^{-10}$ esu.)
(b) What does the result tell us about the role of gravitation at the atomic level?
(c) Explain why the ratio of the electric and gravitational forces is independent of the distance.

6 Show that Equations 15–7 and 15–8 for the orbital radius r_n and electron energy E_n in hydrogen can be derived from Equations 15–4, 15–5, and 15–6.

7 The equation $\lambda = 12{,}400/E$ can be used to calculate the wavelength (in Å) of the light emitted or absorbed by an electron when it loses or gains a quantum of energy E. Discuss whether or not the use of this equation requires a belief in the existence of photons.

8 Using the energy level diagram for hydrogen shown in Figure 15–6, calculate the wavelength of the light emitted and identify the region of the electromagnetic spectrum where the light will be found for (a) the $3 \longrightarrow 1$ transition, (b) the $4 \longrightarrow 2$ transition, and (c) the $\infty \longrightarrow 2$ transition.

9 Prove that the constant E_0, which appears in Equation 15–12, is related to the Rydberg constant by the equation $E_0 = Rch$.

10 Show that for a one-electron atom with Z positive charge units in its nucleus (a) $r_n = r_0 n^2/Z$ and (b) $E_n = -Z^2E_0/n^2$.

11 Use the equation given in 10(a) to calculate the radii of the first four orbits of singly ionized helium (He^+). Compare these orbits with the corresponding orbits of the hydrogen atom. Can you give a physical explanation of why the He^+ orbits are smaller?

12 (a) Using the equation given in 10(b), construct the energy level diagram for He^+. Compare it with the energy level diagram for hydrogen, and explain whether the electron is more tightly bound in He^+ or in H.
(b) Calculate the wavelength of the light emitted in the $3 \longrightarrow 2$ transition. Where will it appear in the spectrum?
(c) Find a transition in He^+ that will give off visible light.
(d) What is the ionization energy for He^+?

13 What does the spectrum of ionized hydrogen (H^+) look like? Of doubly ionized helium (He^{2+})? What is another name for He^{2+}?

14 Imagine a hypothetical universe in which atoms obey the Bohr theory, but the physical constants (c, h, e, etc.) have slightly different values. In this universe the $4 \longrightarrow 2$ transition in hydrogen results in the emission of a 3-eV photon.
(a) What value does E_0 have in this universe?
(b) What is the energy of the photon emitted in the $3 \longrightarrow 1$ transition?
(c) What is the ionization energy of hydrogen?
(d) Can the wavelengths of the photons emitted in these transitions be calculated from the given data?

15 When the absorption spectrum of hydrogen gas at room temperature is examined, all the absorption lines are found in the ultraviolet.
(a) Explain why this is so.
(b) What must be done to make the hydrogen gas exhibit absorption lines in the visible?

16 Spectral line widths can be attributed to the Doppler shift of the light given off by moving atoms. Explain why spectral line widths tend to increase as the temperature of the gas increases.

17 Explain why the phenomenon of fluorescence is a violation of the law of exchanges.

18 What is the minimum wavelength of light required to make hydrogen gas fluoresce? Explain your answer.

19 Emission nebulae are huge clouds of hydrogen gas in space. (The Orion nebula, found in the sword of the constellation Orion, is the best-known example.) Embedded in these clouds are very hot stars that emit substantial quantities of ultraviolet radiation. These nebulae tend to glow with a distinct red light that has a wavelength of 6563 Å. Explain why this phenomenon is an example of fluorescence.

20 Explain how Young's wave-interference experiment can be simplified if a laser is used as the light source.

THE BOHR THEORY AND THE PERIODIC TABLE

16

Of course I am unhappy!
I cannot understand the anomalous Zeeman effect.

Wolfgang Pauli

THE DECADE from 1915 to 1925 was a period of intense activity in the field of atomic physics. Stimulated by the first successes of the Bohr theory, physicists attempted through a variety of different approaches to make improvements in the theory that would extend its range of applicability. A few of these attempts did in fact lead to a more detailed picture of atomic structure and spectra. Perhaps the most notable advance of this period was the use of a modified form of the Bohr theory to provide a systematic explanation of the properties of the chemical elements. This led eventually to a thorough understanding of the nature of the chemical properties of matter and, in effect, brought the separate sciences of chemistry and physics together under one roof. We will see, however, that in order for this result to be achieved, the Bohr theory had to be buttressed with so many new postulates and empirical rules that it could no longer be taken seriously as a fundamental theory.

16–1

CHEMICAL PROPERTIES AND THE BOHR THEORY

Following his first success with the hydrogen spectrum, Bohr's major concern was to extend his theory to cover more complex atoms. In this area Bohr was blessed with a wealth of experimental information: spectral line data for nearly all the known chemical elements, measurements of energy levels and ionization energies, and all the accumulated knowledge from the field of chemistry on the physical and chemical properties of the elements. We will begin with a summary of how the chemical elements are classified.

THE PERIODIC TABLE OF THE ELEMENTS

In 1800, a time when chemistry had only recently begun to evolve above the level of alchemy, there were only about thirty known chemical elements. Spurred on by such developments as Dalton's atomic theory and the invention of electrolysis, chemistry grew rapidly during the next fifty years. By the 1860s more than sixty elements had been identified, and their properties were being thoroughly studied and classified.

Among the most important chemical properties being measured during this time were the ***atomic weights*** of the elements. The atomic weight is a number assigned to an element that gives the relative weight of one atom of that element as compared to some standard atom.* The first table of atomic weights was proposed by Dalton in 1808. But the early methods for determining atomic weights were full of ambiguities, and the table was changed many times. A consistent method for determining atomic weights was finally established by about 1860.

A few years later it was discovered that when the elements are listed in order of increasing atomic weight, certain regularities appear in the listing. In 1864 John Newlands observed that every successive eighth element has similar properties. He called this discovery the "law of octaves" and likened it to the notes of the musical scale (harking back to the days of Pythagoras and the "music of the spheres"). But Newlands' law was based on rather incomplete information about the elements, and it broke down in the middle of the third octave. It was received, not unjustifiably, with considerable ridicule.

Newlands' seemingly ineffective idea came to fruition in the work of the remarkable Russian chemist Dmitri Mendeleev (1834–1907). The Siberian-born Mendeleev was the youngest of fifteen children. At the age of fourteen he made the arduous journey across Russia to St. Petersburg to further his education. He graduated from the university with a degree in chemistry and eventually rose to the position of professor of chemistry at St. Petersburg. Mendeleev adopted Newlands' idea of listing the elements by atomic weight, but rejected the necessity of a law of octaves. Instead, he simply arranged the elements in a table in such a way that elements having similar chemical properties fall naturally into the same group.

A modern-day version of Mendeleev's table is shown in Figure 16–1. We see that the elements are arranged in rows in order of increasing atomic weight. Each *row* constitutes one ***period.*** The elements in each *column* form a ***family.*** All the elements of one family are rather similar chemically. In particular, they enter into chemical reactions with other elements in the same way and form similar chemical compounds. Mendeleev called this table the ***periodic table*** and stated the ***periodic law:***

The properties of the elements are periodic functions of their atomic weights.

Mendeleev's first table, proposed in 1869, listed sixty-three elements. In order to maintain the periodicity of the table, Mendeleev left gaps that suggested the existence of as yet undiscovered elements. For example, in 1869 the element arsenic (As) followed zinc (Zn) in order of atomic weight and should then have been listed under aluminum (Al) in the table. But the properties of arsenic are much more similar to those of phosphorus (P) than to those of aluminum. Mendeleev therefore listed arsenic under phosphorus, leaving two gaps in the table. This arrangement was vindicated by the discovery of the elements gallium (Ga) in 1874 and germanium (Ge) in 1885. These elements fit perfectly into the places that Mendeleev left for them.

Within each family there is a generally systematic gradation of physical and

*The present-day scale of atomic weights, established in 1961, is based on one form of carbon (C^{12}) as the standard.

figure 16–1 THE PERIODIC TABLE OF THE ELEMENTS

An alphabetical listing of the elements by name and symbol may be found in Appendix B.

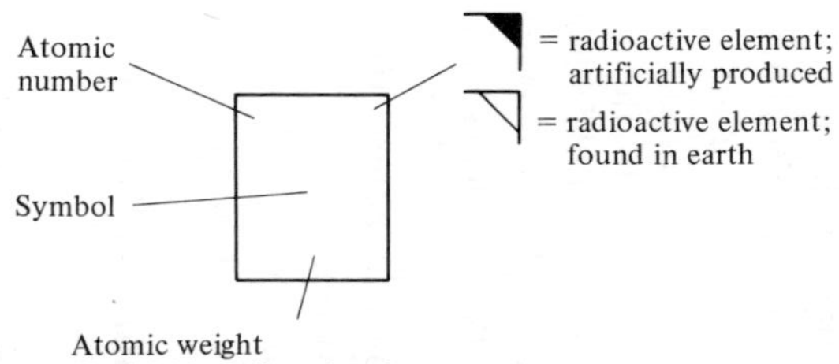

Period																			
1	1 H 1.008																		2 He 4.0026
2	3 Li 6.939	4 Be 9.013												5 B 10.81	6 C 12.011	7 N 14.007	8 O 15.999	9 F 18.998	10 Ne 20.18
3	11 Na 22.99	12 Mg 24.31												13 Al 26.98	14 Si 28.09	15 P 30.97	16 S 32.06	17 Cl 35.45	18 Ar 39.95
4	19 K 39.10	20 Ca 40.08		21 Sc 44.96	22 Ti 47.90	23 V 50.94	24 Cr 52.00	25 Mn 54.94	26 Fe 55.85	27 Co 58.93	28 Ni 58.71	29 Cu 63.54	30 Zn 65.37	31 Ga 69.72	32 Ge 72.59	33 As 74.92	34 Se 78.96	35 Br 79.91	36 Kr 83.80
5	37 Rb 85.47	38 Sr 87.62		39 Y 88.91	40 Zr 91.22	41 Nb 92.91	42 Mo 95.94	43 Tc (99)	44 Ru 101.07	45 Rh 102.91	46 Pd 106.4	47 Ag 107.87	48 Cd 112.40	49 In 114.82	50 Sn 118.69	51 Sb 121.75	52 Te 127.60	53 I 126.90	54 Xe 131.30
6	55 Cs 132.91	56 Ba 137.34	* (57-70)	71 Lu 174.97	72 Hf 178.49	73 Ta 180.95	74 W 183.85	75 Re 186.2	76 Os 190.2	77 Ir 192.2	78 Pt 195.09	79 Au 196.97	80 Hg 200.59	81 Tl 204.37	82 Pb 207.19	83 Bi 208.98	84 Po (210)	85 At (210)	86 Rn (222)
7	87 Fr (223)	88 Ra 226.03	** (89-102)	103 Lw (257)	104 Rf (261)	105 Ha (260)													

*Lanthanide elements:	57 La 138.91	58 Ce 140.12	59 Pr 140.91	60 Nd 144.24	61 Pm (147)	62 Sm 150.35	63 Eu 151.96	64 Gd 157.25	65 Tb 158.92	66 Dy 162.50	67 Ho 164.93	68 Er 167.26	69 Tm 168.93	70 Yb 173.04
**Actinide elements:	89 Ac (227)	90 Th 232.04	91 Pa (231)	92 U 238.03	93 Np (237)	94 Pu (244)	95 Am (243)	96 Cm (247)	97 Bk (247)	98 Cf (251)	99 Es (254)	100 Fm (253)	101 Md (256)	102 No (254)

table 16–1 **PROPERTIES OF THE ALKALI METALS**[a]

Element	Atomic number	Atomic weight	Density at 0°C (g/cm^3)	Melting point (°C)	Boiling point (°C)	Atomic radius (Å)	Ionization energy (eV)
Lithium (Li)	3	6.94	0.53	181	1347	1.61	5.39
Sodium (Na)	11	22.99	0.97	98	883	2.05	5.14
Potassium (K)	19	39.10	0.86	64	774	2.50	4.34
Rubidium (Rb)	37	85.48	1.53	39	688	3.10	4.18
Cesium (Cs)	55	132.9	1.87	28	678	3.40	3.89

[a]The element francium (Fr) is also a member of the alkali family, but it is not listed here. It is a radioactive element that exists in such small quantities that its properties cannot be accurately measured.

chemical properties from the lightest to the heaviest element. Because of this effect, Mendeleev was able to estimate the properties of the undiscovered elements with remarkable accuracy from the positions within a family of the gaps in the table. To illustrate this gradation, the properties of the ***alkali metal*** family (the first column in the periodic table) are listed in Table 16–1.

One family of elements that is of particular interest is the ***inert gas*** family, listed on the extreme right of the periodic table. These elements were discovered after the periodic table was proposed, but it was found that they could easily be added onto the periodic table as an extra column. The inert gases are characterized by chemical unreactivity, very low freezing and boiling points, and high ionization energies. (It was believed for many years that these elements were totally inert and could not react chemically at all. However, since 1962 a number of compounds of Kr, Xe, and Rn have been produced in the laboratory.)

The periodic table contains a wealth of information about the chemical elements, which could not possibly be included in this brief discussion. It has proved to be an extremely useful tool in both physics and chemistry. Yet in spite of the considerable attention focused on the periodic table from its inception, there was virtually no understanding of *why* the chemical elements displayed these regularities. It was the Bohr theory that eventually revealed the physical meaning behind the periodic table.

THE ATOMIC NUMBER AND MOSELEY'S LAW

In the periodic table the elements are listed in order of atomic weight, and each element is assigned an ***atomic number,*** which indicates its place in the listing. The atomic number was considered to be nothing more than a catalog number; it had no physical significance. Rutherford's experiments gave the first indication that the atomic number might be related to some physical property of the atom. Based on the alpha-scattering data of Geiger and Marsden, the Dutch physicist A. Van der Broek proposed in 1913 that *the atomic number of a given element is equal to the number of positive charges in the nucleus* (*and also the number of orbiting electrons*) *of an atom of that element.*

Strong evidence for Van der Broek's hypothesis was provided by the experimental work of Henry Moseley. Moseley, a student of Rutherford's at Manchester, analyzed the spectrum of the x rays produced when a beam of high-speed electrons strikes

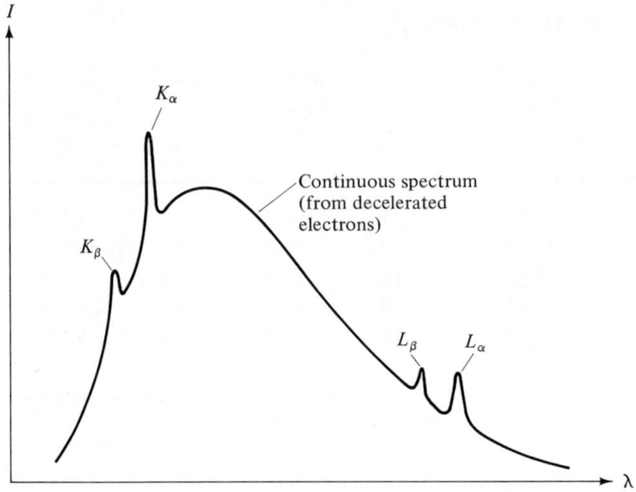

figure 16–2 **X-RAY SPECTRA**

The graph represents a typical spectrum of the x rays produced when high-speed electrons strike a target. The spectral lines K_α, K_β, etc., have wavelengths that are characteristic of the target material. Moseley's law (Equation 16–1) relates the frequency of the K_α line to the atomic number of the element.

a metal target. (Moseley's apparatus was essentially Roentgen's x-ray tube, shown in Figure 10–7.) He observed that the x-ray spectrum was mainly continuous, but also contained several prominent emission lines (Figure 16–2). The continuous spectrum of x rays could be attributed to the energy radiated by the electrons as they are decelerated by the target, but the emission lines were found to be characteristic of the target material itself. Moseley found that the frequency of the so-called K_α line (the most prominent line in the spectrum) varied in a very systematic way from element to element. By a trial-and-error procedure he found that his results could be fit rather closely by the formula

$$f = 2.48 \times 10^{15}(Z - 1)^2 \tag{16–1}$$

where f is the frequency of the K_α line in the x-ray spectrum of an element, and Z is the *atomic number of the element.* (In anticipation of the identity of the atomic number and the nuclear charge number, we have used the same symbol, Z, for both of them.) This equation is known as ***Moseley's law.***

Moseley's discovery did not directly establish any connection between the atomic number and the nuclear charge. It did show that the atomic number was more than just a catalog number; the law established that the atomic number was at least related to some physical property of the atom pertaining to x-ray emission. In the past the atomic number of a given element had been determined only by its location in the periodic table. Moseley's law now provided an unambiguous and accurate method for the *experimental measurement* of the atomic number: the observed frequency of the K_α x-ray spectral line for a particular element is substituted into Equation 16–1, which is then solved to find Z. Since the atomic number locates the position of an element in the periodic table, the measurement of Z by Moseley's law helped to dissolve many of the uncertainties as to the order of the elements in the table, particularly of elements with very similar chemical properties (notably the "transition" elements $_{21}$Sc to $_{30}$Zn and $_{39}$Y to $_{48}$Cd, and the "rare earth" elements $_{57}$La to $_{70}$Yb).

Moseley seemed to be headed for a distinguished career as an experimental physicist. When war broke out in 1914, however, Moseley voluntarily enlisted in the British army, and in 1915 he was killed in action in the abortive Dardanelles campaign.

Moseley's results were published in the same year as the Bohr theory. Soon afterward Moseley's law was explained in terms of the Bohr theory in a way that clearly showed that the atomic number of an element and the number of positive charges in the nucleus of an atom of that element were one and the same. The explanation assumes that a high-speed free electron emitted by the cathode of the x-ray tube collides with a bound electron in the $n = 1$ orbit of a target atom, causing the bound electron to be knocked completely out of the atom. Immediately, an electron from the $n = 2$ level falls into the $n = 1$ level, filling the spot vacated by the knocked-out electron. The characteristic K_α x-ray photon that is emitted carries away the energy lost by the electron in making the $2 \longrightarrow 1$ transition.

The wavelength of this photon can be calculated from Bohr's equation (Equation 15–15)

$$\frac{1}{\lambda} = RZ^2\left(\frac{1}{{n_1}^2} - \frac{1}{{n_2}^2}\right)$$

where Z is the nuclear charge number. For a $2 \longrightarrow 1$ transition,

$$\frac{1}{\lambda} = RZ^2\left(\frac{1}{1^2} - \frac{1}{2^2}\right) = \tfrac{3}{4}RZ^2$$

The frequency of the photon is then

$$f = \frac{c}{\lambda} = \tfrac{3}{4}cRZ^2$$

Substituting $c = 3 \times 10^{10}$ cm/sec and $R = 1.097 \times 10^5$ cm^{-1} gives

$$f = (2.48 \times 10^{15})Z^2 \qquad (16\text{–}2)$$

This equation would be identical to Moseley's law (Equation 16–1) if the Z^2 were replaced by $(Z - 1)^2$. This substitution can be justified by making the following assumption. Suppose the $n = 1$ level contains a *second* electron, which remains after the first electron is knocked out. This single negative charge exerts a repulsive force on any outer electrons; in effect, it cancels or "screens" the attractive force of one of the positive charges in the nucleus. Thus an outer electron experiences the attractive force of an *effective* nuclear charge of $Z - 1$ positive charges. If we then replace Z by $Z - 1$ in Equation 16–2, we arrive at Moseley's law.

By similar analyses the other x-ray emission lines can also be accounted for. It can be shown that certain lines come from the $3 \longrightarrow 1$ or $4 \longrightarrow 1$ transitions, and other lines arise from the $3 \longrightarrow 2$ and $4 \longrightarrow 2$ transitions (assuming that an electron is knocked out of the $n = 2$ level rather than $n = 1$). Thus Moseley's observations could be explained by the Bohr theory. More significantly, Bohr's explanation clearly identified the atomic number with the nuclear charge number and provided the first connection between the atomic theory and the chemical properties of the elements.

THE EVIDENCE FOR CLOSED ENERGY LEVELS

The assumptions needed to derive Moseley's law from the Bohr theory require careful scrutiny. They imply that in an atom with more than one electron *there is a limit to the number of electrons that can occupy a given energy level.* If this were not so, *all* the electrons would lie in the $n = 1$ level and the characteristic x-ray spectral lines would not appear. In other words, the fact that the K_α x ray comes from a $2 \longrightarrow 1$ transition implies that there are electrons in the $n = 2$ level to begin with, and that they cannot fall into $n = 1$ until there is a vacancy in that level. Also, to account for the $Z - 1$ term in Moseley's law, it had to be assumed that the $n = 1$ level holds a maximum of two electrons.

The concept of fully occupied or "closed" energy levels is certainly not inherent in the Bohr theory. In fact, one of the basic principles of the theory of the hydrogen atom is that an electron in a higher energy level will spontaneously drop down to lower energies until it reaches the $n = 1$ level. The evidence seems to indicate, however, that this is not the case for atoms with more than one electron. The idea that an object cannot go to a level of lower energy because that level is "closed" is extremely strange from a classical viewpoint. To draw an analogy, it is as if twelve bricks were thrown off the roof of a ten-story building, but only two reached the ground—six stopped in midair at the fifth floor, and the remaining four could not fall at all!

It might seem that the successful use of the hypothesis of closed energy levels in the explanation of Moseley's law is hardly a rigorous proof of the validity of this concept. However, there is a considerable range of additional experimental evidence pointing to the existence of closed energy levels in the atom.

Let us consider first the ionization energies of the elements, which can be measured experimentally. The ionization of hydrogen can be written symbolically as

$$\mathrm{H} \xrightarrow{13.6\ \mathrm{eV}} \mathrm{H}^+$$

which signifies that it takes 13.6 eV of energy to remove an electron from a neutral hydrogen atom, changing the atom to an ion with a net positive charge of one unit. The measured ionization energy of 13.6 eV agrees exactly with the value predicted by the Bohr theory.

The next element is helium ($Z = 2$), with two electrons:

$$\mathrm{He} \xrightarrow{24.6\ \mathrm{eV}} \mathrm{He}^+ \xrightarrow{54.4\ \mathrm{eV}} \mathrm{He}^{2+}$$

That is, it takes 24.6 eV to remove the first electron from a neutral helium atom and an additional 54.4 eV to remove the second electron. The latter number agrees with the calculated ionization energy of He^+ (which is a one-electron atom). The energy required to remove the first electron, 24.6 eV, is approximately half of 54.4 eV. The reduced ionization energy of the first electron can be attributed to the repulsive force of the second electron, which partially screens the nuclear charge and reduces the attractive force on the first electron. But both electrons are evidently in the $n = 1$ level of the helium atom. (The energy of the $n = 2$ level of the He^+ atom is only 13.6 eV.)

Now we come to lithium ($Z = 3$), with three electrons. The experimental findings are as follows:

$$\text{Li} \xrightarrow{5.4\ \text{eV}} \text{Li}^{+} \xrightarrow{75.6\ \text{eV}} \text{Li}^{2+} \xrightarrow{122.4\ \text{eV}} \text{Li}^{3+}$$

The energy required to remove the first electron (5.4 eV) is much less than that needed to remove the second and third electrons. This effect is much too large to be attributed to screening; it is evident that the *first electron is not in the same energy level as the other two*. Thus the lithium atom has two electrons in the $n = 1$ level and one electron in the $n = 2$ level.

Finally, we look at the data for beryllium ($Z = 4$):

$$\text{Be} \xrightarrow{9.3\ \text{eV}} \text{Be}^{+} \xrightarrow{18.2\ \text{eV}} \text{Be}^{2+} \xrightarrow{153.9\ \text{eV}} \text{Be}^{3+} \xrightarrow{217.6\ \text{eV}} \text{Be}^{4+}$$

The large difference between the energies required to remove the second and the third electrons again indicates that there are only two electrons in the first level.

Examination of the ionization energies for additional elements in turn provides clear evidence for the existence of a sequence of closed energy levels in the atom. This led Bohr to propose that in a many-electron atom the electrons are arranged in successive "shells" of increasing size. Figure 16–3 shows a graph of the ionization energies required to remove the first electron from each of the elements from hydrogen to uranium. Going from one element to the next in order of atomic number, each successive element possesses one more electron than the previous one. The sharp drops that occur periodically indicate the closing of a shell. That is, the low ionization energies for Li, Na, K, Rb, Cs—which indicate that the first electron is easily removed

figure 16–3 **EVIDENCE FOR CLOSED ENERGY LEVELS IN ATOMS**

A graph of the ionization energies of the elements, arranged in order of atomic number, shows periodic peaks and drops that indicate the filling of an energy level. A low ionization energy indicates that the electron is easily removed. Hence the electron is assumed to be in a higher energy level.

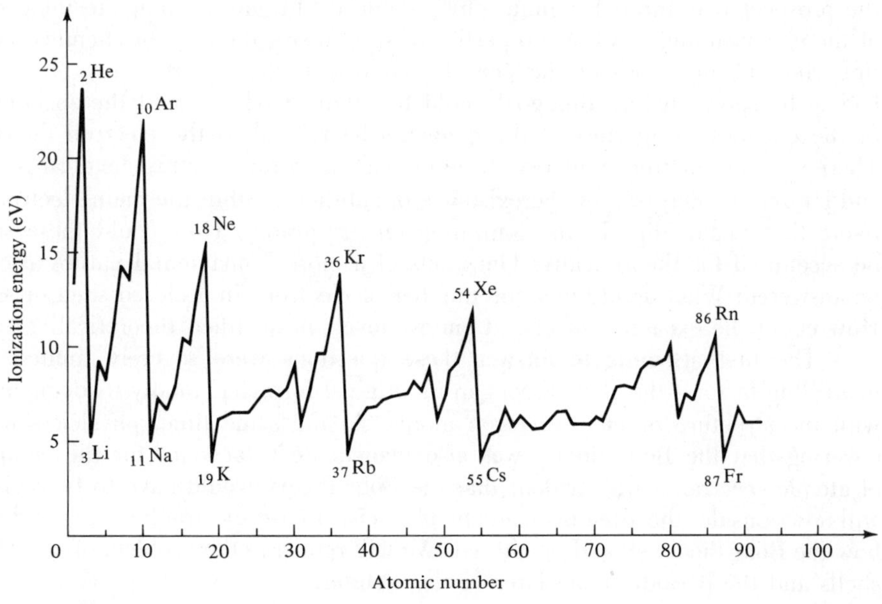

table 16–2 **FILLING OF ELECTRON SHELLS**

Shell number	Shell begins with	Shell closes with	Number of electrons in closed shell
1	$_1H$	$_2He$	2
2	$_3Li$	$_{10}Ne$	8
3	$_{11}Na$	$_{18}Ar$	8
4	$_{19}K$	$_{36}Kr$	18
5	$_{37}Rb$	$_{54}Xe$	18
6	$_{55}Cs$	$_{86}Rn$	32

from these atoms—means that an electron shell was filled with the previous element, and the first electron is in the next higher shell. It is a simple matter to show from the graph that there is a maximum of two electrons in the first shell, eight each in the second and third shells, eighteen each in the fourth and fifth, and thirty-two in the sixth. The details are given in Table 16–2.

In examining these results and comparing them with the periodic table, we find that the elements that open and close each successive shell are the same elements that begin and end each successive period of the periodic table! The atoms that close each shell are the inert gases; hence their chemical unreactivity can be attributed to the fact that all their electrons are in closed energy levels. Each shell begins with an alkali metal, and we can trace the high chemical activity of these elements to the single, easily removed electron that is the first to occupy the new shell. This establishes, for the first time, *a connection between the chemical properties of the elements and the distribution of electrons among the energy levels of the atom.*

The discovery of the relationship between the chemical properties of the elements and the electron configurations within the atom was a momentous one. It held out the prospect that through atomic physics one could gain a complete understanding of the chemical and physical properties of substances, the laws of chemical combination, and the mysteries of the periodic table. But there were many problems that had to be solved before this goal could be attained. How could the specific details of the chemical properties of the elements be related to the electron distribution? There was considerable evidence, from optical and x-ray spectra, chemical properties, and ionization energies, for the existence of subshells within the main electron shells. (Note the smaller dips in the ionization energy graph.) How could these subshells be accounted for theoretically? Questions of a more fundamental nature also had to be answered: What determines the number of electrons in a closed shell or subshell? How could the existence of closed energy levels be justified theoretically?

The first attempts to answer these questions were severely limited by one overriding factor—the Bohr theory in its original form was totally inadequate to deal with the structure of many-electron atoms. At the same time, physicists were discovering that the Bohr theory was also inadequate to account for the complexities of atomic spectra. It was evident that the Bohr theory would have to be revised. We will now consider the attempts made by physicists to use spectral line data to determine how the Bohr theory should be altered. We will return to the problem of closed energy shells and the periodic table later in the chapter.

16–2

THE BOHR–SOMMERFELD THEORY

The first significant improvement in the Bohr theory came about through attempts to account for the fine structure of the hydrogen spectral lines. (See Figure 15–11.) Success was achieved through the inclusion of two new factors in the theory: (1) elliptical electron orbits and (2) the theory of relativity. Originally, Bohr did not use elliptical orbits because he found that their introduction would have no significant effect; all orbits with the same major axis would have the same energy, regardless of eccentricity. Bohr later realized that this would not be the case if relativity were taken into account. If the electron is traveling in an elliptical orbit, its mass should change as its velocity changes. The mass variation would be greatest for the most elliptical orbits; hence the electron energy ought to show a small dependence on the orbital eccentricity.

The magnitude of the relativity correction can be estimated by calculating v/c for an orbiting electron. By combining Equations 15–6 and 15–7, it can be shown that for the electron in the hydrogen atom,

$$v = \frac{2\pi e^2}{nh}$$

For the $n = 1$ orbit, where the electron has its greatest speed, the quantity v/c (which determines the size of the relativity correction) has the value

$$\frac{v}{c} = \frac{2\pi e^2}{hc} = \frac{1}{137.04} = 0.0073$$

This is a rather small number, but because of the accuracy with which spectroscopic observations can be made, the corrections predicted by the use of relativity theory would be measurable.

The relativistic Bohr theory was worked out in detail by the German physicist Arnold Sommerfeld in 1915. In the original form of the Bohr theory all the orbits were circles, and the energy corresponding to a given orbit depended only on the radius of the orbit. A single quantum number, n, was sufficient to define the allowed values of both the electron energy and the orbital radius. But in the relativistic form of the theory, the energy of an orbit depends on both the average radius and the eccentricity. If the eccentricity of an electron orbit were allowed to take on any arbitrary value between 0 and 1, there would no longer be a finite number of electron energy levels; instead there would be a continuum of levels corresponding to all the possible values of the eccentricity. The basic concept of discrete energy levels in the atom would then be lost, and without this concept line spectra cannot be accounted for. So Sommerfeld introduced a *second quantum number* into the theory to restrict the possible values of the orbital eccentricity.

In Sommerfeld's form of the Bohr theory, the two quantum numbers are n, ***the principal quantum number*** (a carry-over from the old Bohr theory) and l, ***the orbital quantum number.**** Together, these two numbers define the size and

* In the original work Sommerfeld used a quantum number k $(= l + 1)$, which was replaced by l many years later. We have maintained the modern usage throughout.

shape of the allowed orbits. The original Bohr radius equation ($r_n = r_0 n^2$) now gives the *average radius* (equal to a, the semimajor axis) of the elliptical orbit, and the shape of the orbit is determined by the relationship

$$\frac{b}{a} = \frac{l+1}{n}$$

where b and a are the lengths of the semiminor and semimajor axes, respectively.

As before, n can have any integral value (1, 2, 3, . . .), but the values of l are restricted by the above equation, since b must be less than or equal to a in an ellipse. The *maximum* value of l occurs for a circular orbit; in this case $b = a$, so that $l + 1 = n$, or $l = n - 1$. The *minimum* value of l is 0, since b cannot be 0 or negative. Hence for a given value of n, l can have any integral value from 0 to $n - 1$. In other words, the allowed values of l for a given n are

$$l = 0, 1, 2, \ldots, n-1$$

For example, if $n = 2$, then l can be 0 or 1; if $n = 3$, then l can be 0, 1, or 2; and so on.

In the Bohr–Sommerfeld theory a *pair* of numbers (n, l) subject to these restrictions defines each possible electron orbit. With two quantum numbers there are considerably more allowed orbits for the electron. For example, in place of the first three orbits (n = 1, 2, and 3) of the original Bohr theory, there are now *six* orbits, as follows:

n	l	Orbit	Average radius (a)
1	0	Circle	r_0
2	0	Ellipse ($b/a = \frac{1}{2}$)	$4r_0$
2	1	Circle	$4r_0$
3	0	Ellipse ($b/a = \frac{1}{3}$)	$9r_0$
3	1	Ellipse ($b/a = \frac{2}{3}$)	$9r_0$
3	2	Circle	$9r_0$

In Sommerfeld's version of the Bohr theory, each of these orbits corresponds to a different energy level. The energy of the electron in the hydrogen atom (or any one-electron atom) now depends on the values of *two* quantum numbers, n and l. To replace the simple Bohr energy equation $E_n = -Z^2E_0/n^2$, Sommerfeld derived a much more complicated formula:

$$E_{n,l} = -\frac{Z^2E_0}{n^2} - \frac{Z^4E_0}{n^4}\alpha^2\left(\frac{n}{l+1} - \frac{3}{4}\right) \quad \begin{matrix}(n = 1, 2, 3 \ldots, \infty) \\ (l = 0, 1, 2 \ldots, n-1)\end{matrix} \tag{16–3}$$

where $\alpha = 2\pi e^2/hc = 0.0073$. The first term of the new energy equation is the simple energy equation from the original Bohr theory, and the second term is the relativistic correction. Since $\alpha^2 \approx 5 \times 10^{-5}$, the second term is quite small compared to the first. Nevertheless, there are now many more energy levels available for the electron, so there should be many more possible energy transitions, each one leading to a different spectral line.

The old and new forms of the hydrogen energy level diagram are compared in Figure 16–4. The former $n = 2$ level now consists of two levels of only slightly different energies; the old $n = 3$ level now consists of three closely spaced levels, and so on. In place of the old $2 \longrightarrow 1$ transition, which accounted for a *single* wavelength in the hydrogen spectrum, there should now be two transitions—$(n, l) = (2, 0) \longrightarrow (1, 0)$ and $(2, 1) \longrightarrow (1, 0)$—which give rise to *two* spectral lines of nearly the same wavelength. In general, the Bohr–Sommerfeld theory shows that all the single spectral lines predicted by the original Bohr theory are actually composed of several closely spaced spectral lines. Hence *the Bohr–Sommerfeld theory accounts for the existence of fine structure.* Sommerfeld's predicted values for the wavelengths of the fine structure lines of the hydrogen spectrum were found to be in rather good agreement with the experimental data available at that time.

One further correction had to be made in this model to make it agree with observations. As shown in Figure 16–4, the $3 \longrightarrow 2$ transition of the old Bohr theory should give rise to six spectral lines in the Bohr–Sommerfeld theory. However, only five are actually seen in the hydrogen spectrum. Bohr later showed on theoretical grounds that a transition cannot take place if l changes by more than one unit. The $(3, 2) \longrightarrow (2, 0)$ transition ($\Delta l = 2$) is forbidden under this rule, hence the missing spectral line can be explained.

Consequently, one of the major weaknesses of the original Bohr theory—the inability to explain the fine structure of the hydrogen spectrum—was apparently solved by the application of relativity theory and the introduction (through the use of elliptical orbits) of a second quantum number. More refined measurements of the hydrogen spectrum made many years later showed that the Bohr–Sommerfeld theory was only approximately correct, but in 1916 the apparent success of the theory was considered

figure 16–4 **THE BOHR–SOMMERFELD THEORY OF THE HYDROGEN ATOM**

A comparison is made between the original Bohr theory and the Bohr-Sommerfeld theory with respect to the electron energies and transitions for the first three levels of the hydrogen atom.

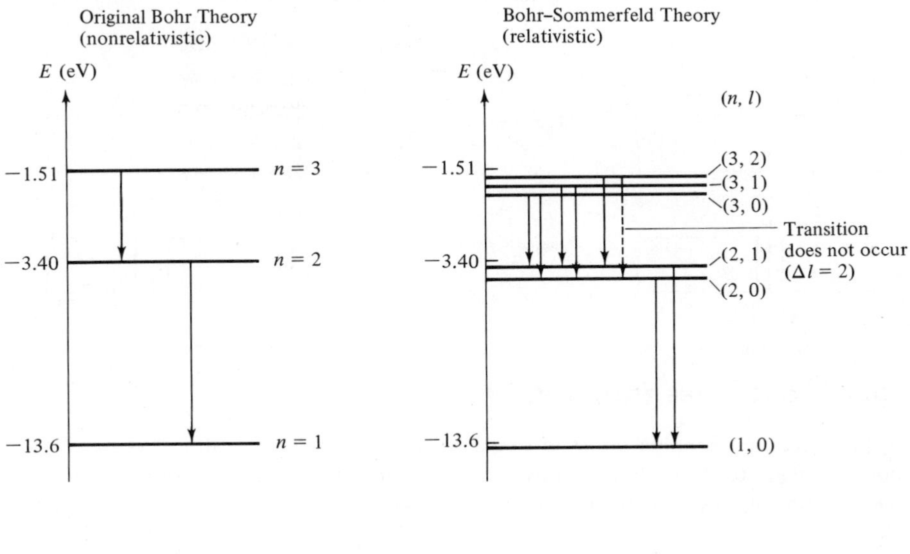

a major triumph for Bohr and his supporters. (The successful application of relativity to atomic physics could also be interpreted as a major triumph for Einstein. In 1915 there were still many physicists who were not convinced of the validity of the theory of relativity.)

The Bohr–Sommerfeld theory, like the original Bohr theory, was applicable only to one-electron atoms. In his attempts to extend it to many-electron atoms, Sommerfeld made little headway at first. From a careful analysis of the spectra of such atoms, Sommerfeld recognized that two quantum numbers (n and l) were still not enough to account for all the observed spectral lines. He then adopted the technique of working backward from the spectral lines to determine, by trial and error, what additional new quantum numbers would be needed. This approach resulted in the discovery of some rather interesting atomic properties.

16–3

MAGNETIC PROPERTIES OF THE ATOM

An orbiting electron is a charged particle moving in a closed path and, as such, constitutes an electric current loop. From this standpoint the orbiting electron can be considered as the fundamental unit of magnetism that Ampère postulated a century before. According to the principles of electromagnetic theory, an orbiting electron should generate its own magnetic field and should be affected by externally applied magnetic fields. It therefore seemed reasonable to expect that the spectrum of an element might be affected by the presence of a magnetic field.

THE ZEEMAN EFFECT

The first attempt to measure the effect of magnetism on spectra was made as far back as 1862 by the resourceful experimenter Michael Faraday. He placed a sodium flame between the poles of a magnet and examined the light with a spectroscope. He found no change in the spectral lines. In 1896 Faraday's experiment was repeated by the Dutch physicist Pieter Zeeman. Employing a more powerful electromagnet and a more sensitive spectroscope, Zeeman observed that *every spectral line splits into several components in the presence of a magnetic field.* This discovery has come to

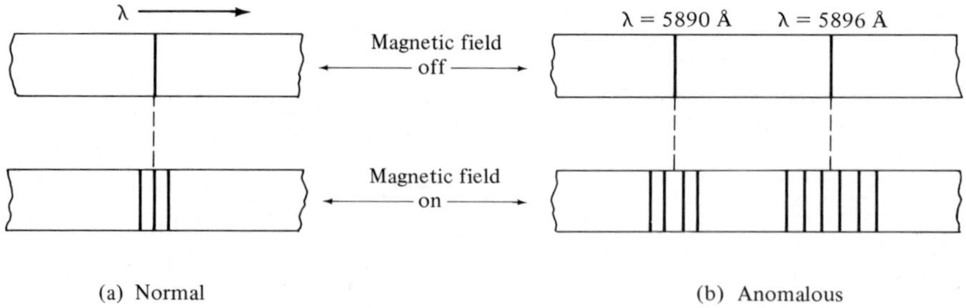

figure 16–5 **THE ZEEMAN EFFECT**

(a) In the normal Zeeman effect, a single spectral line splits into three lines in a magnetic field. (b) In the anomalous Zeeman effect (shown here for the yellow lines of sodium), the patterns are more complicated.

be known as the ***Zeeman effect.*** Some examples of this phenomenon are shown in Figure 16–5. In the ***normal Zeeman effect*** (so called because it conformed with theoretical expectations), a spectral line splits into *three* separate lines (Figure 16–5a). The central line has the same wavelength as the original line, and the two symmetrically spaced outer lines are shifted in wavelength by an amount that is directly proportional to the magnetic field strength. In the ***anomalous Zeeman effect*** (for which there was no explanation at first), a single spectral line splits into a more complex pattern of lines in the presence of a magnetic field. Figure 16–5b shows the anomalous Zeeman effect of the two closely spaced yellow lines of the sodium spectrum. In a magnetic field one splits into four lines, the other into six lines.

In the spectra of most elements, the Zeeman effect is extraordinarily complex. The anomalous effect appears much more often than the normal effect; sometimes both appear in the same spectrum. Generally, the anomalous effect appears when there is a relatively weak magnetic field; as the field gets stronger and stronger, the pattern gradually changes to the normal Zeeman set of three lines.

A quantum theory of the normal Zeeman effect was supplied independently by Sommerfeld and the Dutch physicist Peter Debye in 1916. They found that the normal Zeeman splitting could be accounted for by introducing a *third* quantum number in the Bohr–Sommerfeld theory. This number is called the ***magnetic quantum number*** m_L. The allowed values of m_L depend on the value of l for a given energy level. The rule is

$$m_L = 0, \pm 1, \pm 2, \ldots, \pm l$$

That is, if

$$\begin{aligned} l = 0&: \quad m_L = 0 \\ l = 1&: \quad m_L = -1, 0, \text{ or } 1 \\ l = 2&: \quad m_L = -2, -1, 0, 1, \text{ or } 2 \\ &\qquad \text{etc.} \end{aligned}$$

According to the Sommerfeld–Debye theory, each energy level in the atom is characterized by *three* quantum numbers (n, l, m_L). Additional terms must be added to the energy level equation (Equation 16–3) to account for the change in the electron energy when the atom is in a magnetic field. The introduction of the new quantum number m_L means that there will be more energy levels. Every energy level having $l = 1$ should split into three sublevels (corresponding to $m_L = -1, 0,$ or 1); every $l = 2$ level should split into five sublevels, and so on. (See Figure 16–6.) If the atom is not in a magnetic field, these sublevels all have the same energy, and they cannot be distinguished from one another. But if the atom is in a magnetic field, each sublevel has a different energy, so there is a greater number of distinct energy levels. The extra levels permit more transitions, resulting in more spectral lines. With the introduction of certain restrictions as to the transitions that can or cannot take place,* the *normal* Zeeman effect could be accounted for. However, there was still no way to explain the *anomalous* Zeeman effect within the framework of this theory.

SPACE QUANTIZATION

Because of the failure of the Sommerfeld–Debye theory to explain the anomalous Zeeman effect, the theory's usefulness was extremely limited—in fact, it could not

*The specific restriction is that in any transition between two levels, m_L cannot increase or decrease by more than one unit (i.e., $\Delta m_L = 0, 1,$ or -1).

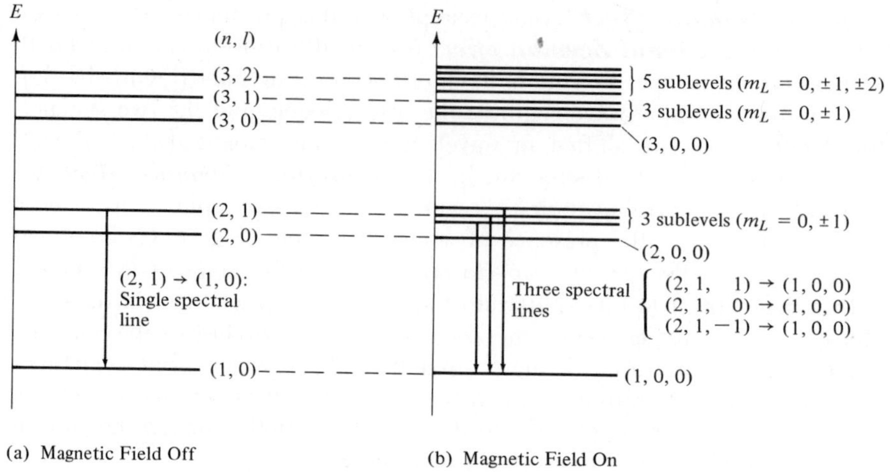

figure 16–6 **SOMMERFELD–DEBYE THEORY OF ENERGY LEVELS IN A MAGNETIC FIELD**

The presence of a magnetic field introduces additional energy levels in the hydrogen atom. Transitions between these levels (subject to the restriction that $\Delta m_L = 0, +1$, or -1) account for the normal Zeeman effect. The splitting of the (2, 1) → (1, 0) spectral line is shown.

even be successfully applied to the hydrogen spectrum! However, to justify the introduction of a third quantum number, Sommerfeld provided a physical picture of the atom that turned out to be an extremely valuable tool in the eventual solution of the problem.

An electron in an atom possesses angular momentum by virtue of its orbital motion. As we learned in Chapter 6, angular momentum is a vector quantity. For an orbiting electron, the direction of the angular momentum vector **L** is perpendicular to the plane of the orbit (Figure 16–7a). Hence the direction of **L** indicates how the orbital plane is oriented in space. In the absence of a magnetic field, the electron orbits of a group of atoms would be randomly oriented in space; there is no reason for **L** to point in any preferred direction as opposed to any other direction.

When a magnetic field is applied, however, the **L** vector will be oriented in a specific way. Because the electron orbit is also a current loop, the motion of the electron will be altered by the magnetic field. It can be shown that the magnetic interaction will cause the orbit to precess or "wobble" around the direction of the magnetic field, simulating the motion of a top, as illustrated in Figure 16–7b.

The wobbling motion that occurs in a magnetic field provides extra energy to the electron. The amount of additional energy depends on the angle θ between **L** and the magnetic field direction. If this angle were allowed to have any arbitrary value (as is expected classically), there would be an *infinite* number (i.e., a continuum) of extra energy levels in a magnetic field, since every value of θ corresponds to a different energy. The Zeeman effect shows, however, that only a few extra energy levels are created in a magnetic field. Hence there can be only a few allowed values of θ. In other words, *the* **L** *vector must be aligned in certain fixed directions with*

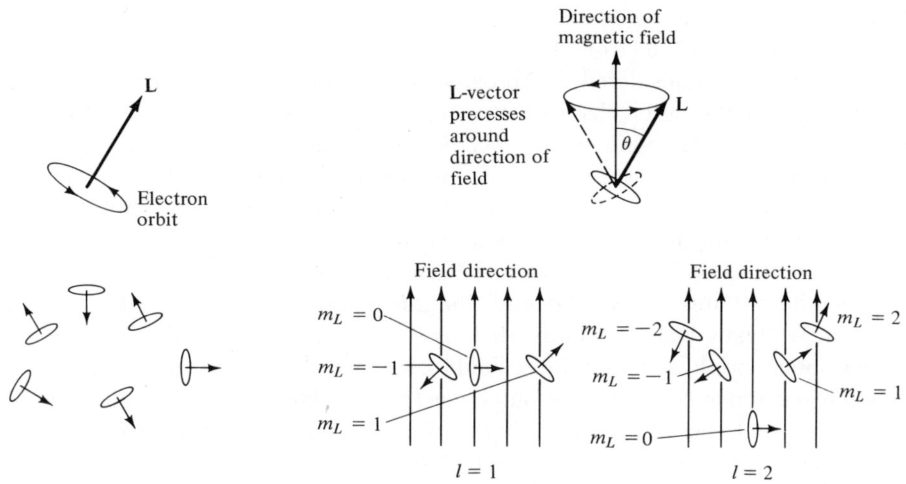

(a) Magnetic Field Off: Electron orbits are randomly oriented.

(b) Magnetic Field On: Orbits are oriented in specific directions

figure 16–7 **SPACE QUANTIZATION**

(a) The angular momentum vector **L** is perpendicular to the plane of the electron orbit. In the absence of a magnetic field, orbits are oriented in random directions in space. (b) In a magnetic field, **L** precesses about the direction of the field. The angle θ can have only certain specific values; that is, **L** must be oriented in certain fixed directions that are determined by the magnetic quantum number m_L.

respect to the direction of the magnetic field. This restriction of the electron orbit to specific fixed orientations in space is known as ***space quantization.***

The allowed orientations are determined by the quantum numbers l and m_L. For $l = 1$ there are three possible orientations ($m_L = -1, 0, 1$), as shown in Figure 16–7b; for $l = 2$ there are five possible orientations ($m_L = -2, -1, 0, 1, 2$); and so on.* The $l = 0$ ($m_L = 0$) state is not affected by a magnetic field. According to the requirements of Sommerfeld's model, an $l = 0$ orbit has *zero angular momentum.* The problem of how an orbiting electron could possess no angular momentum or magnetic effects posed a serious difficulty that Sommerfeld never resolved satisfactorily.

ELECTRON SPIN

Buoyed by his success with the normal Zeeman effect, the indefatigable Sommerfeld continued his battle to understand the mysteries of line spectra. He now focused his attention on the alkali metals. The evidence for closed electron shells in the atom showed that these elements are characterized by a single electron in the next highest energy level after a closed shell. Consequently, these elements are somewhat similar

* The relationship between **L** (the angular momentum vector) and l (the orbital quantum number) is that the magnitude of $\mathbf{L} = \sqrt{l(l+1)}\, h/2\pi$. The allowed orientations are determined from the restriction that the component of **L** along the magnetic field direction is equal to $m_L h/2\pi$.

to one-electron atoms, since they possess a single electron orbiting about a symmetrical inner core of closed electron shells.

Sommerfeld considered the possibility of a magnetic interaction between the inner core and the outer electron. By analyzing the properties of the observed lines in the alkali spectra, Sommerfeld successfully deduced a set of rules to govern the energy levels added to the atom by this interaction. With these rules the alkali spectra could be described in a systematic way, but the apparent success of these discoveries was tainted by the fact that there was virtually no theoretical justification for any of the new rules or formulas that had been introduced.

In an attempt to understand the anomalous Zeeman effect, the Austrian physicist Wolfgang Pauli, who was a former student of Sommerfeld's, began to question some of Sommerfeld's assumptions. Pauli calculated that if there were any interaction between the inner core and the outer electron, it should vary according to the size of the core and should therefore depend on the atomic number. However, no such dependence was observed. Pauli therefore rejected the idea of any interaction caused by the inner core and proposed instead that the additional energy levels arise from a property not of the inner core, *but of the electron itself.* Pauli did not identify this property; he referred to it as a quantum property of the electron "not describable classically."

Pauli reported his findings in 1924. A year later two Dutch physicists, George Uhlenbeck and Samuel Goudsmit, identified the mysterious property as ***electron spin.*** They proposed a picture of the electron as a small sphere of charge that spins on its axis as it orbits the nucleus (in the same way that the earth rotates on its axis as it revolves around the sun). By virtue of its spin, the electron possesses additional angular momentum; that is, an electron in an atom possesses *two* kinds of angular momentum—an orbital angular momentum **L** from its orbital motion and a spin angular momentum **S** from its axial rotation. Agreement with observed spectra could be achieved by assigning to every electron the same fixed amount of spin angular momentum.* In other words, **S** has a constant value that is the same for all electrons—unlike the orbital angular momentum **L,** which can vary from orbit to orbit.

As a spinning electric charge, the electron acts as a tiny magnet. Thus the orbiting electron becomes the effective source of *two* intrinsic magnetic fields, created respectively by the revolution and rotation of the electron. In fact, the modern theory of magnetism suggests that the magnetism created by electron spin is actually a more significant determinant of the magnetic properties of metals than is the magnetism created by the orbital motion of the electron.

Complicated electromagnetic forces are created within the atom by the combined orbital motion and spin of the electron. These effects are known collectively as the ***spin–orbit interaction.*** Because of the spin–orbit interaction, the electron spin can be aligned in only two ways: either **L** and **S** are *parallel,* or they are *antiparallel,* as shown in Figure 16–8a. In the former the total angular momentum is the sum of the two (**L** + **S**); in the latter the total angular momentum is the difference (**L** − **S**). As a consequence, the two alignments have slightly different energies. Thus every energy level associated with a given value of the orbital quantum number l is actually *two* distinct levels, corresponding to the parallel or antiparallel orientations.

*The magnitude of **S** is found to be $\sqrt{3}h/4\pi$. This can be obtained from the equation $S = \sqrt{s(s+1)}h/2\pi$ with $s = \frac{1}{2}$. Hence the electron spin has associated with it a quantum number s, which has a value of $\frac{1}{2}$.

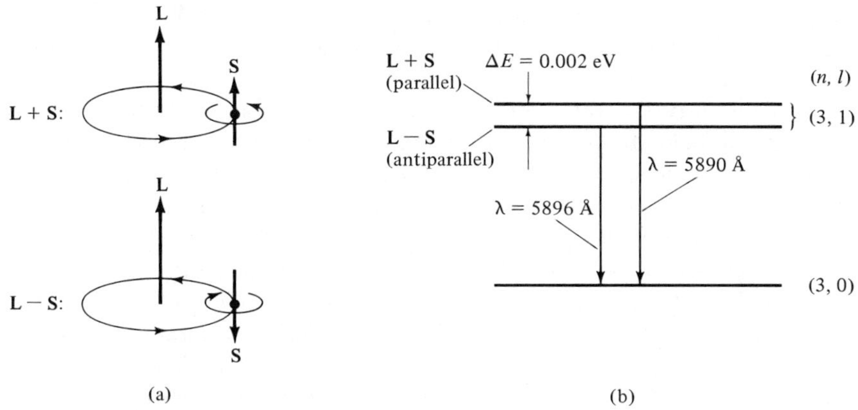

figure 16–8 **SPIN–ORBIT INTERACTIONS**

(a) For a single orbiting electron, the direction of **S** must be either parallel or antiparallel to **L.** (b) The twin yellow lines of the sodium spectrum result from transitions from two closely spaced energy levels of the sodium atom corresponding to the two possible spin-orbit alignments.

(The exception to this rule is the $l = 0$ state, which exhibits no spin–orbit effect.)

All of the rules that Sommerfeld derived to explain spectra could now be explained in terms of the principles of electron spin and the spin–orbit interaction. For instance, it had been known for many years that the main spectral lines of alkali metals were actually double lines of slightly differing wavelengths, the most well-known example being the pair of yellow lines that appear in the sodium spectrum. This effect could now be explained in terms of the two energy levels corresponding to the two spin orientations. As shown in Figure 16–8b, one line (5890 Å) represents a transition to a lower level from an energy level in which **L** and **S** are parallel, the other (5896 Å) a transition from the accompanying antiparallel level. The difference in energy between the two levels is only 0.002 eV, but it is readily observable in spectra.

Perhaps the most significant result of the electron spin hypothesis is that it led to an explanation of the anomalous Zeeman effect. The explanation is rather complicated, so only a brief summary of it will be presented here. As we learned previously, the application of an external magnetic field results in the space quantization or alignment of the orbital angular momentum vector **L** in certain specific directions depending on the value of the magnetic quantum number m_L. It seems reasonable to expect the spin angular momentum also to exhibit space quantization in a magnetic field. Hence there is a corresponding ***spin quantum number*** m_s, which determines the allowed orientations of **S.** It is assumed that there are only two possible orientations of **S** (either in the same direction as the field or in the opposite direction), so that only two values are possible for m_s: $+\frac{1}{2}$ or $-\frac{1}{2}$.

Both the normal and the anomalous Zeeman effects can be understood in terms of the competition between an externally applied magnetic field and the internal spin–orbit interaction (which is always present). If the external magnetic field is strong, it effectively negates the spin–orbit interaction; **L** and **S** act independently of one another, and each obeys its own separate rules for space quantization. The normal Zeeman effect can be accounted for on the basis of this picture. If the external magnetic field is weak, the spin–orbit interaction dominates. The rules of space quantization now apply to the *total* angular momentum, rather than to **L** and **S** separately.* The anomalous Zeeman effect can be explained in terms of the complex energy level structure that results.

In this way the complete details of the Zeeman effect and its dependence on the magnetic field strength could be understood. In 1925 this was considered a most remarkable achievement, and it gave great credence to the electron spin hypothesis. Yet there was considerable doubt as to the validity of the model of the electron as a rotating sphere of charge. Most experiments seemed to indicate that the electron is a point charge; at best, the radius of the electron (calculated from classical electromagnetic theory) is no more than 10^{-13} cm. If so, the speed of rotation of the electron would have to exceed the speed of light in order to generate the observed angular momentum! To the present day the conceptual problems have not been solved. It must simply be *assumed* that an electron possesses an intrinsic angular momentum and magnetic field as fundamental properties, since they cannot be accounted for in classical mechanical terms. The concept of electron spin remains as a convenient visualization of these properties, but it is no longer taken literally.

Thus with the inclusion of the concepts of space quantization and electron spin, the Bohr–Sommerfeld theory could explain, at least qualitatively, virtually all the significant properties of atomic spectra. The price of this success was an increase in the number of postulates and in the overall complexity of the theory. As compared to the single quantum number n, which appeared in the original Bohr theory, it now took four quantum numbers—n, l, m_L, m_s—to account for all the electron energy levels in the atom.

16–4

THE PAULI EXCLUSION PRINCIPLE

The crowning glory—and the final triumph—of the Bohr–Sommerfeld theory came about through the discovery of the principle that determines the filling of energy levels in the atom and explains the structure of the periodic table. By 1925 much information had been obtained on the number of electrons that occupy the various shells and subshells of different atoms. It was Wolfgang Pauli who first recognized that there was a connection between the filling of shells and subshells and the four quantum numbers needed to explain line spectra. In 1925 he proposed what has come to be known as the ***Pauli exclusion principle:***

No two electrons in the same atom can have the same set of quantum numbers n, l, m_L, m_s.

These four quantum numbers are required to account for all the energy levels in the atom. In a magnetic field, where space quantization comes into play, each set of n, l, m_L, m_s corresponds to a single, distinct energy level. The essence of the Pauli exclusion principle is that each of these energy levels can be occupied by no more than one electron. In other words, no two electrons in the same atom can have exactly the same energy in a magnetic field. The "shells" of electrons previously referred to represent groups of electrons occupying closely spaced energy levels that differ slightly in energy.

* Specifically, there is a new quantum number m_J, which governs the space quantization of the total angular momentum **J** (= **L** + **S**).

THE FILLING OF SHELLS AND SUBSHELLS

We will now show how the Pauli exclusion principle can be used to predict the number of electrons in a closed shell or subshell. First let us review the rules governing the allowed values of the four quantum numbers:

n (principal quantum number):	$n = 1, 2, 3, \ldots, \infty$
l (orbital quantum number):	$l = 0, 1, 2, \ldots, n-1$
m_L (magnetic quantum number):	$m_L = 0, \pm 1, \pm 2, \ldots, \pm l$
m_s (spin quantum number):	$m_s = \pm\frac{1}{2}$

The number of different combinations of n, l, m_L, m_s that are allowed under these rules can be determined easily. Let us begin by letting $n = 1$. Since the maximum value of l is always one less than n, the only possible value of l is zero. For $l = 0$, m_L can only be zero. But m_s may be either $+\frac{1}{2}$ or $-\frac{1}{2}$. Consequently, there are only two possible sets of quantum numbers beginning with $n = 1$:

$$(n, l, m_L, m_s) = (1, 0, 0, \tfrac{1}{2}) \text{ and } (1, 0, 0, -\tfrac{1}{2})$$

Thus according to the Pauli exclusion principle, no more than two electrons can occupy the $n = 1$ shell of an atom.

For $n = 2$ there are two possible values of l: 0 and 1. For $l = 0$ we once again obtain two sets of numbers: $(2, 0, 0, \frac{1}{2})$ and $(2, 0, 0, -\frac{1}{2})$. For $l = 1$, m_L can have three values: 1, 0, and -1. For each of these values, m_s may be $\frac{1}{2}$ or $-\frac{1}{2}$. This gives six more different sets of quantum numbers. Thus for the $n = 2$ level the possible combinations of the four quantum numbers (n, l, m_L, m_s) are:

$l = 0$	$l = 1$	
$(2, 0, 0, \frac{1}{2})$	$(2, 1, 1, \frac{1}{2})$	$(2, 1, 1, -\frac{1}{2})$
$(2, 0, 0, -\frac{1}{2})$	$(2, 1, 0, \frac{1}{2})$	$(2, 1, 0, -\frac{1}{2})$
	$(2, 1, -1, \frac{1}{2})$	$(2, 1, -1, -\frac{1}{2})$

Hence eight electrons can occupy the $n = 2$ levels of an atom—two in the $l = 0$ subshell and six in the $l = 1$ subshell.

Before going on, let us familiarize ourselves with the terminology and learn a few rules that will simplify the task of counting up all the different sets of the four quantum numbers. A *subshell* consists of all the energy levels having the same values of n and l, but different values of m_L and m_s. The number of energy levels in a given subshell can be calculated by a simple formula: for a given value of l, there are $2l + 1$ possible values of m_L (corresponding to $m_L = 0, \pm 1, \pm 2, \ldots, \pm l$). Each value of m_L can be paired with two possible values of m_s ($+\frac{1}{2}$ or $-\frac{1}{2}$). Hence there are $2(2l + 1) = 4l + 2$ different combinations of m_L and m_s for a given value of l. In other words, *there are* $4l + 2$ *different energy levels in a subshell.* Therefore according to the Pauli exclusion principle, a maximum of $4l + 2$ electrons can occupy a subshell of a given l. For example, an $l = 0$ subshell holds $4 \times 0 + 2 = 2$ electrons; an $l = 1$ subshell holds $4 \times 1 + 2 = 6$ electrons, and so on.

It is customary to identify the different subshells by specific letters. The letters $s, p, d, f, g, h, \ldots$ designate the $l = 0, 1, 2, 3, 4, 5, \ldots$ subshells, respectively.*

*This peculiar sequence of letters originates from the first letters of the words "sharp," "principal," "diffuse," and "fundamental," used to describe the different series of spectral lines found in alkali spectra. Each series terminates on the designated sublevel. After f the letters follow in alphabetical order.

The rules pertaining to subshells are summarized below:

Subshell designation	l	Maximum number of electrons in subshell $(= 4l + 2)$
s	0	2
p	1	6
d	2	10
f	3	14
g	4	18
h	5	22
⋮	⋮	⋮

Each subshell of a given atom is designated by a number (the n quantum number of the subshell) and a letter (indicating the l quantum number). For example, the $n = 2$, $l = 1$ subshell is the $2p$; the $n = 4$, $l = 2$ is the $4d$; and so on. The number of electrons in a given subshell is written as a superscript next to the subshell designation. For example, the symbol $2p^6$ indicates that there are six electrons in the $2p$ subshell (the maximum number of electrons allowable); the symbol $3d^5$ indicates that only five electrons are occupying the $3d$ subshell of a particular atom (out of a maximum of ten electrons).

The number of subshells having the same n depends on the rule that $l = 0, 1, 2, \ldots, n - 1$. There is a single $n = 1$ shell ($1s$), two $n = 2$ subshells ($2s$, $2p$), three $n = 3$ subshells ($3s$, $3p$, $3d$), and so on. The total number of electrons having a given n can be obtained by adding up the electrons in the filled subshells of the same n, as shown in Table 16–3.

It seems reasonable to expect that subshells will be filled in order of increasing n—that is, in the order $1s$; $2s$, $2p$; $3s$, $3p$, $3d$; $4s$, $4p$, $4d$, $4f$; and so on, with the completion of each n corresponding to a closed shell. But the evidence from chemical properties, ionization energies, and so forth shows that the successive closed shells in the atom (and the successive periods in the periodic table) contain 2, 8, 8, 18, 18, 32, . . . electrons. (See Table 16–2.) Thus the subshells are *not* filled in the expected order. However, agreement with the observed sequence of closed shells can be obtained by assuming that the subshells are grouped into shells in the following order:

Shell	Subshells	Number of electrons
1	$1s^2$	2
2	$2s^2\ 2p^6$	8
3	$3s^2\ 3p^6$	8
4	$4s^2\ 3d^{10}\ 4p^6$	18
5	$5s^2\ 4d^{10}\ 5p^6$	18
6	$6s^2\ 4f^{14}\ 5d^{10}\ 6p^6$	32

We see that there are certain inversions in the order. The $4s$ is filled *before* the $3d$, the $5s$ *before* the $4d$, the $5p$ and $6s$ *before* the $4f$, and so on. Note also that each shell begins with an s subshell and (with the exception of the first shell) closes with a p subshell. Therefore the grouping of subshells into a shell does not depend strictly on n; in other words, all the subshells of a given shell do not necessarily table 16–3 **MAXIMUM NUMBER OF ELECTRONS HAVING THE SAME n**

	Subshell					
n	s $(l = 0)$	p $(l = 1)$	d $(l = 2)$	f $(l = 3)$	g $(l = 4)$	Total number of electrons with same n
1	$1s^2$					2
2	$2s^2$	$2p^6$				8
3	$3s^2$	$3p^6$	$3d^{10}$			18
4	$4s^2$	$4p^6$	$4d^{10}$	$4f^{14}$		32
5	$5s^2$	$5p^6$	$5d^{10}$	$5f^{14}$	$5g^{18}$	50

have the same value of n. With this arrangement, the properties of the periodic table can be fully explained, as we shall learn shortly.

The order in which successive subshells are filled can be easily obtained by the following device. The subshells are arranged in a triangular pattern, as shown:

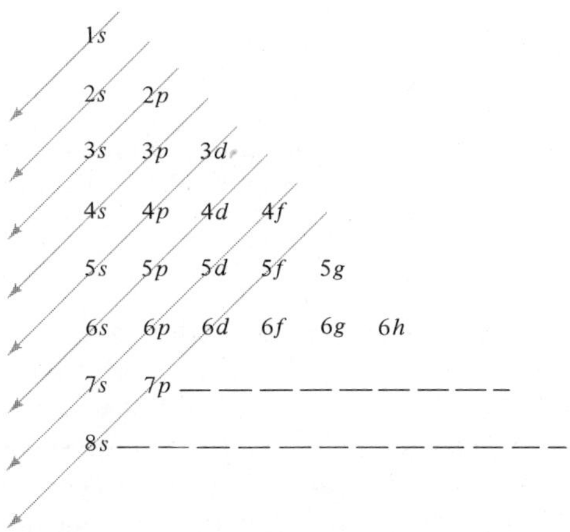

The correct order is obtained by reading down along successive diagonals: $1s$; $2s$; $2p$, $3s$; $3p$, $4s$; $3d$, $4p$, $5s$; and so on.

A simplified explanation for the inversions in this sequence can be given in terms of the Bohr–Sommerfeld theory, according to which each pair of quantum numbers (n, l) defines an orbit of given average radius and eccentricity. For a given n the orbits become less elliptical as l increases. Among any group of subshells, the $l = 0$ orbit is always the most elliptical, and the $l = n - 1$ orbit is always circular. (Thus electrons in the $1s$, $2p$, $3d$, $4f$, $5g$, . . . subshells have circular orbits.)

In a many-electron atom the electrons in the outermost shell are partially shielded by the inner closed electron shells from the full attractive force of the nucleus. The amount of shielding, however, depends on the shape of the orbit. An elliptical orbit will penetrate the inner shells and bring the electron close to the nucleus, so that the electron will experience the full nuclear attraction during at least part of its orbit. However, an electron in a circular orbit will always remain outside the inner shells and will experience the full shielding effect at all times. As a consequence, for a

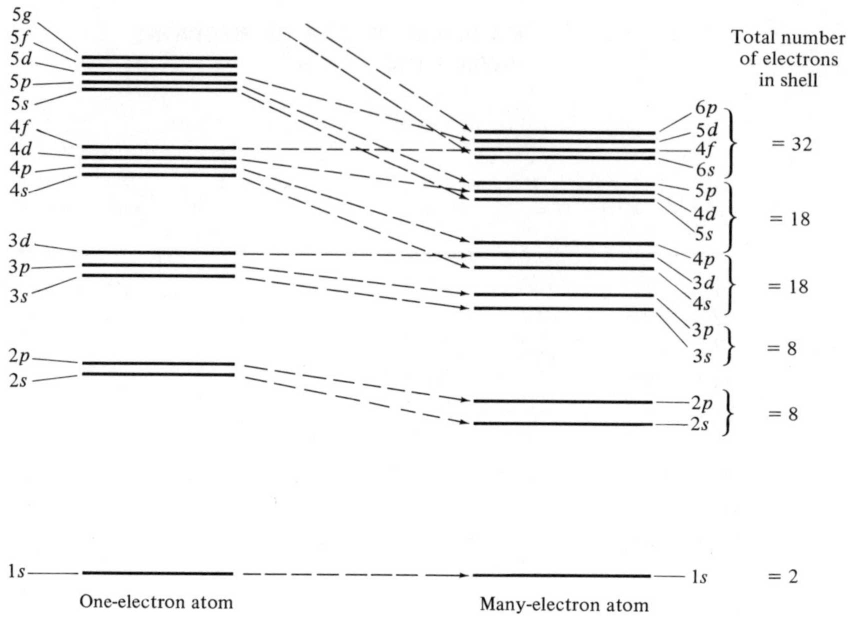

figure 16–9 **GROUPING OF SUBSHELLS IN A MANY-ELECTRON ATOM**

Because of the shielding effects, the order and spacing of subshells are rearranged as shown, leading to the observed sequence of shells and subshells in many-electron atoms.

given n an electron in an $l = 0$ orbit (the most elliptical) will be more tightly bound to the atom than an electron in a higher l orbit; an electron in a circular orbit will be the least tightly bound. Thus for any group of subshells of the same n, the s subshell will have a lower (more negative) energy than the p subshell, which will be lower than the d subshell, and so on. The result is a rearrangement of the energy level diagram, as shown in Figure 16–9. Note that the 4s level drops below the 3d, the 5s below the 4d, and so on. The various subshells tend to group together to produce the observed sequence of shells and subshells. The large gaps between each p subshell and the next successive s subshell are the effective boundaries between closed shells.

The details of this picture are too complex to be worked out mathematically here, so the description is necessarily imprecise. The arrangement of subshells is not fixed and will vary somewhat from element to element. Nevertheless, the properties of the elements and the structure of the periodic table can now be explained rather readily.

ELECTRON DISTRIBUTION AND CHEMICAL PROPERTIES

We will now see how the restrictions enforced by the Pauli exclusion principle on the distribution of electrons among the various shells and subshells lead to an explanation of the chemical properties of the elements. The basic principles are that the chemical activity of an element depends mainly on (1) the number of electrons in the element's outermost shell and (2) the ease with which electrons can be added to or removed from this shell.

The relationship between the filling of successive subshells and the structure of the periodic table is illustrated in Figure 16–10. The first element is hydrogen ($Z = 1$) with its single electron. In the ground state the electron will be in the 1*s* level. Thus we describe the electron distribution of hydrogen as follows:

$$_1\mathrm{H}:\quad 1s^1$$

Hydrogen will combine chemically with other elements either by giving up its single electron or by capturing a second electron to complete the 1*s* subshell.

Helium ($Z = 2$) has two electrons, both of which go into the 1*s* shell. That is, the electron distribution is

$$_2\mathrm{He}:\quad 1s^2$$

One of these electrons will have quantum numbers 1, 0, 0, $\frac{1}{2}$; the other will have 1, 0, 0, $-\frac{1}{2}$. With helium the 1*s* subshell is filled. This fact allows us to explain why helium is an inert gas. The helium atom has a very small affinity for additional electrons (which would go into the next subshell and not be tightly bound); also it is quite difficult to remove either of the helium atom's electrons from the closed shell. Therefore helium will not react chemically with other elements.

The next element, lithium, has the following distribution:

$$_3\mathrm{Li}:\quad 1s^2\,2s^1$$

The third electron is farther from the nucleus than the other two and is also partially shielded by them from the nucleus. Hence this electron can be readily removed (i.e., its ionization energy is low), which explains the high chemical activity of lithium. It can be seen that a single *s* electron is characteristic of all the members of the alkali family:

$$\begin{aligned}
_3\mathrm{Li}&:\quad 1s^2\,\boxed{2s^1}\\
_{11}\mathrm{Na}&:\quad 1s^2\,2s^2\,2p^6\,\boxed{3s^1}\\
_{19}\mathrm{K}&:\quad 1s^2\,2s^2\,2p^6\,3s^2\,3p^6\,\boxed{4s^1}\\
_{37}\mathrm{Rb}&:\quad 1s^2\,2s^2\,2p^6\,3s^2\,3p^6\,4s^2\,3d^{10}\,4p^6\,\boxed{5s^1}\\
_{55}\mathrm{Cs}&:\quad 1s^2\,2s^2\,2p^6\,3s^2\,3p^6\,4s^2\,3d^{10}\,4p^6\,5s^2\,4d^{10}\,5p^6\,\boxed{6s^1}
\end{aligned}$$

This explains why these elements have similar chemical properties. In this family chemical activity increases with atomic number, since a 6*s* electron is less tightly bound than a 5*s*, which in turn is less tightly bound than a 4*s*, and so on. The observed gradation in properties with increasing atomic number (shown in Table 16–1) can be understood in terms of the progressively weaker binding of the *s* electron in higher shells.

After lithium the second period is completed by the filling of the 2*s* and 2*p* shells. (Because the 2*s* and 2*p* subshells are relatively close in energy and size, the filling of the 2*s* level does not constitute a closed shell.) The period closes with element 10—neon—($1s^2\,2s^2\,2p^6$), which is an inert gas. The closed shell structure is typical of all inert gases:

$$\begin{aligned}
_2\mathrm{He}&:\quad 1s^2\\
_{10}\mathrm{Ne}&:\quad 1s^2\,2s^2\,2p^6\\
_{18}\mathrm{Ar}&:\quad 1s^2\,2s^2\,2p^6\,3s^2\,3p^6\\
_{36}\mathrm{Kr}&:\quad 1s^2\,2s^2\,2p^6\,3s^2\,3p^6\,4s^2\,3d^{10}\,4p^6
\end{aligned}$$

etc.

figure 16–10 **THE STRUCTURE OF THE PERIODIC TABLE**

The relationship between the filling of subshells and the periods of the table is shown. The members of each family have the same number of electrons in the outermost subshell.

KEY

1*s*			
2*s*			2*p*
3*s*			3*p*
4*s*		3*d*	4*p*
5*s*		4*d*	5*p*
6*s*	4*f*	5*d*	6*p*
7*s*	5*f*	6*d*	

s		*f*														*d*										*p*					
H 1	He 2																														
Li 3	Be 4																									B 5	C 6	N 7	O 8	F 9	Ne 10
Na 11	Mg 12																									Al 13	Si 14	P 15	S 16	Cl 17	Ar 18
K 19	Ca 20															Sc 21	Ti 22	V 23	Cr 24	Mn 25	Fe 26	Co 27	Ni 28	Cu 29	Zn 30	Ga 31	Ge 32	As 33	Se 34	Br 35	Kr 36
Rb 37	Sr 38															Y 39	Zr 40	Nb 41	Mo 42	Tc 43	Ru 44	Rh 45	Pd 46	Ag 47	Cd 48	In 49	Sn 50	Sb 51	Te 52	I 53	Xe 54
Cs 55	Ba 56	La 57	Ce 58	Pr 59	Nd 60	Pm 61	Sm 62	Eu 63	Gd 64	Tb 65	Dy 66	Ho 67	Er 68	Tm 69	Yb 70	Lu 71	Hf 72	Ta 73	W 74	Re 75	Os 76	Ir 77	Pt 78	Au 79	Hg 80	Tl 81	Pb 82	Bi 83	Po 84	At 85	Rn 86
Fr 87	Ra 88	Ac 89	Th 90	Pa 91	U 92	Np 93	Pu 94	Am 95	Cm 96	Bk 97	Cf 98	Es 99	Fm 100	Md 101	No 102	Lw 103	Rf 104	Ha 105													

The closed shell structure is very stable. With each successive element in a given period, there is an additional positive charge in the nucleus and an additional electron in the outermost shell. The effect of a greater nuclear charge is to increase the electric force on the electrons. Electrons in the same subshell are essentially the same distance from the nucleus, and they do not shield one another to any great extent. Hence the electrons become more and more tightly bound as the subshell fills. Thus considerable energy is required to remove an electron from a closed shell. Furthermore, once the shell is filled, any additional electrons must enter the next shell, where they will be farther from the nucleus, shielded by the inner shells, and therefore rather weakly bound to the atom. This inability either to attract or to give up electrons accounts for the extremely low chemical activities of the inert gases.

The third period, from ${}_{11}$Na to ${}_{18}$Ar, is associated with the filling of the 3*s* and 3*p* subshells. In the fourth period the first two elements (${}_{19}$K and ${}_{20}$Ca) fill the 4*s* subshell. Now the filling of the 3*d* subshell begins, which encompasses the first ***transition elements,*** ${}_{21}$Sc to ${}_{30}$Zn. These ten elements—all metals—have rather similar chemical properties. Even though the 3*d* subshell has the lower energy, it is the two 4*s* electrons that mainly govern the chemistry of these elements; on the average the 4*s* electrons (which have highly elliptical orbits) are farther from the nucleus than the 3*d* electrons. A similar effect occurs with the second set of transition elements (${}_{39}$Y to ${}_{48}$Cd), which correspond to the filling of the 4*d* after the 5*s*.

The rather similar chemical properties of the ***rare earth*** or ***lanthanide elements*** (${}_{57}$La to ${}_{70}$Yb) are attributable to the filling of the 4*f* subshell; the analogous ***actinide elements*** (${}_{89}$Ac to ${}_{102}$No)—nearly all of which are artifically produced—are associated with the filling of the 5*f* subshell. The periodic table presently ends with element 105 (produced artificially in 1970), which has its outermost electron in the 6*d* subshell.

This brief and rather sketchy summary represents but a small fraction of the rich harvest of information that has been reaped through an understanding of the relationship between the filling of electron shells and the regularities of the periodic table. A wide range of physical properties, such as the melting and boiling point, density, crystal structure in the solid state, color, and electric and magnetic activity of any element can be accounted for in terms of the electron distribution for that element. Many of the chemical properties—types of compounds formed, strengths of the bonds within compounds, reaction energies, and so on—can be similarly explained. All in all, the discovery of the principles behind the periodic table has proved to be of immense scientific and practical value to physicists and chemists alike.

THE SIGNIFICANCE OF THE EXCLUSION PRINCIPLE

Let us return to the Pauli exclusion principle, which, as we have seen, provided the key to the periodic table. It is of considerable interest to note that the exclusion principle, which is fundamental to the understanding of atomic structure, is itself unexplained. The principle states that no two electrons in an atom can have the same four quantum numbers, but it does not explain *why* this is so. Our justification for accepting the Pauli exclusion principle is simply that it works. In other words, the filling of energy levels in the atom is merely *described,* but not *explained,* by the exclusion principle. Thus the fundamental question of why there are filled energy levels in the atom remains unanswered.

Moreover, it has been found that the Pauli exclusion principle is not restricted just to the electrons in a single atom, but to any group of electrons in a common force field—such as the free electrons within a metal (the conducting electrons that make up an electric current) or the electrons in the ionized gas in the dense interior of a star. In fact, the principle applies not only to electrons, but to protons and neutrons in the nucleus as well. In general, the Pauli exclusion principle is obeyed by any particle with an intrinsic spin of $\frac{1}{2}$. The Pauli exclusion principle is generally regarded as being a fundamental physical principle of deep significance. At the present time, it cannot be derived or explained in terms of any other physical concepts.

THE FAILURE OF THE BOHR–SOMMERFELD THEORY

It is ironic that the years of the Bohr–Sommerfeld theory's greatest success (1925–1926), in which the complexities of spectra and the periodic table came to be understood, marked the beginning of its decline. It seems remarkable that a theory with which so much had been accomplished should be rejected less than fifteen years after its inception. There was no violent overthrow of the Bohr–Sommerfeld theory; there was no dramatic, critical experiment whose findings clearly contradicted theoretical predictions. Rather, the theory fell mainly from the weight of too many postulates, hypotheses, and rules. In describing the status of the Bohr–Sommerfeld theory as of 1925, science historian Max Jammer referred to it as

> a lamentable hodgepodge of hypotheses, principles, theorems, and computational recipes rather than a logically consistent theory. Every single problem had to be solved first in terms of classical physics; its classical solution had then to pass through the mysterious sieve of the quantum conditions. . . . usually, the process of finding "the correct translation" was a matter of skillful guessing and intuition rather than of deductive and systematic reasoning. . . . In short, the theory still lacked two essential characteristics of a full-fledged scientific theory, conceptual autonomy and logical consistency.[1]

Thus the Bohr–Sommerfeld theory, although a practical success, could only be viewed as a philosophical failure. In the next chapter we will examine the attempts to replace the Bohr–Sommerfeld theory with a more conceptually rigorous theory of the atom.

QUESTIONS FOR REVIEW

1 What does the atomic weight of an element represent?

2 Describe the basic features of the periodic table. What criteria did Mendeleev use in arranging the elements in his table?

3 Before 1900 how was the atomic number of an element determined? What was the significance of this number? According to the present-day theory, what does the atomic number represent?

4 Describe Moseley's experiments on x-ray spectra. What results did he obtain? Discuss the importance of his findings.

5 Explain how Moseley's law could be accounted for in terms of the Bohr theory. What assumptions were made about the arrangement of electrons in the atom?

6 Explain how the data on the ionization energies for the various elements are used as evidence for the existence of closed energy levels in the atom. Describe the relationship between the ionization energy of an element and the element's position in the periodic table.

7 Explain why Bohr did not use elliptical orbits in his original theory, and why they were used in the Bohr–Sommerfeld theory. What role does the theory of relativity play in the Bohr–Sommerfeld theory?

8 Explain why *two* quantum numbers (n and l) are needed in the Bohr–Sommerfeld theory. State the rules that govern the allowed combinations of n and l.

9 Explain why the Bohr–Sommerfeld theory was successful in accounting for the fine structure of the hydrogen atom.

10 Describe the Zeeman effect. What role did it play in the development of atomic theory?

11 State the rule governing the allowed values of the magnetic quantum number, m_L. How was the establishment of this rule justified?

12 Explain what is meant by space quantization. What factors determine the orientation of an electron orbit in a magnetic field?

13 How does the spin angular momentum (**S**) of an orbiting electron differ from the orbital angular momentum (**L**)? How are **L** and **S** aligned with one another?

14 What property of an electron does the spin quantum number m_s describe? Why does m_s have only two possible values?

15 Discuss the value of the electron spin hypothesis. What problems arise when we take the concept of electron spin literally? What is the modern-day view of this concept?

16 Explain why the concepts of space quantization and electron spin were both needed in order to account for the normal and anomalous Zeeman effects.

17 State and explain the Pauli exclusion principle. What is its significance? How do we justify its acceptance as a valid physical principle?

18 Explain in terms of the Pauli exclusion principle why there can be no more than two electrons in the $n = 1$ level of an atom.

19 Explain why a subshell of a given l contains a maximum of $4l + 2$ electrons.

20 List the order in which the subshells are filled with electrons.

21 Explain why a $3d$ electron is less tightly bound to an atom than a $4s$ electron.

22 Discuss the successes and failures of the modified Bohr theory. Why did most physicists believe that it had to be replaced?

EXERCISES

1 The rule that elements are listed in order of increasing atomic weight in the periodic table is broken in a few places in the table. For example, argon ($_{18}$Ar) is listed before potassium ($_{19}$K), even though argon has a larger atomic weight. What is the reason for reversing the order of these elements in the periodic table?

2 The element beryllium ($_4$Be) has two of its four electrons in the $n = 2$ shell. Why does it take only 9.3 eV to remove the first electron and 18.2 eV (nearly twice as much energy) to remove the second electron, if both electrons are in the same shell?

3 The following table gives the ionization energy for each of the five electrons of the element boron ($_5$B):

1st electron:	8.30 eV	($B \longrightarrow B^+$)
2nd electron:	25.15 eV	($B^+ \longrightarrow B^{2+}$)
3rd electron:	37.92 eV	($B^{2+} \longrightarrow B^{3+}$)
4th electron:	259.30 eV	($B^{3+} \longrightarrow B^{4+}$)
5th electron:	340.13 eV	($B^{4+} \longrightarrow B^{5+}$)

(a) How are the electrons distributed by subshell?

(b) Why is the ionization energy for the first electron of boron less than that for the first electron of beryllium (9.3 eV)?

(c) Use the Bohr theory for a one-electron atom to calculate the ionization energy of B^{4+}. How does the theoretical result compare with the experimental value?

4 (a) List all the sublevels of the $n = 2$, $n = 3$, and $n = 4$ levels according to the rules of the Bohr–Sommerfeld theory. Identify each sublevel by its (n, l) combination.
(b) List all the transitions from $n = 3$ to $n = 2$ that obey the rule that l never changes by more than one unit ($\Delta l = 0, \pm 1$).
(c) Do the same for the $n = 4$ to $n = 3$ transitions.

5 Into how many different sublevels should the $(n, l) = (2, 1)$ level split in a magnetic field? The (3, 2) level? The (4, 2) level?

6 By virtue of its revolution and rotation, the earth possesses an orbital angular momentum **L** and a spin angular momentum **S**. Are the **L** and **S** vectors for the earth parallel or antiparallel? Explain your answer.

7 Using the rules governing the four quantum numbers n, l, m_L, and m_s, list all the allowed sets of quantum numbers that begin with $n = 3$.

8 State whether each of the following subshell configurations is permissible under the rules. If so, explain what the symbol signifies. If the configuration is not permissible, explain why it is not.

(a) $3p^4$ (c) $1p^6$ (e) $6d^2$
(b) $7s^3$ (d) $5s^1$ (f) $2d^6$

9 Add a row to the bottom of Table 16–3, listing all the subshells beginning with $n = 6$. What is the maximum number of electrons in all of these subshells?

10 Write the electron distribution by subshell for each of the following elements:
(a) $_8O$ (oxygen).
(b) $_{26}Fe$ (iron).
(c) $_{38}Sr$ (strontium).
(d) $_{54}Xe$ (xenon).
(e) $_{92}U$ (uranium).

11 What feature characterizes the electron distribution of the alkali metals? Refer to the properties of the alkali family in Table 16–1. In terms of the electron distributions, explain (a) why the ionization energy decreases as the atomic number increases, (b) why the atomic radius increases as the atomic number increases, and (c) why the melting and boiling points decrease as the atomic number increases. (Hint: Think about the processes of melting and boiling in terms of the interatomic forces in solids and liquids.)

THE BASIC CONCEPTS OF QUANTUM MECHANICS

17

It was a heroic time. It was not the doing of any one man; it involved the collaboration of scores of scientists from many different lands, though from first to last the deep creative and critical spirit of Niels Bohr guided, restrained, deepened, and finally transmuted the enterprise.

J. Robert Oppenheimer

FOR MORE THAN two hundred years the Newtonian system of mechanics dominated physical thought and shaped the development of all physical theories. In 1905 the first breach of the Newtonian citadel was made by Einstein's theory of relativity. At about the same time, Planck and Bohr revealed further shortcomings of the Newtonian system by demonstrating that on the atomic level discontinuous energy changes or "jumps" appear to take place and that bodies can possess certain mechanical quantities such as energy and angular momentum only in specified units or *quanta*. This was in contradiction to Newtonian principles, which held that mechanical quantities could change continuously by any arbitrary amount. Nevertheless, in the Bohr–Sommerfeld theory the spirit of Newtonian physics was allowed to prevail wherever possible; the theory maintained strict adherence to all conservation laws and contained such patently Newtonian concepts as planetary orbits and spinning electrons. The Bohr–Sommerfeld theory had sought and seemingly achieved a marriage of Newtonian mechanics and quantum principles, but it eventually became clear to everyone that what had really taken place was a shotgun wedding.

Between 1925 and 1928 the model that Bohr, Sommerfeld, and others had so painstakingly constructed out of Newtonian and quantum concepts was swiftly and effectively dismantled. It was replaced by a theory that remarkably reproduced all the successes of the Bohr–Sommerfeld theory while avoiding its conceptual inconsistencies. The new theory was based on a rather abstract, highly formalized mathematical model of atomic phenomena, which denied the validity of virtually all of Newtonian mechanics on the atomic level. The theory professed the impossibility of predicting accurately the positions or motions of electrons within an atom and rejected the idea of even attempting to visualize the atom in terms of any mechanical model.

This new theory of the atom—known variously as ***quantum mechanics*** or ***wave mechanics***—is not easy to grasp on an elementary level. It leads to some disturbing conclusions as to the ultimate nature of matter. It presents rather thorny philosophical problems pertaining to the nature of physical theory. Many questions have been raised about the meaning of certain principles of quantum mechanics—questions that have yet to be answered satisfactorily. Yet the theory has proved to be extremely successful in predicting phenomena on the atomic and subatomic levels to a degree of accuracy unexcelled in any other branch of physics or the other sciences.

The replacement of Newtonian mechanics by quantum mechanics on the atomic level stands as one of the most significant revolutions in the history of physics. Unlike other revolutions, this one did not take many decades to gain legitimacy. The overthrow of the old order was so complete that quantum mechanics became part of the mainstream of physics within five years. Moreover, the development of quantum mechanics was not the work of a single man. It was the creation of a remarkable group of young theoretical physicists, ranging in backgrounds from a French count to an unemployed electrical engineer. Overseeing these developments and providing considerable intellectual guidance and encouragement were the two patron saints of twentieth-century physics: Albert Einstein and Niels Bohr.

17–1

WAVE PROPERTIES OF MATTER

Our study of the development of quantum mechanics begins with the doctoral thesis of Prince Louis Victor de Broglie.* A member of one of the oldest titled families in France, de Broglie had originally been a history major at the Sorbonne. His interest in physics was stimulated by his older brother, Maurice, who was a well-known experimental physicist. After World War I de Broglie returned to the university to study physics. In 1924 de Broglie (then thirty-two years old) submitted his completed thesis to the physics faculty at the Sorbonne for their approval.

De Broglie's thesis, entitled "Researches on the Theory of Quanta," was inspired by Einstein's photon theory of light. In his thesis de Broglie argued for a symmetry between light and matter. He hypothesized that if light—which is described classically as a wave—exhibits particle properties, then material particles ought to exhibit wave properties. In other words, *wave–particle duality should be exhibited by both light and matter.*

Einstein's photon theory proposes that to any electromagnetic wave of a given frequency f can be associated a photon of energy $E = hf$. De Broglie's hypothesis of a light–matter symmetry requires that every material particle should have a wave associated with it. Using the equations of relativity, de Broglie derived a formula relating the wavelength of the associated wave to the mass and speed of the particle. The following is a simplified form of de Broglie's derivation of this formula.

The energy E of a photon is related to the frequency of the associated light wave by the equation $E = hf$. A photon also possesses a momentum p, according to the equation $E = pc$. Combining these two equations gives

$$E = pc = hf = \frac{hc}{\lambda}$$

Therefore

$$p = \frac{h}{\lambda}$$

or

$$\lambda = \frac{h}{p} \tag{17–1}$$

*Pronounced de Broïe, which approximately rhymes with Troy.

This equation was derived specifically for the photon, a massless particle that is associated with a light wave, but de Broglie proposed that Equation 17–1 applies to *any* particle, massless or not. The momentum of a particle of mass m and speed v is $p = mv$. Therefore for a material particle Equation 17–1 becomes

$$\lambda = \frac{h}{mv} \tag{17-2}$$

This equation gives the wavelength λ of the "de Broglie wave" associated with a particle of a given mass and speed.*

We now have two equations, $E = hf$ and $p = h/\lambda$, that relate *particle* properties (energy and momentum) to *wave* properties (frequency and wavelength). These equations apply equally to material particles (electrons, atoms, etc.) and to photons. Photons and material particles also obey equally the relativistic equation $E^2 = p^2c^2 + {m_0}^2c^4$, the main distinction being that $m_0 = 0$ for a photon. Both photons and material particles have associated de Broglie waves. The de Broglie wave of a photon is a light wave (which travels with a speed c), whereas the speeds of the de Broglie waves associated with material particles depend on the speed of the particle.

An obvious question now presents itself: What is the nature of the wave associated with a particle? De Broglie avoided a specific answer to the question. He did not attempt to give any physical explanation of the nature or source of these waves. He concluded his thesis by stating that the nature of the associated wave is "intentionally left rather vague. . . . the present theory ought therefore to be considered as a formal structure whose physical content is not entirely precise rather than as a fully formed, unified doctrine."[1]

De Broglie referred to the associated waves as "pilot waves," which are presumed to guide the particle along a particular path. He made one especially interesting application of this principle with respect to electron orbits in the atom. In the simplified picture presented by the Bohr theory, an electron travels in a circular orbit about the nucleus. Suppose we assume the electron is guided around the orbit by a de Broglie wave. Since the wave is traveling in a circular path, it will catch up with itself, producing interference effects (Figure 17–1). If this interference is not to be destructive—leading to a cancellation of the wave—the de Broglie wave must necessarily be in phase with itself as it travels around the orbit. (In effect, a standing wave pattern must be established.) This will occur only if the circumference of the orbit is equal to an exact number of wavelengths. That is, it must be possible to fit *exactly* one, or two, or three, or more waves around the orbit. If there are any half waves or other fractions left over, cancellation will occur. This requirement is written mathematically as follows:

$$n\lambda = 2\pi r \qquad (n = 1, 2, 3, \ldots, \infty)$$

where n is an integer; λ is the wavelength of the de Broglie wave associated with the orbiting electron; and $2\pi r$ is the circumference of the orbit of radius r. We now substitute for λ from Equation 17–2:

$$n\lambda = n\left(\frac{h}{mv}\right) = 2\pi r$$

* Equation 17–2 is relativistically correct if m represents the *relativistic* mass, rather than the rest mass m_0 ($m = m_0/\sqrt{1 - v^2/c^2}$). The rest mass m_0 can be used only if the speed v is small compared to c.

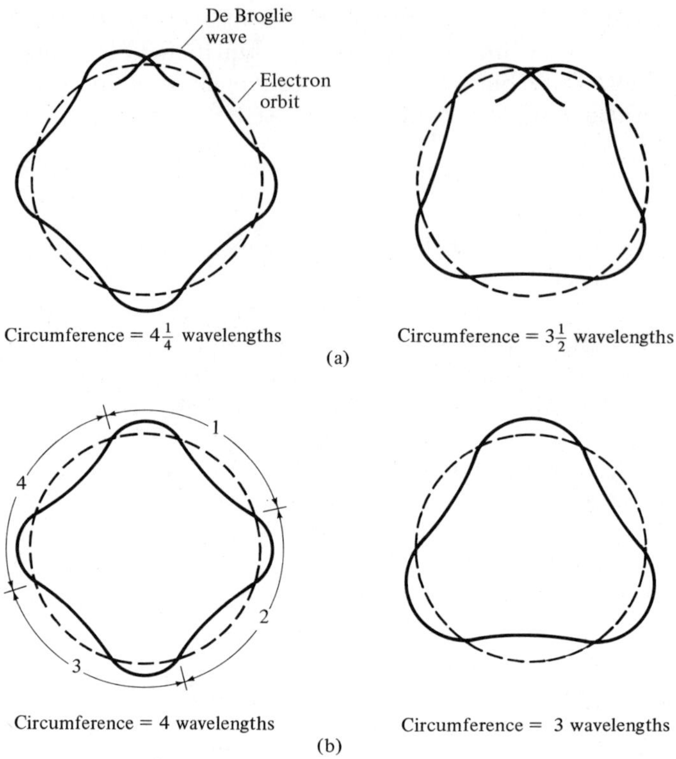

figure 17–1 **DE BROGLIE WAVES AND ELECTRON ORBITS**

(a) An integral number of waves must fit on the circumference of the orbit, or destructive interference will occur. (b) No cancellation occurs when the circumference is exactly equal to an integral number of wavelengths.

which can be rearranged as follows:

$$mvr = \frac{nh}{2\pi} \qquad (n = 1, 2, 3, \ldots, \infty)$$

This equation is none other than Bohr's assumption (Equation 15–6) of the quantization of angular momentum (mvr), which he used to obtain the allowed orbits and energy levels of the hydrogen atom. It will be recalled that Bohr gave no justification for this equation; it was one of the postulates of his theory. We now find a natural explanation for Bohr's assumption in terms of the de Broglie waves accompanying the orbiting electron: *The quantization of the angular momentum follows mathematically from the requirement of noncancellation of de Broglie waves in an electron orbit.*

Other than the above derivation of Bohr's assumption, de Broglie offered no experimental evidence for the existence of these waves. He did not provide any description of their physical nature, nor did he show that their existence follows as a logical necessity from any theory. De Broglie's main argument for their existence was based almost entirely on his belief in the principle of a wave-particle symmetry. Yet in spite of its tentative nature, de Broglie's work was received enthusiastically by many physicists. Einstein read de Broglie's thesis with great interest, then recommended it to a colleague with the comment, "Read it—although it might seem crazy, it is completely sound."[2] In the years that followed, de Broglie's theory proved to be a significant source of many of the new concepts that marked the development of quantum mechanics.

ELECTRON WAVES

In order to provide some experimental basis for the existence of de Broglie waves, one would have to demonstrate that material particles exhibit wave characteristics. This could be done most directly by showing that particles can be made to exhibit interference effects. For example, if Young's double-slit experiment were repeated using a beam of particles instead of a beam of light, and a pattern of alternating dark and bright lines were produced by the particles striking a fluorescent screen, it would provide rather strong evidence for the wave-particle symmetry proposed by de Broglie.

The presence of Planck's constant in the equation for the de Broglie wavelength leads to the result that interference effects will occur only with atomic or subatomic particles. To illustrate this point, let us calculate the de Broglie wavelength of a baseball thrown at a speed of 100 mi/hr. A baseball has a weight of approximately 5 ounces, which is equivalent in CGS units to a mass of nearly 150 grams. In metric units 100 mi/hr = 160.9 km/hr = 4470 cm/sec. Hence the de Broglie wavelength of the baseball is

$$\lambda_{\mathrm{BB}} = \frac{h}{mv} = \frac{6.63 \times 10^{-27}\ \text{erg-sec}}{(150\ \text{grams})(4470\ \text{cm/sec})} \approx 10^{-32}\ \text{cm}$$

A wavelength of 10^{-32} cm is much too small to be detected by any reasonable interference experiment. This extremely minute wavelength results mainly from the small size of h in comparison to the mass of the baseball. (This is actually rather fortunate, for if h were a much larger number, we would have to take into account the possibility that a baseball might diffract around a bat.)

Now suppose we calculate the wavelength of an electron traveling at a speed of 2×10^8 cm/sec (which corresponds to an energy of about 10 eV):

$$\lambda_e = \frac{h}{mv} = \frac{6.63 \times 10^{-27}\ \text{erg-sec}}{(9.11 \times 10^{-28}\ \text{gram})(2 \times 10^8\ \text{cm/sec})} = 3.6 \times 10^{-8}\ \text{cm} = 3.6\ \text{Å}$$

This wavelength is about the same as that of an x-ray photon. Therefore if it is possible to produce an interference pattern with x rays, it should also be possible to produce an interference pattern with electrons.

This comparison between x rays and electrons turned out to be of crucial importance in the demonstration of wave–particle symmetry. Following the discovery of x rays by Roentgen in 1895, a controversy raged for many years as to whether x rays were waves or particles. The earliest evidence seemed to point strongly to the particle theory. The first attempts to produce x-ray interference or diffraction met with little success. The difficulty, from the wave theory point of view, arose from the very short ($\approx 10^{-8}$ cm) wavelengths of x rays. The visibility of an interference or diffraction pattern (as measured by the separation of the bright and dark lines) is proportional to the ratio λ/d, where λ is the wavelength and d is the dimension of the obstacle producing the interference. (In Young's experiment d is the distance between the two slits; in single-slit diffraction d is the width of the slit.) In order for x-ray interference to be observed, d would have to be smaller than 10^{-6} cm. A spacing of such small size could not be produced mechanically with the available technology.

In 1912 the German physicist Max von Laue hit upon the idea of using natural crystals to produce x-ray interference. In a crystal the atoms are arranged in an orderly array of parallel planes of evenly-spaced atoms. The typical interatomic spacing is of the order of 10^{-8} cm, which is exactly the right size for x-ray interference. Thus the crystal atoms themselves act as suitable barriers for the interference and diffraction of x-ray waves. Laue and two associates, Walter Friedrich and Paul Knipping, successfully demonstrated the diffraction of x rays by crystals in 1913. This apparent victory for the proponents of the wave theory was rather short-lived, for the experiments of Compton and others clearly indicated that x rays exhibit particle properties when they interact with electrons. Thus the net result of the various experiments with x rays was to show that they possess both wave *and* particle characteristics.

In defense of his thesis de Broglie suggested that an experiment be performed to see if *electrons* could be diffracted by crystals. As it turned out, evidence for electron diffraction had been reported in 1921—three years *before* de Broglie's theory was published—but no one had recognized its significance at the time. Clinton J. Davisson, an American physicist employed by the Western Electric (now the Bell Telephone) Company in New York, had been engaged in experiments to study the scattering of electron beams by metal surfaces.* Davisson observed that electrons were not always scattered uniformly, as would be expected for particles; under certain conditions they were reflected in a particular direction depending on their speeds. Davisson reported his findings, but he was uncertain how to explain them.

In 1925 the German physicist James Franck learned of de Broglie's work. He remembered Davisson's findings and realized that they could be explained in terms of the diffraction of electron waves by the atoms of the metal crystal. Davisson did not accept Franck's interpretation, but continued his own experiments. At the time, Davisson was using as a target a plate of pure nickel metal. In this form the nickel does not consist of a single crystal, but of innumerable microscopic crystals that are randomly oriented within the metal plate. Because of this, there was no clear-cut indication of an electron diffraction pattern. Providentially, there was a small explosion in Davisson's laboratory, which caused the breakage of the evacuated tube containing the nickel target. The exposure to air completely oxidized the metal surface, and the metal had to be purified by a long, slow heating process—as a result of which a single crystal of nickel was formed. When Davisson and his assistant, Lester H. Germer, began to experiment again with the purified target, the diffraction effects were considerably more pronounced. Davisson and Germer found that their observations could be fully explained in terms of the reflection, refraction, and diffraction of electron waves by the crystal atoms. From the observed diffraction patterns, they could calculate the wavelengths of these waves. The values of these wavelengths for electrons of a given energy turned out to be in exact agreement with the values calculated by the de Broglie formula, $\lambda = h/mv$. Thus Davisson and Germer's results

*The purpose of these experiments was to obtain evidence for a patent suit lodged against Western Electric by the General Electric Company over the design of a vacuum tube. (Western Electric won the suit.)

not only demonstrated the existence of electron diffraction, but also validated de Broglie's equation.

Davisson and Germer's findings were published in April 1927. A few months later George P. Thomson, a physicist at the University of Aberdeen in Scotland, produced electron diffraction by sending a beam of electrons through a thin metal foil. The pattern—consisting of a series of alternating bright and dark concentric circles—was virtually identical to that obtained when x rays are used. This discovery provided an independent verification of electron diffraction. Thomson summarized his findings as follows: "Cathode rays behave as waves of wavelength h/mv according to de Broglie's theory of wave mechanics."[3] This statement is all the more remarkable when we consider that thirty years before, George Thomson's father, J. J. Thomson, had convincingly demonstrated that cathode rays consist of a stream of particles.

Since 1927, the diffraction of hydrogen molecules, helium atoms, and neutrons have all been achieved. Hence there is a substantial amount of experimental evidence to support the contention that material particles possess wave properties.

WAVE–PARTICLE DUALITY

The discovery that material particles exhibit wave characteristics adds a new dimension to the problem of wave–particle duality. The classical descriptions of light as a wave and matter as composed of solid particles no longer seem valid—for both matter and light have been found to display wave *and* particle characteristics. How can we provide a coherent explanation for these extremely puzzling discoveries? One way to resolve the problem of wave–particle duality is to assume that one or the other is more fundamental. There are two possible alternatives:

(1) Light and matter are ultimately composed of particles. Their wave properties derive from the group behavior of a large number of interacting particles. (This approach, it will be recalled, was unsuccessfully adopted by Einstein in an attempt to explain the behavior of light.)
(2) Light and matter are ultimately composed of waves. The particle properties are then derivative. One could hypothesize that "particles" are really concentrations of waves or perhaps stable condensations in an underlying fluid or field.

Unfortunately, neither of these alternative hypotheses has been developed with any great success. Most modern-day physicists generally believe that *neither* particles *nor* waves are more fundamental, but rather that they are two manifestations of some as-yet-unidentified (and possibly unidentifiable) entity.

If the beginning student has trouble understanding how an entity can possess both wave and particle attributes at the same time, he or she may be comforted to learn that most experienced physicists are just as disturbed by this problem. The difficulty seems to stem from the fact that most of us have become accustomed to viewing nature in terms of Newtonian-inspired mechanical models. Before 1900 physicists worked from a model in which all matter was thought to be composed of small, solid, impenetrable particles. Within this framework one could explain wave phenomena as disturbances that propagate *mechanically* through a medium (i.e., from particle to particle, through the action of mutual forces). We recall that electromagnetic phenomena—including light—were originally described by Maxwell in terms of *mechanical* disturbances in the ether.

When the Michelson–Morley experiment showed that the ether could not be depicted as a mechanical medium, our troubles began. We are now forced to think of electromagnetic fields as fundamental entities that can exist in a vacuum, and we have great difficulty in understanding electromagnetic fields because we can no longer visualize them in terms of a simple mechanical model. Nor can we imagine how an electromagnetic wave can propagate and carry energy and momentum from place to place without a medium, since we tend to think of energy and momentum exclusively as properties of individual material particles. The problem of understanding light is especially difficult, because when we think of light as a wave, we tend to categorize it with sound waves, water waves, and the like. We must remember, however, that light is *fundamentally different* from sound and water waves, because light—unlike the others—*is not a mechanical wave.* Although electromagnetic waves and mechanical waves can be described by similar mathematical equations and they have certain appearances in common, they represent *inherently different phenomena.* In other words, part of the confusion engendered by the wave–particle duality of light arises because we tend to think of light in terms of mechanical waves, and since mechanical waves do not possess "particle" properties, we cannot comprehend how light can exhibit particle behavior.

On the other hand, if we adopt a particle model of light by visualizing it as a stream of individual photons, we run into other difficulties—notably, how do we explain interference phenomena? With modern-day techniques it is possible to devise light sources so feeble that they effectively emit one photon at a time. The standard interference experiments—such as Young's double-slit experiment—can be performed with these sources. The results of these experiments seem to indicate that *wave properties must be attributed to individual photons.* For example, the results of the double-slit experiment using single photons can be accounted for only by assuming that the photon passes through both slits at the same time. These findings serve to illustrate that light is not simply *either* a particle *or* a wave, but an entity that exhibits the characteristics of both.

Now let us turn our attention to the electron. We find it rather easy to visualize the electron as a tiny particle in the Newtonian tradition, and we are rather shocked to find that an electron exhibits wave characteristics. We recall the experiments of J. J. Thomson, who supposedly demonstrated the existence of the electron as a particle by showing that a beam of electrons behaves as if it consists of individual units each possessing a fixed charge and mass. As an analogy we have the photoelectric effect in which a beam of light behaves as if it consists of individual units of fixed energy. Yet we know that a light beam also exhibits wave properties. The demonstration of electron diffraction, which shows that an *electron* beam exhibits wave properties, forces us to recognize that the electron is more than just a simple particle. If Young's experiment were to be done with electrons instead of light, we would fully expect to obtain an interference pattern.* This result would be interpreted to mean that each electron passes through both slits at the same time—a rather difficult thing to conceive of if we cling to the notion of the electron as a pure particle.

One of the lessons of the quantum theory is that we must recognize the limitations of using mechanical models to visualize events on the atomic level. Atomic processes simply cannot be described completely in terms of familiar, large-scale phenomena such as billiard balls and water waves.

*This experiment has never actually been performed! It would provide a most interesting test of the concept of the wave nature of electrons.

Thus we have come to appreciate the pitfalls of depicting light or the electron exclusively as either a wave or a particle. In doing so, we have learned what light and electrons *are not*, but we have understood very little about what they *are*. It would be most gratifying at this point to learn that physicists have come up with a theory that accounts for all of the properties of light and matter in terms of some fundamental entity that exhibits both wave and particle characteristics. Unfortunately, at the present time no such theory exists, and wave–particle duality remains as one of the great unsolved mysteries of nature.

17–2

THE SCHRÖDINGER WAVE EQUATION

The discovery of electron diffraction in 1927 naturally provided a strong experimental basis for de Broglie's theory of the wave properties of matter. But in fact, there were several physicists who were so convinced of the validity of de Broglie's ideas that they did not bother to wait for experimental confirmation before developing the theory in greater detail. Early in 1926 Erwin Schrödinger, a professor at the University of Zurich, published a paper that rather dramatically demonstrated the fruitfulness of de Broglie's hypothesis. Working from the principle that an electron in an atom can be described by the properties of its associated de Broglie wave, Schrödinger devised an equation that, when solved for the hydrogen atom, yielded nearly all the results previously obtained with the Bohr–Sommerfeld theory.

The Austrian-born Schrödinger had received a traditional education in both the humanities and the sciences. He studied physics at the University of Vienna, where the influence of Ludwig Boltzmann, one of the last great classical physicists, still prevailed. Paradoxically, although Schrödinger's work was to provide the fundamental equation of the new mechanics, Schrödinger's philosophy of science remained deeply rooted in the classical tradition throughout his life.

FORMULATION OF THE WAVE EQUATION

The mathematics required to understand Schrödinger's equation fully is well beyond the scope of this book. Nevertheless, this equation is of such fundamental importance to twentieth-century physics that the beginning student ought at least to see what it looks like, even if he or she cannot fully comprehend its mathematical meaning. The equation will be presented here in the hope that the student will gain at least some feeling for its significance.

The question that Schrödinger proposed to answer was: How is a de Broglie wave affected by the presence of a force? Schrödinger took as his starting point the principle of conservation of energy, which is concerned with the potential energy associated with a given force, rather than the force itself. The effect of a force on the motion of a body is described by the interplay between kinetic energy and potential energy.

The kinetic energy of a body in the nonrelativistic case can be written in terms of the momentum p, as follows:

$$KE = \frac{1}{2}mv^2 = \frac{(mv)^2}{2m} = \frac{p^2}{2m}$$

Thus the total energy E becomes

$$E = KE + PE = \frac{p^2}{2m} + PE$$

Solving for p, we obtain

$$p = \sqrt{2m(E - PE)}$$

Schrödinger proposed to substitute this equation for p into de Broglie's equation (Equation 17–1):

$$\lambda = \frac{h}{p} = \frac{h}{\sqrt{2m(E - PE)}} \tag{17-3}$$

This equation gives the de Broglie wavelength of a body that is being acted on by a force.

Schrödinger now proposed that a de Broglie wave obeys the same wave equation that applies to mechanical waves. In classical wave theory a wave disturbance is represented mathematically by a *wave function* ψ, which is an equation for the amplitude of the wave at any given point in space and time. (The concept of a wave function was described in the section on the mathematical representation of waves on pages 261–65.) It can be shown that the wave function for *any* mechanical wave—rope waves, sound waves, and so on—satisfies the *classical wave equation*

$$\nabla^2\psi + \left(\frac{2\pi}{\lambda}\right)^2\psi = 0 \tag{17-4}$$

The symbol ∇^2 that appears in this equation is a shorthand notation for a mathematical operation involving second derivatives of ψ. It will be of no concern to us here and can safely be ignored. The λ represents the wavelength of the given wave. In the classical wave equation ψ is the unknown. The purpose of the equation is to solve for the wave function that describes the mechanical wave produced by a given set of conditions. This same equation can be used to describe electromagnetic waves if ψ represents the electric or magnetic field intensity (**E** or **B**).

Schrödinger modified the classical wave equation by substituting for the wavelength λ his own expression for the de Broglie wavelength. If we substitute Equation 17–3 for λ into Equation 17–4, we obtain

$$\nabla^2\psi + \frac{8\pi^2 m(E - PE)}{h^2}\psi = 0 \tag{17-5}$$

Equation 17–5 is the time-independent form of the ***Schrödinger wave equation.**** Schrödinger did *not* obtain this equation by a rigorous theoretical derivation from fundamental principles. He simply assumed that de Broglie waves behave like mechanical waves and modified the classical wave equation accordingly. In doing so, Schrödinger also changed the meaning of the wave function ψ. For mechanical waves ψ represents the wave amplitude (the displacement of the medium), but the wave amplitude of a de Broglie wave does not have a simple physical interpretation. Thus

*The more general (time-dependent) form of the Schrödinger equation, which introduces the variation with time of a wave pattern, is not of interest to us here. We will be concerned mainly with the steady-state (standing wave) solutions, which turn out to correspond to the stationary states of the Bohr theory.

ψ in the Schrödinger equation should be regarded for the moment purely as an abstract mathematical function with no direct physical significance—just as **E** and **B**, the electric and magnetic field intensities, have no physical significance in the etherless electromagnetic theory. Later in the chapter we will examine some of the efforts that have been made to attach a physical meaning to ψ.

SOLVING THE SCHRÖDINGER EQUATION

A student without advanced mathematical training who encounters the Schrödinger equation for the first time will probably find that it conveys little or no meaning. The purpose in introducing the Schrödinger equation here is not to teach the student how to use it, but simply to demonstrate its significance as a basic postulate and a source of some of the fundamental concepts of quantum physics.

The purpose of the Schrödinger equation is to determine the form of the de Broglie wave (or waves) associated with a given particle under specific conditions. The wave function ψ is, in effect, the "unknown" in the equation, so that to solve the Schrödinger equation is to find a mathematical formula for ψ that satisfies the equation. The form of the wave function will naturally depend on the forces acting on the particle; these forces affect the solution because they determine the potential energy term (PE) that appears in the Schrödinger equation. In addition, we expect the solution to be a "well-behaved" mathematical function—that is, it should have one and only one value at every point in space; it should be a smooth, continuous function, and so on.

The Schrödinger equation is one example of a type of mathematical equation known as a ***partial differential equation,*** and mathematicians have developed standardized methods for solving such equations. Generally, there will be more than just one solution to the Schrödinger equation for a given set of conditions; the equation will usually be satisfied by a whole family of mathematically related wave functions. Each one of these wave functions represents a permissible pattern for the de Broglie wave associated with the particle. Corresponding to each of these wave functions will be a distinct value of the total energy E of the given particle. To summarize: One substitutes into the Schrödinger equation the specific forces and other conditions imposed on a given particle, and one solves the equation for the wave functions (and the corresponding total energies) of the de Broglie waves associated with the particle.

To illustrate the procedure, let us consider the simple example of a freely moving particle confined between two impenetrable walls a distance L apart—the so-called particle-in-a-box problem (Figure 17-2). Within the walls there are no forces acting on the particle; therefore we set $PE = 0$ in the Schrödinger equation. Also, since the walls are impenetrable, the particle (and its associated wave) must be confined to the space between the walls. Therefore the wave function must have a value of zero outside the walls.

Under these conditions the solutions of the Schrödinger equation turn out to be a family of wave functions

$$\psi_n = A \sin \frac{n\pi x}{L} \qquad (n = 1, 2, 3, \ldots, \infty; \quad 0 \leq x \leq L)$$

where A is a constant. In other words, there is an infinite number of solutions, each characterized by a different value of the integer n, as follows:*

* In addition, every sum of two or more of these solutions is also a solution.

$$\psi_1 = A \sin \frac{\pi x}{L}$$

$$\psi_2 = A \sin \frac{2\pi x}{L}$$

$$\psi_3 = A \sin \frac{3\pi x}{L}$$

$$\vdots$$

Each of these wave functions can be thought of as a permissible de Broglie wave for the particle. A more important result is that the solution of the Schrödinger equation assigns to each of these wave functions a total energy

$$E_n = \frac{n^2h^2}{8mL^2} \qquad (n = 1, 2, 3, \ldots, \infty)$$

This equation describes the permissible energies that the particle can possess. We see that the energy of the particle is *quantized;* that is, it can only be $h^2/8mL^2$, $4h^2/8mL^2$, $9h^2/8mL^2$, and so on. The quantum number n that appears in the energy equation arises naturally from the mathematical solution of the Schrödinger equation. In the Bohr theory, on the other hand, quantum numbers were introduced without theoretical justification because they were needed to make the theory fit the observational data obtained from spectra.

In Figure 17–2 the first three wave functions have been drawn. We see that they are rather similar in form to the standing wave patterns for a vibrating string, shown in Figure 8–23. This result comes about as a consequence of the similarity between the Schrödinger equation and the classical wave equation—"the particle in the box" and "standing waves on a string" are mathematically the same problem. From this vantage point, the appearance of quantum numbers is not at all surprising. In analyzing standing waves, we found that standing wave patterns appear only for certain wavelengths—those obeying the equation (8–6a)

$$\lambda_n = \frac{2L}{n} \qquad (n = 1, 2, 3, \ldots, \infty)$$

In this equation the integer n is, in effect, a quantum number that restricts the possible values of the wavelengths for standing waves on a string. Thus it is wrong to think that quantum numbers appear only on the atomic level. In fact, they are characteristic of many wave phenomena.

This simple hypothetical problem of the particle in the box demonstrates an important aspect of Schrödinger's approach. If we assume that a particle is governed by the Schrödinger wave equation and then solve the equation for the specific conditions imposed on the particle, we obtain wave functions that are supposed to describe the de Broglie waves associated with the particle. We are also able to calculate the energy of the particle, and we find that the energy turns out to be quantized. This is a significant result, because the quantization of energy is a known feature on the atomic level. The fact that quantization arises naturally from the Schrödinger equation clearly indicates that we are on the right track.

THE HYDROGEN ATOM

Now let us turn to a more significant application of the Schrödinger equation: the solution of the equation for the wave functions and energy states of the hydrogen figure 17-2 **THE SCHRÖDINGER WAVE FUNCTIONS FOR THE PARTICLE IN A BOX**

The particle is confined to move in the x direction between two impenetrable walls located at $x = 0$ and $x = L$. The wave functions representing the first three solutions of the Schrödinger equation are shown. Corresponding to each wave function ψ_n is a specific value of E_n, the energy of the particle.

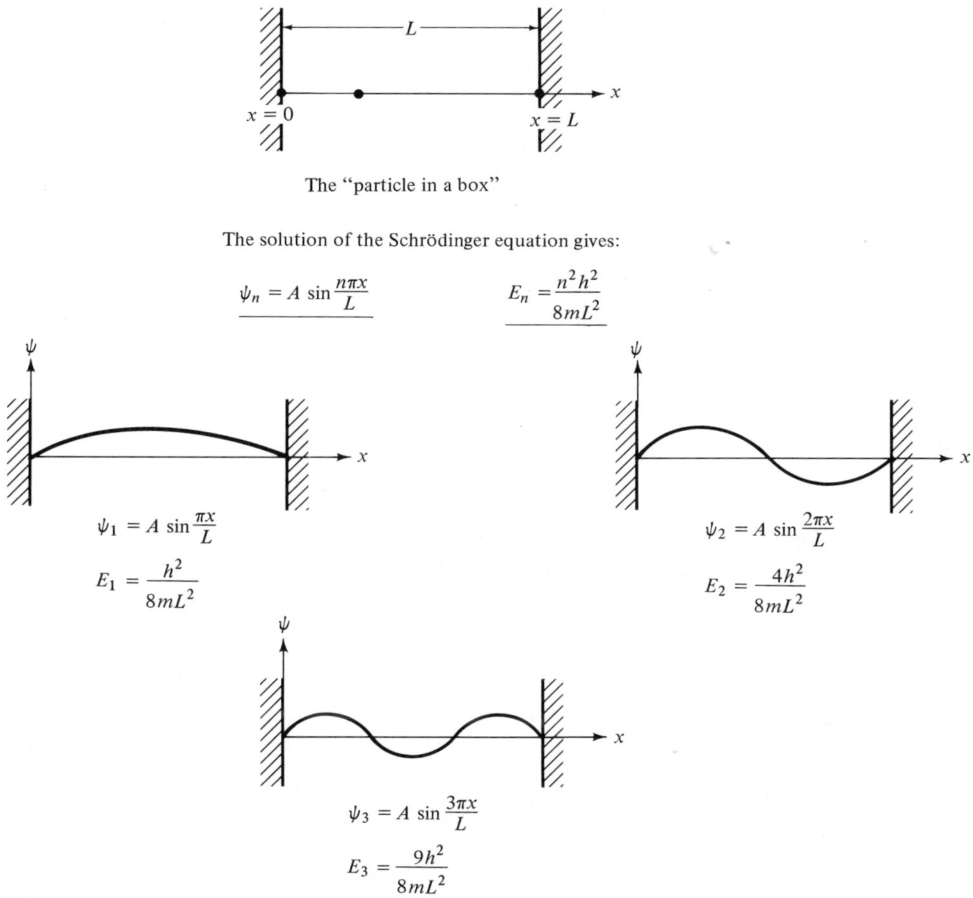

atom. As was the case with the Bohr theory, the Schrödinger equation can be solved exactly only if there is a single electron—so the results obtained are applicable to any one-electron atom.

The electron in such an atom is acted on by a single force—the electrical attraction of the positively charged nucleus. If there are Z positive charges in the nucleus, the electron has an electric potential energy $PE_e = -Ze^2/r$. This quantity is substituted into the Schrödinger equation. The methods used to solve the equation in this case are extremely complex, but an exact solution can be obtained nevertheless. The hydrogen atom is three-dimensional, so the Schrödinger equation has to be written in terms of three coordinates. What is done is to split the Schrödinger equation into three separate equations—one for each coordinate—and the equations are solved one by one. The solution of each one of these equations results in a different quantum number. The three solutions are then recombined to give three-dimensional wave functions. These wave functions contain three quantum numbers in all, *and they turn* *out to correspond exactly to the three quantum numbers n, l, m_L of the Bohr–Sommerfeld theory.* The rules governing the values of these quantum numbers

$$\begin{aligned} n &= 1, 2, 3, \ldots, \infty \\ l &= 0, 1, 2, \ldots, n-1 \\ m_L &= 0, \pm 1, \pm 2, \ldots, \pm l \end{aligned}$$

evolve automatically from the mathematics of the solution. Thus the solution of the Schrödinger equation for the one-electron atom consists of a family of three-dimensional wave functions, each one characterized by a different set of quantum numbers n, l, m_L.

Now we turn to the question of the total energy E of the electron. The solution of the Schrödinger equation leads to two different possibilities. If E is taken to be a negative number, the electron energy is found to be quantized according to the following equation:

$$E_n = -\frac{2\pi^2 m Z^2 e^4}{n^2 h^2} \qquad (n = 1, 2, 3, \ldots, \infty)$$

which is identical to the energy level equation of the Bohr theory (Equation 15–8).

However, if E is taken to be a positive number ($E \geq 0$), there is no restriction on the value of E; it can be *any* positive number. This result corresponds to the *continuum* of the Bohr theory, representing the energies of free electrons. (See Figure 15–6.)

Thus we find that the Schrödinger equation for the one-electron atom leads to the same energy levels that are in the original Bohr theory and generates three of the four quantum numbers that appear in the Bohr–Sommerfeld theory. Schrödinger's method achieves this with fewer arbitrary postulates, and in this respect it is considered superior to the Bohr–Sommerfeld theory. Some of the other main accomplishments of Schrödinger's theory are as follows:

(1) The relative intensities of the different spectral lines can be calculated.
(2) The rules determining which transitions between energy levels can or cannot occur can be derived theoretically.* In the Bohr–Sommerfeld theory these rules could only be inferred from spectra.
(3) When a term is added to the Schrödinger equation that represents the potential energy of an electron in a magnetic field, the solution of the equation accounts for the normal Zeeman effect. By introducing an additional term representing the potential energy of a spinning electric charge, one can derive the anomalous Zeeman effect.

At the same time Schrödinger's theory of the atom has some rather notable shortcomings:

(1) The Schrödinger equation cannot be solved exactly for anything more complicated than a one-electron atom.
(2) The Schrödinger equation does not take the theory of relativity into account. Schrödinger had originally tried to construct a relativistic wave equation, but the solutions of this equation did not agree with observations. Finally, he had to settle for a nonrelativistic formulation.

*The rules are that no transition can occur between two levels unless $\Delta l = \pm 1$ and $\Delta m_L = 0, \pm 1$ in the transition.

(3) The Schrödinger equation (like the original Bohr theory) does not account for the fine structure of the hydrogen spectrum. This failure can be attributed to the fact that the equation is nonrelativistic. (We recall that the fine structure was explained by Sommerfeld when he included relativity in the Bohr theory.)
(4) The phenomenon of electron spin cannot be derived from Schrödinger's theory. (The fourth quantum number, m_s, does not appear in the solution of the Schrödinger equation.) To account for electron spin, additional terms must be postulated and added on to the Schrödinger equation.
(5) The Pauli exclusion principle cannot be derived from Schrödinger's theory.

THE RELATIVISTIC WAVE EQUATION

The difficulties raised by items 2, 3, and 4 were dissolved in 1928, when a successful *relativistic* wave equation was derived by the British mathematical physicist Paul Dirac. Dirac studied to be an electrical engineer, but he could not find employment in this profession, so he decided to return to school and take up the study of physics. Dirac soon established himself as a brilliant mathematician and made many important contributions to the theory of quantum mechanics.

Of all Dirac's contributions, none is more magnificent than the relativistic wave equation, which unites the two great physical theories of the twentieth century, relativity and quantum mechanics. The solution of this equation leads naturally to a fourth quantum number with two values, $+\frac{1}{2}$ and $-\frac{1}{2}$, which is identical to the spin quantum number m_s. The energy terms connected with this quantum number can be interpreted as assigning to the electron an intrinsic spin and magnetism identical to that postulated for the electron by Uhlenbeck and Goudsmit. Thus the phenomenon of electron spin can be accounted for as a mathematical consequence of the Dirac equation; it does not have to be introduced as a separate postulate.*

When the Dirac equation is solved for the hydrogen atom—in itself a brilliant mathematical feat—Sommerfeld's equation for the fine structure of the hydrogen spectrum (Equation 16–3) is obtained. The presence of an electron spin term in the Dirac equation leads to a spin–orbit interaction for the hydrogen electron, which in turn accounts for the anomalous Zeeman effect.

The Dirac equation, as we have seen, accounts for all the phenomena described by the Bohr–Sommerfeld theory and does so in a conceptually superior (albeit mathematically more difficult) way. To the present day relativistic quantum mechanics remains as the supreme formulation of quantum theory; it has been used to predict certain properties of the electron correct to five and six decimal places. However, the Schrödinger equation is quite adequate for the solution of most problems, so it is the more widely used of the two equations. It has proved to be extremely successful in the solution of a wide range of problems in atomic and nuclear physics and has contributed greatly to our understanding of heat, magnetism, electrical conductivity, crystal structure, and other properties of matter.

In view of the vast superiority of Schrödinger's approach as compared to the Bohr–Sommerfeld theory, why has so much space been devoted to the latter? Although conceptually ill-founded, the Bohr–Sommerfeld theory nevertheless gives a fairly accurate (though incomplete) *description* of atomic properties. In this sense we can consider

*Perhaps the most remarkable aspect of the Dirac equation is that it can be interpreted to predict the existence of ***antimatter.*** We will return to this topic in Chapter 20.

the Bohr–Sommerfeld model to contain the *laws* of atomic behavior, whereas Schrödinger's model gives the best *theory* from which these laws can be derived. The Bohr–Sommerfeld model provides a visual representation of the atom, which, although it is oversimplified, serves as a useful pedagogical tool for introducing the student to some of the fundamental quantum concepts.

SCHRÖDINGER'S MODEL OF THE ATOM

Much of the difficulty in understanding the Schrödinger wave equation centers on the mysterious wave function ψ. Schrödinger was naturally most concerned that it be given a clear-cut physical meaning in his theory. Basically, the wave function is supposed to represent the de Broglie wave or waves associated with a particle. De Broglie's original depiction of "pilot waves" that guide particles along specific paths was generally considered a much too simplistic model; what was sought was a more specific relationship between the de Broglie waves and the associated particle.

In Chapter 8 we learned that a wave characterized by a single wavelength appears as an infinitely long wave train (see Figure 8–19), whereas a wave train of finite length can be constructed by adding together a group of waves of slightly different wavelengths. A finite wave train, sometimes known as a ***wave packet,*** has a beginning and an end and is therefore localized in a particular region of space. If a particle is associated with a single de Broglie wave (an infinitely long wave train), the wave does not locate the position of the particle in space, but if the particle is associated with a wave packet (constructed from a group of waves), the position of the particle can be related to the location of the wave packet (Figure 17–3). The wave packet travels with the same speed as the particle, and the motion of the wave packet through space corresponds precisely to the motion of the particle. Schrödinger proposed that the family of wave functions obtained from the solution of the Schrödinger equation describes a group of de Broglie waves that, when added together, form a wave packet that locates the particle in space.

According to Schrödinger's picture, the electron is not a point particle, but is "smeared out" over the whole wave packet. For a free electron traveling through space, the wave packet is very small, so the electron appears to be almost a point particle. If the electron is bound to an atom, however, the wave packet completely surrounds the nucleus. Thus the electron itself surrounds the nucleus; in other words, the electron is smeared into an "electron cloud" around the nucleus. Hence Schrödinger discarded the old picture of the point electron traveling in a distinct orbit around the nucleus and assumed instead that the electron is spread out over the entire orbit. Since the electron is a charged particle, it follows that the electron charge must also be distributed over the entire wave packet. To complete this model, Schrödinger proposed a physical interpretation for the magnitude of the wave function ψ. (Remember that in classical physics the magnitude of ψ is equal to the wave amplitude.) Schrödinger hypothesized that the quantity ψ^2 denotes the *density of the smeared-out electron at any point in space.* (The quantity ψ^2 is used instead of ψ because ψ can be negative, and a negative density is physically meaningless. However, ψ^2 is always a positive quantity.) That is, the electron mass and charge are concentrated wherever ψ^2 is large and are thinly distributed wherever ψ^2 is small.

Schrödinger's interpretation gained little support. Few physicists could accept the idea of a smeared-out electron. It was generally felt that Schrödinger had placed too much emphasis on the wave aspects of the electron and had ignored the electron's particle properties. In June 1926 a more acceptable interpretation was put forth by

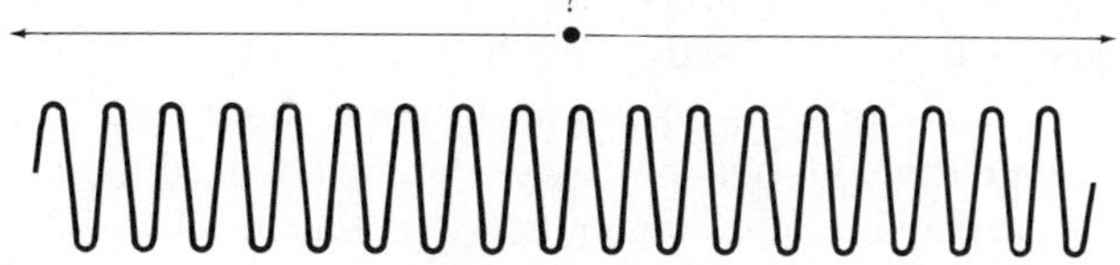

(a)

Infinite wave train (single wavelength). Where is the particle?

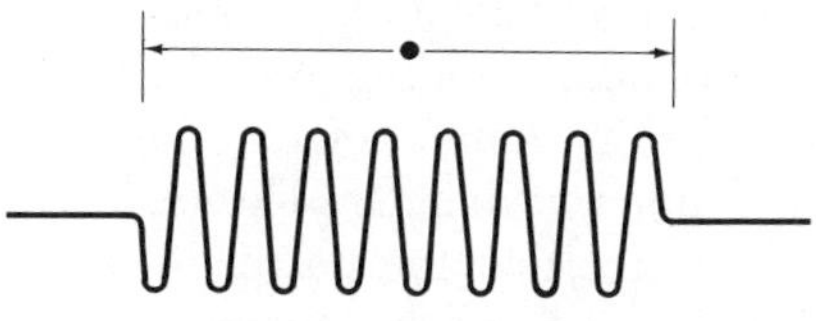

(b)

Finite wave train formed as sum of group of de Broglie waves. Position of particle coincides with position of wave packet.

figure 17–3 **WAVE PACKET REPRESENTATION OF PARTICLE**

An individual particle cannot be represented by a de Broglie wave of a single wavelength, since the wave does not locate the particle at a given point in space. Instead, a wave packet constructed out of a group of waves of different wavelengths must be used.

Max Born, a professor at the University of Göttingen in Germany* and one of the prime contributors to the development of quantum mechanics.

As were so many aspects of the quantum theory, Born's proposal was inspired by Einstein's work. In his attempt to construct a photon theory of the wave properties of light, Einstein had theorized that the square of the amplitude of a light wave at any point is proportional to the probability of finding a photon at that point. In analogy with Einstein's suggestion, Born hypothesized that if ψ is the wave function representing some particle, the value of ψ^2 at a given point in space represents the *probability that the particle is located at that point.* Thus the particle is no longer considered to be distributed over the entire wave packet. Instead, it is assumed only that the particle is located somewhere within the confines of the wave packet, and the *probability* that the particle can be found at various points within the wave packet can be calculated from the wave function (more specifically, from ψ^2). The idea of an electron cloud within the atom is consequently rejected. In Born's interpretation the electron cloud represents not the electron itself but the probability distribution for the location of the electron. *The electron is most likely to be found where ψ^2 has its largest value and least likely to be found where ψ^2 is smallest.* Thus Born's

* In the 1920s Göttingen and Copenhagen were the two major centers for the development of the quantum theory, but the rise of Nazism in the 1930s reduced Göttingen to a shambles. Purged of its non-Aryan professors, the university never again attained its former greatness.

interpretation introduces the notion of probability into quantum mechanics—a matter that will be discussed in depth later in the chapter.

It is rather ironic that while Schrödinger's *mathematical* theory won great acclaim, the physical meanings that he attached to his findings were generally rejected. In particular, Schrödinger believed that he had done away with the idea of discontinuous quantum "jumps" within the atom—an idea that had been particularly disturbing to the more classically oriented physicists. Schrödinger visualized the atom as a nucleus surrounded by three-dimensional standing wave patterns with fixed vibration frequencies. (The waves are de Broglie waves, and the standing wave pattern comprises the electron cloud.) He pictured energy transitions in the atom in terms of changes from one standing wave pattern to another, comparable to the way in which the different harmonics of a violin string can be produced by bowing or plucking the string in a different manner. Through this analogy with classical wave phenomena, Schrödinger thought that the postulate of quantum jumps could be discarded.

In the fall of 1926 Schrödinger was invited to Copenhagen to discuss his ideas with Bohr and his students. The discussion turned into a marathon debate, which lasted several days. Schrödinger was not nearly as energetic or as forceful a debater as Bohr, and he gradually gave ground. Eventually, Schrödinger had to admit the inadequacy of his model. In his despair (and exhaustion) he shouted at Bohr, "If we are going to stick to this damned quantum jumping, then I regret having ever been involved in this thing."—to which Bohr replied, "But the rest of us are thankful that you did, because you have done so much to promote the quantum theory."[4]

Schrödinger never brought himself to accept the idea of quantum jumps. As time progressed, he drifted further and further from the main lines of thought in quantum mechanics. Yet the Schrödinger equation remains as the single most important equation of this theory.*

17–3

THE UNCERTAINTY PRINCIPLE

One of the spectators at the Bohr–Schrödinger debate was a brilliant young German physicist, Werner Heisenberg. Heisenberg studied physics under Sommerfeld and received his Ph.D. at the age of twenty-two. For several years afterward, he divided his time between working with Bohr in Copenhagen and with Born in Göttingen. In 1925 Heisenberg—with the aid of Born and the mathematician Pascual Jordan—constructed a highly abstract mathematical formulation of the quantum theory known as ***matrix mechanics.*** A year later Schrödinger's theory was published, and although it seemed at first that the two theories were fundamentally different, it was soon demonstrated that Heisenberg's matrix mechanics and Schrödinger's wave mechanics were mathematically equivalent. Thus Heisenberg and Schrödinger are generally given equal credit for constructing the mathematical foundations of the quantum theory, although Schrödinger's formulation is more widely used because it is generally easier to work with.

Heisenberg realized that any arguments as to the location of the electron in the atom and the physical interpretation of the wave function were meaningless in

* A most interesting summary of Schrödinger's own thoughts as to the nature of the atom and the validity of the modern interpretation of quantum mechanics can be found in Schrödinger's article "What Is Matter?" See Supplementary Reading, Appendix E.

the absence of relevant experimental data. "If one wants to clarify what is meant by 'the position of an electron,'" Heisenberg argued, "one has to describe an experiment by which the 'position of an electron' can be measured; otherwise this term has no meaning at all."[5] Heisenberg therefore proposed that the quantum theory should pay particular attention to the *measurement* of physical quantities, and he began to investigate this problem in detail.

A fundamental aspect of Heisenberg's approach is the idea that *in making a measurement, we necessarily disturb the object being measured.* In order to measure the position of some object, we must "see" the object,* and in order to see an object, we must shine light on it. This hardly poses any problem on the laboratory or astronomical scales; when shining light on a moving pendulum or bouncing radar beams off Venus, we do not alter the motions of these bodies by any detectable amount. On the atomic scale, however, any act of measurement must bring about significant changes in at least one of the properties of the object under observation.

In order to "see" (and locate the position) of an electron, one must receive at least one photon of light that has been either reflected or emitted by the electron. But *in reflecting or emitting a photon, the electron must recoil, thereby altering its motion significantly*. The electron recoil will be relatively large because an electron and a photon are objects of comparable energy and momentum. We do not expect the motion of a pendulum to be changed when a photon strikes the pendulum, because the momentum of the pendulum is enormously larger than the momentum of the photon. But the momentum of an electron and that of a photon are generally of the same order of magnitude, so that the motion of the electron should be altered considerably in an interaction with a photon. By analogy, if we chose to detect the position of a moving pendulum by firing precisely aimed machine-gun bullets at it, the effect would be quite similar. A bullet striking the pendulum would accurately locate the pendulum's position but would seriously disturb its motion.

As an alternative we could try to reduce the electron recoil by using low-energy photons. Conceivably, we could make the photon energy so low compared to that of the electron that the electron would hardly recoil at all, but the use of lower-energy photons reduces the precision to which the position of the electron can be measured. This effect is a consequence of the *wave nature* of photons, for as the energy of a photon decreases, the wavelength of the photon increases. In effect, the photon's "size" increases (it becomes less concentrated) as its energy decreases, and if the photon becomes much larger than the electron itself, the position of the electron cannot be ascertained to any high degree of accuracy.

Consequently, we find that we cannot determine *both* the position and the motion of the electron as precisely as we wish. If we determine the *position* precisely (by using high-energy photons), we disturb the electron's *motion* appreciably. If we determine the electron's *motion* precisely (through a series of nearly recoil-free observations with low-energy light), its *position* cannot be determined accurately.

Heisenberg was able to express this concept in mathematical form. Let Δx represent the uncertainty† in the *position* x of the object being observed. Let Δp

*The word "see" is being used in its most general sense here; it is not restricted to the visual sense. With the aid of instruments we can see an object by receiving infrared, ultraviolet, or some other form of radiation from it.

†Up to now we have used the Δ symbol to indicate the *change* in a quantity. Here the *uncertainty* Δq denotes the precision of an experimentally determined value of q, where q is any quantity. For example, if the result of a series of measurements is given as $q = 10.5 \pm 0.3$, then ± 0.3 is the uncertainty Δq in any single measurement of q.

be the uncertainty in the *momentum* p ($= mv$) of the object. (The momentum is a suitable measure of the motion of the object.) In analyzing a number of idealized experiments designed to measure the position or momentum of a particle, Heisenberg found that in every case

$$(\Delta x)(\Delta p) \geq h/2\pi \qquad (17\text{–}6)$$

In other words, the *product* of the two uncertainties must always be greater than $h/2\pi$, where h is Planck's constant. This equation, known as the ***uncertainty principle,*** was announced by Heisenberg in 1927.

According to the uncertainty principle, if the precision of a position measurement is improved (Δx is made smaller), there must be a comparable *decrease* in the precision of the momentum measurement (Δp gets larger), and vice versa. It is possible to measure any *one* quantity to arbitrarily high precision (in principle, one could approach $\Delta p = 0$), but then no meaningful measurement could be made of the other quantity ($\Delta x \longrightarrow \infty$). Hence the uncertainty principle can also be stated as follows:

It is impossible to measure simultaneously both the position and the momentum of a particle to an arbitrarily high degree of precision.

Whereas Heisenberg thought that the uncertainty principle was inherent in the measurement process, Bohr—with whom Heisenberg was working closely at the time—felt that the uncertainty principle was a natural consequence of wave–particle duality. In fact, the principle can be shown to follow directly from the wave packet representation introduced by de Broglie and Schrödinger. As illustrated in Figure 17–3, a wave train defined by a single wavelength is infinitely long, so the position of the associated particle is undefined ($\Delta x \longrightarrow \infty$). But from de Broglie's equation $\lambda = h/p$, we see that a precisely defined wavelength implies a precisely known momentum ($\Delta p = 0$). Now suppose the particle is associated with a finite wave packet. Then the particle is presumed to be located somewhere within the range of the wave packet, so the uncertainty Δx in the position of the particle is equal to the width of the wave packet. But a finite wave packet is constructed from a group of waves of slightly different wavelengths; hence the particle does not have a precisely defined wavelength. Through the equation $\lambda = h/p$, there will be an uncertainty Δp in the momentum of the particle associated with the spread of the wavelengths ($\Delta\lambda$) needed to construct the wave packet. Thus the following chain of reasoning is established:

(1) A smaller uncertainty Δx in the position of a particle requires a smaller wave packet.
(2) Therefore more waves are needed to construct the wave packet, which requires a greater range of wavelengths $\Delta\lambda$.
(3) This greater range of $\Delta\lambda$ leads to a greater uncertainty Δp in the momentum of the particle.

The uncertainty principle can be derived mathematically from an analysis of the relationship between Δx (the size of the wave packet) and $\Delta\lambda$ (the wavelength range needed to construct the wave packet), which is in turn related to Δp through the equation $\lambda = h/p$.

To understand the limitations imposed by the uncertainty principle, let us now consider some examples. First let us see how the uncertainty principle applies to a baseball thrown at 100 mi/hr (for which m = 152 grams, v = 4470 cm/sec). How accurately can we determine both the position and the momentum of the baseball? Presumably, with suitable equipment we could measure the position of the baseball at any given instant to a precision of, say 0.1 inch; but let us assume that we know only that the ball is somewhere between the pitcher's mound and home plate (a distance of 60.5 feet). Thus $\Delta x = 60.5$ feet ≈ 2000 cm. Now suppose that while the ball is in flight, we attempt to measure its momentum. According to Equation 17–6, this measurement will have an inherent uncertainty of

$$\Delta p \geq \frac{h/2\pi}{\Delta x} = \frac{1.05 \times 10^{-27} \text{ erg-sec}}{2 \times 10^{3} \text{ cm}} = 5.3 \times 10^{-31} \text{ g-cm/sec}$$

Here Δp represents the *maximum precision attainable by any means.* That is, there is no way to measure the momentum of the baseball in this circumstance that would give an uncertainty in p *smaller* than 5.3×10^{-31}. Since the actual momentum of the baseball is $p = mv = (150 \text{ grams})(4470 \text{ cm/sec}) = 6.71 \times 10^{5}$ g-cm/sec, the uncertainty principle tells us that p could conceivably be measured to an accuracy of $\Delta p/p \approx 1$ part in 10^{36}, which is beyond the wildest dreams of even the most talented experimental physicist. Generally, for large-scale objects the uncertainty principle does not impose any serious limitations on the precision of our best experiments.

On the atomic level, however, the situation is quite different. Consider an orbiting electron in a hydrogen atom. Assuming that for the electron $m = 9.11 \times 10^{-28}$ gram and $v = 2 \times 10^{8}$ cm/sec (the approximate speed of an electron in the $n = 1$ orbit of the hydrogen atom), we calculate that the momentum of the electron should be $p = mv = 1.8 \times 10^{-19}$ g-cm/sec. Now suppose we wish to check this calculation by making a measurement of the momentum of an orbiting electron. Since we know that the electron is somewhere in the atom, the uncertainty in the electron's position is $\Delta x = 1.06 \times 10^{-8}$ cm (the diameter of the $n = 1$ orbit of the hydrogen atom). Therefore

$$\Delta p \geq \frac{h/2\pi}{\Delta x} = \frac{1.05 \times 10^{-27} \text{ erg-sec}}{1.06 \times 10^{-8} \text{ cm}} \approx 1 \times 10^{-19} \text{ g-cm/sec}$$

Note that Δp is nearly as large as p itself! The best result we could hope for under ideal experimental conditions would be $p = (1.8 \pm 1.0) \times 10^{-19}$ g-cm/sec, which represents an error of more than 55 percent! Obviously, we cannot pretend to be able to measure the momentum of an orbiting electron to within acceptable standards of accuracy. On the other hand, if we could design an experiment to measure p to within 10 percent ($\Delta p/p = 0.1$, or $\Delta p = 1.8 \times 10^{-20}$), the uncertainty Δx in the electron's position created by the experiment would become so large that we could no longer be certain that the electron was still in the atom. Physically, this means that any experiment to measure p to 10 percent accuracy could easily impart enough energy to the electron to knock it out of the atom.

To summarize: The uncertainty principle does not prevent us from measuring any *single* quantity as precisely as we wish. It does place a limit, however, on the accuracy with which one can *simultaneously* measure *two* quantities; that is, the position and momentum of a particle. This limit is not one that could be exceeded by devising better experimental equipment; it represents the limit obtainable by *ideal* measuring instruments. This limit is inherent in the nature of matter rather than in the present state of experimental technique; it is *theoretically* impossible to exceed it.

In the measurement of objects large enough to be seen with the naked eye, the uncertainty principle has little or no significance because the theoretical uncertainties are extremely small compared to the sizes of the quantities being measured. It is only on the atomic and subatomic levels that the uncertainties become relatively large. But since the uncertainty principle deals only with the *simultaneous* measurement of two quantities, we are not prevented from making extremely precise measurements of single quantities on the atomic level. Why, then, is the uncertainty principle considered to be one of the most significant concepts of quantum mechanics? As we shall learn in the next section, it is the *philosophical* consequences of the uncertainty principle, rather than any practical limitations it imposes, that makes it so important.

17–4

THE PHILOSOPHICAL IMPLICATIONS OF QUANTUM THEORY

The modern-day physicist is naturally pleased to know that the mathematical equations of quantum mechanics can be counted on to yield highly accurate predictions of a wide variety of atomic phenomena. Yet few physicists today are concerned with the wider significance and underlying meaning of these equations. They may simply adopt the pragmatic attitude that as long as the equations work, there is no need to speculate about their meaning. They may defend their attitude on the ground that a mathematical theory that successfully predicts the values of measurable physical quantities through appropriate equations is entirely sufficient in itself and requires no further interpretation.

It is perhaps characteristic of the greatness of such thinkers as Einstein, Bohr, Heisenberg, and Schrödinger that they devoted so much of their efforts to clarifying the physical meaning behind the mathematical structure of the quantum theory. All of these men wrote extensively on the philosophical significance of the quantum theory and its status with respect to physics as a whole. Yet in spite of their efforts, there remains substantial disagreement as to the meaning of a number of the basic concepts of quantum mechanics.

THE SIGNIFICANCE OF THE UNCERTAINTY PRINCIPLE

Heisenberg's uncertainty principle, which expresses the fundamental inability to obtain certain information on the atomic level, is a veritable Pandora's box of difficult philosophical problems. Heisenberg and Bohr both recognized that the limitations imposed by the uncertainty principle are subject to the same kind of interpretation that Einstein employed in the theory of relativity. We recall that Einstein adopted the principle that if such entities as absolute time, absolute space, and the ether are inherently unmeasurable, it is meaningless to talk about them, and they should be eliminated from physical theory. This is the essence of the philosophical doctrine of positivism, which states that science should deal only with quantities and entities that can actually be measured.

Because of the uncertainty principle, this same doctrine has considerable relevance for atomic theory. In Newtonian mechanics if the position and velocity of a body are known at any given instant and the forces acting on the body are also known, the future path of the body can be predicted exactly. However, on the atomic scale we *cannot* precisely determine both the position and velocity of an electron at a given instant; therefore it is *impossible* to predict accurately the future path of the electron. Furthermore, it is impossible to measure the path of the electron, because in the act of observing the electron we significantly disturb its motion. For that matter, we have no right to assume that the electron was on any particular path *before* it was disturbed, since we have no information as to what the electron was doing previous to our observations.

Since we are unable to determine the path of the electron—either past or future—the idea of an electron orbit becomes meaningless. Applying the positivist principle, we must reject entirely the notion of orbiting electrons. The solar system model of the atom may be a convenient way to visualize the atom, but is experimentally unverifiable and—like the ether—is therefore nothing more than a metaphysical speculation. In this respect *the uncertainty principle emphasizes the futility of constructing classical models of atomic phenomena.*

Having lost the right to construct easily visualized models of the atom, our understanding of the nature of the atom is consequently obscured. As a consequence of the uncertainty principle, we are ultimately unable to determine what the atom is "really" like. Because the act of measurement disturbs the object being measured, we are able to determine only how an atom behaves *when it is being observed,* and we can learn nothing about the properties of an atom when it is *not* being disturbed. In Heisenberg's words, "What we observe is not nature itself, but nature exposed to our method of questioning."[6] According to the Heisenberg–Bohr interpretation,* *statements made with respect to how atoms behave when they are not being observed are meaningless.*

This approach seems to present a picture of the world that conflicts with our fundamental assumptions about the orderliness of the universe. We believe that the properties of an object do not change capriciously, and we assume that the properties of an object when it *is not* being observed are the same as when the object *is* being observed. Therefore it should be possible to make meaningful predictions (based on observations) as to the past or future behavior of an object. In terms of the uncertainty principle, this attitude is quite acceptable for large-scale objects, because their properties are not necessarily altered to any great extent by an act of measurement. But on the atomic scale an observation *necessarily* changes the properties of a system to the extent that we cannot infer anything about the properties of the system before or after the observation.

Thus Heisenberg and Bohr asserted that *on the atomic level* one cannot ascribe a reality to objects independent of our observations. According to Heisenberg, "The atoms themselves are not as real as the objects of daily life; they form a world of potentialities or possibilities rather than one of things or facts." The physicist Victor F. Weisskopf has criticized this view on the ground that "This restriction of reality may have the disadvantage that it sets atomic phenomena in a frame of unreality or subjectivity which they do not deserve. The fact that atomic phenomena are hard to describe does not make them less real."[7] But from Heisenberg's point of view the "reality" of atomic phenomena is permanently beyond our reach, and it is therefore meaningless to talk about it.

Bohr believed that the mystery of wave–particle duality could be understood through the uncertainty principle. The properties exhibited by an object are presumed to be characteristic of the interaction between the object and the measuring instrument, rather than of the object itself. Hence it follows that "wave properties" and "particle properties" should not be attributed to the object alone, but *to the combination of*

* Usually referred to in the literature as the ***Copenhagen interpretation,*** after its place of birth.

the object and the measuring instrument. In other words, an object will exhibit either wave or particle properties depending on what kind of experiment is being performed. Noting that an object can only be described in terms of its interactions with other objects, and that in no experiment are wave and particle properties exhibited *simultaneously,* Bohr proposed that "wave properties" and "particle properties" represent complementary (i.e., opposite and mutually exclusive) aspects equally necessary for a complete description of phenomena.

Bohr suggested that a "principle of complementarity" extends throughout the atomic domain and that two kinds of description are needed to account for many phenomena. He traced this dualism to the uncertainty principle, where position and momentum appear as complementary quantities. Thus Bohr attempted to dissolve the problem of wave–particle duality by declaring that this duality is fundamental; in essence, the dilemma is solved simply by accepting both horns. Although Bohr's interpretation has gained considerable acceptance, it has also been strongly criticized by some scientists and philosophers. If nothing else, Bohr's complementarity principle represents an interesting and provocative philosophical speculation.*

THE ROLE OF PROBABILITY

Newtonian mechanics comprises a completely deterministic system; given adequate knowledge of the present state of an object, it is possible to predict its future course. It is this feature that enables astronomers to calculate the motions of the planets, the times of lunar and solar eclipses, and so on, with considerable accuracy. However, the uncertainty principle severely limits our ability to make accurate predictions of future events at the atomic level. The more precisely we determine the state of an object at some given instant, the more we disturb its motion, and the more we reduce the accuracy with which we can predict the future position of the object. On this basis Bohr proposed that "knowledge of present state" and "prediction of future state" represent complementary aspects of the description of an object under the complementarity principle.

As a rule when scientists reach a point at which they cannot accurately predict the future, they take refuge in statistics. Because it is too difficult to identify all the forces acting on a coin when it is flipped, it is a practical impossibility to predict with certainty the result of a single coin flip. However, we can predict reasonably well what the net outcome of a thousand coin flips will be. Similarly, although it is much too complicated to trace the individual paths of all the molecules of a gas, it is possible to determine the *average* properties of a large aggregate of molecules by using statistical methods.

Given the limitations imposed by the uncertainty principle, it is clear that a statistical approach is dictated at the atomic level. But there is a fundamental distinction to be made. There is no *theoretical* reason why we cannot describe precisely the motion of a single coin or gas molecule; we use statistics merely as a matter of practicality. In other words, we agree that there are fundamental deterministic laws underlying the statistics of coins and gas molecules. But at the atomic level we are presumably prevented *in theory* from identifying any underlying laws relating present

*Students of philosophy will be interested to learn that Bohr admitted to being strongly influenced by the writings of the Danish philosopher Sören Kierkegaard, who wrote on the dualism of thought and reality, and the psychologist William James, who examined the complementary aspects of human consciousness.

to future; according to Heisenberg, it is therefore meaningless to talk about such laws. Heisenberg went one step further, arguing that it is not just that underlying laws are undetectable; *they do not even exist.* Accordingly, the behavior of atoms is *inherently* statistical; their actions are governed by chance—an "absolute" chance not controlled by hidden deterministic laws. In simplified terms the laws that govern the behavior of atoms are the laws of probability. This is the interpretation tacitly accepted by most physicists, and it represents the "orthodox" view.

In this context Born's interpretation of ψ^2 as denoting the probability of finding a particle in a given position takes on added force. According to the orthodox view, the results of a single measurement of some property of a particle cannot be predicted accurately—just as we cannot predict the outcome of a single coin flip. Instead, the equations of quantum mechanics predict the *average* value of a *series* of measurements on a particle, based on a probability distribution (mathematically related to ψ^2)—just as the average result of a sequence of coin flips can be predicted on the basis of probability principles.

The orthodox view has been advanced and defended most notably by Heisenberg, Bohr, and Born. It has been strongly opposed by Planck, de Broglie, Schrödinger, and Einstein. Of these, Einstein was most vehement in his disagreement. At first Einstein tried to discredit the uncertainty principle by inventing idealized experiments in which pairs of quantities supposedly could be measured simultaneously to arbitrary precision. In every instance Bohr was able to show that these experiments were in fact limited by the uncertainty principle. Eventually, Einstein accepted the validity of the uncertainty principle and admitted that quantum mechanics contained no logical inconsistencies.

However, Einstein never accepted the orthodox interpretation of quantum mechanics. He rejected the positivist contention that scientific assertions are meaningless if they cannot be directly verified by experiment. He felt that such an attitude would stifle scientific imagination. For example, the concept of the atom proved to be extremely valuable to the development of physics and chemistry in the nineteenth century. Yet at the time there was no way to verify experimentally the existence of atoms. How would science have progressed if, at the outset, all statements regarding the existence of atoms had been declared meaningless?

Ironically, it was Einstein himself who had first introduced positivist notions into physics with the theory of relativity. In later years he completely abandoned his early positivist attitudes. (When Einstein was reminded by a friend that he had initiated the use of positivism in science, Einstein replied, "A good joke should not be repeated too often.") He believed in the existence of an external world independent of our observations and measuring instruments, and felt that it was the goal of science to discover the nature of this external world, using metaphysical concepts if necessary. The major requirement was that theories based on these concepts had to be in harmony with observable facts. Thus Einstein believed that our observations held the *key* to physical reality, but they are not the *sum total* of reality.

Einstein could not accept the idea that atomic behavior is fundamentally probabilistic, for it would mean that the universe—which seems to be highly ordered and deterministic in the large—is ultimately chaotic in the small. The dependence of quantum mechanics on statistical principles clearly indicated to Einstein that the theory gives an incomplete description of atomic behavior. He rejected the contention that the wave function describes the properties of a single particle; he claimed instead that the wave function describes the *average properties of a large group of particles,* rather than the random fluctuations of the properties of a single particle. Einstein firmly believed that underlying the quantum theory—perhaps on a subquantum level—there had to be fully deterministic laws. In a letter to Max Born written in 1926, Einstein summarized his position:

> Quantum mechanics is very impressive. But an inner voice tells me that it is not the real thing. The theory has much to offer . . . but I am convinced that God does not throw dice.[8]

At the present time, nearly fifty years after the birth of quantum mechanics, the argument has not been settled. There is no evidence whatsoever that a deterministic subquantum level exists. Nor is there any convincing evidence to support the orthodox interpretation. For example, no experiments have ever been performed on a single atom or electron to test the orthodox contention that the wave function describes the properties of a single particle rather than of a group of particles. Although the orthodox interpretation is generally accepted, there remain a few who, like Einstein, believe that the mathematical formalism of quantum mechanics is not the final answer. This attitude has been most eloquently described by Erwin Schrödinger:

> Many maintain that no objective picture of reality is possible. However, the optimists among us (of whom I consider myself one) look upon this view as a philosophical extravagance born of despair. We hope that the present fluctuations of thinking are only indications of an upheaval of old beliefs which in the end will lead to something better than the mess of formulas which today surrounds our subject.[9]

It is easy for the beginning student of physics to sympathize with Schrödinger's view. The mathematical complexity of quantum mechanics alienates many students. One would like the laws of atomic behavior to be as simple and as compact as Newton's laws of motion. We must realize, however, that the criterion of simplicity is a human aesthetic preference that the universe is not compelled to follow. Regardless of its mathematical complexity and its unseemly philosophical implications, the quantum theory has thus far been remarkably successful in accounting for atomic behavior. Whether it is to be replaced by "something better" is a matter for future generations of physicists to decide.

QUESTIONS FOR REVIEW

1 Explain what is meant by the de Broglie wavelength of a particle. State the equation for the de Broglie wavelength of a particle of mass m and speed v.

2 Explain how de Broglie derived Bohr's postulate of the quantization of the angular momentum of an orbiting electron. What basic assumptions did de Broglie make in his derivation?

3 Explain why natural crystals must be used to demonstrate x-ray diffraction.

4 How did Davisson and Germer demonstrate the wave properties of the electron?

5 Discuss the problem of wave–particle duality. Summarize the experimental evidence for (a) the wave nature of light and matter and (b) the particle nature of light and matter. How does the modern-day physicist deal with this problem?

6 What is the expected outcome of Young's experiment if electrons are used instead of light? How would the result be explained?

7 What does a wave function represent in classical wave theory? What is its role in quantum mechanics?

8 Summarize the basic assumptions that Schrödinger made in the derivation of his wave equation.

9 What is the purpose of the Schrödinger wave equation? For what quantity or quantities is the equation solved? On what factors does the solution depend?

10 What is the significance of the fact that the solution of the Schrödinger wave equation leads to the appearance of quantum numbers and the quantization of energy?

11 Summarize the merits and shortcomings of the solution of the Schrödinger equation for the one-electron atom. Specifically, what aspects of the Bohr–Sommerfeld theory can and cannot be accounted for?

12 What theoretical improvements were made through the application of Dirac's relativistic wave equation to the problem of the one-electron atom?

13 Compare Schrödinger's interpretation of the meaning of the wave function ψ with Born's interpretation.

14 What is a wave packet? How is it related to the particle it describes?

15 Explain why the act of measurement must necessarily disturb the object being measured on the atomic level. Does this problem exist with large-scale objects?

16 State and explain Heisenberg's uncertainty principle. What is meant by the *uncertainty* (in position, momentum, or any other quantity) of a measurement?

17 Outline Bohr's interpretation of the uncertainty principle in terms of the wave packet representation of a particle.

18 If the uncertainty principle places severe limits on our ability to make measurements on the atomic level, how do we account for the extreme accuracy of predictions made using quantum mechanics? What precisely are the limitations imposed by the uncertainty principle?

19 Explain what Heisenberg meant by the statement "What we observe is not nature itself, but nature exposed to our method of questioning." What are the implications of Heisenberg's statement with respect to the construction of realistic models of the atom?

20 Discuss the differences between the use of probability and statistics on the large scale and on the atomic level. What does it mean to say that the behavior of atoms is *inherently* statistical?

21 What is the orthodox interpretation of what the equations of quantum mechanics predict with respect to the properties of objects on the atomic level?

22 Discuss Einstein's objections to the orthodox interpretation. What did he mean by his statement "I am convinced that God does not throw dice"?

EXERCISES

1 Calculate the de Broglie wavelength of each of the following:
(a) A 4400-lb automobile ($m = 2000$ kg) traveling at a speed of 60 mi/hr (approximately 30 m/sec).
(b) A tennis ball ($m = 50$ grams) traveling at a speed of 1000 cm/sec.
(c) A hydrogen atom ($m = 1.67 \times 10^{-24}$ gram) at room temperature ($v = 10^5$ cm/sec).

2 In general, how does the de Broglie wavelength of a particle change (a) as the particle's speed increases? (b) as the particle's mass increases?

3 Prove that in the nonrelativistic case the de Broglie wavelength of a particle is related to the particle's energy by the equation $\lambda = h/\sqrt{2mE}$.

4 At what speed must an electron ($m = 9.11 \times 10^{-28}$ gram) be traveling in order to have a wavelength of 5000 Å (the wavelength of visible light)? What is the energy of the electron in eV?

5 In their electron diffraction experiment Davisson and Germer used electrons with kinetic energies of 54 eV. What is the de Broglie wavelength of a 54-eV electron? How does this wavelength compare with the typical interatomic distance of about 1 Å in a crystal?

6 High-speed electrons are used experimentally to determine the forces within the nucleus. In order for these experiments to be successful, the electron wavelength must be no larger than the size of the nucleus, or about 10^{-12} cm.
(a) Using the nonrelativistic equation $\lambda = h/mv$, calculate the speed of an electron having a wavelength of 10^{-12} cm.
(b) According to your answer, is the use of a nonrelativistic equation justifiable? Explain how you would redo the problem taking relativity into account.

7 Show that the Schrödinger wave equation (Equation 17–5) follows mathematically from Equations 17–3 and 17–4.

8 The equation for the allowed energies of the "particle in a box" is $E_n = n^2h^2/8mL^2$.
(a) Using the equation $E = p^2/2m$, find an equation for the momentum of the particle.
(b) Using the result of (a), show that the de Broglie wavelength of the particle is $\lambda = 2L/n$. Why is this equation the same as that for the wavelengths of standing waves on a string?

9 Discuss the solution of the particle-in-a-box problem in terms of *classical* physics. What limitations would be expected (if any) on the energy of the particle? In particular, should the energy of the particle be quantized?

10 Suppose that a 5-gram marble is confined to a box 20 cm across.
(a) Calculate the *minimum* energy that the marble can have, assuming that it obeys the energy equation $E_n = n^2h^2/8mL^2$.
(b) What is the next highest allowed energy value?
(c) Do you think that the quantization of the energy of the marble could be verified experimentally? Explain your answer.

11 In what respects does Schrödinger's interpretation of ψ place too much emphasis on the wave aspects of electrons?

12 The momentum of a proton inside the nucleus of an atom is estimated to be 3×10^{-15} g-cm/sec. An experiment is devised to measure the momentum of the proton to 10 percent accuracy (so that $\Delta p = 3 \times 10^{-16}$ g-cm/sec).
(a) What will be the uncertainty in the position of the proton as a consequence of this experiment?
(b) Can we be certain that the proton is still inside the nucleus (diameter $= 10^{-12}$ cm) after the experiment? Explain your answer.

13 The path of a beam of electrons is located by allowing the beam to pass through a small hole 10^{-5} cm in diameter in a screen.
(a) What uncertainty in the momentum of an electron is produced by this process?
(b) If the electron has a speed of 3×10^9 cm/sec, to what percent error can the momentum be measured?

14 An equivalent form of the uncertainty principle links the simultaneous measurement of energy (E) and time (t) by the relationship $(\Delta E)(\Delta t) \geq h/2\pi$.
(a) A metastable state of a particular atom has an average lifetime of 100 sec. (That is, an electron raised to a metastable energy level will remain there for 100 sec on the average before falling to a lower energy level.) What is the uncertainty in the energy of the metastable state?
(b) The spectral line emitted when an electron leaves the metastable state tends to have a narrower than normal line width. How can this phenomenon be explained in terms of the uncertainty principle?

15 An electron is fired horizontally from an electron gun. Does the electron travel in a straight line? Analyze this question from the positivist point of view in the light of the uncertainty principle. Is the question meaningless?

16 If the uncertainty principle makes the idea of an electron orbit meaningless, does the principle require that we reject the idea of a planetary orbit in the solar system? Explain your answer.

17 The basic concepts behind the uncertainty principle can be applied to fields outside of atomic physics. Explain, for example, how a kind of "uncertainty principle" might enter into the relationship between a psychoanalyst and a patient.

18 Einstein believed that quantum mechanics was an incomplete theory, but he agreed that it was logically consistent. Are these two viewpoints contradictory? Explain your answer.

Six

THE NUCLEAR FRONTIER

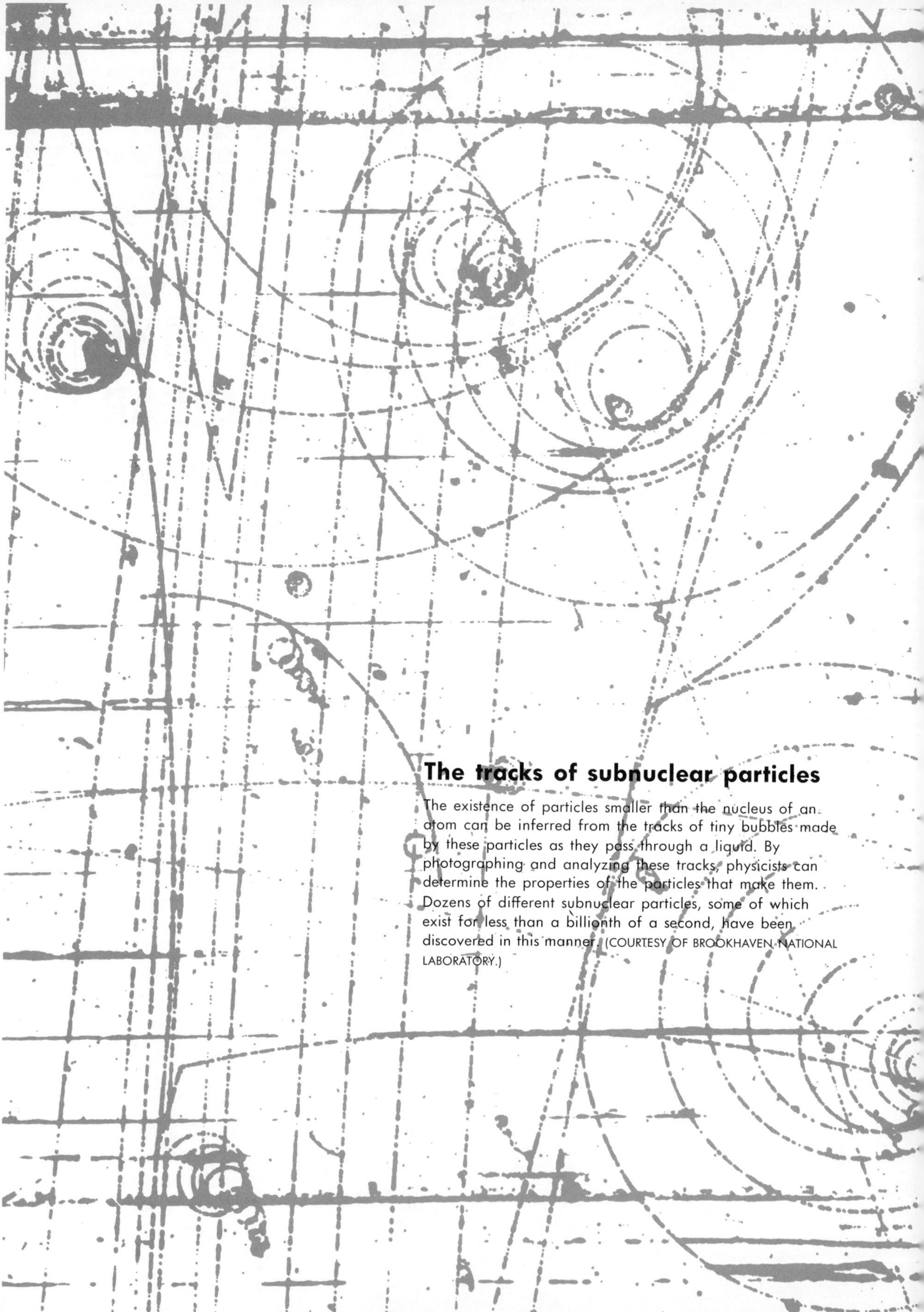

The tracks of subnuclear particles

The existence of particles smaller than the nucleus of an atom can be inferred from the tracks of tiny bubbles made by these particles as they pass through a liquid. By photographing and analyzing these tracks, physicists can determine the properties of the particles that make them. Dozens of different subnuclear particles, some of which exist for less than a billionth of a second, have been discovered in this manner. (COURTESY OF BROOKHAVEN NATIONAL LABORATORY.)

THE NATURE OF THE NUCLEUS

18

It seems probable to me that God in the beginning formed Matter in solid, massy, hard, impenetrable, moveable Particles . . . and that these primitive Particles being Solids are incomparably harder than any porous Bodies compounded of them; even so very hard as never to wear or break in pieces; no ordinary Power being able to divide what God himself made in the first Creation.

Isaac Newton

BY 1930 virtually all of the major questions about atomic structure had been answered. Attention now shifted to a new frontier of knowledge: the atomic nucleus. The existence of the nucleus had been discovered by Rutherford in 1911, but it had been pretty well ignored by most physicists in their efforts to understand atomic structure. As we learned in the preceding chapters, atomic physics is concerned mainly with the behavior of the electrons within the atom, and the nucleus is of rather peripheral interest. One needs to know only that it is the source of positive charge and mass in the atom; any details about the internal structure of the nucleus are of little value in atomic theory. Although the nucleus was generally ignored in the attempt to construct a theory of the atom, it was not entirely forgotten, and it rather naturally became the center of interest once a successful atomic theory had been forged.

Since the early 1930s the nucleus has occupied center stage. At the present time nuclear physics is generally the most active and popular area of physics research. In view of the relatively large number of men and women who are presently at work in physics, it can be reasonably estimated that throughout history more man-hours have been devoted to the study of the nucleus than to any other area of physics. Yet with this concentration of effort physicists have not been able to duplicate the spectacular achievements of the few dozen physicists who constructed a successful atomic theory in fifteen years—beginning with the Bohr theory in 1913 and culminating with the Dirac wave equation in 1928. Today, more than forty years after the study of the nucleus was begun in earnest, many of the major problems of nuclear physics remain unsolved.

The difficulties inherent in the study of the nucleus can be appreciated when one considers the sizes of the objects being dealt with. A *large* nucleus has a radius of less than 10^{-12} cm and occupies one-trillionth the volume of the atom. The jump in scale from the nuclear domain to the atomic domain is nearly as wide as the jump from the atomic domain to the macroscopic world. In our study of atomic theory, we learned that classical physics could not bridge the gap between the macroscopic and atomic levels, and a new physics—quantum mechanics—had to be developed. It would not have been surprising to find that another "new physics" was needed to describe events at the nuclear level. To the present time, however, a new physics has not emerged, and there has been no clear-cut demonstration that one is needed at the nuclear level. Many aspects of nuclear behavior can be explained quite adequately through the use of quantum mechanics. There have been no out-and-out failures of quantum theory at the nuclear level comparable to the failures of classical theory discovered at the atomic level by Planck, Einstein, Bohr, and others. Most modern-day physicists believe that the remaining unsolved problems of nuclear physics will eventually be solved through the application of quantum mechanics. The difficulties that have arisen are believed to be a consequence of the intrinsic complexity of the nucleus rather than of any shortcomings of the quantum theory.

18–1

THE BEGINNINGS OF NUCLEAR PHYSICS

Although the early 1930s mark the beginning of nuclear physics as a full-fledged discipline, quite a bit of nuclear research was done before 1930—nearly all of it by Rutherford and his associates. A brief survey of these early developments follows.

The history of nuclear physics begins with the discovery of natural radioactivity by Henri Becquerel of France in 1896. Becquerel found that the element uranium emitted an unknown radiation that was capable of penetrating solid materials and of ionizing gases. In 1898 Pierre and Marie Curie began to search for other elements possessing the same property. They found that the element thorium is also radioactive, and they discovered in the mineral pitchblende (uranium ore) two previously unknown elements, polonium and radium, of even higher activity than uranium or thorium. It was found that the radioactivity of these elements was unaffected by any physical or chemical changes in their environment; hence it was concluded that radioactivity is an intrinsic property of individual atoms. The discovery of radium evoked considerable interest because of its exceptional properties. The amount of energy continually given off by a sample of pure radium is unusually high, much more than that given off in a chemical reaction; yet the radium showed no evidence of any chemical change. It was believed by some that a violation of conservation of energy had been found. In contrast, the Curies believed that radioactivity is stimulated by the absorption of subtle unknown radiations from the ether—radiations that could be absorbed only by heavy elements such as uranium, thorium, and radium.

In 1899 Rutherford discovered the alpha and beta radiation given off by radioactive substances. As we have already learned, alpha particles are helium nuclei, and beta particles are high-speed electrons. In 1900 Paul Villard discovered a third kind of radiation, which he called **γ *(gamma) radiation.*** Later it was found that gamma rays are electromagnetic radiation (or, in modern terminology, photons) of extremely high energy. Typically, the energies of alpha, beta, and gamma particles are on the order of 0.1 to 10 MeV (1 MeV $=$ 1 million eV $= 10^6$ eV $= 1.6 \times 10^{-6}$ erg).

TRANSMUTATION OF CHEMICAL ELEMENTS

In 1900 Rutherford, then at McGill University, was joined by the chemist Frederick Soddy in an investigation of the radioactivity of thorium. It had been discovered previously that both radium and thorium give off a gaseous emanation that is also radioactive. Rutherford and Soddy found that the emanation consisted of a new chemical element (radon), which has the properties of an inert gas. They were a bit awed by what they had discovered; according to Soddy's account, he exclaimed "Rutherford, this is transmutation: the thorium is disintegrating and transmuting itself into an inert gas."

Rutherford replied, "For Mike's sake, Soddy, don't call it *transmutation*. They'll have our heads off as alchemists."[1]

The complete dissimilarity between the chemical properties of thorium and radon convinced Rutherford and Soddy that they had indeed observed the transmutation of one element (thorium) to another (radon). (Actually, thorium does not transform directly to radon, but to an intermediate element—a short-lived form of radium—which in turn transforms to radon.) In 1902 they proposed a theory of radioactivity in which they hypothesized that the atoms of radioactive elements spontaneously disintegrate to form atoms of different elements, releasing large amounts of energy in the process. In 1905 Albert Einstein suggested that the energy released by radioactive elements might result from the conversion of mass to energy, and that radioactivity might provide a test of the mass–energy relationship, $E = mc^2$. Einstein was, of course, quite correct in this assertion.

The Rutherford–Soddy theory challenged long-held ideas about the immutability of the atom and the impossibility of the "alchemist's dream" of transforming one element into another. But the theory explained the phenomena of radioactivity so well that it received little opposition. Yet Rutherford and Soddy had given no account of the exact mechanism of radioactive decay. At the time the theory was proposed, the nucleus had not yet been discovered, so it was assumed that radioactivity was a property of the atom as a whole.

With the discovery of the nucleus in 1911 and Bohr's elucidation of atomic structure in 1913, it became evident that radioactivity originates in the *nucleus* of the atom. An alpha particle possesses two units of positive charge and a mass four times that of the hydrogen atom. Since the positive charge and nearly all of the mass of an atom reside in the nucleus, it follows that an alpha particle emitted by a radioactive atom is being ejected directly from the nucleus of the atom. Also, the energies of the alpha, beta, and gamma particles emitted by radioactive atoms are of the order of millions of electron volts—much too large to be related to the orbiting electrons, which have energies typically of the order of a few eV. Therefore one must conclude that a nucleus is capable of disintegrating and emitting high-energy charged particles or photons in the process.

The next important development was the *artificial* transmutation of an element, achieved by Rutherford in 1919. Rutherford placed a source of alpha particles at one end of a container filled with nitrogen gas. At the other end was a screen coated with zinc sulfide. When the container was completely empty, alpha particles struck the screen and produced light flashes. One would expect to see fewer flashes when nitrogen gas is in the container, because the alpha particles would be stopped by collisions with the nitrogen atoms, but Rutherford observed to the contrary that the number of flashes *increased*. He determined experimentally (using electric and magnetic deflection) that the flashes were being produced not by alphas or nitrogen atoms, but by high-speed hydrogen nuclei striking the screen. He concluded that a hydrogen nucleus is ejected from the nitrogen nucleus when the latter is struck by an alpha particle. Later experiments showed that in the reaction the alpha particle is completely captured by the nitrogen nucleus, and a hydrogen nucleus is immediately ejected. Since the alpha particle contains two positive charges and the emitted hydrogen nucleus carries away one positive charge, it follows that the charge of the target nucleus has increased by one unit from seven to eight. Consequently, nitrogen ($Z = 7$) is transformed to oxygen ($Z = 8$) by the interaction. Rutherford was able to produce similar transmutations of other light elements, including boron, fluorine, sodium, aluminum, and phosphorus.

Thus Rutherford became the first successful alchemist. Why did he succeed where countless others before him (including Newton) had failed? The explanation is a simple one. The old-time alchemists were trying to change one element to another by *chemical* means. We now know that chemical reactions involve the outer electrons of an atom. The charge of the nucleus (which is what distinguishes one element from another) is unaffected in chemical changes; the amount of energy required to alter the nucleus of an atom is about a million times greater than the energy involved in chemical processes. Consequently, any attempt to transmutate an atom by *chemical* means is automatically doomed to failure.

THE PROTON–ELECTRON HYPOTHESIS

The accumulated evidence from Rutherford's experiments on radioactivity and nuclear transmutation clearly showed that the nucleus is a complex structure. In various circumstances a nucleus was found to emit a hydrogen nucleus, a helium nucleus (alpha particle), or electron (beta particle). Evidently, the nucleus must *contain* these particles in some way. Furthermore, the fact that one element could be changed to another indicated that the nuclei of different elements must contain some fundamental unit in common.

The most likely candidate for this fundamental unit is the hydrogen nucleus. It is the nucleus of the lightest known atom, possesses a single unit of positive charge, and presumably cannot be subdivided in any way. In 1920 Rutherford coined the name ***proton*** for the hydrogen nucleus. The hydrogen atom therefore consists of a single proton as the nucleus, with a single orbital electron. The proton and electron have *exactly* the same quantity of charge but are opposite in sign, so that the hydrogen *atom* is electrically neutral. However, the mass of the proton is found to be 1836 times the mass of the electron.

In 1920 it was believed that all matter was composed of combinations of protons and electrons—which were the only two particles known at the time. Each element, as we know, is characterized by two numbers: the *atomic number,* which gives the number of positive charges contained in the nucleus, and the *atomic weight,* which gives the mass of the atom (as compared to a standard atom). In general, the atomic weight is not a round number, so let us define a quantity called the ***mass number*** A, which is the atomic weight rounded off to the nearest integer. For example, hydrogen has an atomic weight of 1.008; therefore its mass number is 1. We will represent hydrogen by the symbol ${}_1H^1$, where the subscript at the left represents the atomic number (Z), and the superscript to the right is the mass number (A). The next element, helium ($Z = 2$), has an atomic weight of 4.0026; therefore its mass number is 4, and the appropriate symbol is ${}_2He^4$.

The question that now arises is: What is the helium nucleus composed of? Rutherford assumed that the alpha particle is not a fundamental unit but is composed of protons and electrons. The helium nucleus is approximately four times heavier than a proton; therefore it must contain four protons. However, four protons have a total charge of $+4$, and the helium nucleus has a charge of only $+2$. Therefore there must also be two electrons in the helium nucleus. Thus if the helium nucleus consists of four protons and two electrons, it will have a *net* positive charge of 2 and a mass number of 4, in agreement with observations. Of course, the helium nucleus does not have a mass exactly equal to that of four protons, but the difference can be readily attributed to interactions between the protons and electrons in the nucleus.

Thus it was believed that *every nucleus is composed of a combination of protons and electrons.* The mass number gives the total number of protons in the nucleus, and the atomic number (which is the net nuclear charge) is equal to the difference between the number of protons and the number of electrons in the nucleus. According to this model, the nucleus of fluorine ($_9F^{19}$) should contain 19 protons and $(19 - 9 =)$ 10 electrons; the nucleus of iron ($_{26}Fe^{56}$) should contain 56 protons and $(56 - 26 =)$ 30 electrons; and the nucleus of uranium ($_{92}U^{238}$) should consist of 238 protons and 146 electrons.

In 1920 the proton–electron hypothesis seemed perfectly reasonable. It accounted for all that was known at that time about the nucleus. The assumed existence of electrons in the nucleus agreed with the observed emission of high-speed electrons (beta particles) by certain radioactive nuclei, and it could be presumed that the electrical attraction between the positive and negative charges provides the force that holds the nucleus together. With the development of quantum mechanics, however, certain inconsistencies in the proton–electron model became evident. These inconsistencies were revealed by the discovery of an even finer subdivision of atomic spectral lines known as ***hyperfine structure*** (in effect, a fine structure of the fine structure). In 1925 Wolfgang Pauli demonstrated that the hyperfine structure could be explained by assuming that the nucleus possesses an ***intrinsic angular momentum*** (or "nuclear spin") and an ***intrinsic magnetism.**** In other words, the atom has *three* types of angular momentum: (1) the orbital angular momentum of the electrons, (2) the spin of the orbiting electrons, and (3) the spin of the nucleus. Each of these has a magnetic field associated with it. Hyperfine structure can be explained in terms of the interactions of all three. By analyzing the hydrogen spectrum, one can show that the spin of the proton has exactly the same magnitude as the spin of the electron; that is, the proton has an intrinsic spin of $\frac{1}{2}$. (To be precise, the intrinsic angular momentum of the proton is $\sqrt{3}h/4\pi$, as indicated in the footnote on page 566.)

However, the intrinsic magnetism of the proton is much smaller that that of the electron. According to the quantum theory, the intrinsic magnetism of a charged particle is measured in terms of $eh/4\pi mc$, where m is the mass of the particle; for the electron, $eh/4\pi mc = 0.927 \times 10^{-20}$ erg/gauss. Since the proton mass is 1836 times the electron mass, the proton's intrinsic magnetism should be 1/1836 *smaller* than the electron's, or 5.05×10^{-24} erg/gauss. Thus if there are protons and electrons in the nucleus, the intrinsic magnetism of the electrons ought to dominate completely the total magnetism of the nucleus, and the measured intrinsic magnetism of the nucleus ought to be on the order of 10^{-20} erg/gauss. However, experimental measurements clearly show that *the intrinsic magnetism of any nucleus is very small—on the order of* 10^{-24} *erg/gauss.* This means that if the proton–electron hypothesis is correct, the electron's magnetism is somehow canceled when it is inside the nucleus.

A second, more damaging, argument against the presence of electrons in the nucleus has to do with the spin of the nucleus. For certain nuclei the spins determined from measurement of the hyperfine spectra simply do not agree with the spins calculated on the assumption that the nucleus contains protons and electrons. In addition to the difficulties with intrinsic magnetism and spin, a number of other

* Physicists generally use the term ***magnetic moment*** to denote the intrinsic magnetism of a particle.

discrepancies were found that clearly showed that the proton–electron hypothesis was incompatible with quantum mechanics. Yet there seemed to be no other way to account for the composition of the nucleus. The only alternative was to believe that somehow the electron loses certain of its properties (i.e., its spin and intrinsic magnetism, but not its charge) when it is in the nucleus. All the evidence pointed to the conclusion that the electron could not exist as an independent particle within the nucleus.

THE DISCOVERY OF THE NEUTRON

As early as 1920 Rutherford had proposed that within the nucleus a proton and an electron could become so closely bound together that they form a neutral entity. Rutherford suggested the name ***neutron*** for this object. Over the next decade several unsuccessful attempts were made to demonstrate the existence of neutrons. Finally, in 1932 James Chadwick, a longtime associate of Rutherford's, provided convincing experimental evidence for the existence of this particle.

Chadwick's investigations were inspired by the earlier experiments of Walther Bothe of Germany, who found in 1930 that the bombardment of the light elements beryllium and boron by alpha particles produced high-energy, uncharged radiation able to penetrate through relatively large thicknesses of absorbing material. Early in 1932 these experiments were repeated by Irène and Frédéric Joliot-Curie, the daughter and son-in-law of Marie and Pierre Curie. They discovered that this radiation was able to eject high-energy (5.7 MeV) protons from paraffin (a solid hydrocarbon compound rich in hydrogen). Both groups of experimenters assumed that the radiation consisted of gamma-ray photons. The Joliot-Curies calculated that the gamma rays must have energies of about 55 MeV in order to eject 5.7-MeV protons from paraffin. In analyzing these results, Chadwick noted that the number of protons ejected from the paraffin was much higher than the theoretical predictions based on the assumption that the protons are knocked out by 55-MeV gamma rays. Furthermore, the maximum possible energy for a gamma ray emitted in the alpha–beryllium reaction is about 10 MeV if mass–energy is conserved.

Because of these inconsistencies, Chadwick decided to repeat the experiment in greater detail. (See Figure 18–1.) He found that the mysterious radiation could eject particles from other materials besides paraffin. In particular, Chadwick determined that nitrogen atoms gained 1.6 MeV when struck by the radiation. Calculations showed, however, that if the radiation consists of 55-MeV gamma rays, the maximum amount of energy that the nitrogen atom could gain when struck is only 0.45 MeV! Chadwick was able to demonstrate that his experimental findings were completely incompatible with the assumption that the radiation consists of gamma-ray photons—unless the principles of conservation of energy and momentum are rejected.

Chadwick then showed that his findings were entirely consistent with the assumption that particles are ejected from the absorber in a "billiard-ball" collision (one in which energy and momentum are conserved) with a particle having a mass approximately equal to that of a proton. Since this particle is also uncharged, it does not correspond to any previously known particle, so Chadwick proposed that it was the long-sought-after neutron. Chadwick concluded that the bombardment of a beryllium or boron nucleus with an alpha particle leads to the emission of a neutron, a fundamental particle with a charge of zero and a mass approximately equal to the proton mass.

Since 1932 more precise measurements of the neutron mass have been made, which show that the mass of the neutron is slightly (about 0.13 percent) greater than

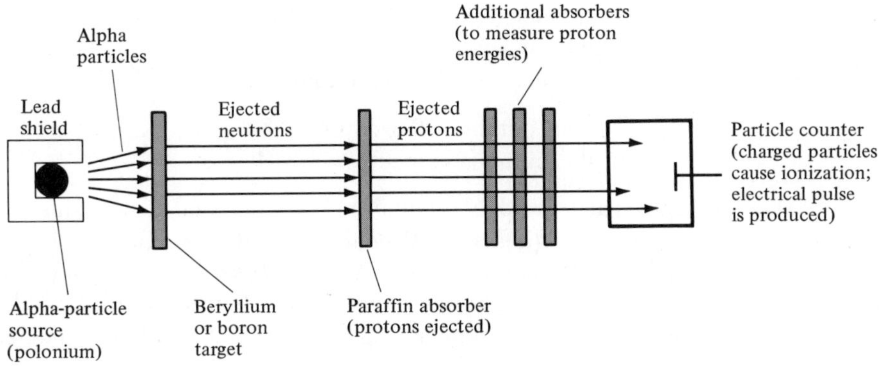

figure 18–1 **CHADWICK'S EXPERIMENT FOR THE DISCOVERY OF THE NEUTRON**

Alpha particles from a radioactive source strike a target and eject uncharged particles. These particles are capable of ejecting protons of high energy from paraffin. On the basis of conservation of mass-energy, Chadwick was able to prove that the uncharged particles are neutrons.

the mass of the proton. Moreover, the mass of a neutron is greater than the combined mass of a proton and an electron, so that a neutron cannot be a simple combination of a proton and an electron, as originally proposed by Rutherford. The neutron possesses an intrinsic angular momentum; like the proton and electron, its spin is $\frac{1}{2}$. The neutron also has an intrinsic magnetism that is about 0.7 as strong as that of the proton. The existence of an intrinsic magnetism for the neutron points to the conclusion that although the neutron is electrically neutral, it is not entirely devoid of charge. The zero charge of the neutron is generally interpreted to mean that the neutron has no *net* charge, but—as is the case with any neutral atom—there is an internal distribution of equal amounts of positive and negative charge. Thus there are strong indications that the neutron has an internal structure, but at the present time there is no clear picture of what this structure is like.

18–2

THE CONSTITUTION OF THE NUCLEUS

The discovery of the neutron in 1932 marked a turning point in the development of nuclear physics. It was recognized rather quickly that the difficulties created by the proton–electron hypothesis could be dissolved by assuming that the nucleus is composed of protons and neutrons, rather than protons and electrons. The credit for proposing the ***proton–neutron hypothesis*** is generally given to Werner Heisenberg.

Let us now see how the properties of the nucleus can be accounted for in terms of protons and neutrons. Since neutrons are uncharged, the positive charge on the nucleus is equal to the total charge of all the protons. Therefore *the atomic number Z is equal to the total number of protons in the nucleus*. Since neutrons and protons have approximately the same mass, it follows that the atomic weight of a nucleus depends on the combined masses of all the protons and neutrons in the nucleus. Hence *the mass number A* (the atomic weight rounded off to the nearest integer) *is equal to the total number of protons and neutrons in the nucleus.* It is easy to see that for a nucleus of atomic number Z and mass number A,

$$\text{the number of protons} = Z$$
$$\text{the number of neutrons} = A - Z$$

Hence an alpha particle ($_2He^4$) contains 2 protons and 2 neutrons; a fluorine nucleus ($_9F^{19}$) contains 9 protons and 10 neutrons, and a uranium nucleus ($_{92}U^{238}$) contains 92 protons and 146 neutrons. The word ***nucleon*** has been coined to denote the particles found in the nucleus. Hence there are two kinds of nucleon—the proton and the neutron—and the mass number A gives the total number of nucleons in a given nucleus.

The problem of the nuclear spin—a serious failure of the proton–electron hypothesis—was immediately resolved by the assumption of the proton–neutron hypothesis. When the nuclear spin is calculated on this basis (protons and neutrons both have a spin of $\frac{1}{2}$), no disagreements between the predicted and measured spins are found for any known nucleus. In addition, the problem of the observed magnitude of the intrinsic nuclear magnetism is also cleared up. Since electrons are no longer presumed to exist in the nucleus, the intrinsic magnetism of the nucleus will be determined by the combined magnetic properties of the protons and neutrons in the nucleus. Since the intrinsic magnetism of the neutron is slightly smaller than that of the proton, the small magnitude of nuclear magnetism can be accounted for.

One difficulty appears to be raised by the banishment of the electron from the nucleus. It is known that high-speed electrons (beta particles) are ejected from certain radioactive nuclei. Where do these electrons come from if they are not in the nucleus beforehand? To resolve this apparent contradiction, it must be assumed that the electron is *created* at the moment of nuclear disintegration and is immediately ejected. This process will be discussed later in the chapter.

THE NUCLEAR FORCE

In the proton–electron model of the nucleus, it was assumed that the nucleus was held together by electric forces of attraction. This assumption is clearly invalid if the nucleus contains protons and neutrons. The electric force between two protons is a *repulsive* force, and the neutron—which is electrically neutral—should experience no electric force at all. What, then, binds protons and neutrons together in the nucleus?

It can be shown that the gravitational force of attraction between nucleons is of no consequence. Let us compare the relative strengths of the gravitational attraction and the electrostatic repulsion between two protons:

$$\frac{F_{\text{grav}}}{F_{\text{elec}}} = \frac{Gm_{\text{p}}^2/r^2}{e^2/r^2} = \frac{Gm_{\text{p}}^2}{e^2}$$

where m_{p} is the proton mass, G the gravitational constant, and e the fundamental unit of charge. In CGS units

$$\frac{F_{\text{grav}}}{F_{\text{elec}}} = \frac{Gm_{\text{p}}^2}{e^2} = \frac{(6.67 \times 10^{-8}\ \text{cm}^3/\text{g-sec}^2)(1.66 \times 10^{-24}\ \text{gram})^2}{(4.8 \times 10^{-10}\ \text{esu})^2} = 0.8 \times 10^{-36}$$

Thus the gravitational attraction is about 10^{36} *times weaker* than the electrostatic repulsive force! It is evident that neither electric nor gravitational forces can hold the nucleus together. There must then be another kind of force, neither electric nor gravitational, that nucleons exert on one another. This force is called the ***nuclear force.***

Hence there are at least *three* fundamental forces in nature: gravitational, electromagnetic, and nuclear. The nuclear force is much more complex than the other two. No one has as yet been able to obtain a simple mathematical formula, similar to the law of gravitation or Coulomb's law, that adequately describes the nuclear force. It is known to have the following properties:

(1) *The nuclear force has an extremely short range.* Its strength is completely negligible beyond a distance of about 10^{-12} cm (roughly one ten-thousandth of the radius of an atom). Thus its effect does not extend very far beyond the nucleus.
(2) *The nuclear force is charge-independent.* Experimental studies of the interactions among protons and neutrons indicate that the nuclear forces between two protons, between two neutrons, and between a proton and a neutron are essentially the same. Thus the charge of a nucleon has no effect on the strength of the nuclear force.
(3) *The nuclear force is spin-dependent.* The strength of the force between two nucleons depends on how their spins are aligned. In general, the force is greatest when the spins of the nucleons point in the same direction.
(4) *The nuclear force is the strongest force known.* For two protons 10^{-13} cm apart, the nuclear force is about ten times stronger than the electric force between them. In general, it is somewhat difficult to make a direct comparison between these forces because of the extremely short range of the nuclear force.
(5) *At very close range the nuclear force is repulsive.* Experiments indicate that the nuclear force between two nucleons is *repulsive* when the nucleons are less than about 0.5×10^{-13} cm apart. For greater distances the nuclear force is attractive.

The picture that we get of the nuclear force is one of a very complex force. The repulsive–attractive nature of the force described in (5) above resembles the force between two atoms, where the mutual electric repulsion of the two positively charged nuclei competes with the attraction of the nuclei for each other's electrons. It seems quite possible that the nuclear force is not in itself a fundamental force but a combination of two or more forces. At the present time the lack of a precise mathematical description of the nuclear force is one of the major unsolved problems of nuclear physics and represents a significant barrier to a more complete understanding of nuclear structure.

INTERNAL STRUCTURE

How are protons and neutrons arranged within the nucleus? One of the most effective ways to learn about the internal structure of the nucleus is to bombard it with charged particles. The bombarding particles must have high enough energies (and small enough de Broglie wavelengths) so that they will interact with the nucleus, rather than with the atom as a whole. Generally, electrons that have been accelerated up to energies of hundreds of MeV are best suited for this purpose. By observing how high-energy electrons interact with nuclei, physicists can determine how the nuclear charge is distributed within the nucleus. In effect, Rutherford's scattering experiment is repeated on the nuclear level. In Rutherford's experiment alpha particles with energies of 5 to 10 MeV penetrated the atom and were deflected by the nucleus as a whole. If electrons with energies on the order of 500 MeV are used, they will penetrate the nucleus and be deflected by the charges *inside* the nucleus.

The results of such experiments show that the electric charge is distributed rather uniformly through the nucleus. This gives us a picture of the nucleus that bears somewhat of a resemblance to Thomson's (rather than to Rutherford's) model of the atom; that is, the charge is not concentrated at the center of the nucleus, but is spread out fairly evenly over the whole nucleus.

With electron-scattering experiments it is possible to determine the size and shape of a nucleus, and it turns out that nuclei are generally spherical or ellipsoidal in shape. Since both the radius and the mass of a given nucleus can be measured, it is possible to calculate the nuclear density (= mass/volume). These calculations show that all nuclei have about the same average density, regardless of their mass or size. This is interpreted to mean that within the nucleus the protons and neutrons are packed together as closely as possible, so that the nucleus cannot be further compressed by the addition of more nucleons. This indicates that there is little or no empty space within the nucleus. In fact, the average density of nuclear matter is on the order of 10^{14} g/cm^3 (100 trillion times the density of water!). This figure would seem to represent the "true" density of solid matter, when all the space between and within atoms is taken into account.

NUCLEAR MASSES

The technique of electric and magnetic deflection devised by J. J. Thomson to measure the q/m ratio of the electron can be applied to any charged particle. In particular, the q/m ratio for any ionized atom can be measured. Since the charge on the ionized atom is known (remember than q stands for the *net* charge on the ion and not the total number of electrons), the mass of the *ionized atom* can be determined. The mass of the *nucleus* is then calculated by subtracting off the total mass of all the electrons in the atom.

Since Thomson's time the methods for measuring atomic and nuclear masses have been greatly refined. Special devices called ***mass spectrometers*** have been designed for the expressed purpose of making precise mass measurements by the electromagnetic deflection method. Today it is possible to measure atomic and nuclear masses accurately to one part in a million. The use of mass spectrometers has made almost obsolete the use of chemical procedures to measure atomic weights.

Nuclear masses can also be determined spectroscopically. We recall that in the Bohr theory of the atom the electrons and nucleus both rotate about the center of mass of the atom. The position of the center of mass, and thus the allowed orbital radii and electron energy levels, will therefore depend on the mass of the nucleus. Consequently, from precise measurement of spectral line wavelengths it is possible to determine the nuclear mass.

Atomic and nuclear masses are extremely small—on the order of 10^{-24} gram—so a more convenient unit of mass has been defined. This is the ***atomic mass unit (amu)***:

$$1 \text{ amu} = 1.660 \times 10^{-24} \text{ gram}$$

This definition of the amu was made by selecting the atom $_6C^{12}$ (the carbon atom with a nucleus containing six protons and six neutrons) as a standard.* By definition, the mass of the C^{12} atom is precisely 12 amu (i.e., 1 amu = 1/12 the mass of one C^{12} atom). All the atomic weights in the periodic table (Figure 16–1) are given in amu on the C^{12} scale.

By comparison to C^{12}, the mass of a hydrogen atom (proton plus electron) is

$$m({}_1H^1) = 1.007825 \text{ amu}$$

The mass of an electron is

$$m_e = 0.000549 \text{ amu}$$

Thus the proton mass is 1.007825 − 0.000549, or

$$m_p = 1.007276 \text{ amu}$$

The measured value of the neutron mass is found to be

$$m_n = 1.008665 \text{ amu}$$

Until the early 1900s it was generally believed that all the atoms of a given element have the same weight. Then the chemists and physicists doing research on radioactivity began to discover different "elements" that turned out to be chemically identical to other known elements, but had different atomic weights. For example, as of 1907 there were at least three substances—called uranium X, ionium, and radiothorium—that were found to be chemically and spectroscopically inseparable from ordinary thorium. However, they had different atomic weights—234, 230, and 228 respectively, as compared to 232 for ordinary thorium—and quite different rates of radioactive decay. Although these were considered to be different elements, they all belonged in exactly the same place in the periodic table!

In 1913 Frederick Soddy proposed that the same element can have different forms with different atomic weights, and he coined the word ***isotope*** to signify one of the forms of a given element. In terms of the proton–neutron picture of the nucleus, *the isotopes of a given element have the same number of protons, but a different number of neutrons.* As we know, the proton number is the same as the atomic number (Z), which determines the position in the periodic table and the chemical properties of an element. However, the atomic weight is related to the total number of nucleons (protons and neutrons). Thus the isotopes of an element have the same Z but different values of A—which means they must have different numbers of neutrons.

Consequently, the belief that the atoms of a given element all have the same weight has been quite thoroughly disproved. It has been found, in fact, that most of the elements have two or more stable isotopes. There are, for example, two stable isotopes of carbon (${}_6C^{12}$ and ${}_6C^{13}$), three stable isotopes of oxygen (${}_8O^{16}$, ${}_8O^{17}$, and ${}_8O^{18}$), and ten stable isotopes of tin (${}_{50}Sn$). Hence any chemical sample of an element found in nature can contain a mixture of stable isotopes. As a consequence, the chemical atomic weight of an element represents an *average* of the relative amounts of the different isotopes. This effect helps to explain some of the peculiar fractional atomic weights found in the periodic table. For example, the chemical atomic weight of chlorine—35.45—comes about because natural chlorine contains 75.5 percent Cl^{35}

*The C^{12} atom was selected because it is the most convenient atom with which to calibrate a mass spectrometer. Atomic masses can then be measured relative to C^{12} most easily. The choice of a C^{12} standard in 1961 settled a long-standing despute between physicists and chemists over an appropriate basis for the scale of atomic weights.

and 24.5 percent Cl^{37}. The relative isotopic abundances of most elements as they are found in the earth tend to be remarkably constant from sample to sample. However, measurements of relative isotopic abundances of the elements found in moon rocks, meteorites, and stellar spectra indicate that the isotopic composition of a chemical element can vary considerably throughout the universe.

One isotope that is of particular interest in nuclear physics is the stable isotope of hydrogen known as ***deuterium*** or ***heavy hydrogen,*** which was discovered spectroscopically (because of its different center-of-mass motion as compared to ordinary hydrogen) by Harold C. Urey in 1932. The deuterium nucleus contains one proton and one neutron; therefore its symbol is ${}_1H^2$. It is the simplest nucleus containing more than one particle; it is bound together by the nuclear force between a single proton and a single neutron. It therefore provides a useful "laboratory" for the testing of theories on nuclear forces and structure. Deuterium, which has several important uses in nuclear technology, comprises only 0.015 percent of natural hydrogen. There is also a third isotopic form of hydrogen known as ***tritium***—${}_1H^3$—which is naturally radioactive.

NUCLEAR BINDING ENERGY

Measurements of nuclear masses reveal a most interesting property: the mass of a nucleus is always *less* than the total masses of all the protons and neutrons in that nucleus—that is, the whole is less than the sum of its parts! To illustrate this phenomenon, let us compare the experimentally measured mass of the deuterium nucleus

$$m_d = 2.013553 \text{ amu}$$

with the combined masses of a proton and a neutron:

$$\begin{aligned} m_p &= 1.007276 \\ m_n &= 1.008665 \\ \hline m_p + m_n &= 2.015941 \text{ amu} \end{aligned}$$

Note that the mass of a deuterium nucleus is less than the total mass of its constituent parts. The difference is known as the ***mass defect*** Δm, which for deuterium is $2.015941 - 2.013553 = 0.002388$ amu. In general, for any nucleus containing Z protons and $(A - Z)$ neutrons, the mass defect is

$$\Delta m = Zm_p + (A - Z)m_n - m_{nuc} \tag{18–1}$$

where m_{nuc} is the measured mass of the nucleus.

The disappearance of mass in the formation of a nucleus from protons and neutrons can be explained in terms of the relativistic principle of the conservation of mass–energy. The "missing" mass (the mass defect) is converted to energy in the process of formation. The energy equivalent of the mass defect is known as the ***nuclear binding energy*** B. From the equation $E = mc^2$ it follows that

$$B = c^2\,\Delta m \tag{18–2}$$

where Δm is the mass defect for the given nucleus.

The binding energy of deuterium can be readily calculated. First we convert the mass defect (0.002388 amu) to grams:

$$\Delta m = (0.002388 \text{ amu})(1.66 \times 10^{-24} \text{ gram/amu}) = 3.96 \times 10^{-27} \text{ gram}$$

We then calculate the binding energy using Equation 18–2:

$$B = c^2\,\Delta m = (3 \times 10^{10}\ \text{cm/sec})^2(3.96 \times 10^{-27}\ \text{gram}) = 3.57 \times 10^{-6}\ \text{erg}$$

Since 1 MeV $= 1.6 \times 10^{-6}$ erg,

$$B = \frac{3.57 \times 10^{-6}\ \text{erg}}{1.6 \times 10^{-6}\ \text{erg/MeV}} = \underline{2.23\ \text{MeV}}$$

Therefore the binding energy of the deuterium nucleus is 2.23 MeV.

The nuclear binding energy should not be thought of as energy that is "stored" inside the nucleus. It represents the amount by which the mass–energy of a nucleus is *less* than the mass–energy of its constituent parts. In other words, a bound nucleus has a lower total energy than a collection of unbound particles, just as an atom has a lower energy when its electron is in the $n = 1$ state as compared to the $n = \infty$ (unbound) state. Thus when a proton and neutron come together to form deuterium, 2.23 MeV must be *lost* (by emission of a photon or transfer of kinetic energy to some other particle), just as an electron must lose 13.6 eV when it falls from $n = \infty$ to $n = 1$ in the hydrogen atom. By the same token, the binding energy can be viewed as the amount of energy that must be supplied by an outside source in order to break up a nucleus into its component protons and neutrons. In this sense the nuclear binding energy is akin to the ionization energy required to remove an electron from an atom. The comparison of the energy diagrams for the deuterium nucleus and the hydrogen atom in Figure 18–2 may help the student visualize this concept.

The analogy between the nuclear binding energy and the atomic electron binding energy is not at all accidental. When particles come together under the influence of an attractive force (such as gravitation or the electrical attraction between opposite charges), there is always a *decrease* in their potential energy. Thus a bound state always has a lower potential energy than an unbound state. The nuclear binding energy can similarly be thought of as the decrease in the potential energy associated with nucleons coming together under the influence of the nuclear force.

figure 18–2 **COMPARISON OF NUCLEAR BINDING ENERGY AND ELECTRON BINDING ENERGY IN THE ATOM**

(a) The binding energy of the deuterium nucleus ($B = 2.23$ MeV) is the amount of energy that must be supplied to decompose deuterium into a proton and neutron. This is analogous to (b) the ionization energy required to remove the electron from a hydrogen atom.

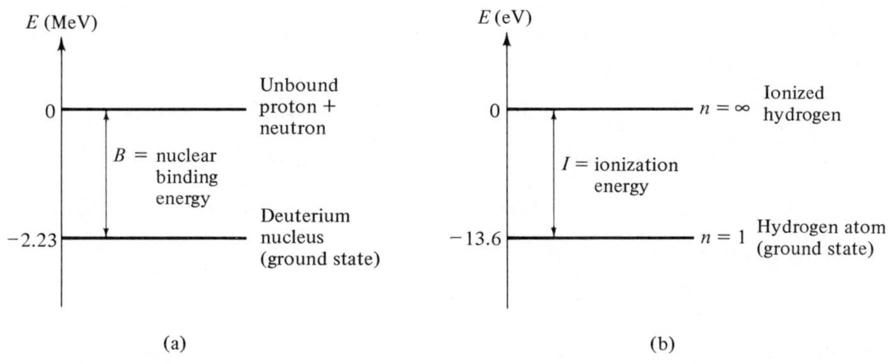

The concept of nuclear binding energy has been given extensive experimental validation. In the case of deuterium we can cite two experimental measurements of the calculated binding energy of 2.23 MeV. The first involves the capture of slow-moving neutrons by hydrogen atoms to form deuterium atoms. It is found that neutron capture is accompanied by the emission of 2.23-MeV gamma-ray photons, which carry away the energy equivalent of the mass defect:

$$\mathrm{p} + \mathrm{n} \longrightarrow \mathrm{d} + \gamma(2.23\ \mathrm{MeV})$$

The reverse of this process is also observed. Deuterium can be disintegrated by bombarding it with high-energy photons (a process known as ***photodisintegration***). Experiments show that the reaction

$$\gamma + \mathrm{d} \longrightarrow \mathrm{p} + \mathrm{n}$$

will not take place unless the photon has an energy of 2.23 MeV or more. That is, the photon must supply the energy equivalent of the mass difference between the deuterium nucleus and the proton plus neutron.

Similar experiments have been performed with many other nuclei. The verification of Equations 18–1 and 18–2 by these experiments represents further evidence for the validity of Einstein's mass–energy relationship.

In calculating nuclear binding energies, it proves convenient to be able to convert directly from the mass defect in amu to the binding energy in MeV. Let us calculate the energy equivalent of 1 amu:

$$E = mc^2 = (1.66 \times 10^{-24}\ \mathrm{gram})(3 \times 10^{10}\ \mathrm{cm/sec})^2$$
$$= 1.49 \times 10^{-3}\ \mathrm{erg}$$
$$\frac{1.49 \times 10^{-3}\ \mathrm{erg}}{1.6 \times 10^{-6}\ \mathrm{erg/MeV}} = 931\ \mathrm{MeV}$$

Thus the mass–energy relationship can be expressed as follows:*

$$1\ \mathrm{amu\ of\ mass} = 931\ \mathrm{MeV\ of\ energy}$$

With this factor the binding energy in MeV is obtained directly by multiplying the mass defect in amu by 931 MeV/amu. For example, the deuterium mass defect is 0.002388 amu; therefore the binding energy is (0.002388 amu)(931 MeV/amu) = 2.23 MeV.

EXAMPLE 18–1: What is the binding energy of the alpha particle if it has a mass of 4.001505 amu?

SOLUTION: An alpha particle is composed of two protons and two neutrons. Their combined mass is

$$2m_\mathrm{p} + 2m_\mathrm{n} = 2(1.007276) + 2(1.008665) = 4.031882\ \mathrm{amu}$$

The mass defect for an alpha particle ($m_\alpha = 4.001505$ amu) is:

$$\Delta m = (2m_\mathrm{p} + 2m_\mathrm{n}) - m_\alpha = 4.031882 - 4.001505 = 0.030377\ \mathrm{amu}$$

The binding energy is then

$$(0.030377\ \mathrm{amu})(931\ \mathrm{MeV/amu}) = \underline{28.3\ \mathrm{MeV}}$$

*Another way to express this relationship is to state that if the energy is measured in MeV and the mass in amu, the c^2 in $E = mc^2$ has the value of 931 MeV/amu.

NUCLEAR ENERGY LEVELS

There is a rather large body of experimental evidence to show that nucleons occupy distinct, quantized energy levels within the nucleus—just as electrons occupy distinct energy levels in the atom. As Neils Bohr showed in 1913, the fact that atoms exhibit line spectra can be explained in terms of quantum jumps between allowed electron energy levels accompanied by the absorption or emission of photons. It is found experimentally that *nuclei* also exhibit absorption and emission line spectra. The difference between atomic spectra and nuclear spectra is mainly one of energy. *Electron* energy levels are a few electron volts apart, and the photons emitted or absorbed in transitions are generally in the visible or ultraviolet range. *Nuclear* energy levels are on the order of a few MeV apart, so that transitions between these levels result in the emission or absorption of gamma-ray photons.

Hence when a nucleus is raised to an excited state, it returns to its lowest energy or ground state by emitting one or more gamma rays with specific wavelengths characteristic of the particular nucleus. A nucleus can be raised to an excited state by bombarding it with gamma rays; the nucleus will absorb only those gamma rays whose energies correspond exactly to the differences between nuclear energy levels.

The existence of nuclear energy levels can be further demonstrated by bombarding nuclei with high-energy electrons. It is found that the electrons will not lose any energy unless they have sufficient energy (usually about an MeV or so) to raise the nucleus to an excited state. The electrons lose a specific quantum of energy; immediately afterward the nucleus returns to its ground state by emitting a gamma-ray photon of the same energy as that lost by the electrons. Through the techniques of fast-electron bombardment and gamma-ray absorption, nuclear physicists can map out the main energy levels of different nuclei. Figure 18–3 shows an energy level diagram for the first few levels of the sodium nucleus; for comparison the electron energy levels of the sodium *atom* have also been shown.

The existence of quantized nuclear energy levels is strong evidence that quantum mechanical principles are applicable at the nuclear level. In principle, we ought to be able to use the Schrödinger equation to calculate the allowed nuclear energy levels for a given nucleus. However, in order to solve the Schrödinger equation, we must be able to identify the forces acting on the particles in question. For the electron in a hydrogen atom, there is a single force—the Coulomb attraction between the electron and the nucleus—and an exact solution of the Schrödinger equation can be obtained. But the significant force inside the nucleus is the nuclear force, which cannot be exactly described mathematically. Moreover, each nucleon presumably has acting on it the nuclear forces exerted by all the other nucleons in the particular nucleus. Thus the force on a nucleon is so complicated that the Schrödinger equation cannot be written, much less solved, for the particles inside a nucleus.

Under the circumstances the best that can be hoped for is an approximate solution. One approach, known as the ***independent-particle*** model, assumes that each nucleon moves in the nucleus independently of all the other nucleons, under the influence of a single average force which is the same for all nucleons. In this way the Schrödinger equation can be written as if it described a single particle being acted on by a single force. On the presumption that all quantum-mechanical principles are applicable at the nuclear level, it is assumed that each of the energy levels described by the independent-particle Schrödinger equation splits into two sublevels, according to whether a nucleon has its spin "up" ($m_s = \frac{1}{2}$) or "down" ($m_s = -\frac{1}{2}$). It is also assumed that the Pauli exclusion principle is valid in the nucleus: no two protons

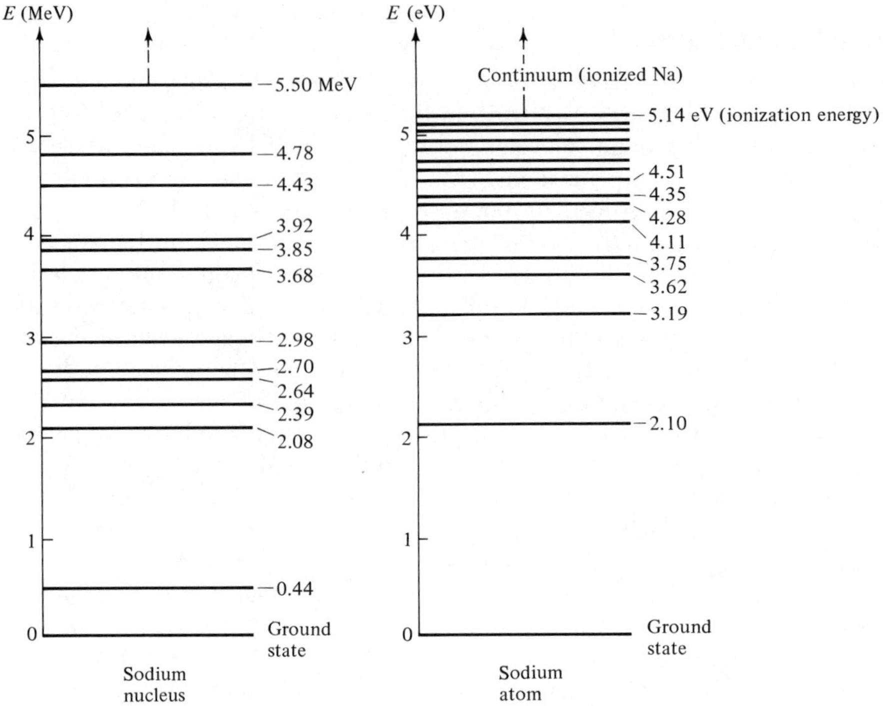

figure 18–3 **ATOMIC AND NUCLEAR ENERGY LEVELS**

A comparison can be made between the energy levels of the sodium nucleus (measured in MeV) and the electron energy levels of the sodium atom (measured in eV). In both cases, energies are measured relative to the ground state.

(or neutrons) can have the same set of quantum numbers. It follows that each energy level can be occupied by no more than two protons (with opposite spins) and two neutrons (also with opposite spins).

NUCLEAR SHELL STRUCTURE

In the simplest form of the independent-particle model, the solution of the Schrödinger equation gives a succession of equally spaced energy levels. This result does not agree with observations; no nuclei are found to exhibit equally spaced energy levels. (Note, for example, the energy level diagram for the sodium nucleus in Figure 18–3.) In fact, there is considerable evidence to show that proton and neutron energy levels tend to bunch together (as do the electron energy levels in the atom) to form shells containing several closely spaced subshells.

In the atom the bunching together of energy levels is responsible for the closed shell structure of the inert gases. There is parallel evidence for closed shells in the *nucleus,* as well. It is found that if the number of protons and/or neutrons in the nucleus is equal to 2, 8, 20, 28, 50, 82, or 126, the nucleus will be particularly stable. These numbers have come to be known among nuclear physicists as the ***magic numbers.*** Nuclei having a magic number of protons or neutrons tend to have properties on the nuclear level that resemble the properties of the inert gases on the atomic level, such as high binding energy and low affinity for capturing additional particles. For example, a nucleus with a magic number of neutrons ($_{38}Rb^{88}$ or $_{56}Ba^{138}$) will have a very low affinity for capturing an additional neutron, but nuclei that are one neutron short of a magic number of neutrons ($_{38}Rb^{87}$ or $_{56}Ba^{137}$) will have an extremely high probability for neutron capture so as to complete the closed shell.

Of course, there is nothing magic about the magic numbers; they simply represent the number of nucleons needed to fill successive groups of bunched energy levels. The problem for theoretical nuclear physicists is to explain how these particular numbers come to represent the number of nucleons in a closed shell. In 1949 a theoretical explanation was provided by Maria Goeppert Mayer in the United States and independently by J. H. Jensen, Otto Haxel, and Hans Suess in Germany. The theory is too complicated to go into here, but by modifying the independent-particle model appropriately, they were able to show that the proton and neutron levels should bunch together in such a way as to yield the magic numbers at the closed shells. In the shell model of the atom, chemical properties are determined by the electrons outside the last closed electron shell. Similarly, the properties of nuclei can be related to the nucleons outside the last closed nucleon shell. In this manner the independent-particle model has proved to be quite successful in accounting for many of the properties of nuclei. This model has its shortcomings, however. There are certain nuclear phenomena—such as nuclear fission, to be described in the next chapter—that cannot be accounted for by the independent-particle model. In recent years physicists have formulated more complex theories of the nucleus in an effort to remedy the weaknesses of the independent-particle model by taking into account the overall properties of the entire nucleus and de-emphasizing the properties of individual nucleons. These "collective" models are still being improved upon at the present time. Fittingly enough, one of the more successful collective models of the nucleus was developed by Aage Bohr and Ben Mottelson of Denmark. Aage Bohr is the son of Niels Bohr and the present head of the Bohr Institute.

18–3

NUCLEAR REACTIONS

When the nucleus of an atom collides with another nucleus or is struck by some high-energy particle, the result is often a ***nuclear reaction,*** which can lead to the transmutation of one element to another or to the formation of a different isotope of the same element. In any nuclear reaction certain physical quantities must be conserved: mass–energy, linear momentum, angular momentum, spin, charge. Another quantity that is conserved in nuclear reactions is the *nucleon number.* That is, the total number of nucleons (protons plus neutrons) is the same before and after a reaction. (Note that if a neutron changes into a proton or vice versa, the total number of *nucleons* is unchanged.) By taking all of these conserved quantities into account, physicists are able to make reasonably good predictions as to the outcome of any proposed nuclear reaction.

Nuclear reactions are usually written just like chemical reactions, with reactants on the left and products on the right. Each particle or nucleus taking part in the reaction is represented by a term in the form $_Z\mathrm{X}^A$, where Z is the atomic number or nuclear charge, A is the mass number, and X is the symbol for the particle or nucleus.* Every nuclear reaction must be "balanced"; that is, the net charge and the total number of nucleons must be the same on both sides of the reaction.

For illustrative purposes let us write the complete reaction for the transmutation of nitrogen by alpha-particle bombardment, first performed by Rutherford in 1919. Rutherford found that protons are emitted in this reaction; that is,

$$\text{alpha} + \text{nitrogen} \longrightarrow \text{proton} + ?$$

The alpha particle and proton are helium and hydrogen nuclei, respectively, and the stable form of nitrogen is ${}_7N^{14}$. The question mark represents the residual nucleus created by the reaction. Thus we write

$${}_2He^4 + {}_7N^{14} \longrightarrow {}_1H^1 + ?$$

In balancing this reaction, we see that the nucleus produced in the reaction must have a charge of 8 ($= 2 + 7 - 1$) and a mass number of 17 ($= 4 + 14 - 1$). Looking at the periodic table, we find that the element with atomic number $Z = 8$ is oxygen. The complete reaction is therefore

$${}_2He^4 + {}_7N^{14} \longrightarrow {}_1H^1 + {}_8O^{17}$$

The nuclear reaction with which Chadwick discovered the neutron (symbol: ${}_0n^1$) involved the bombardment of beryllium (${}_4Be^9$) by alpha particles:

$${}_2He^4 + {}_4Be^9 \longrightarrow {}_0n^1 + {}_6C^{12}$$

Thus one should also expect to find C^{12} produced by the reaction. The reaction describing the photodisintegration of deuterium into a proton and a neutron is as follows:

$$\gamma + {}_1H^2 \longrightarrow {}_1H^1 + {}_0n^1$$

In this reaction the symbol γ represents the bombarding gamma-ray photon. A photon, of course, has neither charge nor mass. (To be precise, its symbol should be ${}_0\gamma^0$.)

THE REACTIONS OF RADIOACTIVE DECAY

We will now write the nuclear reactions describing the different types of radioactive decay. We will begin with ***gamma decay,*** which is the simplest type. When a radioactive nucleus emits a gamma-ray photon, the emitted photon carries away neither charge nor mass. Hence there is no nuclear transmutation; the effect of gamma decay is simply to remove excess energy from the nucleus. The general reaction for gamma decay thus has the form

$${}_ZX^A \longrightarrow \gamma + {}_ZX^A$$

Alpha decay, on the other hand, involves the emission of an alpha particle (${}_2He^4$); for example, the alpha decay of radium (${}_{88}Ra^{226}$) is

$${}_{88}Ra^{226} \longrightarrow {}_2He^4 + {}_{86}Rn^{222}$$

The alpha decay of any radioactive element must result in a decrease of two

* For identification purposes, it is generally not necessary to put both the atomic number *and* the symbol, since the symbol automatically identifies the atomic number. In other words, the 6 in ${}_6C^{12}$ is superfluous. However, in writing nuclear reactions, it is good practice to include the atomic number.

units in Z and four units in A. Thus the general reaction for alpha decay is

$$_{Z}X^{A} \longrightarrow {}_{2}He^{4} + {}_{Z-2}Y^{A-4}$$

where X is the radioactive nucleus, and Y is the decay product.

In ***beta decay*** an electron is emitted from the nucleus. An electron has a negative charge and is not a nucleon; hence its charge number is -1, its mass number is 0, and its symbol is $_{-1}e^{0}$. The beta decay of Th^{234} (a naturally occurring beta emitter) proceeds as follows:

$$_{90}Th^{234} \longrightarrow {}_{-1}e^{0} + {}_{91}Pa^{234}$$

Note that there is an *increase* in the charge of the nucleus upon beta decay; by losing a negative charge, the nucleus becomes more positive. The original $_{90}Th^{234}$ nucleus contains 90 protons and 144 neutrons; the resulting $_{91}Pa^{234}$ nucleus contains 91 protons and 143 neutrons. The effect of beta decay is that a *neutron changes to a proton inside the nucleus.* Thus the general reaction for beta decay is

$$_{Z}X^{A} \longrightarrow {}_{-1}e^{0} + {}_{Z+1}Y^{A}$$

In 1934 another type of radioactive decay was discovered. Irène and Frédéric Joliot-Curie were studying the emission of neutrons by the alpha bombardment of boron:

$$_{2}He^{4} + {}_{5}B^{10} \longrightarrow {}_{0}n^{1} + {}_{7}N^{13}$$

They discovered that the nitrogen nucleus produced by this reaction is radioactive. The N^{13} isotope does not exist in nature; thus the Joliot-Curies produced for the first time *induced* or *artificial radioactivity*. Furthermore, they found that the radioactive N^{13} nucleus decays by emitting a ***positron.*** This is a particle, discovered in 1932, that has the same mass as an electron but possesses one unit of *positive* charge. (The extraordinary history of the discovery of the positron will be taken up in Chapter 20.) Generally, the symbol used for the positron is e^{+} or β^{+}, as compared to e^{-} or β^{-} for the electron. In nuclear reactions the appropriate symbol for the positron is $_{1}e^{0}$. Thus the positron decay of N^{13} proceeds as follows:

$$_{7}N^{13} \longrightarrow {}_{1}e^{0} + {}_{6}C^{13}$$

Note that positron emission carries away a positive charge from the nucleus, without changing the nucleon number. In general, positron emission proceeds as follows:

$$_{Z}X^{A} \longrightarrow {}_{1}e^{0} + {}_{Z-1}Y^{A}$$

The effect of positron emission is that *a proton changes to a neutron in the nucleus.* (In the preceding example, $_{7}N^{13}$, with seven protons and six neutrons, changes to $_{6}C^{13}$ with six protons and seven neutrons).

Thus positron emission is analogous to normal beta decay and is, in fact, considered a form of beta decay. In other words, the process of beta decay has been expanded to encompass two types of decay: electron (β^{-}) emission and positron (β^{+}) emission. Both of these processes play important roles in the establishment of nuclear stability; a nucleus that is unstable because of an imbalance in the number of protons or neutrons can become stable by converting a proton to a neutron (or vice versa) through the process of beta decay. The phenomenon of beta decay is considerably more complicated than described here; it will be analyzed more fully in Chapter 20.

In writing balanced nuclear reactions, we use the conservation of charge and nucleon number to identify all the participants in a given reaction. However, these two conservation principles alone cannot tell us whether a particular reaction will or will not occur. Up to this point, then, we are not able to predict in advance whether the bombardment of, say, a nitrogen nucleus with an alpha particle will lead to the emission of a neutron, a proton, or both—or possibly some other combination of particles. In order to predict the outcome of a nuclear reaction under given conditions, we obviously need more information. Some of this information is provided by other conservation principles.

NUCLEAR REACTION ENERGIES

It is a fundamental assumption that mass–energy is conserved in all nuclear reactions. That is, in any nuclear reaction the total mass–energy of the reactants must be equal to the total mass–energy of the products. In general, in a nuclear reaction mass by itself is not conserved; any changes in mass are compensated by an increase or decrease of energy in other forms. There are two possibilities:

(1) *The total mass of the reactants is greater than the total mass of the products.* In this case there is a *decrease* in mass as a result of the reaction; mass has been converted to energy. The energy created by the reaction may appear as the kinetic energy of the products, or it may be transformed into electromagnetic energy in the form of a gamma-ray photon. A reaction of this type, in which energy is created at the expense of mass, is known as an ***exoergic reaction.***

(2) *The total mass of the reactants is less than the total mass of the products.* In this case there is an *increase* in mass as a result of the reaction; energy has been converted to mass. Thus energy must be supplied to the reactants in order for the reaction to take place. This energy may come from the kinetic energy of the reactants (if they are moving at high enough speeds) or from the absorption of a photon (as in the photodisintegration of deuterium, described previously). A reaction in which energy must be supplied—that is, in which energy is absorbed in the reaction to create mass—is known as an ***endoergic reaction.***

The amount of energy absorbed or created in a nuclear reaction is known as the ***Q value*** of the reaction. The Q value is defined as follows:

$$Q = c^2[(\text{total mass of reactants}) - (\text{total mass of products})]$$

It follows that if

(1) Q is *positive,* mass has been destroyed, and the reaction is *exoergic.*
(2) Q is *negative,* mass has been created, and the reaction is *endoergic.*

Now let us consider some examples. Again we will use Rutherford's reaction:

$${}_2\text{He}^4 + {}_7\text{N}^{14} \longrightarrow {}_1\text{H}^1 + {}_8\text{O}^{17}$$

Consulting a table of experimentally measured nuclear masses, we find the following values in amu:

$${}_1\text{H}^1 = 1.007276 \qquad {}_7\text{N}^{14} = 13.999231$$
$${}_2\text{He}^4 = 4.001505 \qquad {}_8\text{O}^{17} = 16.994741$$

Adding reactants and products, we find

Total Mass of Reactants	Total Mass of Products
${}_2He^4 = 4.001505$	${}_1H^1 = 1.007276$
${}_7N^{14} = 13.999231$	${}_8O^{17} = 16.994741$
Total = 18.000736 amu	Total = 18.002017 amu

The total mass *increases* as a result of the reaction; therefore the reaction is endoergic. The Q value is (using $c^2 = 931$ MeV/amu)

$$\begin{aligned} Q &= 931(18.000736 - 18.002017) \\ &= 931(-0.001281) = -1.19 \text{ MeV} \end{aligned}$$

Therefore 1.19 MeV of energy must be supplied in order for the reaction to proceed. The required energy can be provided by the kinetic energy of the alpha particle. In Rutherford's original experiment the alphas were emitted from a radioactive source and had kinetic energies of about 5.5 MeV, well above the 1.19 MeV needed to trigger the reaction.

When conservation of momentum is taken into account, it is found that the minimum or ***threshold energy*** needed to trigger an endoergic reaction is slightly higher than the calculated Q value for the reaction. Momentum conservation requires the total momentum of the products to be equal to the total momentum of the reactants. Therefore some additional energy, above and beyond that needed to make up the mass increase, must be supplied to provide for the motion of the products after the reaction. Any extra energy beyond what is required for energy and momentum conservation will simply add to the kinetic energy of the reaction products.

In many experiments the energies and particles needed to trigger endoergic reactions cannot be obtained from any naturally occurring radioactive sources. Heating the reactants to raise their average kinetic energies (a procedure that is used successfully to trigger *chemical* reactions) is highly impractical; in order for a group of atoms to have an average kinetic energy of 1 MeV, they must be at a temperature of more than 10 billion °K! Hence for experimental purposes particles are raised to higher energies by accelerating them in electromagnetic fields using such devices as the cyclotron.

As an example of an exoergic reaction, we will examine the radioactive decay of the alpha emitter ${}_{88}Ra^{226}$. The reaction is as follows:

$${}_{88}Ra^{226} \longrightarrow {}_2He^4 + {}_{86}Rn^{222}$$

The nuclear masses are:

$$\begin{aligned} {}_{88}Ra^{226} &= 225.977048 \text{ amu} \\ {}_2He^4 &= 4.001505 \text{ amu} \\ {}_{86}Rn^{222} &= 221.970317 \text{ amu} \end{aligned}$$

We find that the total mass of the products is 225.971822 amu, which is less than the mass of the initial Ra^{226} nucleus. The reaction is therefore exoergic, with

$$Q = 931(225.977048 - 225.971822) = 4.87 \text{ MeV}$$

This energy appears as the kinetic energy of the product nuclei. In order to maintain conservation of momentum, the kinetic energy must be shared by the two products

(He and Rn) in inverse proportion to their masses. The alpha particle will have 222/226 of the available energy (222/226 × 4.87 = 4.78 MeV), and the radon nucleus obtains the remaining 4/226 of the available energy. The kinetic energy of the products of an exoergic reaction can be transferred to the surroundings by collisions, causing an increase in temperature and possibly the emission of light.

THE THEORY OF NUCLEAR REACTIONS

The application of the conservation principles of mass–energy and momentum enables us to predict whether a given nuclear reaction will be endoergic or exoergic. If it is exoergic, we can calculate how much energy will be released in the reaction, and if it is endoergic, we can determine how much energy must be supplied if the reaction is to take place. This is, of course, extremely useful information, but it is still not sufficient to allow us to predict fully the outcome of a given experiment. Suppose, for example, we set up an experiment in which alpha particles of a given energy strike a carbon target. How do we determine what the products of this reaction will be? There is a rather large number of possible outcomes, such as

$$
{}_2\text{He}^4 + {}_6\text{C}^{12} \rightarrow
\begin{cases}
{}_8\text{O}^{16} + \gamma \\
{}_8\text{O}^{15} + {}_0\text{n}^1 \\
{}_7\text{N}^{15} + {}_1\text{H}^1 \\
{}_7\text{N}^{14} + {}_1\text{H}^2 \\
\text{etc.}
\end{cases}
$$

Using mass–energy conservation, we might be able to rule out some of these possibilities—a reaction might be endoergic, and the alpha particle energy might not be large enough to trigger the reaction—but in general, a number of the reactions will be energetically possible. If so, which of the reactions will be most likely to occur? Will the same reaction occur every time, or will there be some sort of statistical distribution among all the possible outcomes? If we actually perform the experiment, we find that some protons, some neutrons, some deuterium nuclei, and so on are observed. In other words, there is a competition between the different reactions; some collisions of alpha particles and carbon nuclei lead to one set of reaction products, and others lead to different outcomes.

What we would like to know, then, is the frequency with which each of the possible reaction outcomes takes place. Suppose alpha particles of a given energy strike a carbon target (say a block of solid carbon) at a given rate. How many of these alpha particles will collide and react with carbon nuclei? What fraction of these reactions will produce protons, neutrons, gammas, and so on? This is what we expect a good theory of nuclear reactions to be able to predict.

In the present stage of development of theoretical nuclear physics, there is no single theory that can be used to predict nuclear reaction outcomes under all conditions. Different theoretical models of the nuclear reaction process are used, depending on the type and energy of the bombarding particle. These models have met with varying degrees of success, but none of them can be counted on to provide the accuracy one has come to expect of physical theories. No theory can predict the outcome of any *single* reaction; rather, they are used to calculate the *probability* that any given reaction will occur. Thus nuclear reaction theories are necessarily statistical—they are an attempt to predict the distribution of results of a large number of reactions. Some of these theories are extremely complicated and can be used only with the aid of a high-speed electronic computer.

In general, the present-day theory (or theories) of nuclear reactions can make adequate, but not especially accurate, predictions of nuclear reaction rates. (These predictions are nevertheless quite useful in the design of nuclear physics experiments and applications.) The difficulty in obtaining a good general theory of nuclear reactions can be attributed to the complexity of the nucleus and to the incomplete knowledge of the nature of the nuclear force.

The present state of nuclear physics is strongly reminiscent of the state of atomic physics in 1925. A great deal of experimental data has been collected, and some theoretical models have been developed that explain many, but not all, nuclear properties. What is lacking is a theoretical understanding of the fundamental phenomena at work inside the nucleus. In particular, a more precise knowledge of the nature of the nuclear force and of the interactions between protons and neutrons in the nucleus must be attained if the properties of the nucleus are to be understood fully. Thus today the study of the nucleus represents one of the frontiers of physics, well worthy of the effort and attention of the present generation of physicists.

QUESTIONS FOR REVIEW

1 Describe the three types of radiation given off by naturally radioactive substances.

2 Discuss the evidence that radioactivity originates in the nucleus of the atom. Explain why radioactive decay constitutes the transmutation of one chemical element into another.

3 What do the atomic number (Z) and mass number (A) of a nucleus represent? Explain how these numbers are related to the constituents of the nucleus according to (a) the proton–electron hypothesis and (b) the proton–neutron hypothesis.

4 What is hyperfine structure? What information about the nucleus and the particles it contains can be obtained from an analysis of hyperfine structure?

5 Explain why the assumption that electrons exist in the nucleus conflicts with the observed magnitude of the intrinsic magnetism of the nucleus. Why was this problem dissolved by the discovery of the neutron?

6 Describe the nuclear reactions that Chadwick used to prove the existence of the neutron. What inconsistencies arise if it is assumed that a gamma ray is emitted in these reactions?

7 What are the properties of the neutron? How do they compare with the properties of the proton and the electron?

8 Explain why it is necessary to postulate the existence of a nuclear force that is neither gravitational nor electromagnetic in nature. Summarize the main features of the nuclear force.

9 What picture of the internal structure of the nucleus has been obtained from high-energy electron scattering experiments?

10 What is the atomic mass unit (amu)? How is 1 amu defined?

11 What is an isotope? Explain how the isotopes of the same element differ from one another.

12 Define and explain the terms mass defect and nuclear binding energy, and tell how they are related.

13 What is the evidence for the existence of nuclear energy levels? How do these energy levels differ from the electron energy levels of the atom?

14 Discuss the difficulties involved in the application of the Schrödinger equation to a particle in the nucleus.

15 What are the "magic numbers" of nuclear physics? What do they represent? How can their existence be accounted for?

16 List the physical quantities that are conserved in a nuclear reaction.

17 Discuss in general terms the nuclear transmutations that take place when each of the following particles is emitted:
(a) Alpha particle.
(b) Electron.
(c) Positron.
(d) Gamma ray.

18 What is the difference between an endoergic reaction and an exoergic reaction? How can we determine whether a given reaction will be endoergic or exoergic? What is meant by the Q value of a given reaction?

19 Explain (a) the ways in which energy can be supplied to trigger an endoergic reaction and (b) the forms in which energy can be released in an exoergic reaction.

EXERCISES

1 Rutherford was able to induce artificial transmutations by alpha-particle bombardment of the elements B, F, Na, Al, and P. What do these elements have in common? If a proton is ejected in each case, to what elements are they transmuted by the reaction?

2 Show that according to the proton–electron hypothesis
(a) the total number of protons in the nucleus $= A$.
(b) the total number of electrons in the nucleus $= A - Z$.
(c) the total number of particles in the nucleus $= 2A - Z$.

3 The nuclear spin is equal to the vector sum of all the intrinsic spins of the individual particles in the nucleus. (Protons, neutrons, and electrons all have intrinsic spins of $\frac{1}{2}$.) The individual spins must line up in the same direction, either parallel or antiparallel to one another.
(a) Explain why the nuclear spin of a nucleus containing an *even* number of particles must be an integer (0, 1, 2, 3, . . .).
(b) Explain why the nuclear spin of a nucleus containing an *odd* number of particles must be an odd half-integer (1/2, 3/2, 5/2 . . .).

4 For each of the following nuclei, predict whether the nuclear spin should be an integer or a half-integer according to (a) the proton–electron hypothesis and (b) the proton–neutron hypothesis. (The total number of particles in the nucleus will be different according to each theory.) Compare the predicted spin with the observed value obtained from spectra. For which nuclei does the proton–electron hypothesis predict the wrong spin?

Nucleus	Observed spin
$_2He^4$	0
$_3Li^7$	$\frac{3}{2}$
$_5B^{10}$	3
$_6C^{12}$	0
$_7N^{14}$	1

5 Why is a high-energy electron a suitable particle to probe the internal structure of the nucleus? What would happen if protons or alpha particles of the same energy were used instead of electrons?

6 Using the data given in the text, calculate the mass in *grams* of (a) a C^{12} atom, (b) a proton, and (c) an alpha particle.

7 The element copper ($_{29}Cu$) found in nature contains two isotopes, Cu^{63} and Cu^{65}. Natural copper on the earth is 69 percent Cu^{63} and 31 percent Cu^{65}. Estimate the average atomic weight of natural copper. Compare your result with the value given in the periodic table.

8 Explain why the mass of a nucleus is always *less* than the sum of the masses of its protons and neutrons. Where does the missing mass go?

9 Calculate the nuclear binding energy for each of the following nuclei:
(a) ${}_1H^3$ (m_{nuc} = 3.015501 amu).
(b) ${}_2He^3$ (m_{nuc} = 3.014932 amu).

10 Discuss some experimental methods with which one can measure the binding energy of a given nucleus.

11 From the energy level diagram given for the sodium nucleus in Figure 18–3, predict some of the energies of the gamma rays that could be emitted or absorbed by the sodium nucleus.

12 Explain in terms of the magic numbers denoting closed shells in the nucleus
(a) why each of the following nuclei is especially stable: ${}_{28}Ni^{58}$, ${}_{36}Kr^{86}$, ${}_{40}Zr^{90}$, and ${}_{54}Xe^{136}$.
(b) why ${}_{23}V^{50}$ and ${}_{56}Ba^{137}$ are good neutron absorbers.
(c) why ${}_{82}Pb^{208}$ (the heaviest stable isotope) is "doubly magic."

13 Give the atomic number, mass number, and name of the missing nucleus in each of the following reactions:
(a) ${}_2He^4 + {}_{15}P^{31} \longrightarrow {}_{16}S^{34} + ?$
(b) ${}_1H^3 \longrightarrow {}_2He^3 + ?$
(c) ${}_2He^4 + {}_{10}Ne^{22} \longrightarrow {}_{12}Mg^{25} + ?$
(d) ${}_{29}Cu^{64} \longrightarrow {}_{28}Ni^{64} + ?$
(e) ${}_1H^2 + {}_8O^{16} \longrightarrow {}_0n^1 + ?$

14 Propose a nuclear reaction by which one can transmute mercury (${}_{80}Hg^{200}$) into gold (${}_{79}Au^{197}$).

15 Calculate the Q value for the reaction ${}_{26}Fe^{56} + {}_2He^4 \longrightarrow {}_{28}Ni^{59} + {}_0n^1$. The masses of the Fe^{56} and the Ni^{59} nuclei are 55.920662 amu and 58.918970 amu, respectively. Is this reaction endoergic or exoergic?

16 The first nuclear reaction used to test the validity of $E = mc^2$ was the following (studied by Cockcroft and Walton in 1932): ${}_1H^1 + {}_3Li^7 \longrightarrow {}_2He^4 + {}_2He^4$. Calculate the Q value for this reaction. (The mass of a Li^7 nucleus is 7.014357 amu.) What is the kinetic energy in MeV of each of the two alpha particles produced in this reaction?

17 Explain why all naturally occurring radioactive decay reactions *must* be exoergic.

APPLICATIONS OF NUCLEAR PHYSICS

19

The energy produced by breaking down the atom is a very poor kind of thing. Anyone who expects a source of power from the transformation of these atoms is talking moonshine.

Ernest Rutherford

IN THE NATURAL ENVIRONMENT of the earth, nuclear processes are of little consequence. The temperatures that occur on the earth are much too low to trigger nuclear reactions. The processes that define the geological and biological regime on earth are *chemical* processes; that is, they involve transformations of the outer atomic electrons, which have no effect on the atomic nucleus. With the exception of naturally occurring radioactivity—which is a rather minor phenomenon over the earth as a whole—nuclear reactions were not observed on the earth until they were produced in scientific laboratories in the 1930s. Since that time the situation has changed quite drastically. Nuclear processes initiated by man have come to exert a profound effect on the conditions and future of life on the earth, for the study of the nucleus—once regarded as a highly esoteric area of physics, dealing as it did with an unseen entity supposedly less than 10^{-12} cm in diameter—has led to several discoveries of far-reaching significance.

In this chapter we will be concerned with two of these discoveries—artificial radioactivity and nuclear fission—and their practical applications. We will learn how the first of these discoveries, artificial radioactivity, has been put to use in a way that has led to significant advances in knowledge in a wide variety of areas, including astronomy, geology, chemistry, biology, archaeology, and art history. We will also learn how the discovery of nuclear fission has led to the unlocking of a vast source of energy that may prove to be enormously beneficial or ultimately destructive to life on the earth.

19–1

RADIOACTIVITY: THEORY AND APPLICATIONS

The phenomenon of radioactivity has played a central role in the development of nuclear physics. The early investigations of radioactivity by the Curies and by Rutherford and Soddy gave the first indications of the complexity of the nucleus. For many years (before the first particle accelerators were constructed), radioactive nuclei provided the only source of high-energy particles with which nuclear reactions could be triggered. On the other hand, the technological uses of radioactivity were rather limited at first. There were only about fifty known naturally occurring radioactive isotopes, confined mainly to the heavy-element region ($Z > 82$) of the periodic table. These elements exist on the earth in very low abundance and are quite expensive to extract in commercially usable quantities. However, in recent years methods have been developed for the artificial production of radioactive isotopes of nearly all the elements. This advance has widely increased the technological value of radioactivity. In other to understand why radioactive isotopes have proved to be so valuable to science and industry, let us now examine the principles of radioactive decay in detail.

THE RADIOACTIVE DECAY LAW

The law that describes the rate of disintegration of the atoms of a radioactive isotope was discovered by Rutherford and Soddy in 1902. They found that *the number of atoms of a radioactive sample that disintegrate in a given interval of time is always proportional to the number of atoms in the sample.* In other words, the disintegration rate—the number of atoms that disintegrate per second—is directly proportional to the number of atoms present. Thus in a given time interval a fixed fraction of the atoms in a sample will disintegrate. As time progresses, the number of atoms of the radioactive isotope steadily decreases, so the rate of disintegration decreases as well. That is, the same *fraction* of the atoms present will disintegrate, but since there are progressively fewer atoms remaining, there will be a smaller number of disintegrations. In this manner the number of radioactive atoms—and the number of them that disintegrate in a given time interval—diminish gradually to zero.

The radioactive decay law is best explained in terms of a quantity known as the ***half-life,*** $T_{1/2}$. The half-life of a radioactive isotope is defined as the *time interval during which half of the atoms of the isotope initially present disintegrate.* To illustrate this concept, let us suppose that initially (at $t = 0$) there are N_0 atoms of a radioactive isotope that has a half-life of 5 hours. After 5 hours have elapsed, half of the atoms will have disintegrated; $\frac{1}{2}N_0$ remain. During the next 5 hours half of the remaining atoms will decay, so that at the end of 10 hours $\frac{1}{2}(\frac{1}{2}N_0) = \frac{1}{4}N_0$ atoms remain. After 15 hours, $\frac{1}{2}(\frac{1}{4}N_0) = \frac{1}{8}N_0$ remain, and so on. Thus during each 5-hour interval the number of atoms decreases by a factor of 2. The results are summarized in Table 19–1.

table 19–1 **STATISTICS OF RADIOACTIVE DECAY FOR A HYPOTHETICAL ELEMENT WITH A 5-HOUR HALF-LIFE**

Elapsed time (hours)	Total number of atoms remaining	Number that disintegrated in previous time interval	Total number of atoms of decay product
0	N_0	—	0
5	$\frac{1}{2}N_0$	$\frac{1}{2}N_0$	$\frac{1}{2}N_0$
10	$\frac{1}{4}N_0$	$\frac{1}{4}N_0$	$\frac{3}{4}N_0$
15	$\frac{1}{8}N_0$	$\frac{1}{8}N_0$	$\frac{7}{8}N_0$
20	$\frac{1}{16}N_0$	$\frac{1}{16}N_0$	$\frac{15}{16}N_0$
25	$\frac{1}{32}N_0$	$\frac{1}{32}N_0$	$\frac{31}{32}N_0$
⋮	⋮	⋮	⋮

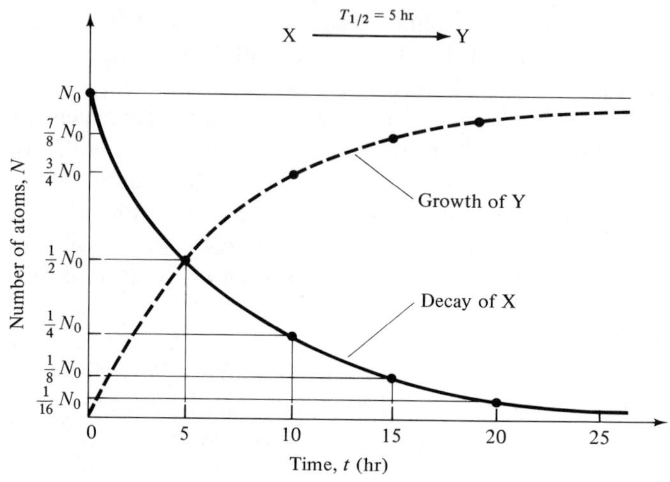

figure 19–1 **THE RADIOACTIVE DECAY LAW**

The curves show the decay of a radioactive isotope X having a half-life of 5 hours and the associated growth of the stable decay product Y.

Note that as time progresses, the number of atoms of the radioactive isotope gradually decreases, and the rate of disintegration also decreases. At the same time there is a gradual buildup of the number of atoms of the ***decay product***—the atom that is formed as a result of the decay.

These results are shown graphically in Figure 19–1. The ***radioactive decay curve*** shown in this figure clearly indicates the gradual decrease toward zero in the number of radioactive atoms present as time progresses.* This type of curve is known as an ***exponential decay curve,*** since it can be shown mathematically that the equation of this curve is

$$\frac{N}{N_0} = e^{-0.693t/T_{1/2}}$$

where N_0 is the number of atoms initially present; N is the number of atoms present after a time t has elapsed; $T_{1/2}$ is the half-life; and e is the base of natural logarithms. (The number 0.693 in the exponent is equal to the natural logarithm of 2.)

The half-life of a particular radioactive isotope is a fundamental and constant property of the nuclei of that isotope. The half-life can be determined experimentally by observing the rate of decay of an isotope over a period of time and plotting the radioactive decay curve. The half-life of an isotope is evidently independent of any external conditions; experimental attempts to change the half-life of an isotope by altering the temperature, pressure, and so on have thus far been unsuccessful. The half-life can be considered a reliable measure of the stability of an isotope. The shorter the half-life of an isotope, the more rapidly it decays; hence a short half-life is associated with a relatively unstable nucleus. The half-lives of the known radioactive

* It is assumed here that the decay product is a stable isotope. If the decay product is itself radioactive, as is often the case, its decay curve will be considerably more complicated.

isotopes (naturally occurring and artificially produced) range from microseconds to trillions of years.

LIMITATIONS OF THE RADIOACTIVE DECAY LAW

It is important to recognize that radioactive decay is described in terms of a radioactive decay *law*. Rutherford and Soddy discovered this law from their measurements of the decay of naturally occurring radioactive elements; they did not attempt to explain *why* radioactive decay takes place. To this day, the mechanism by which a nucleus disintegrates is still not completely understood.

The *statistical* nature of the radioactive decay law should also be stressed. The law *cannot* predict when the nucleus of a single radioactive atom will disintegrate. For a very large number of atoms, the law accurately predicts how many of these atoms will decay in a given time interval, but it cannot be used to determine precisely which atoms will decay and which will not. A strong analogy can be drawn to the process of coin flipping. We can predict with good accuracy that if 1000 coins are flipped, 500 of them will be heads, and if 500 coins are flipped, 250 will be heads. That is, we are able to predict the *fraction* ($\frac{1}{2}$) of the coins, out of a large number of coin flips, that will turn up heads—but we cannot predict the outcome of any individual coin flip. Similarly, given a large number of radioactive atoms, we are able to predict the fraction of these that will decay in a given time interval, but we cannot predict when any individual atom will decay. Hence the radioactive decay law is accurate for large numbers of atoms; it becomes less and less reliable as the number of atoms decreases. In the extreme limit, where there are two atoms of an isotope whose measured half-life is 5 hours, the prediction that only one of these atoms will decay within 5 hours is highly unreliable; and if there is only one atom, no meaningful prediction can be made.

In view of the statistical nature of the radioactive decay law, we find that we can state nothing more than the *probability* that a given atom will decay in a certain time interval. (This probability is inversely proportional to the half-life; the shorter the half-life, the more likely the atom is to decay.) This conclusion provides some fuel for the argument about the role of probability on the atomic level. Is the probability associated with radioactive decay an example of the "absolute chance" that is supposed to operate at this level? Is the disintegration of a radioactive nucleus an intrinsically unpredictable and chaotic event? Or is there some underlying, deterministic mechanism at work in the nucleus, which—if discovered—would enable us to predict the precise time and conditions for a given nucleus to disintegrate? At the present time there is no evidence for the existence of such a mechanism within the nucleus. Should a deterministic law of radioactive decay ever be found, it would be an exceedingly important discovery.

Although the theoretical interpretation of the radioactive decay law is still open to question, the law itself is quite reliable when applied to large numbers of atoms. Let us now consider some of the implications of the radioactive decay law.

NATURAL AND ARTIFICIAL RADIOACTIVITY

We have learned that the number of atoms of a radioactive isotope decreases by a factor of two over a period of one half-life. As shown in Table 19–1, if there are N_0 atoms at $t = 0$, then after one half-life has elapsed, $N_0/2$ atoms remain; after two half-lives $\frac{1}{2}(\frac{1}{2}N_0) = N_0/2^2$ atoms remain; after three half-lives $N_0/2^3$ atoms remain; and in general, after n half-lives $N_0/2^n$ atoms remain. The factor 2^n has the following approximate values:

$$\begin{aligned} 2^{10} &= 1024 \approx 10^3 \\ 2^{20} &\approx 10^6 \\ 2^{30} &\approx 10^9 \\ 2^{40} &\approx 10^{12} \\ 2^{50} &\approx 10^{15} \\ 2^{100} &\approx 10^{30} \end{aligned}$$

Accordingly, the number of atoms of a radioactive isotope decreases by a factor of about a thousand in ten half-lives, a million after twenty half-lives, a trillion after forty half-lives, and so on. Thus if there is initially 1 gram of a radioactive isotope at $t = 0$, after forty half-lives only one trillionth of a gram of that isotope will remain.

On this basis we can begin to understand why some radioactive isotopes occur naturally in the earth, whereas others were not known to exist until they were produced artificially. Let us suppose for the sake of argument that a billion years ago many radioactive isotopes were found naturally in the earth, and that since that time there has been no additional production of these isotopes. Obviously, any isotopes with half-lives of a year or less would not now be found in the earth. More than a billion half-lives would have elapsed for these isotopes, and it would be highly unlikely that more than a few atoms still remain. It is also unlikely that an isotope with a half-life of as much as a million years would be found in measurable quantities, for over a period of a billion years, the isotope would have gone through a thousand half-lives. Thus any radioactive isotopes still found on the earth after a billion years would fall into two categories: (1) isotopes with half-lives of about 0.1 billion (10^8) years or more (for these isotopes less than ten half-lives have elapsed, and they still can be found in sufficient abundance to be detected) and (2) isotopes that have been freshly created during the billion-year period.

Now let us consider what is actually found in the earth. The two heaviest naturally occurring elements, uranium and thorium, are both radioactive. Thorium has only one naturally occurring isotope, Th^{232}, with a half-life of 1.39×10^{10} years. Uranium has two isotopes: U^{238} (99.3 percent of natural uranium), with $T_{1/2} = 4.5 \times 10^9$ years, and U^{235} (0.7 percent of natural uranium), which has $T_{1/2} = 7.1 \times 10^8$ years. Each of these three isotopes decays to form an isotope that is also radioactive, which in turn decays to form another isotope, and so on. In each case there is a ***decay chain*** that continues until a stable isotope of lead ($Z = 82$) is reached. For example, U^{238} decays through a chain of seventeen short-lived isotopes until Pb^{206} (which is stable) is reached. In similar fashion the decay chain that begins with U^{235} terminates at the stable isotope Pb^{207}, and the Th^{232} decay chain ends at Pb^{208}.

Over a period of many thousands of years, an equilibrium has been established between the production and decay of all the isotopes in the three decay chains. Hence these radioactive isotopes are found in the earth, even though nearly all of them have relatively short half-lives. The short-lived naturally occurring radioactive elements thus belong to category 2; they are found in the earth because they are being continually created by the decay of uranium and thorium. Uranium and thorium, on the other hand, belong to category 1; they appear in the earth because they have very long half-lives.

As time progresses, all the uranium and thorium, as well as all the elements in their decay chains, should eventually decay to lead. Thus the abundance of lead in the earth will steadily increase, and the abundance of the naturally occurring radioactive elements will gradually decrease until they can no longer be found in the earth. The fact that uranium and thorium are still found in some abundance in the earth can be used to estimate the age of the earth itself, as we shall learn later in the chapter.

Thus the distinction between "natural" and "artificial" isotopes is not really based on any significant physical differences. The artificial radioactivities may very well have existed at one time on the earth, but they have since become "extinct" because of their relatively short half-lives and the absence of any source for further production. In time, elements such as radium, polonium, thorium, and uranium will also become extinct and will no longer be classified as "natural" radioactive elements.

As mentioned in the previous chapter, the existence of additional radioactive isotopes, beyond those found naturally in the earth, became known in 1934 through the experiments of the Joliot-Curies. Since that time more than a thousand different radioactive isotopes have been produced artificially by controlled nuclear reactions. The production of these radioactive isotopes to supplement those already found in nature has proved to be of significant theoretical and practical value to science. By analyzing the systematics of the decay of radioactive isotopes, physicists have learned a lot about the properties of the nucleus. In addition, radioactive isotopes have been used extensively in chemical, biological, and medical research.

ARTIFICIAL ELEMENTS

As an illustration of how radioactive isotopes are produced in the laboratory, we will now consider how the ***transuranic elements*** (elements 93 to 105) were produced. None of these elements is found naturally in the earth; they have been created artificially by suitably chosen nuclear reactions. Nuclear reaction theory is extremely useful for predicting the most suitable targets and bombarding particle energies for the most efficient production of these elements.

The first artificial element, neptunium (${}_{93}Np$), was produced by Edwin McMillan and Philip Abelson at the University of California in 1940.* Their method was to bombard U^{238} with low-energy neutrons, leading to the reactions

$$ {}_0n^1 + {}_{92}U^{238} \longrightarrow {}_{92}U^{239} \longrightarrow {}_{93}Np^{239} + {}_{-1}e^0 $$

Neutron capture by U^{238} produces the unstable isotope U^{239}, which then beta-decays (with a 23.5-min half-life) to Np^{239}. This work was continued by Glenn Seaborg, who found in the same year that Np^{239} beta-decays (with a half-life of 2.33 days) to another new element, which he called plutonium:

$$ {}_{93}Np^{239} \longrightarrow {}_{94}Pu^{239} + {}_{-1}e^0 $$

Since 1944 successively heavier elements have been created by using quantities of the transuranic elements as targets for further reactions. For example,

$$ {}_2He^4 + {}_{94}Pu^{239} \longrightarrow {}_{96}Cm^{242} + {}_0n^1 $$
$$ {}_2He^4 + {}_{96}Cm^{242} \longrightarrow {}_{98}Cf^{245} + {}_0n^1 $$

*Nearly all of the transuranic elements were first produced by scientists at the University of California—as the names of some of them (americium, berkelium, californium) indicate. Elements 99 (einsteinium) and 100 (fermium) were discovered in the debris from the first hydrogen bomb explosion in 1952. The names chosen for the most recently discovered elements—rutherfordium (104) and hahnium (105)—have not yet been given official acceptance.

The nuclei of light elements, accelerated to high energies in a cyclotron, have also been used successfully as bombarding particles:

$$
\begin{aligned}
{}_{6}C^{12} + {}_{96}Cm^{246} &\longrightarrow {}_{102}No^{254} + 4\,{}_{0}n^{1} \\
{}_{5}B^{10} + {}_{98}Cf^{252} &\longrightarrow {}_{103}Lw^{257} + 5\,{}_{0}n^{1} \\
{}_{6}C^{12} + {}_{98}Cf^{249} &\longrightarrow {}_{104}Rf^{257} + 4\,{}_{0}n^{1} \\
{}_{7}N^{15} + {}_{98}Cf^{249} &\longrightarrow {}_{105}Ha^{260} + 4\,{}_{0}n^{1}
\end{aligned}
$$

The heaviest elements tend to be highly unstable with half-lives no longer than a few seconds. Nevertheless, they can be identified by the type and energy of the radiation given off when they disintegrate. The theoretical indications are that identifiable isotopes of elements up to $Z = 126$ could be produced with the existing technology.

The creation of the transuranic elements has not been pursued merely for the pleasure of adding new symbols to the periodic table. Much has been learned by studying the nuclear and chemical properties of these new elements. The practical uses of the transuranic elements are relatively limited at the present time because most of the elements exist in extremely minute quantities. The exception is plutonium, which is manufactured by the ton for use in nuclear weapons. Currently, californium is being produced at the rate of 1 gram per year and einsteinium at the rate of about 0.1 gram per year.

Given that it is possible to produce virtually any isotope of any element by nuclear reactions, one might wonder whether some modern-day alchemists are engaged in the conversion of base metals into gold. One could easily produce gold from mercury by the reaction

$$ {}_{1}H^{1} + {}_{80}Hg^{200} \longrightarrow {}_{79}Au^{197} + {}_{2}He^{4} $$

Unfortunately, the expenses involved in the construction and maintenance of a cyclotron and in the chemical processing needed to separate the reaction products would far exceed the value of the gold produced.

RADIATION CHEMISTRY AND BIOLOGY

The usefulness of radioactive isotopes in chemical and biological research derives mainly from the fact that the chemical behavior of a radioactive isotope of some element is virtually the same as that of any stable isotope of the same element. As we know, chemical reactions involve only the outer electrons of the atom—so that even though their nuclei are radioactive and have slightly different masses, the atoms of radioactive isotopes will enter into the same chemical reactions at the same rates as their stable counterparts and will also be absorbed and used by biological systems in the same way as the stable isotopes of the same element. The advantage of radioactive isotopes is that they can be "traced"; that is, the path of a radioactive isotope in a chemical reaction or in a biological system can be followed by detecting the radiation given off by the isotope. The uses of radioactive tracers in chemical and biological research are so diverse that only a brief summary of them can be given here.

Most elements have a radioactive isotope with a half-life that is neither too short (so that it decays away before its path can be traced) nor too long (so that the decay rate is too small to be useful). Such isotopes can be produced in usable quantities by bombarding a suitable target with high-speed particles from a nuclear reactor or particle accelerator. Radioactive tracers have great value in biological and medical research. The rate at which plants absorb nutrients from soil can be studied by chemically synthesizing the nutrients so that they contain radioactive isotopes. The way in which certain minerals are used by plants or animals can be examined by using tracers containing radioactive isotopes of calcium, phosphorus, magnesium, iron, sodium, and so on. A great deal has been learned about thyroid function through the use of radioactive iodine (I^{123} or I^{131}). The importance to the human body of vitamin B_{12}, which contains cobalt, has been examined by tagging molecules of the vitamin with radioactive Co^{60}. The list of research applications is almost endless. It has been said that radioactive isotopes represent the most useful tool for biological research since the invention of the microscope.

Another factor in the application of radioactivity is the effect of radiation on matter. The alpha, beta, and gamma particles emitted by radioactive isotopes have considerable energy. When they travel through matter, they transmit their energy to surrounding atoms and molecules by electromagnetic interactions or direct collisions. This results in the excitation, ionization, and dissociation of atoms and molecules along the path of the particle. Chemical reactions are often triggered as a consequence; for instance, the passage of alpha particles through water will dissociate water molecules and bring about the production of some hydrogen gas (H_2) and hydrogen peroxide (H_2O_2).

Radiation passing through a biological cell can cause considerable damage by triggering ionization and chemical reactions that break up the molecules. For this reason radiation poses a considerable health hazard to man. Many of the early researchers in radioactivity, unaware of the dangers of radiation, died from various radiation-induced diseases. Marie Curie suffered from skin lesions, cataracts, and leukemia (which caused her death). Her daughter Irène Joliot-Curie also died from radiation-induced leukemia.* Many of the workers who were employed to paint radium dials on watches contracted cancer of the mouth or stomach from accidentally ingesting bits of radium.

In the past few decades, as scientists have become more aware of the dangers of radiation, more rigid procedures have been instituted for the handling of radioactive materials. However, there is still considerable question as to what constitutes a "tolerable" exposure to radiation. Humans are constantly exposed to radiation caused by man-made sources (from dental x rays to fallout from nuclear bomb tests) and natural sources (naturally occurring radioactivity and cosmic rays). It has been suggested that sudden increases in cosmic-ray intensity in the past may have significantly affected the evolution of life on earth by increasing the rate of genetic mutations. (Most radiation-induced mutations are "bad" mutations, so that an increase in cosmic-ray intensity would have been more likely to wipe out less resistant species than to produce forward mutations.)

The damaging effects of radiation have, nevertheless, been put to good use, for it has been found that certain types of cancer and tumorous growths can be retarded or destroyed by intense dosages of radiation. As more is learned about the effects of radiation on biological processes, the value of radioactivity to man may eventually come to far outweigh its dangers.

* Pierre Curie, on the other hand, was killed when he was run over by a horse-drawn wagon in a Paris street.

19–2

RADIOACTIVITY AND HISTORY

We now turn to an entirely different use of radioactivity: the analysis of the abundances of naturally occurring radioactive isotopes and their decay products as a means of determining the ages of objects on the archaeological, geological, and astronomical time scales. Let us first examine some of the basic principles underlying the methods of radioactive age determination.

There are roughly thirty known radioactive isotopes that have half-lives greater than a million years. Roughly half of them are found to exist naturally in the earth; the other half have been produced artificially. The isotopes that are found naturally all have half-lives of 700 million years or more. None of the long-lived isotopes not found in the earth have half-lives greater than 100 million (10^8) years. It seems reasonable to assume that at least some of these isotopes did exist naturally at some time in the past and have since decayed into extinction. There is some real, although indirect, evidence to support this contention: the decay products of a number of these isotopes have a relatively high abundance in the earth. For example, the most abundant isotope of xenon is the stable isotope ${}_{54}Xe^{129}$. This isotope is formed by the decay of the artificially produced radioactive isotope ${}_{53}I^{129}$, which has a half-life of 16 million years. Although it cannot be established with complete certainty, it seems highly likely that I^{129} was once present in the earth but all of it has since decayed to Xe^{129}.

How did a now-extinct radioactive isotope like I^{129} come to be present in the first place? Presumably, there must have been a time in the distant past when these isotopes were being created. The conditions for their creation ceased to exist at some point, and with no further production, the isotopes have decayed out of existence. In recent years a theory of the origin of the elements has been developed that seems to explain how the different elements were formed and how they came to have their present isotopic abundances. The theory assumes that at one time extremely high temperature conditions caused extensive nuclear reactions to take place in the medium out of which the solar system eventually condensed. These nuclear reactions brought about transmutations that built up the abundances of some isotopes and depleted others. Eventually, the medium cooled down, and the nuclear reactions stopped. At this point the isotopic abundances became fixed. Shortly afterward the solar system formed out of this material. Since that time there has been no significant element production by nuclear reactions. Any radioactive isotopes that existed at that point continued to decay, and their abundances steadily decreased. The shorter-lived isotopes have since become extinct; the longer-lived isotopes can still be found in the earth. (Exceptions are the radioactive isotopes in the uranium and thorium decay chains, which are still being produced.)

Among the radioactive isotopes *not* found in nature, there are several with half-lives close to 10^8 years. Let us assume that these isotopes were initially present and have since decayed into extinction. In order for these isotopes to have become extinct, they must have been decaying for at least twenty to thirty half-lives. This means that the nuclear reactions that formed these isotopes must have terminated several billion years ago.

According to this theory, the solar system must be at least a few billion years old. There are methods by which the age of the solar system can be determined more precisely, as we shall learn later. For the purposes of radioactive age determinations, however, the key point of the theory is the following: *The abundances of the isotopes were determined in the past by the effects of nuclear reactions that have long since terminated. Since that time any changes in isotopic abundances have been due primarily to the decay of any radioactive isotopes that existed when the nuclear reactions stopped.* With these basic ideas in mind, we will now learn how the age of an object can be determined from the radioactive isotopes the object contains.

THE AGE OF A ROCK

The decay of a radioactive isotope is accompanied by a corresponding increase in the isotope's decay products, as shown in Figure 19–1. By comparing the relative amounts of the isotope and its product in an object that contains both, one can determine what fraction of the isotope has decayed since the object was chemically isolated from its surroundings. Using the decay curve and the known half-life of the radioactive isotope, one can then easily determine the age of the object.

Let us suppose the object under consideration is a rock that contains measurable quantities of a long-lived radioactive isotope (such as uranium) and its decay product (lead). From a comparison of the relative amounts of uranium and lead in the rock, the age of the rock can be determined. By the "age of the rock," we mean the time since the rock last solidified, for when the material from which the rock crystallized was in the liquid or gaseous state, the decay products could have easily escaped or been chemically separated from the parent isotope. Once a rock solidifies, the decay products stay locked in place and remain in close proximity to the remaining parent isotope.

The validity of this method appears to depend on the assumption that all the lead found in a rock has come from the decay of uranium. How do we know that there wasn't some lead originally present in the rock? How is it possible to distinguish between naturally occurring lead and decay-product lead? This distinction *can* be made because uranium decays to a *specific isotope* of lead: U^{238} decays to Pb^{206}, and the considerably less abundant U^{235} decays to Pb^{207}. Thus any rock that contains uranium decay-product lead will be enriched in Pb^{206} and (to a lesser extent) in Pb^{207}. Hence the *excess* Pb^{206} and Pb^{207} in a rock, as compared with their isotopic abundances in samples of normal lead not found in association with uranium, is assumed to be derived from uranium decay. Thus the abundance ratios of U^{238}/Pb^{206} or U^{235}/Pb^{207} can be used to determine the age of a rock. The major remaining uncertainty in this method is whether the samples of "normal" lead used as a basis of comparison are truly pure and have never been contaminated with uranium decay-product lead.

There are several long-lived naturally occurring isotopes suitable for age determinations. In addition to the method using the two uranium isotopes, there are methods based on the naturally occurring radioactive isotopes Th^{232}, K^{40}, and Rb^{87}. These different methods can often be used on the same rock sample (or an different rocks from the same geological region) to give independent age measurements. The measured ages usually turn out to be in good agreement with one another, thereby attesting to the reliability of these methods.

THE AGE OF THE SOLAR SYSTEM

The oldest rocks that have been found so far turn out not to be rocks originating on the earth, but meteorites that have landed on the earth after surviving the plunge through the earth's atmosphere. Several of these meteorites have measured ages (according to several different methods) of about 4.5 billion years. It is generally believed that these objects represent the debris from the formation of the solar system and are as old as the solar system itself. Presumably, they had been traveling in orbits around the sun since the beginning of the solar system until they happened to intersect the earth's orbit and fall into the earth's atmosphere.

The oldest rocks found on the earth are about 4 billion years old. This does not mean that the earth itself is 4 billion years old; it means only that the oldest known available rocks solidified about 4 billion years ago. Remember that if a rock melts or enters into chemical reactions, the decay products can be separated from the parent isotope—thereby setting the "radioactive clock" back to zero. Nevertheless, it is still possible to determine the age of the earth from the relative isotopic abundances of Pb^{207} and Pb^{206} in different rocks. We know that Pb^{207} comes from U^{235}, and Pb^{206} from U^{238}. Since U^{235} and U^{238} do not decay at the same rate (U^{235} decays more rapidly), the relative isotopic abundances of Pb^{207} and Pb^{206} will change with time in rocks that contain uranium. However, the Pb^{207}/Pb^{206} ratio in normal lead remains contant in time and presumably represents the "primeval" value—the abundance ratio that existed when the nuclear reactions terminated. Thus by comparing the Pb^{207}/Pb^{206} abundance ratio in normal lead with the ratio in uranium-bearing rocks of various ages, it is possible to determine how much time has elapsed since the termination of the nuclear reactions that created U^{238} and U^{235}. This measurement is unaffected by chemical reactions, melting, and so on because Pb^{207} and Pb^{206}, being isotopes of the same element, *will have the same chemistry* and will not be separated from each other. The time determined by this procedure turns out to be 4.55 ± 0.10 billion years—the same age as the oldest known meteorites. Modern astronomers generally believe that the earth formed a short time after the nuclear reactions stopped.* Hence the above figure is taken to represent the age of the earth. The oldest meteorites and the earth apparently formed at about the same time, more than 4.5 billion years ago.

The age of another body in the solar system was determined when the rocks and soil that the Apollo astronauts brought back from the moon were examined. Scientists analyzing these objects by the Pb^{207}–Pb^{206} method have obtained an age of about 4.6 billion years for the moon. This means that the moon is as old as, or slightly older than, the earth. It also supports the theory that all the bodies in the solar system formed at essentially the same time. The ages of individual moon rocks have also been measured, and several rocks with ages in excess of 4 billion years have been found. By comparing the ages of rocks brought back from different regions of the moon, astronomers have been able to get some idea of the history and evolution of the moon's surface. Thus the study of radioactive decay has provided the modern-day astronomer with some valuable data about the solar system.

ARCHAEOLOGICAL AGE DETERMINATIONS

The radioactive dating methods described thus far are based on the decay of isotopes that have half-lives on the order of billions of years. These methods are useful for determining the ages of objects that are several hundred million years old or more. However, because these isotopes decay so slowly, it is extremely difficult to determine the age of much younger objects. The amount of lead produced by the decay of uranium over a period of a thousand years is virtually undetectable.

*A "short time" means about 50 million years, which is a rather brief time span on the scale of astronomical events. Note that the *uncertainty* in the age of the earth is 0.1×10^9, or 100 million years!

In 1949 the American chemist Willard F. Libby announced a new radioactive dating method capable of establishing the ages of objects less than 30,000 years old. This method is based on the decay of the radioactive isotope $_6C^{14}$, which has a half-life of 5600 years and is found in the earth's atmosphere in small amounts.

Since C^{14} has such a short half-life, its presence on the earth can be explained only if this isotope is being continually produced by some nuclear reaction or radioactive decay. In fact, it has been well established that C^{14} is produced by the action of cosmic rays on the atoms of the earth's atmosphere. Cosmic rays are extremely high energy particles (mostly protons and alphas) that bombard the earth from beyond our solar system. The cosmic-ray particles react with the nuclei of atmospheric atoms and produce, among other particles, high-energy neutrons. These neutrons collide with the nuclei of the nitrogen atoms in the atmosphere and produce C^{14} by the reaction

$$_0n^1 + {}_7N^{14} \longrightarrow {}_6C^{14} + {}_1H^1$$

The C^{14} is unstable and beta-decays (with a 5600-year half-life) back to nitrogen:

$$_6C^{14} \longrightarrow {}_7N^{14} + {}_{-1}e^0$$

Equilibrium has been reached between the production and decay of C^{14}, so that the C^{14} content of the atmosphere remains relatively constant. It has been estimated that only one out of every trillion carbon atoms is a C^{14} atom—an extremely small, but measurable, amount. On the average there should be about fifteen disintegrations per minute in a gram of natural carbon because of the C^{14} contained therein.

The radioactive C^{14} atoms will have the same chemical behavior as the stable C^{12} and C^{13} atoms. They will react to form CO_2 in the atmosphere and will be assimilated into all living plants and animals on the earth. Because of the steady intake and outgo of carbon compounds, living organisms will have a uniform content of C^{14}. Thus all human beings are slightly radioactive because of the C^{14} in their bodies, which disintegrates at the equilibrium rate of 15 C^{14} atoms/min per gram of carbon.

When an organism dies—say a tree is cut down—its intake of C^{14} ceases. The C^{14} in the organism at the time of its demise simply decays away according to the radioactive decay law. After 5600 years (one half-life) have elapsed, the disintegration rate will be half of the equilibrium rate, or 7.5 atoms/min/g. After 11,200 years the rate will have decreased again by a half to 3.75 atoms/min/g, and so on. Hence by measuring the present decay rate in a formerly living object, one can determine the time that has elapsed since its death. With sensitive detection equipment, the C^{14} method can be extended to date objects up to 30,000 years old.

The validity of the C^{14} method rests on the assumption that the C^{14} content of the atmosphere has remained constant in the past. To test this assumption, Libby and his associates determined the C^{14} ages of objects whose ages were already known from archaeological data—wood from Egyptian tombs, bread from the ruins of Pompeii, linen wrapping from the Dead Sea scrolls—and found that the C^{14} ages were in good agreement with the historically established ages.

More precise determinations have shown that the C^{14} content has fluctuated somewhat in time. These fluctuations may be due to variations in solar activity (sun spots), cosmic-ray intensity, or the strength of the earth's magnetic field (which shields out much of the cosmic rays reaching the earth). Other fluctuations are definitely man-made. Since the 1870s the industrial burning of coal (which is millions of years old and contains no C^{14}) has reduced the relative abundance of C^{14} in the atmosphere; however, the nuclear bomb tests in the 1950s and 1960s tended to increase it. Nevertheless, by carefully correlating C^{14} ages with the known ages of objects, these fluctuations can be taken into account. One procedure that has proved useful in this respect is to measure the C^{14} content of successive rings of very old trees.

The C^{14} dating technique has become an indispensable tool of the modern archaeologist. Any organically derived artifact, such as a piece of wood or bone, can be used to determine the age of an archaeological site. Caves have been found in northern Europe and North America that were occupied by man 10,500 years ago, just after the end of the last ice age. All art lovers know of the beautiful prehistoric paintings found on the walls of the caves of Lascaux in central France. The carbon dating of charcoal found in these caves indicates that these earliest-known works of art are 15,000 years old. Thus the record of man's creativity has been extended back in time through modern nuclear physics.

19–3

NUCLEAR FISSION AND FUSION

Few scientific discoveries have changed the course of history so dramatically as the discovery of nuclear fission. Since 1945 world politics has been dominated by an arms race between the great powers, involving the amassing of great stockpiles of nuclear weapons capable of wiping out entire cities if not all civilization. Yet if man's proclivities for self-destruction can be contained, the controlled use of nuclear reactions holds the prospect of a source of energy sufficient to meet the needs of mankind for the foreseeable future.

Since one of the aims of this book is to show how science and scientists are related to their social and political environment, the events that led to the development of the atomic and hydrogen bombs will be described in some detail. Much criticism has been directed against the scientists who contributed to the invention of these terrible weapons. The thrust of this criticism is that these scientists operated in a moral vacuum, that they created the bombs simply because it was within the power of science to do so, that they had no concern for the consequences of their actions, and that the use of pure science for the creation of weapons is a perversion of the aims and purposes of science. In the following sections we shall consider some of the reasons why scientists chose to develop nuclear weapons, and how they reacted to the use of these weapons. The student can then make his own judgment with respect to the validity of these criticisms and the extent to which scientists are responsible for the threat to civilization posed by the existence of nuclear weapons.

THE DISCOVERY OF NUCLEAR FISSION

The accidental discovery of the fission of uranium, set against the background of Hitler's rise to power and the outbreak of World War II, is one of the great dramas in the history of science. The story begins in 1934 with the researches of a group of Italian physicists led by Enrico Fermi. Fermi was exceptionally adept at both experimental and theoretical physics, and was one of the few physicists to make important discoveries in both of these areas since the days of Galileo and Newton. Fermi had already made significant contributions to the quantum theory and to the theory of beta decay before he turned his attention to experimental nuclear physics in 1934.

Fermi's "conversion" was inspired by Chadwick's discovery of the neutron and the Joliot-Curies' production of artificial radioactivity. He reasoned that the neutron, being uncharged, could penetrate the nucleus very easily and would provide an ideal particle with which to trigger nuclear reactions. Fermi and his co-workers systematically subjected the elements to neutron bombardment one by one and in doing so discovered many new radioactive isotopes. When they finally reached the heaviest element, uranium, they found that the radioactive isotopes produced by neutron bombardment of uranium could not be associated with any of the known elements at the end of the periodic table. Fermi assumed that he had produced isotopes of transuranic elements. However, a German chemist, Ida Noddack, disagreed with Fermi's chemical identification; the new element did not have the right chemical properties to be a transuranic element. Noddack proposed that Fermi had actually split the nucleus into two much lighter elements.*

Most physicists, including Fermi, considered this proposal ridiculous. The uranium nucleus has a very large total binding energy. The average binding energy of a nucleon in the uranium nucleus is 7.59 MeV. The neutrons used in Fermi's experiments had energies of less than 1 eV. How could a single low-energy neutron possibly split the uranium nucleus?

Between 1935 and 1938 Fermi's experiments were repeated by two highly respected chemists, Otto Hahn and Lise Meitner of Germany. They concluded that the radioactive isotopes produced by the neutron bombardment of uranium were isotopes of radium, actinium, and thorium ($Z = 88, 89, 90$, respectively), which were presumed to be formed from the alpha decay of uranium isotopes. However, no alpha particles were observed among the reaction products.

In 1938 the Third Reich decided that it could do without the services of Lise Meitner in view of her Jewish ancestry, despite a personal appeal that Max Planck made to Hitler. After Meitner escaped to Sweden, Fritz Strassmann became Hahn's assistant. Later in the year Strassmann learned that Irène Joliot-Curie had repeated the uranium experiment and found an element closely resembling lanthanum ($Z = 57$) among the reaction products. At Strassmann's urging, Hahn repeated the experiment. This time Hahn and Strassmann found without any doubt that the uranium nucleus had been split into smaller fragments. The elements that they had mistakenly identified as Ra, Ac, and Th were actually found to be isotopes of barium and the rare earth elements. A glance at the periodic chart will explain the mistaken identification. Ba, La, and Ce are directly below Ra, Ac, and Th and therefore have quite similar chemical properties. In early 1939 Hahn and Strassmann reported:

> As chemists we should, from the evidence, replace Ra, Ac, Th by the symbols Ba, La, Ce in our scheme. . . . As nuclear chemists, closely associated with physics, we cannot as yet decide to take this step in contradiction to the prevailing knowledge in nuclear physics. Our results could possibly have been simulated through a series of peculiar occurrences.[1]

Hahn passed his findings along to Lise Meitner in Sweden. On the day she received Hahn's letter, Meitner was being visited by her nephew, Otto Frisch, a physicist at Bohr's institute. Together they worked out a rough picture of how the uranium nucleus, upon absorbing a neutron, distorts and splits into two roughly equal parts. They calculated that the fission of uranium should be a highly exoergic reaction

*Actually, the reaction that Fermi *thought* had taken place—neutron capture by U^{238}, followed by beta decay to Np^{239}—was the one eventually used by McMillan and Abelson in 1940 to demonstrate the existence of neptunium.

in which about 200 MeV of energy are released—about ten times more energy than that of any previously known nuclear reaction.

When Frisch returned to Copenhagen, he informed Niels Bohr about what had happened. Shortly thereafter Bohr went on a visit to the United States, where he and John Wheeler of Princeton University worked out a detailed theoretical explanation of the fission process.

THE PHYSICS OF NUCLEAR FISSION

The Bohr–Wheeler theory of nuclear fission assumes that the nucleus behaves like an incompressible drop of liquid. Nucleons are packed closely together in the nucleus under the influence of the strong nuclear forces. These forces cause the nucleus to assume a roughly spherical shape, just as a drop of liquid assumes a spherical shape to minimize surface tension. In other words, in the analogy between a nucleus and a liquid drop, the nuclear forces play the role of surface tension and bring the nucleus into a minimum-energy spherical shape.

When the nucleus absorbs a neutron, it is deformed into an ellipsoidal shape. But the nuclear "surface tension" tends to restore the nucleus back to a spherical shape, and oscillations set in (Figure 19–2a). If these oscillations are large enough, a "neck" will form, and the nucleus will assume a dumbbell shape (Figure 19–2b). A separation of charge is created, and the positive charges in the two halves of the dumbbell repel each other. If the shape of the nucleus is too deformed, the nuclear force will not be strong enough to overcome this repulsion (remember that the nuclear force has a much shorter range than the electrostatic force), and the nucleus will split into two roughly equal parts (Figure 19–2c).

With the Bohr–Wheeler theory it is possible to estimate the ***activation energy*** needed to deform a nucleus so that it splits. A small activation energy means that

figure 19–2 NUCLEAR DEFORMATION AND FISSION

(a) When a nucleus is excited (by capturing a neutron, for example), it is deformed from its normal spherical shape and begins to oscillate. (b) If the deformation is large enough, a neck forms in the nucleus, but surface tension brings the nucleus back to a spherical shape. (c) For larger nuclei, the repulsion between the positive charges in the two halves of the dumbbell exceeds the surface tension and fission occurs.

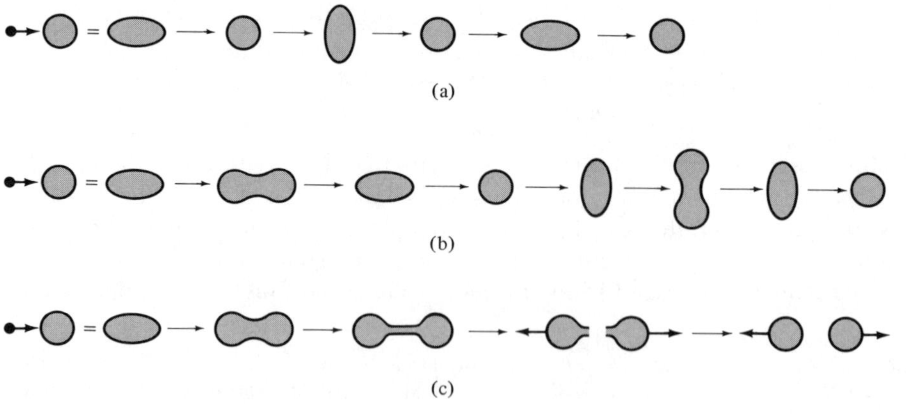

the nucleus is relatively easy to split. In general, the larger the nucleus, the smaller the activation energy needed to achieve fission. For the heaviest nuclei the activation energy for fission is estimated to be about 6 MeV.

The activation energy for fission can be supplied in two ways when a heavy nucleus captures a neutron: (1) by the energy released in the neutron reaction and (2) by the kinetic energy brought in by the captured neutron. If the sum of these two energies is greater than the activation energy, the nucleus will be very likely to undergo fission. There is also a small probability that a nucleus will accidentally absorb sufficient energy through collisions or other random processes and spontaneously split without being hit by a neutron. The likelihood of ***spontaneous fission*** is generally rather small.

Shortly after the discovery of fission, it was confirmed experimentally that it is the less abundant uranium isotope, U^{235}, that undergoes fission with slow neutrons, rather than the more abundant U^{238}. This result was predicted by the Bohr–Wheeler theory. For U^{235} the neutron-capture reaction ${}_0n^1 + {}_{92}U^{235} \longrightarrow {}_{92}U^{236}$ has a Q value of 6.47 MeV, which is more than the activation energy for fission of this nucleus. Hence the capture of a neutron alone provides sufficient energy for fission of U^{235}, regardless of the kinetic energy of the neutron. By contrast, neutron capture by U^{238} releases only 4.78 MeV, which is not enough for fission. Hence the neutron must have a kinetic energy of more than 1 MeV to exceed the activation energy needed to trigger fission of U^{238}. Thus slow neutrons will cause fission of U^{235} but not of U^{238}. Only "fast" fission (by neutrons with energies greater than 1 MeV) is possible for U^{238}.

By making similar calculations for other heavy nuclei, one can determine whether any of them are likely to undergo fission with slow neutrons. Such calculations show that the isotopes ${}_{92}U^{233}$, ${}_{92}U^{235}$, and ${}_{94}Pu^{239}$ are most likely to undergo fission with slow neutrons. Of these, only U^{235} is found in nature. The two heaviest naturally occurring isotopes, ${}_{90}Th^{232}$ and ${}_{92}U^{238}$, both require fast neutrons for fission.

Thus the only naturally occurring isotope that is fissionable by slow neutrons is U^{235}, which comprises only 0.7 percent of natural uranium. All other fissionable isotopes must be produced artificially. Fast fission (of Th^{232} or U^{238}) is a considerably less efficient process; the slow neutron fission rate for U^{235} is about 600 times greater than the fast neutron fission rate for U^{238}. On all accounts—except its availability—U^{235} is the most suitable isotope for nuclear fission.

The fission of U^{235} does not always proceed in exactly the same way. In particular, the ***fission products*** (the two nuclei created when the nucleus splits) are not the same in every reaction. The most likely division is into one nucleus with $A = 141$ and another with $A = 95$, such as

$$ {}_0n^1 + {}_{92}U^{235} \longrightarrow {}_{56}Ba^{141} + {}_{36}Kr^{95} \qquad (Q = 185 \text{ MeV}) $$

The *least* likely split is into two equal parts of 118 nucleons each. It occurs only once in every 10,000 fissions. This is a peculiarity that is not readily explained by the Bohr–Wheeler theory.

The fission product nuclei will invariably be highly unstable and will decay further by emitting neutrons, alphas, betas, and gamma rays. These reactions release an extra 15 to 20 MeV of energy, so that the average energy release in fission is about 200 ± 5 MeV per reaction. The bulk of this energy appears as the kinetic energy of the two fission product nuclei.

THE CHAIN REACTION

The fact that U^{235} fission is such a highly exoergic reaction is interesting, but what is even more significant than the quantity of energy given off is that a number of neutrons are released in the reaction. (To be precise, neutrons are emitted by the unstable fission products within 10^{-14} sec after fission.) The emitted neutrons can trigger more fissions, releasing more neutrons, and so on. The result is a ***chain reaction*** that continues very rapidly until either there are no more neutrons or no more fissionable uranium atoms left.

In a chain reaction the most important factor is the multiplication of neutrons from one generation to the next. If each successive generation has fewer neutrons than the previous one (i.e., the multiplication factor is less than 1), the chain reaction will eventually die out. If the multiplication factor is exactly equal to 1, the chain reaction will proceed along at a constant rate, with a steady production of energy. If the multiplication factor is greater than 1, the chain reaction will escalate rapidly, releasing energy at an extremely rapid rate, until no more fissionable nuclei remain. The amount of energy that would be released if 1 kg of U^{235} (about the size of a golf ball) were consumed in a chain reaction can be calculated as follows. The mass of one U^{235} atom $\approx$ 235 amu $= 3.90 \times 10^{-22}$ gram; therefore 1 kg ($=$ 1000 grams) of U^{235} contains

$$\frac{1000 \text{ grams}}{3.90 \times 10^{-22} \text{ gram/atom}} = 2.56 \times 10^{24}\ U^{235} \text{ atoms}$$

Each fissioning atom releases 200 MeV; therefore the total energy released is

$$(2.56 \times 10^{24} \text{ atoms})(200 \text{ MeV/atom})(1.6 \times 10^{-6} \text{ erg/MeV}) = 8.2 \times 10^{20} \text{ ergs}$$

This is equivalent to approximately 20 trillion calories of heat, or roughly 23 million kilowatt-hr of electric energy, or the energy released by the explosion of 20,000 tons of TNT.

When the discovery of uranium fission was announced, few scientists believed that a large-scale chain reaction was possible, but the discovery that neutrons are released in fission made the possibility more real. Not all of the released neutrons trigger new fissions; some are captured by nonfissioning atoms (in U^{235}, 17 percent of the slow neutron captures *do not* lead to fission), and some escape from the uranium without being captured at all. This "leakage" of neutrons could certainly be large enough to prevent the attainment of a multiplication factor ≥ 1 needed for a self-sustaining chain reaction. Early in 1939 Enrico Fermi (who had escaped from Fascist Italy to the United States in 1938) and his co-workers at Columbia University discovered that on the average 2.5 neutrons are emitted per U^{235} fission—more than enough to counteract the neutron leakage. Shortly afterward Fermi and Leo Szilard—a refugee from Nazi Germany—calculated that it would be theoretically possible to produce a uranium bomb of enormous explosive power.

THE DEVELOPMENT OF THE ATOMIC BOMB

One can only speculate how the discovery of nuclear fission would have affected the course of history if Adolf Hitler had not been in power at the time—or for that matter, if Hitler in his mania for "racial purity" had not forced so many of the great German scientists to leave the country. In 1939, with a world war imminent, the refugee scientists were well aware that Nazi Germany was quite capable of building a uranium bomb and that uranium research had already begun in a few German laboratories. Obsessed by the fear that Germany would develop and use a nuclear weapon, a number of scientists, led by Szilard, tried to inform the United States government of the need to counteract this threat. Albert Einstein—who had little interest in nuclear physics and had no knowledge of the possibility of a uranium chain reaction—was asked to use his prestige to convince the government to support uranium weapons research. In October 1939 a letter signed by Einstein was given to President Franklin D. Roosevelt. With Roosevelt's support a research program to develop an "atomic" bomb was begun.

The first significant breakthrough came in December 1942, when the first self-sustaining chain reaction was achieved under Fermi's leadership. This first nuclear reactor—which generated only a few hundred watts of power—was constructed out of small lumps of natural uranium embedded in graphite blocks. The carbon atoms in the graphite served to slow down the fission-emitted neutrons so that they would be captured by U^{235} instead of U^{238}. The chain reaction was maintained at a constant level by the use of ***control rods*** made of cadmium (a good absorber of neutrons). When fully inserted in the reactor, the control rods absorbed enough neutrons to prevent a chain reaction, but when the control rods were slowly removed, the reaction became self-sustaining. Fermi's reactor—which was constructed under the stands of the football field at the University of Chicago—provided much of the experimental data needed to construct a bomb and also served as the prototype for more advanced nuclear reactors designed to produce electric power.

In 1943 a laboratory was set up at Los Alamos, New Mexico, for the construction of the bomb itself. The scientific director of the new laboratory was J. Robert Oppenheimer, a highly respected physicist who had been a student of Max Born's at Göttingen. The scientists who worked at Los Alamos and the other installations of the "Manhattan Project" were subject to strict military control and operated under rigid rules of security and censorship. This was quite contrary to the traditions of science, but the specter of a German atomic bomb provided the only motivation necessary. Under Oppenheimer's supervision, the Los Alamos installation was organized, and the atomic bomb was designed and constructed in less than two years. On the morning of July 16, 1945, the first atomic bomb was exploded in the New Mexico desert.

In April 1945, when the war in Europe was coming to an end, several German scientists were located and taken into custody.* It was then that the scientists in the United States learned with great relief that there had never been a serious effort to produce a German atomic bomb. Many scientists strongly urged the government not to use the bomb on Japan. A committee of scientists, led by James Franck and Leo Szilard, urged that a demonstration of the bomb be given on a barren island, and that after the war international control of atomic energy should be established. After the successful test of the bomb in July, these efforts were intensified, but the decision was out of the scientists' hands. It was not certain whether the Japanese would surrender or an invasion of Japan would be necessary. The idea of a demonstration was rejected, apparently on the ground that there was no certainty that the bomb would work a second time and also because there was only enough material

*The mission to locate and capture the German scientists was led by none other than Samuel Goudsmit, the co-inventor of electron spin and an amateur criminologist. Goudsmit's description of his adventures appear in his book *Alsos*. (See the Supplementary Reading in Appendix E.)

available for two more bombs. The fact that two billion dollars had been spent in the development of the weapon was also taken into consideration. The final decision was made by President Harry S Truman. On August 6 the city of Hiroshima was leveled by a single uranium bomb. Three days later Nagasaki was devastated by a plutonium bomb. On August 14 Japan surrendered unconditionally.

When Otto Hahn (who had been interned in England with the other German scientists) learned of the use to which his discovery had been put, he became deeply depressed and considered committing suicide. Einstein was particularly anguished; many years later he said, "If I had known that the Germans would not succeed in constructing the atom bomb, I would never have lifted a finger."[2] Einstein, a lifetime pacifist, was deeply disturbed by the belief that in signing the letter to President Roosevelt, he had been "the one who pushed the button."

Today, several decades after the fact and blessed with the gift of hindsight, we may find it easy to take scientists to task for the development of the atomic bomb and its use on the Japanese cities. We must recognize, however, the moral dilemma faced by scientists in 1939, when they had to choose between the use of science to create a terrible weapon and the prospect of a Nazi holocaust. Certainly, the men like Hitler and his supporters, who forced this choice, and those like Roosevelt, Churchill, Truman, and their advisers, who made the ultimate decisions, must bear their share of the terrible burden.

NUCLEAR FUSION: THE HYDROGEN BOMB

After the war ended, many of the scientists who had contributed to the development of the atomic bomb devoted great energy to the task of bringing atomic weapons under international control. Tragically, their efforts did not succeed. The Soviet government was unwilling to cooperate and refused to let its scientists participate in any conferences with non-Communist scientists. At the time, the Russians were busily developing their own atomic bomb, which they exploded in 1949.

It had been known for several years that another type of nuclear weapon was theoretically possible. This new weapon was to be based on the energy released by the fusion of hydrogen nuclei to form helium. It had been calculated that the hydrogen bomb would be thousands of times more powerful than the atomic bomb.

The source of energy in a hydrogen bomb is the nuclear reaction between two isotopes of hydrogen—deuterium and tritium:

$$_1H^2 + {_1H^3} \longrightarrow {_2He^4} + {_0n^1} \qquad (Q = 17.6 \text{ MeV})$$

The energy released in this reaction is considerably less than the 200 MeV given off in uranium fission. However, there is a limit to the size of a uranium bomb but no limit to the size of a hydrogen bomb. A fission bomb depends on the multiplication of neutrons in a chain reaction. As a consequence a certain ***critical mass*** of uranium is needed for an explosion; below the critical mass the neutron leakage is too large to permit rapid neutron multiplication. In a uranium bomb an explosion is created by bringing together two smaller masses of uranium to form a critical mass. The smaller masses must, of course, each have masses less than the critical mass, and this restriction places an upper limit on the amount of uranium that can be placed in a bomb. The upper limit on the energy released by a fission bomb is about 25,000 tons of TNT.

A hydrogen bomb has no such restriction, since it does not depend so critically on the establishment of a chain reaction. The more hydrogen used, the larger the explosion. A temperature of more than 100 million °K is required to trigger the reaction (i.e., to overcome the repulsive electrostatic forces between the nuclei); in a hydrogen bomb this temperature is provided by exploding a uranium bomb, which surrounds the hydrogen core.

At the end of World War II few scientists were willing to work on the development of such a weapon. For many years the only advocate of the hydrogen bomb was Edward Teller, a gifted nuclear physicist who had studied with Heisenberg, Born, and Bohr. Like Szilard (who had argued for the development of the atomic bomb in 1939), Teller was born in Hungary. Both men had emigrated to Germany in the 1920s to escape anti-Semitism. With the rise of Hitler they left Germany and eventually came to the United States, where they both contributed to the atomic bomb project. In the postwar period, when Szilard was working for international controls and cooperation with the Russians, Teller took another tack. To Teller, the Stalinist government of Russia was no better than Hitler's Germany, and there was every reason to believe that Russian scientists would develop a hydrogen bomb. Yet Teller was no warmonger. According to historian Robert Jungk:

> Teller also took every opportunity to advance the cause of world government; it seemed to him to be the only hope of preserving peace. He used to say to other scientists: "It won't be until the bombs get so big that they can annihilate everything that people will really become terrified and begin to take a reasonable line in politics. Those who oppose the hydrogen bomb are behaving like ostriches if they think they are going to promote peace in that way."[3]

With the heightening of cold war tensions, the surprise of the Russian atomic bomb, and finally the discovery that a top nuclear physicist, Klaus Fuchs, had been passing secrets to the Russians, Teller's view came to prevail. In January 1950 President Truman authorized a crash program to develop the hydrogen bomb.

In November 1952 the first hydrogen bomb was exploded on an island in the Pacific. The bomb, which had a power of about 5 million tons of TNT, completely destroyed the island, leaving a hole 1 mile in diameter and 175 feet deep. Less than a year later the Russians exploded a hydrogen bomb that was considerably more advanced than its American counterpart. Nuclear bomb tests continued into the 1960s, culminating in the explosion of a 50-megaton bomb by the Russians in 1961. Great Britain, France, China and India have also exploded atomic or hydrogen bombs, and numerous other countries now have the capability to develop nuclear weapons.

As early as 1944 some scientists foresaw the inevitability of a nuclear arms race; their predictions have turned out to be frighteningly accurate. The problem of stopping the arms race is certainly one of the most difficult ever faced by mankind.

PEACEFUL USES OF NUCLEAR ENERGY

On the other side of the coin, we find the prospect that nuclear fission and/or fusion reactions can be used to provide virtually unlimited sources of cheap, clean, and efficiently produced energy. Unfortunately, these prospects have not as yet been fully realized. It is perhaps indicative of some perversity of man and/or nature that the full potentialities of the destructive aspects of nuclear energy were attained with great speed, whereas the development of the peaceful applications has not progressed as rapidly as had been expected, despite the best efforts of many scientists and engineers.

The large-scale generation of electric power by both nuclear fission and nuclear fusion is theoretically and technologically possible. Nuclear fission reactors are in a much more advanced state of development at the present time. In a nuclear reactor designed for electric power production, the heat generated by fission is used to produce the steam needed to turn an electric generator. In this way electricity is generated without burning coal or oil, and air pollution is avoided. However, there is a problem of the disposal of the radioactive waste material produced in reactors. In addition, there is the question of reactor safety. Although it is impossible for a nuclear reactor to explode with the force of a bomb, a reactor accident could conceivably release radioactive matter into the environment. Until convincingly safe reactors are designed, there is little likelihood that large nuclear reactors will be built close enough to large cities to produce electric power economically. These problems are not insurmountable, however, and the conversion from conventional (coal- and oil-burning) to nuclear-power generation of electricity is gradually taking place.*

A more long-range aspect of nuclear reactors is the matter of fuel consumption. The worldwide supply of uranium is rather limited and may not last for more than a hundred years. However, there is the possibility of using nuclear reactors to create more fuel. As we have already learned, the artificially produced isotope $_{94}Pu^{239}$ is fissionable by slow neutrons. This isotope, which has a half-life of 24,400 years, can be produced artificially in a nuclear reactor from slow-neutron capture by U^{238}. (The reactions leading to the production of Pu^{239} were described earlier in the chapter.) A "breeder" reactor containing U^{235} or Pu^{239} surrounded by a blanket of U^{238} could conceivably produce more fuel (Pu^{239}) than it uses. If this could be achieved on a large scale, the world's fuel problems could be greatly alleviated in the long run.

The prospects for controlled nuclear fusion are presently clouded by technological problems. In a nuclear fusion reactor, deuterium and tritium would be made to react at a controlled rate. The reaction products (helium and neutrons) would transfer their kinetic energy to a circulating coolant, and the heat would be used to generate electricity. A fusion reactor would be quite safe and would not release pollutants into the atmosphere. One advantage of great importance is that one of the fuels is in plentiful supply; deuterium can be extracted from the water in the oceans. Tritium, on the other hand, is radioactive (with a 12-year half-life) and is not found in nature. However, it can be produced as a by-product of the fusion process itself. If the reactor is surrounded with lithium to capture the emitted neutrons, tritium is produced by the reactions

$$_3Li^6 + {_0n^1} \longrightarrow {_2He^4} + {_1H^3}$$
$$_3Li^7 + {_0n^1} \longrightarrow {_2He^4} + {_1H^3} + {_0n^1}$$

In order for the deuterium–tritium reaction to occur, the reacting nuclei must have energies on the order of 10,000 eV—equivalent to a temperature of about 100 million °K. At this temperature the reactants will be completely ionized and will constitute a ***plasma***—an ionized gas of charged particles (nuclei and electrons). It is not too difficult to accelerate the plasma particles to the required temperature; the major problem is to maintain the temperature for a long enough period of time (about 0.1 sec). If the plasma particles strike the walls of the container, they will lose much of their energy, and the temperature will drop. Hence the plasma particles must be prevented from colliding with the walls. This is done by imposing a magnetic field

*As of January 1974 there were 42 nuclear power plants in operation in the United States, and about 160 were under construction or being planned. It is estimated that by 1980 one-fourth of the electricity generated in the United States will be provided by nuclear power.

strong enough to confine the plasma to an interior region away from the walls. The difficulty with this procedure is that instabilities occur in the plasma, and these instabilities distort the magnetic field and cause the particles to hit the walls and lose energy. At the present stage of development, this problem has not been fully overcome, and a sustained controlled fusion reaction has yet to be achieved. A procedure that would avoid the use of magnetic fields entirely has been receiving serious attention since 1970. The idea is to trigger fusion by igniting pellets containing a frozen deuterium–tritium mixture with pulses from a high-intensity laser. This very promising approach is now in the early stages of development.

In general, the problems involved in the development of an efficient fusion reactor are not considered to be insuperable; many experts feel that commercial fusion power will be available by the year 2000. One hopes that by that time the political tensions of the world, which have been unbearably heightened by the dangers of nuclear weapons, will have been relieved so that humanity can enjoy the benefits of nuclear energy.

QUESTIONS FOR REVIEW

1 State and explain the law of radioactive decay. Discuss the statistical nature of this law.

2 Explain what is meant by the half-life of a radioactive isotope. How is the half-life related to the decay rate?

3 How can we account for the existence of naturally occurring radioactive isotopes in the earth?

4 What is a decay chain? What will eventually happen to all the elements in a given decay chain?

5 What are the transuranic elements? How are they produced?

6 Discuss the basic principles underlying the use of a radioactive isotope as a tracer in a chemical or biological process.

7 What kind of evidence could be used to show that a radioactive isotope not now found in nature did exist at one time in the past?

8 What is the evidence that the nuclear reactions that formed the elements terminated several billion years ago?

9 Outline the uranium–lead method for obtaining the age of a rock. Explain how it is possible to distinguish between the lead produced by uranium decay and the lead initially present in the rock.

10 What is the evidence to support the theory that the bodies of the solar system all formed at the same time?

11 Explain why the ratio of the abundances of Pb^{207} and Pb^{206} changes with time in a uranium-bearing rock. Why is it possible to use this ratio to obtain the age of the solar system, rather than just the age of the rock?

12 Discuss the radioactive dating method based on C^{14}. For what kinds of objects (and what range of ages) is this method particularly suitable?

13 Explain how C^{14} is produced in the earth's atmosphere. What effects might cause the C^{14} content of the atmosphere to fluctuate? Discuss the techniques used to determine past variations in the C^{14} content.

14 Describe how the fission of uranium was discovered. Why does the identification of barium, lanthanum, and so on as elements produced by neutron bombardment of uranium indicate that fission has taken place?

15 Summarize the Bohr–Wheeler theory of uranium fission. What is meant by the activation energy for fission? How is the activation energy supplied when a neutron is captured?

16 Explain why U^{235} will undergo fission with slow neutrons, whereas with U^{238} fast neutrons are required to induce fission.

17 Explain why a chain reaction is possible in uranium fission. What must be done to achieve a self-sustaining chain reaction?

18 State the nuclear reaction used in the hydrogen bomb. Why is a hydrogen bomb a more powerful weapon than a uranium bomb?

19 Discuss the basic principles behind the breeder reactor. Write the reactions for the production of Pu^{239} from U^{238}.

20 Outline the major difficulties to be overcome in the achievement of controlled nuclear fusion. What are the main advantages of a fusion reactor as compared to a fission reactor?

EXERCISES

1 A 10-gram sample of the unstable isotope Ca^{47} ($T_{1/2} = 5$ days) is produced artificially. After 20 days how much of the original sample remains?

2 A 32-gram sample of a radioactive isotope is isolated from its surroundings. After 30 days only 1 gram of the original isotope remains.
(a) What is the half-life of this isotope?
(b) What happened to the other 31 grams of matter?

3 The element radium is radioactive. Its most stable isotope has a half-life of only 1622 years. How can we explain the fact that radium occurs naturally in the earth?

4 Natural uranium is 99.3 percent U^{238} and 0.7 percent U^{235}. Have these percentages increased, decreased, or remained constant during the past billion years? Explain your answer.

5 How can the existence of U^{235} ($T_{1/2} = 7.1 \times 10^8$ years) in the earth be used to establish an upper limit to the age of the earth? Make an estimate of this upper limit.

6 The naturally occurring isotope $_{49}In^{115}$ decays to $_{50}Sn^{115}$ with a half-life of 5×10^{14} years. Explain why a radioactive dating method based on In^{115} would not be especially useful.

7 At the present time the oldest known rocks found on the earth are 4 billion years old. Yet the age of the earth is estimated at 4.55 billion years. How do you account for this discrepancy?

8 Explain why the C^{14} activity has a virtually constant value (of 15 disintegrations/min/g of carbon) in all living organisms and begins to decrease only when the organism dies.

9 A 10-gram piece of wood taken from an ancient tomb is found to have a C^{14} activity of 37.5 disintegrations/min. When was the tomb constructed?

10 A small meteorite crater is found in a forest. Surrounding the crater are broken trees, which were knocked down by the impact. Explain how one could determine when the meteorite struck the earth.

11 Explain physically why the activation energy for fission is smallest for the largest nuclei (i.e., why it is easier to split a large nucleus than a small one).

12 Calculate the Q value for the reaction $n + Th^{232} \longrightarrow Th^{233}$. (The nuclear masses of Th^{232} and Th^{233} are 231.988714 and 232.992059 amu, respectively). Can Th^{232} undergo fission with slow neutrons if the activation energy is 6 MeV? If not, how much kinetic energy must the incoming neutron have in order to induce fission?

13 Explain why chain reactions do not occur in the natural uranium found in the earth.

14 Show that in uranium fission less than 0.1 percent of the mass of the uranium nucleus is converted to energy. (Hint: Compare the mass equivalent of 200 MeV to the mass of the uranium nucleus.)

15 What is the percent conversion of mass to energy in the hydrogen fusion reaction? Is the efficiency of this reaction greater or less than that of uranium fission?

16 Discuss whether or not you believe history has vindicated Teller's viewpoint on the construction of the hydrogen bomb.

17 Considering the advantages and disadvantages of nuclear power plants versus conventional coal- or oil-burning power plants, discuss whether or not you would object to a nuclear power plant being located in your community.

SUBNUCLEAR PARTICLES

20

In this century the effort of physicists to comprehend the fundamental building blocks of nature has been a story of bafflement followed by clarification followed by bafflement.

Victor F. Weisskopf

ONE OF THE MAJOR GOALS of physics throughout the centuries has been the identification of the indivisible, impenetrable particle or particles that represent the ultimate or fundamental constituents of all matter. For hundreds of years the atom was thought to be the fundamental particle, but this belief had to be abandoned at the end of the nineteenth century, when the divisibility of the atom was firmly established. The discovery of the nucleus in 1911 did not solve the problem, for it was immediately evident (in view of such phenomena as radioactive decay) that the nucleus was a composite structure constructed from more fundamental entities. The assumption that these fundamental entities were protons and electrons led to a number of inconsistencies, as we learned in Chapter 18.

The discovery of the neutron in 1932 provided what seemed to be a complete picture of the nature of matter. The proton, neutron, and electron were believed to be the fundamental entities out of which all matter was constructed (although it was not certain whether the neutron was an indivisible or a composite particle). One could also include in this list the photon, which could be viewed as the "particle of light" or simply as the carrier of a quantum of energy. However one chose to classify them, these four entities were all that was needed to account for the observed properties of matter as they were known in 1932. All atomic properties could be explained in terms of electrons and photons, and all nuclear properties could be understood in terms of protons and neutrons.

Unfortunately, this remarkably simple and economical picture of matter proved rather quickly to be inadequate. The 1930s and 1940s were marked by the discovery of an almost bewildering array of additional subnuclear particles. Some of these particles were discovered quite by accident; others were found through an interesting interplay of theory and experiment. Some of the new subnuclear particles fit into an available theoretical niche. That is, their existence was needed to explain some apparent theoretical inconsistency. However, a number of the newly discovered particles seemed to serve no function whatsoever.

In this chapter we shall learn about some of the important theoretical and experimental advances that led to the discovery of the additional subnuclear particles. First, the techniques that have been developed to detect and study the properties of these particles will be described.

20–1

THE DETECTION OF SUBNUCLEAR PARTICLES

Subnuclear particles appear as the products of very high energy nuclear reactions. Since these particles appear to come out of the nucleus, their dimensions must be extremely small—certainly no greater than 10^{-13} cm in diameter. Most of them are highly unstable and exist for the tiniest fraction of a second before they disintegrate into other particles. Under the circumstances one of the most remarkable aspects of fundamental particle physics is the fact that such particles can be detected in the first place! It is a tribute to the ingenuity of twentieth-century experimental physicists that there are several methods with which individual subnuclear particles can be detected and their basic properties precisely measured.

The earliest device capable of detecting a single high-energy particle was the ***Geiger counter,*** invented by Hans Geiger in 1908 when he was working in Rutherford's laboratory. A Geiger counter measures the ionization produced when a high-speed particle passes through a gas. A Geiger counter is an extremely valuable device for detecting sources of radiation and measuring the intensity and energy of the particles being emitted. However, it is of little use in determining particle properties such as mass, charge, spin, and lifetime. For these purposes physicists have turned to a different class of particle detector by which one can see and/or photograph the path or track of a particle as it passes through some medium.

The best-known device of this type is the ***cloud chamber,*** invented by C. T. R. Wilson of England. Wilson constructed the first cloud chamber in 1896 because he was interested in the formation of clouds in the atmosphere. At a given pressure and temperature there is an upper limit to the amount of water vapor (water in the gaseous form) that can be contained in a given volume of air. When this limit is reached, the air is said to be ***saturated*** (corresponding to 100 percent humidity), and the addition of any further water vapor will cause water droplets to condense. A cloud is an aggregate of water droplets, which form when the surrounding air becomes saturated with water vapor. The saturation limit depends on the temperature of the air; the lower the temperature, the smaller the amount of water vapor needed to saturate the air. Hence if there is a fixed amount of water vapor in the air, a lowering of the temperature will bring the air closer and closer to saturation. Eventually, a temperature (known as the ***dew point***) is reached at which the amount of water vapor in the air is equal to the saturation limit at that temperature. When the air temperature drops below the dew point, clouds can form in the atmosphere.

One other factor is necessary for cloud formation—small dust particles that act as "seeds" on which the water can condense. In studying the formation of clouds, Wilson found that if the air is completely free of dust particles or other impurities, the air temperature can be lowered several degrees *below* the dew point without condensation taking place. That is, water droplets do not form, even though the air is saturated, because they have nothing to condense on. Under these conditions the air is said to be ***supersaturated;*** that is, it contains *more* water vapor than the saturation limit at the given temperature.

Wilson demonstrated this effect with the cloud chamber, shown in Figure 20–1. The air in the chamber is first saturated with water. When the piston is pulled down rapidly, there is a sudden drop in pressure and temperature, and the air in the chamber

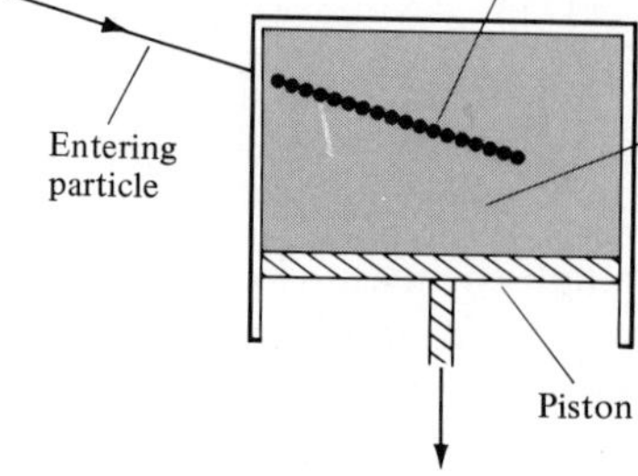

figure 20–1 **THE WILSON CLOUD CHAMBER**

When the piston is pulled down, the pressure drops and the air in the chamber becomes supersaturated. Ions produced by a charged particle passing through the chamber act as centers of condensation, and a track of water droplets is seen.

is cooled below its dew point. If the chamber has been swept free of any impurities, however, no condensation of water occurs.

In the course of his experiments, Wilson found that if ultraviolet radiation or x rays enter the chamber when the air is supersaturated, clouds begin to form. The explanation for this effect is that the entering radiation ionizes some of the air molecules in the chamber, and the ions act as centers of condensation for the water droplets. In 1911, fifteen years after he built the first cloud chamber, Wilson recognized the value of the cloud chamber as a research tool in atomic and nuclear physics. He modified the device so that individual particles entering the chamber would leave clearly visible tracks of water droplets (which condense on the ions formed by the particle). These tracks could be photographed before they dissipated; then the chamber could be swept clear and recycled to produce supersaturation again.

Photographs of cloud chamber tracks yield a great deal of information about subnuclear particles. The particle energy can be deduced from the length and thickness of the track. If the cloud chamber is situated in an electric or magnetic field, charged particles will be deflected along curved paths, and the charge and momentum of the particle can be deduced from the curvature of the path. With a cloud chamber one can observe not only single particles but also collisions between a high-speed particle and an individual nucleus of some absorbing matter placed inside the chamber. Thus single nuclear reactions can be examined. The energies, momenta, and so on of reactants and products can be carefully measured from the tracks they produce.

Throughout the brief history of nuclear physics, no research tool has played a more important role than the cloud chamber. Most of the discoveries that Rutherford and his associates made—including the discovery of the neutron*—were subsequently verified through analysis of cloud chamber photographs. The cloud chamber was also instrumental in the identification and the determination of the properties and interactions of many of the subnuclear particles to be studied in this chapter.

In 1952 Donald Glaser, a physicist at the California Institute of Technology, invented a device known as the ***bubble chamber,*** which is based on a principle

*The neutron, being uncharged, does not produce significant ionization and therefore leaves no track in a cloud chamber. It can, however, be identified by its effects—such as the recoil of protons or other nuclei when they are struck by neutrons.

similar to that of the cloud chamber but is considerably more efficient. In a cloud chamber a high-speed particle causes condensation droplets in a gas; in a bubble chamber a particle causes evaporation bubbles in a liquid. A bubble chamber contains a liquid maintained at a temperature very near its boiling point. The boiling point of a liquid depends on the external pressure; as the pressure decreases, the boiling point decreases. For example, if water is heated to a temperature of 200°F—12° below the boiling point at normal atmospheric pressure—and then the pressure is lowered, the waver will eventually begin to boil.

The boiling of a liquid begins with the formation of tiny gas bubbles in the liquid. In order to form, these bubbles need "triggering centers"—impurities in the liquid or rough points on the surface of the container will serve the purpose. Just as a gas free of impurities can be supersaturated, a pure liquid can be "superheated" —that is, maintained for a short time at a temperature above its boiling point without any boiling actually taking place. Glaser got the idea that the ionization produced by a high-speed particle passing through a superheated liquid might trigger bubble formation along its path, leaving a trail of bubbles that could be photographed.

Glaser's first experiments showed that ether (the well-known anesthetic) could be superheated by bringing it close to boiling and then suddenly dropping the pressure, and it could be maintained in the superheated state for several minutes, during which time numerous particle tracks could be seen. The first bubble chambers that were built were no more than a few inches in diameter. (The largest cloud chambers are about a foot in diameter.) In 1959 work was completed on a 6-foot-long bubble chamber at Berkeley. This device used liquid hydrogen, which must be maintained at a temperature of about 25°K (≈ -415°F). The largest bubble chamber now in operation is the 15-foot-long chamber at the National Accelerator Laboratory in Batavia, Illinois.

The major advantage of a bubble chamber over a cloud chamber has to do with the much greater density of a liquid as compared to a gas. As a consequence, a high-speed particle will collide more frequently and produce more ionization in a bubble chamber than in a cloud chamber. Thus in a bubble chamber there are many more tracks to be seen, and the tracks are more sharply defined. A bubble chamber can also be recycled more quickly than a cloud chamber. However, bubble chambers are quite expensive to build and operate.

Perhaps the most inexpensive way to track subnuclear particles is with photographic film. A high-speed charged particle passing through an undeveloped photographic emulsion will ionize the sensitive silver bromide grains, and when the film is developed, the grains are found to be blackened. Around 1947 special types of high-density emulsions were devised in which a high-speed particle could produce a well-defined track. Because of the high density of the emulsion, particle tracks are extremely short—usually no more than a few millimeters long—and the developed film must be examined under a microscope. From the track thickness and length it is possible to determine the mass, speed, and charge of the particle.

20–2

SOURCES OF SUBNUCLEAR PARTICLES

We turn now to the question of how subnuclear particles can be produced in sufficient quantities to be studied. As has been noted, these particles are produced in high-energy nuclear reactions. Although such reactions do not *commonly* occur in the environment of the earth, there are natural phenomena that lead to the creation of subnuclear particles on a small, but detectable, scale.

One such source is natural radioactivity, which provides alpha, beta, and gamma particles with energies up to 10 MeV. During the early years of nuclear physics, naturally radioactive materials represented the only available source of high-energy particles with which artificial nuclear reactions could be triggered. However, the 10-MeV limit on the energy of the available bombarding particles naturally restricted the amount of progress that could be made with this approach.

COSMIC RAYS

A second and considerably more fruitful natural source of high-energy particles is the cosmic radiation striking the earth's atmosphere. The extraterrestrial origin of this radiation was established early in the twentieth century. It had been known for some time that the air at the earth's surface is slightly ionized, and it had been assumed all along that the ionization was caused by the particles emitted by naturally radioactive elements in the earth. In 1912 Victor F. Hess of Austria, by flying an electroscope to high altitudes in a balloon, discovered that the amount of ionization in the air *increases* with altitude. Hess reasoned that the ionization must be caused by particles coming from outside the earth rather than from within. This theory was given further confirmation by Robert Millikan, who found in 1926 that the intensity of the ionization decreases as one goes below the earth's surface. On the mistaken assumption that the ionization is caused by high-energy gamma rays, Millikan coined the name *cosmic rays.*

Subsequent research has shown that the cosmic rays are not electromagnetic radiation, but mainly charged particles of extremely high energy. There are two groups of particles, the ***primaries*** and the ***secondaries.*** The primary radiation is of extraterrestrial origin and consists of about 90 percent protons, 9 percent alpha particles, and 1 percent heavier nuclei. The secondary particles (consisting of electrons, positrons, neutrons, gamma rays, etc.) are produced by the nuclear reactions that occur when the primary particles collide with the atoms and molecules of the earth's atmosphere. Most of these collisions take place at an altitude of about 60,000 feet ($\approx$12 miles) above the earth, where the cosmic-ray intensity reaches its maximum. Above this altitude the atmosphere is so rarefied that few collisions occur, and the primary particles predominate. Below 60,000 feet the cosmic radiation consists almost entirely of secondary particles; few primaries can penetrate to lower altitudes without colliding with an air molecule. Most of the secondary particles are absorbed in the atmosphere, but about one in a hundred reaches the earth's surface. A person standing on the earth's surface is struck by about one to five cosmic-ray particles each minute.

The energies of the primary particles are enormous. Their *average* energy is about 1 GeV (1 GeV = 1000 MeV = 10^9 eV).* Primaries with energies as high as 10^{11} GeV have been observed. Primary particles strike the earth's atmosphere at a rate of about 2 particles/sec per square inch (a total of about 10^{18} particles/sec over the whole atmosphere). It has been estimated that the total energy of the cosmic rays entering the atmosphere is about equal to the total energy of all the starlight reaching the earth.

The source of the primary cosmic radiation is unknown. The most commonly

*The G in GeV stands for the prefix "giga" (as in "gigantic"). This unit used to be known as the BeV, where B stands for billion. The change was made to avoid a confusion of definitions. In the United States a billion equals a thousand million (10^9), but in Europe a billion is a million million (10^{12}).

accepted theory (proposed by Enrico Fermi) is that these particles gain energy by being accelerated in interstellar magnetic fields for hundreds of millions of years. But it is not certain how the particles get to be injected into these magnetic fields. The orgin of cosmic rays is one of the many interesting unsolved problems in astrophysics.

From the nuclear physicist's point of view, cosmic rays represent a free and unending source of high-energy nuclear reactions. The energies of cosmic-ray primaries are many orders of magnitude larger than the energies that can be attained artificially with the largest particle accelerators. Even the secondary particles produced by the higher-energy primaries can have energies of several hundred GeV. Hence the nuclear reactions triggered by primary and secondary cosmic-ray particles are of great interest, especially since many of these reactions cannot be duplicated in the laboratory. In the 1930s and 1940s much nuclear particle research was conducted by taking a cloud chamber or photographic film package to the top of a mountain and studying the particle tracks produced by cosmic-ray particles. Later balloons, rockets, and even artificial satellites were used to take equipment high above the earth to study the primary particles and their interactions. Overall, the cosmic radiation has turned out to be the most important source of high-energy nuclear reaction data available to nuclear physicists.

PARTICLE ACCELERATORS

The study of cosmic rays and their interactions has its limitations. A physicist is constrained to work with whatever reactions happen to appear in the photographs; he does not have the option of performing a controlled experiment with a chosen target and a bombarding particle of a selected energy. Thus in cosmic-ray research luck plays a significant role. Great discoveries are made when an experimenter is fortunate enough to catch the collision of a high-energy primary with a nucleus on his photographic plate.

For more detailed studies of high-energy nuclear reactions, physicists have turned to particle accelerators. Virtually all particle accelerators use an electric field to supply energy to charged particles. The earliest accelerators created this electric field electrostatically by the separation of positive and negative charges to produce a large voltage difference. The first electrostatic accelerators invented by Van de Graaff in 1929 and Cockcroft and Walton in 1932 were capable of accelerating particles to energies of 1 to 2 MeV. The modern-day descendant of these devices is the ***linear accelerator*** (Figure 20–2), in which charged particles are accelerated in a straight line through a series of cylindrical electrodes. The voltages on these electrodes are alternated in such a way that as a particle passes through each electrode, it is acceler-

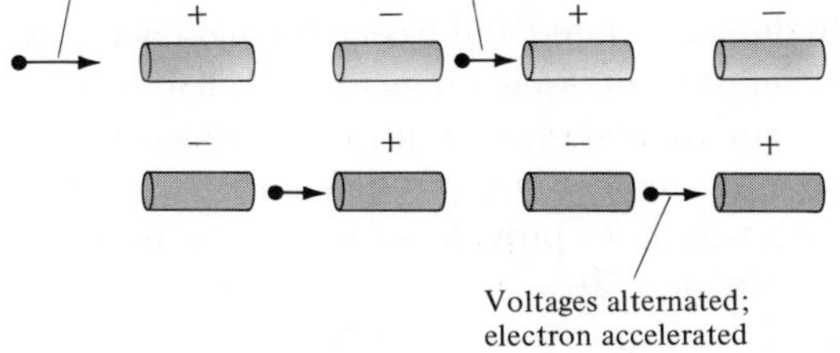

figure 20–2

THE LINEAR ACCELERATOR

Particles are accelerated by a series of hollow cylindrical electrodes arrayed in a straight line. The voltages on successive electrodes are alternated so that the particle is always accelerated to the next electrode.

ated toward the next one. The largest linear accelerator now in operation is at Stanford University in California. It is 2 miles long and can accelerate electrons to a maximum energy of about 20 GeV.

The main disadvantage of a linear accelerator is its length; not many laboratories have space available to build a 1- or 2-mile-long tube! Fortunately, a particle accelerator can be made into a much more compact instrument by using a magnetic field to bend the path of a particle into a curve. This is the principle of the cyclotron, which we studied in Chapter 10. In a cyclotron a particle is accelerated by an electric field; the magnetic field does not increase the particle energy but simply makes the particle travel in a spiral path.

The simple cyclotron, with a constant magnetic field and a uniformly alternating electric field, has a useful upper limit of about 20 MeV. At around this energy the relativistic mass increase of the accelerated particle becomes significant. In the cyclotron the electric field is alternated so that the particle is always accelerated when it crosses the gap from one side of the cyclotron to the other. As the mass of the particle increases, however, the particle begins to fall out of synchronization with the alternations of the electric field. To overcome this difficulty, the frequency of alternation of the electric field must be changed in relation to the relativistic mass increase. With this modification, the cyclotron becomes the ***synchrocyclotron*** (first constructed in 1947), with which particle energies of up to 700 MeV can be attained.

In a cyclotron or synchrocyclotron the accelerated particles spiral outward in a uniform field. Thus a large electromagnet must be constructed; in the first synchrocyclotron the magnet was over 15 feet in diameter. An electromagnet of this size is extremely heavy and very expensive to build and operate. In the next generation of particle accelerators, called ***synchrotrons,*** this problem was resolved. In a synchrotron (Figure 20–3) the particles do not spiral outward but are maintained on a roughly circular path of *constant radius*. This is done by increasing the magnetic field at a rate that corresponds to the increase in the speed of the particle. In other words, as the particle accelerates, the magnetic field is increased in strength so that the particle is bent onto the same path every time (instead of spiraling outward). For this procedure much smaller magnets are needed, and larger energies can be attained economically.

Particle energies well over 1 GeV are attainable with the synchrotron. Beginning in the 1950s, bigger and bigger synchrotrons were constructed in rapid succession. The large synchrotrons that have been constructed since 1952 are listed in chronological order in Table 20–1. The largest synchrotron presently in operation, at Batavia, Ill. (a suburb of Chicago), is 0.62 mile in diameter. It can produce a beam of approximately 10^{13} protons/sec with energies up to 400 GeV.

The construction of more powerful synchrotrons has taken on all the aspects of a contest between different countries. The construction and operating costs are enormous and can only be supported by government funding. (For example, the Batavia accelerator cost about 300 million dollars to build and has an annual operating budget of about 60 million.) Legislative support for such projects is probably more strongly motivated by national prestige than by a desire to advance knowledge in nuclear and subnuclear physics. Whatever the motivation, physicists in these areas are certainly thankful that governments are willing to provide such massive financial assistance for high-energy accelerator construction. It is a safe bet that experiments performed with the Batavia accelerator and the large accelerators being planned will

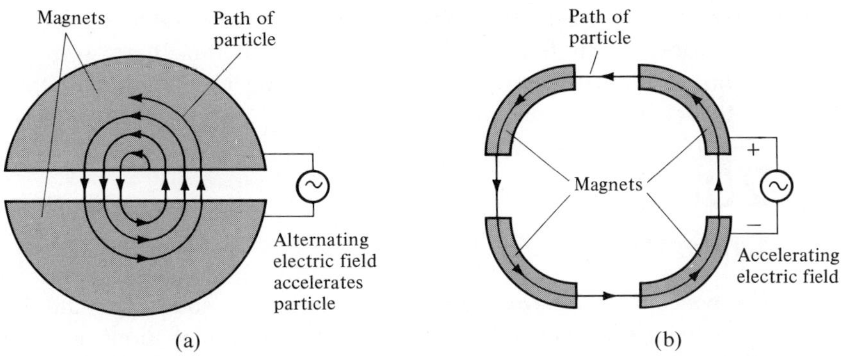

Cyclotron. Particles spiral outward in steady magnetic field.

Synchrotron. Particles travel on constant path. With each acceleration the magnetic field is increased so that particles are bent onto same path.

figure 20–3 **COMPARISON OF CYCLOTRON AND SYNCHROTRON**

Both devices use an electric field to accelerate the particles and a magnetic field to bend the particles onto a curved path. (a) In a cyclotron the magnetic field is constant and the particles spiral outward. (b) In a synchrotron the magnetic field is gradually increased so as to keep the particles on a constant path.

lead to discoveries of fundamental importance with regard to the internal structure of protons and neutrons and to the ultimate nature of all elementary particles.

This completes our survey of the sources and methods of detection of subnuclear particles. The circumstances under which the "fundamental" particles came to be discovered will now be described. Most of these particles were discovered unexpectedly in the debris from cosmic-ray or particle accelerator-induced nuclear reactions, but several of them were theoretically predicted to exist many years before they were

table 20–1 **SYNCHROTONS WITH $E > 1$ GEV IN OPERATION OR BEING PLANNED**

Location	Maximum particle energy (GeV)	Year completed
Brookhaven, N. Y. (U. S.)	3	1952
Berkeley, Calif. (U. S.)	6	1954
Dubna (USSR)	10	1957
Geneva, Switz. (CERN)[a]	28	1959
Brookhaven, N. Y. (U. S.)	33	1960
Serpukhov (USSR)	76	1968
Batavia, Ill. (U. S.)	400	1972
Geneva, Switz. (CERN)	400–1000	Being planned
USSR	1000?	Being planned

[a]CERN is an acronym (in French) for the European Center for Nuclear Research, which is supported by several western European nations.

actually observed. In some cases these predictions were made on the basis of seemingly bizarre theories, which few people accepted at the time they were proposed. The verifications of these theories represent interesting case studies of the relationship between theory and experiment in physics.

20–3

ANTIMATTER

The concept of ***antimatter***—a substance that completely annihilates matter whenever it comes in contact with matter—is the sort of thing one expects to find in a collection of science fantasy stories. Surprisingly, there is a firm theoretical and experimental basis for the existence of antimatter in the real world.

The theoretical basis for antimatter arises from Dirac's relativistic wave equation, which was discussed in Chapter 17. There we learned that the solution of the Dirac wave equation yields a spin quantum number with two values, $m_s = \frac{1}{2}$ or $-\frac{1}{2}$, thus accounting for the phenomenon of electron spin. Now we will examine another aspect of the solution of the Dirac wave equation. Let us consider the case of a free electron—one that is not attached to or otherwise affected by an atom. Such an electron possesses kinetic energy of motion and an energy $E_0 = m_0c^2$ associated with its rest mass, but no potential energy of any sort (since there are presumed to be no forces acting on it). In this case the solution of the Dirac equation yields, for the energy of the free electron, the relativistic formula (seen before as Equation 12–18):

$$E^2 = (m_0c^2)^2 + (pc)^2 \tag{20–1}$$

This equation gives the allowed energy states or levels of the free electron. Notice that no quantum numbers appear in this equation. Thus the momentum and energy of a free electron are *not* quantized; p can take on any value from zero up. Hence there is an infinite number of possible values of p, and corresponding to each of them is a different total energy E (obtained by substituting p into the energy equation). It follows that there is an *infinite* number of allowed energy levels for the free electron—in the language of quantum mechanics, the energy levels form a continuum. Corresponding to each energy level, there are two possible spins, $m_s = \frac{1}{2}$ or $-\frac{1}{2}$. Furthermore, the electron is presumed to obey the Pauli exclusion principle. Therefore each energy level is *filled* when it contains two electrons (with opposite spins).

Now let us consider the range of energy values that the free electron can have. Suppose we take the simplest case of an electron at rest. Then its momentum $p = 0$, and the energy equation reduces to

$$E^2 = (m_0c^2)^2$$

To solve this equation, we take the square root of both sides. One is tempted to write down simply $E = m_0c^2$ as the solution, but we must remember that this equation has *two* mathematical solutions:

$$E = m_0c^2 \qquad \text{and} \qquad E = -m_0c^2$$

The first solution, $E = m_0c^2$, is the relativistic rest energy of the electron and is exactly what we expect to find for the energy of a stationary free electron. The second solution, $E = -m_0c^2$, gives a *negative energy*. This seems nonsensical; the most reasonable thing to do is simply to reject the negative solution as a mathematically correct but physically meaningless solution.

If we now consider the more general case of a moving free electron, we find that the solution of Equation 20–1 gives two possible values:

$$E = +\sqrt{(m_0c^2)^2 + (pc)^2} \qquad \text{and} \qquad E = -\sqrt{(m_0c^2)^2 + (pc)^2}$$

We consider the negative solution to be physically meaningless. It can be shown algebraically that in the nonrelativistic limit of small velocities, the square root term has the approximate value

$$\sqrt{(m_0c^2)^2 + (pc)^2} \approx m_0c^2 + \tfrac{1}{2}m_0v^2$$

That is, the total energy is approximately equal (at low velocities) to the sum of the rest energy and nonrelativistic kinetic energy, as we would expect. Thus in the low-velocity limit the positive (physically meaningful) solution is

$$E = m_0c^2 + \tfrac{1}{2}m_0v^2 \tag{20–2a}$$

and the negative (rejected) solution becomes

$$E = -m_0c^2 - \tfrac{1}{2}m_0v^2 \tag{20–2b}$$

figure 20–4 **ALLOWED ENERGY LEVELS OF A FREE ELECTRON**

(a) Positive energy solution only: the lowest possible energy of a free electron is its rest energy, m_0c^2. (b) According to the Dirac theory, the free electron may have negative as well as positive energy levels.

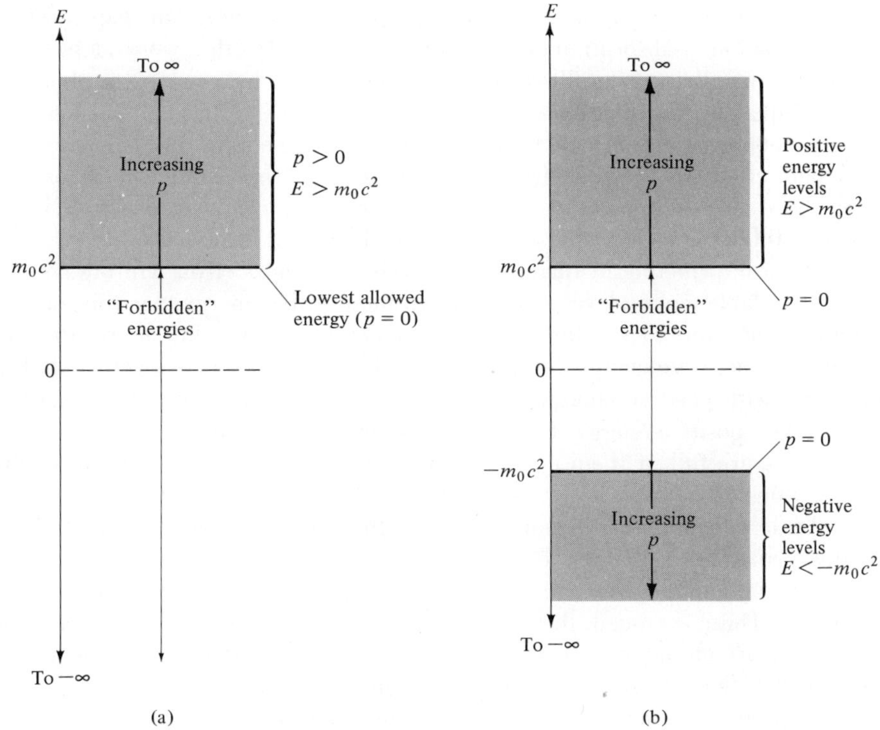

We see that the total energy in Equation 20–2b is equal to the sum of a *negative* rest energy and a *negative* kinetic energy—neither of which makes any physical sense. If the negative energy solutions are discarded, the free electron can have any energy greater than or equal to m_0c^2 (its rest energy), as shown in Figure 20–4a.

The rejection of the negative energy solutions of the Dirac equation seemed reasonable to virtually everyone—except Dirac. He pointed out that there was no theoretical reason why the negative energy solutions *had* to be rejected; he showed, moreover, that these solutions were in fact *required* to maintain the logical and mathematical consistency of the relativistic wave theory and to make certain predictions of the theory agree with observations. Dirac therefore proposed a theory of the electron in which the negative energy solutions are given a real, physical status.

THE DIRAC ELECTRON SEA

Dirac's electron theory begins with the assumption that the negative energy solutions are real, and that it is physically possible for an electron to have a negative rest energy and a negative kinetic energy. According to Equation 20–2b, as the speed of a negative-energy electron *increases,* the electron's energy *decreases* (becomes more negative). Does a negative kinetic energy imply a *negative* mass? If so, does a negative-energy electron exert or experience a gravitational *repulsion* and fall *up* in a gravitational field? Dirac did not attempt to answer these questions—and in fact to the present day they remain unanswered. However, the answers to these questions are not essential to the development of Dirac's theory, and we can ignore them.

The allowed energy states of the free electron, with the negative energies included, are shown in Figure 20–4b. The positive-energy electrons have a *minimum* energy of $+m_0c^2$; the negative-energy electrons have a *maximum* energy of $-m_0c^2$, so there is an energy gap of $2m_0c^2$ between the two regions. This gap can be bridged by the emission or absorption of a quantum of energy. In other words, a positive-energy electron can fall to an unfilled negative-energy level by *emitting* a photon with an energy equal to the difference in energy between the initial and final states, and a negative-energy electron can be raised to a positive-energy level by the *absorption* of a photon. It can be seen from the diagram that the emitted or absorbed photon must have an energy of *at least* $2m_0c^2$. For an electron ($m_0 = 9.11 \times 10^{-28}$ gram $= 0.000549$ amu), this energy is equal to 1.02 MeV.

A basic principle of quantum theory is that an electron will always fall to the lowest available energy level. Therefore all the positive-energy electrons ought to emit photons and drop to the unfilled negative-energy levels—but there are an infinite number of negative-energy levels! We should therefore not expect to find any free electrons with positive energies; they should all have fallen into the negative-energy levels. Yet positive-energy free electrons are quite common in nature. (Electric conductors are full of them.) Thus Dirac's theory would seem to be in conflict with observations.

Dirac's theory is saved, however, by the Pauli exclusion principle. According to this principle, each level can be occupied by only one electron (two, if spin is included), and an electron cannot fall to a lower energy level if that level is occupied. Therefore, Dirac reasoned, the appearance in nature of positive-energy electrons must mean that *all the negative-energy states are occupied.* In other words, there are no unfilled negative-energy levels for positive-energy electrons to fall into, so they must remain in positive-energy levels. A region of space that we consider to be a vacuum must actually be filled with an infinite number of negative-energy electrons; thus in Dirac's view empty space is a "sea of negative-energy electrons."

It follows that a negative-energy electron cannot jump from its own energy state to another negative-energy state, because all the other negative-energy states are filled. Thus a negative-energy electron cannot *lose* energy, and it cannot *gain* energy—unless it gains enough energy (more than $2m_0c^2$) to jump to a positive-energy state. Consequently, *it is impossible to detect a negative-energy electron,* since it cannot exchange energy with any detection device or with any other negative-energy electrons. (A negative-energy electron can exchange energy only by absorbing more than $2m_0c^2$—but then it is no longer a negative-energy electron.) A negative-energy electron cannot emit photons and cannot absorb photons unless they have energies greater than 1.02 MeV $= 2m_0c^2$. Hence the "sea" of filled negative-energy states is indistinguishable from a vacuum; it is completely unobservable by physical means because the negative-energy electrons cannot manifest their existence in any way. Even though the negative-electron sea theoretically has an infinite mass, an infinite negative charge, and an infinite negative energy, the impossibility of energy exchange among negative-energy electrons makes these properties undetectable.

Thus in Dirac's theory the negative-electron sea represents the normal state of affairs in empty space. The sea itself is not detectable; we can only hope to be able to observe *deviations* from the normal state of the sea. Specifically, when a negative-energy electron absorbs a quantum of energy sufficient to raise it to a positive-energy state, an unfilled negative-energy level will result. This unfilled level is, in effect, a "hole" in the negative-energy sea. Thus what we observe is the following: (1) the disappearance of a quantum of energy (greater than 1.02 MeV) and (2) the sudden appearance of a positive-energy electron coupled with (3) a "hole"—that is, an unfilled energy level—in the negative-electron sea:

$$\text{photon } (E \geq 1.02 \text{ MeV}) \longrightarrow \text{electron} + \text{hole}$$

According to Dirac, the hole should have all the outward characteristics of a particle. It can exchange energy with its surroundings—that is, other negative-energy electrons can fall into it, gaining or losing small amounts of energy as they jump from another negative-energy level into the hole. But when another negative-energy electron fills the hole, it must leave a hole in the level it came from—so the hole appears to move from place to place. Moreover, since the hole represents the *absence* of *negative* charge and mass, it gives the appearance of a *positive* charge and a *positive* mass. Thus the hole exhibits all the properties of an electron, except that it has a positive charge.

It seems ridiculous, at first, to think that a hole could be confused with a particle. We must recognize, however, that we do not see the hole itself; we observe its *effects*—that is, its interactions with its surroundings. Since the interactions of the hole are identical to those of a positively charged electron, *the hole will be totally indistinguishable from a positively charged electron.* In this respect the hole becomes a visible, tangible entity. A good analogy can be drawn with a bubble moving through a liquid: to an underwater observer, a bubble has the appearance of a solid object. We do not generally think of a bubble in terms of what it really is—the absence of water! In the same way the holes, or "bubbles," in the negative-electron sea take on the appearance of tangible positively charged electrons. The positively charged electron, or *positron,* can therefore be considered as a real particle; in view of its figure 20–5 **PRODUCTION AND ANNIHILATION OF A POSITRON–ELECTRON PAIR**

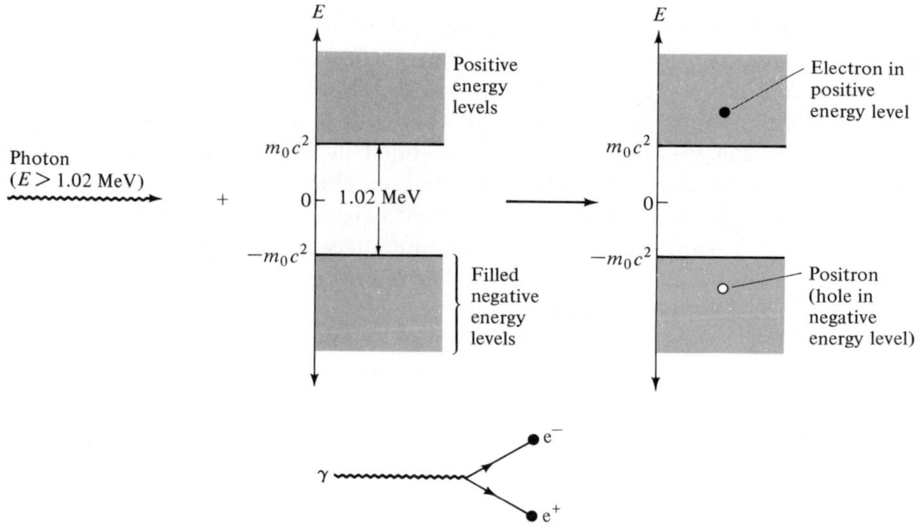

(a) Pair Production

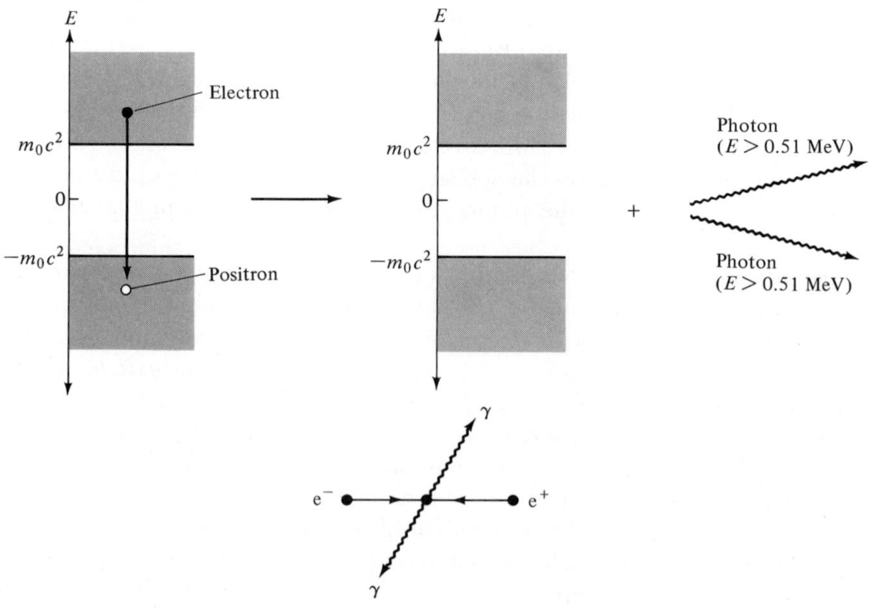

(b) Annihilation

relationship to the electron, it will be represented by the symbol e^+. (The role of the positron in beta decay was described in Chapter 18.)

Thus the process in which a photon is absorbed by a negative-energy electron to form a hole plus a positive-energy electron (see Figure 20–5a) can be described as follows:

$$\gamma \longrightarrow e^- + e^+ \qquad (E_\gamma \geq 1.02 \text{ MeV})$$

This process is known as ***pair production,*** because an electron–positron pair is created. (Note that charge is conserved in the reaction.)

ANNIHILATION: THE CONCEPT OF ANTIMATTER

Once a hole—or positron—is created, it will remain in existence until a positive-energy electron comes along. The hole represents an unfilled lower energy state, so a passing electron can fall into the hole, losing at least 1.02 MeV in the process. With the filling of the hole, the positron disappears, and since the electron is now a part of the filled negative-electron sea, it also disappears. Consequently, in the process both an electron and a positron disappear, and energy appears in the form of electromagnetic radiation. In order for both mass–energy and momentum to be conserved, the energy must be divided between *two* photons of 0.51 MeV each:

$$e^- + e^+ \longrightarrow \gamma + \gamma \qquad (E_\gamma \geq 0.51 \text{ MeV})$$

This is the process of ***annihilation.*** Two particles disappear, and two photons are created; the total energy of the two photons is equivalent to the mass of the two particles (Figure 20–5b). Thus the annihilation process represents the complete conversion of mass to energy.

From the annihilation process arises the concept of antimatter. That which annihilates matter is called antimatter; a particle is annihilated by its ***antiparticle.*** Since an electron and positron annihilate each other to create energy, it follows that the positron is the antiparticle of the electron. In other words, *the positron is the antielectron.* According to Dirac's theory, an antiparticle is a hole—the absence of a negative-energy particle—in the sea of negative-energy states.

THE DISCOVERY OF THE POSITRON

The Dirac theory of the electron certainly must rank among the most bizarre theories ever proposed by a serious physicist. When Dirac announced his theory in 1930, it was greeted with skepticism, if not outright derision. In the original form of his theory, Dirac proposed that the *proton* is the antiparticle of the electron. In this way Dirac thought that the mystery of the exact equality (except for sign) of the proton and electron charge could be explained. (The much larger mass of the proton was attributed to interactions within the negative-energy sea.) But then protons and electrons should annihilate one another! Oppenheimer calculated that ordinary matter ought to annihilate itself within 10^{-10} sec. Wolfgang Pauli (whose sarcasm was legendary among physicists) remarked that if Dirac's theory were correct, Dirac himself should have been annihilated into high-energy photons before he had the time to publish his theory.

In the face of such devastating criticism, Dirac had to retract:

> It thus appears that we must abandon the identification of holes with protons and must find some other interpretation for them. . . . A hole, if there were one, would

> be a new kind of particle, unknown to experimental physics, having the same mass and opposite charge to an electron. We may call such a particle an antielectron. . . . The protons on the above view are quite unconnected with electrons. Presumably the protons will have their own negative-energy states, all of which normally are occupied, an unoccupied one appearing as an antiproton.[1]

Dirac announced this revision of his theory in 1931. In August 1932 the American physicist Carl D. Anderson, who was studying cosmic rays under the direction of Robert Millikan, discovered the track of a positron in a cloud chamber photograph. From the thickness of the positron's cloud chamber track and its deflection in a magnetic field, Anderson was able to show that the mass of the positron is equal to the mass of the electron, and that positrons and electrons have equal but opposite charges. The properties of the positron were exactly as predicted by the Dirac theory. However, the relationship between Dirac's theory and Anderson's discovery is not as direct as it might appear, as Anderson himself has explained:

> It has often been stated in the literature that the discovery of the positron was a consequence of its theoretical prediction by Dirac, but this is not true. The discovery of the positron was wholly accidental. Despite the fact that Dirac's relativistic theory of the electron was an adequate theory of the positron, and despite the fact that the existence of this theory was well known to nearly all physicists, it played no part whatsoever in the discovery of the positron.[2]

Anderson was unaware of Dirac's theory and did not learn about it until 1933. What is even more remarkable is that several observers as early as 1927 saw positron tracks in cloud chambers, but since they could not bring themselves to accept the existence of such a particle, they offered alternative interpretations. As we learned

figure 20–6 **DEFLECTION OF CHARGED PARTICLES IN A MAGNETIC FIELD**

Positive and negative charges initially moving in the same direction are deflected in opposite directions by a magnetic field.

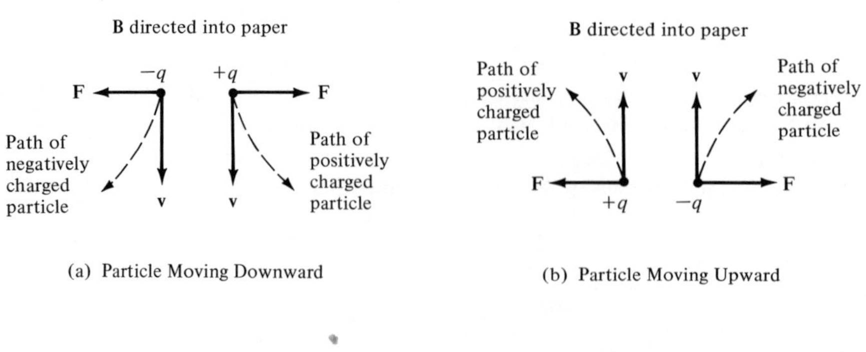

in Chapter 10, when a charged particle moves through a magnetic field, it experiences a force ($F = qvB \sin \theta$) perpendicular to its motion, which deflects the particle sideways. The direction of the deflecting force depends not only on the direction of the particle's motion, but also on the sign of its charge. (A negative q in the force equation changes the sign and hence the direction of the force.) As shown in Figure 20–6, a particle moving downward (in the diagram the magnetic field **B** points into the paper) is deflected to the *right* if it has a positive charge and to the *left* if it has a negative charge. A positively charged particle moving *upward* in the same magnetic field is deflected to the *left*, and a negatively charged particle moving upward is deflected to the *right*. Consequently, a positron moving in one direction will have the same curvature in a magnetic field as an electron of the same energy moving in the opposite direction.

Figure 20–7a illustrates a path made by a cosmic-ray particle passing through a cloud chamber. From the known magnetic field strength and the track curvature

figure 20–7 **THE DISCOVERY OF THE POSITRON**

The particle track shown in (a) could have been made by a positron traveling downward or by an electron traveling upward. (b) By inserting a lead plate in the cloud chamber, Anderson showed that the particle was a positron moving downward. (c) Pair production. A photon decays into a positron and electron, which have equal and opposite curvatures in the magnetic field.

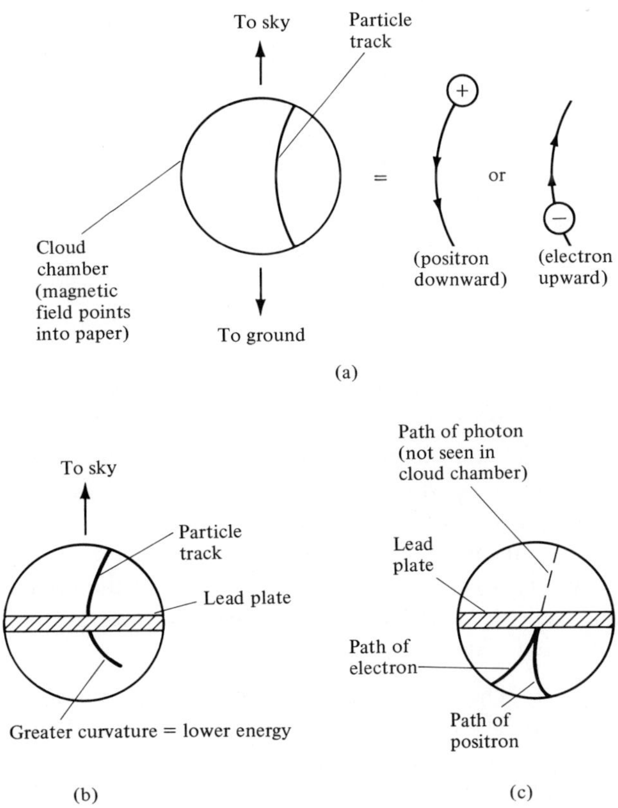

and thickness, one can show that the particle has a mass and charge equal in magnitude to that of an electron. However, there is no way to determine the direction of motion from the picture alone, so it is not known for certain whether the particle was moving downward or upward when it made the track. As a presumed high-energy cosmic-ray secondary, the particle ought to be moving downward from the sky—but an electron moving downward should be deflected to the left! The particle must be a cosmic-ray positron (which would be deflected to the right). A number of observers rejected this interpretation, however; instead, they referred to the particles as "electrons coming up from the floor"! Tracks of positrons emitted in beta decay were also observed before 1930, but they were identified as "electrons moving backward" or "electrons falling back into the source."

Anderson—even though he had no knowledge of Dirac's theory—took a more careful approach. He decided to determine experimentally the direction of the particle's motion. He did this by inserting a lead plate inside the cloud chamber to slow up the particle. As shown in Figure 20–7b, the particle has a greater curvature—which means a smaller speed—*below* the plate; obviously, it must have come from above and lost energy in passing through the plate. Anderson noted that the only other possibility is that the particle was an electron that entered the plate from below with an energy of 20 MeV and left with a much greater energy of 60 MeV. Anderson rejected this as totally unreasonable. Therefore, he concluded, the particle must be a positron.

Why were the earlier observers of positron tracks so reluctant to accept the consequences of their own evidence? Anderson has suggested the following explanation:

> The Dirac theory . . . carried with it so many novel and seemingly unphysical ideas, such as negative mass, negative energy, infinite charge density, etc. Its highly esoteric character was not in tune with most of the scientific thinking of that day. Furthermore, positive electrons were not needed to explain any other observations. Clearly the proton was the fundamental unit of positive charge, and the electron the corresponding unit of negative charge. This kind of thinking prevented most experimenters from accepting the Dirac theory wholeheartedly.[3]

Shortly after Anderson's discovery, photographs of pair production (see Figure 20–7c) were obtained by Blackett and Occhialini in England. They later found that pair production will occur only with gamma rays of energy greater than 1.02 MeV, as predicted by Dirac's theory.

MATTER AND ANTIMATTER

After the discovery of the positron, Dirac's theory had to be taken seriously. All of the properties that Dirac predicted for the positron have since been confirmed, and the concept of matter and antimatter has come to be an accepted part of physical theory. According to the theory, any particle that obeys the Pauli exclusion principle should have an antiparticle. (This excludes the photon; there is no antiphoton distinguishable from the photon.) A particle and antiparticle should be related in the following ways:

(1) The particle and antiparticle have the same mass and intrinsic spin.
(2) The electric charges of the particle and antiparticle are equal in magnitude, but of opposite sign.
(3) The intrinsic magnetisms of the particle and antiparticle are equal in

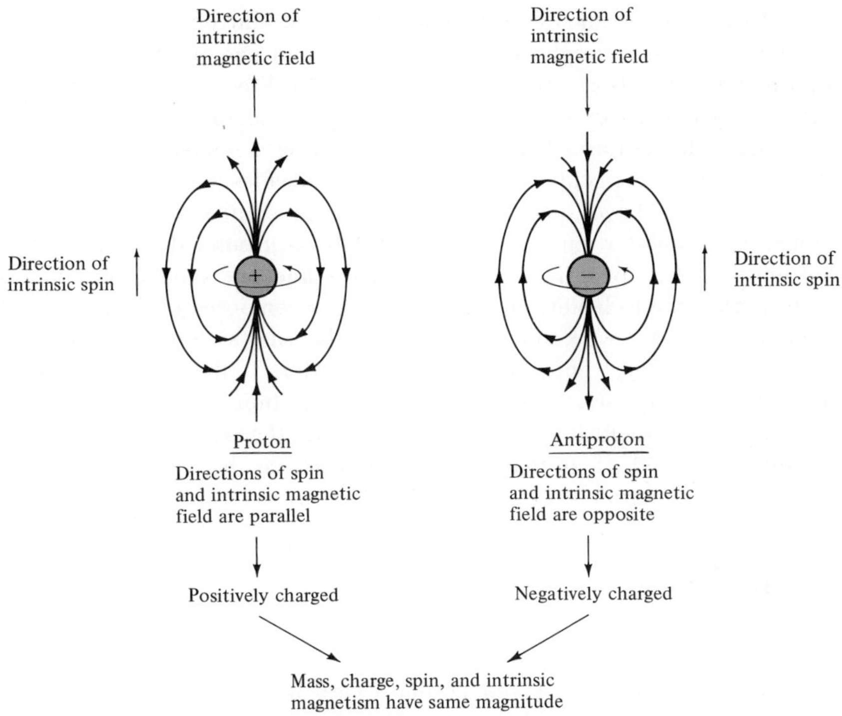

figure 20–8 **COMPARISON OF PROPERTIES OF PROTON AND ANTIPROTON**

strength, but opposite in direction. The intrinsic magnetic field is aligned either parallel or opposite to the spin direction. (See Figure 20–8.)

(4) Particle and antiparticle are formed in pair production from a quantum of energy equal to or greater than $2m_0c^2$, where m_0 is the rest mass of the particle (or antiparticle).
(5) Particle and antiparticle annihilate each other in a direct collision, releasing $2m_0c^2$ of energy in the form of radiation or other particles.
(6) If a particle is stable, so is its antiparticle; if the particle is unstable, the particle and antiparticle will have the same half-life.

The next antiparticle to be looked for after the positron was the antiproton. By the 1950s no antiprotons had turned up in cosmic-ray photographs, so physicists decided to try to produce them with a particle accelerator. A large accelerator was needed, because the energy required for production of a proton–antiproton pair would be greater than 1.88 GeV. The 6-GeV beam of the Berkeley synchrotron was used to trigger nuclear reactions of sufficiently high energy to produce antiprotons, and in 1955 the antiproton was detected. In 1956 the existence of antineutrons was established.* The antineutron, like the neutron, has no charge and leaves no tracks in a cloud or bubble chamber. Its existence can be demonstrated only by observing the products of its annihilation.

*The neutron and antineutron differ in the orientation of their intrinsic magnetism. The direction of the neutron's magnetic field is opposite to the direction of its intrinsic spin; in the antineutron the two directions are parallel.

Antiatoms are theoretically possible and would be as stable as their ordinary counterparts. For example, an antihydrogen atom has an antiproton nucleus and an orbiting positron. Its spectrum is virtually identical to that of a normal hydrogen atom (although subtle differences are detectable in a magnetic field). In fact, any stable antiatom could be formed with positrons orbiting a nucleus of antiprotons and antineutrons. Antiatoms could combine to form molecules and even living things. In fact, many physicists have speculated about the possibility of "antiworlds" composed entirely of antimatter. Such an idea is not entirely grounded in fantasy. Since particles and antiparticles are always created in pairs and are symmetrical in every way, it seems reasonable to believe that the universe is composed of equal amounts of matter and antimatter. Yet we know that the earth, the solar system, and probably our galaxy are composed almost entirely of ordinary matter. If they were composed of matter and antimatter equally, a lot of annihilation reactions would be observed—but they are not! Where, then, is the antimatter located? It may well be that there are entire galaxies composed of antimatter—or even, as has been suggested, a whole other universe of antimatter separated from our own.

20–4

THE NEUTRINO

We will now take up our second "case history" of the relationship between theory and experiment in elementary particle physics. Once again we will deal with a particle—called the ***neutrino***—whose existence was predicted theoretically before it was actually observed in nature.

The theoretical prediction of the existence of the neutrino came about because of a serious discrepancy involving the phenomenon of beta decay. As we learned in Chapter 18, beta decay is an internal nuclear process by which a neutron changes to a proton and electron (or a proton changes to a neutron and positron) and the electron (or positron) is expelled with considerable energy from the nucleus. We expect mass–energy to be conserved in these reactions; any mass defect between the initial and final states should appear as the sum of the kinetic energies of the ejected electron (or positron) and the residual nucleus. For example, in the beta decay of P^{32}

$$_{15}P^{32} \longrightarrow {}_{16}S^{32} + {}_{-1}e^{0}$$

one can calculate from the known nuclear masses that the reaction is exoergic by 1.71 MeV. Hence we expect to find that the kinetic energy of the electron is very nearly 1.71 MeV. (The kinetic energy of the recoiling S^{32} nucleus should be very small because of its much greater mass as compared to the electron.) Yet experiments show that the kinetic energy of the electron is always noticeably *less* than 1.71 MeV—on the average, it is about 0.7 MeV. As shown in Figure 20–9, the electrons have a continuous spectrum of energies, all less than 1.71 MeV. For mass–energy conservation, *all* electrons should have a kinetic energy of 1.71 MeV. Therefore it appears that there is an unaccountable loss of mass–energy in the reaction; in other words, *the principle of conservation of mass–energy seems to be violated.* This is found to be the case in all instances of β^- and β^+ emission.

Further analysis shows that conservation of linear momentum also appears to be violated in these reactions. In addition, nuclear angular momentum (intrinsic spin) is not conserved. The emitted electron or positron has a spin of $\frac{1}{2}$. If spin is to be

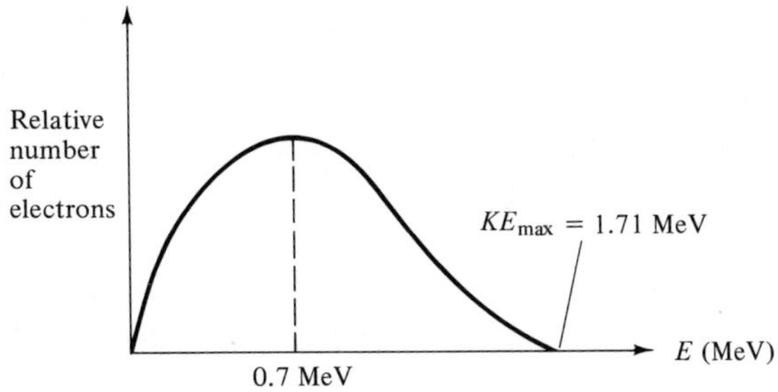

figure 20–9 **ELECTRON ENERGY SPECTRUM FROM BETA DECAY OF P^{32}**

If mass–energy is conserved, every emitted electron should have a kinetic energy of 1.71 MeV. Observations show that there is a continuous distribution of electron energies, all less than 1.71 MeV. In order to preserve conservation of mass–energy, the existence of the neutrino was postulated.

conserved, the spin of the nucleus must change in beta decay; specifically, the spin of the residual nucleus must differ from that of the original nucleus by a half integer. But in beta decay the mass number of the nucleus remains the same, so that it is not possible for the nuclear spin to change by a half integer. For example, both P^{32} and S^{32} contain an even number of nucleons, and their respective spins are 1 and 0. Thus in the reaction

$$\begin{array}{lcccccc} & {}_{15}P^{32} & \longrightarrow & {}_{16}S^{32} & + & {}_{-1}e^{0} \\ \text{spin:} & 1 & \neq & 0 & + & \frac{1}{2} \end{array}$$

spin is evidently not conserved.

THE NEUTRINO HYPOTHESIS

By 1930 a considerable amount of experimental evidence had been amassed that indicated that several conservation principles were not upheld in beta decay. Physicists at the time were not especially shocked by this state of affairs, considering the violence that had already been done to so many classical principles since the beginning of the century. So most physicists were willing to accept the fact that mass–energy and momentum were not conserved in beta decay.

This view was opposed by Wolfgang Pauli. He proposed instead that all of the conservation principles could be preserved if it was assumed that in beta decay the emitted electron is always accompanied by a second particle that carries away the missing energy, momentum, and spin. The new particle would have no charge, a negligible mass, and a spin of $\frac{1}{2}$—just the properties needed to maintain all the conservation principles.

Pauli's proposal was made informally around 1930. In 1934, after the discovery of the positron and the vindication of Dirac's theory of the electron, Enrico Fermi developed Pauli's hypothesis in considerable detail. Fermi proposed that the new particle be called the *neutrino* (Italian for "little neutron") and be represented by the symbol ν (the Greek letter nu). The neutrino must have an antiparticle, the antineutrino ($\bar{\nu}$). Originally, it was not certain whether there was any difference between the neutrino and the antineutrino. In 1957 it was shown that a distinction could be made between them, as we shall learn later. In the present-day form of the Fermi theory of beta decay, it is assumed that the emitted electron is always accompanied by an antineutrino, and the emission of a positron is always coupled to the emission of a neutrino. Thus in the beta-decay process the mass–energy and momentum of the interaction are *shared* between the product nucleus and the two members of the e^-–$\bar{\nu}$ or e^+–ν pair. Consequently, the continuous spectrum of electron (or positron) kinetic energies can be accounted for. In each reaction, part of the kinetic energy is carried off by the neutrino (or antineutrino), so that the *combined* mass–energy of the electron plus antineutrino (or positron plus neutrino) and the product nucleus is always equal to the maximum available mass–energy.

In order for spin to be conserved in a nuclear reaction, the total spin must remain an integer or half integer on both sides of the reaction. By assigning the neutrino (and antineutrino) a spin of $\frac{1}{2}$, spin conservation is assured. The total spin of the e^-–$\bar{\nu}$ (or e^+–ν) pair is either 0 or 1, so that a change in the total spin from an integer to a half integer (or vice versa) is impossible. Thus in the decay of P^{32}

$$\begin{array}{lccccccc} & {}_{15}P^{32} & \longrightarrow & {}_{16}S^{32} & + & {}_{-1}e^{0} & + & \bar{\nu} \\ \text{spin:} & 1 & = & 0 & + & \frac{1}{2} & + & \frac{1}{2} \end{array}$$

the total spin stays an integer on both sides of the reaction.

According to the neutrino hypothesis, the basic reactions of beta decay should be written as follows:

$$\begin{array}{ll} \textit{electron } (\beta^-) \textit{ emission:} & n \longrightarrow p + e^- + \bar{\nu} \\ \textit{positron } (\beta^+) \textit{ emission:} & p \longrightarrow n + e^+ + \nu \end{array}$$

The first of these reactions (electron emission) is an *exoergic* reaction ($Q = 0.78$ MeV) and therefore should occur spontaneously outside the nucleus. In fact, a free neutron (outside the nucleus) is unstable and decays with a 15-min half-life. On the other hand, proton decay (by positron emission) is *endoergic* ($Q = -1.80$ MeV). Hence a free proton is stable, but inside the nucleus the energy needed for this reaction can be obtained from other nucleons. The transformation of a proton to a neutron in the nucleus can also proceed via the reaction

$$p + e^- \longrightarrow n + \nu$$

This process is known as ***electron capture,*** since it involves the capture of an electron by the nucleus.

Fermi worked out an elaborate theory of beta decay, with which he was able to account for the observed distribution of energies in the continuous electron spectrum. In Fermi's theory the energy distribution is theoretically dependent on the mass of the neutrino, and by comparing the theoretical and experimental distributions, it could be shown that the neutrino mass is immeasurably small (certainly no greater than 1/5000 of the electron mass). If the mass of a neutrino—like that of a photon—is truly zero, then the neutrino travels at the speed of light. Although massless, the neutrino has both energy and momentum. From the relativistic equation (20–1) it follows that the momentum of a neutrino, like that of a photon, is described by the equation $p = E/c$.

Fermi was also able to use his theory to relate the half-life of a beta-unstable nucleus to the decay energy. (In general, the greater the energy available, the shorter the half-life.) The relatively long half-lives of beta emitters indicated to Fermi that the force involved in beta decay is much weaker than either the electromagnetic or the nuclear force. For example, the neutron, which decays exoergically with an energy of 0.78 MeV, has a half-life of 15 min. For the same decay energy, orbital electron jumps (governed by the electromagnetic force) and gamma-ray emission (governed by the nuclear force) would take place in about 10^{-10} to 10^{-12} sec. Fermi therefore postulated that *the beta decay interaction must depend on a different fundamental force,* which he called the ***weak interaction.*** The strength of the weak interaction ranks between the gravitational and electromagnetic forces. The relative strengths of the four fundamental forces are approximately as follows:

Force	Relative strength
nuclear	10^{38}
electromagnetic	10^{36}
weak	10^{25}
gravitational	1

That is, the weak interaction is 10^{25} times stronger than the gravitational force, but 10^{13} times weaker than the nuclear force. Like the nuclear force, the exact nature of the weak interaction is not completely known at the present time.

DETECTING THE NEUTRINO

According to the preceding table, the strength of the weak interaction is 10^{11} (100 billion) times weaker than the electromagnetic force. Consequently, the likelihood of a neutrino being attracted to and captured by a neutron or proton is extremely small. It can be calculated from the Fermi theory that a neutrino has a 50–50 chance of passing through a slab of lead *one light year thick* without being captured! A neutrino could easily pass through the earth without interacting with a single nucleon; in fact, only one neutrino out of 10 billion would be stopped by a body the size of the earth. Neutrinos and antineutrinos are produced in great numbers by nuclear reactions in the interiors of the sun and other stars, and are also produced in cosmic-ray interactions in the atmosphere. Thus the surface of the earth receives a rather high intensity neutrino bombardment. As a conservative estimate, a person on the earth will have about 10^{23} neutrinos pass through his body in his lifetime—out of which perhaps one or two will be absorbed.

Under the circumstances it seemed quite unlikly in 1934 that a successful experiment could ever be set up to demonstrate the existence of the neutrino. Fermi's theory, which was firmly based on the assumption that neutrinos exist, was remarkably successful in accounting for the details of beta decay, but this in itself was only indirect and not completely convincing proof that neutrinos really exist and have the properties predicted for them.

In 1956 two American physicists, Clyde Cowan and Frederick Reines, succeeded in providing experimental evidence for the existence of the antineutrino. The antineutrino obviously has an extremely low probability of interacting with matter, but Cowan and Reines hoped that by placing a large, sensitive detection device close to an extremely intense antineutrino source, a few antineutrino captures might be observed. The reaction that they aimed to observe was

$$\bar{\nu} + \mathrm{p} \longrightarrow \mathrm{n} + \mathrm{e}^+$$

For an antineutrino source Cowan and Reines used the large nuclear reactor at Savannah River, South Carolina. In nuclear fission the fission product nuclei tend to have an excess of neutrons, and a neutron-rich unstable isotope will decay by electron (β^-) emission to a more stable nucleus. These ejected electrons should always be accompanied by an antineutrino. Thus a nuclear reactor should emit antineutrinos (but not neutrinos) in great numbers. It was estimated that the Savannah River reactor gave off about 10^{18} antineutrinos/sec.

To capture antineutrinos, Cowan and Reines used a large tank of water (the protons in the hydrogen atoms served as the capturing nuclei) surrounded by suitable detecting and counting devices. Detection of a $\bar{\nu}$–p interaction was based on the observation of the simultaneous appearance of the neutron and positron produced by the reaction. The positron will be annihilated by an electron, creating two 0.51-MeV gamma-ray photons, and the neutron will be captured by a nearby nucleus. (Cadmium—a good neutron absorber—was added to the water for this purpose.) Thus the observation of the characteristic radiations given off almost simultaneously (within 10^{-6} sec) by neutron capture and positron annihilation constitutes "observation" of antineutrino capture. Cowan and Reines very carefully checked against the possible production of a neutron and positron by any process other than antineutrino capture. With the experiment in full operation, Cowan and Reines were able to detect, on the average, about three antineutrino captures per hour—which was within 5 percent of the predicted figure. The Cowan–Reines experiment thus provided strong evidence for the existence of the antineutrino and for the validity of Fermi's theory of the weak interaction.

The detection of the neutrino itself proved to be a bit more difficult. There are no intense artificial sources of neutrinos. However, the sun serves as a readily available neutrino source. The basic reaction for neutrino capture is

$$\nu + \mathrm{n} \longrightarrow \mathrm{p} + \mathrm{e}^-$$

It is quite difficult to observe this reaction with free neutrons, so the neutrons inside a nucleus must be used. A highly suitable reaction of this type is

$$\nu + {}_{17}\mathrm{Cl}^{37} \longrightarrow {}_{18}\mathrm{Ar}^{37} + \mathrm{e}^-$$

The Ar^{37} produced by the reaction is radioactive and can be detected relatively easily. This reaction was observed in 1964 by the American physicist Raymond Davis. Using a tank containing several thousand gallons of cleaning fluid (C_2Cl_4) as a detector, Davis observed an average of one neutrino capture every two days. Thus both the neutrino and antineutrino have been observed and found to have the properties predicted for them by Pauli and Fermi.

THE LEFT-HANDED NEUTRINO

The modern form of the Fermi theory of beta decay assumes that the neutrino and antineutrino are different from one another. Yet the neutrino has neither charge nor mass. It is not affected by the electromagnetic force, which means that the neutrino does not have an intrinsic magnetism. On what basis, then, can a distinction be made between the neutrino and the antineutrino?

The distinguishing factor turns out to be the intrinsic spin. Both particles have an intrinsic spin of $\frac{1}{2}$, but their spin directions have different orientations. We learned previously that intrinsic spin is a fundamental quantum mechanical property of a particle that cannot be fully explained simply as an actual rotation of the particle.

Nevertheless, it is still convenient to visualize the intrinsic spin in terms of a rotating particle. The relationship between the direction of the intrinsic spin vector and the direction of rotation is shown in Figure 20–10a.

As a result of a number of experiments (most of them performed in 1957), it has been established that *the direction of the spin of a neutrino is always opposite to its direction of motion,* and *the spin direction of an antineutrino is parallel to its direction of motion* (Figure 20–10b). Thus if one observes a neutrino as it moves

figure 20–10 **NEUTRINO AND ANTINEUTRINO SPINS**

Neutrinos are emitted with counterclockwise spins, antineutrinos with clockwise spins. Thus the neutrino and antineutrino are mirror images of one another.

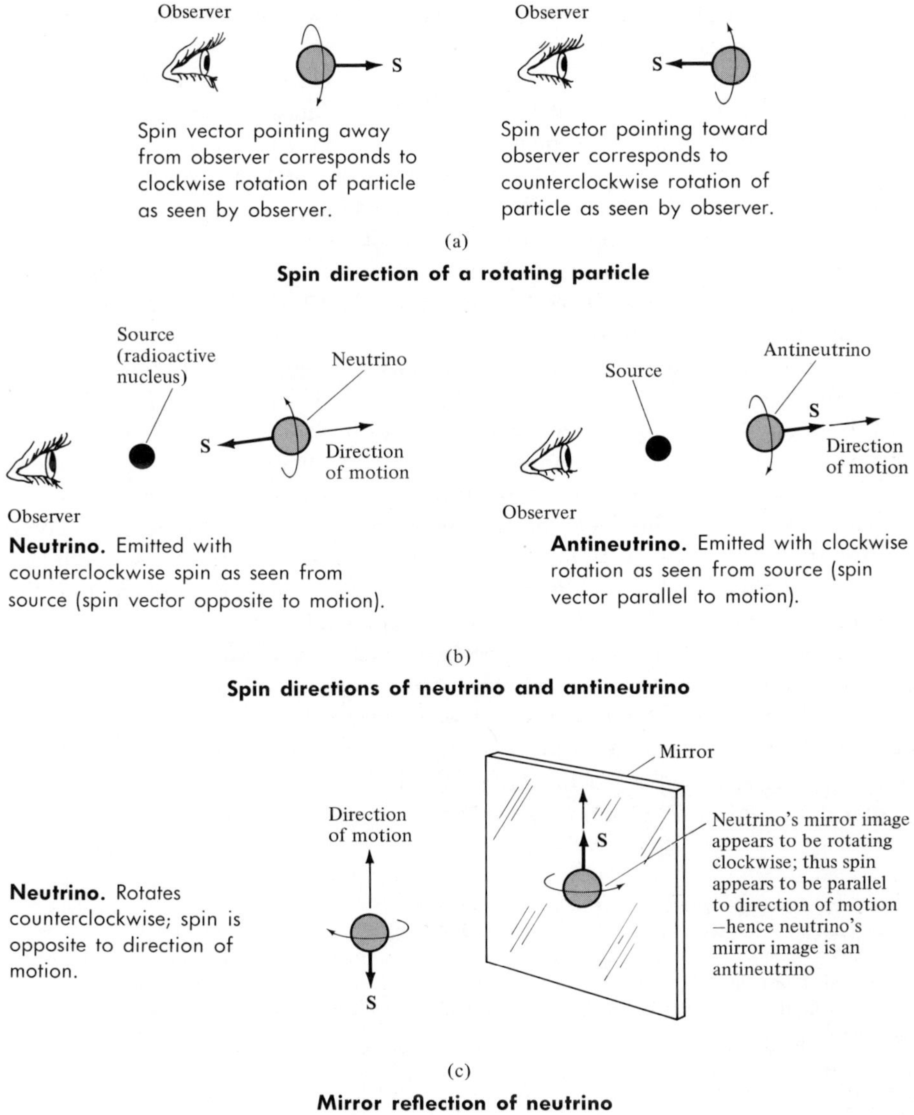

away from the emitting nucleus (the observer being located at the position of the nucleus), one sees the neutrino rotating *counterclockwise,* but under the same circumstances the antineutrino would appear to rotate *clockwise.* Of course, one cannot actually "see" the rotation of the neutrino (or antineutrino); the neutrino spin cannot even be measured directly. However, the spins of the emitted electron (or positron) and the residual nucleus in beta decay can be measured experimentally, and by assuming that the conservation of spin holds, one can infer the direction of the neutrino spin.

Thus the difference in spin direction distinguishes the neutrino from the antineutrino. Because of its parallel spin direction (equivalent to a clockwise rotation), the antineutrino is referred to as "right-handed," whereas the neutrino, which is always created with a counterclockwise spin, is "left-handed."*

The distinction between the "spin sense" (the physicist uses the word "helicity") of the neutrino and the antineutrino may seem a bit unusual to the student, but to the physicist it is extraordinarily puzzling. The physicist believes—for mathematical as well as for aesthetic reasons—that nature ought to be symmetrical and not show any preferences for either hand. On this basis, there ought to be as many "right-handed" neutrinos as "left-handed" neutrinos—but all neutrinos are apparently left-handed!

An even more difficult problem arises when a neutrino emission process is viewed in a mirror. A mirror has the property of reversing left and right; thus the mirror reflection of a right hand is a left hand (try shaking hands with your image in a mirror), and the mirror image of a clockwise rotation is a counterclockwise rotation. (For example, the hands of a clock held up to a mirror appear to move counterclockwise in the mirror.) Hence, as shown in Figure 20–10c, the mirror image of a neutrino is an antineutrino!

Now let us imagine an unstable nucleus that emits a positron–neutrino pair. When this decay is viewed in a mirror, it appears as if a positron and an *antineutrino* have been emitted—a process never seen in nature. Thus when looking at the mirror image of a real process—e^+–ν emission—one sees in the mirror an *impossible* process (e^+–$\overline{\nu}$ emission). It is a bit like looking in a mirror and seeing someone else's face staring back at you. This violation of mirror symmetry—which seems to be characteristic of all processes involving the weak interaction—is extremely puzzling to the modern-day physicist.† It might perhaps be less confusing if the neutrino, like the legendary Count Dracula, had no mirror image at all.

Once again the adoption of a certain "mental set" by physicists delayed the discovery of an unusual physical phenomenon. The principle of mirror symmetry had been tacitly assumed by physicists and was never held in doubt. Then in 1956 two theoretical physicists, C. N. Yang and T. D. Lee, noted that the principle had, in fact, never been tested experimentally. Within a few months three different experimental observations of violations of mirror symmetry were made.

*The human arm is constructed so that it is easier to make a clockwise rotation with the right hand and a counterclockwise rotation with the left hand. An antineutrino can be compared to a football thrown by a right-handed quarterback; the ball is spinning *clockwise* (as seen by the quarterback) when it is released. However, a left-handed quarterback throws a football with a *counterclockwise* spin; so the neutrino's motion is analogous to that of a football thrown by a left-handed passer.

† The physicist uses the term "conservation of parity" to signify that the mirror image of a real process should also be a real process.

20–5

THE MESON THEORY

The early 1930s saw the development of a new theoretical model of the nature of the electromagnetic force. It will be recalled that Maxwell's theory described the force between two charged particles in terms of an electromagnetic field that represented a state of disturbance of the ether. When the ether was done away with by Einstein, the electromagnetic field remained, but it took on a sort of ghostly existence. Physicists still tried to describe electromagnetic phenomena in terms of electromagnetic fields and waves, but they were forced to define them as unexplainable entities able to exist and propagate in a vacuum.

The discovery of wave–particle duality and the quantization of energy provided a new way to visualize electromagnetic interactions. Electromagnetic waves could now be interpreted as a stream of photons, and it seemed reasonable to believe that electromagnetic fields themselves could be quantized in some way. From this starting point Dirac, Pauli, Heisenberg, and others began the development of a ***quantum field theory*** in which the basic electromagnetic interactions are described in terms of a rapid exchange of photons between the interacting bodies. For instance, the electric force between two charged particles was seen to arise from the rapid emission and absorption of photons by the two particles (Figure 20–11). Thus in the quantum field theory the idea that energy is transmitted in some unexplained way through a field is replaced by the assumption that the energy is carried directly from particle to particle by the photons as they shuttle rapidly back and forth.

One key feature of the quantum field theory is that the photon exchange occurs so rapidly that it cannot possibly be observed; hence it is known as a ***virtual process.*** Moreover, in a virtual process mass–energy need not be conserved. The role of a virtual process in the theory is simply that of a mathematical device; a "real" process—one that *is* observed in nature—is depicted mathematically as the sum of a large number of virtual processes. Although mass–energy is not necessarily conserved in any single virtual process, it is conserved in a sum of many virtual processes. Although this model of electromagnetic interactions may seem unnecessarily complex and "unphysical," it is nevertheless quite successful in accounting for many subtle features of a variety of electromagnetic phenomena.

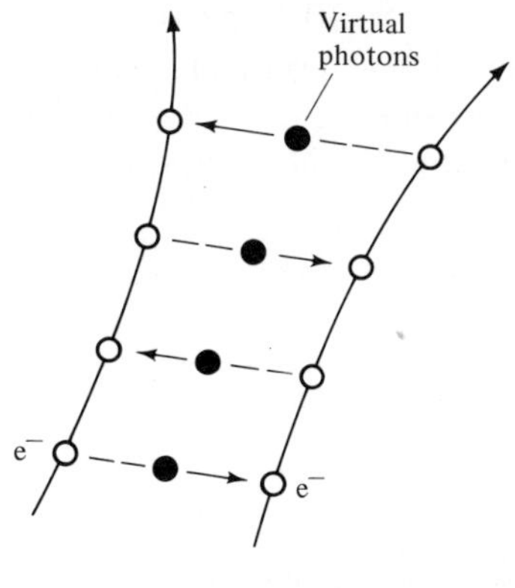

figure 20–11

QUANTUM FIELD THEORY OF THE ELECTROMAGNETIC FORCE

The force of electric repulsion between two electrons is transmitted through the rapid exchange of photons. This exchange is undetectable; hence it is referred to as a virtual process.

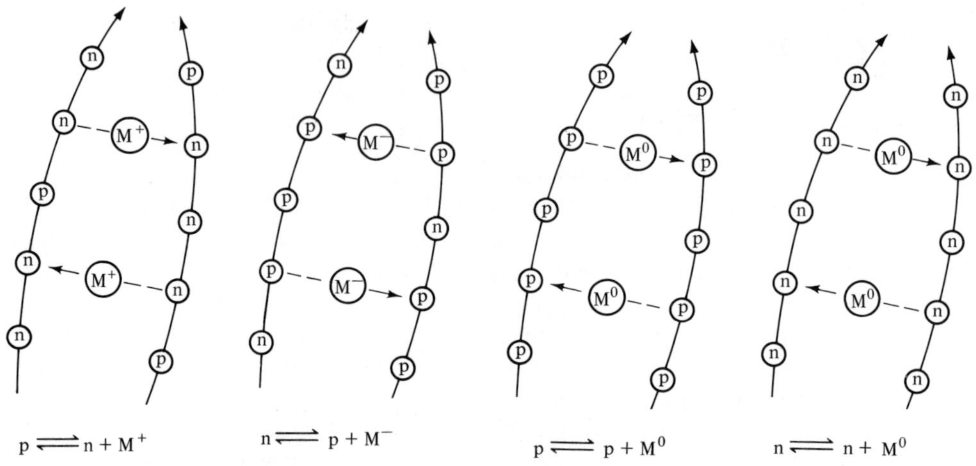

figure 20–12 **MESON THEORY OF NUCLEAR FORCES**

In Yukawa's quantum field theory of nuclear forces, the attractive force between nucleons is transmitted through the rapid exchange of mesons.

When the nuclear force was postulated (upon the discovery of the neutron in 1932), it was suggested that this force could also be described by a quantum field theory. That is, the nuclear force could in principle be depicted in terms of a rapid exchange of virtual particles between nucleons. In 1935 the Japanese physicist Hideki Yukawa noted that none of the then-known particles could possibly be capable of carrying the nuclear force. He therefore proposed the existence of a new family of particles that act as the quanta of the nuclear force field—just as photons are the quanta of the electromagnetic field. Yukawa showed that if a particle is to transmit the nuclear force, it cannot be massless (like the photon), and he estimated that the rest mass of the particle would be about 250 times the mass of the electron. Since the mass of Yukawa's particle falls between that of the proton (1836 electron masses) and the electron, the name ***meson*** (from the Greek *mesos*, meaning "intermediate") was later proposed for the particle.*

Thus in Yukawa's theory the nuclear force is transmitted by the rapid exchange of mesons between protons and neutrons. Since the range of the nuclear force is about 10^{-12} cm and the particles can travel with a maximum speed of $c = 3 \times 10^{10}$ cm/sec, the exchange takes place in a time interval of $d/c = 10^{-12}/(3 \times 10^{10}) \approx 10^{-23}$ sec. To account for all possible interactions, one positively charged, one negatively charged, and one neutral meson are needed. We will represent the mesons here by the symbols M^+, M^-, and M^0. The force between like nucleons is mediated by the neutral meson:

$$p \rightleftharpoons p + M^0$$
$$n \rightleftharpoons n + M^0$$

The force between protons and neutrons can be described in two ways:

$$p \rightleftharpoons n + M^+$$
$$n \rightleftharpoons p + M^-$$

* The name meson appears to have been suggested by Heisenberg's father, who was a professor of classical languages.

These four processes are illustrated in Figure 20–12. Note that in the latter two processes the proton and neutron change rapidly (within 10^{-23} sec) back and forth into one another through the emission and absorption of charged mesons. Furthermore, each meson exchange in itself violates the principle of conservation of mass–energy; roughly 150 MeV (the approximate rest mass of a meson) is required to create a meson—much more energy than is normally available to protons and neutrons. However, we must remember that these are *virtual processes* that cannot be detected by experiment. In order to create an *observable* meson, more than 150 MeV of energy would have to be supplied to a nucleon from an outside source.

THE DISCOVERY OF MESONS

The first mesons to be observed were produced by cosmic-ray interactions and were seen in cloud chamber photographs. The discovery was announced in 1937 by Carl Anderson (the discoverer of the positron) and his co-worker Seth Neddermeyer. Once again, Anderson was unaware that the particle he discovered had been anticipated theoretically. (Yukawa's paper had been published in a Japanese journal, and Anderson was unaware of its existence.) Anderson and Neddermeyer began to observe both positive and negative meson tracks in 1934, but it took them several years to establish the fact that these particles had rest masses intermediate between those of the proton and the electron.

The particles that Anderson and Neddermeyer observed were immediately identified with Yukawa's mesons, and the discovery was hailed as a major theoretical triumph. Further studies revealed, however, that these particles did not have the properties they were supposed to have. In particular, it was expected that if mesons transmit the nuclear force, they should interact very strongly with nuclei—so that if cosmic-ray mesons were allowed to strike a block of solid matter, they would be rapidly captured by the protons and neutrons in the material. However, experiments showed just the opposite; cosmic-ray mesons were rarely captured by nuclei and appeared to interact *weakly* (that is, via the weak interaction proposed by Fermi) with nucleons! By 1947 it had become quite evident that the cosmic-ray mesons discovered by Anderson and Neddermeyer were *not* the same as Yukawa's mesons.

Fortunately, the mystery was cleared up by the discovery in 1947 of another kind of meson. This discovery was made by the British experimental physicist Cecil F. Powell, who was mainly responsible for the development of the use of photographic emulsions in particle physics. Powell left some photographic plates on a mountaintop for several months, and upon retrieving and developing these plates, he discovered the track of a heavy meson that had decayed rapidly into a lighter meson. The heavy meson had never been observed before, but the lighter meson was found to be the same as the one discovered by Anderson and Neddermeyer. The two particles were given the names ***pi-meson*** or ***pion*** (symbol π) and ***mu-meson*** or ***muon*** (symbol μ), respectively. Further experimentation revealed that it is the pion, and not the muon, that has the properties predicted by Yukawa's theory.

THE PROPERTIES OF PIONS AND MUONS

The properties of pions and muons, as they are presently known, are summarized in Table 20–2. The pion can be identified with the theoretical Yukawa meson, which we previously identified by the symbol M. Thus there is a positively charged pion (π^+), a neutral pion (π^0), and a negatively charged pion (π^-). Hence the virtual processes described in the quantum field theory of nuclear forces should be rewritten as follows:

$$\mathrm{p} \rightleftharpoons \mathrm{p} + \pi^0$$
$$\mathrm{n} \rightleftharpoons \mathrm{n} + \pi^0$$
$$\mathrm{p} \rightleftharpoons \mathrm{n} + \pi^+$$
$$\mathrm{n} \rightleftharpoons \mathrm{p} + \pi^-$$

There is considerable evidence that pions are involved in nuclear forces. High-energy nuclear reactions with energies in the high MeV or GeV range result in the production of several pions. Notably, the annihilation of a proton–antiproton (or neutron–antineutron) pair creates pions. (The number of pions released depends on the initial energies of the particles.)

$$\left.\begin{matrix}\mathrm{p} + \overline{\mathrm{p}} \\ \mathrm{n} + \overline{\mathrm{n}}\end{matrix}\right\} \longrightarrow \pi^+ + \pi^0 + \pi^- + \pi^+ + \pi^- + \cdots$$

Note that the annihilation of an electron–positron pair ($e^- + e^+ \longrightarrow \gamma + \gamma$) creates photons (the quanta of the electromagnetic field), whereas the annihilation of a nucleon–antinucleon pair creates pions—the quanta of the nuclear field.

Free pions are unstable. Charged pions decay with a half-life of 2.5×10^{-8} sec to muons (accompanied by a neutrino to conserve mass–energy, momentum, and spin). Cosmic-ray pions are produced as secondaries in nuclear reactions in the upper atmosphere, but because of their short half-lives, they rarely reach the earth's surface. However, the decay product muons, whose half-lives are 100 times longer than those of pions, are able to reach sea level (aided by the relativistic time-dilation effect described in Chapter 12). This explains why muons, but not pions, are observed in laboratory cloud chambers. The higher one goes in the atmosphere, the more pions one sees.

The neutral pion has a slightly lower mass and a much shorter half-life than a charged pion. The mass difference can be attributed to electromagnetic effects. The shorter half-life can be attributed to the fact that a π^0 decays by an *electromagnetic* interaction ($\pi^0 \longrightarrow \gamma + \gamma$), whereas charged pions decay through a *weak* interaction. Since the latter is many orders of magnitude weaker than the electromagnetic interaction, the decay of the π^+ or π^- is expected to proceed much more slowly than the decay of the π^0. The π^0 is its own antiparticle. (That is, it cannot be distinguished from its antiparticle.) The π^+ and π^- are antiparticles of one another. Generally, the π^+ is classified as the particle and the π^- as the antiparticle, but this is not a significant distinction, as we shall learn later on.

We see that the pion family fits nicely into the niche created for it by Yukawa's

table 20–2 PROPERTIES OF PIONS AND MUONS

Particle	Mass (in electron masses)	Spin	Half-life (sec)	Mode of decay
π^+	273	0	2.5×10^{-8}	$\pi^+ \longrightarrow \mu^+ + \nu_\mu$
π^-	273	0	2.5×10^{-8}	$\pi^- \longrightarrow \mu^- + \overline{\nu}_\mu$
π^0	264	0	1.0×10^{-16}	$\pi^0 \longrightarrow \gamma + \gamma$
μ^-	207	$\frac{1}{2}$	2.2×10^{-6}	$\mu^- \longrightarrow e^- + \overline{\nu}_e + \nu_\mu$
μ^+	207	$\frac{1}{2}$	2.2×10^{-6}	$\mu^+ \longrightarrow e^+ + \nu_e + \overline{\nu}_\mu$

theory, but what about the muon? The existence of this particle is not accounted for by any theory; nor does it seem to serve any purpose whatsoever in nature. Moreover, the muon is found to be identical to the electron in every way (charge, spin, intrinsic magnetism, etc.) *except* for its much greater mass ($m_\mu = 207m_e$) and its instability. The μ^- decays to an e^-; its antiparticle, the μ^+, decays to the e^+. (Both decays must be accompanied by a neutrino *and* an antineutrino if spin is to be conserved.) It has been suggested that the μ^- may simply represent an "excited state" of the electron, but the large mass difference between the muon and electron is difficult to explain.

TWO NEUTRINOS?

The difficulty of trying to account for the muon's existence is compounded by the fact that *the muon appears to have its own neutrino.* That is, the neutrino that accompanies muon production and decay *is not the same* as the neutrino that is paired with the electron. Hence there are now believed to be two kinds of neutrinos (and antineutrinos): the ***electron-neutrino*** (ν_e, $\overline{\nu}_e$) and the ***muon-neutrino*** (ν_μ, $\overline{\nu}_\mu$).

The evidence for the existence of two different neutrinos stems from the observed decay of the muon:

$$\mu^\pm \longrightarrow e^\pm + \nu + \overline{\nu}$$

If the neutrino and antineutrino emitted in this reaction are antiparticles of one another, they should occasionally annihilate each other. The observed reaction would then be

$$\mu^\pm \longrightarrow e^\pm + \gamma$$

However, *this reaction has never been observed.* This seems to indicate that the neutrino and antineutrino given off in muon decay are not antiparticles of one another. Instead, the reactions presumably proceed as follows:

$$\begin{aligned} \mu^+ &\longrightarrow e^+ + \nu_e + \overline{\nu}_\mu \\ \mu^- &\longrightarrow e^- + \overline{\nu}_e + \nu_\mu \end{aligned}$$

In 1962 a group of physicists at Columbia University provided substantial evidence for the two-neutrino hypothesis by demonstrating that the muon-neutrino and electron-neutrino trigger different reactions. In the two-neutrino theory the following reactions are predicted:

$$\begin{aligned} &(1a){:}\ \nu_e + n \longrightarrow p + e^- && && (2a){:}\ \nu_e + n \not\longrightarrow p + \mu^- \\ &(1b){:}\ \nu_\mu + n \longrightarrow p + \mu^- && \textit{but} && (2b){:}\ \nu_\mu + n \not\longrightarrow p + e^- \\ &(1c){:}\ \overline{\nu}_e + p \longrightarrow n + e^+ && && (2c){:}\ \overline{\nu}_e + p \not\longrightarrow n + \mu^+ \\ &(1d){:}\ \overline{\nu}_\mu + p \longrightarrow n + \mu^+ && && (2d){:}\ \overline{\nu}_\mu + p \not\longrightarrow n + e^+ \end{aligned}$$

If the two neutrinos are distinct, then muon-neutrino capture by a nucleon should *always* result in muon emission, never electron emission. The Columbia group decided to test this hypothesis. The Brookhaven synchrotron was used to produce a beam of pions, which decay to muons and (presumably) muon-neutrinos:

$$\pi^+ \longrightarrow \mu^+ + \nu_\mu \qquad \text{and} \qquad \pi^- \longrightarrow \mu^- + \overline{\nu}_\mu$$

The pions and muons were shielded out (by a 45-foot-thick wall of stacked steel plates obtained from the dismantled battleship *Missouri*), and the muon-neutrinos were allowed to enter a detector, where they might react with protons and neutrons. Since only muon-neutrinos are produced by pion decay, neutrino capture (via reactions 1b and 1d) should lead to *only* muon production. But if the muon-neutrino and electron-neutrino are identical, *all* of the reactions listed in the two columns can take place, producing both muons and electrons. Hence the appearance of equal numbers of muons and electrons would disprove the two-neutrino hypothesis.

In approximately 300 hours of total observation time, during which an estimated 10^{14} neutrinos entered the detector, nearly thirty muons were observed—in agreement with theoretical predictions—and only one or two electrons, which could readily be attributed to the background radiation (cosmic rays and other particles produced by the synchrotron). The experimental results thus provided strong evidence for the two-neutrino hypothesis.

The muon-neutrino and electron-neutrino seem to be alike in every particular, except that they are created and captured in different reactions. Muon-neutrinos are left-handed (and muon-antineutrinos are right-handed), so that reactions involving muon-neutrinos also violate mirror symmetry. Coupled with the muon-electron puzzle, the existence of two kinds of neutrino represents an extravagance on the part of nature that modern-day physicists are totally at a loss to explain.

QUESTIONS FOR REVIEW

1 Discuss the basic principles of (a) the cloud chamber and (b) the bubble chamber. In each case, explain why a charged subnuclear particle makes a visible track.

2 What information about a particle can be obtained by studying its track in a cloud or bubble chamber? In particular, how can the mass, charge, and energy of a particle be determined?

3 Discuss the basic properties of cosmic rays. What is the distinction between primary and secondary cosmic rays? What are the main constituents of each type?

4 Summarize the basic principles of operation of each of the following types of particle accelerators. Discuss the advantages and disadvantages of each type:
(a) Linear accelerator.
(b) Cyclotron.
(c) Synchrocyclotron.
(d) Synchrotron.

5 Explain why the relativistic energy equation has negative energy solutions. Why were these solutions rejected as meaningless by most physicists?

6 What are the allowed ranges of energies for positive-energy electrons and for negative-energy electrons? What range of energy is forbidden to both? How is it possible for an electron to get from a positive-energy state to a negative-energy state or vice versa?

7 Summarize the properties of the "negative-electron sea." Why did Dirac postulate that all negative-energy states are filled? Why is the negative-electron sea indistinguishable from a vacuum?

8 According to Dirac's theory, what is antimatter? How is it created and destroyed?

9 Explain why the path of a downward-moving positron in a magnetic field is the same as that of an upward-moving electron. How did Anderson overcome this difficulty and establish the existence of the positron?

10 List the basic similarities and differences between the properties of a particle and those of its antiparticle.

11 Explain why it was necessary to postulate the existence of the neutrino in order to explain beta decay. In particular, show that (a) conservation of mass–energy and (b) conservation of intrinsic spin are violated if the electron (or positron) is the only particle emitted in beta decay.

12 Summarize the basic properties of the neutrino. How were these properties established theoretically?

13 Explain why it was necessary to postulate the existence of another fundamental force to explain beta decay. How do we know that this force is neither nuclear, electromagnetic, nor gravitational? What are the relative strengths of the four fundamental forces?

14 Describe the experiments by which the neutrino and the antineutrino were detected. Outline the difficulties involved in detecting neutrino interactions with matter.

15 Explain how the neutrino and antineutrino differ from one another. Why do we refer to the neutrino as "left-handed" and the antineutrino as "right-handed"?

16 Explain why the left-handedness of the neutrino violates basic ideas of mirror symmetry.

17 Outline the basic concepts of the quantum field theory of electromagnetic forces. What is a virtual process?

18 On what basis did Yukawa predict the existence of the meson? Why was it necessary to postulate three types of meson?

19 How did physicists know that the particle that Anderson and Neddermeyer discovered in cosmic rays was not the Yukawa meson?

20 Compare the basic properties of pions and muons.

21 Discuss the experimental evidence for the existence of two types of neutrino. How do the electron-neutrino and muon-neutrino differ?

EXERCISES

1 Explain how clouds form in the atmosphere.

2 Explain why sudden rainstorms occur on humid summer days just around sundown.

3 The Batavia accelerator is designed to produce a beam containing 10^{13} protons/sec with energies of 400 GeV. What is the power output in watts of this beam?

4 Calculate the minimum energy required to create a proton–antiproton pair.

5 Discuss the problems that would arise in the design of an engine that would run on the energy released by matter–antimatter annihilation.

6 What are the main differences between (a) an electron and a positron? (b) a proton and an antiproton? (c) a neutron and an antineutron?

7 It is possible to have a stable "positronium" atom consisting of a single electron orbiting a positron. How would the energy levels of positronium differ from those of hydrogen?

8 Explain why the spectra of hydrogen and antihydrogen would be virtually identical, except in a magnetic field.

9 What are the major similarities and differences between the photon and the neutrino?

10 Complete the following reactions:
(a) ${}_{19}K^{40} \longrightarrow {}_{-1}e^{0} + ?$
(b) ${}_{7}N^{13} \longrightarrow {}_{6}C^{13} + ?$
(c) ${}_{36}Kr^{79} + {}_{-1}e^{0} \longrightarrow ?$
(d) $\nu_e + {}_{16}S^{32} \longrightarrow ?$
(e) $\bar{\nu}_e + {}_{11}Na^{23} \longrightarrow ?$

11 Consider the pion decay reaction $\pi^+ \longrightarrow \mu^+ + \nu_\mu$. Assume that the pion is at rest when it decays.
(a) Explain why the muon and its neutrino must go in exactly opposite directions.
(b) The pion has an intrinsic spin of zero. Since the muon-neutrino is left-handed, the muon emitted in pion decay must also be left-handed. Explain why this is so.

THE FUNDAMENTAL PARTICLES

21

We shall not cease from exploration
And the end of all our exploring
Will be to arrive where we started
And know the place for the first time.

T. S. Eliot
(from *Four Quartets*)

THE DISCOVERY of pions in 1947 stimulated great interest in subnuclear particle physics. With cosmic rays providing a free and unlimited supply of high-energy reactions and photographic film as an inexpensive detection device, research could be conducted with relative ease. One simply had to leave some film on a mountaintop or send it to high altitudes with a balloon or rocket, and then analyze the particle tracks found in the exposed film. The discoveries made with these techniques stimulated an even greater interest in subnuclear particles, and as a consequence high-energy accelerators were designed specifically for the purpose of producing and studying particle interactions.

Because of these efforts, new particles were being discovered at an ever-increasing rate. Between 1947 and 1957 the number of known subnuclear particles more than tripled. The problem faced by theoretical physicists was to find some unifying relationship between these particles. Each of the new particles has a distinct mass, charge, spin, and so on. Are they all equally "fundamental," or is there some hierarchy among them? Could there be yet a deeper level of matter out of which all of these particles are constructed?

The confusion created by the puzzling array of subnuclear particles recalls to mind the bewilderment displayed by turn-of-the-century physicists in the face of the great variety of atomic spectra. At that time physicists had little more than Balmer's and Rydberg's formulas to go by. In comparison, the theories of fundamental particles that modern-day physicists have to work with, although they are mathematically rather sophisticated and complex, are somewhat like the Balmer and Rydberg formulas. They bring some semblance of order to the data, but provide little in the way of a basic understanding of the underlying phenomena.

21–1

PARTICLES IN PROFUSION

One way in which some of these new particles came to be discovered is illustrated in Figure 21–1. A piece of photographic film exposed to cosmic rays exhibited a pattern of tracks consisting of a single track and two sets of "V tracks" (Figure 21–1a).

Careful examination of the track thicknesses (which depend on the amount of ionization produced by the particles) indicated that the tracks were produced by pions and protons, as shown in Figure 21–1b. A cosmic-ray π^-, entering from the left, apparently struck a stationary particle at point A; a π^- and π^+ are seen to emanate from point B, and a π^- and proton are seen to emanate from point C.

The most reasonable explanation for the sudden appearance of particles emanating from B and C is that they were created by the decay of unstable, uncharged particles. (An uncharged particle leaves no tracks in a photographic emulsion.) Presumably, these particles were created at A in the collision between a high-energy π^- and a proton at rest in the emulsion. One of the neutral particles decayed at B, the other at C. By assuming that mass–energy and momentum are conserved in all interactions, physicists could calculate the masses of the neutral particles from the known masses and energies of the product pions and proton. The particle that decayed at B was found to have a rest mass of $966m_e$ (m_e = rest mass of the electron), which makes the particle heavier than a pion ($m = 273m_e$) but lighter than a proton ($m = 1836m_e$). This particle was called the ***K⁰-meson.*** The particle that decayed at C has a calculated rest mass of $2183m_e$, which is heavier than that of a proton. It was named the ***lambda particle*** (symbol Λ^0).

Thus the photograph shows the following series of reactions. (See Figure 21–1c.) A negatively charged pion collided with a positively charged proton at A, producing two neutral particles:

$$\pi^- + p \longrightarrow K^0 + \Lambda^0$$

figure 21–1 **ANALYSIS OF PARTICLE TRACKS**

An exposed photographic emulsion exhibits the pattern of tracks shown in (a). Examination of the tracks (b and c) shows that a K^0 and Λ^0 were created at A by the collision of a π^- with a proton. The K^0 and Λ^0 subsequently decayed at B and C, respectively.

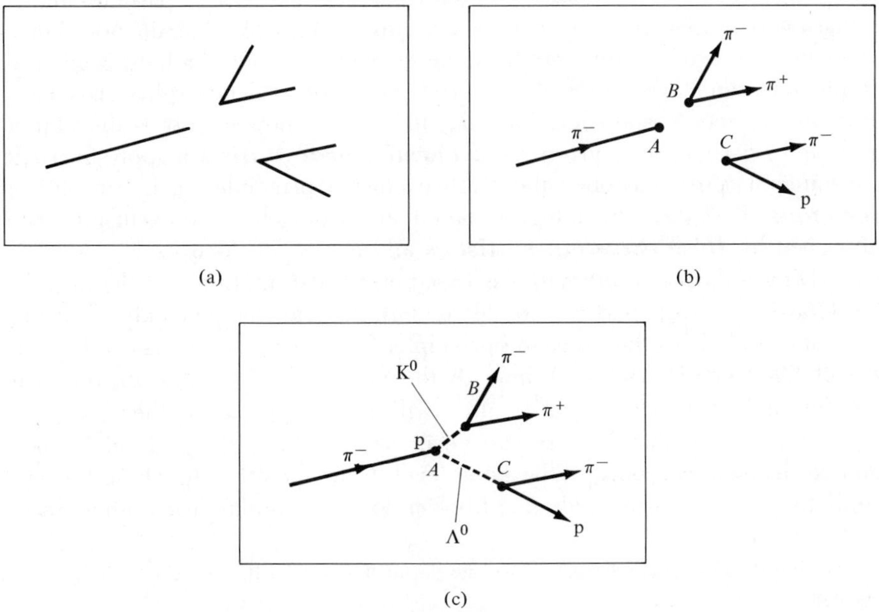

Each of these particles is unstable and decays into two oppositely charged particles:

$$\text{At B:} \quad K^0 \longrightarrow \pi^+ + \pi^-$$
$$\text{At C:} \quad \Lambda^0 \longrightarrow p + \pi^-$$

Finally, from the path lengths *AB* and *AC* of the two particles and their calculated speeds (generally very close to the speed of light), it is possible to estimate the lifetimes of the particles. Both the K^0 and the Λ^0 are found to have lifetimes on the order of 10^{-10} sec.

In addition to the K^0 and Λ^0, several other new short-lived particles were discovered by similar techniques: ***charged K-mesons*** (K^+, K^-), three ***sigma particles*** (Σ^+, Σ^0, Σ^-) with masses of about $2330m_e$, and two ***xi particles*** (Ξ^-, Ξ^0) with masses of approximately $2580m_e$. The antiparticles of most of these particles were also observed. Because of their peculiar properties, these particles were referred to as the ***strange particles.*** Physicists now set about trying to find some sense in this seemingly inexplicable and unnecessary proliferation of particles.

PARTICLE FAMILIES

Given a large amount of observational data, the first step that a scientist takes is usually to organize or classify the data in various ways in the hope that some recognizable pattern might emerge. Let us see what happens when we try to classify the subnuclear particle data according to a logical scheme. The broadest classification scheme used by particle physicists divides all particles into two groups, according to whether their intrinsic spins are half integers ($s = \frac{1}{2}, \frac{3}{2}, \frac{5}{2}, \ldots$) or integers ($s = 0, 1, 2, \ldots$). For some unexplained reason, particles with half-integral spins (electrons, protons, etc.) obey the Pauli exclusion principle, and particles with integral spins (photons, pions, etc.) do not.

If the Pauli exclusion principle is *not* obeyed, any number of particles can occupy the same energy level, so there is no such thing as a "filled" energy level. It follows that energy will be distributed differently among a group of particles that obey the exclusion principle as compared to a group of particles that do not. For example, in an atom no two electrons can have the same energy, but in a light beam any number of photons can have identical energies! As a consequence, physicists have had to develop two sets of statistical formulas to describe how energy is distributed among a group of particles of a given kind. ***Fermi–Dirac statistics*** apply to particles with half-integral spins that obey the Pauli exclusion principle; such particles are called ***fermions.*** Particles with integral spins that do not obey the exclusion principle are described by ***Bose–Einstein statistics*** and are called ***bosons.****

Dirac's theory of antimatter is based on the assumption that the negative energy levels available to a particle are filled, but this can happen only if the particle is a fermion and obeys the exclusion principle. Hence only fermions can have a negative sea of filled energy states. A hole in the negative sea creates an antifermion; any fermion and its associated antifermion will annihilate one another.

On the other hand, there can never be a "filled" energy level for bosons; any number of them can occupy the same level. Consequently, the Dirac theory does not apply to bosons; a boson and an antiboson do not annihilate each other. In fact, there

* Satyendranath Bose, an Indian physicist, developed the original theory, which was later modified by Einstein.

is no clear way to distinguish a boson from an antiboson, and it is questionable whether the concept of antimatter has any meaning at all with respect to bosons. For example, the π^+ and π^- have the same properties except for the opposite charge, and appear to constitute a particle–antiparticle pair. But pions are bosons, and in this respect it is meaningless to try to identify one of them as a particle and the other as an antiparticle. That is, each is a particle in its own right; neither one is a "hole" in a negative energy level.

Keeping in mind the distinction between bosons and fermions, let us now look at Table 21–1, which lists the properties of all of the subnuclear particles theoretically predicted or known to exist as of 1959. Originally, particle masses were given in terms of the mass of the electron (the lightest known particle with nonzero mass), but it is now considered more useful to give the energy equivalent in MeV of the particle rest mass ($E_0 = m_0c^2$). In the table, both values are listed.

When the known particles are arranged in order of increasing mass, they are found to fall neatly into four groups, or "families." At the bottom of the list we find two ***massless bosons.*** One of them is the familiar photon. The second is a hypothetical particle, the ***graviton,*** which has never been observed. Its existence has been postulated on the assumption that gravitational fields are quantized. The graviton is the quantum that transmits the gravitational force, just as the photon is the quantum that transmits the electromagnetic force. Because of the extreme weakness of the gravitational interaction, the graviton should be considerably more difficult to detect than even the neutrino. At the present time there is no experimental evidence that gravitons exist, but at the same time their existence cannot be ruled out.

The second group of particles is the ***lepton*** family, consisting of the muon, the electron, and their associated neutrinos.* Since these particles are all fermions, each one has a distinct antiparticle with which it annihilates. The nature of leptons and the relationship between the four members of the family are not understood. Particle physicists are extremely puzzled by the large mass difference between the muon and electron and the existence of *two* neutrinos. At the present time the existence of the lepton family is almost a complete mystery.

The third particle family is the ***meson*** family, a group of bosons with masses between the leptons and the proton. The family consists of three pions (π^+, π^0, π^-) and four K-mesons or ***kaons*** (K^+, K^0, $\overline{K}^0$, K^-). The meson family is the only one without a stable member. Most mesons decay eventually to leptons, except for the π^0, which decays to photons. The π^+–π^-, K^+–K^-, and K^0–$\overline{K}^0$ pairs are sometimes classified as antiparticles of one another (note the similar masses, decay schemes, etc.), even though they are bosons and do not obey the Dirac theory. It has proved more useful, however, to drop the particle–antiparticle distinction and list them as separate particles.

Among the mesons the two neutral kaons (K^0, $\overline{K}^0$) have the most confusing properties. Each of these particles appears to have *two* different half-lives, depending on its mode of decay. In a beam of otherwise identical kaons, about half will decay with a half-life of 5.7×10^{-8} sec into two particles ($\pi^+ + \pi^-$, $\pi^0 + \pi^0$). The others will decay with a much shorter half-life of 10^{-10} sec into *three* particles ($\pi^+ + \pi^0 + \pi^-$; $3\pi^0$; $\pi^+ + e^- + \overline{\nu}_e$; $\pi^- + e^+ + \nu_e$; $\pi^+ + \mu^- + \overline{\nu}_\mu$; $\pi^- + \mu^+ + \nu_\mu$; etc.). The K^0 is sometimes classified as two particles, ${K_L}^0$ and ${K_S}^0$

* Although the muon-neutrino was not discovered until 1962, it has been included in Table 21–1 to avoid confusion in describing the various decay reactions and properties of the other leptons.

table 21–1 **PROPERTIES OF KNOWN SUBNUCLEAR PARTICLES (1959)**

Family	Particle name	Symbol	Rest mass ($m_e = 1$)	Rest mass (MeV)	Charge	Spin	Anti-particle	Lifetime (sec)	Common modes of decay
Baryons (heavy fermions)	Xi	Ξ^-	2586	1321.2	−1	$\frac{1}{2}$	$\overline{\Xi}^+$	1.7×10^{-10}	$\Lambda^0 + \pi^-$
		Ξ^0	2573	1314.7	0	$\frac{1}{2}$	$\overline{\Xi}^0$	3.0×10^{-10}	$\Lambda^0 + \pi^0$
	Sigma	Σ^-	2343	1197.4	−1	$\frac{1}{2}$	$\overline{\Sigma}^+$	1.7×10^{-10}	$n + \pi^-$
		Σ^0	2338	1192.6	0	$\frac{1}{2}$	$\overline{\Sigma}^0$	$<10^{-14}$	$\Lambda^0 + \gamma$
		Σ^+	2328	1189.5	1	$\frac{1}{2}$	$\overline{\Sigma}^-$	0.8×10^{-10}	$p + \pi^0$; $n + \pi^+$
	Lambda	Λ^0	2183	1115.6	0	$\frac{1}{2}$	$\overline{\Lambda}^0$	2.5×10^{-10}	$p + \pi^-$; $n + \pi^0$
	Nucleon	n	1839	939.6	0	$\frac{1}{2}$	$\overline{n}$	≈ 1000	$p + e^- + \overline{\nu}_e$
		p	1836	938.3	1	$\frac{1}{2}$	$\overline{p}$	Stable	
Mesons (intermediate bosons)	Kaon	K^0	974	497.9	0	0		$K_S{}^0$: 0.9×10^{-10}	$\pi^+ + \pi^-$; $2\pi^0$
		$\overline{K}^0$	974	497.9	0	0		$K_L{}^0$: 5.7×10^{-8}	$\pi + \mu + \nu_\mu$; $\pi + e + \nu_e$; 3π
		K^+	966	493.8	1	0		1.2×10^{-8}	$\mu^+ + \nu_\mu$; $\pi^+ + \pi^0$
		K^-	966	493.8	−1	0		1.2×10^{-8}	$\mu^- + \overline{\nu}_\mu$; $\pi^- + \pi^0$
	Pion	π^+	273	139.6	1	0		2.5×10^{-8}	$\mu^+ + \nu_\mu$
		π^-	273	139.6	−1	0		2.5×10^{-8}	$\mu^- + \overline{\nu}_\mu$
		π^0	264	135.0	0	0		1.0×10^{-16}	2γ
Leptons (light fermions)	Muon	μ^-	207	105.7	−1	$\frac{1}{2}$	μ^+	2.2×10^{-6}	$e^- + \overline{\nu}_e + \nu_\mu$
	Electron	e^-	1	0.51	−1	$\frac{1}{2}$	e^+	Stable	
	Muon-neutrino	ν_μ	0	0	0	$\frac{1}{2}$	$\overline{\nu}_\mu$	Stable	
	Electron-neutrino	ν_e	0	0	0	$\frac{1}{2}$	$\overline{\nu}_e$	Stable	
Massless bosons	Photon	γ	0	0	0	1		Stable	
	Graviton	G	0	0	0	2		Stable	

(which decay with long and short half-lives, respectively), but except for their decay schemes, the $K_L{}^0$ and $K_S{}^0$ are identical. Understanding the nature of the K^0 is one of the many problems faced by the modern-day particle physicist.

The ***baryon*** family of fermions includes the familiar neutron and proton (classified together as two forms of the nucleon) and the lambda, sigma, and xi particles. (The lambda, sigma, and xi particles, all of which are heavier than nucleons, are sometimes referred to as ***hyperons.***) The proton is the lightest member of the family and the only stable baryon; all the other baryons decay in one or more steps to the proton. Each baryon has a distinct antibaryon, with which it annihilates. The uncharged baryons (n, Λ^0, Σ^0, Ξ^0) differ from their antiparticles in the relative orientations of the intrinsic spin and intrinsic magnetic field directions. (See Figure 20–8.) The Σ^+ and $\overline{\Sigma}^+$ and the Σ^- and $\overline{\Sigma}^-$ are distinguishable from one another in the same way.

Altogether, Table 21–1 contains twenty-one distinct subnuclear particles—thirty-three if antiparticles are counted. Are all of these particles to be considered independent and equally "fundamental," or is there some underlying unity through which they are all related?

From this point on, we shall focus our interest on two of the four particle families—the baryons and mesons. The two massless bosons (the photon and the hypothetical graviton) are theoretically understood to be the quanta of their respective fields. The leptons have thus far defied all attempts to describe them theoretically, and there is little more that can be said about them. Fortunately, physicists have had more success with the baryon and meson families.

21–2

PARTICLE INTERACTIONS AND CONSERVATION LAWS

Important keys to the properties of the various particles are provided when the different particle reactions are studied in detail. We expect every reaction to obey four conservation principles: mass–energy, linear momentum, angular momentum (including intrinsic spin), and charge. In fact, no reactions have ever been observed to violate any of these conservation principles. Yet one can propose reactions that *theoretically* ought to occur—that is, they do not violate any of the above conservation principles—but have never been observed. For example, such reactions as

$$\begin{aligned} p &\longrightarrow e^+ + \gamma \\ n &\longrightarrow e^+ + e^- + \nu_e \\ p + n &\longrightarrow K^+ + \gamma \\ p + e^- &\longrightarrow \pi^0 + K^0 \\ \Lambda^0 + \Lambda^0 &\longrightarrow K^+ + K^- \end{aligned}$$

are never seen to occur, even though they obey the four conservation principles.

There is a famous unwritten principle known to physicists that "anything that is not forbidden is compulsory," or in simpler terms, "anything that can happen, will happen." From this premise—which has been applied with great success in many areas of physics—one can conclude that if reactions such as those listed above do not occur, they must be forbidden by some as-yet-undiscovered principle. In other words, there must be some other conservation principle or principles according to which these reactions cannot occur.

CONSERVATION OF BARYONS

One such principle is relatively easy to identify. We note that (1) all unstable baryons decay eventually to the proton, which is itself a stable particle, and (2) any baryon (including the proton) can be annihilated in an interaction with its antibaryon. In more general terms, one can say that a baryon always transforms to another baryon and can be destroyed only by annihilation with its antibaryon. If we now define a ***baryon number*** B such that

$$\begin{aligned} B &= 1 \text{ for all baryons} \\ &= -1 \text{ for all antibaryons} \\ &= 0 \text{ for all nonbaryons (mesons, leptons, etc.)} \end{aligned}$$

then we can formulate a ***principle of baryon conservation:***

The total baryon number is constant in all reactions.

As far as is known, there are no particle reactions that violate this principle. This new principle "explains" why the reactions listed previously do not occur; in none of them does B remain constant! For example, consider the reaction

$$\begin{array}{ccc} \mathrm{p} & \longrightarrow & \mathrm{e}^+ + \gamma \\ B = 1 & \neq & 0 + 0 \end{array}$$

This reaction is forbidden because B is not conserved. However, such reactions as

$$\begin{array}{ccc} \mathrm{p} + \overline{\mathrm{p}} & \longrightarrow & \pi^+ + \pi^0 + \pi^- \\ B = 1 + (-1) & = & 0 + 0 + 0 \end{array}$$

or

$$\begin{array}{ccc} \mathrm{p} + \overline{\mathrm{p}} & \longrightarrow & \mathrm{n} + \overline{\mathrm{n}} \\ B = 1 + (-1) & = & 1 + (-1) \end{array}$$

or

$$\begin{array}{ccc} \mathrm{p} + \mathrm{p} & \longrightarrow & \Sigma^+ + \mathrm{K}^0 + \mathrm{p} \\ B = 1 + 1 & = & 1 + 0 + 1 \end{array}$$

do *not* violate the baryon conservation principle and are, in fact, observed to occur in nature.

However, the baryon conservation principle does not account for all "forbidden" reactions. For example, the reaction

$$\pi^- + \mathrm{p} \longrightarrow \Sigma^- + \mathrm{K}^+$$

is observed to occur in nature, but the reaction

$$\pi^- + \mathrm{p} \longrightarrow \Sigma^+ + \mathrm{K}^-$$

has never been seen, even though it does not violate any of the conservation principles (including baryon conservation) previously mentioned. Some similar examples are listed in Table 21–2.

None of the reactions in the right-hand column of Table 21–2 is forbidden by any of the known conservation principles, so it is reasonable to hypothesize that there is yet another conservation principle to be discovered—one that would be violated by the unobserved reactions.

table 21–2 **SOME OBSERVED AND "MISSING" HIGH-ENERGY REACTIONS**

Observed	Not observed
$\pi^- + p \longrightarrow \Lambda^0 + K^0$	$\pi^- + n \longrightarrow \Lambda^0 + K^-$
$K^- + p \longrightarrow \Sigma^- + \pi^+$	$K^+ + p \longrightarrow \Sigma^+ + \pi^+$
$K^0 + p \longrightarrow n + K^+$	$K^0 + n \longrightarrow p + K^-$
$p + p \longrightarrow p + n + \pi^+$	$p + p \longrightarrow p + n + K^+$
$n + n \longrightarrow \Sigma^0 + K^0 + n$	$n + n \longrightarrow \Sigma^+ + K^- + n$

CONSERVATION OF STRANGENESS

The discovery of a new conservation principle was made in 1953 by the American physicist Murray Gell-Mann and independently by Kazuhiko Nishijima of Japan. Of all the physicists who have worked on the problems of particle physics, none has made more significant contributions than Gell-Mann. Born and raised in New York, he entered college when he was fifteen years old and received his Ph.D. at twenty-one. He was twenty-three years old when he derived the missing conservation principle needed to solve the puzzle of the nonappearing reactions. Gell-Mann proposed the existence of a new property, which he called ***strangeness,*** because the so-called strange particles are involved in these reactions. He defined a new quantum number called the ***strangeness quantum number,*** S. He found that each particle could be assigned a unique value of S such that strangeness is conserved in the observed reactions, but not in the unobserved reactions. In other words, the unobserved reactions are forbidden because they do not conserve strangeness! The values of S for the baryons and mesons listed in Table 21–1 are as follows:*

$$\begin{aligned} S &= 0: \quad \pi^+, \pi^-, \pi^0; \text{n}, \text{p} \\ S &= 1: \quad \text{K}^+, \text{K}^0 \\ S &= -1: \quad \Sigma^+, \Sigma^0, \Sigma^-; \Lambda^0; \overline{\text{K}}^0, \text{K}^- \\ S &= -2: \quad \Xi^-, \Xi^0 \end{aligned}$$

Thus in the reaction (which is observed in nature)

$$\begin{array}{rcccc} & \pi^- + \text{p} & \longrightarrow & \Sigma^- + \text{K}^+ \\ S = & 0 + 0 & = & -1 + 1 \end{array}$$

strangeness is conserved (the total S is the same on both sides), whereas the "missing" reaction

$$\begin{array}{rcccc} & \pi^- + \text{p} & \longrightarrow & \Sigma^+ + \text{K}^- \\ S = & 0 + 0 & \neq & -1 - 1 \end{array}$$

is one in which strangeness is *not* conserved. The student can readily verify that S is conserved in all of the observed reactions in Table 21–2 but not in any of the unobserved reactions listed in the table.

The principle of strangeness conservation does not have the universality of the other conservation principles, however, for there is a whole group of reactions in which

*For the antibaryons S has the opposite sign. Thus $S = 1$ for the $\overline{\Lambda}^0$ and $\overline{\Sigma}$; $S = 2$ for the $\overline{\Xi}$.

strangeness is *not* conserved. Consider some of the reactions by which the unstable baryons and mesons are observed to decay (see Table 21–1):

$$\begin{array}{ccccc} & \Xi^- & \longrightarrow & \Lambda^0 + \pi^- \\ S = & -2 & \neq & -1 + 0 \end{array}$$

or

$$\begin{array}{ccccc} & \Lambda^0 & \longrightarrow & p + \pi^- \\ S = & -1 & \neq & 0 + 0 \end{array}$$

We find that *strangeness is not conserved in these reactions.* In fact, the K, Ξ, Σ, and Λ particles all are found to decay by reactions in which strangeness is not conserved.

What, then, is the value of the strangeness conservation principle if it is found to be violated by these reactions? Actually, one can make a clear distinction between "strangeness-conserving" and "strangeness-nonconserving" reactions. We have identified four fundamental forces or interactions that exist in nature: *nuclear, electromagnetic, weak,* and *gravitational.* The nuclear force is the strongest of the four (it is often referred to simply as the ***strong interaction***), and we expect a reaction mediated by this strong force to occur very rapidly with a characteristic time of about 10^{-23} sec. The weak interaction is 10^{13} times weaker than the strong nuclear force; theoretically, the characteristic time should be 10^{13} times longer, or about 10^{-10} sec. Thus an unstable particle should decay in about 10^{-23} sec if it decays by a strong interaction and in about 10^{-10} sec if the decay is a consequence of a weak interaction. The characteristic time for an electromagnetic interaction is about 10^{-20} sec.

Looking at the half-lives of the unstable particles in Table 21–1, we see that most baryons and mesons decay in about 10^{-10} sec or longer—indicating that they are decaying by a weak interaction. In all these reactions *strangeness is not conserved.* The Σ^0 and π^0 have much shorter half-lives, but these particles decay by electromagnetic interactions. (Note that photons are emitted when they decay.) In these reactions *strangeness is conserved.* Furthermore, in every known reaction in which two high-energy particles collide and interact strongly to produce two or more different particles, strangeness is found to be conserved. Hence *strangeness is conserved in strong and electromagnetic interactions, but not in weak interactions.*

Why do the various baryons and mesons decay by weak rather than strong or electromagnetic interactions? For example, the Σ^- could decay by either of the following strong, strangeness-conserving reactions:

$$\Sigma^- \longrightarrow \Lambda^0 + \pi^- \qquad \text{or} \qquad \Sigma^- \longrightarrow n + K^-$$

These are, in fact, the *only* possible decay reactions available to the Σ^- that conserve strangeness—but both of them are forbidden by a different principle: mass–energy conservation. In each case the total rest mass of the products is greater than the rest mass of the Σ^-; therefore each reaction is endoergic and cannot occur as a *spontaneous* decay process. Hence there are *no* strangeness-conserving reactions by which the Σ^- can decay. Thus if strangeness were *always* conserved, the Σ^- would be a stable particle. Instead, the Σ^- decays by a weak interaction, in which strangeness is not conserved:*

$$\Sigma^- \longrightarrow n + \pi^- \qquad (T_{1/2} = 1.7 \times 10^{-10} \text{ sec})$$

* We identify the reaction as a weak interaction by the decay time (10^{-10} sec is characteristic of weak interactions) rather than by the particles that are produced in the reaction.

Nearly all of the unstable particles listed in Table 21–1 decay by weak interactions because there are no strong, strangeness-conserving decay paths available to them. The Σ^0 and π^0 are the exceptions to this rule—they decay rapidly via strangeness-conserving electromagnetic interactions.

RESONANCES

The observation that the strange particles are relatively long-lived because they can decay only by weak interactions raises the possibility that there may be additional undiscovered particles that are able to decay by strong interactions. Such particles would have half-lives on the order of 10^{-23} sec—much too short a time to leave an observable track in a photographic emulsion or bubble chamber.

But in physics everything that can happen will happen. Beginning in 1960 short-lived particles that decayed in 10^{-23} sec began to be discovered. These particles could not be seen directly, but their existence could be inferred from careful analysis of particle accelerator-induced reactions. For example, in the reaction

$$K^- + p \longrightarrow \Lambda^0 + \pi^+ + \pi^-$$

the energy released in the reaction should be distributed among the three products in a certain specific way, defined by the principles of energy and momentum conservation. However, when the energy distribution for a large number of the reactions was statistically analyzed, it was found that the three product particles could not have been produced simultaneously. Instead, the reaction proceeds in two steps, which involve the production and rapid decay of a new particle, the Y_1^*:

$$K^- + p \longrightarrow Y_1^* + \pi^+$$
$$Y_1^* \longrightarrow \Lambda^0 + \pi^-$$

In other words, the energy released in the reaction is first shared by *two* particles—the Y_1^* and the π^+—followed by the rapid decay of the Y_1^* into a Λ^0 and a π^-. But the track of the Y_1^* is not seen; its lifetime of 10^{-23} sec is too short for it to leave an observable track (Figure 21–2). Nevertheless, it can be determined

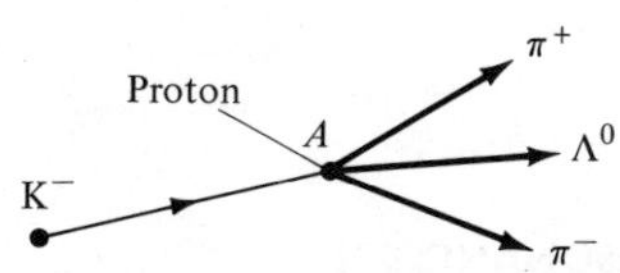

(a) Observed Reaction: $K^- + p \longrightarrow \Lambda^0 + \pi^+ + \pi^-$

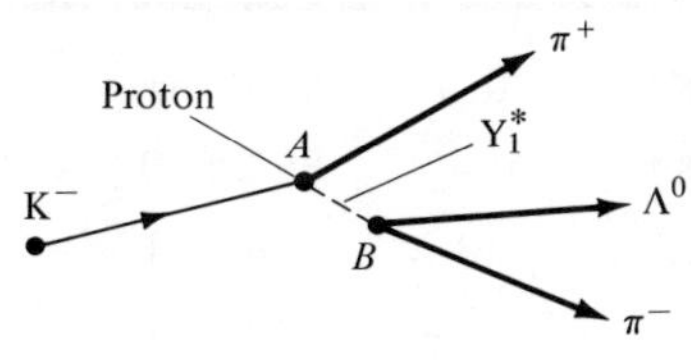

(b) Actual Reaction: $K^- + p \longrightarrow Y_1^* + \pi^+$
$Y_1^* \longrightarrow \Lambda^0 + \pi^-$

figure 21–2

THE DISCOVERY OF THE Y_1^* RESONANCE

(a) A K^- strikes a proton at *A*, and three particles are created. (b) An analysis of the distribution of energy among the three particles indicates that the reaction proceeds in two steps. The path of the Y_1^* (*AB*) is too short to be observed because of the very short lifetime of the Y_1^*.

that the Y_1^* has a definite mass (1385 MeV), spin ($\frac{3}{2}$), charge ($Q = 1$), and strangeness ($S = -1$), and it can therefore be classified as a distinct particle. Subsequent experiments showed that there are actually *three* Y_1^* particles, with charges of 1, 0, and -1.

The discovery of the Y_1^* was followed rapidly by the discovery of numerous short-lived baryons and mesons. (See Table 21–3.) Each of these new particles is distinguished by a very short lifetime ($\approx 10^{-23}$ sec), indicative of decay via a strong interaction. However, many physicists preferred to refer to these new entities not as particles but as ***resonances*** or temporary associations of particles. A case in point is provided by the four ***delta particles***—Δ^{++}, Δ^+, Δ^0, and Δ^-—which are created in pion–nucleon interactions:

$$\pi^+ + p \longrightarrow \Delta^{++}$$

$$\left.\begin{matrix}\pi^0 + p \\ \pi^+ + n\end{matrix}\right\} \longrightarrow \Delta^+$$

$$\left.\begin{matrix}\pi^- + p \\ \pi^0 + n\end{matrix}\right\} \longrightarrow \Delta^0$$

$$\pi^- + n \longrightarrow \Delta^-$$

Each Δ particle has a mass of 1238 MeV, a strangeness of 0, and an intrinsic spin of $\frac{3}{2}$. Since pions and nucleons have intrinsic spins of 0 and $\frac{1}{2}$, respectively, it is difficult to understand how the Δ can have a spin of $\frac{3}{2}$, and it would appear that spin has not been conserved in the above reactions. However, the missing unit of spin can be attributed to *orbital* angular momentum—as if the Δ consisted of a pion temporarily in orbit around a nucleon.

The short-lived resonances have all the properties normally attributed to a particle, and they could reasonably be classified as such. By 1962 the total number of subnuclear particles, resonances, and antiparticles had reached nearly eighty, and it seemed quite likely that many more would be discovered. Fortunately, some underlying relationships were discovered, and these relationships helped to simplify the chaos created by this profusion of particles.

table 21–3 **SOME SHORT-LIVED PARTICLES ("RESONANCES")**

Baryons					Mesons				
Symbol	Charge	Mass (MeV)	Spin	Strangeness	Symbol	Charge	Mass (MeV)	Spin	Strangeness
Y_0^{**}	0	1520	$\frac{3}{2}$	-1	K*	0, 1	890	1	1
N*	1, 0	1512	$\frac{1}{2}$	0	$\overline{K}$*	0, -1	890	0	-1
Y_0^*	0	1405	$\frac{1}{2}$	-1	ω^0	0	780	1	0
Y_1^*	1, 0, -1	1385	$\frac{3}{2}$	-1	ρ	1, 0, -1	760	1	0
Δ	2, 1, 0, -1	1238	$\frac{3}{2}$	0	η^0	0	548	0	0

21–3

ISOSPIN AND HYPERCHARGE

Thus far we have classified all known particles into four families, and we have concentrated our attention on the baryons and mesons. Upon examination of each of these families, we see that the family members tend to be clustered in ***charge groups*** or ***multiplets*** of one, two, or three particles of nearly the same mass. The pions and sigma particles represent ***charge triplets*** ($Q = -1, 0, 1$). The xi particles ($Q = -1, 0$), nucleons ($Q = 0, 1$), and the two kaon groups—K^+, K^0 ($Q = 1, 0$) and $\overline{K}^0$, K^- ($Q = 0, -1$)—each form a ***charge doublet.*** The lambda particle ($Q = 0$) stands by itself and is regarded as a ***charge singlet.***

It proves quite useful to define a new quantum number, called the ***isospin*** I, to describe this charge-grouping effect. A charge multiplet is assigned an isospin quantum number of $I = 0, \frac{1}{2}$, or 1, if it is a singlet, doublet, or triplet, respectively. Specifically, I is defined such that the number of members of the charge multiplet is equal to $2I + 1$. (That is, for $I = 0, \frac{1}{2}$, or 1, $2I + 1 = 1, 2$, or 3.) In addition, we define an ***isospin component*** I_Z, which is related to I in the same way that m_L is related to l (or m_s to s) in atomic theory. Thus

(1) If $I = 0$ (charge singlet), $I_Z = 0$ (one value).
(2) If $I = \frac{1}{2}$ (charge doublet), $I_Z = \frac{1}{2}, -\frac{1}{2}$ (two values).
(3) If $I = 1$ (charge triplet), $I_Z = 1, 0, -1$ (three values).

Hence each member of a charge multiplet can be assigned a unique value of I_Z. In the pion triplet ($I = 1$), for example, the π^-, π^0, and π^+ have $I_Z = -1, 0$, and 1, respectively. In the nucleon doublet ($I = \frac{1}{2}$) the proton is assigned $I_Z = \frac{1}{2}$, and the neutron $I_Z = -\frac{1}{2}$.

Historically, the concept of isospin (or isotopic spin, as it was originally called) is a relatively old one. It was first suggested by Heisenberg in 1932 (just after the neutron was discovered) and was used to describe the nucleon doublet. Isospin is *mathematically* similar to electron spin and angular momentum, but it does not correspond to any *physical* spin.

In Table 21–1 each particle is distinguished by a specific mass, charge, and intrinsic spin. We have now added four new quantum numbers: baryon number (B), strangeness (S), isospin (I), and isospin component (I_Z). Table 21–4 is a modified version of Table 21–1 that includes the new quantum numbers. In Table 21–4 a single (average) mass has been assigned to each charge multiplet; the slight mass differences between the members of the multiplet have been ignored. (For simplicity, antibaryons have not been included in the table. For an antibaryon B, S, and I_Z will have the opposite sign as compared to the associated baryon. However, I has the same value for a baryon and its corresponding antibaryon.)

In 1953 Gell-Mann and Nishijima showed that the four quantum numbers Q, I_Z, S, and B for each particle are related by the equation

$$Q = I_Z + \tfrac{1}{2}(S + B) \qquad (21\text{–}1)$$

This relationship can be verified by consulting the data in Table 21–4. In order to maintain the consistency of this equation, Gell-Mann and Nishijima predicted the table 21–4 **QUANTUM NUMBERS FOR CHARGE MULTIPLETS**

	Charge multiplet	Mass (MeV)	Isospin, I	Isospin component, I_z	Charge number, Q	Strangeness, S	Baryon number, B	Hypercharge, Y
Baryons	Ξ	1318	$\frac{1}{2}$	$\Xi^0: \frac{1}{2}$ $\Xi^-: -\frac{1}{2}$	0 -1	-2	1	-1
	Σ	1193	1	$\Sigma^+: 1$ $\Sigma^0: 0$ $\Sigma^-: -1$	1 0 -1	-1	1	0
	Λ	1115	0	$\Lambda^0: 0$	0	-1	1	0
	N	939	$\frac{1}{2}$	p: $\frac{1}{2}$ n: $-\frac{1}{2}$	1 0	0	1	1
Mesons	K	496	$\frac{1}{2}$	$K^+: \frac{1}{2}$ $K^0: -\frac{1}{2}$	1 0	1	0	1
	$\overline{K}$	496	$\frac{1}{2}$	$\overline{K}^0: \frac{1}{2}$ $K^-: -\frac{1}{2}$	0 -1	-1	0	-1
	π	138	1	$\pi^+: 1$ $\pi^0: 0$ $\pi^-: -1$	1 0 -1	0	0	0

existence of the Σ^0 and Ξ^0, which had not yet been discovered in 1953. (These particles were needed to complete the sigma triplet and xi doublet, respectively.) The subsequent discovery of the Σ^0 and Ξ^0 was a notable triumph for the Gell-Mann–Nishijima theory.

Particle physicists have also found it convenient to combine S and B into a new quantum number, Y:

$$Y = S + B \tag{21–2}$$

Thus $Y = S + 1$ for baryons and $Y = S$ for mesons. This new quantity is named the ***hypercharge*** (so-called because it is equal to twice the average charge of all the members of a given multiplet). Hypercharge, like strangeness, is conserved in strong and electromagnetic interactions, but *not* in weak interactions. The value of Y for each of the multiplets is also listed in Table 21–4. Equation 21–1 can now be written as follows:

$$Q = I_z + \frac{Y}{2} \tag{21–3}$$

The new quantum numbers I and Y are the basis of an important scheme for classifying all particles. This scheme includes the short-lived resonances, which are also grouped into charge multiplets and can be assigned values of I and Y. For example, the four Δ particles represent a charge ***quartet;*** thus $I = \frac{3}{2}$ for this multiplet (so that $2I + 1 = 4$). Each Δ has a strangeness of 0 (obtained by observing its decay reactions) and a hypercharge $Y = S + B = 0 + 1 = 1$. Thus the Δ multiplet is characterized by $(I, Y) = (\frac{3}{2}, 1)$. All of the known resonances can be assigned values of I and Y in a similar way.

Most of the resonances are found to have the same (I, Y) combinations as the longer-lived particles. This suggests that the baryons and mesons be grouped according to their (I, Y) values. The result is shown in Figures 21–3 and 21–4, in which particles of the same I and Y are listed in order of increasing rest mass. The figures have the appearance of a series of energy level diagrams, and the relationship between

figure 21–3 **BARYON STATES**

The baryon subfamilies and a few of their related lower-energy resonances of the same I and Y are arranged in order of increasing mass. For each state the mass in MeV and the intrinsic spin are given.

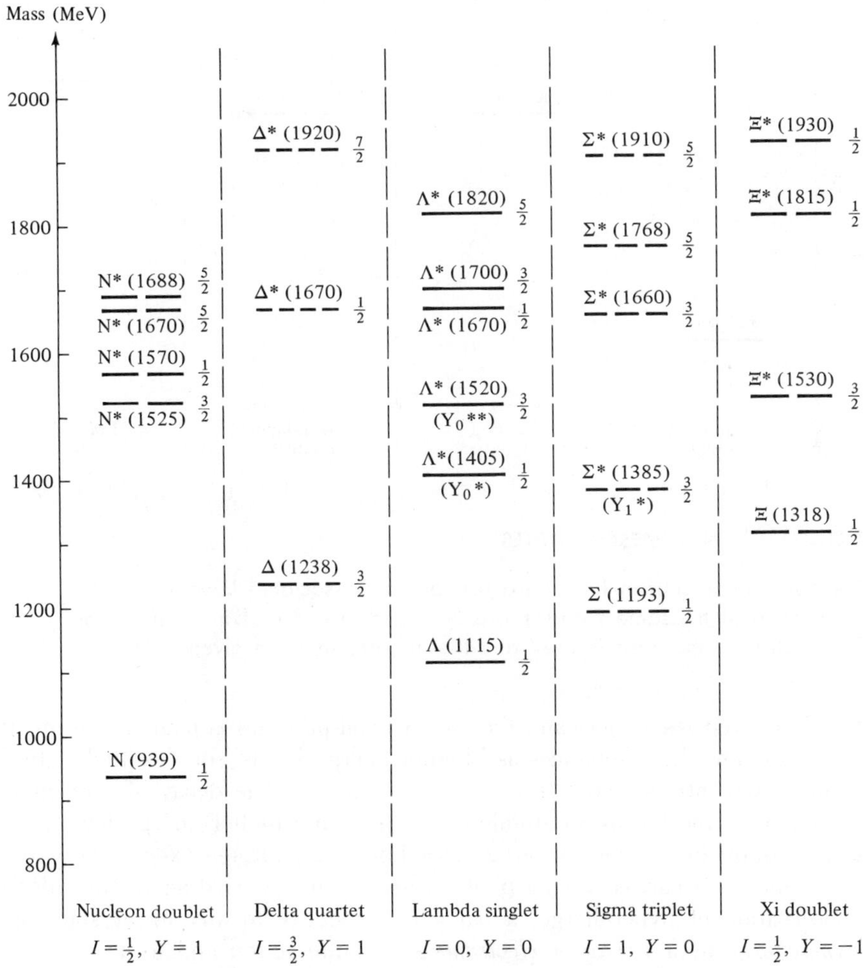

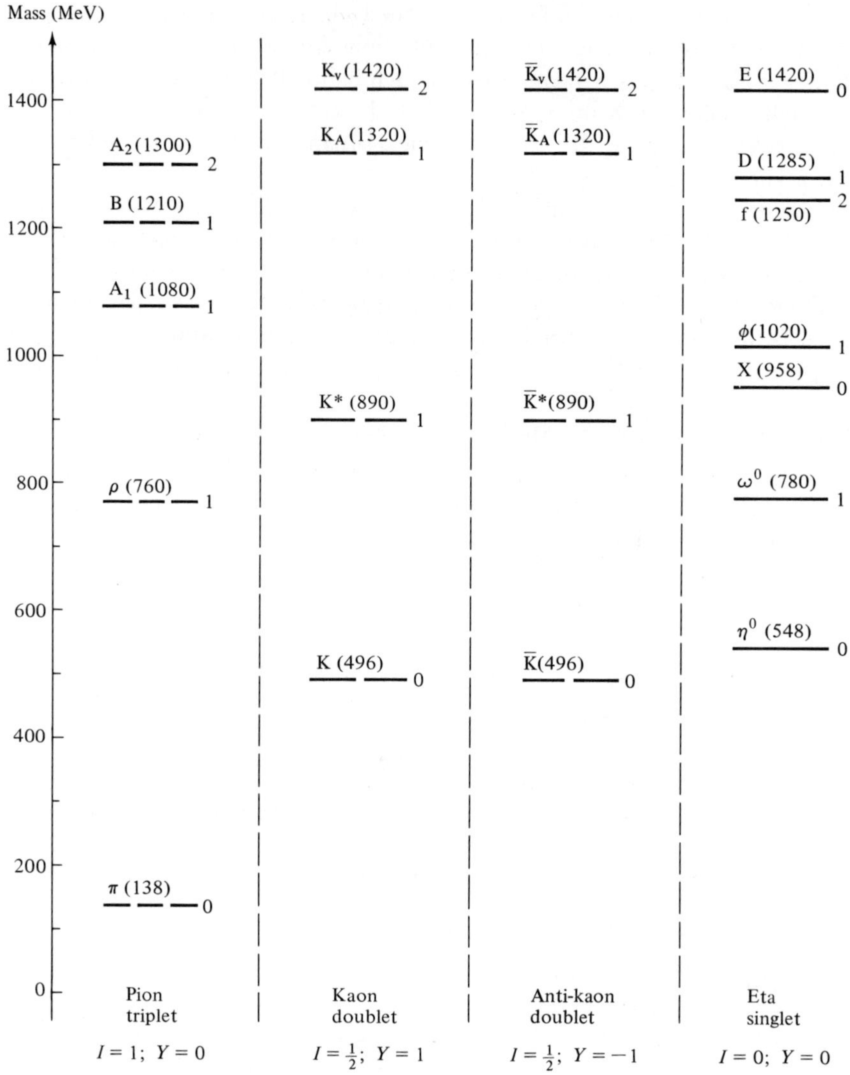

figure 21–4 **MESON STATES**

The four meson subfamilies and a few of their associated lower-energy resonances of the same I and Y are arranged in order of increasing mass. For each state the mass in MeV and the intrinsic spin are given.

the short-lived resonances and the longer-lived particles is made apparent. It becomes useful to view the resonances as "higher energy levels" or "excited states" differing in mass and intrinsic spin from a "ground state." The decay of a resonance is thus analogous to the fall of an atomic electron from a higher energy level to the ground state with the excess energy being carried off by a photon—except that here the excess mass–energy is carried off by photons or mesons. According to the rules governing conservation of hypercharge, a particle can decay to any lower state of the *same hypercharge* by a *strong* or *electromagnetic* interaction and to a state of a *different hypercharge* by a *weak* interaction.

This classification scheme also brings about a reduction in nomenclature. We see that the Y_1^* originally listed in Table 21–3 is actually an excited state of the Σ, so it becomes the $\Sigma^*(1385)$ triplet. Similarly, the Y_0^* and Y_0^{**} in the same table become the $\Lambda^*(1405)$ and $\Lambda^*(1520)$, respectively. Thus we find that the profusion of particles is greatly simplified; there are now only five distinct baryon subfamilies (the N, Δ, Λ, Σ, and Ξ multiplets) and four meson subfamilies (the K, η, π, and $\bar{K}$ multiplets). Each subfamily terminates at a "ground state." The higher-mass "excited states" decay to the various ground states via strong interactions (in which hypercharge and strangeness are conserved). The different ground states also communicate with one another. The Δ ground state decays to the N ground state (the proton–neutron doublet) via a strong interaction, and the Ξ, Σ, and Λ ground states decay by weak interactions through one another to the N ground state. (All baryon decays eventually terminate at the proton, which is the only stable baryon.) The meson ground states, however, are all unstable. The π, K, and $\bar{K}$ all decay through weak interactions to members of the lepton family. The η(eta) singlet has as its ground state the η^0 (previously listed as a resonance), which decays via an electromagnetic interaction ($\eta^0 \longrightarrow 2\gamma$) with a lifetime of about 10^{-20} sec.

21–4

THE EIGHTFOLD WAY

A great deal of order has been created by classifying the baryons and mesons according to their isospin and hypercharge. We now recognize only five baryon and four meson subfamilies, instead of eighty or more different particles. Yet there is much that remains to be explained. How do we account for the pattern of mass differences within a subfamily? How are the various subfamilies related? Since all the members of a subfamily have the same I and Y, could there be still more undiscovered subfamilies, corresponding to other (I, Y) combinations?

A significant theoretical advance was made in 1961 by Gell-Mann and Yuval Ne'eman, an Israeli physicist studying at the University of London. Gell-Mann and Ne'eman independently proposed equivalent theories defining the relationship between the (I, Y) values of the various baryon and meson subfamilies. They examined the conservation principles governing the particle quantum numbers (I, I_z, Q, S, B, and Y). Of these, Q and B are conserved in *all* interactions; Y, S, and I_z are conserved in strong and electromagnetic interactions; and I is conserved only in strong interactions. In addition mass–energy, linear momentum, and angular momentum are conserved in all reactions. Furthermore, Q, S, B, and Y are always integers $(= 0, 1, 2, 3\ldots)$, whereas I and I_z may be integers or half integers $(= \frac{1}{2}, 1, \frac{3}{2}, 2, \frac{5}{2}, \ldots)$. Finally, S, B, Y, Q, and I_z are related through Equations 21–2 and 21–3.

Altogether, Gell-Mann and Ne'eman identified eight mathematical relationships that describe the various absolute and partial conservation principles. Gell-Mann named this group of relationships the ***eightfold way.**** These relationships severely limit the possible combinations of I and Y that a real particle can possess. For example,

* Gell-Mann borrowed this name from Buddhist philosophy. The eightfold way consists of eight principles or practices that lead to enlightenment: right views, right intention, right speech, right action, right living, right effort, right mindfulness, right concentration.

table 21–5 **THE EIGHTFOLD WAY: BARYON AND MESON OCTETS**

(I, Y)	Associated multiplet	Baryons: Spin $= \frac{1}{2}$ (mass in MeV)	Mesons: Spin $= 0$ (mass in MeV)
(0, 0)	singlet ($Q = 0$)	Λ (1115)	η (548)
($\frac{1}{2}$, 1)	doublet ($Q = 0, 1$)	N (939)	K (496)
($\frac{1}{2}$, −1)	doublet ($Q = -1, 0$)	Ξ (1315)	$\overline{\text{K}}$ (496)
(1, 0)	triplet ($Q = -1, 0, 1$)	Σ (1193)	π (138)

a combination such as ($I = 1$, $Y = 1$) corresponds to a charge triplet ($I_Z = -1, 0, 1$) with $Q = -\frac{1}{2}, \frac{1}{2}, \frac{3}{2}$ (according to Equation 21–3). But no particles with fractional charges have ever been observed in nature, so this combination is ruled out by the mathematics of the eightfold way.

Besides limiting the possible (I, Y) pairs, the eightfold way combines them into a number of related groups that are the only ones allowed by the various conservation laws. One such group predicted by the mathematics of the eightfold way consists of the four combinations $(I, Y) = (0, 0), (\frac{1}{2}, 1), (\frac{1}{2}, -1)$, and $(1, 0)$, which represent four multiplets—a charge singlet, two doublets, and a triplet—for a total of $1 + 2 + 2 + 3 = 8$ particles. Thus these four related (I, Y) pairs comprise a ***supermultiplet,*** which breaks down into four related multiplets comprising an ***octet*** of particles.

Now let us re-examine the various subfamilies in Figures 21–3 and 21–4. We find that the four (I, Y) combinations $(0, 0), (\frac{1}{2}, 1), (\frac{1}{2}, -1), (1, 0)$ correspond exactly to the (I, Y) values of the Λ, N, Ξ, and Σ ground states, as shown in Table 21–5. These four ground states include eight baryons (all with intrinsic spin $\frac{1}{2}$), which comprise the ***baryon octet.*** (Note that the Δ state, with $s = \frac{3}{2}$, is not included). A similar octet groups together the corresponding antibaryons. The four meson ground states (η, K, $\overline{\text{K}}$, π) also represent eight particles—the ***meson octet***—which fit into the same scheme. Thus the eightfold way has provided a unifying mathematical principle by which the various baryon and meson ground states can be related.

When the octet representation was proposed by Gell-Mann and Ne'eman in 1961, the η^0 meson had not yet been discovered. In order to extend the octet scheme to the mesons, both men predicted the existence of a new meson having the properties required by the eightfold way. When the η^0 was discovered in 1962 with exactly the predicted properties, the Gell-Mann–Ne'eman theory gained recognition as a significant key to the understanding of the fundamental particles.

THE OMEGA-MINUS

An even more spectacular prediction grew out of Gell-Mann's attempts to bring the Δ quartet into a supermultiplet. The Δ multiplet is characterized by $I = \frac{3}{2}$, $Y = 1$, and an intrinsic spin of $\frac{3}{2}$, and is the only one of the five baryon subfamilies that does not fit into the octet scheme. However, the mathematical relationships of the eightfold way lead to a second group of related (I, Y) combinations:

$$(I, Y) = (\tfrac{3}{2}, 1), (1, 0), (\tfrac{1}{2}, -1), (0, 2)$$

corresponding to a quartet, triplet, doublet, and singlet, respectively, for a combined ***decuplet*** of ten particles $(4 + 3 + 2 + 1)$ altogether. The Δ quartet corresponds table 21–6 **THE BARYON DECUPLET (SPIN = $\frac{3}{2}$)**

(I, Y)	Associated multiplet	Particle	Mass (MeV)	ΔM
$(0, -2)$	singlet $(Q = -1)$	Ω^-	1679 (predicted)	147
$(\frac{1}{2}, -1)$	doublet $(Q = -1, 0)$	Ξ^*	1532	147
$(1, 0)$	triplet $(Q = -1, 0, 1)$	Σ^*	1385	147
$(\frac{3}{2}, 1)$	quartet $(Q = -1, 0, 1, 2)$	Δ	1238	

to the $(\frac{3}{2}, 1)$ combination. The $(1, 0)$ triplet can be associated with the Σ^* triplet (formerly called the Y_1^*) with a mass of 1385 MeV, which is the "first excited state" of the Σ family. Then in 1962 the first Ξ^* resonance (at 1532 MeV) was discovered, and Gell-Mann incorporated it into the new group as the $(\frac{1}{2}, -1)$ doublet. This left only the $(0, -2)$ singlet to be identified.

A particle with $I = 0$, $Y = -2$ must have a negative charge ($Q = I_Z + Y/2 = -1$) and a strangeness of -3 ($S = Y - B = -2 - 1 = -3$). Like the other members of the group (Δ, Σ^*, and Ξ^*), the particle should have an intrinsic spin of $\frac{3}{2}$. In 1962 there was no particle known to have such properties, but Gell-Mann predicted that such a particle could be produced in high-energy reactions. He named it the ***omega-minus*** (Ω^-), omega being the last letter of the Greek alphabet.

Gell-Mann also estimated the mass of the hypothetical particle from the pattern established by the other members of the decuplet. As shown in Table 21–6, the mass of the Σ^* is greater than that of the Δ by $1385 - 1238 = 147$ MeV. Similarly, the Ξ^* mass exceeds that of the Σ^* by $1532 - 1385 = 147$ MeV. Gell-Mann therefore predicted that the mass of the Ω^- would be 147 MeV greater than that of the Ξ^*, or 1679 MeV.

Gell-Mann's predictions were strikingly confirmed in 1964. After a three-month search in which more than 50,000 bubble chamber photographs were analyzed, a group of physicists* at Brookhaven National Laboratory finally observed the track of an Ω^-. (See Figure 21–5.) The Brookhaven synchrotron was used to produce a beam of 5 GeV K^- mesons, and the Ω^- was created in a reaction between a K^- and a proton in the liquid hydrogen bubble chamber:

$$K^- + p \longrightarrow \Omega^- + K^+ + K^0$$

An analysis of the track made by the Ω^- gave a value of 1675 MeV for the rest mass—remarkably close to Gell-Mann's prediction of 1679 MeV. The lifetime of the Ω^- was found to be 10^{-10} sec, indicating that it decays by a weak interaction in which strangeness is not conserved. This was expected, because the Ω^- is the only known particle with $S = -3$, so it must decay to a baryon of different strangeness. The reactions

$$\Omega^-\ (S = -3) \begin{cases} \longrightarrow \Xi^0\ (S = -2) + K^-\ (S = -1) \\ \longrightarrow \Xi^-\ (S = -2) + \bar{K}^0\ (S = -1) \end{cases}$$

conserve strangeness, but not mass–energy (the Ξ mass plus the K mass exceeds the Ω^- mass), so they cannot occur. Instead the Ω^- decays via the following weak

*The $2\frac{1}{2}$-page paper announcing the discovery of the Ω^- listed thirty-three coauthors.

interactions:

$$\Omega^- \begin{array}{l} \nearrow \Xi^- + \pi^0 \\ \longrightarrow \Xi^0 + \pi^- \\ \searrow \Lambda^0 + K^- \end{array}$$

Thus in 1964 the Ω^- was added to the family of particles. Because of its unique (I, Y) combination, the Ω^- cannot be classified as a higher energy level of any of the other known baryons and must be considered the ground state of a new baryon subfamily. Thus we now recognize *six* baryon subfamilies: N, Δ, Λ, Σ, Ξ, and Ω.

As a philosophical aside, let us take note of the pattern obtained when the members of the baryon decuplet are plotted on a graph of I_z and Y (Figure 21–6). There appears the mystical *tetractys*, worshiped for its symmetry by the Pythagoreans 2500 years ago. (See footnote on page 35.) The Pythagoreans were the first to believe

figure 21–5 **THE DISCOVERY OF THE OMEGA-MINUS**

The Ω^- was first observed in the bubble chamber photograph shown at the left. A K^- entered from the lower left and struck a stationary proton, producing a K^0, K^+, and Ω^-. The Ω^- subsequently decayed into a Ξ^0 and a π^-. The Ξ^0 then decayed to a Λ^- and a π^0 (not shown) that immediately decayed to two photons (γ_1 and γ_2). Each photon then converted to an electron-positron pair. The dashed lines in the diagram represent the paths of neutral particles, which leave no tracks in a bubble chamber. The other tracks in the photograph were made by charged particles not involved in the reaction.

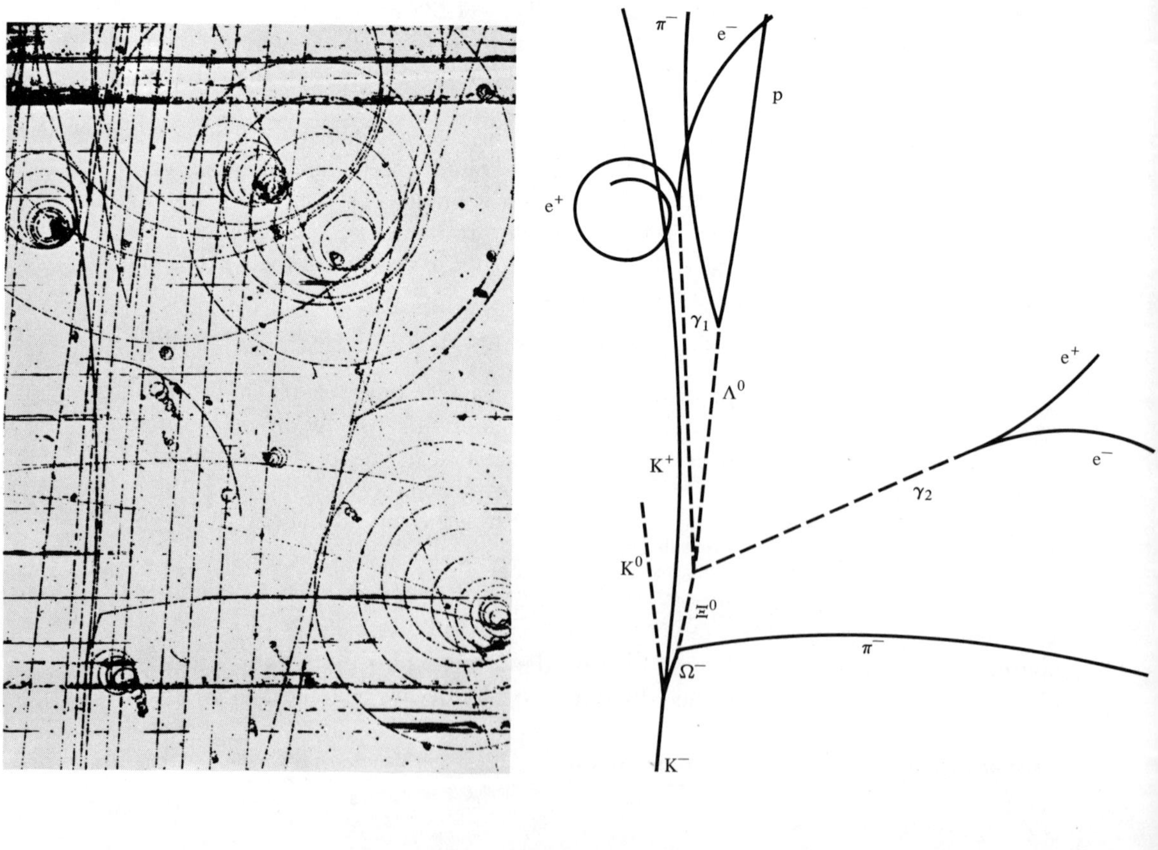

COURTESY BROOKHAVEN NATIONAL LABORATORY

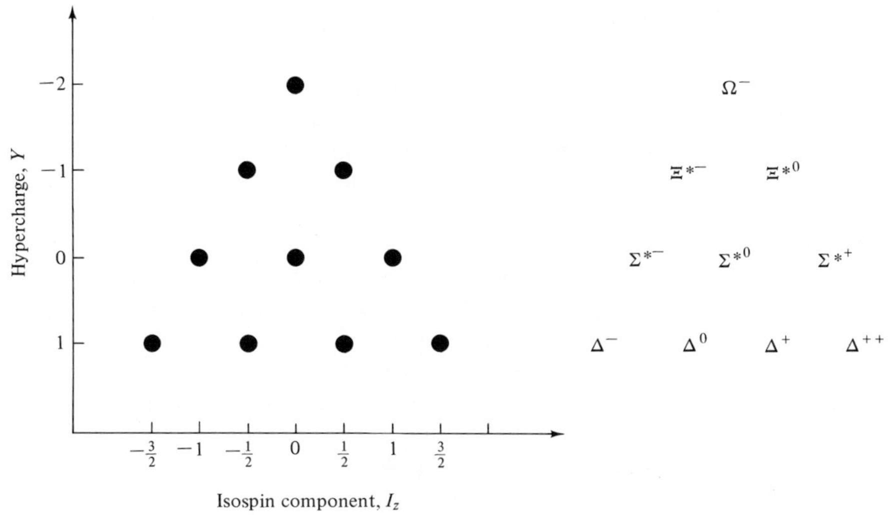

figure 21–6 **PYTHAGOREAN MYSTICISM AND MODERN PHYSICS**

When the particles of the baryon decuplet are plotted according to their hypercharge Y and isospin component I_z, they reveal a familiar pattern (see footnote on page 35).

in the underlying symmetry and mathematical harmony of the universe—an idea that re-emerges with Kepler and proceeds in an unswerving line through the works of Newton, Maxwell, Einstein, and Bohr. The sudden reappearance of the Pythagorean mystical symbol among the abstruse mathematical relationships of particle physics is no accident; it serves to remind us that the modern-day scientist still adheres to the Pythagorean faith.

21–5

QUARKS

The discovery of the Ω^- naturally represented a great triumph for the theory of the eightfold way. As we have learned, the various conservation principles impose mathematical relationships on the particle quantum numbers in such a way that particles of the same intrinsic spin are bound together in groups of eight (the octet) or ten (the decuplet). We have already identified a baryon ($s = \frac{1}{2}$) octet and baryon ($s = \frac{3}{2}$) decuplet, and a meson ($s = 0$) octet. In addition, one can identify in Figure 21–4 another meson octet of excited states with $s = 1$ (ω^0, ρ, K^*, $\bar{K}^*$). However, the other known excited baryon and meson states have not all been definitely identified as members of octets, decuplets, or higher supermultiplets. Thus the eightfold way has brought considerable—but by no means complete—order to the table of particles.

In spite of its notable success, the eightfold way leaves much to be desired as a physical theory. The various conservation laws on which the theory is based are unexplained; the forces underlying these conservation laws have not been identified. Nor is it understood why certain quantities (I, Y, S) are not conserved in weak interactions. There seems to be some connection between the mass differences between multiplets and the partial or "broken" conservation laws, but at the present time the connection is obscure.

Thus one is left with nothing more than a series of abstract mathematical relationships that make up the eightfold way. It is quite difficult to understand these relationships in simple terms and to explain how and why the various (I, Y) pairs are linked together. Thus to the student—and even to the physicist who has not specialized in particle physics—the eightfold way is every bit as mysterious as the Pythagorean numerology must have been to the local villagers in 500 B.C.

Gell-Mann made an attempt to give a physically understandable interpretation of the eightfold way in 1964. It had been recognized earlier that the mathematical relationships of the eightfold way and the observed properties of the baryon and meson families could be interpreted to mean that all baryons and mesons are *composite structures,* consisting of various combinations of *three* distinct fundamental subparticles. Gell-Mann adopted a scheme in which the three subparticles have *fractional charges.* He proposed to call these particles ***quarks***—a word that appears in the sentence "Three quarks for Muster Mark," found on p. 383 of James Joyce's *Finnegan's Wake.* One of these particles—the ***p-quark*** (q_p)—has a charge number $Q = \frac{2}{3}$; the other two—the ***n-quark*** (q_n) and ***lambda-quark*** (q_λ)—each have charge numbers of $Q = -\frac{1}{3}$. The p- and n-quarks form a doublet $(I = \frac{1}{2})$ of strangeness $S = 0$; the λ-quark forms a singlet $(I = 0)$ with $S = 1$. Each quark is stable and possesses an intrinsic spin of $\frac{1}{2}$ and a fractional baryon number of $B = \frac{1}{3}$. Corresponding to each quark is an antiquark of the opposite charge, strangeness, and so on. The properties of the three quarks and their antiquarks are listed in Table 21–7.

According to the quark hypothesis, each *baryon* is composed of *three quarks;* each *antibaryon* is composed of *three antiquarks;* and each *meson* consists of a *quark–antiquark pair.* Let us begin with the baryons. Since each quark has a baryon number $B = \frac{1}{3}$, three quarks together will have a total B of 1 (and three antiquarks will total $B = -1$). Each quark has an intrinsic spin of $\frac{1}{2}$, so that the net spin for a three-quark system must be $\frac{3}{2}$ or $\frac{1}{2}$, according to whether the spins are aligned parallel $(\uparrow\uparrow\uparrow)$ or antiparallel $(\uparrow\downarrow\uparrow)$. It is also assumed that a three-quark system is ***spin-symmetric,*** which means that *in any quark triplet identical quarks must have parallel spins.*

For example, let us consider a baryon constructed from three λ-quarks. Adding up their quantum numbers, we obtain for the resultant baryon

$$Q = -1, S = -3, Y = -2, I = 0, I_z = 0$$

table 21–7 **PROPERTIES OF QUARKS AND ANTIQUARKS**

Particle	Intrinsic spin	Q	B	I	I_z	S	Y
p-quark (q_p)	$\frac{1}{2}$	$\frac{2}{3}$	$\frac{1}{3}$	$\frac{1}{2}$	$\frac{1}{2}$	0	$\frac{1}{3}$
n-quark (q_n)	$\frac{1}{2}$	$-\frac{1}{3}$	$\frac{1}{3}$	$\frac{1}{2}$	$-\frac{1}{2}$	0	$\frac{1}{3}$
λ-quark (q_λ)	$\frac{1}{2}$	$-\frac{1}{3}$	$\frac{1}{3}$	0	0	−1	$-\frac{2}{3}$
anti-p-quark $(\bar{q}_p)$	$\frac{1}{2}$	$-\frac{2}{3}$	$-\frac{1}{3}$	$\frac{1}{2}$	$-\frac{1}{2}$	0	$-\frac{1}{3}$
anti-n-quark $(\bar{q}_n)$	$\frac{1}{2}$	$\frac{1}{3}$	$-\frac{1}{3}$	$\frac{1}{2}$	$\frac{1}{2}$	0	$-\frac{1}{3}$
anti-λ-quark $(\bar{q}_\lambda)$	$\frac{1}{2}$	$\frac{1}{3}$	$-\frac{1}{3}$	0	0	1	$\frac{2}{3}$

table 21–8 CONSTRUCTION OF BARYONS FROM QUARK TRIPLETS

Strange-ness	Quark triplet	Spin alignment	Intrinsic spin	Q	I	I_z	Particle: Spin-$\frac{1}{2}$ octet	Particle: Spin-$\frac{3}{2}$ decuplet
−3	λλλ	$\uparrow\uparrow\uparrow$ λ λ λ	$\frac{3}{2}$	−1	0	0		Ω^-
−2	pλλ	$\uparrow\uparrow\uparrow$ p λ λ	$\frac{3}{2}$	0	$\frac{1}{2}$	$\frac{1}{2}$		Ξ^{*0}
		$\downarrow\uparrow\uparrow$ p λ λ	$\frac{1}{2}$	0	$\frac{1}{2}$	$\frac{1}{2}$	Ξ^0	
	nλλ	$\uparrow\uparrow\uparrow$ n λ λ	$\frac{3}{2}$	−1	$\frac{1}{2}$	$-\frac{1}{2}$		Ξ^{*-}
		$\downarrow\uparrow\uparrow$ n λ λ	$\frac{1}{2}$	−1	$\frac{1}{2}$	$-\frac{1}{2}$	Ξ^-	
−1	ppλ	$\uparrow\uparrow\uparrow$ p p λ	$\frac{3}{2}$	1	1	1		Σ^{*+}
		$\uparrow\uparrow\downarrow$ p p λ	$\frac{1}{2}$	1	1	1	Σ^+	
	pnλ	$\uparrow\uparrow\uparrow$ p n λ	$\frac{3}{2}$	0	1	0		Σ^{*0}
		$\uparrow\uparrow\downarrow$ p n λ	$\frac{1}{2}$	0	1	0	Σ^0	
		$\uparrow\downarrow\uparrow$ p n λ or $\uparrow\downarrow\downarrow$ p n λ	$\frac{1}{2}$	0	0	0	Λ^0	
	nnλ	$\uparrow\uparrow\uparrow$ n n λ	$\frac{3}{2}$	−1	1	−1		Σ^{*-}
		$\uparrow\uparrow\downarrow$ n n λ	$\frac{1}{2}$	−1	1	−1	Σ^-	
0	ppp	$\uparrow\uparrow\uparrow$ p p p	$\frac{3}{2}$	2	$\frac{3}{2}$	$\frac{3}{2}$		Δ^{++}
	ppn	$\uparrow\uparrow\uparrow$ p p n	$\frac{3}{2}$	1	$\frac{3}{2}$	$\frac{1}{2}$		Δ^+
		$\uparrow\uparrow\downarrow$ p p n	$\frac{1}{2}$	1	$\frac{1}{2}$	$\frac{1}{2}$	p	
	pnn	$\uparrow\uparrow\uparrow$ p n n	$\frac{3}{2}$	0	$\frac{3}{2}$	$-\frac{1}{2}$		Δ^0
		$\downarrow\uparrow\uparrow$ p n n	$\frac{1}{2}$	0	$\frac{1}{2}$	$-\frac{1}{2}$	n	
	nnn	$\uparrow\uparrow\uparrow$ n n n	$\frac{3}{2}$	−1	$\frac{3}{2}$	$-\frac{3}{2}$		Δ^-

table 21–9 **CONSTRUCTION OF MESONS FROM QUARK–ANTIQUARK PAIRS**

Strangeness	Quark–antiquark	Q	I	I_z	Particle: Spin 0 (↑↓)	Particle: Spin 1 (↑↑)
0	$p\bar{n}$	1	1	1	π^+	ρ^+
	$p\bar{p}$ or $n\bar{n}$	0	1	0	π^0	ρ^0
	$n\bar{p}$	-1	1	-1	π^-	ρ^-
	$\lambda\bar{\lambda}$	0	0	0	η^0	ω^0
1	$p\bar{\lambda}$	1	$\frac{1}{2}$	$\frac{1}{2}$	K^+	K^{*+}
	$n\bar{\lambda}$	0	$\frac{1}{2}$	$-\frac{1}{2}$	K^0	K^{*0}
-1	$\lambda\bar{n}$	0	$\frac{1}{2}$	$\frac{1}{2}$	$\bar{K}^0$	$\bar{K}^{*0}$
	$\lambda\bar{p}$	-1	$\frac{1}{2}$	$-\frac{1}{2}$	K^-	K^{*-}

This is the Ω^-. Note that it is the only possible particle with strangeness -3 that can be constructed out of three quarks. The intrinsic spins of the three λ-quarks must be aligned parallel to one another (↑ ↑ ↑); hence the Ω^- has a spin of $\frac{3}{2}$. The alignment (↑ ↓ ↑) is ruled out because two λ-quarks would have opposite spins, in violation of the spin-symmetry requirement. Therefore a spin-$\frac{1}{2}$ Ω^- is not possible according to the quark model, and in fact no spin-$\frac{1}{2}$ Ω^- is known to exist. On the other hand, a combination of one p-quark and two λ-quarks (a combination that has the same total quantum numbers as the Ξ^0) can have two separate spin alignments (↑ ↑ ↑ or ↓ ↑ ↑). The two λ-quarks must have parallel spins, but the p-quark can be aligned parallel or antiparallel to them. Thus there are two Ξ^0 states, spin $\frac{1}{2}$ and spin $\frac{3}{2}$, corresponding to the Ξ^0 and the Ξ^{*0}, respectively.

All the possible distinguishable quark triplets have been listed in Table 21–8. Each one of the eight members of the baryon octet and the ten members of the baryon decuplet can be matched to a quark triplet. The corresponding antibaryons can be constructed from similar combinations of three antiquarks. Table 21–9 shows how the eight spin-0 mesons and eight spin-1 mesons correspond to the various quark–antiquark combinations. Higher-order baryon and meson states (i.e., the higher energy levels not included in the octet or decuplet schemes) can be constructed out of more complicated combinations of more than three quarks and antiquarks. In this manner the relationships among the various baryons and mesons described mathematically by the eightfold way can be understood by assuming that baryons and mesons are composed of various combinations of three fundamental particles and their antiparticles. Thus the eightfold way reduces to the fundamental quark triad.*

The quark model has been successful in accounting for many of the properties of baryons and mesons and in predicting the outcomes of high-energy particle reactions, but there are a number of unresolved difficulties. It is not understood how quarks are arranged inside a baryon; apparently, they would have to be bound together by an extremely strong or "superstrong" force, which is at least several hundred times stronger than the known "strong" force. At present there is no evidence pertaining

* Coincidentally, the eight principles of the Buddhist eightfold way are arranged into three classes: higher morality, higher thought, and higher insight.

to the nature of this superstrong force. Furthermore, there is considerable uncertainty as to the masses of the three quarks. It is not completely understood how the observed baryon and meson masses can be related to the fundamental quark masses. The data seem to indicate that the λ-quark has a greater mass than the other two. but otherwise the quark masses cannot be reliably determined.

Perhaps the most damaging evidence against the quark hypothesis is the fact that no quarks have ever been observed. Quarks would be easily detectable in cloud or bubble chamber photographs because of their fractional charges. But intensive searches of cosmic-ray and particle-accelerator reaction products have failed to yield definite evidence for the existence of a free quark. Proponents of the quark model believe that the binding force between quarks may be so large that the energy supplied in the particle reactions presently being observed is insufficient to release free quarks. Other physicists argue that quarks simply do not exist in nature and that the quark model is nothing more than a hypothetical scheme with which the mathematical relationships of the eightfold way can be visualized.

There is some experimental evidence that provides rather indirect support for the physical reality of quarks. Electrons produced by the 20-GeV linear accelerator at Stanford University have been used to probe the internal structure of protons and neutrons. The experiment is quite analogous to Rutherford's alpha-scattering experiment, in which the pattern of alpha-particle scattering by atoms revealed the internal structure of the atom. In the Stanford experiments the high-energy electrons penetrate the nucleon and are scattered by interactions inside the particle. The available evidence to date seems to indicate that protons and neutrons have a complicated internal structure. Electrons appear to be scattered by interactions with point particles distributed within the nucleon. The scattering data are consistent with the idea that these point particles (which have been called ***partons***) have fractional charges and an intrinsic spin of $\frac{1}{2}$. It is possible that partons may very well be quarks bound together in some unspecified way, although this identification is by no means certain or even widely accepted at the present time.

Thus the question of whether or not quarks are physically real entities remains unanswered to date. Particle physicists are hopeful that the new or proposed 400 to 1000 GeV synchrotrons will provide the experimental data needed to solve this crucial problem.

WHAT IS A FUNDAMENTAL PARTICLE?

Progress in the field of fundamental particle physics has followed the classic pattern embodied in the scientific method. At the beginning *observations* were made (the discoveries of pions, muons, and the various strange particles) and data on particle masses, spins, decay processes, and so on were collected. Next came the *classification* of particles into families according to their properties, followed by the formulation of a number of *laws* (conservation of baryons, strangeness, etc.) based on the classification schemes and on particle reaction data. Then a comprehensive *theory* (the eightfold way) was constructed in which the various laws were unified in various mathematical relationships. This theory was then used to *predict* the existence and properties of new particles (the η^0 and Ω^-) and new particle interactions.

The picture, however, is by no means complete. A good Pythagorean might believe that the mathematical relationships of the eightfold way represent the ultimate explanation for the existence of the baryon and meson families. But from another point of view, one might argue that the eightfold way does not explain the existence of particles any more than Balmer's formula explains the pattern of lines in the hydrogen spectrum. It seems more reasonable to believe that the mathematical relationships of the eightfold way are merely manifestations of the properties of an underlying stratum of subparticles—possibly quarks. Until this argument is settled, a clear understanding of the nature of the fundamental particles of matter cannot be achieved.

It would be helpful at this point to remember just what it is we are searching for. A fundamental particle should be an *indivisible* concentration of mass–energy that is localized at a point in space (subject to the restrictions of the uncertainty principle). A fundamental particle may or may not possess other properties, such as charge, intrinsic spin, or intrinsic magnetism. A fundamental particle should not have an internal structure, for an internal structure implies a differentiation between two or more parts, which must then be regarded as more fundamental than the particle itself. Similarly, a fundamental particle should be stable against disintegration, for decay implies divisibility, which contradicts our idea of fundamentality.

If we now examine the list of known particles in the light of these considerations, we find ourselves in a rather confused situation. All the known baryons and mesons seem to be composite structures; with the sole exception of the proton, they are all unstable. The proton is rather puzzling to us because it is entirely stable—hence "indivisible"—yet it has an internal structure. The same sort of things were said about atoms in the mid-nineteenth century, but when suitable energy sources became available, the atom was found to be decomposable into smaller structures. Perhaps at ultra-high energies much greater than what is presently attainable it will be possible to decompose a proton into its constituents. If it turns out that the proton and the other baryons and mesons are truly composed of quarks, obviously the three quarks may justifiably be classified as fundamental particles. But we are not at all certain that quarks exist.

In the lepton family we find three stable particles—the electron and the two neutrinos—and all three seem to qualify as fundamental particles. Yet we know very little about the properties and interactions of neutrinos and are at a total loss to explain why there are two of them. The electron has the best credentials as a fundamental particle. It does not, as far as we know, have any internal structure or compositeness. But its relationship to the muon has yet to be explained. The exact equality of the proton and electron charges is also a deep mystery. If the proton is composed of quarks possessing *fractional* positive and negative charges, then the *integral* negative charge of the electron becomes a puzzle.

The massless photon (and the graviton, if it exists) is difficult to classify; one generally thinks of it as "energy" rather than "matter." Whether or not it should be considered a fundamental particle becomes a matter of definition. One problem is that under certain conditions a photon can disintegrate or transform into a particle and an antiparticle. This phenomenon in turn brings us to the question of antimatter. Do we consider antiparticles as particle "holes" or as independent particles in their own right?

Considering all of these difficulties, it should be evident that the age-old question "What is the ultimate nature of matter?" is still far from being answered. In this century physicists have proceeded from the atom to the nucleus to the subnuclear level, and the end is still not in sight. This progress—or lack of it, depending on one's perspective—serves to remind us that the field of fundamental particle physics is still in its infancy. It represents one of the many frontiers of human knowledge to be crossed by future generations.

QUESTIONS FOR REVIEW

1 Discuss the classification of particles into fermions and bosons. What are the general properties of each class? Why are different statistical formulas needed for each class?

2 Identify the four major particle families.

3 What is a graviton? Why is it so difficult to detect? What properties does a graviton have in common with a photon?

4 List the members of the lepton family. What are some of the difficulties involved in attempting to understand the properties of this family?

5 State and compare the general properties of the baryon and the meson families.

6 State and explain the principle of conservation of baryons. What are the rules for determining the baryon number of a given particle?

7 Discuss the principle of conservation of strangeness. List the strangeness numbers for the baryons, antibaryons, and mesons listed in Table 21–1.

8 What criteria do we use to determine whether a reaction is controlled by the strong, electromagnetic, or weak interaction? Identify the type of interaction that governs the decay of each of the unstable particles listed in Table 21–1.

9 Discuss the difficulties involved in detecting the existence of particles that decay by a strong interaction. How are such particles identified?

10 Discuss the grouping of particles into charge multiplets (singlets, doublets, and triplets). How do the members of a multiplet differ from one another? Explain how an isospin quantum number (I) and an isospin component (I_Z) are assigned to the members of a multiplet.

11 Define hypercharge (Y). How is the hypercharge related to strangeness for baryons, antibaryons, and mesons? In what reactions is hypercharge conserved?

12 Explain how the classification of particles according to I and Y enables us to relate resonances to the longer-lived baryons and mesons. How can we explain the much shorter lifetimes of the resonance particles?

13 How many different (I, Y) combinations are needed to classify all baryons? All mesons? What are these combinations?

14 In general terms, what is the eightfold way? What restrictions does the eightfold way place on the possible combinations of I and Y?

15 Show that the members of (a) the baryon octet, (b) the antibaryon octet, and (c) the meson octet are all linked by the same supermultiplet of related (I, Y) pairs.

16 Identify the decuplet scheme that links the spin-$\frac{3}{2}$ particles. How did this grouping lead to the prediction of the existence of the Ω^-?

17 Describe the properties of the Ω^-. Why can't it decay by a strangeness-conserving reaction?

18 Discuss the quark hypothesis. What is the purpose of the concept of quarks?

19 Identify the properties of the three kinds of quarks and antiquarks.

20 According to the quark hypothesis, how are baryons, antibaryons, and mesons constructed? What is the spin-symmetry rule?

21 Show that every possible quark–antiquark combination leads to a known meson.

22 Discuss the present status of fundamental particle physics. What are the major unsolved problems?

EXERCISES

1 What properties of the track made by a particle in a photographic emulsion or a cloud or bubble chamber are used to determine the particle's (a) energy, (b) lifetime, (c) charge, and (d) mass?

2 List the properties of each of the following antiparticles, and write the reaction by which it decays:
(a) The charged anti-xi.
(b) The negatively charged anti-sigma.
(c) The anti-lambda.
(d) The antineutron.

3 Explain why kaons are never emitted by baryons when the baryons decay.

4 Verify that strangeness is conserved in all of the observed reactions in Table 21–2 and not in the nonobserved reactions.

5 Discuss the merits of the concept of strangeness. Does the concept have any physical meaning? What is the value of this concept?

6 A principle of conservation of leptons can be formulated by assigning a lepton number of $L = 1$ to the e^-, μ^-, ν_e, and ν_μ and $L = -1$ to the e^+, μ^+, $\overline{\nu}_e$, and $\overline{\nu}_\mu$. Show that the total lepton number is conserved in each of the following reactions:
(a) $\mu^- \longrightarrow e^- + \nu_\mu + \overline{\nu}_e$
(b) $e^- + e^+ \longrightarrow 2\gamma$
(c) $n \longrightarrow p + e^- + \overline{\nu}_e$
(d) $\pi^+ \longrightarrow \mu^+ + \nu_\mu$
(e) $\overline{\nu}_\mu + p \longrightarrow n + \mu^+$
(f) $K^0 \longrightarrow \pi^+ + \mu^- + \overline{\nu}_\mu$

7 Determine whether or not any of the following reactions is forbidden by the principles of conservation of baryons, strangeness, or charge. (It is assumed in each case that the reacting particles have sufficient kinetic energy to satisfy conservation of mass–energy.)
(a) $\pi^- + p \longrightarrow \Lambda^0 + \pi^0$
(b) $\mu^- + p \longrightarrow \pi^+ + \pi^0 + \pi^-$
(c) $p + \overline{p} \longrightarrow K^0 + \overline{K}^0$
(d) $K^+ + p \longrightarrow \Lambda^0 + \pi^+ + \pi^+$
(e) $\gamma + p \longrightarrow K^+ + \Lambda^0$
(f) $\overline{p} + n \longrightarrow K^+ + \overline{K}^0$
(g) $p + n \longrightarrow \Sigma^0 + K^0 + \pi^+$
(h) $p + \overline{n} \longrightarrow n + \overline{p}$

8 Explain why the decay reaction $\Sigma^0 \longrightarrow \Lambda^0 + \gamma$ is permissible, whereas the following reactions are forbidden:
(a) $\Xi^0 \longrightarrow \Sigma^0 + \gamma$
(b) $\Sigma^0 \longrightarrow \Lambda^0 + \pi^0$

9 In each of the following reactions, identify the conservation principle that is violated:
(a) $n \longrightarrow \mu^+ + e^- + \nu_\mu + \overline{\nu}_e$
(b) $K^- \longrightarrow \mu^- + \overline{\nu}_\mu$
(c) $\overline{n} \longrightarrow \overline{p} + e^- + \overline{\nu}_e$
(d) $\Xi^0 \longrightarrow \Sigma^0 + \overline{K}^0$
(e) $\Xi^0 \longrightarrow \Lambda^0 + \pi^0$
(f) $\Lambda^0 \longrightarrow n + \overline{K}^0$
(g) $\overline{\Lambda}^0 \longrightarrow \overline{p} + \pi^+$
(h) $\eta^0 \longrightarrow \pi^+ + \pi^+ + \pi^-$
(i) $\Sigma^- \longrightarrow \pi^0 + \pi^0 + K^-$
(j) $\Sigma^+ \longrightarrow \Lambda^0 + \mu^+ + \nu_\mu$

10 Show that the reaction $K^- + p \longrightarrow \Lambda^0 + \pi^+ + \pi^-$ may occur as a two-step process involving the production of the negatively charged Y_1^{*-} resonance particle.

11 The Y_0^* resonance particle decays by the reaction $Y_0^* \longrightarrow \Lambda^0 + \pi^0$. What is the strangeness number of the Y_0^*? Explain your answer.

12 Verify the fact that each of the particles listed in Table 21–4 satisfies Equation 21–1.

13 Consider a hypothetical multiplet with $I = 2$:
(a) How many particles are included in the multiplet?
(b) What are the isospin components of each of these particles?
(c) If the particles are baryons with a strangeness number $= 1$, what are the charges on each of the particles?

14 Explain why the hypercharge of an antiparticle is always equal to the negative of the hypercharge of the corresponding particle.

15 Using the data given in Table 21–3, assign an isospin (I) and hypercharge (Y) to each of the following resonance multiplets:
(a) Y_1^*
(b) N^*
(c) η^0
(d) ρ

16 Explain why the following (I, Y) combinations correspond to multiplets that cannot exist in nature. (Hint: Use Equation 21–3 to calculate the charges of the members of the multiplet.)
(a) $I = 0,\ Y = 1$
(b) $I = \frac{1}{2},\ Y = 0$
(c) $I = 1,\ Y = -1$
(d) $I = \frac{3}{2},\ Y = 2$

MATHEMATICAL AIDS AND FORMULAS

APPENDIX A

A–1

EXPONENTS AND POWERS-OF-TEN NOTATION

The use of exponents in mathematics provides a useful shorthand notation which greatly simplifies many calculations. We will first review the basic definitions and rules governing the use of exponents.

DEFINITIONS

The mathematical expression x^n signifies that the number x is multiplied by itself n times (n is the exponent of x). We say that "x is raised to the nth power." That is,

$$x^n = x \cdot x \cdot x \cdot x \cdot \cdots \quad (n \text{ times})$$

For example,

$$2^4 = 2 \cdot 2 \cdot 2 \cdot 2 = 16$$
$$3^3 = 3 \cdot 3 \cdot 3 = 27$$
$$5^2 = 5 \cdot 5 = 25$$

A *negative exponent* signifies a fraction, or more specifically the reciprocal of the positive exponent:

$$x^{-n} = \frac{1}{x^n} = \frac{1}{x \cdot x \cdot x \cdot x \cdot \cdots (n \text{ times})}$$

Thus

$$2^{-4} = \frac{1}{2^4} = \frac{1}{2 \cdot 2 \cdot 2 \cdot 2} = \frac{1}{16}$$

$$3^{-3} = \frac{1}{3^3} = \frac{1}{3 \cdot 3 \cdot 3} = \frac{1}{27}$$

A *fractional exponent* represents the *root* of a number; that is, $x^{1/n} = \sqrt[n]{x}$. Thus $x^{1/2}$ = square root of x; $x^{1/3}$ = cube root of x; and so on. For example,

$$
\begin{aligned}
2^{1/2} &= \sqrt{2} = 1.414 \\
9^{1/2} &= \sqrt{9} = 3 \\
27^{1/3} &= \sqrt[3]{27} = 3 \\
4^{3/2} &= \sqrt{(4)^3} = \sqrt{64} = 8 \\
8^{2/3} &= \sqrt[3]{(8)^2} = \sqrt[3]{64} = 4
\end{aligned}
$$

MULTIPLICATION

The multiplication of two or more numbers containing the same base number raised to different exponents proceeds as follows:

$$x^m \cdot x^n = x^{m+n}$$

In other words, the rule is to *add the exponents*. Note that

$$x^2 \cdot x^3 = (x \cdot x)(x \cdot x \cdot x) = (x \cdot x \cdot x \cdot x \cdot x) = x^5$$

Thus

$$
\begin{aligned}
2^2 \cdot 2^3 &= 2^5 \quad (\text{i.e., } 2^2 = 4,\ 2^3 = 8;\ 2^2 \cdot 2^3 = 4 \cdot 8 = 32 = 2^5) \\
4^3 \cdot 4^6 &= 4^9 \\
10^3 \cdot 10^4 \cdot 10^6 &= 10^{13}
\end{aligned}
$$

The rule also applies to negative and fractional exponents:

$$
\begin{aligned}
2^6 \cdot 2^{-2} &= 2^{6+(-2)} = 2^4 \\
5^3 \cdot 5^{-4} &= 5^{3+(-4)} = 5^{-1} \\
4^{3/2} \cdot 4^{1/2} &= 4^{3/2+1/2} = 4^2 \\
10^7 \cdot 10^{-5} \cdot 10^4 &= 10^6
\end{aligned}
$$

Note also that

$$
\begin{aligned}
(10^2)^3 &= 10^2 \cdot 10^2 \cdot 10^2 = 10^6 \\
(5^2)^4 &= 5^2 \cdot 5^2 \cdot 5^2 \cdot 5^2 = 5^8
\end{aligned}
$$

Or, in general terms,

$$(x^m)^n = x^{m \cdot n}$$

DIVISION

To divide two numbers containing the same base raised to different exponents, we *subtract* the exponents:

$$\frac{x^m}{x^n} = x^{m-n}$$

For example,

$$\frac{x^5}{x^3} = \frac{x \cdot x \cdot x \cdot x \cdot x}{x \cdot x \cdot x} = x \cdot x = x^2$$

Thus

$$\frac{2^6}{2^2} = 2^{6-2} = 2^4 \quad (2^6 = 64;\ 2^2 = 4;\ 2^6/2^2 = 64/4 = 16 = 2^4)$$

$$\frac{3^8}{3^6} = 3^2$$

$$\frac{10^5}{10^{-3}} = 10^{5-(-3)} = 10^8$$

$$\frac{4^{-6}}{4^3} = 4^{-6-3} = 4^{-9}$$

$$\frac{5^2 \cdot 5^{-3}}{5^{-1} \cdot 5^4} = \frac{5^{-1}}{5^3} = 5^{-4}$$

Note the following peculiarity:

$$\frac{x^n}{x^n} = x^{n-n} = x^0$$

But $x^n/x^n = 1$ (a number divided by itself equals 1); therefore $x^0 = 1$. That is, *any number raised to the zero power equals 1.*

EXERCISES

Evaluate the following:

1 3^4

2 $3^2 \cdot 3^3$

3 $4^{-2} \cdot 4^{-1}$

4 $10^5 \cdot 10^{-3}$

5 $10^5/10^{-3}$

6 $\dfrac{5^{-3}}{5^{-5}}$

7 $(4^{-3})^2$

8 $(125)^{1/3}$

9 $\left(\frac{1}{4}\right)^3$

10 $\left(\frac{1}{2}\right)^5\left(\frac{1}{2}\right)^{-3/2}$

11 $\dfrac{2^6 \cdot 2^4 \cdot 2^{-3}}{2^{-5} \cdot 2^7}$

12 $\dfrac{4^2 \cdot 4^{-3} \cdot (4^{-1/2})^3}{4^{1/2} \cdot 4^{-2}}$

EXPONENT ARITHMETIC OF BASE 10

Because of the way our number system is defined, the exponent arithmetic of the number 10 is particularly easy to work with. It is rather simple to determine the value of 10^n, where n is any integer. Note that

$$\begin{aligned} 10^0 &= 1 \\ 10^1 &= 10 \\ 10^2 &= 100 \\ 10^3 &= 1000 \\ 10^4 &= 10{,}000 \\ 10^5 &= 100{,}000 \\ &\text{etc.} \end{aligned}$$

We see that the numbers follow the simple rule that *the number of zeros is equal to the exponent.* Thus, for example, we know immediately that 10^{14} is equal to 1 followed by 14 zeros. Equivalent numbers using other bases, like 2^{14} or 5^9, are much more difficult to evaluate.

A simple rule can also be developed for negative exponents of 10:

$$10^{-1} = \frac{1}{10} = 0.1$$

$$10^{-2} = \frac{1}{10^2} = \frac{1}{100} = 0.01$$

$$10^{-3} = \frac{1}{10^3} = \frac{1}{1000} = 0.001$$

$$10^{-4} = \frac{1}{10^4} = \frac{1}{10{,}000} = 0.0001$$

etc.

We find that *the number of places to the left of the 1 is equal to the exponent* (as an alternative rule, we can say that the number of zeros plus 1 equals the exponent).

The rules for multiplication and division are the same as for any base.

$$10^m \cdot 10^n = 10^{m+n}$$

$$10^m/10^n = 10^{m-n}$$

$$(10^m)^n = 10^{m \cdot n}$$

etc.

POWERS-OF-TEN REPRESENTATION OF A NUMBER

The simplicity of the powers-of-ten arithmetic enables us to develop a shorthand for writing very large or very small numbers in an abbreviated form. By factoring out enough zeros (powers of ten), it is possible to write any number as a coefficient (with a value between 1 and 10) times some power of ten. For example

$$3{,}790{,}000 = 3.79 \times 1{,}000{,}000 = 3.79 \times 10^6$$

or

$$0.000322 = 3.22 \times 0.0001 = 3.22 \times 10^{-4}$$

The rules for converting a number to powers-of-ten form are as follows:

If the number is *greater than* 1, the decimal place is moved to the *left*, and the exponent of 10 is equal to the number of places moved. Thus:

$$93{,}000{,}000 = 9.3 \times 10^7$$

$$177.6 = 1.776 \times 10^2$$

$$15.4 = 1.54 \times 10^1$$

If the number is a *decimal fraction*, the decimal place is moved to the *right*, and the exponent of 10 is equal to *minus* the number of places moved. For example,

$$0.00917 = 9.17 \times 10^{-3}$$

$$0.127 = 1.27 \times 10^{-1}$$

$$0.0000026 = 2.6 \times 10^{-6}$$

EXERCISES

Write the following numbers in powers-of-ten form:

13 238,800

14 0.0162

15 26.3

16 0.2

17 15,000,000,000

18 0.0000000166

19 3461.5

20 0.0040

MULTIPLICATION AND DIVISION

The multiplication and division of large or small numbers is greatly simplified by the use of powers-of-ten notation. Suppose we wish to multiply $46{,}000{,}000 \times 0.0000052$; the first step is to put each number in powers-of-ten form.

$$46{,}000{,}000 \times 0.0000052 = (4.6 \times 10^7)(5.2 \times 10^{-6})$$

Next we multiply the coefficients separately, then add the exponents:

$$(4.6 \times 10^7)(5.2 \times 10^{-6}) = (4.6)(5.2) \times 10^7 \times 10^{-6} = 23.92 \times 10^1$$

To get the answer in standard form, we move the decimal point one place to the left (increasing the exponent by one):

$$23.92 \times 10^1 = (2.392 \times 10^1) \times 10^1 = \underline{2.392 \times 10^2}$$

Division is performed in similar fashion:

$$\frac{300{,}000{,}000}{0.00006} = \frac{3 \times 10^8}{6 \times 10^{-5}} = \frac{3}{6} \times 10^{8+5} = 0.5 \times 10^{13}$$
$$0.5 \times 10^{13} = (5 \times 10^{-1}) \times 10^{13} = \underline{5 \times 10^{12}}$$

Here, of course, we *divide* the coefficients separately, and *subtract* the exponents. These techniques greatly simplify complicated computations involving multiplication or division of large or small numbers; for example,

$$\begin{aligned}\frac{(1{,}600{,}000)(0.05)^2}{(250)(0.00006)} &= \frac{(1.6 \times 10^6)(5 \times 10^{-2})^2}{(2.5 \times 10^2)(6 \times 10^{-5})} \\ &= \frac{(1.6)(5)^2}{(2.5)(6)} \times \frac{10^6 \times 10^{-4}}{10^2 \times 10^{-5}} \\ &= \frac{40}{15} \times \frac{10^2}{10^{-3}} = \underline{2.67 \times 10^5}\end{aligned}$$

EXERCISES

Perform the following calculations using the powers-of-ten format:

21 $(0.000056)(15{,}000)$

22 $(200{,}000{,}000)(900{,}000)$

23 $(0.0000069)(0.00028)$

24 $(27{,}000)/(0.0054)$

25 $(0.0066)/(0.02)^2$

26 $(160)/(300{,}000)$

27 $(500)^2(630)/(0.0031)$

28 $(0.036)/(2000)^3(0.05)$

29 $(8000)(0.00067)/(15)(16{,}000)$

30 $(0.056)(960)/(0.28)(0.0001)^4$

A–2

GEOMETRIC FORMULAS

Circle (radius $= r$)

$$\text{Circumference} = 2\pi r$$
$$\text{Area} = \pi r^2$$

Rectangle

$$\text{Perimeter} = 2(l + w)$$
$$\text{Area} = l \cdot w$$

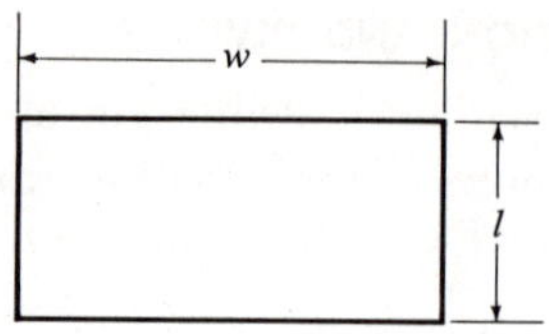

Triangle

$$\text{Area} = \tfrac{1}{2}(\text{base}) \times (\text{altitude}) = \tfrac{1}{2}bh$$

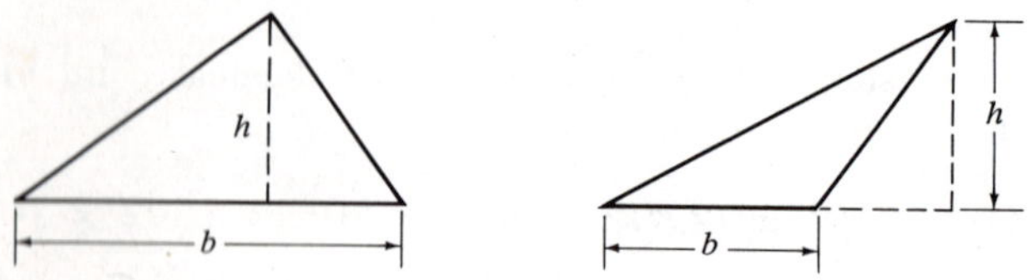

Right Triangle

$$a^2 + b^2 = c^2 \quad \text{(Pythagorean theorem)}$$

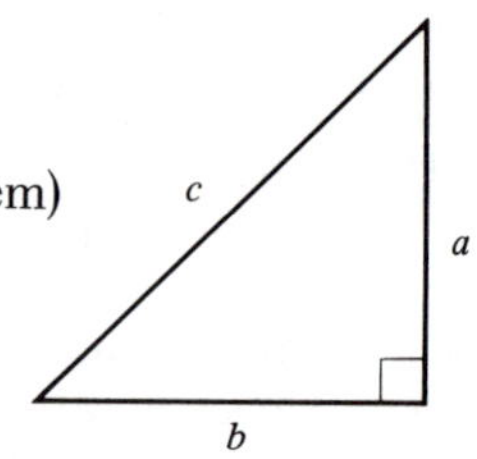

Special right triangles:

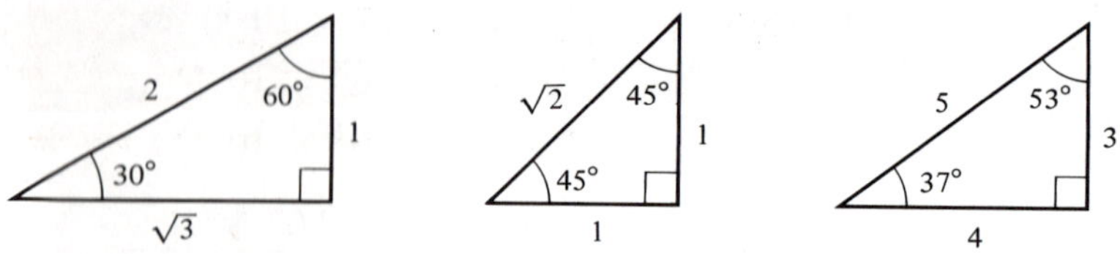

Cylinder (radius $= r$, height $= h$)

$$\text{Volume} = \pi r^2 h$$

Sphere (radius $= r$)

$$\text{Surface area} = 4\pi r^2$$
$$\text{Volume} = \frac{4\pi r^3}{3}$$

A–3

ANGULAR MEASURE

Degrees, Minutes, and Seconds of Arc

A circle contains 360°
One degree of arc = 60 minutes of arc ($1° = 60'$)
One minute of arc = 60 seconds of arc ($1' = 60''$)
$1° = 3600''$

Radians

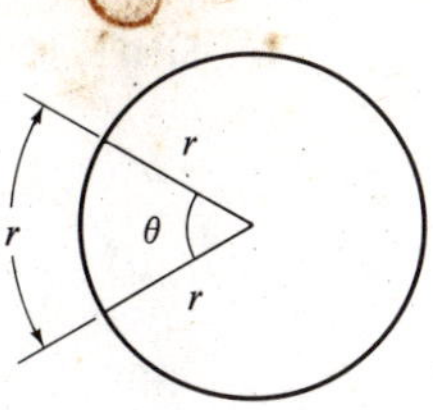

$\theta = 1$ radian

One *radian* is defined as the angle of the segment of a circle such that the *arc length* is equal to the *radius*. Since the circumference of a circle $= 2\pi r$, a circle contains 2π radians. Therefore

$$2\pi \text{ radians} = 360°$$
$$1 \text{ radian} = \frac{360°}{2\pi} = 57°\ 18' \approx 57.3°$$
$$1° = 0.01745 \text{ radian}$$

CONVERSION TABLE: DEGREES VERSUS RADIANS

Degrees	Radians	Degrees	Radians	Degrees	Radians
30°	$\frac{\pi}{6}$	150°	$\frac{5\pi}{6}$	270°	$\frac{3\pi}{2}$
60°	$\frac{\pi}{3}$	180°	π	300°	$\frac{5\pi}{3}$
90°	$\frac{\pi}{2}$	210°	$\frac{7\pi}{6}$	330°	$\frac{11\pi}{6}$
120°	$\frac{2\pi}{3}$	240°	$\frac{4\pi}{3}$	360°	2π

A–4

TRIGONOMETRIC FORMULAS

$$\sin\theta = \frac{\text{opposite}}{\text{hypotenuse}} = \frac{BC}{AC}$$
$$\cos\theta = \frac{\text{adjacent}}{\text{hypotenuse}} = \frac{AB}{AC}$$
$$\tan\theta = \frac{\text{opposite}}{\text{adjacent}} = \frac{BC}{AB}$$

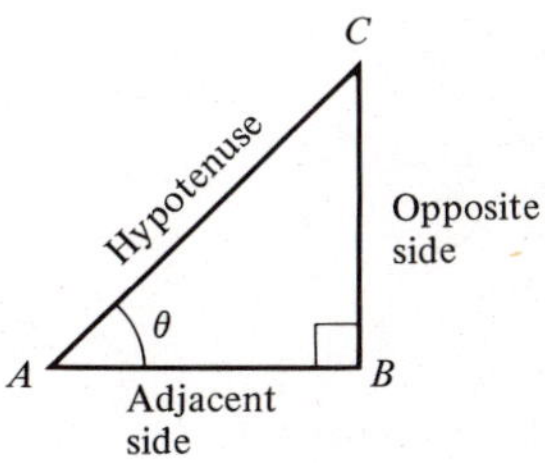

$$\tan\theta = \frac{\sin\theta}{\cos\theta} \qquad \sin^2\theta + \cos^2\theta = 1$$

$$\sin(x \pm y) = (\sin x)(\cos y) \pm (\cos x)(\sin y)$$
$$\cos(x \pm y) = (\cos x)(\cos y) \mp (\sin x)(\sin y)$$

SINES, COSINES AND TANGENTS OF SELECTED ANGLES

θ	$\sin\theta$	$\cos\theta$	$\tan\theta$
0°	0	1	0
15°	0.259	0.966	0.268
30°	0.500	0.866	0.577
45°	0.707	0.707	1
60°	0.866	0.500	1.732
75°	0.966	0.259	3.732
90°	1	0	∞
180°	0	-1	0
270°	-1	0	$-\infty$

ANGLES GREATER THAN 90°

	θ between 90° and 180°	θ between 180° and 270°	θ between 270° and 360°
$\sin\theta$	$\sin(180° - \theta)$	$-\sin(\theta - 180°)$	$-\sin(360° - \theta)$
$\cos\theta$	$-\cos(180° - \theta)$	$-\cos(\theta - 180°)$	$\cos(360° - \theta)$
$\tan\theta$	$-\tan(180° - \theta)$	$\tan(\theta - 180°)$	$-\tan(360° - \theta)$

USEFUL INFORMATION

APPENDIX B

B–1

THE GREEK ALPHABET

Alpha	A	α	Nu	N	ν
Beta	B	β	Xi	Ξ	ξ
Gamma	Γ	γ	Omicron	O	o
Delta	Δ	δ	Pi	Π	π
Epsilon	E	ϵ	Rho	P	ρ
Zeta	Z	ζ	Sigma	Σ	σ
Eta	H	η	Tau	T	τ
Theta	Θ	θ	Upsilon	Υ	υ
Iota	I	ι	Phi	Φ	ϕ
Kappa	K	κ	Chi	X	χ
Lambda	Λ	λ	Psi	Ψ	ψ
Mu	M	μ	Omega	Ω	ω

B–2

UNITS AND CONVERSION FACTORS

FUNDAMENTAL UNITS FOR PHYSICAL QUANTITIES FOUND IN THE TEXT

Physical quantity	MKS unit	CGS unit	British unit
Distance	meter (m)	centimeter (cm)	foot (ft)
Time	sec	sec	sec
Speed	m/sec	cm/sec	ft/sec
Acceleration	m/sec^2	cm/sec^2	ft/sec^2
Mass	kilogram (kg)	gram (g)	slug

Physical quantity	MKS unit	CGS unit	British unit
Force	newton (N)	dyne	pound (lb)
Pressure	N/m^2	$dyne/cm^2$	lb/ft^2
Mass density	kg/m^3	g/cm^3	$slug/ft^3$
Work and energy	joule	erg	ft-lb
Heat	calorie (cal)	calorie	Btu
Temperature	°C	°C	°F
Power	watt	erg/sec	horsepower
Electric charge	coulomb	esu	
Electric potential	volt		
Electric current	ampere (amp)	emu	
Electric resistance	ohm		
Electric field intensity	N/coulomb		
Magnetic field strength	tesla	gauss	

OTHER UNITS AND CONVERSION FACTORS

Numerical Prefixes

giga (G) = 10^9 (billion)
mega (M) = 10^6 (million)
kilo (k) = 10^3 (thousand)
centi (c) = 10^{-2} (hundredth)
milli (m) = 10^{-3} (thousandth)
micro (μ) = 10^{-6} (millionth)

Kinematics

1 inch = 2.54 cm
1 mile = 1.609 km
1 meter = 3.281 feet
1 km = 0.6214 mile

1 micron (μ) = 10^{-6} m = 10^{-4} cm
1 Ångström (Å) = 10^{-8} cm

1 mi/hr = 44.7 cm/sec = 1.467 ft/sec (15 mi/hr = 22 ft/sec)
1 year = 3.156×10^7 sec

Mechanics

1 slug = 14.59 kg
1 lb = 4.448 newtons
1 g/cm³ = 1.94 slug/ft³

1 newton = 10^5 dynes = 0.2248 lb
1 dyne = 2.248×10^{-6} lb

Energy and Heat

1 joule = 10^7 ergs = 0.2389 calorie
1 Btu = 252 calories = 778 ft-lb = 1055 joules
1 kilowatt-hour (kwhr) = 3413 Btu = 3.6×10^6 joules = 8.60×10^5 calories
1 electron volt (eV) = 1.6×10^{-12} erg = 1.6×10^{-19} joule
1 MeV = 1.6×10^{-6} erg
1 GeV = 1.6×10^{-3} erg
1 ton of TNT = 4.2×10^{16} ergs
1 horsepower (hp) = 550 ft-lb/sec = 746 watts
1 kilowatt = 1.341 hp

Mass–Energy Equivalents

1 atomic mass unit (amu) = 1.66×10^{-24} gram = 931 MeV = 1.492×10^{-3} erg
1 gram = 8.987×10^{20} ergs

Electromagnetism

1 coulomb = 3×10^{9} esu
1 emu = 10 amp = 3×10^{10} esu/sec
1 tesla = 10^{4} gauss

B–3

MATHEMATICAL AND PHYSICAL CONSTANTS

$\pi = 3.14159$ $\qquad \sqrt{2} = 1.41421$
$\sqrt{\pi} = 1.77245$ $\qquad \sqrt{3} = 1.73205$
$\pi^2 = 9.86960$ $\qquad \sqrt{5} = 2.23607$
$\sqrt{10} = 3.16228$

Base of natural logarithms: $e = 2.71828$

Speed of light: $c = 2.9979 \times 10^{10}$ cm/sec = 186,282 mi/sec
Universal constant of gravitation: $G = 6.668 \times 10^{-8}$ cm^3/g-sec^2
Planck's constant: $h = 6.6262 \times 10^{-27}$ erg-sec
Boltzmann constant: $k = 1.3805 \times 10^{-16}$ erg/°K
Loschmidt number: $n_0 = 2.687 \times 10^{19}$ gas molecules/cm^3 at 0°C and 1 atm
Mechanical equivalent of heat: $J = 4.185$ joules/calorie
Stefan–Boltzmann constant: $\sigma = 5.6692 \times 10^{-5}$ erg/cm^2-sec-°K^4
Rydberg constant: $R = 1.0974 \times 10^{5}$ cm^{-1}
Electrostatic constant: K_e $(= 1/4\pi\epsilon_0) = 1$ dyne-cm^2/esu
$= 8.9874 \times 10^{9}$ N-m^2/coulomb^2
Magnetic constant: K_m $(= \mu_0/4\pi) = 1$ dyne/emu^2 = 10^{-7} N/amp^2
Electron charge: $e = 1.602 \times 10^{-19}$ coulomb = 4.803×10^{-10} esu
Mass of electron: $m_e = 9.1096 \times 10^{-28}$ gram
Mass of hydrogen atom: $m_H = 1.673 \times 10^{-24}$ gram
q/m (electron) = 1.76×10^{11} coulomb/kg
q/m (proton) = 9.65×10^{7} coulomb/kg

Particle Masses in amu

electron: $m_e = 0.000549$
proton: $m_p = 1.007276$
H atom: $m_H = 1.007825$
neutron: $m_n = 1.008665$
deuterium nucleus: $m_d = 2.013553$
helium nucleus: $m_\alpha = 4.001505$
C^{12}: $m_C = 12.000000$

Sea Level Pressure

$p = 1$ atm = 14.7 lb/in.^{2} = 76 cm Hg = 1.013×10^{6} dynes/cm^2

Acceleration of Gravity at Surface of Earth

equator: $g = 32.09\ \text{ft/sec}^2 = 978.1\ \text{cm/sec}^2$
poles: $g = 32.26\ \text{ft/sec}^2 = 983.3\ \text{cm/sec}^2$

B–4

ASTRONOMICAL DATA

mass of sun $= 1.989 \times 10^{33}$ grams
mass of earth $= 5.977 \times 10^{27}$ grams
mass of moon $= 7.349 \times 10^{25}$ grams $(= M_E/81.3)$
equatorial radius of earth $= 6378$ km $= 3963$ miles
radius of moon $= 1738$ km $= 1080$ miles $(= R_E/3.67)$
g at lunar surface $= 5.33\ \text{ft/sec}^2$ $(= g_E/6.04)$
1 astronomical unit (AU) $= 92{,}956{,}200 \pm 300$ miles $= 149{,}598{,}500 \pm 500$ km
1 light year (LY) $= 9.46 \times 10^{17}$ cm $= 5.88 \times 10^{12}$ miles

PLANETARY DATA

Planet	Synodic period (days)	Sidereal period (d = days; y = years)	Orbital radius AU	Orbital radius millions of miles	Orbital eccentricity	Planetary radius (compared to earth)	Mass (compared to earth)	Surface gravity (ft/sec^2)	Escape speed (mi/sec)	Rotation period	Oblateness	No. of known satellites
Mercury	116	88 d	0.387	36	0.206	0.38	0.05	11.8	2.6	59 d	0	0
Venus	584	225 d	0.723	67	0.007	0.96	0.82	28.5	6.4	244 d	0	0
Earth	—	1 y	1.00	93	0.017	1.00	1.00	32.2	6.94	23^h56^m	1/298	1
Mars	780	1.88 y	1.52	141	0.093	0.53	0.11	12.3	3.2	24^h37^m	1/192	2
Jupiter	399	11.86 y	5.20	483	0.048	11.19	317.8	85.3	37.0	9^h51^m	1/16	13
Saturn	378	29.46 y	9.54	886	0.056	9.47	95.2	36.7	22.0	10^h14^m	1/10	10
Uranus	370	84.01 y	19.18	1,783	0.047	3.73	14.5	30.8	13.9	10^h49^m	1/16	5
Neptune	367	165 y	30.07	2,794	0.009	3.49	17.2	49.2	15.4	15^h	1/50	2
Pluto	367	248 y	39.44	3,670	0.249	0.5	0.11	14.2	3.3	6.4 d	0	0

B–5

THERMAL PROPERTIES OF COMMON SUBSTANCES

Substance	Specific heat at 0°C (cal/g-°C)	Melting point (°C)	Latent heat of fusion (cal/g)	Boiling point (°C)	Latent heat of vaporization (cal/g)
Aluminum	0.21	660	94	2057	2000
Copper	0.09	1083	49	2336	1760
Iron	0.10	1535	65	3000	1620
Bromine	0.11	−7	16	59	44
Ethyl alcohol	0.60	−130	25	79	204
Mercury	0.03	−39	2.7	357	71
Water	1.00	0	80	100	539
Ammonia	1.08	−75	108	−28	327
Hydrogen	3.40	−259	14	−253	108
Nitrogen	0.25	−210	6.1	−196	48
Oxygen	0.22	−218	3.3	−183	51

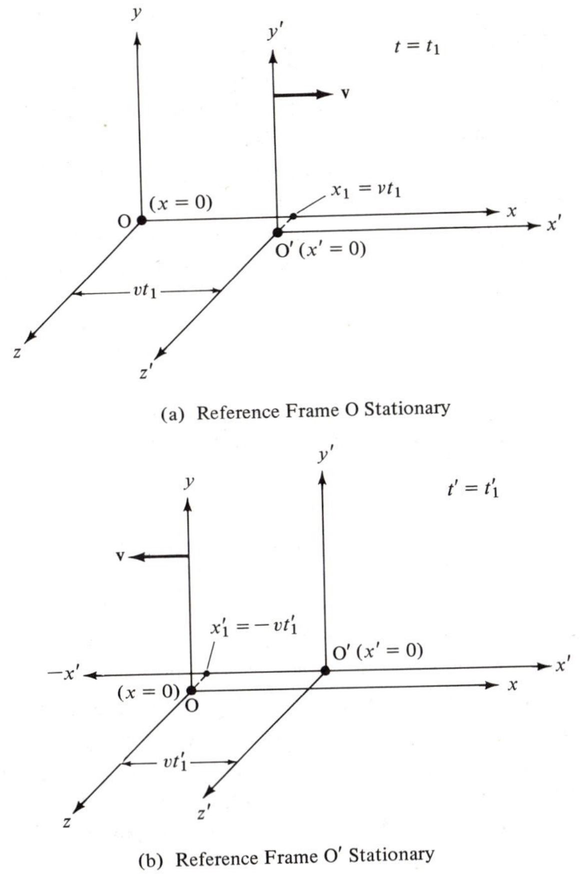

figure C–2

DIAGRAM FOR DERIVATION OF LORENTZ TRANSFORMATION

(a) The O frame is assumed stationary; O′ is moving to the right with a speed v along the positive x direction. (b) The O′ frame is assumed stationary; O is moving along the negative x' direction with a speed v.

condition that $x' = 0$ corresponds to $x_1 = vt_1$ at $t = t_1$. Substituting $x' = 0$, $x_1 = vt_1$, and $t = t_1$ into Equation C–1a gives

$$0 = a_1(vt_1) + a_2t_1$$

Eliminating t_1, we obtain $a_2 = -a_1v$. Substituting this result back into Equation C–1a gives

$$x' = a_1(x - vt) \quad \text{(C–2a)}$$
$$t' = a_3x + a_4t \quad \text{(C–2b)}$$

From the point of view of the O′ reference frame, the origin of the O frame is moving to the left along the negative x' axis (Figure C–2b). Therefore after a time t'_1 the point $x = 0$ should have the coordinate $x'_1 = -vt'_1$ in the O′ system. We substitute $x = 0$, $x'_1 = -vt'_1$, and $t' = t'_1$ into Equations C–2:

$$-vt'_1 = a_1(0 - vt_1)$$
$$t'_1 = a_3(0) + a_4t_1$$

Solving these equations simultaneously, we obtain $a_4 = a_1$. We can now eliminate a_4 from the transformation equations by substituting $a_4 = a_1$ in Equation C–2b:

$$x' = a_1(x - vt) \quad \text{(C–3a)}$$
$$t' = a_3x + a_1t \quad \text{(C–3b)}$$

B–6

TABLE OF THE ELEMENTS

Element	Symbol	Atomic number	Element	Symbol	Atomic number
Actinium	Ac	89	Mercury	Hg	80
Aluminum	Al	13	Molybdenum	Mo	42
Americium	Am	95	Neodymium	Nd	60
Antimony	Sb	51			
Argon	Ar	18	Neon	Ne	10
			Neptunium	Np	93
Arsenic	As	33	Nickel	Ni	28
Astatine	At	85	Niobium	Nb	41
Barium	Ba	56	Nitrogen	N	7
Berkelium	Bk	97			
Beryllium	Be	4	Nobelium	No	102
			Osmium	Os	76
Bismuth	Bi	83	Oxygen	O	8
Boron	B	5	Palladium	Pd	46
Bromine	Br	35	Phosphorus	P	15
Cadmium	Cd	48			
Calcium	Ca	20	Platinum	Pt	78
			Plutonium	Pu	94
Californium	Cf	98	Polonium	Po	84
Carbon	C	6	Potassium	K	19
Cerium	Ce	58	Praseodymium	Pr	59
Cesium	Cs	55			
Chlorine	Cl	17	Promethium	Pm	61
			Protactinium	Pa	91
Chromium	Cr	24	Radium	Ra	88
Cobalt	Co	27	Radon	Rn	86
Copper	Cu	29	Rhenium	Re	75
Curium	Cm	96			
Dysprosium	Dy	66	Rhodium	Rh	45
			Rubidium	Rb	37
Einsteinium	E	99	Ruthenium	Ru	44
Erbium	Er	68	Samarium	Sm	62
Europium	Eu	63	Scandium	Sc	21
Fermium	Fm	100			
Fluorine	F	9	Selenium	Se	34
			Silicon	Si	14
Francium	Fr	87	Silver	Ag	47
Gadolinium	Gd	64	Sodium	Na	11
Gallium	Ga	31	Strontium	Sr	38
Germanium	Ge	32			
Gold	Au	79	Sulfur	S	16
			Tantalum	Ta	73
Hafnium	Hf	72	Technetium	Tc	43
Helium	He	2	Tellurium	Te	52
Holmium	Ho	67	Terbium	Tb	65
Hydrogen	H	1			
Indium	In	49	Thallium	Tl	81
			Thorium	Th	90
Iodine	I	53	Thulium	Tm	69
Iridium	Ir	77	Tin	Sn	50
Iron	Fe	26	Titanium	Ti	22
Krypton	Kr	36			
Lanthanum	La	57	Tungsten	W	74
			Uranium	U	92
Lawrencium	Lw	103	Vanadium	V	23
Lead	Pb	82	Xenon	Xe	54
Lithium	Li	3	Ytterbium	Yb	70
Lutetium	Lu	71			
Magnesium	Mg	12	Yttrium	Y	39
			Zinc	Zn	30
Manganese	Mn	25	Zirconium	Zr	40
Mendelevium	Mv	101			

DERIVATIONS

APPENDIX C

C–1

GRAVITATIONAL POTENTIAL ENERGY

(Section 6–4)

Mathematical proof to show that $PE_G = mgh$ is a valid approximation to $PE_G = -GMm/d$ for a body near the surface of the earth.

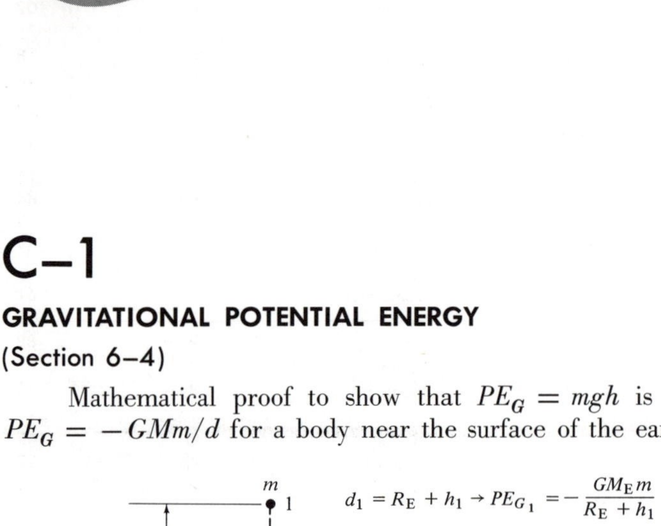

figure C–1

Assume that a body of mass m at point 1, a distance h_1 from the earth's surface, falls to point 2, a distance h_2 above the surface (see Figure C–1). The gravitational potential energy is $PE_G = -GMm/d$, where $M = M_E$ (the mass of the earth) and $d =$ the distance to the center of the earth. Hence $d_1 = R_E + h_1$, and $d_2 = R_E + h_2$ (R_E = radius of the earth). The decrease in the gravitational potential energy of the body as it falls is:

$$PE_{G_1} - PE_{G_2} = -\frac{GMm}{d_1} - \left(-\frac{GMm}{d_2}\right) = -\frac{GM_E m}{R_E + h_1} - \left(-\frac{GM_E m}{R_E + h_2}\right)$$

$$= GM_E m\left(-\frac{1}{R_E + h_1} + \frac{1}{R_E + h_2}\right) = \frac{GM_E m(h_1 - h_2)}{(R_E + h_1)(R_E + h_2)}$$

If the body is very close to the surface of the earth, h_1 and h_2 will be very small compared to the radius of the earth. We can then make the following approximations:

$$R_E + h_1 \approx R_E \qquad \text{and} \qquad R_E + h_2 \approx R_E$$

Continuing with the calculation, we have

$$PE_{G_1} - PE_{G_2} = \frac{GM_E m(h_1 - h_2)}{(R_E + h_1)(R_E + h_2)} \approx m\left(\frac{GM_E}{R_E{}^2}\right)(h_1 - h_2)$$

But at the earth's surface the acceleration of gravity $g = GM_E/R_E{}^2$; therefore

$$PE_{G_1} - PE_{G_2} \approx mg(h_1 - h_2) = mgh_1 - mgh_2$$

from which the equation $PE_G = mgh$ is seen to follow.

C–2

THE LORENTZ TRANSFORMATION

(Section 12–2)

The following is a simplified version of Einstein's derivation of the Lorentz transformation as the proper coordinate transformation of special relativity. The derivation is based on the two axioms of special relativity described in Chapter 12:

(1) All laws of physics have the same form in all inertial reference frames.
(2) Light propagates in free space with a speed of c relative to all observers.

We will consider two inertial reference frames: O, with x-y-z coordinate axes, and O′, with x'-y'-z' coordinate axes. (See Figure C–2.) It is assumed that O′ is moving to the right along the positive x axis with a speed v relative to O. We wish to determine the following: Given a point with fixed space–time coordinates (x, y, z, t) in reference frame O, what are its space–time coordinates (x', y', z', t') with respect to reference frame O′?

If we align the y and z axes parallel to the y' and z' axes, respectively, then we can assume that these coordinates will be the same in both reference frames; that is, $y' = y$ and $z' = z$. Therefore we have only to be concerned with the transformation from the (x, t) coordinates to the (x', t') coordinates. The most general form of the transformation equations* is as follows:

$$x' = a_1 x + a_2 t \qquad \text{(C–1a)}$$
$$t' = a_3 x + a_4 t \qquad \text{(C–1b)}$$

where a_1, a_2, a_3, and a_4 are constants whose values have to be determined.

These constants can be evaluated by considering certain situations in which the axioms of relativity require a specific relationship between the coordinates. For example, let us assume that the origins initially coincide ($x = x' = 0$ at $t = t' = 0$). As O′ moves to the right, it travels in a time t_1 a distance vt_1 with respect to O. In other words, at $t = t_1$ the origin of O′ ($x' = 0$) has the coordinate $x_1 = vt_1$ relative to the O reference frame (Figure C–2a). The equations should therefore satisfy the

*To ensure that one point in one reference frame transforms into only one point in the other reference frame, the equations must be linear. That is, terms containing x^2, t^2 or higher powers must be excluded.

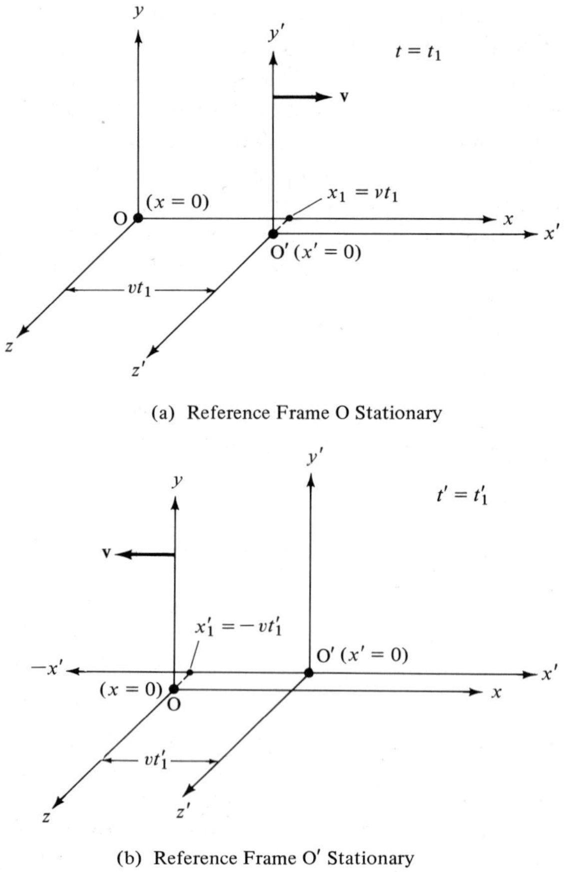

figure C–2

DIAGRAM FOR DERIVATION OF LORENTZ TRANSFORMATION

(a) The O frame is assumed stationary; O′ is moving to the right with a speed v along the positive x direction. (b) The O′ frame is assumed stationary; O is moving along the negative x' direction with a speed v.

condition that $x' = 0$ corresponds to $x_1 = vt_1$ at $t = t_1$. Substituting $x' = 0$, $x_1 = vt_1$, and $t = t_1$ into Equation C–1a gives

$$0 = a_1(vt_1) + a_2 t_1$$

Eliminating t_1, we obtain $a_2 = -a_1 v$. Substituting this result back into Equation C–1a gives

$$x' = a_1(x - vt) \tag{C–2a}$$
$$t' = a_3 x + a_4 t \tag{C–2b}$$

From the point of view of the O′ reference frame, the origin of the O frame is moving to the left along the negative x' axis (Figure C–2b). Therefore after a time t'_1 the point $x = 0$ should have the coordinate $x'_1 = -vt'_1$ in the O′ system. We substitute $x = 0$, $x'_1 = -vt'_1$, and $t' = t'_1$ into Equations C–2:

$$-vt'_1 = a_1(0 - vt_1)$$
$$t'_1 = a_3(0) + a_4 t_1$$

Solving these equations simultaneously, we obtain $a_4 = a_1$. We can now eliminate a_4 from the transformation equations by substituting $a_4 = a_1$ in Equation C–2b:

$$x' = a_1(x - vt) \tag{C–3a}$$
$$t' = a_3 x + a_1 t \tag{C–3b}$$

The next condition to be required of the transformation equations is that they obey the second axiom of relativity. That is, a light pulse traveling through space must have a speed c with respect to *both* reference frames. Suppose that a light pulse is emitted from the origin of reference frame O along the x axis at the time $t = 0$ (when both origins coincide). After a time t_1 has elapsed, the pulse will have traveled a distance $x_1 = ct_1$ with respect to reference frame O. But the same light pulse when viewed from reference frame O′ must also travel with a speed c; hence the pulse will have traveled a distance $x_1' = ct_1'$ relative to reference frame O′. We therefore substitute $x_1 = ct_1$, $x_1' = ct_1'$ into Equations C–3:

$$ct_1' = a_1(ct_1 - vt_1)$$
$$t_1' = a_3(ct_1 + a_1t_1)$$

Solving these equations simultaneously will give $a_3 = -a_1v/c^2$. With a_3 substituted for, the transformation equations now have one remaining unknown:

$$x' = a_1(x - vt) \tag{C–4a}$$

$$t' = a_1\left(-\frac{v}{c^2}x + t\right) \tag{C–4b}$$

To evaluate a_1, we will now invoke the first axiom, which states that the laws of physics must have the same form in all inertial reference frames. The "laws of physics" include the coordinate transformation that we are in the process of deriving. We require the equations for transforming from (x', t') coordinates to (x, t) to have the same form as the equations used to transform from (x, t) to (x', t'). The only difference is that we must replace v by $-v$, since O is moving with a negative velocity in the positive x' direction relative to O′. Replacing x by x', t by t', and v by $-v$ in Equations C–4, we obtain the inverse transformation equations in the form

$$x = a_1(x' + vt') \tag{C–5a}$$

$$t = a_1\left(\frac{v}{c^2}x' + t'\right) \tag{C–5b}$$

Let us substitute into Equation C–5a the equations given for x' and t' by Equations C-4:

$$x = a_1(x' + vt') = a_1\left\{a_1(x - vt) + v\left[a_1\left(-\frac{v}{c^2}x + t\right)\right]\right\}$$

Collecting terms on the right side, we have

$$x = a_1{}^2x - a_1{}^2\frac{v^2}{c^2}x = a_1{}^2x\left(1 - \frac{v^2}{c^2}\right)$$

Therefore

$$1 = a_1{}^2\left(1 - \frac{v^2}{c^2}\right)$$

so that

$$a_1 = \frac{1}{\sqrt{1 - v^2/c^2}}$$

Now we can write the complete transformation equations:

$$x' = \frac{1}{\sqrt{1 - v^2/c^2}}(x - vt) \tag{C–6a}$$

$$y' = y \tag{C–6b}$$

$$z' = z \tag{C–6c}$$

$$t' = \frac{1}{\sqrt{1 - v^2/c^2}}\left(t - \frac{v}{c^2}x\right) \tag{C–6d}$$

C–3

RELATIVISTIC KINETIC ENERGY

(Section 12–4)

Mathematical derivation of $KE = \frac{1}{2}m_0v^2$ as a low-velocity approximation to the relativistic kinetic energy equation:

The relativistic kinetic energy is given by the equation

$$KE = m_0c^2\left(\frac{1}{\sqrt{1 - v^2/c^2}} - 1\right) \tag{12–14}$$

To obtain the low-velocity approximation to this equation, we first derive an expression for the factor $1/\sqrt{1 - v^2/c^2}$ when v is small compared to c. For this purpose, we make use of the mathematical formula known as the *binomial expansion:*

$$(a + b)^n = a^n + na^{n-1}b + \frac{n(n-1)}{2}a^{n-2}b^2 + \frac{n(n-1)(n-2)}{(3)(2)}a^{n-3}b^3 + \cdots$$

We now write the factor in the form

$$\frac{1}{\sqrt{1 - v^2/c^2}} = (1 - v^2/c^2)^{-1/2}$$

Identifying $a = 1$, $b = -(v^2/c^2)$, and $n = -\frac{1}{2}$, we apply the binomial expansion to obtain

$$(1 - v^2/c^2)^{-1/2} = 1 + \frac{1}{2}\frac{v^2}{c^2} + \frac{3}{8}\frac{v^4}{c^4} + \frac{5}{16}\frac{v^6}{c^6} + \cdots$$

When we substitute this expression back into Equation 12–14, the 1 cancels out, and we obtain

$$KE = m_0c^2\left(\frac{1}{2}\frac{v^2}{c^2} + \frac{3}{8}\frac{v^4}{c^4} + \frac{5}{16}\frac{v^6}{c^6} + \cdots\right)$$

If v is much smaller than c, v^2/c^2 will be a small fraction, and the higher-order terms, v^4/c^4, v^6/c^6, etc., will be even smaller by comparison to v^2/c^2 and can be neglected. Thus as an approximation we drop all terms except the first:

$$KE \approx m_0c^2\left(\frac{1}{2}\frac{v^2}{c^2}\right) = \frac{1}{2}m_0v^2$$

and the familiar Newtonian equation for kinetic energy emerges as the low-velocity approximation to the relativistic formula.

REFERENCES

APPENDIX D

CHAPTER 2
ASTRONOMY: THE FIRST SCIENCE

1. Koestler, Arthur, *The Sleepwalkers,* Macmillan Publishing Co., Inc., New York, 1959, p. 20.
2. Kepler, Johannes, *The Harmony of the World,* quoted in Koestler, *op. cit.*, p. 262.
3. Kepler, Johannes, quoted in Koestler, *op. cit.*, p. 345.

CHAPTER 3
MOTION

1. Galilei, Galileo, *Dialogues Concerning Two New Sciences* (translated by Henry Crew and Alfonso de Salvio), Macmillan Publishing Co., Inc., New York, 1914; (reprint), Dover Publications, New York, p. 166.
2. *Ibid.*, p. 178.
3. *Ibid.*, p. 215.

CHAPTER 4
NEWTON'S PRINCIPLES OF MOTION

1. Newton, Isaac, quoted in F. E. Manuel, *A Portrait of Isaac Newton,* Belknap Press, Cambridge, Mass., 1968, p. 80.
2. Newton, Isaac, *Sir Isaac Newton's Mathematical Principles of Natural Philosophy and His System of the World* (translated by Andrew Motte, revised by Florian Cajori), University of California Press, Berkeley, 1934, p. 2.

CHAPTER 5
THE SYSTEM OF THE WORLD

1. Stukeley, William, *Memoirs of Sir Isaac Newton's Life,* quoted in F. E. Manuel, *A Portrait of Isaac Newton,* Belknap Press, Cambridge, Mass., 1968, p. 82.
2. Newton, Isaac, quoted in E. A. Burtt, *The Metaphysical Foundations of Modern Science* (Rev. Ed.), Doubleday & Co., Inc., New York, 1954, p. 291.
3. *Ibid.*, p. 266.

CHAPTER 7
HEAT AND THE NATURE OF MATTER

1. Wilson, Mitchell, "Count Rumford," *Scientific American,* Vol. 203 (October 1960), p. 158.
2. *Ibid.*, p. 164.
3. Durant, Will, *The Life of Greece,* Simon & Schuster, New York, 1939, p. 353.

CHAPTER 8
LIGHT AS A WAVE

1. Burtt, E. A., *The Metaphysical Foundations of Modern Science* (Rev. Ed.), Doubleday & Co., Inc., New York, 1954, p. 217.

CHAPTER 9
ELECTRICITY AND MAGNETISM

1. Gilbert, William, *On the Magnet* (translated by Sylvanus P. Thompson), Basic Books, Inc., New York, 1958, p. 116.
2. Cajori, Florian, *A History of Physics*, Macmillan Publishing Co., Inc., New York, 1929; (reprint) Dover Publications, New York, 1962, p. 126.
3. White, A. B., *A History of the Warfare of Science with Theology*, Vol. I, George Braziller, New York, 1955, p. 367.
4. Whittaker, Sir Edmund, *A History of the Theories of Ether and Electricity*, Vol. I, Harper & Row, New York, 1960, p. 67.
5. De Santillana, Giorgio, "Alessandro Volta," *Scientific American*, Vol. 212 (January 1965), p. 90.

CHAPTER 10
THE ELECTROMAGNETIC FIELD THEORY

1. *The Scientific Papers of James Clerk Maxwell* (edited by W. D. Niven), Dover Publications, New York, 1890, p. 527.
2. *Ibid.*, p. 535.
3. Whittaker, Sir Edmund, *A History of the Theories of Ether and Electricity*, Vol. I, Harper & Row, New York, 1960, p. 354.

CHAPTER 11
THE ORIGINS OF RELATIVITY

1. Newton, Isaac, *Sir Isaac Newton's Mathematical Principles of Natural Philosophy and His System of the World* (translated by Andrew Motte, revised by Florian Cajori), University of California Press, Berkeley, 1934, p. 6.
2. *Ibid.*, p. 545.
3. Einstein, Albert, "Autobiographical Notes," *Albert Einstein, Philosopher–Scientist* (edited by P. A. Schilpp), Tudor Publishing Co., New York, 1951, p. 17.
4. Frank, Philipp, *Einstein: His Life and Times*, Alfred A. Knopf, New York, 1947, p. 144.
5. Lenard, Philipp, *German Physics*, quoted in Frank, *op. cit.*, p. 252.

CHAPTER 12
THE SPECIAL THEORY OF RELATIVITY

1. Einstein, Albert, "Does the Inertia of a Body Depend upon Its Energy Content?" *The Principle of Relativity* (translated by W. Perrett and G. B. Jeffrey), Dover Publications, New York, 1923, p. 71.
2. Minkowski, Hermann, "Space and Time," *The Principle of Relativity, op. cit.*, p. 75.

CHAPTER 13
THE GENERAL THEORY OF RELATIVITY

1. Einstein, Albert, "On the Influence of Gravitation on the Propagation of Light," *The Principle of Relativity* (translated by W. Perrett and G. B. Jeffrey), Dover Publications, New York, 1923, p. 100.
2. Einstein, Albert, letter to Paul Ehrenfest, quoted in Banesh Hoffman, *Albert Einstein, Creator and Rebel*, The Viking Press, New York, 1972, p. 124.
3. McVittie, G. C., *General Relativity and Cosmology*, University of Illinois Press, Urbana, 1965, p. 7.

CHAPTER 14
THE ORIGINS OF THE QUANTUM THEORY

1. Jammer, Max, *The Conceptual Development of Quantum Mechanics*, McGraw-Hill Book Co., New York, 1966, p. 6.
2. *Ibid.*, p. 44.
3. Cropper, William H., *The Quantum*

Physicists, Oxford University Press, New York, 1970, p. 28.

4. Einstein, Albert, Letter to M. Besso, quoted in Martin J. Klein, "Thermodynamics in Einstein's Thought," *Science,* Vol. 157 (August 4, 1967), p. 514.

CHAPTER 15
THE STRUCTURE OF THE ATOM

1. Andrade, E. N. Da C., *Rutherford and the Nature of the Atom,* Anchor Books, New York, 1964, p. 111.
2. Gamow, George, *Thirty Years That Shook Physics,* Anchor Books, New York, 1966, p. 55.

CHAPTER 16
THE BOHR THEORY AND THE PERIODIC TABLE

1. Jammer, Max, *The Conceptual Development of Quantum Mechanics,* McGraw-Hill Book Co., New York, 1966, p. 196.

CHAPTER 17
THE BASIC CONCEPTS OF QUANTUM MECHANICS

1. Jammer, Max, *The Conceptual Development of Quantum Mechanics,* McGraw-Hill Book Co., New York, 1966, p. 247.
2. *Ibid.*, p. 249.
3. *Ibid.*, p. 253.
4. *Ibid.*, p. 324.
5. *Ibid.*, p. 328.
6. Heisenberg, Werner, *Physics and Philosophy: The Revolution in Modern Science,* quoted in a review by Victor F. Weisskopf, *Scientific American,* Vol. 199 (September 1958), p. 215.
7. *Ibid.*, p. 216.
8. Adapted from *The Born–Einstein Letters,* Walker & Company, New York, 1971, p. 91.
9. Schrödinger, Erwin, "What Is Matter?" *Scientific American,* Vol. 189 (September 1953), p. 52.

CHAPTER 18
THE NATURE OF THE NUCLEUS

1. Badash, Lawrence, "How the 'Newer Alchemy' Was Received," *Scientific American,* Vol. 215 (August 1966), p. 91.

CHAPTER 19
APPLICATIONS OF NUCLEAR PHYSICS

1. Hahn, Otto, and F. Strassmann, "Uber den Nachweis und das Verhalten der bei Bestrahlung des Urans mittels Neutronen entstehenden Erdalkalimetalle," *Die Naturwissenschaften,* Vol. 27 (1939), p. 15 (translated by Peter J. Brancazio).
2. Jungk, Robert, *Brighter Than a Thousand Suns,* Harcourt Brace Jovanovich, Inc., New York, 1958, p. 87.
3. *Ibid.*, p. 273.

CHAPTER 20
SUBNUCLEAR PARTICLES

1. Hanson, Norwood Russell, *The Concept of the Positron,* Cambridge University Press, New York, 1963, p. 151.
2. Anderson, C. D. "Early Work on the Positron and Muon," *American Journal of Physics,* Vol. 29 (December 1961), p. 825.
3. *Ibid.*, p. 827.

SUPPLEMENTARY READING

APPENDIX E

The following texts are recommended as source books pertaining to the historical and/or philosophical development of physics as a whole:

Burtt, E. A., *The Metaphysical Foundations of Modern Science* (Rev. Ed.), Doubleday & Co., Inc., New York, 1954.

Cajori, Florian, *A History of Physics,* Macmillan Publishing Co., Inc., New York, 1929 (reprinted by Dover Publications, New York).

Dampier, Sir William C. *A History of Science* (4th Ed.) Cambridge University Press, New York, 1961.

Einstein, Albert, and Leopold Infeld, *The Evolution of Physics,* Simon & Schuster, Inc., New York, 1954.

Shamos, Morris H. (Ed.), *Great Experiments in Physics,* Holt, Rinehart & Winston, Inc., New York, 1960.

Whittaker, Sir Edmund, *A History of the Theories of Ether and Electricity,* Harper & Row, New York, 1960.

The following texts and articles are recommended as source material or further readings pertaining to specific topics covered in this book:

PART I

FROM THE CAVE TO THE RENAISSANCE

(Chapters 2–3)

DeSantillana, Giorgio, *The Crime of Galileo,* University of Chicago Press, Chicago, 1955.

Galilei, Galileo, *Dialogue Concerning the Two Chief World Systems* (translated by Stillman Drake), University of California Press, Berkeley and Los Angeles, 1953.

Galilei, Galileo, *Dialogues Concerning Two New Sciences* (translated by Henry Crew and Alfonso de Salvio), Macmillan Publishing Co., Inc., New York, 1914 (reprinted by Dover Publications, New York).

Hawkins, Gerald S., *Stonehenge Decoded,* Doubleday & Co., Inc., New York, 1965.

Koestler, Arthur, *The Sleepwalkers,* Macmillan Publishing Co., Inc., New York, 1959.

Kuhn, Thomas S., *The Copernican Revolution,* Random House, Inc., New York, 1959.

Rosen, E. (Ed.), *Three Copernican Treatises,* Dover Publications, New York, 1959.

Taton, Rene (Ed.), *History of Science: Ancient and Medieval Science*, Basic Books, Inc., New York, 1963.

Toulmin, Stephen, and June Goodfield, *The Fabric of the Heavens*, Harper & Row, New York, 1961.

PART II

THE NEWTONIAN UNIVERSE

(Chapters 4–7)

Carnot, Sadi, *Reflections on the Motive Power of Fire* (edited by E. Mendoza), Dover Publications, New York, 1960.

Gamow, George, "Gravity," *Scientific American*, Vol. 204 (March 1961), p. 94.

Grosser, Morton, *The Discovery of Neptune*, Harvard University Press, Cambridge, Mass., 1963.

Jammer, Max, *Concepts of Force*, Harvard University Press, Cambridge, Mass., 1957.

Jammer, Max, *Concepts of Mass*, Harvard University Press, Cambridge, Mass., 1961.

Lounasmaa, O. V., "New Methods for Approaching Absolute Zero," *Scientific American*, Vol. 221 (December 1969), p. 26.

Manuel, Frank E., *A Portrait of Isaac Newton*, Belknap Press, Cambridge, Mass., 1968.

Newton, Isaac, *Sir Isaac Newton's Mathematical Principles of Natural Philosophy and His System of the World* (translated by Andrew Motte, revised by Florian Cajori), University of California Press, Berkeley, 1934.

Sciama, Dennis, "Inertia," *Scientific American*, Vol. 196 (February 1957), p. 99.

Stewart, Albert B., "The Discovery of Stellar Aberration," *Scientific American*, Vol. 210 (March 1964), p. 100.

Struve, Otto, "The First Determinations of Stellar Parallax," *Sky and Telescope*, Vol. 22 (November 1956), p. 9; (December 1956), p. 69.

Wilson, Mitchell, "Count Rumford," *Scientific American*, Vol. 203 (October 1960), p. 158.

PART III

LIGHT, ELECTROMAGNETISM, AND THE ETHER HYPOTHESIS

(Chapters 8–10)

Bascom, Willard, "Ocean Waves," *Scientific American*, Vol. 201 (August 1959), p. 75.

Blackham, E. Donnell, "The Physics of the Piano," *Scientific American*, Vol. 213 (December 1965), p. 88.

DeSantillana, Giorgio, "Alessandro Volta," *Scientific American*, Vol. 212 (January 1965), p. 82.

Ford, Kenneth W., "Magnetic Monopoles," *Scientific American*, Vol. 209 (December 1963), p. 122.

Gilbert, William, *On the Magnet* (translated by Silvanus P. Thompson), Basic Books, Inc., New York, 1958.

Hesse, Mary B., *Forces and Fields*, Philosophical Library, Inc., New York, 1961.

Josephson, Matthew, "The Invention of the Electric Light," *Scientific American*, Vol. 201 (November 1959), p. 99.

Kolm, Henry H., and Arthur J. Freeman, "Intense Magnetic Fields," *Scientific American*, Vol. 212 (April 1965), p. 66.

Sharlin, Harold I., "From Faraday to the Dynamo," *Scientific American*, Vol. 204 (May 1961), p. 107.

Thomson, Sir George, "The Septuagenarian Electron," *Physics Today*, Vol. 20 (May 1967), p. 55.

Williams, L. Pearce, "Michael Faraday and the Physics of 100 Years Ago," *Science*, Vol. 156 (June 9, 1967), p. 1335.

PART IV

THE EINSTEINIAN REVOLUTION

(Chapters 11–13)

Bonola, Robert, *Non-Euclidean Geometry*, Dover Publications, New York, 1955.

DeBenedetti, Sergio, "The Mossbauer Effect," *Scientific American*, Vol. 202 (April 1960), p. 73.

Einstein, Albert, *The Meaning of Relativity*, Princeton University Press, Princeton, N. J., 1950.

Einstein, Albert, "Autobiographical Notes," *Albert Einstein, Philosopher–Scientist* (edited by P. A. Schilpp), Tudor Publishing Co., New York, 1951.

Frank, Philipp, *Einstein: His Life and Times,* Alfred A. Knopf, New York, 1947.

Hoffman, Banesh, *Albert Einstein, Creator and Rebel,* The Viking Press, New York, 1972.

Holcomb, Robert W., "Experimental Tests of Relativity," *Science,* Vol. 169 (July 3, 1970), p. 40.

Klein, Martin, "Einstein and Some Civilized Discontents," *Physics Today,* Vol. 18 (January 1965), p. 38.

Perrett, W., and G. B. Jeffrey, (translators), *The Principle of Relativity,* Dover Publications, New York, 1923.

Resnick, Robert, *Introduction to Special Relativity,* John Wiley and Sons, Inc., New York, 1968.

Schmeidler, F., "The Einstein Shift—An Unsettled Problem," *Sky and Telescope,* Vol. 27 (April 1964), p. 217.

Sciama, Dennis, "Inertia," *Scientific American,* Vol. 196 (February 1957), p. 99.

Shankland, R. S., "The Michelson–Morley Experiment," *Scientific American,* Vol. 211 (November 1964), p. 107.

Toth, Imre, "Non-Euclidean Geometry Before Euclid," *Scientific American,* Vol. 221 (November 1969), p. 87.

PART V

LIGHT AND THE NATURE OF THE ATOM

(Chapters 14–17)

Andrade, E. N. Da C., *Rutherford and the Nature of the Atom,* Anchor Books, New York, 1964.

Behrens, Carl E., "Atomic Theory from 1904 to 1913," *American Journal of Physics,* Vol. 11 (1943), p. 61.

———, "The Early Development of the Bohr Theory," *American Journal of Physics,* Vol. 11 (1943), p. 135.

———, "Further Developments of Bohr's Early Atomic Theory," *American Journal of Physics,* Vol. 11 (1943), p. 272.

Bunge, Mario, "Survey of the Interpretations of Quantum Mechanics," *American Journal of Physics,* Vol. 24 (1956), p. 272.

Cline, Barbara Lovett, *The Questioners,* Thomas Y. Crowell Co., New York, 1965.

Cropper, William H., *The Quantum Physicists,* Oxford University Press, New York, 1970.

Darrow, Karl K., "Davisson and Germer," *Scientific American,* Vol. 178 (May 1948), p. 51.

Feinberg, Gerald, "Light," *Scientific American,* Vol. 219 (September 1968), p. 50.

Gamow, George, *Thirty Years That Shook Physics,* Anchor Books, New York, 1966.

———, "The Principle of Uncertainty," *Scientific American,* Vol. 198 (January 1958), p. 2.

Jammer, Max, *The Conceptual Development of Quantum Mechanics,* McGraw-Hill Book Co., New York, 1966.

Klein, Martin, "Thermodynamics in Einstein's Thought," *Science,* Vol. 157 (August 4, 1967), p. 509.

Parker, Sherwood, "Single-Photon Double-Slit Interference—A Demonstration," *American Journal of Physics,* Vol. 40 (July 1972), p. 1003.

Pauli, Wolfgang, "Remarks on the History of the Exclusion Principle," *Science,* Vol. 103 (February 22, 1946), p. 213.

Schawlow, Arthur L., "Optical Masers," *Scientific American,* Vol. 204 (June 1961), p. 54.

Schrödinger, Erwin, "What Is Matter?" *Scientific American,* Vol. 189 (September 1953), p. 52.

PART VI

THE NUCLEAR FRONTIER

(Chapters 18–21)

Anderson, C. D., "Early Work on the Positron and Muon," *American Journal of Physics,* Vol. 29 (1961), p. 825.

Badash, Lawrence, "How the 'Newer Alchemy' Was Received," *Scientific American,* Vol. 215 (August 1966), p. 89.

Brown, Laurie M., "Quarkways to Particle

Symmetry," *Physics Today,* Vol. 19 (February 1966), p. 44.

Chew, Geoffrey F., Murray Gell-Mann, and Arthur H. Rosenfeld, "Strongly Interacting Particles," *Scientific American,* Vol. 210 (February 1964), p. 74.

Fowler, William B., and Nicholas P. Samios, "The Omega-Minus Experiment," *Scientific American,* Vol. 211 (October 1964), p. 36.

Gell-Mann, Murray, and E. P. Rosenbaum, "Elementary Particles," *Scientific American,* Vol. 197 (July 1957), p. 72.

Glaser, Donald, "The Bubble Chamber," *Scientific American,* Vol. 192 (February 1955), p. 46.

Goudsmit, Samuel A., *Alsos,* Sigma Books Ltd., London, 1947.

Hammond, Allen, "Fission: The Pros and Cons of Nuclear Power," *Science,* Vol. 178 (October 13, 1972), p. 147.

Hanson, Norwood Russell, *The Concept of the Positron,* Cambridge University Press, New York, 1963.

Jones, Lawrence W., "The Continuing Search for Quarks," *Physics Today,* Vol. 26 (May 1973), p. 30.

Jungk, Robert, *Brighter Than a Thousand Suns,* Harcourt Brace Jovanovich, Inc., New York, 1958.

Leachman, R. B., "Nuclear Fission," *Scientific American,* Vol. 213 (August 1965), p. 49.

Lederman, Leon M., "The Two-Neutrino Experiment," *Scientific American,* Vol. 208 (March 1963), p. 60.

Libby, Willard F., "Radiocarbon Dating," *Science,* Vol. 133 (1961), p. 621.

Mattauch, J. H. E., W. Thiele, and A. H. Wapstra, "1964 Atomic Mass Table," *Nuclear Physics,* Vol. 67 (1965), p. 1.

Metz, William D., "Laser Fusion: A New Approach to Thermonuclear Power," *Science,* Vol. 177 (September 29, 1972), p. 1180.

Patterson, Claire, "Ages of Meteorites and the Earth," *Geochimica et Cosmochimica Acta,* Vol. 10 (1956), p. 230.

Rose, David J., "Controlled Nuclear Fusion: Status and Outlook," *Science,* Vol. 172 (May 21, 1971), p. 797.

Seaborg, Glenn T., and Justin L. Bloom, "The Synthetic Elements: IV," *Scientific American,* Vol. 220 (April 1969), p. 57.

——— and Arnold R. Fritsch, "The Synthetic Elements: III," *Scientific American,* Vol. 208 (April 1963), p. 68.

——— and Evans G. Valens, *Elements of the Universe,* E. P. Dutton & Co., New York, 1965.

Weisskopf, Victor F., "The Three Spectroscopies," *Scientific American,* Vol. 218 (May 1968), p. 15.

Wilson, Robert R., "Particle Accelerators," *Scientific American,* Vol. 198 (March 1958), p. 64.

Yang, Chen-Ning, *Elementary Particles,* Princeton University Press, Princeton, N. J., 1961.

ANSWERS TO NUMERICAL EXERCISES

APPENDIX F

CHAPTER 2

1 maximum altitude = (a) $53\frac{1}{2}°$; (b) $90°$; (c) $63\frac{1}{2}°$
9 $R = 0.866$ AU
10 $e_{max} = 30°$
14 $d_{peri} = 3.6$ AU; $d_{ap} = 4.4$ AU
15 $d_{ap} = 38$ AU
16 $e = 0.5$
17 $P = 1000$ years
18 $P = 64$ years
21 $R = \frac{5}{7}$ AU

CHAPTER 3

1 (a) 7.87 in.; (b) 1.88 meters; (c) 109.4 yards; (d) 1.496×10^8 km
3 (a) 7 west; (b) 1 south; (c) 10 northwest; (d) 13 northeast; (e) 1 east
4 (a) 13 north; (b) 13 southwest; (c) 12.4 southeast; (d) 23.9 northwest
5 (a) $\mathbf{D} = 3.86$ west (along negative x axis)
(b) $\mathbf{D} = 23.06$ southeast (4° below positive x axis)
(c) $\mathbf{D} = 21.49$ northwest (58° above negative x axis)
(d) $\mathbf{D} = 27.7$ east (along positive x axis)
6 $v_{av} = 42$ mi/hr
7 $v_{av} = 56$ mi/hr
8 (a) $v_{av} = 40$ ft/sec; (b) $\mathbf{v}_{av} = 0$; (c) $v_{av} = 40$ ft/sec; (d) $\mathbf{v}_{av} = 25.5$ ft/sec
9 $d = 5.88 \times 10^{12}$ mi
10 $v_f = 60$ ft/sec
11 $d = 400$ ft
12 (a) $a = -4$ ft/sec^2; (b) $v_{av} = 30$ ft/sec; (c) $d = 450$ ft
14 $\mathbf{a} = 16$ ft/sec^2 directed toward center of circle
16 (a) $d = 1600$ ft; (b) $v_f = 320$ ft/sec
17 (a) $a = 24$ ft/sec^2; (b) $d = 48$ ft, $v_f = 48$ ft/sec; (c) $d = 108$ ft, $v_f = 72$ ft/sec
18 $v_i = 80$ ft/sec
19 (a) $d = 256$ ft; (b) $d = 240$ ft, $\mathbf{v}_f = 32$ ft/sec downward, $\mathbf{a} = 32$ ft/sec^2 downward
20 $d = 1536$ ft
21 $v = 2800$ ft/sec
22 height of center = 2 ft

CHAPTER 4

4 $a = 4$ m/sec^2
5 $F = 2$ N
6 $\mathbf{a} = 0.33$ m/sec^2 northwest
7 $\mathbf{F} = 8000$ dynes west
8 (a) $m = 5$ slugs; (b) $w = 294{,}000$ dynes; (c) $w = 588$ N
9 $F = 750$ lb
10 $a = 68$ ft/sec^2
11 $F = 20$ lb
14 $v = 10$ cm/sec
15 $F = 1800$ lb
16 $w =$ (a) 62.1 lb; (b) 0.08 lb; (c) 1200 lb
17 $P = 4420$ lb/in.^{2}
18 $H =$ (a) 42.5 ft; (b) 3.4 ft
19 $H =$ (a) 253 ft; (b) 0.39 in.

CHAPTER 5

1 $F = Q/4$; $Q/16$; $4Q$; $25Q$
2 $F =$ (a) 8.58×10^{-7} lb; (b) 6.67×10^{-7} N
4 $g =$ (a) 26.4 ft/sec^2; (b) 8 ft/sec^2; (c) 0.26 ft/sec^2
5 12,000 mi above the surface
6 (a) $g = 24$ ft/sec^2; (b) $w = 120$ lb
7 $g_J = 81.3$ ft/sec^2
8 $R = 6\ R_E$
9 $M_V = M_E/12$
10 (a) $R_K = 3000$ mi; (b) $g_K = 48$ ft/sec^2
11 $a = 0.0356$ ft/sec^2
17 $v_{esc} =$ (a) 1.47 mi/sec (5310 mi/hr); (b) 37.0 mi/sec (133,000 mi/hr); (c) 4.01 mi/sec (14,400 mi/hr)
18 4 AU from center of more massive star
19 $v = 26{,}000$ ft/sec (17,700 mi/hr); $P = 1.42$ hr
21 (a) $v = 18.5$ mi/sec; (b) $c = 185{,}000$ mi/sec

CHAPTER 6

1 $\Delta\mathbf{p} =$ (a) 100 kg-m/sec east; (b) 1000 g-cm/sec west; (c) 5000 g-cm/sec northwest
2 $\mathbf{F} =$ (a) 10 N east; (b) 100 dynes west; (c) 500 dynes northwest
3 $\mathbf{v}_B = 15$ m/sec to the right
4 $\mathbf{v}_2 = 23.3$ cm/sec in the same direction
5 $v = 5.88$ m/sec
6 $\mathbf{T} = 1000$ ft-lb inward
8 $L = 4 \times 10^6$ kg-m^2/sec
9 (a) $L = 60{,}000$ g-cm^2/sec; (b) $F = 2667$ dynes; (c) $T = 0$
10 (a) $T = 30$ N-m; (b) $t = 200$ sec
16 $W = 0$
17 $W = 848.4$ ft-lb
18 $W = 588$ joules
19 (a) $W = 1200$ ft-lb; (b) $W = -800$ ft-lb; (c) $W_{net} = 400$ ft-lb
20 (a) $W = 4000$ ergs; (b) $KE_f = 4000$ ergs; (c) $v_f = 20$ cm/sec
21 $KE_{tot} =$ (a) 4250 joules; (b) 3610 joules
22 (a) $PE_{G_i} = 5.88 \times 10^6$ ergs; (b) $PE_{G_f} = 2.352 \times 10^6$ ergs; (c) $\Delta PE = -3.528 \times 10^6$ ergs
23 $v =$ (a) 112 ft/sec; (b) 64 ft/sec
24 (a) $H = 75$ ft; (b) $v = 62$ ft/sec
26 (a) $v = 16.6$ m/sec; (b) $W = -1372$ joules; (c) $F = 27.44$ N

CHAPTER 7

1 $T_C = -17.8°$C; 10°C; 37°C; 82.2°C
2 $T_F = 14°$F; 68°F; 95°F; −40°F
3 (a) $T_R = \frac{4}{5}T_C$; (b) $T_R = \frac{4}{9}(T_F - 32)$
7 $V = 13.3$ liters
8 $T = 427°$C
9 $T = -223°$C
10 $T = 15.2°$C
11 $m = 66.7$ grams
12 $Q = 3750$ calories
13 $m = 30$ grams
14 $W = 12{,}000$ ft-lb $= 15.42$ Btu $=$ 3886 calories
15 (a) $W = 1.98 \times 10^8$ ft-lb; (b) $Q = 2.54 \times 10^5$ Btu
16 $Q = 80.3$ Btu
18 $W = 2.334 \times 10^6$ ft-lb
19 $t = 11.52$ min
21 $v_{av} =$ (a) 4.6×10^4 cm/sec; (b) 2.60×10^5 cm/sec

CHAPTER 8

2 $r = 19.5°$
3 $i = 41.8°$
6 $v = 100$ cm/sec
8 $\lambda =$ (a) 0.217 ft; (b) 12.5 cm
10 $\lambda =$ (a) 40 cm; (b) 20 cm
11 (a) $f_1 = 100$ Hz; (b) $f_2 = 200$ Hz; $f_3 = 300$ Hz; $f_4 = 400$ Hz
14 (a) $\lambda = 4 \times 10^{-5}$ cm; (b) $x_{B_2} = 0.6$ cm
17 $i = 56.3°$
18 $f =$ (a) 5×10^{14} Hz; (b) 10^{15} Hz

CHAPTER 9

3 $F =$ (a) 2.5 dynes attraction; (b) 0.2 N repulsion
4 $F_{elec} = 2.3 \times 10^{-3}$ dynes; $F_{grav} = 5.54 \times 10^{-46}$ dynes ($F_{elec} = 4.15 \times 10^{42}\ F_{grav}$)
7 (a) $PE_e = -8$ ergs, -4 ergs, -2 ergs; (b) $\Delta PE_e = -4$ ergs, -2 ergs; (c) $v = 2$ cm/sec.
8 $W = 0.225$ joules
10 (a) $U = 180$ volts, 45 volts; (b) $V = 135$ volts; (c) $W = 1.35 \times 10^{-3}$ joule
12 $I = 10$ amp; $R = 12$ ohms
13 $P = 480$ watts
14 (a) $P = 6600$ watts; (b) $Q = 1577$ cal/sec; (c) cost $= \$1.32$

18 $F = 10^{-4}$ N repulsion
19 $F = 4$ N inward on side BC, 4 N outward on side AD, 0 on AB and CD

CHAPTER 10

10 $f =$ (a) 1.5×10^9 Hz; (b) 1.5×10^{13} Hz; (c) 3×10^{16} Hz
12 (a) $\mathbf{F} = 4.8 \times 10^{-9}$ N perpendicular to $\mathbf{v}$ and $\mathbf{B}$; (b) $r = 3.13 \times 10^{-4}$ meter

CHAPTER 11

6 $\mathbf{v} =$ (a) 7 mi/sec westward; (b) 5 mi/sec northwest
9 A measures $3c/2$; B measures $c/2$.
10 $\mathbf{c}' = c\sqrt{5}/2$ southeast

CHAPTER 12

2 (a) A exploded 300 years after B; (b) observer is 400 LY farther from B than from A; observer is 100 LY farther from B than from A.
5 $y' = y + 4$, $x' = x - 3$
7 $\Delta t =$ (a) 1.67 flashes/min; (b) 1.67 flashes/min
8 $v = 0.866c$
9 (a) $v = 0.6c$
11 $u = 0.988c$
13 $u = 0.357c$
14 $u = 0.994c$
16 (a) $m = 4.33 \times 10^{12}$ grams/sec; (b) $t = 1.46 \times 10^{13}$ years
17 $m =$ (a) 16.7 grams; (b) 10 grams
18 $KE_{rel} = 3.76 \times 10^{-4}$ ergs; $KE_{nonrel} = 2.71 \times 10^{-4}$ ergs
19 $v = 0.866c$
20 $Q = 2.15 \times 10^{10}$ Calories
21 (a) $\Delta t = 3.125 \times 10^{-8}$ sec; (b) $d = 562.5$ cm

CHAPTER 13

7 $\Delta f/f_o = 3.32 \times 10^{-15}$
18 $P =$ (a) 3.84″/century; (b) 8.73″/century

CHAPTER 14

9 (a) 16 times; (b) 81 times
10 $E = 7.56 \times 10^3$ ergs
13 $T =$ (a) 20,000°K; (b) 500°K; (c) 0.01°K
14 $\lambda_{max} = 9.67$ microns (infrared)
15 (a) $E = 1.99 \times 10^{-12}$ erg; (b) $E = 3.98 \times 10^{-12}$ erg; (c) $\lambda = 10^{-4}$ cm
16 $E =$ (a) 1.99×10^{-10} erg $= 124$ eV; (b) 1.99×10^{-13} erg $= 0.124$ eV; (c) 9.93×10^{-18} erg $= 6.20 \times 10^{-6}$ eV
18 $\lambda =$ (a) 912 Å (UV); (b) 20.7 microns (IR); (c) 0.41 Å (x ray); (d) 0.0021 Å (gamma ray)
20 (a) $W = 4.0$ eV; (c) $\lambda = 1770$ Å; (d) $v_{max} = 8.8 \times 10^7$ cm/sec

CHAPTER 15

3 $q/m = 4.83 \times 10^7$ coulombs/kg
4 (a) increases by 10^8; (b) decreases by 47/79
5 $F_{elec}/F_{grav} = 2.27 \times 10^{39}$
8 $\lambda =$ (a) 1026 Å (UV); (b) 4863 Å (visible); (c) 3646 Å (UV)
12 (b) $\lambda = 1641$ Å (UV); (d) $I = 54.4$ eV
14 (a) $E_0 = 16$ eV; (b) $E = 14.22$ eV; (c) $I = 16$ eV

CHAPTER 17

1 $\lambda =$ (a) 1.11×10^{-36} cm; (b) 1.33×10^{-31} cm; (c) 3.97×10^{-8} cm
4 $v = 1.46 \times 10^5$ cm/sec; $E = 6.07 \times 10^{-6}$ eV
5 $\lambda = 1.67$ Å
6 $v = 7.28 \times 10^{12}$ cm/sec
10 (a) $E_1 = 2.75 \times 10^{-57}$ erg; (b) $E_2 = 5.50 \times 10^{-57}$ erg
12 (a) $\Delta x = 3.51 \times 10^{-12}$ cm
13 (a) $\Delta p = 1.05 \times 10^{-22}$ g-cm/sec; (b) percent error = 0.00385%
14 (a) $\Delta E = 1.05 \times 10^{-29}$ erg

CHAPTER 18

6 $m =$ (a) 1.992×10^{-23} gram; (b) 1.672×10^{-24} gram; (c) 6.642×10^{-24} gram
7 $A_{av} = 63.62$
9 $B =$ (a) 8.48 MeV; (b) 7.71 MeV
13 (a) proton ($_1H^1$); (b) electron ($_{-1}e^0$); (c) neutron ($_0n^1$); (d) positron ($_1e^0$); (e) fluorine ($_9F^{17}$).
15 $Q = -5.09$ MeV (endoergic)
16 $Q = 17.34$ MeV; $KE_\alpha = 8.67$ MeV

CHAPTER 19

1 $m = 0.625$ gram
2 (a) $T_{1/2} = 6$ days
9 11,200 years ago
12 $Q = 4.95$ MeV (neutron must have at least 1.05 MeV kinetic energy)
15 0.38% of mass converted

CHAPTER 20

3 $P = 6.4 \times 10^5$ watts
4 $E = 1876$ MeV
10 (a) ${}_{20}Ca^{40} + \bar{\nu}_e$; (b) $e^+ + \nu_e$; (c) ${}_{35}Br^{79} + \nu_e$; (d) ${}_{17}Cl^{32} + e^-$; (e) ${}_{10}Ne^{23} + e^+$

CHAPTER 21

7 (b),(g)—baryons; (a),(d)—strangeness; (f),(h)—charge
9 (a),(i)—baryons; (b),(e),(g)—strangeness; (c),(h)—charge; (d),(f),(j)—mass-energy
13 (a) 5 particles; (b) $I_Z = 2, 1, 0, -1, -2$; (c) $Q = 3, 2, 1, 0, -1$
15 $(I, Y) =$ (a) $1, 0$; (b) $\frac{1}{2}, 1$; (c) $0, 0$; (d) $1, 0$

APPENDIX A-1

2 243
4 100
6 25
8 5
10 $\sqrt{2}/16$
12 $\frac{1}{4}$
14 1.62×10^{-2}
16 2×10^{-1}
18 1.66×10^{-8}
20 4.0×10^{-3}
22 1.8×10^{14}
24 5×10^{6}
26 5.33×10^{-4}
28 9×10^{-11}
30 1.92×10^{18}

INDEX

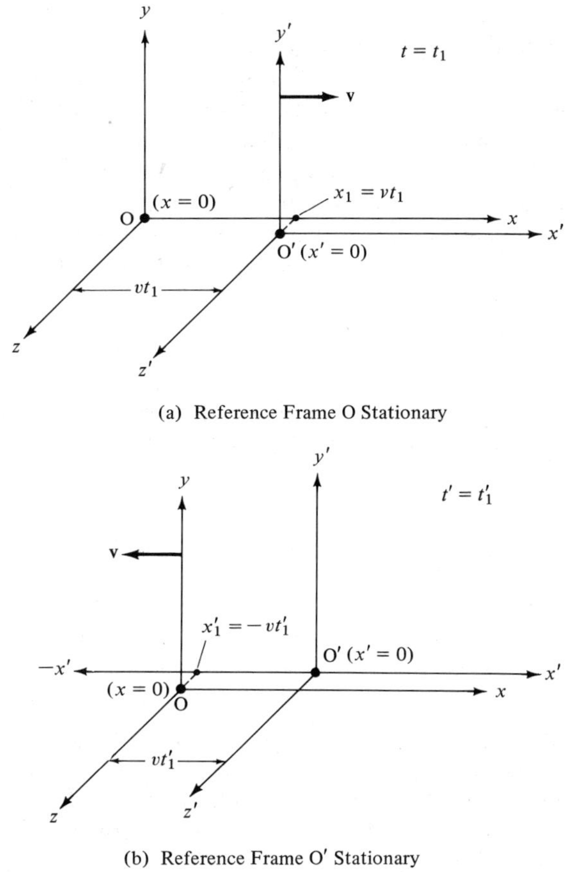

figure C–2

DIAGRAM FOR DERIVATION OF LORENTZ TRANSFORMATION

(a) The O frame is assumed stationary; O′ is moving to the right with a speed v along the positive x direction. (b) The O′ frame is assumed stationary; O is moving along the negative x' direction with a speed v.

condition that $x' = 0$ corresponds to $x_1 = vt_1$ at $t = t_1$. Substituting $x' = 0$, $x_1 = vt_1$, and $t = t_1$ into Equation C–1a gives

$$0 = a_1(vt_1) + a_2 t_1$$

Eliminating t_1, we obtain $a_2 = -a_1 v$. Substituting this result back into Equation C–1a gives

$$x' = a_1(x - vt) \qquad \text{(C–2a)}$$
$$t' = a_3 x + a_4 t \qquad \text{(C–2b)}$$

From the point of view of the O′ reference frame, the origin of the O frame is moving to the left along the negative x' axis (Figure C–2b). Therefore after a time t_1' the point $x = 0$ should have the coordinate $x_1' = -vt_1'$ in the O′ system. We substitute $x = 0$, $x_1' = -vt_1'$, and $t' = t_1'$ into Equations C–2:

$$-vt_1' = a_1(0 - vt_1)$$
$$t_1' = a_3(0) + a_4 t_1$$

Solving these equations simultaneously, we obtain $a_4 = a_1$. We can now eliminate a_4 from the transformation equations by substituting $a_4 = a_1$ in Equation C–2b:

$$x' = a_1(x - vt) \qquad \text{(C–3a)}$$
$$t' = a_3 x + a_1 t \qquad \text{(C–3b)}$$

If the body is very close to the surface of the earth, h_1 and h_2 will be very small compared to the radius of the earth. We can then make the following approximations:

$$R_{\mathrm{E}} + h_1 \approx R_{\mathrm{E}} \qquad \text{and} \qquad R_{\mathrm{E}} + h_2 \approx R_{\mathrm{E}}$$

Continuing with the calculation, we have

$$PE_{G_1} - PE_{G_2} = \frac{GM_{\mathrm{E}}m(h_1 - h_2)}{(R_{\mathrm{E}} + h_1)(R_{\mathrm{E}} + h_2)} \approx m\left(\frac{GM_{\mathrm{E}}}{R_{\mathrm{E}}^2}\right)(h_1 - h_2)$$

But at the earth's surface the acceleration of gravity $g = GM_{\mathrm{E}}/R_{\mathrm{E}}^2$; therefore

$$PE_{G_1} - PE_{G_2} \approx mg(h_1 - h_2) = mgh_1 - mgh_2$$

from which the equation $PE_G = mgh$ is seen to follow.

C–2

THE LORENTZ TRANSFORMATION

(Section 12–2)

The following is a simplified version of Einstein's derivation of the Lorentz transformation as the proper coordinate transformation of special relativity. The derivation is based on the two axioms of special relativity described in Chapter 12:

(1) All laws of physics have the same form in all inertial reference frames.
(2) Light propagates in free space with a speed of c relative to all observers.

We will consider two inertial reference frames: O, with x-y-z coordinate axes, and O′, with x'-y'-z' coordinate axes. (See Figure C–2.) It is assumed that O′ is moving to the right along the positive x axis with a speed v relative to O. We wish to determine the following: Given a point with fixed space–time coordinates (x, y, z, t) in reference frame O, what are its space–time coordinates (x', y', z', t') with respect to reference frame O′?

If we align the y and z axes parallel to the y' and z' axes, respectively, then we can assume that these coordinates will be the same in both reference frames; that is, $y' = y$ and $z' = z$. Therefore we have only to be concerned with the transformation from the (x, t) coordinates to the (x', t') coordinates. The most general form of the transformation equations* is as follows:

$$x' = a_1 x + a_2 t \qquad \text{(C–1a)}$$
$$t' = a_3 x + a_4 t \qquad \text{(C–1b)}$$

where a_1, a_2, a_3, and a_4 are constants whose values have to be determined.

These constants can be evaluated by considering certain situations in which the axioms of relativity require a specific relationship between the coordinates. For example, let us assume that the origins initially coincide ($x = x' = 0$ at $t = t' = 0$). As O′ moves to the right, it travels in a time t_1 a distance vt_1 with respect to O. In other words, at $t = t_1$ the origin of O′ ($x' = 0$) has the coordinate $x_1 = vt_1$ relative to the O reference frame (Figure C–2a). The equations should therefore satisfy the

*To ensure that one point in one reference frame transforms into only one point in the other reference frame, the equations must be linear. That is, terms containing x^2, t^2 or higher powers must be excluded.

DERIVATIONS

APPENDIX C

C–1

GRAVITATIONAL POTENTIAL ENERGY

(Section 6–4)

Mathematical proof to show that $PE_G = mgh$ is a valid approximation to $PE_G = -GMm/d$ for a body near the surface of the earth.

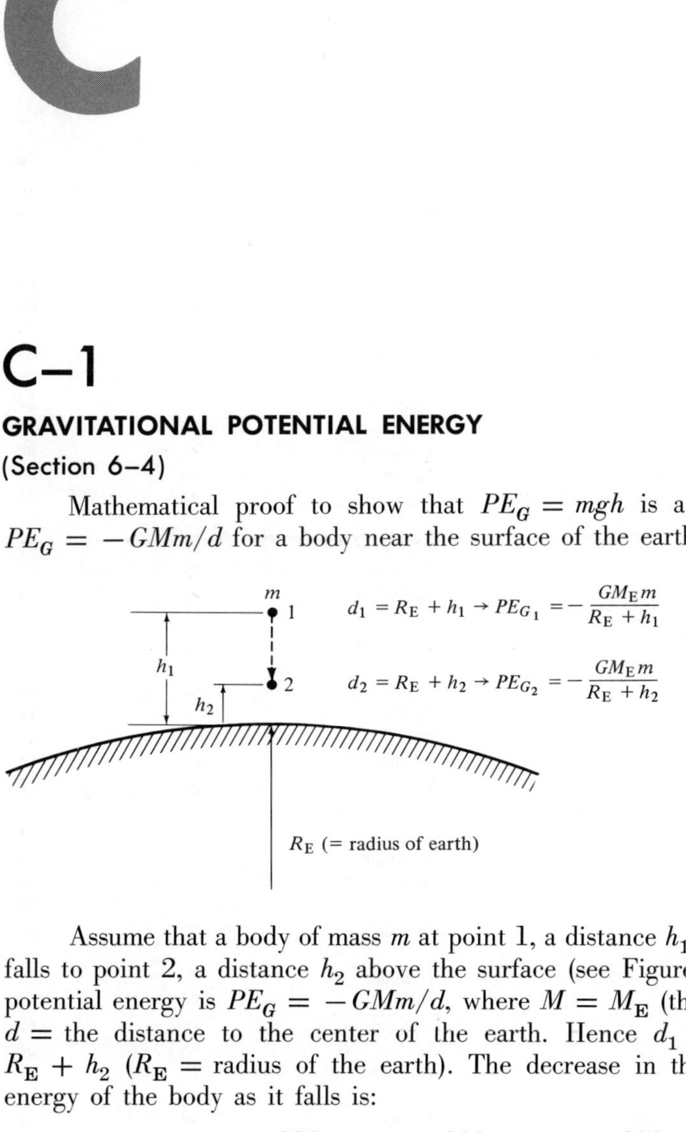

figure C–1

Assume that a body of mass m at point 1, a distance h_1 from the earth's surface, falls to point 2, a distance h_2 above the surface (see Figure C–1). The gravitational potential energy is $PE_G = -GMm/d$, where $M = M_E$ (the mass of the earth) and $d =$ the distance to the center of the earth. Hence $d_1 = R_E + h_1$, and $d_2 = R_E + h_2$ ($R_E =$ radius of the earth). The decrease in the gravitational potential energy of the body as it falls is:

$$PE_{G_1} - PE_{G_2} = -\frac{GMm}{d_1} - \left(-\frac{GMm}{d_2}\right) = -\frac{GM_E m}{R_E + h_1} - \left(-\frac{GM_E m}{R_E + h_2}\right)$$

$$= GM_E m\left(-\frac{1}{R_E + h_1} + \frac{1}{R_E + h_2}\right) = \frac{GM_E m(h_1 - h_2)}{(R_E + h_1)(R_E + h_2)}$$

B–6

TABLE OF THE ELEMENTS

Element	Symbol	Atomic number
Actinium	Ac	89
Aluminum	Al	13
Americium	Am	95
Antimony	Sb	51
Argon	Ar	18
Arsenic	As	33
Astatine	At	85
Barium	Ba	56
Berkelium	Bk	97
Beryllium	Be	4
Bismuth	Bi	83
Boron	B	5
Bromine	Br	35
Cadmium	Cd	48
Calcium	Ca	20
Californium	Cf	98
Carbon	C	6
Cerium	Ce	58
Cesium	Cs	55
Chlorine	Cl	17
Chromium	Cr	24
Cobalt	Co	27
Copper	Cu	29
Curium	Cm	96
Dysprosium	Dy	66
Einsteinium	E	99
Erbium	Er	68
Europium	Eu	63
Fermium	Fm	100
Fluorine	F	9
Francium	Fr	87
Gadolinium	Gd	64
Gallium	Ga	31
Germanium	Ge	32
Gold	Au	79
Hafnium	Hf	72
Helium	He	2
Holmium	Ho	67
Hydrogen	H	1
Indium	In	49
Iodine	I	53
Iridium	Ir	77
Iron	Fe	26
Krypton	Kr	36
Lanthanum	La	57
Lawrencium	Lw	103
Lead	Pb	82
Lithium	Li	3
Lutetium	Lu	71
Magnesium	Mg	12
Manganese	Mn	25
Mendelevium	Mv	101
Mercury	Hg	80
Molybdenum	Mo	42
Neodymium	Nd	60
Neon	Ne	10
Neptunium	Np	93
Nickel	Ni	28
Niobium	Nb	41
Nitrogen	N	7
Nobelium	No	102
Osmium	Os	76
Oxygen	O	8
Palladium	Pd	46
Phosphorus	P	15
Platinum	Pt	78
Plutonium	Pu	94
Polonium	Po	84
Potassium	K	19
Praseodymium	Pr	59
Promethium	Pm	61
Protactinium	Pa	91
Radium	Ra	88
Radon	Rn	86
Rhenium	Re	75
Rhodium	Rh	45
Rubidium	Rb	37
Ruthenium	Ru	44
Samarium	Sm	62
Scandium	Sc	21
Selenium	Se	34
Silicon	Si	14
Silver	Ag	47
Sodium	Na	11
Strontium	Sr	38
Sulfur	S	16
Tantalum	Ta	73
Technetium	Tc	43
Tellurium	Te	52
Terbium	Tb	65
Thallium	Tl	81
Thorium	Th	90
Thulium	Tm	69
Tin	Sn	50
Titanium	Ti	22
Tungsten	W	74
Uranium	U	92
Vanadium	V	23
Xenon	Xe	54
Ytterbium	Yb	70
Yttrium	Y	39
Zinc	Zn	30
Zirconium	Zr	40